T5-CUT-211

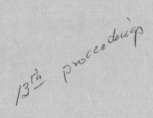
13th proceedings

TJ	International
1185	Machine Tool ...
.A1	Proceedings of
I53	the Thirteenth
1972	International ...

THE TIMKEN COMPANY

Research Library

Canton, Ohio 44706

PROCEEDINGS OF THE

THIRTEENTH INTERNATIONAL
MACHINE TOOL DESIGN AND RESEARCH
CONFERENCE

PROCEEDINGS OF THE

THIRTEENTH INTERNATIONAL
MACHINE TOOL DESIGN AND RESEARCH
CONFERENCE

held in Birmingham
18–22 September 1972

Edited by

S. A. TOBIAS

Chance Professor and Head of Department
Department of Mechanical Engineering
University of Birmingham

and

F. KOENIGSBERGER

Professor of Machine Tool Engineering
University of Manchester Institute of Science and Technology

TIMKEN

OCT 2 1974

RESEARCH LIBRARY

TJ
1185
.A1
I53
CONF. 13
1972

7409263   £22.40

MACMILLAN

All rights reserved. No part of this
publication may be reproduced or transmitted,
in any form or by any means, without permission.

First published 1973 by
THE MACMILLAN PRESS LTD

London and Basingstoke
Associated companies in New York
Dublin Melbourne Johannesburg and Madras

SBN 333 14747 2

© The Macmillan Press Limited 1973

Text set in 10/11 pt. IBM Press Roman, printed by photolithography,
and bound in Great Britain at The Pitman Press, Bath

CONTENTS

METAL CUTTING AND TOOL WEAR

GRINDING AND SURFACE PROPERTIES

ELECTROCHEMICAL GRINDING AND ELECTRO-DISCHARGE MACHINING

HIGH VELOCITY FORMING

EXPLOSIVE AND ELECTROHYDRAULIC FORMING

POWDER COMPACTION AND SINTER FORGING

METAL FORMING: PROCESSES AND MACHINES

OPENING ADDRESS:

DESIGNING FOR THE CUSTOMER

by

VISCOUNT CALDECOTE*

Machine tools are truly the basis of modern industry and, indeed, of today's civilisation itself, and the potential importance of an international gathering such as this is self-evident and needs no emphasis from me. The programme of papers indicates very clearly too the enormous knowledge and experience of machine tools brought by the delegates to this conference from all over the world, and you will not expect me to try to add to it.

Instead I would like to put before you one or two more general issues which I hope you feel are relevant to the theme of this conference and to your industry.

First, there is the old issue of the relationship between the academic world and the world of industry. There are very different worlds and it is misleading and unhelpful to pretend they are not, even where they appear to be in close contact as they are in the fields which this conference is covering. Universities and colleges are devoted to imparting and increasing knowledge; industry is concerned with producing wealth and profit. To the academic, time and cost have little significance, whereas in industry they are fundamental factors affecting success or failure. But the two worlds have a common interest where new designs and the achievement of higher performance are involved, and they can do much to help and stimulate each other.

Academics can and do make valuable contributions to the development of new products and the improvement of existing ones by analysing the basic principles and providing essential knowledge from research to support development. Industry can provide the resources which stimulate work in universities and colleges and help to maintain a sensible balance between the theoretical and practical. I have seen examples of both here in Birmingham yesterday. Why then does it so often appear that the two worlds are in some way antagonistic or, at best, mutually unsympathetic?

The principal reason, I believe, is that too often neither party takes the trouble to understand the other party's problems and objectives with sufficient care. The results obtained from cooperation are then disappointing and the work ends in recrimination, providing more ammunition for the sceptics. I very much hope, therefore, that in presenting and discussing the papers, and in all the informal but valuable contacts that are made in a conference like this, the academics and the industrialists will seek to bridge this gap and will take the opportunity to gain a better understanding of each other's point of view. In particular

I hope the academics will show that they appreciate industry's interest in the practical outcome of new ideas and knowledge within a realistic time, and at a cost which will be acceptable in the market place in relation to the benefits obtained. On the other side, I hope industrialists will make clear what their problems are not only in general terms, but more precisely in a way which links performance, time and costs, so that the academic who is serious in his wish to contribute may do so with a full appreciation of the whole problem. Thus will greater confidence be established between the two worlds, and this conference will make a valuable contribution to removing the misunderstandings which all too often seem to exist.

Secondly, I would like to make a few comments on the environment in which industry, and particularly your industry, works. Recently machine tool industries in many countries have been through a difficult time because of the low demand for their products. Some of the comments made by those who are concerned for the welfare of your industry give the impression that the rest of manufacturing industry is under some obligation to place orders for machine tools. They seem to forget there are simple reasons why potential customers don't buy. The price—performance relationship is wrong, the delivery unacceptable, the marketing inferior. To be fair to machine tool manufacturers, the economic climate is also an important factor, but too often it seems to me that this is assumed to be the principal, even the only, reason for falling sales, and the cry goes up for government action. Experience in many places suggests that this is seldom effective if the other factors, for which the industry is responsible, are unfavourable; and it is often the beginning of serious long term trouble, as has happened in the world's shipbuilding industry.

So I would like to spend a few minutes looking at those industry factors. At the risk of being accused of stating the obvious, I suggest that the most important factor in any industry is customer relations, commonly called marketing. I hope no-one in these enlightened days confuses this with selling, which is only one element in marketing and will only be successful if it is preceded and followed by other essential marketing activities.

Some delegates to this conference may still believe that the industry's prosperity and successful marketing starts with having the right products; some may even equate 'right' with 'technically most advanced and complex'. The right choice of product is of course

* Executive director of the Delta Metal Co. Ltd.

essential, but it is the end of a long process, and determination of the required performance must be made in relation to price and the timing of introduction to the market, for this is the basis of real prosperity. The process involves discussions with potential customers on their requirements and an understanding of their problems, detailed analysis of the problems and another round of visits to confirm that the deductions are correct. Only then can the specification be firmly settled and the design work go full speed ahead, but even then the new products will not be a success if the time and cost parameters set by the marketing investigations are not maintained. Unfortunately, market conditions and customer requirements do not stand still during the design and development period, so the longer the time between specification and having the new product available for delivery, the greater the risk of failure.

So I suggest that in your work and discussions you would do well to attach at least as great importance to the influence of time and cost as to perfection in design and performance. For it is only when all these are right that the customer will be fully satisfied and the essential element of customer—supplier loyalty be established.

I have been particularly interested to see that the subject of this conference is 'Machine Tool Design and Research' and that there is no mention of Development — that difficult process of turning a design into a fully proved and reliable product. But this is no criticism, for just as good marketing is the basis for a correct specification, so competent research is an important element in good design and forms a basis from which sound development can go forward within the planned time and cost. As I indicated earlier, this is the area where the specialised knowledge of research workers in universities and colleges can be of the utmost value in supplementing the knowledge of experienced designers, and in helping them to gain a complete understanding of the basic principles affecting new design features, the lack of which can cause much trouble during development. But even though there is no direct reference to development in this conference, design itself cannot and must not take place without consideration of cost, and it is therefore surprising that amongst the ninety-odd papers being presented during the next three days, only one mentions cost in its title. Perhaps others will bring in this factor, but I must admit that the titles of the papers do not encourage me to expect that many will do so. This is I think unfortunate, partly because it is unrealistic, and partly because it gives ammunition to those who seek to emphasise the gap in understanding between the academic world and industry.

Another aspect of marketing which is directly connected with design relates to reliable and efficient performance, not only over the working life of the product, but particularly in the initial period after installation. Industry invests in machine tools in order to meet an identified production requirement, and the moment from which reliable production starts is of the utmost importance. There is nothing more infuriating than when the installation of a new machine tool is followed by a long period of faulty operation. This is usually attributed to so-called teething trouble, which is a euphemism for bad installation arrangements. A good design should be easy to install, with a high degree of confidence that, provided instructions have been followed and prescribed tests made — and these should be as straightforward as possible — the machine will start its production work without further delay. It is wise to remember that every new feature added to a design increases the chance of trouble after installation, and special attention must be given to these problems if customer satisfaction is to be maintained.

But let us assume that you have done your market survey efficiently and produced a product which can be sold at an attractive price; still the selling has to be effective too. A good time to buy new machine tools is in a period of recession, to be ready for the upturn when it comes, as surely it will. This suits the machine tool industry, too, because not every company has resources to be able to buy, and so there will always be plenty of orders when the boom comes. The problem is to even out the flow of orders. I will leave aside the ingenious schemes that have been put to governments, such as the Swedish scheme for tax-free credits which can subsequently be released for buying machine tools. These have their place and are important, but how much better for the industry to stand on its own feet and sell its products on their merits. To do that in times of low activity it is useless for salesmen to peddle their wares to production engineers and works managers who will send them on their way empty handed because existing equipment can already produce more than is needed. The place to sell in these circumstances is the managing director's office, but the economic case will have to be convincing, and it will be the design that gives the best value for money which will win and not the one with the most advanced features. Similar considerations apply to the university research worker who seeks to persuade industry to develop and market the design which he has already proved. In my experience this approach to the top management is hardly ever made, and I would recommend that you, industry and university alike, do all you can to get in to put your case at the top, but concentrate on the financial advantages which your novel design features can bring rather than the ingenuity of the features themselves. Of course, you must not omit first to make sure that your product is well known and that its good features are understood by the customer's experts, the works managers and production engineers; but my plea is that you do more than win that battle, for the war is to be won higher up.

Finally, I would like to end on a more philosophical note. The machines which you design will form part of a production complex made up of people and equipment in widely varying ways. At present, all over the free world there are signs that people are becoming alarmed at the impact on their lives of new machines and advancing technology. It is becoming increasingly obvious that prosperity is inadvisable and that is is not practicable — certainly not in a free society and probably not on a world scale — for affluence and satisfaction with life to exist alongside poverty and frustration. So the benefits of increasing output per person employed, through the use of ever more ingenious and productive machines, is beginning to follow the law of diminishing returns. It is therefore necessary when designing new products and production systems to consider not only the machine output but the output of the man—machine complex, taking into account job satisfaction and all that goes with it.

You may feel that this is outside your sphere, that your task is to do research and design better machines, and that it is someone else's job to deal with these wider human problems. My own view is that none of us has the right to opt out of these problems. Already we are reaching a situation in which specialists in many fields have the knowledge and skill to alter our lives by making it possible to do things now which have been impossible in the past. But those who determine how this new knowledge is to be used — the politicians and industrialists — are often sadly lacking in understanding of the basic principles behind the new knowledge of the specialists. So, I believe, it is the duty of all of us — not least, of you, who are involved in designing the basic tools of modern industry — to do what we can to take account of the effect of our work on the community around us, and not to dig our heads into the sand of our own back garden and expect others to sort out these important human problems.

Now it is high time for me to stop and allow you to get on with the serious business of the conference. I hope that what I have said may at least remind you that there is more in 'Machine Tool Design and Research' than the acquisition of more and more specialist knowledge.

I hope you have a very successful and happy conference.

MACHINE TOOL DYNAMICS

PREDICTION OF DYNAMIC CUTTING COEFFICIENTS FROM STEADY-STATE CUTTING DATA

by

M. M. NIGM*, M. M. SADEK* and S. A. TOBIAS*

SUMMARY

An analytical method is presented with which the dynamic cutting coefficients can be derived from steady state cutting data. The technique is used for predicting the experimental results of previous investigators, covering wave cutting and removing tests as well as two actual stability charts, and good correspondence is noted.

NOTATION

A_s = shear plane area
C = force ratio = F_t/F_c
dC = effective force ratio increment
F_c = tangential cutting force along the mean cutting direction
F_t = thrust force normal to the mean cutting direction
dF_c = tangential force increment along the mean cutting direction
dF_t = thrust force increment normal to the mean cutting direction
F_s = shearing force along the shear plane
j = $\sqrt{-1}$
K_s = machining shear stress along the shear plane area
m_1, m_2 = effective force ratio incremental coefficients
n_1, n_2, n_3 = effective shear plane angle incremental coefficients
N = workpiece rotational speed (rev/min)
p = chip thickness ratio = s/s'
R = workpiece radius (in)
s' = mean deformed chip thickness (in)
s = mean undeformed chip thickness (in)
s_i = effective undeformed chip thickness normal to the instantaneous cutting direction
s_1 = tool displacement normal to the mean cutting direction
s_2 = surface undulation at the free end of the shear plane normal to the mean cutting direction
ds = effective undeformed chip thickness variation normal to the instantaneous cutting direction
v = mean cutting speed (ft/min)
dv = cutting speed increment
w = width of cut (in)
$P_r + jQ_r$ = direct receptance of the machine tool structure normal to the mean cutting direction
$G_r + jH_r$ = cross receptance of the machine tool structure between the normal and the tangent to the mean cutting direction
α = rake angle

$d\alpha$ = effective rake angle increment
γ = tool clearance angle
$d\gamma$ = effective clearance angle increment
δ = angle of action enclosed between the resultant cutting force and the mean cutting direction
ϵ = geometric lead of the free end of shear plane relative to tool edge
θ = phase angle between successive waves cut on the workpiece surface
λ = attenuation factor of force ratio dynamic variation
ϕ = mean shear plane angle measured from the mean cutting direction
ϕ_i = effective shear plane angle measured from the instantaneous cutting direction
$d\phi$ = effective shear plane angle increment
$d\psi$ = effective free surface slope increment relative to the instantaneous cutting direction
$d\psi_1$ = incremental angle between instantaneous and mean cutting directions
μ = overlap factor of successive cuts
ω = frequency of oscillation (rad/s)

INTRODUCTION

It has been held for a considerable time that the dynamic cutting coefficients required for the prediction of machine tool stability can be obtained only from specially designed experiments in which the cutting tool is made to vibrate in the manner simulating conditions arising when the machine is chattering. A method has been developed by Das and Tobias[1] which showed that such experiments are unnecessary since these coefficients can be derived directly from steady-state cutting data. Although correlation between their theoretical predictions and experiments carried out independently by Shumsheruddin[2] was on the whole satisfactory, when applying their method for establishing the unconditional threshold of stability of machine tools it was found that the maximum unconditionally stable cut was 50–70% of that measured experimentally[3, 4].

Re-examining the basis of their method, Kainth[5] concluded that the discrepancies may well be due to

*Mechanical Engineering Department, University of Birmingham, England.

one of their assumptions, this being that the shear plane does not respond to the chip thickness and rake angle variations. However, this modified theory led to a predicted unconditional threshold of stability level which was above that measured experimentally.

In the present paper the original model of the dynamic cutting of metals is further refined. This development is based on dimensional analysis of the steady-state cutting data which forms the basis of the dynamic theory.

The theoretical argument leading to equations with which the dynamic cutting coefficients can be predicted from steady-state cutting data being very complex and laborious; the paper confines itself to a summary of the assumptions and the argument which leads to the final conclusions. These are then applied to the experimental data reported by Shumsheruddin[2], and it is found that correlation over the whole range of conditions is a very great deal closer than was the case with the previous two investigations. In addition to this, the technique is also applied for the prediction of the unconditional threshold of stability measured in a particular case by Knight[3] and also in a series of tests performed within the framework of a C.I.R.P. International Co-operative Research Project[4]. It is concluded that the method evolved permits both a qualitative and quantitative prediction of the unconditional threshold of stability to an acceptable level of accuracy.

STATIC AND DYNAMIC CUTTING MODELS

Steady-state cutting[6]

Consider the conventional idealised shear plane model for orthogonal cutting shown in figure 1(a). The geometry of this can be specified by the rake angle α as an independent parameter and the shear plane angle ϕ and the angle of action of the resultant cutting force δ as two dependent parameters. As a scale factor the width of cut w and the depth of cut s must be specified as two independent parameters whilst the machining shear stress K_s is a dependent parameter.

The three dependent parameters are determined as follows:

$$\phi = \tan^{-1}\left(\frac{p\cos\alpha}{1 - p\sin\alpha}\right) \tag{1}$$

$$\delta = \tan^{-1} C \tag{2}$$

and K_s = the slope of the F_s versus A_s relationship.

It is well known that ϕ and δ are dependent on the cutting conditions in a manner that has hitherto not been predictable from these. However, by carrying out a dimensional analysis of the cutting process, in which dimensionless groups characterising strain rate and temperature effects have been introduced, it is possible to derive explicitly the dependence of the chip thickness ratio p and the force ratio C as a function of the cutting parameters α, s, v.

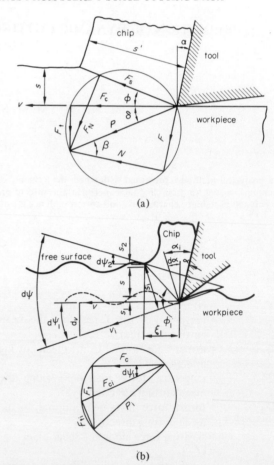

(a)

(b)

Figure 1. Shear plane model for orthogonal cutting: (a) steady state; (b) regenerative.

The process leading to this end result is explained elsewhere[6] where it is concluded that

$$p = e\sin\alpha + (1 - f\sin\alpha) \times$$
$$\left[g + \frac{h}{(10^3 s)k}\ \ln\left(\frac{sv^n}{G}\right)\right] \tag{3}$$

$$C = -e'\sin\alpha + (1 + f'\sin\alpha) \times$$
$$\left[g' - \frac{h'}{(10^3 s)k'}\ \ln\left(\frac{sv^n}{G'}\right)\right] \tag{4}$$

Both equations are valid in the absence of the built-up edge, this being the case when the following condition is satisfied:

$$sv^n \geqslant A + B\sin\alpha \tag{5}$$

The coefficients contained in equations (3) to (5), listed in table 1, are determined from steady-state cutting tests as explained in the appendix: for a particular case, when machining in orthogonal cut, mild steel with a carbide tipped tool without using any lubricant, the coefficients were found to have the values presented in table 1.

The form of equations (3) to (5) is of general validity as was established by analysing data published

TABLE 1

e	f	g	h	k	G	e'	f'	g'	h'	k'	G'	n	A	B
0.56	0.83	0.0	0.147	0.10	0.10	0.92	0.73	0.075	0.142	0.10	100.0	0.90	0.61	1.2

by previous investigators[5, 8] covering a wide range of steels. The coefficients contained in these equations, however, are dependent on the workpiece and tool materials used.

As far as the machining shear stress K_s is concerned, for the particular case for which the coefficients are presented in table 1, this was found to be independent of the chip thickness s and the rake angle α but varied slightly with the cutting speed in a manner shown in figure 2.

Having obtained the chip thickness ratio p and the force ratio C with the help of equations (3) and (4),

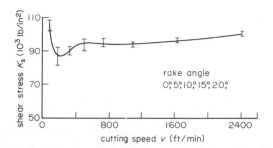

Figure 2. Variation of machining shear stress K_s with cutting speed for hot finished M.S. tubes. Rake angle: $0-20°$, chip thickness: $0.001\ 83-0.009\ 33$ in.

which in turn determine the shear angle ϕ and the angle of action δ through equations (1) and (2), and bearing in mind that K_s is independent of the chip thickness, the cutting force components can now be found as

$$F_s = wK_s \left(\frac{s}{\sin\phi}\right) \tag{6}$$

$$\left.\begin{aligned} F_c &= wK_s \left[\frac{s}{\sin\phi\,(\cos\phi - C\sin\phi)}\right] \\[2mm] F_t &= wK_s \left[\frac{Cs}{\sin\phi\,(\cos\phi - C\sin\phi)}\right] \end{aligned}\right\} \tag{7}$$

Dynamic cutting[7]

Assume that the relative oscillation between the tool and the workpiece is confined to the normal to the mean cutting direction; it will result in a cyclical variation of the independent cutting parameters about their steady-state values as shown diagrammatically in figure 1(b). These incremental variations can be represented by the following.

A. variation in the instantaneous cutting direction

$$d\psi_1 \approx \left(\frac{\dot{s}_1}{v}\right) \tag{8}$$

B. variation in the instantaneous cutting speed

$$dv \approx \frac{\dot{s}_1}{2}\left(\frac{\dot{s}_1}{v}\right) \tag{9}$$

C. variation in the instantaneous clearance angle

$$d\gamma \approx -\left(\frac{\dot{s}_1}{v}\right) \tag{10}$$

D. variation in the instantaneous rake angle

$$d\alpha \approx \left(\frac{\dot{s}_1}{v}\right) \tag{11}$$

E. variation in the free surface slope relative to the instantaneous cutting direction

$$d\psi \approx \left(\frac{\dot{s}_1 - \dot{s}_2}{v}\right) \tag{12}$$

F. variation in the effective undeformed chip thickness normal to the instantaneous cutting direction

$$ds \approx s_1 - s_2 + \frac{\epsilon\dot{s}_1}{v} \tag{13}$$

It will now be assumed that the following hold.

(1) The instantaneous cutting force components can be expressed by equations (6) and (7), these being instantaneously valid also in the case of oscillatory cutting.

(2) The instantaneous shear plane angle ϕ_i and the instantaneous force ratio C_i acquire dynamic variations about their steady-state values which can be expressed as

$$\phi_i = \phi + d\phi \tag{14}$$

$$C_i = C + dC \tag{15}$$

(3) The incremental variation in the effective shear plane angle is given by

$$d\phi = n_1\,ds + n_2\,d\alpha + n_3\,d\psi \tag{16}$$

where n_1, n_2 and n_3 are obtained from steady-state cutting data as

$$n_1 = \frac{\partial\phi}{\partial s},\ n_2 = \frac{\partial\phi}{\partial\alpha}\text{ and }n_3 = n_2 \tag{17}$$

showing the effects of the variation of the shear plane angle due to the chip thickness variation, the rake angle variation and the upper surface slope.

(4) The incremental variation in the effective force ratio is given by

$$dC = m_1\,ds + m_2\,d\alpha \tag{18}$$

where m_1 and m_2 are obtainable from steady state cutting data as

$$m_1 = \lambda\frac{\partial C}{\partial s}\text{ and }m_2 = \lambda\frac{\partial C}{\partial\alpha} \tag{19}$$

where $0 < \lambda < 1$ is a dynamic factor arising from the attenuation of the dynamic response of the chip–tool interface temperature T_{int}, assumed to be a decreasing function of T_{int}. Since the parameter $\ln(sv^n)$ in equations (3) and (4) has a dominant effect on the chip formation process, T_{int} can also be characterised by the same parameter. Hence the following form was introduced to represent λ:

$$\lambda = 1 - A'\ln\left(\frac{sv^n}{B'}\right) \tag{20}$$

where A' and B', are constants to be determined from dynamic tests.

(5) The dynamic variation in the clearance angle has a negligible effect on the dynamic cutting force provided that rubbing does not arise between the flank face and the workpiece surface.

(6) The cutting speed variation dv is also neglected as being of second-order magnitude. This permits K_s to be considered as a static variable dependent only on the mean cutting speed.

The dynamic cutting force components along the mean tangential and thrust directions resulting from the incremental changes in the chip thickness ds, shear angle $d\phi$, force ratio dC, and the cutting direction $d\psi_1$ can now be expressed as follows:

$$dF_c = k_{lc}(ds + e_1\, d\phi + e_2\, dC - Cs\, d\psi_1)$$
$$dF_t = k_{lc}(C\, ds + Ce_1\, d\phi + e_2\cot\phi\, dC + s\, d\psi_1) \quad (21)$$

where the first three terms have been obtained by partial differentiation of equation (7) and the last term has been added to account for the change in the instantaneous cutting direction.

Assuming that the tool oscillates s_1 and that workpiece upper surface has been modulated s_2 both varying harmonically with a frequency ω rad/s, then a substitution of equations (8), (11) to (13), (16) and (18) into equation (21) leads to

$$
\left.
\begin{aligned}
dF_c &= k_{lc}\left[C_1(s_1 - s_2)\right.\\
&\qquad \left. + j\left(\frac{\omega}{v}\right)\left\{C_2 s_1 + C_3(s_1 - s_2)\right\}\right]\\
dF_t &= k_{lc}\left[T_1(s_1 - s_2)\right.\\
&\qquad \left. + j\left(\frac{\omega}{v}\right)\left\{T_2 s_1 + T_3(s_1 - s_2)\right\}\right]
\end{aligned}
\right\} \quad (22)
$$

where

$$k_{lc} = wK_s/\{\sin\phi\,(\cos\phi - C\sin\phi)\}$$

$$C_1 = 1 + n_1 e_1 + m_1 e_2$$

$$C_2 = \epsilon + (n_1\epsilon + n_2)e_1 + (m_1\epsilon + m_2)e_2 - Cs$$

$$C_3 = n_3 e_1$$

$$T_1 = (1 + n_1 e_1)C + m_1 e_2\cot\phi$$

$$T_2 = [\epsilon + (n_1\epsilon + n_2)e_1]C + (m_1\epsilon + m_2)e_2\cot\phi + s$$

$$T_3 = n_3 e_1 C$$

$$e_1 = \left(\frac{\sin\phi + C\cos\phi}{\cos\phi - C\sin\phi} - \cot\phi\right)s$$

$$e_2 = \left(\frac{\sin\phi}{\cos\phi - C\sin\phi}\right)s$$

$$\epsilon = s\cot\phi$$

The coefficients n_1, n_2, n_3, m_1 and m_2 as determined from the steady-state cutting data represented in the form of equations (1), (3) and (4) are

given as follows:

$$
\left.
\begin{aligned}
n_1 &= \frac{\cos\alpha(1 - f\sin\alpha)}{1 + p^2 - 2p\sin\alpha}\left(\frac{m}{s}\right)\left[1 - k\ln\cdot\frac{sv^n}{G}\right]\\[2mm]
m &= \frac{h}{(10^3 s)^k}\\[2mm]
n_2 &= \\
&\frac{p^2 - p\sin\alpha + [e - f\{g + m\ln(sv^n/G)\}]\cos^2\alpha}{1 + p^2 - 2p\sin\alpha}\\[2mm]
n_3 &= n_2\\[2mm]
m_1 &= -\lambda(1 + f'\sin\alpha)\left(\frac{m'}{s}\right)\left[1 - k'\ln\left(\frac{sv^n}{G'}\right)\right]\\[2mm]
m' &= \frac{h'}{(10^3 s)^{k'}}\\[2mm]
m_2 &= -\lambda\left[e' - f'\left\{g' - m'\ln\left(\frac{sv^n}{G'}\right)\right\}\right]\cos\alpha
\end{aligned}
\right\} \quad (23)
$$

Except for the attenuation factor λ all the above-mentioned parameters required for the prediction of the incremental cutting force components are thus obtainable from steady-state cutting data.

VERIFICATION OF THEORETICAL RESULTS

Equations (22) with the associated expressions (23) will now be verified by reference to previous experimental results relating to both dynamic cutting and experimentally determined stability charts.

Dynamic experiments

The experimental results due to Shumsheruddin[2] relating to wave cutting and wave removing are presented in figure 3. In these experiments the workpiece material was mild steel with a machining shear stress of $K_s = 30$ ton/in^2. The results presented in the figure were obtained at the cutting speeds of 454 ft/min and 118 ft/min respectively. The figure shows the incremental thrust and tangential component amplitude dF_t and dF_c respectively as well as thier phase in relation to the chip thickness variation for a range of frequencies up to 500 Hz. Experimentally determined values are marked in these figures by $\circ$ and $\square$ for the tangential component dF_c and x and + for the thrust component dF_t.

Using equations (1), (3) and (4) together with table 1 defines the steady-state cutting data in the form of the shear angle ϕ and the force ratio C. From equations (23) the shear plane angle coefficients n_1, n_2, n_3, as well as the force ratio coefficients m_1 and m_2 are found. These data are summarised in table 2.

The parameters required for predicting the dynamic cutting force coefficients k_{lc}, C_1, C_2 and C_3, as well as T_1, T_2, T_3, as contained in equations (22) were thus found using the data presented in table 2, in conjunction with the value of K_s already specified. The attentuation factor λ was determined by a matching process between the predicted and experimentally obtained phase angles of the thrust force dF_t for the case of wave cutting. The appropriate

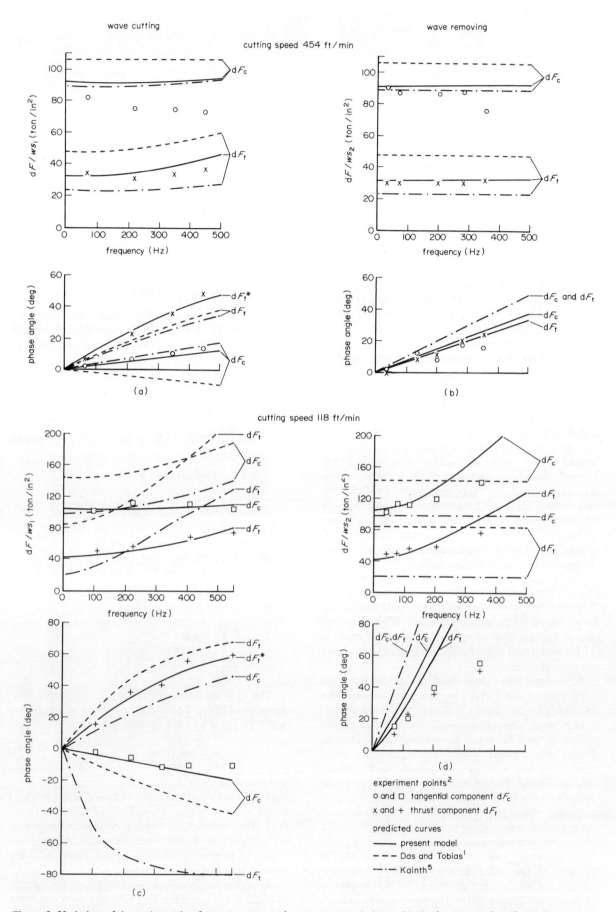

Figure 3. Variation of dynamic cutting force components in magnitude and phase with the frequency of oscillation in both wave cutting and wave removing: work material, M.S. 0.25% C; tool, carbide 5° rake and 5° clearance; mean chip thickness, 0.010 in.

TABLE 2

v (ft/min)	θ (deg)	C	n_1 (rad/in)	n_2	n_3	m_1/λ (in^{-1})	m_2/λ	λ
118	15.0	0.59	8.40	0.40	0.40	−17.9	−0.46	0.84
454	23.0	0.45	6.75	0.34	0.34	−16.4	−0.55	0.53

curves used are marked in figures 3(a) and (c) by dF_t^*. The following expression for the attenuation factor λ is derived by substituting the values of λ presented in table 2 into equation (20):

$$\lambda = 1 - 0.28 \ln\left(\frac{sv^{0.9}}{0.4}\right) \qquad (24)$$

Substituting the above-mentioned parameters into equations (22) leads to the variation of the incremental force component dF_c and dF_t as a function of the oscillation frequency. The appropriate curves are drawn in full in figure 3.

Comparing these curves with the experimental points it can be seen that on the whole correspondence between theory and experiment is satisfactory. An exception to this is the phase angle for the case of wave removing at a cutting speed of 118 ft/min (figure 3(d)) which shows a marked discrepancy at high frequencies.

Figure 3 contains also the theoretical prediction arrived at by using the method due to Das and Tobias[1], these being shown by the interrupted curves. As can be seen, the force amplitudes predicted by this method are always above those obtained experimentally. Moreover, in the case of wave cutting shown in figure 3(a) this method predicts a negative phase angle for the tangential force dF_c in contradiction to the experimental results.

As far as the method due to Kainth[5] is concerned, the predicted curves for these are also shown in the figure by the chain-dotted lines. This predicts generally lower thrust force amplitudes than those measured except for the case of wave cutting at the speed of 118 ft/min (figure 3(c)) where the experimental and theoretical curves cross.

Comparing the three sets of curves with the experimental points, it can be seen that the prediction of the present theory follows the experimental results in all cases qualitatively and in most quantitatively as well. However, the method due to Das and Tobias as well as that due to Kainth, though in some cases coinciding with the experiments, show appreciable qualitative discrepancies, for instance, in figure 3(c). In this, a lagging thrust force dF_t and a leading tangential force dF_c is predicted by Kainth whereas experimental results show exactly the opposite trend.

Similar remarks can be made concerning the 'wave-removing' case. At the cutting speed of 118 ft/min both Das and Tobias and Kainth predict force amplitudes which are independent of the frequency of oscillation (figure 3(d)). The experiments indicate that the force amplitude arises with the frequency, as is predicted by the present method.

Prediction of stability chart

For the prediction of the stability chart of a cutting process, the vibration characteristics of the machine tool structure must be provided in the form of two harmonic response loci, these being the direct receptance between the thrust force and the normal to the machined surface and the cross-receptance between the tangential force and the normal. These data are given in the form of

direct receptance $(P_r + jQ_r)$
cross-receptance $(G_r + jH_r)$

These are determined from forced vibration tests in the normal way as already explained in reference 9.

Given the harmonic response data of the structure, this is linked with the dynamic cutting force components and as a result the stability conditions of the process are derived as follows:

$$0 = A_1(1 - \mu\cos\theta') + A_2\mu\sin\theta'$$
$$+ \frac{\theta}{2\pi R}(T_2 P_r + C_2 G_r) \qquad (25)$$

$$\left(\frac{1}{k_{1c}}\right) = -A_2(1 - \mu\cos\theta')$$
$$+ A_1\mu\sin\theta' + \frac{\theta}{2\pi R}(T_2 Q_r + C_2 H_r) \qquad (26)$$

where

$$A_1 = T_1 Q_r + C_1 H_r + \frac{\theta}{2\pi R}(T_3 P_r + C_3 G_r)$$

$$A_2 = T_1 P_r + C_1 G_r - \frac{\theta}{2\pi R}(T_3 Q_r + C_3 H_r)$$

$$\theta' = \theta\left(1 - \frac{\epsilon}{2\pi R}\right)$$

$$\theta = \frac{\omega}{N/60}$$

The chip thickness coefficient k_{1c} calculated from equations (25) and (26) can be related to the critical width of cut w by the following expression:

$$w = \sin\phi\,(\cos\phi - C\sin\phi)\frac{k_{1c}}{K_s} \qquad (27)$$

Equations (25) to (27) have been applied for predicting an experimentally obtained stability chart taken from the work of Knight[3] and shown in figure 4(a). This refers to a turning operation performed with a specially designed test rig which had only one degree of freedom, this being in the direction of the normal to the machined surface.

Applying the present method to this particular case in which $\mu = 1$, using the data provided by Knight, the stability chart shown in figure 4(b) is obtained. Comparing this with figure 4(a) it is seen that the theory predicted the 'hump' in the boundary between the unconditionally and the conditionally stable regions (the envelope of the unstable lobes). However, this by itself is not significant since this hump was also predicted by Knight using the method due to Das and Tobias. However, that prediction, though qualitatively reasonable, was by no means sufficiently accurate quantitatively.

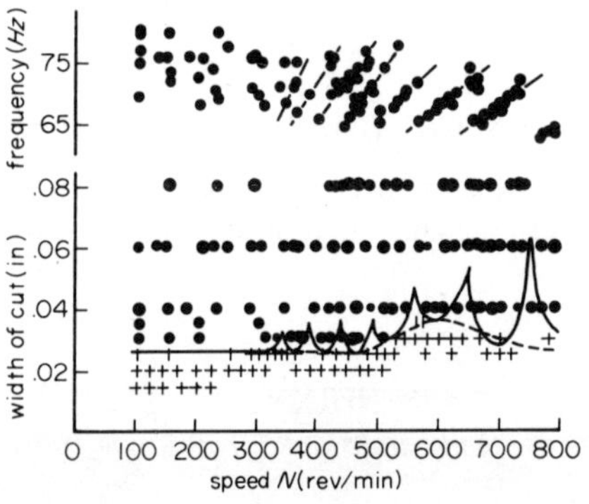

(a)

Figure 4(a). Experimental stability chart[3]: workpieces, hot finished M.S. tubes; tool, carbide, 5° rake, 5° clearance; feed, 0.0056 in/rev; mean radius, 2.9 in.

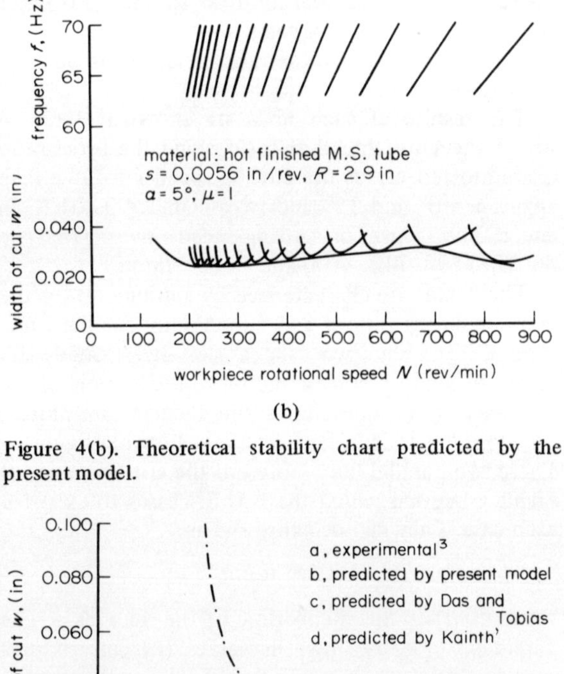

(b)

Figure 4(b). Theoretical stability chart predicted by the present model.

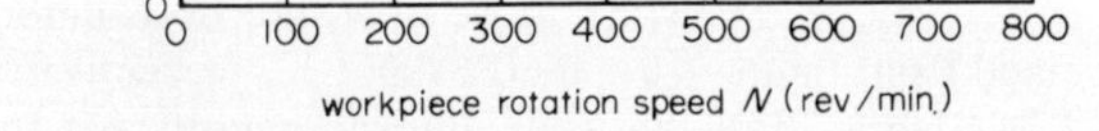

(c)

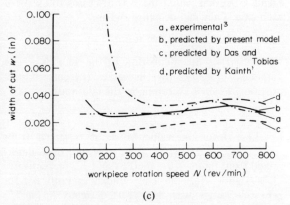

Figure 4(c). Envelopes of experimental and predicted stability charts.

The quantitative aspect becomes clearer by considering solely the envelope of the unstable regions as predicted by the three methods and comparing these with the experimental results, as is done in figure 4(c). In this, the experimentally determined envelope (curve a) is shown dotted. The envelope as predicted by Das and Tobias (interrupted line) curve c, lies substantially below the experimental one. The curve d as predicted by Kainth shows exactly the opposite tendency, lying considerably above this. Curve b has been calculated by the present method. It not only lies between the two extreme curves c and d but it does not deviate appreciably from the experimental results over a wide range of workpiece speeds.

The present technique has also been applied in a similar manner for the prediction of the stability chart of a single-degree-of-freedom cutting rig which was used within the C.I.R.P. Co-operative Research Project on the susceptibility of machine tool systems to chatter[4]. This differed from the test rig used by Knight[3] by the orientation of the principle mode of vibration. In the case of Knight this coincided with the normal to the cut surface but for the C.I.R.P. rig the common normal and the principle axis formed an angle of 30°. This necessitated the determination of the direct and cross receptances, as explained in connection with equations (25) and (26).

The experimental stability chart of this test rig, for a particular machining condition as specified in the figure caption, is shown in figure 5(a). The predicted stability chart is presented in figure 5(b).

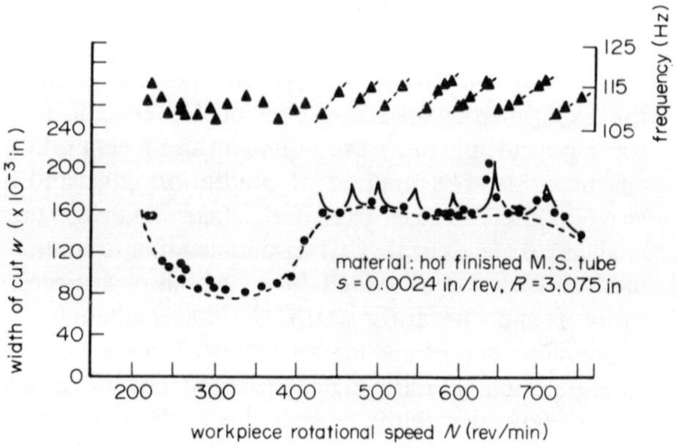

(a)

Figure 5(a). Experimental stability chart of the C.I.R.P. calibrating rig[4].

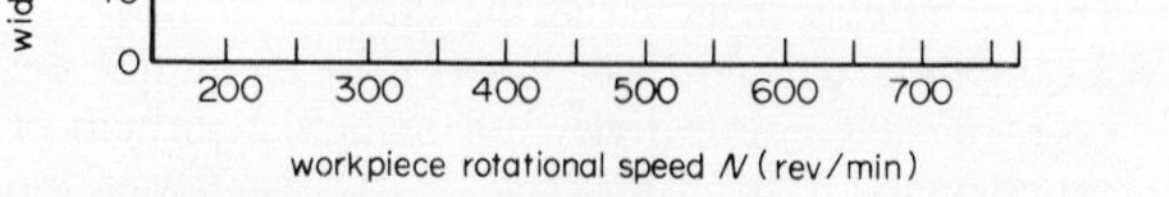

(b)

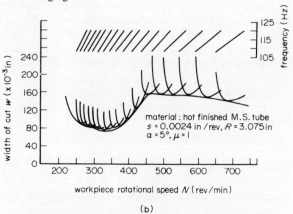

Figure 5(b). Theoretical stability chart of the C.I.R.P. calibrating rig as predicted by the present model.

The degree of approximation that can be achieved by the present method and also the progress made in relation to previous work will be fully appreciated from figure 5(c). This shows the envelope to the unstable lobes of the stability chart shown in figure

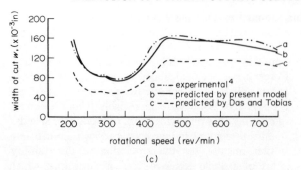

Figure 5(c). Envelopes of the experimental and predicted stability charts of the C.I.R.P. calibrating rig.

5(a), the same curve as predicted by the present method, as plotted in figure 5(b), as well as the prediction by the method due to Das and Tobias[1].

Figure 5(c) shows very clearly that, whereas the method of Das and Tobias was able to predict, for this particular case, the qualitative variation of the envelope, when it comes to a quantitative prediction then it has proved itself to be inadequate. In that case it is necessary for the dynamic coefficients of the cutting process to be determined by the present, more complex method.

CONCLUSION

The method presented predicts with good agreement the experimental results of previous investigators[2-4] for wave cutting and wave removing and regenerative chatter tests. The method of prediction consists of two distinct stages. The first stage involves the analysis of steady-state cutting data leading to mathematical expressions which give the chip thickness ratio p and the force ratio C. The mathematical expressions describing the variation of these factors as a function of rake angle, feed and cutting speed are of general validity though the coefficients contained within these are dependent on the workpiece and tool materials.

In the second stage, the steady-state cutting data as expressed by the equations giving p and C is used to determine the incremental force components dF_t and dF_c as a function of the cutting conditions. The derivatives of ϕ and C involved in the analysis are obtained by mathematical differentiation, thus avoiding the errors arising from graphical or numerical techniques.

Even from this brief summary of the steps leading to the final result it is clear that the method is more complex than that originally proposed by Das and Tobias. This is so as far as the theoretical aspect of the method is concerned, but it is certainly not the case from the point of view of the actual amount of experimental effort required, as is clear from the appendix.

The degree of approximation with which the method yielded the envelope of the unstable regions of the two stability charts presented is remarkable indeed. However, from this it should not be concluded that the problem of the prediction of chatter is completely solved. Both stability charts were obtained with idealised cutting rigs. In both the damping of the

oscillatory system was highly controlled and consequently the response data were obtained with a high degree of reliability. The accuracy of the predictions made depends, of course, not only on the reliability of the dynamic cutting model but also on the accuracy with which the dynamic vibration characteristics of the structure can be determined, from tests which do not necessarily simulate the friction conditions that arise in slides when cutting.

ACKNOWLEDGEMENTS

The first author wishes to express his thanks to the University of Birmingham for granting him a research studentship.

APPENDIX
Determination of steady-state and dynamic coefficients

The following steady-state orthogonal cutting tests are necessary to determine the coefficients appearing in equations (3) to (5).

(a) Cutting force component measurement for two rake angles, two feeds and about six speeds. Altogether amounting to 24 cutting tests.

(b) The chip thickness ratio p arising in each of these tests.

The results of such tests are shown in figure 6 where the chip thickness ratio p and the force ratio C are plotted versus the cutting speed v for two rake angles $\alpha = 0°$ and $15°$ and two feeds $s = 0.001\ 83$ in and $0.009\ 33$ in. The cutting speeds investigated are 85, 183, 500, 740, 1100 and 2420 ft/min.

The curves are characterised by a minimum for the chip thickness ratio p and a maximum for the force ratio C. For the two rake angles investigated, the values of s and v corresponding to the minima of the p curves or the maxima of the C curves are plotted on a log–log scale as straight lines, shown in figure 7a. These two parallel lines represent the critical values of s and v beyond which the B.U.E. ceases to exist for each case. They can be expressed as

$$sv^n = E$$

where $(-n)$ is the inclination to the $\ln v$ axis. The parameter E is, as shown in figure 7(b), linearly proportional to $\sin \alpha$ according to the relationship

$$E = A + B \sin\alpha$$

In these tests the value of n, as obtained from figure 7(a), is 0.9 while those of A and B, as determined from figure 7(b), are 0.61 and 1.2 respectively.

In figure 7(c) the chip thickness ratio p_0 for zero rake angle was thus plotted versus $\ln (sv^{0.9})$ for values of $sv^n \geq A$, that is free of B.U.E. This figure shows that the experimental results corresponding to the two investigated feeds can be represented by the linear relationship

$$p_0 = g + m \ln(sv^n/G)$$

where (G,g) are the co-ordinates of the point of intersection of the two lines and m is the slope which

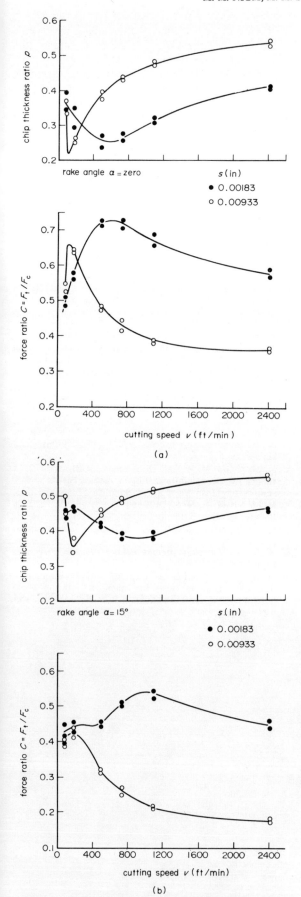

Figure 6. Variation of chip thickness ratio p and force ratio C with cutting speed v for various chip thickness s: (a) zero rake angle; (b) 15° rake angle; work material, hot finished M.S. tubes; tool, carbide 5° clearance; width of cut, 0.15 in.

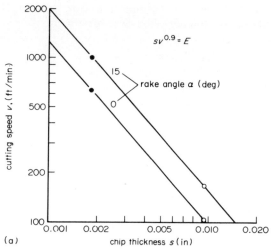

Figure 7(a). Cutting conditions characterising the termination of the B.U.E.

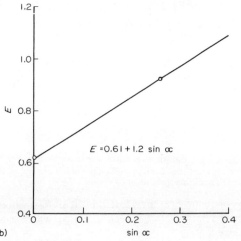

Figure 7(b). Variation of the B.U.E. characteristic parameter E with rake angle.

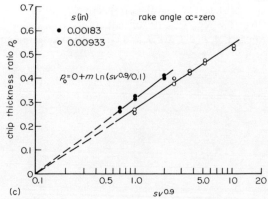

Figure 7(c). Variation of chip thickness ratio p_0 at zero rake angle with sv^n.

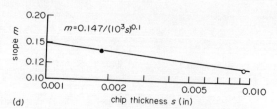

Figure 7(d). Relation between the slope m and the chip thickness s.

varies with the feed s as shown in figure 7(d). This is characterised by the following expression:

$$m = \frac{h}{(10^3 s)^k}$$

where h is the intercept of the m axis and $(-k)$ is the slope with the abscissa. Figure 7(c) shows that, in this investigation, $g = 0$ and $G = 0.1$ while the values of $h = 0.147$ and $k = 0.1$ are obtained from figure 7(d).

The variation of the chip thickness ratio p with the rake angle α is governed by the following expression:

$$p = p_0 + d \sin\alpha$$

This can be seen from figure 8a in which p is plotted versus $\sin\alpha$ for a particular s and v satisfying equation (5). The slope d and the intercept p_0 of the straight

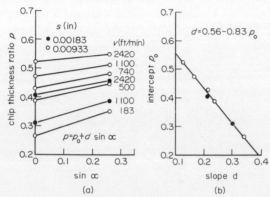

Figure 8. Variation of chip thickness ratio p with rake angle for various cutting speeds v and chip thickness s: (a) p versus $\sin\alpha$; (b) slope d versus intercept p_0.

lines of figure 8(a) are plotted in figure 8(b) showing a linear relationship between d and p_0 which can be expressed as

$$d = e - f p_0$$

where the coefficients e and f are found from figure 8(b) as 0.56 and 0.83 respectively.

All the coefficients of equations (3) and (5), namely n, A, B, e, f, g, h, k and G, are thus determined.

Similar procedure is repeated for the curves of the force ratio C to determine the remaining coefficients of equation (4) namely e', f', g', h', k' and G', the values of which are given in table 1.

As a result of this analysis, the steady-state cutting data of a particular work and tool materials is established for a wide range of cutting conditions from a limited number of cutting tests.

The dynamic attenuation factor λ given by equation (20) is determined from oscillatory cutting tests either wave cutting or wave removing, in which either the cutting force variations or their phase are measured with respect to the chip thickness variation. Only two of such tests are required either at two different speeds or at two different feeds.

REFERENCES

1. M. K. DAS and S. A. TOBIAS, 1967. 'The relation between the static and the dynamic cutting of metals', *Intern. J. Machine Tool Design Res.*, 7, 63.
2. A. A. SHUMSHERUDDIN, 1964. 'Dynamic orthogonal metal cutting', *Ph.D. Thesis,* Birmingham University.
3. W. A. KNIGHT, 1968. 'Application of the universal machinability chart to the prediction of machine tool stability,' *Intern. J. Machine Tool Design Res.,* 8, 1.
4. M. I. CHOWDHURY, M. M. SADEK and S. A. TOBIAS, 1973. 'The prediction of the chatter behaviour of the C.I.R.P. cutting rig from steady-state cutting tests', p. 19 of the present volume.
5. G. S. KAINTH, 1969. 'Investigation into the dynamics of the metal cutting process', *Ph.D. Thesis,* Birmingham University.
6. M. M. NIGM, M. M. SADEK and S. A. TOBIAS, 1972. 'A theory of the dynamic cutting of metals, Part I; steady-state cutting',*Inst. Mech. Engrs.,* submitted for publication.
7. M. M. NIGM, M. M. SADEK and S. A. TOBIAS, 1972. 'A theory of the dynamic cutting of metals, Part II: dynamic cutting', *Inst. Mech. Engrs.,* submitted for publication.
8. N. N. ZOREV, 1966. *Metal Cutting Mechanics,* Pergamon Press, Oxford.
9. M. M. SADEK and S. A. TOBIAS, 1971. 'Comparative dynamic acceptance tests for machine tools applied to horizontal milling machines',*Proc. Inst. Mech. Engrs.,* 185, 319.

THE INTERRELATIONSHIP OF SHEAR AND FRICTION PROCESSES IN MACHINING UNDER REGENERATIVE CHATTER CONDITIONS

by

V. A. STEWART* and R. H. BROWN†

SUMMARY

An experimental investigation, using high-speed cine photography, of the variation in rake-face friction and the length and inclination of the shear plane in orthogonal machining under regenerative chatter conditions. It is shown that fluctuations in the rake-face coefficient of friction are $180°$ out-of-phase with fluctuations in the length of the shear plane and with fluctuations in the toolface normal force. The coefficient of friction is approximately proportional to the toolface normal force, but the dependence of friction on normal force is considerably greater than that in steady-state cutting. The shear angle, which in steady-state cutting is approximately linearly related to the difference between friction angle and rake-angle, is shown to vary cyclically with this quantity under dynamic conditions. Differences between the steady-state and dynamic cutting processes are discussed and a model of the dynamic relationship between shear geometry and tool friction is presented. An important factor in dynamic cutting is the finite time required for the toolface stress distribution to reach its steady-state configuration.

INTRODUCTION

In an earlier paper[1], the authors concluded that available models of the mechanics of orthogonal cutting under regenerative chatter conditions inadequately described the process. To solve practical vibration problems in machining there is a need for the formulation of an adequate model. In particular this requires an understanding of the behaviour of the shear zone and the nature of the rake-face friction under vibration conditions. The previous review[1] also indicated a serious lack of fundamental experimental data on these topics. The gathering of such data has formed an important part of the study reported here.

In this paper the term *dynamic cutting* will be used to describe cutting when the tool is vibrating relative to the tool. *Static cutting* will be used to describe cutting when the tool is essentially free from vibration. In both cases the discussion will be restricted to situations in which a continuous chip is formed.

SOME ASPECTS OF THE STATIC CUTTING MODEL

It is known from static cutting data that Amonton's friction law (the coefficient of friction μ is constant and independent of the contact area) does not hold over most of the chip–tool contact region. The coefficient of friction averaged over the contact depends on cutting conditions, particularly the rake-angle. This dependence can be explained by considering the stress distribution on the rake-face. From photoelastic studies, Zorev[2] has proposed the idealised stress distribution shown in figure 1. The region AB of the tool–chip contact is known as the *sliding* region and displays the characteristics of normal sliding with constant coefficient of friction. OA is known as the plastic contact or *sticking* region. In this zone the normal stress is high and the chip adheres to the toolface causing plastic flow in

the chip material close to the interface. The tangential force is thus governed by the force required to shear the contacting layers. Neglecting workhardening, the shear stress at the interface will be constant and equal to τ_s'. The coefficient of friction will thus not be constant in the sticking region, but will be proportional to the normal stress, σ, at any point.

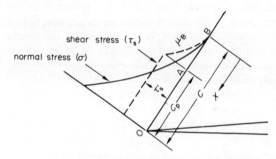

Figure 1. Distribution of normal and shear stresses on the rake-face of the tool (after Zorev[2]).

The coefficient of friction, μ, measured in a cutting test is an average based on the friction processes in both the sticking and the sliding regions. Where the sticking region is large, the measured value of μ will be sensitive to changes in the normal force of the rake-face. Any alterations in the cutting condition which affect the ratio of the length of sticking region to that of the sliding region, or the value of the shear stress τ_s, will be expected to influence μ[2,3,4].

A feature of static cutting (which has been the subject of much discussion) is the relation between the shear angle ϕ, the friction angle β and the rake-angle α. An approximately linear relationship between ϕ and $(\beta - \alpha)$ has been observed by many investigators. Zorev[2] was able to explain this dependence by considering moment and force equilibrium between the shear zone and rake-face.

* Department of Supply, Ammunition Factory, Victoria, Australia.
† Department of Mechanical Engineering, Monash University, Victoria, Australia.

CONSIDERATION OF DYNAMIC CUTTING IN RELATION TO STATIC CUTTING

Previously there has been controversy about the behaviour of the shear zone in dynamic-cutting[5]. An experiment described in the authors' earlier paper[1] has demonstrated that the orientation of the shear zone oscillates over an appreciable angle at high frequency. If ϕ is linearly related to $(\beta - \alpha)$ in dynamic-cutting, as it is in static cutting, this would imply a fluctuation in β. However, there may be inherent dynamic effects which exert an influence on the instantaneous friction angle. For example several authors, including Doi and Kato[6], have suggested a lag in rake-face friction processes and there is evidence[7,8] indicating a limited frequency response of contact length fluctuations. Thus one feature of dynamic-cutting requiring clarification is whether or not the shear and friction processes are interdependent. Because of a lack of published data, the experimental techniques described later were developed to study this question.

THEORETICAL MODEL OF FRICTION IN DYNAMIC CUTTING

It is assumed that under the most general dynamic conditions, the primary deformation zone can be described by a thin shear plane of length L_{si}, inclined at an angle ϕ_i to the instantaneous cutting velocity vector. Although it is recognised that the thin shear plane model is not an exact description of the physical process, it is considered a reasonable assumption for the cutting speeds applying during regenerative chatter and has been adopted so that analysis of the model is possible. It is further assumed that, instantaneously, the stress distribution is of the shape shown in figure 1, and the normal stress distribution has the form

$$\sigma = x^n \qquad (1)$$

where n may be defined as the normal stress index. C_i defines the instantaneous contact length.

Force equilibrium is considered in the manner proposed by Merchant[9]. This assumes that the resultant forces on the shear plane and rake-face, R and R' in figure 2, are equal, opposite and collinear.

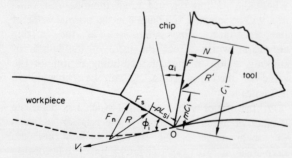

Figure 2. Representation of the instantaneous forces acting on the shear plane and the rake-face.

Hence the normal force (N) and tangential force (F) on the rake-face may be expressed in terms of the shear and normal forces on the shear plane F_s and F_n.

$$N = F_s\cos(\phi_i - \alpha_i) + F_n\sin(\phi_i - \alpha_i) \qquad (2)$$

$$F = F_n\cos(\phi_i - \alpha_i) - F_s\sin(\phi_i - \alpha_i) \qquad (3)$$

In addition to force equilibrium, Zorev[2] also considered moment equilibrium. Taking moments about 0 in figure 2 gives

$$pL_{si}F_n = mC_iN \qquad (4)$$

where the moment arms (pL_{si}) and (mC_i) are fractions of the instantaneous shear plane and chip contact lengths respectively. Assuming the normal stress distribution of equation (1),

$$m = \frac{1}{n + 2} \qquad (5)$$

It might be noted that for a uniform normal stress on the shear plane $p = \frac{1}{2}$.

Eliminating F_n from (2) to (4) gives

$$N = F_s\left[\frac{\cos(\phi_i - \alpha_i)}{1 - \dfrac{mC_i}{pL_{si}}\sin(\phi_i - \alpha_i)}\right] \qquad (6)$$

and

$$F = F_s\left[\frac{mC_i}{pL_{si}}\frac{\cos^2(\phi_i - \alpha_i)}{1 - \dfrac{mC_i}{pL_{si}}\sin(\phi_i - \alpha_i)} - \sin(\phi_i - \alpha_i)\right] \qquad (7)$$

Instantaneously, the coefficient of friction averaged over the rake-face is

$$\mu = \frac{F}{N} = \frac{mC_i}{pL_{si}}\left[\cos(\phi_i - \alpha_i) + \sin(\phi_i - \alpha_i)\tan(\phi_i - \alpha_i)\right]$$

$$- \tan(\phi_i - \alpha_i) \qquad (8)$$

Thus μ depends on the instantaneous shear plane geometry, defined by ϕ_i and L_{si}, the instantaneous rake angle α_i, and the quantities p, m and C_i which characterise the shear plane and rake-face stress distributions. It is known that ϕ_i, L_{si} and α_i all fluctuate in dynamic-cutting. The variation in p, m and C_i is not so well known. Tashlitskii[7] has suggested a limited frequency response for changes in contact length as the uncut chip thickness varies. MacManus and Pearce[8] demonstrated this effect by noting that contact length responded to changes in uncut thickness at low frequency, but there was almost no variation in the contact length at high frequencies.

There is no data available on the variation of p and m in dynamic cutting. For static cutting, Zorev[2] demonstrated that the ratio p/m was approximately constant for the conditions he tested. In dynamic cutting it is clear that there must be changes in the index n to accommodate the typical variations in the rake-face normal load N under the extreme dynamic conditions where contact length does not vary. However, from equation (5) it follows that changes in n result only in small changes in m. It is reasonable to assume that the normal stress distribution on the shear plane does not alter significantly as ϕ_i and L_{si} oscillate. Thus we might expect p/m to remain sensibly constant in dynamic cutting. Consequently fluctuations in coefficient of friction in dynamic cutting should be expected to depend primarily on the length and orientation of the shear plane. Experiments to study this dependence are described later.

The length of the shear plane may be written in terms of the instantaneous uncut chip thickness t_1.

$$L_{si} = \frac{t_1}{\sin \phi_i} \qquad (9)$$

Substituting into (8),

$$\mu = \frac{mC_i}{pt_1} \sin \phi_i [\cos(\phi_i - \alpha_i) + \sin(\phi_i - \alpha_i) \tan(\phi_i - \alpha_i)]$$
$$- \tan(\phi_i - \alpha_i) \qquad (10)$$

For static cutting, Zorev[2] has shown that C_i is proportional to t_1 and that the ratio m/p is constant, so (10) reduces to

$$\beta = \tan^{-1} \mu = f_1(\phi, \alpha) \qquad (11)$$

This is a particular form of the more general function describing the interdependence of shear and friction processes in static cutting

$$f_2(\phi, \beta, \alpha) = 0 \qquad (12)$$

The most commonly used function is the linear relationship between ϕ and $(\beta - \alpha)$. The constants of proportionality for this have been discussed at great length. Pugh[10] experimentally verified the linearity and has given values of the proportionality constants for a range of work materials. Bailey and Boothroyd[11] have argued that it is more meaningful to consider β as a function of $(\phi - \alpha)$ as in (10) or (11), rather than ϕ as a function of $(\beta - \alpha)$. In either case, it may be concluded that the model described by (8) or (10) is consistent with the observed static cutting behaviour. Its application to dynamic cutting is considered below.

EXPERIMENTAL STUDY OF INTERRELATIONSHIP OF SHEAR AND FRICTION IN DYNAMIC CUTTING

Experimental data is required to test the validity of the model described by equation (8), which postulates that μ fluctuates dynamically as a function of ϕ_i, L_{si} and α_i. A high-speed cine camera (framing speed 8000 frames a second) was used to record the instantaneous shear geometry in an orthogonal cut. Forces were measured by a two-component strain gauge dynamometer. A second lens system on the camera enabled the dynamometer output, displayed on an oscilloscope, to be recorded on the film simultaneously with the cutting geometry. The output of the dynamometer was calibrated for dynamic force amplitude and phase, using the method described in an earlier paper[12].

It has been shown[13] that forces acting on the nose and flank-faces of the tool may vary significantly during dynamic cutting. The contribution of these forces could lead to erroneous conclusions regarding the variation of μ. Realising this difficulty, the tool geometry was chosen to minimise the nose and flank

forces: a large clearance angle ($20°$) was used and the cutting-edge radius kept as small as possible (less than 0.001 in).

The experimental cuts were taken on the end of a 4 in diameter tube in the spindle of an infinitely variable speed lathe. Both the dynamometer and the camera were mounted on the lathe cross-slide. Clear regenerative chatter with a frequency of 1100 Hz resulted when a cut was taken under the following test conditions:

tool material	Stellite 100
work material	1010 cold drawn steel
rake-angle	$7°$
clearance angle	$20.°$
cutting-edge radius	less than 0.001 in
cutting speed	730 s ft/min
uncut chip thickness	0.0095 in
width of cut	0.230 in

The forces tangential and normal to the mean cutting direction, the shear angle, shear plane length, the chip contact length and the instantaneous rake-angle were recorded on each frame of the high-speed film. Figure 3 shows some of this data obtained from a sequence of frames: the forces shown are calculated values of the components along and normal to the rake-face, that is, F and N. These were determined by resolving the measured force components of the resultant minus and the mean values of the ploughing force components. These latter values were estimated from force measurements at several uncut chip thicknesses by extrapolating to zero uncut chip thickness. The friction angle β_i was calculated from the ratio F/N.

Examining figure 3 reveals that the normal force N is in phase with the shear plane length L_{si} and that the frictional force F is $180°$ out-of-phase with both N and L_{si}. As a consequence of the large phase difference between F and N, there is a large cyclic

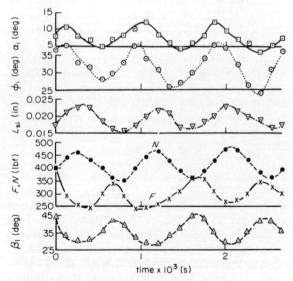

Figure 3. Experimentally-determined variations in some parameters in dynamic cutting.

variation in the magnitude of β_i. The variation in β_i is $180°$ out-of-phase with N, indicating the dependence of β_i on N, as observed in static cutting. If Amonton's law held, F and N would be expected to be in phase and μ would remain constant. It may further be noted that β_i (and therefore μ) is $180°$ out-of-phase with L_{si}. This is in agreement with the inverse relationship between these two quantities postulated by equation (8).

The chip produced in the test was examined to see if it bore evidence of the large fluctuations in μ apparent from the force observations. Figure 4

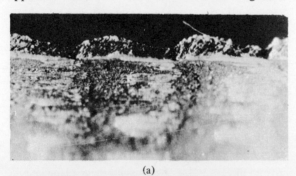

(a)

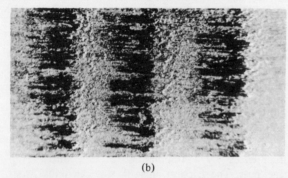

(b)

Figure 4. Photographs of the chip produced in dynamic cutting: (a) the side and bottom (sliding) surface of the chip; (b) nature of the sliding surface.

shows (a) an angled side photograph of the chip and (b), in line under this, a photograph taken normal to the surface which had contacted the tool. Under the thickest sections of the chip (that is, large L_{si}) the chip shows evidence of build-up associated with a plastic flow or a low coefficient of friction condition. Under thin sections of the chip a sliding condition is revealed. Micrographs of sections through the chip showed large variations in the direction of metalflow in the secondary deformation zone. The flow changes corresponded to the variations noted on the sliding surface of the chip and to the friction angle fluctuations of figure 3.

It is interesting to plot the values of β_i against N for dynamic cutting and compare them with corresponding values for static cutting. This is shown in figure 5 where the static cutting data was obtained by testing at a speed of 730 s ft/min with four different rake-angles. The steeper slope in the dynamic case indicates a stronger dependence of β_i on N than that in static cutting. This difference can be attributed to an attenuated response of the contact length at the high-frequency of geometry fluctuations in the shear zone. For the same changes in the uncut chip thickness, in both static and dynamic cutting, the size of the shear zone and the magnitude of the normal force N would be expected to change by

similar amounts in the two cases. In static cutting a change of uncut chip thickness will result in a change of contact length. If, in dynamic cutting, the change in contact length is reduced by lack of response

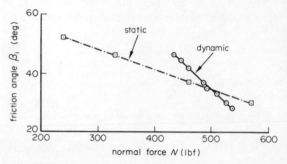

Figure 5. A comparison of measured values of friction angle plotted against normal force on the rake-face for static and dynamic cutting.

time, then β_i will show a greater sensitivity to the normal load variation than that for static cutting.

The variation in contact length was examined by measuring the observed contact at the side of the cut. A comparison with the contact indicated by rubbing marks on the tool, showed that the contact length near the centre of the cut was slightly greater than that at the side of the cut. The value at the side has to be used because it is the only one in which fluctuations can be measured. Figure 6 shows

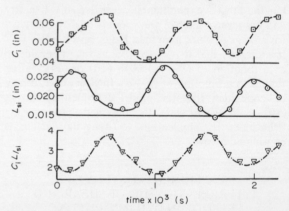

Figure 6. Measured variations in contact length C_i and shear plane length L_{si} for dynamic cutting.

measured values of the contact length C_i and corresponding shear plane lengths L_{si} taken from a sequence of the high-speed film. The contact length does not remain constant, but is seen to fluctuate with a cycle lag of approximately $60°$ behind the shear plane length fluctuation. The ratio C_i/L_{si} is also plotted in the figure: this is seen to be at a phase of almost $180°$ to the L_{si} variation. The apparent domination exerted over this ratio by the instantaneous length of the shear plane explains the inverse dependence of μ on L_{si} observed in figure 3.

From the data of figure 3, values of ϕ_i can be plotted against $(\beta_i - \alpha_i)$ for dynamic cutting. This is shown in figure 7, which also gives similar data obtained with the present apparatus under static cutting conditions (cutting speed 730 s ft/min, rake-angles from $0°$ to $30°$, depth of cut 0.0095 in). Over a vibration cycle the dynamic curve has the form of a closed-loop, which contrasts with the straight line of the static characteristic. Treating the dynamic loop as a Lissajous figure, there appears to be a phase lag of approximately 60° between ϕ_i and $(\beta_i - \alpha_i)$. From static behaviour, ϕ_i and $(\beta_i - \alpha_i)$ would be expected to be in phase with each other. The observed

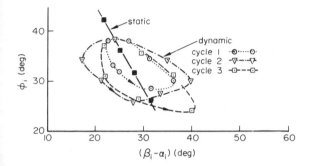

Figure 7. Experimental results comparing the static and dynamic relationships of ϕ_i to $(\beta_i - \alpha_i)$.

difference can be explained by reconsidering equation (8). In dynamic cutting this does not reduce to the form of equation (11) as it does in static cutting.

Tests reported in the earlier paper[1] have indicated that in dynamic cutting, the angle between the shear plane and a tangent to the workpiece surface at the free end of the shear plane is constant. This dependence of shear angle on the slope of approaching work surface had been postulated by Shaw and Holken[14]. Additional dynamic cutting tests under different cutting conditions have confirmed this dependence.

DISCUSSION OF THE SHEAR AND FRICTION RELATIONSHIP

One significant experimental result is that the variation in coefficient of friction μ is 180° out-of-phase with the shear plane length L_{si}. This behaviour corresponds with that predicted by equation (8) taking account of the further experimental observation that the ratio C_i/L_{si} is primarily a function of L_{si}. This finding is in line with the suggestion of Tashlitskii[7] and the evidence of MacManus and Pearce[8], that the response of the contact length is attenuated as the vibration frequency increases. In the extreme case C_i may be considered constant.

The correlation observed between the instantaneous values of μ and L_{si} suggests that a simplified model of equation (8) may suffice to describe the interrelationship between shear and friction processes for any one given dynamic cutting situation. The following simplifying assumptions may be made:

• The ratio m/p remains constant. As discussed previously, it is likely that this will be approximately correct.

• The contact length C_i remains constant as the cutting geometry changes. This assumption appears realistic at high frequencies.

• The dynamic fluctuation in the difference $(\phi_i - \alpha_i)$ is small and may be neglected. This will not be strictly correct. $(\phi_i - \alpha_i)$ will vary as a result of changes in the slope of the incoming work surface, but it will not be affected by changes in the slope of the cutting velocity vector since changes in ϕ_i and α_i will cancel. In general the fluctuation in $(\phi_i - \alpha_i)$ is very small.

With these assumptions, equation (8) can be written

$$\mu = \frac{G}{L_{si}} \qquad (13)$$

where G is a proportionality factor, the value of which might be approximately taken as the product of μ and L_{si} obtained in a static cutting test at the same rake-angle, uncut chip thickness and a speed close to that of the dynamic cut.

The model represented by equation (13) may be considered as one extreme for dynamic frictional behaviour: the case of very-high-frequency fluctuations in the shear zone size. Equation (12) which describes the static cutting process may be taken as the other extreme.

The observed dependence of shear angle on the approaching workpiece surface slope may be explained by the limited contact length response time. A slope in the incoming work surface will give a change in the uncut chip thickness and hence the rake-face normal force. It will also tend to change the contact length, but if there is a lag in contact length response, the rake-face friction angle may change significantly and so alter the shear angle. For example, when the incoming surface has an upward slope so that the uncut chip thickness is increasing, the normal force on the contact zone will increase. If the length of contact remains about constant, the rake-face normal stress will increase, giving a decrease in friction angle and a corresponding increase in shear angle. With a downward sloped incoming surface this argument indicates a shear angle decrease.

CONCLUSIONS

(1) Large cyclic variations in rake-face friction have been observed for dynamic cutting.

(2) There is a limitation in dynamic response of the rake-face contact length to fluctuations in the shear zone geometry. This will affect the stress distribution on the rake-face and can explain observed differences in the frictional processes for dynamic and static cutting.

(3) In dynamic cutting the coefficient of friction is more dependent on the rake-face normal load than is the case in static cutting. This indicates a more direct relationship between the friction and primary shear processes in dynamic cutting.

(4) A simplified model, in which the coefficient of friction is inversely proportional to the instantaneous shear plane length, gives an adequate description of the observed interrelationship between the friction and shear processes.

(5) The linear relationship between ϕ and $(\beta - \alpha)$ generally accepted for static cutting, is quite unacceptable for dynamic cutting.

(6) The observed dependence of shear angle on the slope of incoming work material can be qualitatively explained in terms of the limitation in dynamic response of rake-face contact length.

REFERENCES

1. V. A. STEWART and R. H. BROWN, 1969. A study of the metal cutting process under dynamic conditions. *Mech. and Chem. Eng. Trans., Inst. Engrs. Australia, MC5,* 51.

2. N. N. ZOREV, 1963. Interrelationship between the shear processes occurring along the tool face and on the shear plane in metal cutting. *International Research in Production Engineering.* Proc. of the Int. Prod. Conf., Pittsburgh, 42.

3. R. G. FENTON and P. L. B. OXLEY, 1968/9. Mechanics of orthogonal machining, allowing for the effects of strain rate and temperature on tool–chip friction. *Proc. Inst. Mech. Engrs.,* **183,** 417.

4. E. J. A. ARMAREGO and R. H. BROWN, 1969. The machining of metals, Prentice-Hall, New Jersey.

5. M. K. DAS and S. A. TOBIAS, 1967. The relationship between static and dynamic cutting of metals, *Int. J. Mech. Tool Des. Res.,* 7, 63.

6. S. DOI and S. KATO, 1956. Chatter vibrations in lathe tools, *Trans. Amer. Soc. Mech. Engrs.,* 78, 1127.

7. N. I. TASHLITSKII, 1960. Primary source of the energy inducing self-oscillations when metal cutting. *Vestnik-Mashinostroeniya,* **2,** 36. (Translation by PERA.)

8. B. R. MACMANUS and D. F. PEARCE, 1970. The dynamic response of the effective shear angle in wave removal. *Proc. 11th Int. Mach. Tool Des. Res. Conf.,* 119.

9. M. E. MERCHANT, 1944. Basic mechanics of the metal cutting process. *J. of Appl. Mech.,* **15,** A-168.

10. H. L. D. PUGH, 1958. Mechanics of the cutting process. *Proc. Conf. on Technology and Engineering Manufacture.* Inst. of Mech. Engrs., London, 237.

11. J. A. BAILEY and G. BOOTHROYD, 1968. Critical review of some previous work on the mechanics of the metal cutting process. *Trans. Amer. Soc. Mech. Engrs., J. of Eng. for Ind.,* **90** (Series B), 54.

12. R. KOMANDURI, V. A. STEWART and R. H. BROWN, 1971. Dynamic calibration of dynamometers. *Annals C.I.R.P.,* **19,** 231.

13. V. A. STEWART, 1971. The mechanics of continuous chip formation under regenerative chatter conditions. Ph.D. Thesis, Monash University.

14. M. C. SHAW and W. HOLKEN, 1957. Uber selbsterregte Schwingungen bei der spanenden Bearbeitung. *Industrie-Anzieger,* **79,** 185.

THE RELIABILITY OF THE C.I.R.P. CALIBRATING RIG FOR TESTING MACHINE TOOLS

by

M. I. CHOWDHURY,* M. M. SADEK* and S. A. TOBIAS*

SUMMARY

Techniques proposed for assessing the dynamic performance of machine tools are based on either the harmonic excitation or the actual cutting. The C.I.R.P. calibrating test rig technique falls into the latter category. This has been proposed with the aim of calibrating the material susceptibility to chatter.

This paper deals with the chatter behaviour of this rig, from both the experimental and the theoretical aspects. The reliability and the limitations of the rig as a calibrator are discussed.

INTRODUCTION

For the dynamic evaluation of machine tools, three distinct techniques have been proposed.

(1) Coefficient of merit technique[1,2].
(2) Cutting test technique[3,4].
(3) C.I.R.P. calibrating rig technique[5].

The coefficient of merit technique utilises vibration tests[6] for determining the so-called operative receptance locus of the machine tool structure in question and then uses a geometric feature of this for measuring its resistance to chatter in terms of a coefficient of merit (CoM). The operative receptance locus is obtained by exciting the machine in the direction of the mean cutting force P and measuring the response in the direction of the mean normal to the machined surface X, these two directions being illustrated in figure 1(a), which refers to milling.

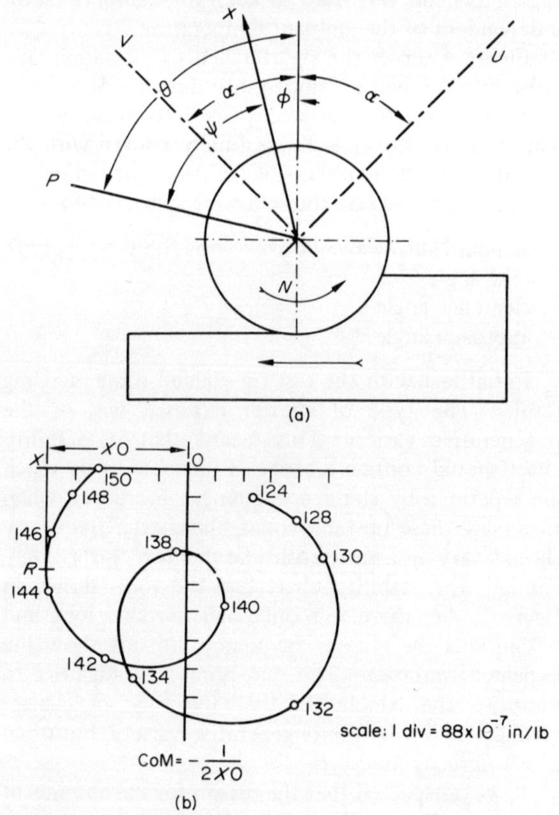

The operative receptance locus can be found by computation from a set of direct and cross receptance locii measured in two orthogonal directions, U and V, shown in figure 1(a). Once the operative receptance locus is given, the CoM for the particular cutting conditions, characterised by the orientation of the mean cutting thrust P and the normal to the machined surface X, is obtained as the inverse of the maximum negative in-phase component, as shown in figure 1(b).

The CoM has been proposed as a measure for the dynamic quality of machine tools on the ground that it is directly proportional to the maximum width of cut (or drill diameter) which, for the particular cutting operation, is stable at all speeds. In the stability chart of the process it is represented by the horizontal asymptote of the envelope of the unstable region, denoted by h_m in figure 2. That chart was

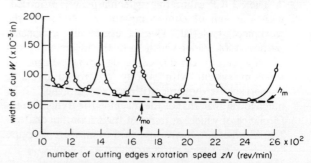

Figure 2. Experimental stability chart.

obtained experimentally but it can also be predicted from the operative receptance locus, a graph of this type being shown in figure 3. Thus, by performing

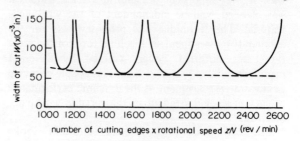

Figure 3. Theoretical stability chart, $C = 0.001$.

Figure 1. (a) Schematic representation of a milling operation.
(b) Operative receptance locus.

*Department of Mechanical Engineering, University of Birmingham.

vibration tests, it is not only possible to determine a measure of the dynamic quality of the machine in relation to some other design, but if necessary the complete stability chart of the process can be obtained. This can be done, in principle, for any material, tool design or cutting condition, provided only that the appropriate dynamic cutting coefficients are known. With recent developments in computer-aided design, it is even possible to predict the effect of design changes on the CoM or the influence of a change of the foundation, of the physical dimensions of the workpiece, etc.

The cutting test technique aims at determining directly the width of cut (or drill diameter) at which the machine starts to chatter, for standardised cutting conditions, covering feeds, speeds, tool geometry, workpiece material, etc. In order to avoid an exorbitant amount of experimental work, the range of conditions that can be covered by such a test is necessarily limited and this throws grave doubt on its usefulness. The width of cut at which the machine starts to chatter depends very much on the cutting speed, as is clear from figure 2, and hence a few isolated tests may well give completely misleading results. In addition to this, it is practically impossible to standardise the workpiece material. It is well known that the machinability of the nominally same material can vary between wide limits and it can only be expected that such scatter will also be found as far as its effect on the threshold of chatter is concerned. Finally, even with parameters which can be controlled, such as tool angles, only a limited range of conditions can be covered and from these it cannot be predicted how the system will behave for others not included in the tests.

The C.I.R.P. calibrating rig technique was proposed[7] with the aim of eliminating the variability of the workpiece material. This is essentially a special-purpose tool holder which from the mechanics point of view can be considered to be a single degree of freedom system with strictly controlled dynamic properties. Its dynamic stiffness was designed to be much lower than that of the machine the dynamic performance of which, in terms of the maximum chatter-free width of cut, it was supposed to measure. Because of this, when used on a machine as a tool support, the maximum width of cut at which chatter arose was also much smaller than when the tool was held in a normal tool post. Thus, the ratio of the two widths of cut, that is, the width of cut obtained without the calibrating rig divided by that arising with it, was supposed to establish a scale which, it was claimed, will be independent of the workpiece material. The technique was proposed also for the comparative assessment of machine tools of the same type, but different design, offered by different manufacturers. It was held that it is capable of giving a comparative assessment of the dynamic performance of such cases, independent of the variable workpiece material.

The basic idea of a 'chatter calibrator' looks attractive, promising some obvious advantages. However, the whole concept of the C.I.R.P. rig was based on a number of assumptions and unless these are satisfied the scheme cannot work in practice. The assumptions in question are as follows.

(1) For a particular machining set-up, the only factor which influences the critical width of cut is the dynamic characteristics of the machine tool structure. The influence of the cutting parameters must be assumed to be constant since otherwise their effect cannot be eliminated.

(2) The cutting process is characterised by a single material coefficient, which is independent of the cutting speed.

(3) The scatter of machinability characteristics of a standardised test material does not affect the ratio of the cutting force components.

It will be shown that these assumptions are not justified and hence the C.I.R.P. calibrating test rig is not suitable for the purpose for which it was intended, as has already been pointed out by Peters[5]. The chatter behaviour of this rig shows some peculiarities which have not been explained hitherto and these will be discussed. From this kind of behaviour some general conclusions can be drawn in relation to dynamic acceptance tests in general.

C.I.R.P. CALIBRATING TEST RIG

The rig, shown in figure 4, consists of a mass M, connected to the base plate B by means of two inclined springs S. A viscous damper is inserted between the mass and the base plate, acting in the direction of the principle mode of vibration of the rig. Silicon oil was used to keep the damping factor independent of the ambient temperature.

Figure 4 shows the rig attached to the saddle of a lathe, a tool T being clamped into the mass M and this is machining a tubular workpiece in orthogonal cut. This type of workpiece was finally chosen with the aim of avoiding the effects of the tool nose radius[8].

The tools used had the following specification:

standard throw-away carbide tips (I.S.O. standard)
rake angle 5°
clearance angle 6°
approach angle 90°.

Initial test with the test rig yielded some puzzling results. The type of chatter exposed was of the re-generative variety. This meant that its stability chart should contain a series of unstable lobes which are separated by stable speed ranges. Moreover, when traversing these unstable lobes, the chatter frequency should vary in a saw-tooth like manner. Surprisingly enough, the stability chart had the form shown in figure 5, i.e. there was only one unstable lobe and within this the chatter frequency did not show the expected variation. This led some investigators to jump to the conclusion that the type of chatter arising was not of the re-generative variety, but of an as yet unrecognised type.

It was suspected that the reason for the absence of the saw-tooth variation of the chatter frequency was that the natural frequency of the test rig being very high (142 Hz)[8] at the relatively low speeds the unstable bands overlapped. To separate them, the natural

Figure 4. C.I.R.P. calibrating test rig.

frequency of the test rig was lowered to 105 Hz by attaching an additional mass to the main mass M in figure 4. With this modified rig the stability chart shown in figure 6(a) was obtained. This shows that at the higher speeds the unstable lobes, though overlapping, are separated by stable regions and, even more

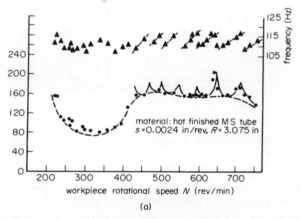

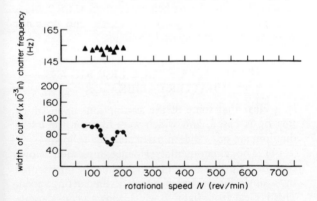

Figure 5. Experimental stability chart of the C.I.R.P. rig (without extra mass).

significantly, the saw-tooth variation of the chatter frequency. However, what was still puzzling was the shape of the envelope of the unstable region. This is generally hyperbolic, as it was originally proposed by Tobias and Fishwick[9]. However, envelopes of the type shown in figure 6(a) have been noted by Knight[10] while experimenting with an idealised machine tool system and it was shown that they can also arise on a centre lathe. Such envelopes can be explained by using a more refined dynamic cutting force model due to Das and Tobias[11] and subsequently improved upon by Knight[10]

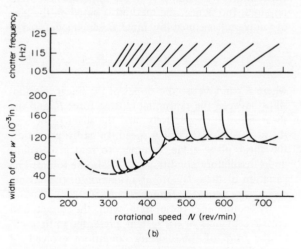

Figure 6. (a) Experimental stability chart of the C.I.R.P. rig with added mass. (b) Theoretical stability chart of the C.I.R.P. rig with added mass.

PREDICTION OF STABILITY CHART

The stability chart of the C.I.R.P. test rig, for the conditions stated, was predicted by using the method due to Knight[10]. On the structural side, this utilised the direct receptances in the directions of the normal to the machined surface and the cutting speed, as well as the cross-receptance between these two, this information being presented in figure 7. For representing the dynamic cutting process, the method due to Das and Tobias is applied. This permits the prediction of the dynamic cutting coefficient from steady state cutting data.

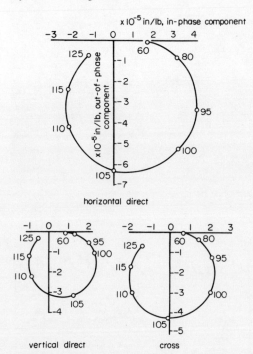

Figure 7. Harmonic receptance loci of the C.I.R.P. rig with added mass.

The method due to Das and Tobias[11], as modified by Knight[10], has been explained elsewhere and in view of this it will be sufficient to confine the present exposition to a summary of the experimental steps required. In essence, the method is based on the idea of a universal machinability index D which is defined as

$$D = \frac{K_c}{K_s}$$

where K_c and K_s are the slopes of the linear relationships between the tangential cutting force F_c and the shear force F_s, respectively, and the shear plane area A_s, at a particular cutting speed. In addition to this, the shear plane angle, assumed to remain constant under oscillatory conditions, must also be found. The variation of these factors, as a function of the cutting speed, is presented in figures 8(a) to 8(c).

By combining these data with the direct and cross-receptances of the system, presented in figure 7, in the form of the stability conditions evolved by Knight[10], the theoretical stability chart is obtained, this being shown in figure 6(b).

A comparison of this with the experimental chart

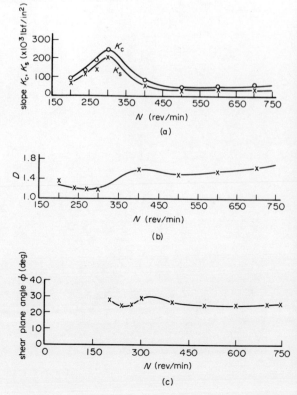

Figure 8. (a) Variation of K_c and K_s with speed. (b) Variation of ratio, D, with speed. (c) Variation of shear plane angle, ϕ_m, with speed.

of figure 6(a) shows good qualitative correspondence. From the quantitative point of view, the level of stability predicted is substantially lower than that obtained experimentally. Nevertheless, it can be concluded that the chatter behaviour of the C.I.R.P. rig is essentially of the re-generative variety and does not raise any new fundamental problems.

SUITABILITY OF THE C.I.R.P. RIG AS A CHATTER CALIBRATOR

It is clear that some of the assumptions upon which this rig is based, and which were listed in the introduction, are not valid, in particular, the following.

(1) The experimental evidence presented shows that the influence of the material cannot be represented by a single factor, independent of the cutting speed. This is clear from figure 8 which shows a considerable variation of the coefficients K_c, K_s and ϕ_m over the wide speed range investigated. These variations resulted in the irregularity in the envelope to the unstable lobes of both the experimental and the theoretical stability charts shown in figures 6(a) and 6(b).

(2) Within the framework of the C.I.R.P. co-operative work, the stability charts obtained for materials of the same specification, but from different origins, showed considerable variation in the critical width of cut from one batch of material to the other.

(3) It could be argued that the rig can be used to calibrate the material over the entire speed range. However, the variability of the material gives rise to

changes in the cutting force direction in addition to those of the cutting coefficients.

Every cutting force direction requires a different operative response locus of the rig and any machine tool structure it is proposed to test. The actual relationship is a complex trigonometric function involving the direction of the cutting force and the principal vibration directions of the machine tool structure.

Thus the rig cannot be used as a calibrator of the material because the effects of variations in the cutting force direction on the structural response cannot be separated from those due to changes in the cutting force coefficients.

CONCLUSION

Although the present paper is very largely concerned with the suitability of the C.I.R.P. test rig as a means of evaluating the chatter responses of machine tools, from the experimental results conclusions can be drawn also relating to the other two techniques.

It is clear that the C.I.R.P. calibrating rig technique did not prove to be feasible simply because chatter is a very great deal more complex phenomena than was assumed. The experimental stability charts presented in figures 3, 5 and 6(a) indicate that cutting tests performed at a limited number of speeds may lead to highly misleading conclusions. This, coupled with the variability of the machining characteristics of the material, effectively kills any proposal for the evaluation of machine tools solely on the basis of actual cutting tests.

The coefficient of merit technique, utilising the vibration characteristics of the machine tool for its dynamic evaluation, therefore appears to be the only feasible basis for a dynamic acceptance test. It therefore follows that future research effort must be directed at the development of reliable and repeatable techniques for the determination of response data under the actual machining conditions. This by itself gives the information required for evaluating machine tools. However, since production engineers will continue to enquire what these data actually mean in terms of machining capacity, it is also necessary that methods for the prediction of the dynamic cutting coefficients be improved. In this connection it ought to be added that the method by Nigm, Sadek and Tobias, which represents the latest development of the method due to Das and Tobias, is capable of predicting these coefficients to a very high degree of accuracy, as has been shown elsewhere[12].

REFERENCES

1. S. A. TOBIAS, 1962. 'Dynamic acceptance tests for machine tools', *Intern. J. Machine Tool Design Res.*, **2**, 267.

2. M. M. SADEK and S. A. TOBIAS, 1971. 'Comparative dynamic acceptance tests for machine tools applied to horizontal milling machines', *Proc. Inst. Mech. Engrs.*, **189**, 319.

3. M. BURDEKIN, A. COWLEY and J. TLUSTY, 1971. 'Establishing standard cutting conditions for performance testing of universal metal cutting machine tools', *12th Intern. M.T.D.R. Conf.*, Macmillan, 289–98.

4. B. J. STONE, 1971. 'The development of a dynamic performance test in lathes', *12th Intern. M.T.D.R. Conf.*, Macmillan, 299–308

5. J. PETERS, P. VANHERCK and H. V. BRUSSEL, 1971. 'The measurement of the dynamic cutting coefficients', *C.I.R.P. Gen. Assem.*, Warsaw.

6. I. CHOWDHURY, M. M. SADEK and S. A. TOBIAS, 1969–70. 'Determination of the dynamic characteristics of machine tool structures', *Proc. Inst. Mech. Engrs.*, **70**, Vol. 184, pt. 1.

7. J. PETERS and P. VANHERCK, 1968. 'Machine tool stability and the incremental stiffness', *Rept. C.I.R.P. Ma Group*, Leuven.

8. M. I. CHOWDHURY, M. M. SADEK and S. A. TOBIAS, 1969. 'Provisional report on C.I.R.P. Co-operative work on susceptibility of materials to chatter', *C.I.R.P. Gen. Meet.*, Geneva.

9. S. A. TOBIAS and W. FISHWICK, 1958. 'Theory of regenerative machine tool chatter', *The Engineer*, **205**, 199.

10. W. A. KNIGHT, 1968. 'Application of universal machinability chart to the prediction of machine tool stability', *Intern. Machine Tool Design Res.*, **8**.

11. M. K. DAS and S. A. TOBIAS, 1965. 'The basis of a universal machinability index', *Proc. 5th Intern. M.T.D.R. Conf.*, Birmingham 1964, Pergamon Press, Oxford, p. 183.

12. M. M. NIGM, M. M. SADEK and S. A. TOBIAS, 1972. 'Prediction of dynamic cutting coefficients from steady-state cutting data', p. 3 in the present volume.

APPLICATION OF APERIODIC TEST SIGNALS TO THE MEASUREMENT OF THE DYNAMIC COMPLIANCE OF MACHINE TOOLS

by

H. OPITZ* and M. WECK*

SUMMARY

Apart from the well-known test procedures using harmonic and random exciting force signals a further possibility is given by applying normalised aperiodic test signals. The importance of this new test procedure can be seen in the fact that a relatively small expenditure of the test equipment is necessary. The theoretical background is briefly described. Based on a theoretical estimation of the functional and statistical errors a generator of aperiodic test signals is developed which fulfils the given requirements. Finally a representative result of the application of this new test method is presented and compared with the measurement using harmonic test signals.

NOTATION

D	=	damping factor
f	=	frequency (c/s or Hz)
f_0	=	natural frequency (c/s or Hz)
F	=	force (kp)
$F(if)$	=	frequency response (μm/kp)
F_1	=	pulse force (kp)
F_2	=	step force (kp)
G	=	weighting function (μm/kp)
GU	=	response of pulse step function (μm/kp)
i	=	$\sqrt{-1}$
Im	=	imaginary part of
k	=	current integer number
m	=	scale factor (μm/mm rec. paper or μm/voltage tape rec.)
P	=	confidence limit (%)
Re	=	real part of
t	=	time (s)
Δt	=	equidistant time distance (s)
Δt_0	=	digitising time distance (s)
T	=	time constant (s)
T_0	=	observation time (s)
T^*	=	duration of the pulse (s)
U	=	step response (μm/kp)
V	=	number of digitised points per period
σ	=	standard deviation

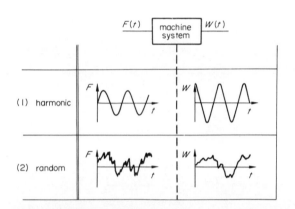

Figure 1. Hitherto used force signals for the analysis of the dynamic compliance behaviour.

1.   INTRODUCTION

The hitherto existing experimental investigation into the dynamic compliance behaviour of machine tools is carried out by two typical shapes of force signals. Figure 1 illustrates the harmonic and random force excitation.

As is well known for the measurement of the frequency response by means of harmonic signals the frequency of the exciting force is changed continuously or stepwise. The ratio of displacement and force amplitudes, as well as the phase shift, are measured in the steady-state condition.

The other method of the frequency response determination is performed by random or stochastic exciting forces[1,2]. Hereby the random force signals as well as the resulting random displacement signals are simultaneously recorded. The analysis of the frequency response is carried out by means of a spectral density or Fourier analyser. The advantage of this method consists in the possibility that the dynamic behaviour can be measured under machining conditions. That means the machine elements − for example, slides and tables − can move during the measurement, because the recording time is very short. Only a few seconds are necessary for getting complete information.

The disadvantage of both mentioned procedures is the expensive test equipment. Therefore a further test procedure was developed. The accuracy of this new method is comparable with the method using harmonic test signals. Against that the expenditure of the necessary instruments is relatively small. The new test procedure depends on the use of aperiodic exciting force signals.

2.   TEST PROCEDURE BY APERIODIC TEST SIGNALS

For a long time use of aperiodic test signals has been known in the aircraft industry. Bollinger and Bonesko first applied pulse testing in machine tool dynamic analysis[3].

Figure 2 describes the test equipment and the data flow of the test procedure. A force signal in the form of a pulse or step function is generated by means of a

* Laboratorium für Werkzeugmaschinen und Betriebslehre der Technischen Hochschule Aachen.

special force generator. The run of the force signal as well as that of the machine displacement due to the force signal are measured and digitised. The data of the force and of the displacement signals are Fourier transformed and the desired compliance frequency response of the machine tool is computed.

Without doubt aperiodic *standard* test signals with infinite edge steepness are ideal concerning the expenditure of evaluation. Figure 3 shows three

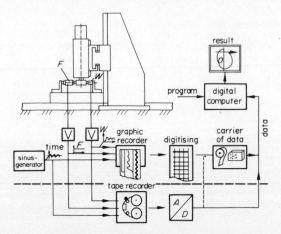

Figure 2. Flow diagram of signals and data.

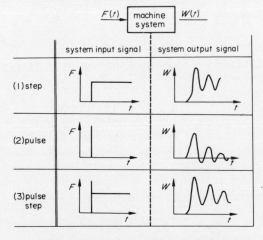

Figure 3. Aperiodic standard test signals.

typical shapes of exciting signals. The advantage of these signals depends on the fact that the standardised aperiodic force signals have not to be registered and digitised. The illustrated exciting signals are the step function, the Dirac delta function and a combination of the two mentioned signals, the pulse step function.

Naturally the ideal shape of the curves cannot be realised because an infinite power would be necessary. Therefore theoretical consideration will explain the dependency of the errors on the difference between the ideal and real aperiodic standard test signals. Furthermore, the accuracy of the digital Fourier transformation must be identified.

2.1. Theoretical relations
The determination of the frequency response $F(if)$ from the step response is carried out from the following relationship[4]:

$$F(if) = \int_0^\infty G(t)\, e^{-i2\pi ft}\, dt \qquad (1)$$

$$F(if) = \int_0^\infty \frac{d}{dt}\Big\{U(t)\Big\}\, e^{-i2\pi ft}\, dt$$

$$F(if) = \int_0^\infty d\Big\{U(t)\Big\}\, e^{-i2\pi ft} \qquad (2)$$

where $G(t)$ = weighting function
$U(t)$ = step response.

The analytical form of Equation (2) is substituted by a difference equation.

$$F(if) = \sum_{k=0}^\infty \Delta U(k\Delta t)\, e^{-i2\pi f(k\Delta t)} \qquad (3)$$

with $\Delta U(k\Delta t) = U(k\Delta t) - U((k-1)\Delta t)$

$$= \sum_{k=0}^\infty \Delta U(k\Delta t)\cos(2\pi f(k\Delta t))$$

$$- i\sum_{k=0}^\infty \Delta U(k\Delta t)\sin(2\pi f(k\Delta t)) \qquad (4)$$

$$F(if) = \mathrm{Re}\Big\{F(if)\Big\} + i\,\mathrm{Im}\Big\{F(if)\Big\} \qquad (5)$$

The integration is compensated by a summation. using a pulse step function equation (2), (3) or (4) is Equation (2), (3) or (4) must be multiplied by $1/(i2\pi ft)$ or $1/(i2\pi f(\Delta tk))$ respectively and finally using a pulse step function equation (2), (3) or (4) is multiplied by a correction frequency response[4]:

$$\frac{1}{1 + i\,\dfrac{F_1 T^*}{F_2}\, 2\pi f} \qquad (6)$$

where $F_1 T^*$ means the pulse integral and F_2 the step force.

Assuming that the exciting signal can be observed as an ideal aperiodic standard input signal, in all cases only the response signal has to be digitised and Fourier transformed.

In practice the real shape of the aperiodic signals differs from the ideal shape. The influence of this real shape, the finite observation time and the digitising distance of the system response on the accuracy of the results will be regarded. Furthermore the statistical errors are of interest.

2.2 Influence of the real shape of aperiodic standard test signals on the accuracy
The real shape of the step function is approximated by a step response of a first-order system. The behaviour of this real function is sufficiently described by the time constant T as illustrated in figure 4. With the limiting value $\lim T \to 0$ s the shape of the real step function is identical with the ideal.

Figure 5 shows the dependency of the relative error of the frequency response on the product of analysis frequency and the time constant of the step function. For example the relative error of the frequency response must not be greater than 10%; the product of analysis frequency and time constant has to be less than 0.015. This means that the time constant of the

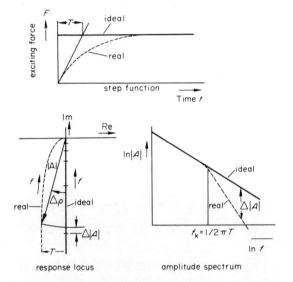

Figure 4. Shape of ideal and real step function and associated response locus and amplitude spectrum.

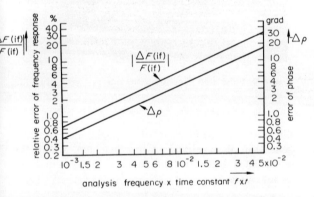

Figure 5. Relative error of the frequency response and phase as function of the analysis frequency and the time constant of the step response.

step function must be less than 0.1 ms, if the analysis frequencies of greater than 150 Hz are of interest. This time constant requires a special constructional lay-out of the force generator. The maximum necessary time constant depends on the allowable error and on the maximum interesting frequency of the analysis.

2.3 Influence of the finite observation time and the digitising distance

Theoretically the building-up time of the system response is infinite. Really only a finite duration of the response per period V of the computed frequency is which has to be estimated. Figure 6 illustrates the behaviour of error. The error relates to a standardised observation time which results from the product of the observation time T_0, the natural frequency of the system f_0 and the damping factor D. Simultaneously the dependency of the number of digitised points of the response per period V of the computed frequency is taken into consideration. As we see by figure 6 the

error of the computed frequency response decreases with increasing number of digitised points per period of the computed frequency and approaches a limit value asymptotically. This limit value depends on the standardised observation time. This means that the standardised observation time has to be greater than 0.8 assuming that the number of digitised points per period of the highest interesting natural frequency is about 10. Under these conditions the relative error will be less than 10%. The exact values can be read from the diagram (figure 6).

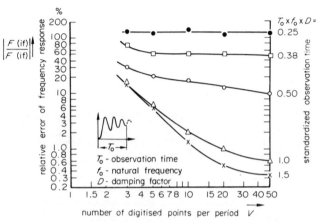

Figure 6. Maximum relative error of frequency response as a function of the observation time and the number of digitised points per period (frequency range: $0 < f_0 < 1.5 f_0$).

In other words the necessary observation time is determined by the value of rest amplitude of the dying-out response oscillation. In order that the error is less than 10% the rest amplitude of the step response in relation to the static displacement has to be less than 3%, assuming that the number of digitised points per period of the highest interesting natural frequency is about 10.

2.4 Statistical errors

The errors of the incorrect digitising, or the noise in the displacement signals, which is not based on the exciting force, can be sufficiently described by the standard deviation 'σ'. The absolute error decreases with increasing exiting force 'F' and decreasing scale factor 'm' of the recorder (m in μm/mm$_{rec. paper}$ or in μm/voltage$_{tape rec.}$). Therefore the statistical error of the frequency response due to incorrect digitising can be standardised by the following relationship:

$$\frac{|\Delta F(if)|}{\sigma m / F_2}$$

Figure 7 shows this standardised statistical error of the frequency response as a function of the product of the analysis frequency 'f' and digitising time distance 'Δt_0'. The upper and lower borders of the statistical errors are drawn into the diagram where the two aperiodic exciting signals, the step and the pulse step function are taken into consideration. As can be seen from figure 7 the errors increase with increasing frequency of analysis or increasing digitising time distance respectively. Regarding the shape of the curves for the step function, the statistical error decreases if the product $f\Delta t_0$ is greater than 0.35. This behaviour is based on the frequency spectrum of the noise signal caused by the incorrect digitising. This noise frequency spectrum decreases above $f\Delta t = 0.35$.

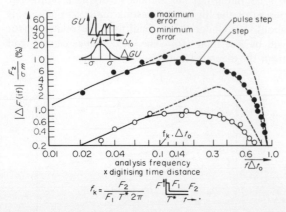

$$f_k = \frac{F_2}{F_1 \; T^* \; 2\pi} \qquad F \uparrow \boxed{F_1 \atop T^* \; t \rightarrow} F_2$$

Figure 7. Range of statistical errors of the frequency response using step or pulse step force signals ($P = 95\%$).

The advantage of the pulse step function in comparison with the pure step function consists of the difference of the frequency spectrum. As is well known the frequency spectrum of the step function is inversely proportional to the frequency. On the other hand, the frequency spectrum of a pulse with a large edge steepness is constant over a wide frequency range. The combination of both signals lifts the hyperbolic spectrum of the step function. Thus the statistical error of the frequency response decreases with the increasing part of the spectrum due to the pulse. Figure 7 shows this relationship[4]. The cut-off frequency 'f_k' is determined by the quotient of the step force F_2 over the pulse integral '$F_1 T^*$'.

2.5 Generator of standardised aperiodic test signals

As previously derived in Section 2.2 the force generator has to realise a time constant of the step function or the pulse step function of less than 0.1 ms. This is the most important supposition of the generation of standardised aperiodic test signals. Further characteristics of the force generator have to be low stiffness, definable and variable static and dynamic force adjustment and a small built-in dimension.

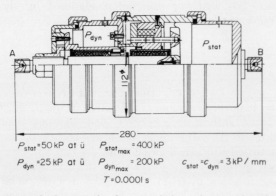

$P_{stat} = 50 \text{ kP at ü} \qquad P_{stat_{max}} = 400 \text{ kP}$

$P_{dyn} = 25 \text{ kP at ü} \qquad P_{dyn_{max}} = 200 \text{ kP} \qquad c_{stat} = c_{dyn} = 3 \text{ kP / mm}$

$T = 0.0001 \text{ s}$

Figure 8. Constructive lay-out of the pulse step function force generator.

These requirements are sufficiently fulfilled by a physical procedure basing on a pneumatic prestressed spring[4]. Figure 8 shows the constructive lay-out of the pulse step function force generator. There are two separate chambers of compressed air. The right chamber produces the static force and the left the dynamic force. The movable piston of the dynamic force is locked by an electromagnet. After interruption of the induction current the piston snaps against the collar of the nut. During the acceleration the piston gets kinetic energy and the rebound produces a definable force pulse which is superimposed by a definable step function.

As was previously (Section 2.4) described, the pulse improves the behaviour of the statistical errors. For the illustrated generator an important improvement exists for frequencies greater than 60 Hz. Figure 9

Figure 9. Pulse step function generator at a vertical milling machine.

illustrates the application of the pulse step function generator at a vertical milling machine, where the compliance behaviour of the spindle system is to be measured.

3. Application to the dynamic investigation of machine tools

Now a representative example of a machine tool investigation is presented. Here different test procedures are used and compared.

Figure 10 demonstrates the results of a measurement of a vertical milling machine which was tested by harmonic signals, by pulse step function signals and by pure pulse signals (hammer). As can be seen from this figure, the amplitude characteristic as well as the phase characteristic agrees sufficiently. The difference of the phase behaviour within the lower range of frequencies depends on the different dynamic behaviour of the pick-ups of the displacement that are used.

The agreement is especially satisfactory within a frequency band near the resonance frequencies. The deviations of the amplitudes are absolutely small and are negligible in the frequency band of small amplitudes, although the differences obviously appear great on the logarithmic scale.

A presupposition of the application of aperiodic test signals is the linear behaviour of the system to be

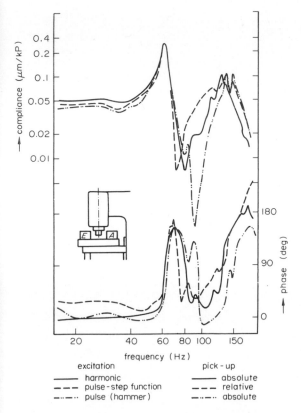

excitation
——— harmonic
— — — pulse - step function
—·—··— pulse (hammer)

pick - up
——— absolute
— — — relative
—··—··— absolute

Figure 10. Comparison of the results of the different test procedures.

analysed. This condition is fulfilled in most cases. However, it should be controlled by a variation of the static preload if the linearity is sufficient.

CONCLUSIONS

By the side of the hitherto known test procedures using harmonic and stochastic exciting signals the standardised aperiodic test signals represent a further possibility of the dynamic machine tool investigations. As a lot of tests have shown, this method is equivalent to the tests with harmonic signals. Three kinds of signal shapes are presented: the pure step function, the pulse function and the combination of both, the pulse step function. The standardisation of the shapes was realised by a sufficient edge steepness. Thus the amplitude spectrum allows the analysis up to relative high frequencies and the reproducibility of the exciting signals enables us to renounce the record and the digitising of the exciting signals.

An advantage of this method is the relatively small expenditure of the test equipment. The prime cost of a transfer function analyser with the associated exciter and amplifiers is more significant than the prime cost of the instruments for the analysis with aperiodic test signals. The expense of the use of the digital computer is negligible. The computation of

100 points of the response locus from the digitised system response is carried out in four seconds only (Control Data Computer 6400).

Figure 11 shows the advantages and disadvantages of the single test procedures. Only an analysis with stochastic test signals allows the determination of a

Criterion		Shape of the exciting signals		
		Harmonic	Stochastic	Aperiodic
		standstill	under machining conditions	standstill
results	(1)	frequency response ↓ analysis of stability in the frequency domain	frequency response under machining conditions ↓ analysis of stability in the frequency domain	step response ↓ frequency response ↓ analysis of stability in the time and frequency domain
	(2)	mode shape	—	—
	(3)	discovery of dynamic and static weak points	—	discovery of static weak points
	(4)	—	dermination of machining parameters on the dynamic behavior	—
time of standstill due to investigation		great	small	small
expenditure of test equipment		great	great	small

Figure 11. Comparison of the advantages and the disadvantages of the different test procedures.

frequency response economically under machining conditions. This means the slides and tables can move during the test, because only a very short recording time of the test signals is necessary.

The registration of the mode shape can only be carried out by means of harmonic test signals. Thus the dynamic and static weak points can be found out by sinusoidal excitation. The necessary expenditure of the test equipment for the different procedures was previously mentioned. As it can be seen from figure 11 each test procedure has advantages and disadvantages regarding the different criteria. The developed test method with aperiodic test signals completes the hitherto existing test procedures in a suitable manner.

REFERENCES

1. M. WECK, 1969. 'Analyse linearer Systeme mit Hilfe der Spektraldichtemessung und ihre Anwendung bei dynamischen Werkzeugmaschinenuntersuchungen unter Arbeitsbedingungen', *Thesis,* TH Aachen.
2. H. OPITZ and M. WECK, 1969. 'Determination of the transfer function by means of spectral density measurement and its application to the dynamic investigations of machine tools under machining conditions', *Proc. 10th M.T.D.R. Conf.,* Manchester, Pergamon Press, Oxford.
3. I. G. BOLLINGER and I. A. BONESKO, 1965. 'Pulse testing in machine tool dynamic analysis', *Intern. J. Machine Tool Design Res.,* 5, 167–81.
4. H. OPITZ and M. WECK, 1971. 'Anwendung aperiodischer Testsignale zur Bestimmung des dynamischen Nachgiebigkeitsverhaltens von Werkzeugmaschinen. Forschungsbericht des Landes Nordrhein-Westfalen, No. 2209, Westdeutscher Verlag Opladen, Germany.
5. J. O. HOUGEN and R. A. WALSH, 1961. 'Pulse testing method', *Chemical Engng. Progr.,* **57,** No. 3, March.

A RESPONSE PREDICTION AND OPTIMISATION OF A FRICTIONALLY DAMPED STRUCTURE

by

S. W. E. EARLES* and N. MOTT*

SUMMARY

Theoretical analysis assuming coulombic friction is, even for simple structures, lengthy in computation time. This has led to the development of an analysis using an equivalent sinusoidal force. The normal force is determined that, when applied across a frictional damper attached to a beam, minimises the displacement response at a given point over a limited excitation frequency range. The experimental and theoretical response patterns obtained are similar, and such differences that do exist can be explained by comparing the assumed and measured characteristics of the damper interface surfaces. The normal force, frictional force and coefficient of friction varied by only 5% over the first 10^7 cycles and tended to constant values. The coefficient of friction was found to increase with the amplitude of slip of the damper interface. It is shown theoretically that the optimum response using a frictional damper is comparable with that using a viscous damper. The essential requirement of the frictional damper system is that it shall traverse the free resonant peaks in a 'locked' state and the clamped resonant peaks in a 'slipping' state.

NOTATION

C	viscous damping coefficient
F	frictional force amplitude
$F(t)$	frictional force at time t
I	invariant point
M	magnification ratio (dynamic/static response)
N	normal force on damper
P	applied force amplitude
Q	F/P
R	frequency ratio (excitation frequency/first 'free' resonant frequency)
X	displacement response amplitude
f	frequency
p	pulsatance ($= 2\pi$ x excitation frequency)
$\hat{s}$	slip amplitude
t	time
$x(t)$	displacement at time t
α_{ij}	receptance
β, θ, ϕ	phase angles
μ_s	static coefficient of friction
μ_d	dynamic coefficient of friction

Subscripts

F	free state
I	invariant point state
L	locked state
s	slipping state

INTRODUCTION

A difficulty in the theoretical prediction of the response of a vibrating structure lies in obtaining a mathematical expression for the damping forces that is both reasonable from a physical viewpoint and amenable to analysis. The essential criterion of the practical usefulness of any expression adopted is whether it enables a prediction of the actual pattern of behaviour to be made within acceptable limits.

Assuming coulombic friction and that the frictional force at the point of slipping was equal to that occurring during slipping, Den Hartog[1] has analysed a single-degree of freedom frictionally damped system by (i) an 'exact' method, obtained by solving the relevant differential equations, and (ii) an 'approximate' method, obtained by linearising the frictional force.

For large slip amplitudes he found that his 'approximate' method gave responses which compared favourably with those produced using his 'exact' method, but that the similarity in the results decreased with decreasing slip amplitude.

The exact method has been extended by Yeh[2] to a two-degree of freedom system; his conclusions were the same as those of Den Hartog. For the two-degree of freedom system Sawaragi et al.[3] have obtained some fair experimental correlation.

Within the forcing frequency range of a practical continuous structure it is likely that several of the structure's vibrational modes will be excited. Thus the useful application of the exact method would seem unlikely, for even if formulated it would be most tedious to apply to a multi-degree of freedom system.

The work to be presented is concerned with a continuous system subjected at one point to a sinusoidal force excitation having a range of magnitudes and frequencies. An attempt is made to justify a linearisation of the frictional force; this enables a theoretical optimisation of the damper force and its support flexibility to be made.

THEORETICAL ANALYSIS

Damper linearisation

When a damped system is excited by a sinusoidally varying force then, provided the damping forces are linear functions, their magnitudes also vary sinusoidally

* Department of Mechanical Engineering, Queen Mary College, London.

and with a frequency equal to that of the excitation frequency. Under certain experimental conditions a frictional damped structure has been observed to have a periodic response of frequency equal to that of the excitation frequency[4]. This, at least, indicates the possibility of representing the damping force in terms of a Fourier series. Thus, if the exciting force is $P \sin pt$ the damping force $F(t)$ may be represented by the series

$$F(t) = \sum_{n=1}^{\infty} F_n \sin n(pt + \phi)$$

Coulombic friction is characterised by the frictional force being

(a)   dependent on the materials in contact,
(b)   proportional to the normal force across the interface,
(c)   substantially independent of the sliding speed and the apparent area of contact.

It is further observed that the magnitude of the frictional force just prior to relative motion occurring is greater than that during uniform relative motion.

Consider the system shown in figure 1 in which a subsystem (A) is connected to a subsystem (B) via a frictional damper D. Individually subsystems (A) and (B) both have linear characteristics. Let subsystem (A) be excited by a force $P \sin pt$ at point 1.

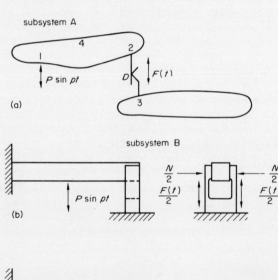

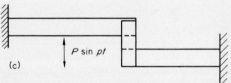

Figure 1. System assemblies (a) general; (b) damper rigidly supported; (c) damper flexibly supported.

Now if the damper has an infinitely *large* normal force across its interface (in such a state the system will be locked), the response characteristic of any point on the system will contain a series of resonant peaks. Likewise if the damper has an infinitely *small* normal force across its interface (in such a state the

system will be free), the response characteristic of any point on subsystem (A) will contain a series of resonant peaks. In the latter case subsystem (B) will be effectively stationary.

The practical problem therefore is that, given the physical details of subsystem (A), to arrange the damping force and the physical details of subsystem (B) so that over a range of excitation frequencies the system response will be minimised. This means in effect attempting to arrange for the system to pass through the free resonant conditions in a locked state and through the locked resonant conditions in an apparent free state, that is, with slip occurring.

The analytical problem is how to make an acceptable prediction of the physical requirements to produce such a response state. The limiting conditions of slip are

(i)   the frictional force on the damper is a maximum but no slip occurs
(ii)   the frictional force on the damper is very small and slip is occurring for most of each cycle.

It is well known that the frictional force necessary to initiate motion is greater than that required to maintain uniform motion. Thus the variation of the frictional force on the damper for (i) will be represented by a sinewave having an amplitude of $\mu_s N$, where μ_s is the static coefficient of friction, and the variation for (ii) represented in the limit by a rectangular wave of amplitude $\mu_d N$, where μ_d is the dynamic coefficient of friction. Thus for

(i) $F(t) = \mu_s N \sin(pt + \phi)$ (1)

and for (ii)

$$F(t) = \sum_{n=1}^{\infty} A_n \sin n(pt + \phi) \qquad (2)$$

where ϕ is a phase angle.

When the waveform in (ii) is rectangular

$$F(t) = \frac{4\mu_d N}{\pi} \sum_{n=1}^{\infty} \frac{1}{n} \sin n(pt + \phi)$$

$$= \frac{4\mu_d N}{\pi} [\sin(pt + \phi) + \tfrac{1}{3}\sin 3(pt + \phi) + \ldots] \quad (3)$$

Now it would very much simplify the subsequent analysis if it could be shown reasonable to neglect all but the first term in the series expression for $F(t)$. To do so for a rectangular waveform would result in an error of about 35%. However, it is to be expected that the frictional forces of interest will lie approximately midway between the frictional forces of the limiting cases (i) and (ii). Hence there is an error of much less than 35% if only the first term in the series is considered. Some supporting evidence for an expected near-sinusoidal frictional force waveform is found in reference 5.

Let it be assumed that the variation in the frictional force at the damper during slipping is given by

$$F(t) = \frac{4\mu_d N}{\pi} \sin(pt + \phi) \qquad (4)$$

It is to be expected that the displacement of any point on the beam will be a continuous function of the forcing frequency, which requires that the forces producing the displacement are continuous functions of the forcing frequency. This implies that the magnitude of the frictional force just before slip occurs is the same as that when slip is occurring; that is, $\mu_s = 4\mu_d/\pi$ if equation (4) is assumed to apply. Such a ratio of static to dynamic coefficients of friction is similar to that generally found in practice.

Application to the system

Assuming that coulombic friction obtains, that the frictional force always opposes relative motion, and that equation (4) applies, the structure shown in figure 1(a) is dynamically analysed.

The displacement response at any point (4) on subsystem (A) is given by

$$x_4(t) = \alpha_{14} P \sin pt + \alpha_{24} F \sin(pt + \phi) \qquad (5)$$

where α_{14} is a typical cross-receptance[6], $F = \mu_s N = 4\mu_d N/\pi$ when slip occurs, and $F < \mu_s N$ when no slip occurs.

(i) For no slip at the damper

For no slip occurring $x_2(t) = x_3(t)$ and

$$x_2(t) = \alpha_{12} P \sin pt + \alpha_{22} F \sin(pt + \phi)$$

$$x_3(t) = \alpha_{33} F \sin(pt + \phi)$$

hence

$$F\sin(pt + \phi) = -\frac{\alpha_{12}}{\alpha_{22} + \alpha_{33}} P \sin pt$$

Thus the response at any point (4) on subsystem (a) is given by

$$x_4(t) = \alpha_{14} P \sin pt - \frac{\alpha_{12}\alpha_{24}}{\alpha_{22} + \alpha_{33}} P \sin pt$$

$$= \left[\alpha_{14} - \frac{\alpha_{12}\alpha_{24}}{\alpha_{22} + \alpha_{33}}\right] P \sin pt \qquad (6)$$

(ii) For slip at the damper

The magnitude of the slip is $x_2(t) - x_3(t)$ and

$$x_2(t) - x_3(t) = \alpha_{12} P \sin pt + (\alpha_{22} + \alpha_{33})\mu_s N \sin(pt + \phi)$$

$$= [\alpha_{12} P + (\alpha_{22} + \alpha_{33})\mu_s N \cos\phi]\sin pt$$

$$+ (\alpha_{22} + \alpha_{33})\mu_s N \sin\phi \cos pt$$

Using

$$\sin(pt + \beta) = \sin pt \cos\beta + \cos pt \sin\beta$$

$$\tan\beta = \frac{(\alpha_{22} + \alpha_{33})\mu_s N \sin\phi}{\alpha_{12} P + (\alpha_{22} + \alpha_{33})\mu_s N \cos\phi}$$

and

$$x_2(t) - x_3(t) = [\{\alpha_{12} P + (\alpha_{22} + \alpha_{33})\mu_s N \cos\phi\}^2$$

$$+ \{(\alpha_{22} + \alpha_{33})\mu_s N \sin\phi\}^2]^{\frac{1}{2}} \sin(pt + \beta)$$

$$= [\alpha_{12}^2 P^2 + (\alpha_{22} + \alpha_{33})^2 \mu_s^2 N^2$$

$$+ 2\alpha_{12}(\alpha_{22} + \alpha_{33})P\mu_s N$$

$$\times \sin\phi \cos\phi]^{\frac{1}{2}} \sin(pt + \beta)$$

Now let $x_2(t) - x_3(t) = \hat{s} \sin(pt + \beta)$, where $\hat{s}$ is the slip amplitude. Since, when slipping, the damper must always dissipate energy

$$\mu_s N \sin(pt + \phi) \frac{d}{dt}[x_2(t) - x_3(t)] < 0$$

That is

$$\mu_s N \sin(pt + \phi)\,\hat{s}p\cos(pt + \beta) < 0$$

and

$$\mu_s N \hat{s}p\sin(pt + \phi)\sin(pt + \beta + \frac{\pi}{2}) < 0$$

As the above must be true for all $t > 0$, the two sinusoidals are exactly out of phase, so $\beta = (\phi + \pi/2)$.

Thus

$$\tan\beta = \frac{\sin\left(\phi + \frac{\pi}{2}\right)}{\cos\left(\phi + \frac{\pi}{2}\right)} = \frac{(\alpha_{22} + \alpha_{33})\mu_s N \sin\phi}{\alpha_{12} P + (\alpha_{22} + \alpha_{33})\mu_s N \cos\phi}$$

which results in

$$-(\alpha_{22} + \alpha_{33})\mu_s N \sin^2\phi = \alpha_{12} P \cos\phi$$

$$+ (\alpha_{22} + \alpha_{33})\mu_s N \cos^2\phi$$

and

$$\cos\phi = -\frac{(\alpha_{22} + \alpha_{33})\mu_s N}{\alpha_{12}} \frac{1}{P} \qquad (7)$$

Hence the response at any point (4) on subsystem (a) is given by

$$x_4(t) = \alpha_{14} P \sin pt + \alpha_{24}\mu_s N \sin(pt + \phi) \qquad (8)$$

where ϕ is determined from equation (7)
Rewriting

$$x_4(t) = \alpha_{14} P \sin pt + \alpha_{24}\mu_s N(\sin pt \cos\phi + \cos pt \sin\phi)$$

$$= (\alpha_{14} P + \alpha_{24}\mu_s N \cos\phi)\sin pt + \alpha_{24}\mu_s N \sin\phi \cos pt$$

$$= X_{4s} \sin(pt + \phi)$$

where

$$X_{4s}^2 = \alpha_{14}^2 P^2 + \alpha_{24}^2 \mu_s^2 N^2 + 2\alpha_{14}\alpha_{24} P\mu_s N \cos\phi \qquad (9)$$

and

$$\tan\theta = \frac{\alpha_{24}\mu_s N \sin\phi}{(\alpha_{14} P + \alpha_{24}\mu_s N \cos\phi)} \qquad (10)$$

Substituting for $\cos\phi$ from equation (7) in equation (9) gives

$$X_{4s}^2 = \alpha_{14}^2 P^2 + \left[\alpha_{24}^2 - \frac{2\alpha_{14}\alpha_{24}(\alpha_{22} + \alpha_{33})}{\alpha_{12}}\right]\mu_s^2 N^2 \qquad (11)$$

Equivalent energy dissipations

The energy dissipation by the damper/cycle assuming coulombic friction will be $4\mu_d N\hat{s}$, and that by the assumed frictional force $-\int_0^\tau \mu_s N \sin(pt + \phi)(\dot{x}_2 - \dot{x}_3)dt$,

where τ is the periodic time.

Equating these expressions

$$4\mu_d N\hat{s} = -\int_0^{2\pi/p} \mu_s N\hat{s}\, P\sin(pt + \phi)(pt + \beta)dt$$

$$= \mu_s N\hat{s}\, \pi$$

and hence

$$\mu_s = 4\mu_d/\pi.$$

This shows that the theoretical damper, assuming a sinusoidally varying frictional force of amplitude $\mu_s N$, will dissipate per cycle the same amount of energy as an actual damper which is assumed to have coulombic friction characteristics.

Invariant points

Den Hartog[1] has shown that for a system with viscous damping there exist excitation frequencies at which the amplitude of the response at a specified point in the system is independent of the damping force. Such positions on a response-frequency plot are termed 'invariant points'.

When the normal force N across the damper is high and no slip occurs that is, the damper is in a locked condition, the amplitude of the response at point (4) can be found from equation (6),

$$\frac{X_{4L}}{P} = \alpha_{14} - \frac{\alpha_{12}\alpha_{24}}{\alpha_{22} + \alpha_{33}} \qquad (12)$$

and when the normal force N is substantially zero for the damper is in a free condition, the amplitude of the response at point (4) can be found from equation (8)

$$\frac{X_{4F}}{P} = \alpha_{14} \qquad (13)$$

Equations (12) and (13) are essentially frequency equations as the receptances are functions of frequency. Thus the frequencies at which the responses of these two conditions have equal amplitudes is given by

$$|\alpha_{14}| = |\alpha_{14} - \frac{\alpha_{12}\alpha_{24}}{\alpha_{22} + \alpha_{33}}| \qquad (14)$$

that is, for $\alpha_{24} = 0$

or

$$2\alpha_{14} = \frac{\alpha_{12}\alpha_{24}}{\alpha_{22} + \alpha_{33}} \qquad (15)$$

It is seen that $\alpha_{12}/(\alpha_{22} + \alpha_{33}) \neq 0$ as this would lead to $\cos\phi = \infty$ in equation (7).

The response at point (4) during slipping at the damper is given by equation (11). Substituting this in either equation (14) or (15) leads to $X_{4s} = \alpha_{14}P$.

Hence, for frequencies that satisfy either equation (14) or (15), the response at point (4) is independent of the damper, that is, invariant points exist on the response plots.

Response optimisation

It is evident that irrespective of the normal force applied across the damper the minimum displacement response possible is that which has the magnitudes of the invariant points as its maxima.

Den Hartog[1] has proposed for a two-degrees of freedom, viscously damped system that optimum damping occurs when the two invariant points are on the same level and the response curve passes with a horizontal tangent through one of them. To obtain such an optimum requires being able to vary both the damping force and stiffness of one element of the system (tuned damping).

The above definition of optimum response is in general insufficient, for while one maximum occurs at an invariant point, the other maximum (unless it occurs at the second invariant point) must be higher. Usually optimum response is that which, over a given frequency range, produces the minimum amplitude, but it is considered that the differences to be observed between these two definitions are likely to be small in practice.

The essential realisation is that the complete process of response optimisation involves two variables: (i) the stiffness of the damper support by which the levels of the invariant points may be adjusted, that is, optimum-tuning, and (ii) the damping force by which the response during slipping may be altered, that is, optimum-damping.

THEORY APPLIED TO THE EXPERIMENTAL SYSTEMS

Two systems I and II, shown diagrammatically in figures 1(b) and (c), will be discussed theoretically, although experimentally in this work only system I will be considered in detail. In both systems I and II the subsystem (A) is a beam of uniform rectangular cross-section, 2 x 2.25 in, rigidly clamped at one end with two vertical damping pads attached at the other end to the outer faces of the beam. The distance between the beam root and the centre of the damping pads is 25 in. For system I the pair of interacting damping pads are rigidly supported: in system II they are supported at the end of a cantilever the length of which could be varied between 2 and 20 in. The cross-section of the cantilever was the same as that of the main beam.

System I: rigidly supported damper

In this case $\alpha_{33} = 0$,
hence for no slip,

$$X_{4L} = \left[\alpha_{14} - \frac{\alpha_{12}\alpha_{24}}{\alpha_{22}} \right] P \qquad (16)$$

and

$$\frac{F}{P} = -\frac{\alpha_{12}}{\alpha_{22}} \qquad \text{for } F < \mu_s N \qquad (17)$$

During slipping

$$X_{4S}^2 = \alpha_{14}^2 P^2 + \left[\alpha_{24}^2 - \frac{2\alpha_{14}\alpha_{24}\alpha_{22}}{\alpha_{12}}\right]\mu_s^2 N^2 \quad (18)$$

The response at the invariant points is given by

$$X_{4I} = \alpha_{14}P \quad (19)$$

System II: flexibility mounted damper
For no slip

$$X_{4L} = \left[\alpha_{14} - \frac{\alpha_{12}\alpha_{24}}{\alpha_{22} + \alpha_{33}}\right]P \quad (20)$$

and

$$\frac{F}{P} = -\frac{\alpha_{12}}{\alpha_{22} + \alpha_{33}} \quad \text{for } F < \mu_s N \quad (21)$$

During slipping

$$X_{4S}^2 = \alpha_{14}^2 P^2 + \left[\alpha_{24}^2 - \frac{2\alpha_{14}\alpha_{24}(\alpha_{22} + \alpha_{33})}{\alpha_{12}}\right]\mu_s^2 N^2 \quad (22)$$

As before, the response at the invariant point is given by $X_{4I} = \alpha_{14}P$ but in general the invariant point frequencies, and therefore their amplitudes, will not be the same for the two systems.

EXPERIMENTAL RESULTS

Apparatus
The two arrangements of the experimental rig are shown diagrammatically in figures 1(b) and (c).

An electrodynamic exciter was supported above the main beam, subsystem (A) and connected to the beam via an impedance head. From signals fed back from the impedance head to a vibration control unit, the excitation force amplitude could be maintained constant. Alternatively acceleration, velocity or displacement at the excitation position could have been controlled.

Semiconductor strain gauges in the damper assembly, arranged in a double bridge, were used to measure the normal force and frictional force.

The normal force N and the frictional force F quoted are the total effective forces on the damping system, that is, the sum of the respective forces on the two halves.

Preliminary tests showed that the frictional properties of the damping pads varied considerably with time during the initial 30 min or so of running, but it was found that by adopting a running-in procedure, whereby the rig was operated at a high-slip amplitude for 1 h before measurements were taken, enabled reproducible results to be obtained.

The interface surfaces of the damper were ground and cleaned with acetone before each series of tests, and the surfaces assembled so that the grinding marks were at right angles.

Because it was impossible to ensure similar alignment of the interface surfaces for each test, the damping system was dismantled and reassembled three times during the course of one test and each

time a graph of frictional force against applied force was obtained. A variation of ±15% was found between the three curves produced, although their shapes were similar. The differences were attributed to variations in the effective coefficient of friction resulting from different areas of the interface surfaces being in contact.

Characteristics of the friction damper: system I
In the theoretical analysis it is assumed that after slip occurs the magnitude of the frictional force is proportional to the normal force and is independent of the slip amplitude, excitation frequency and time. Attempts were made to assess the reasonable validity of these assumptions.

A series of tests was performed with excitation frequencies of 200, 300, 400 and 500 Hz, normal forces of 10, 20 and 40 lbf, and applied force amplitudes up to 15 lbf. A small capacitance probe (0.15 in diameter and 0.15 in long), attached to the damper, on subsystem (B), and close to the underside of the beam, was used to measure the damper slip amplitude.

The results obtained are shown in figure 2. For normal forces of 10 and 20 lbf no significant variation

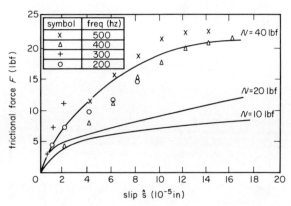

Figure 2. Variation of frictional force with slip for different normal forces: system of figure 1(b).

with frequency was observed and the mean curves only have been presented. At 40 lbf, as indicated by the actual measurements plotted, some dependence on frequency was observed although the variation from a mean line is not large.

The theoretical analysis implies that the slip should be zero for F/N ratios less than μ_s, and when slip occurs the ratio F/N should be constant and equal to μ_s. However, from figure 2 it is seen that F/N initially rises rapidly with increasing slip amplitude and then more slowly as slip increases.

These experimental observations are compatible with a relative movement of the damping surfaces with increasing applied force characteristic, obtainable by an immediate elastic deformation of the contacting asperities in the damper, followed by plastic deformation and eventual gross movement between the interfaces. The experimental results therefore suggest that the theoretical analysis is inadequate in not taking into account elastic deformations of the damper contacting asperities.

The coefficients of friction derivable from figure 2

vary with slip amplitude between 0.3 and 0.6 for normal forces of 20 and 40 lbf, and between 0.5 and 0.7 for a normal force of 10 lbf.

The time dependence of the normal force (possible reduction by wear of the contacting surfaces) and frictional force (due to changes in the normal force and/or μ_s) was examined by carrying out tests over periods of 3 and 9 h at an applied force of 15 lbf and a frequency of 500 Hz. At a nominal normal force of 10 lbf a reduction in actual normal force of 20% was observed over 3 h, although over a similar period at 20 lbf normal force the reduction was only 2%, as was also observed over 9 h at a normal force of 40 lbf. In the last two tests over the same periods a change in the frictional force of 5% and in the slip amplitude of 10% was observed.

Frequency response: system I

The findings of the preceding section suggest that the theoretical representation of the damper system should contain, between the damping surfaces and the damper support, an elastic spring, the stiffness of which depends on the normal force being applied to the damper.

The effect of this on the frequency response is indicated in figure 3. For a spring of infinite stiffness,

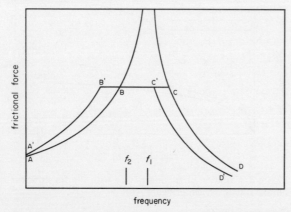

Figure 3. Effect on the response curves of damper interface stiffness.

(infinite normal force), the response is given by ABCD, where curves AB and CD are the lower parts of the response curve for the damper in a locked state, which has a resonant frequency, f_1. Whereas for a finite normal force resulting in a finite spring stiffness the response is given by $A^1B^1C^1D^1$, where curves A^1B^1 and C^1D^1 are as before, the lower parts of the corresponding locked state, which has a resonant frequency f_2.

The effect therefore that may be expected in the experimental results, as a result of elastic deformation of the interlocking interface asperities, is for the response curves to be displaced, towards lower apparent resonant frequencies, by an amount which increases as the magnitude of the normal force across the damper decreases.

Further as a result of observing that μ_s is not independent of the slip amplitude, BC and B^1C^1 in figure 3 will not be found experimentally to be horizontal.

The set of results obtained for $N = 20$ lbf is shown in figure 4. The curves approximate to the shape of figure 3 and, as expected, the frictional force during slipping is not constant. The corresponding locked resonant frequency ratio from figure 4 is about 4.5 while that observed for a similar set of results for $N = 40$ lbf was about 4.8. This to some extent indicates a decrease in resonant frequency with decreasing normal force.

From the set of curves figure 4 it is seen that the apparent coefficient of friction during slipping

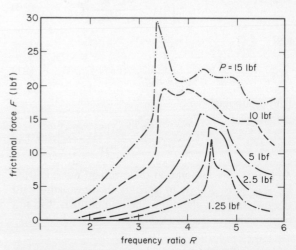

Figure 4. Variation of frictional force with frequency ratio for different applied forces: system of figure 1(b), $N = 20$ lbf.

increases with applied force (and therefore slip amplitude) from about 0.3 to 1.0. Similarly it was observed, for a normal force of 40 lbf, that the variation was from about 0.6 to 0.8: for a normal force of 40 lbf the variation was from about 0.2 to 0.8.

Displacement response: system l

The displacement response was measured at the point where the beam was excited. This position was chosen partly for convenience and partly because in many structures it is the point at which minimisation of the response is required. However, the methods employed are equally applicable to any other point on the beam with little modification.

Measurements were obtained from a piezoelectric transducer which produces a charge signal proportional to the acceleration. The signal, after amplification, was integrated twice to give the displacement. This method was considered satisfactory as the initial signal contained little harmonic distortion.

Typical results are shown in figure 5 for an applied force amplitude of 7.5 lbf. The effect of the elasticity of the contacting asperities can be seen in the displacement of the curves to the left with decreasing normal force magnitude. It is also observed that all the curves pass close to the first invariant-point I_1, but due to the displacement of the curves they do not pass close to the second invariant point I_2. Similar observations were made from two other sets of displacement response curves obtained for applied force amplitudes of 15 and 3.75 lbf.

An estimation of the optimum normal force from

figure 5 is 8 lbf. For these test conditions the effective coefficient of friction is estimated as about 0.6. Thus the ratio Q of frictional force to applied force is about 0.64. Similar estimations, for applied forces of 15 and 3.75 lbf, of the optimum normal force μ_s and Q were 10 and 5 lbf, 0.7 and 0.5, and 0.5 and 0.67 respectively.

The theoretical optimum response curve is shown in figure 6. Other theoretical curves produced were

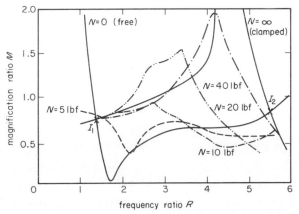

Figure 5. Variation of magnification ratio with frequency ratio for different normal forces: system of figure 1(b), $P = 7.5$ lbf.

generally similar in form to the experimental curves, although differences existed due to the shift of the experimental curves to the left with decreasing compliance between the interlocking asperities (that is, with decreasing normal force) due to material damping in the experimental system, which tends to reduce the magnitude of the responses, and due to the variation of coefficient of friction during slipping, which causes the response curves to reduce after the initiation of slip and not remain constant as assumed theoretically.

OPTIMUM DAMPING

System I
Certain requirements for optimum damping can be interpreted from figure 6. At zero frequency the damper is clamped and hence the initial response will start from A. A clamped condition must be main-

tained until at least the first free resonant frequency has been passed. Breakaway from the clamped (locked) state must occur before the first clamped resonant frequency is reached, say at B, rejoin the clamped curve before the second free resonant frequency is reached, say at D, and pass through the second invariant point I_2.

The theoretical analysis was used to predict the frictional force to applied force ratio Q which would give the optimum response between 20 to 1000 Hz. This optimum response, shown as curve AI_1BI_2DE in figure 6, was obtained by considering the maximum response over the frequency range for various frictional forces and so choosing the frictional force that this response was minimised. The value of Q found was 0.8.

Also included on figure 6 for comparison is the optimum response that could be obtained using a viscous damper in place of the frictional damper.

System II
The locked resonant frequencies of system II depend on the length of the damper beam, whereas the free resonant frequencies are independent of the length of the damper beam, and are in fact the same as for system I.

Thus the essential difference between the two systems is that for the rigidly mounted frictional

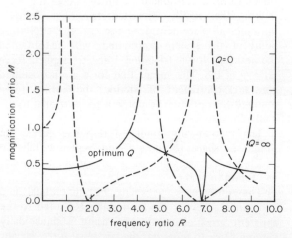

Figure 7. Optimum theoretical response curve for a flexibly mounted frictional damper: system of figure 1(c).

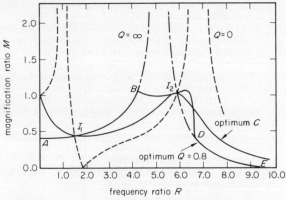

Figure 6. Optimum theoretical response curves: system of figure 1(b). Optimum Q frictional damper, optimum C viscous damper.

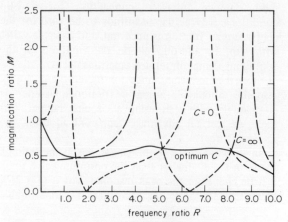

Figure 8. Optimum theoretical response curve for a flexibly mounted viscous damper: system of figure 1(c).

damper system optimum response involves only one variable Q, while for the flexibly mounted frictional damper system there are two variables Q and L. This in effect means that in the latter system the magnitudes of the invariant points can be adjusted.

Experimental results and the process of response optimisation will be dealt with in a subsequent paper. For the present the optimum theoretical response for the frictional damper is shown in figure 7 and the corresponding optimum curve for a viscous damper in figure 8.

For the frictional damper, optimum response was found for $L = 6.41$ in and $Q = 0.664$.

DISCUSSION

This investigation has been concerned with observing experimentally how a frictional damper might be arranged to minimise the response of a sinusoidally excited continuous system, and assessing whether this response can be predicted theoretically using linearised equations.

The frequency range of the excitation has included the first two free resonant frequencies and the first clamped resonant frequency of the system, and it has been shown that a normal force across the damper can be chosen so that these resonant frequencies are traversed without resonant conditions occurring.

For the rigidly mounted frictional damper, optimum displacement response, at the excitation point of the system considered, will be obtained theoretically for a frictional force to applied force ratio Q of 0.8. This means that for a given (assumed constant) coefficient of friction, the normal force required for optimum response is proportional to the applied force.

The theoretically predicted response curves are similar in shape to those found experimentally in which the average value of Q for optimum response was found to be 0.6.

The differences between the theoretical and experimental results is only in part attributed to the apparent gross assumption of using a sinusoidally varying force to represent the frictional force. Having an equal or even greater effect is the assumption in the theory of a constant coefficient of friction and a rigid damper interface connection for nonslip conditions. The actual magnitude to be used for the coefficient of friction is uncertain and its value varies with slip. In addition elastic deformations of the contacting damper interface asperities occur.

Over the frequency range considered the optimum response obtainable using a frictional damper is comparable with that using a viscous damper, but there are certain advantages apparent in using the viscous damper.

In the case of the frictional damper, if the ratio Q, figure 6, is 25% higher than the optimum value there is little change in the peak response, but if Q is 25% lower than the optimum value there is a considerable increase in the peak response. Equivalent changes using a viscous damper produce only small changes in the peak response.

Thus having regard to the possible reduction in normal force due to wear of the damper interface surfaces, thus reducing the effective frictional force, it would be necessary initially to set the frictional damper with a higher-than-optimum normal force, so that during its operating period the value of Q did not fall below the optimum value.

To embody a frictional damper in a structure, it is necessary either to predict the damper's essential features in the design stage or determine them from measurements taken on the actual structure. In either case it is necessary to determine direct and cross-receptances at the positions of the excitation, damper and specific points of interest, before it is possible to design (for the mean estimated applied force) a damper having the right size and a reasonable range of adjustment.

CONCLUSIONS

A theoretical linearised analysis of a frictional damped beam system is presented from which a reasonable prediction of the actual response can be obtained.

The theoretical analysis assumes the coefficient of friction to be constant during slipping, whereas it is found experimentally to increase with slip amplitude.

The frictional damper can be arranged so as to limit the forced excitation response of a beam system over a frequency range, which includes one locked and two free resonant frequencies, to a level comparable with that using a viscous damper.

For a frictional damped system, invariant points exist through which all response curves pass independent of the frictional force.

ACKNOWLEDGEMENT

The authors wish to acknowledge the support given by the Machine Tool Industry Research Association.

REFERENCES

1. J. P. DEN HARTOG, 1931. Forced vibrations with combined coulomb and viscous friction. *Trans. ASME*, **53**, p. APM-107.
2. G. C. K. YEH, 1966. Forced vibrations of a two-degree of freedom system with combined coulomb and viscous friction. *J. Acoust. Soc. Am.*, **39**, p. 14.
3. Y. SAWARAGI, T. FUJII and Y. OKADA, 1959. Forced vibrations of the system with two-degrees of freedom with coulomb damping. *Bull. JSME*, **2**, p. 311.
4. S. W. E. EARLES and M. G. PHILPOT, 1967. Energy dissipation at plane surfaces in contact. *J. Mech. Engng. Sci.*, **9**, p. 86.
5. S. W. E. EARLES and C. F. BEARDS, 1970. Some aspects of frictional damping as applied to vibrating beams. *Int. J. Mech. Tool Des. Res.*, **10**, p. 123.
6. R. E. D. BISHOP and D. C. JOHNSON, 1960. The mechanics of vibrations. Cambridge University Press.

EFFECT OF FEEDMOTION DURING DYNAMIC TESTS ON THE CHATTER PREDICTION OF A LATHE

by

N. H. HANNA* and A. W. KWIATKOWSKI*

SUMMARY

Excitation tests were carried out on a lathe under stationary and moving-carriage conditions. The theoretical stability for both cases was predicted and found to be affected by the presence of feed. The effect is discussed with the aid of a model with two degrees-of-freedom. It is concluded that chatter predictions based on stationary-carriage receptance are liable to be far from accurate.

INTRODUCTION

One of the main objectives of dynamic testing of machine tools is to assess their chatter-resistance characteristics. The most widely-accepted criterion on which this assessment is based, is the maximum unconditionally stable width-of-cut (h_0) which can be determined either theoretically or experimentally. The theoretical h_0 is usually assumed proportional to the maximum unconditionally stable chip thickness coefficient (k_0) or what is known as the coefficient of merit (com)[1]. The larger the com, the higher the resistance of the machine to chatter and vice versa. The com of a particular machine tool is inversely proportional to the maximum negative inphase component (a_0) of its operative receptance, which is calculated from a set of nine receptances experimentally measured, generally in three directions. In some cases, however, such as lathes and probably horizontal milling with straight cutters, the response along the spindle axis is usually small compared to the values in the normal directions, so two-dimensional receptance measurement is sufficient.

In many instances in the past, when theoretically-calculated stability charts (or com's) were compared with those experimentally determined, there were discrepancies, in some cases quite considerable[2]. Most investigators are inclined to attribute these discrepancies entirely to the inaccuracy of the cutting coefficients used in the calculations: this is because, till the present day, no universally recognised, accurate method for their determination has been developed, whereas the method used for receptances identification has reached a great deal of perfection.

The direct relationship between the com and a_0 makes the theoretical assessment of chatter resistance very sensitive to any variations in the operative receptance. Therefore, not only the accuracy of the receptance measurement, but also the conditions under which this receptance was determined, are of considerable importance. Strictly speaking, cutting conditions, that is, spindle rotation, preloads, force level and feedmotion, should be faithfully simulated. Although the effects of the first three factors on the dynamic response of machine tools have been investigated[3,4], in all reported stability investigations

using simulated cutting conditions the feedmotion was neglected. The reason for this was simply that the exciter design or the experimental setup used, did not allow such motion to take place.

It has been reported recently[5] that the feedmotion of a lathe carriage does, in fact, affect the dynamic response of its structure, with the consequence that a difference is observed between receptances measured under zero and finite feed conditions. The investigation was conducted using an electromagnetic exciter and a long workpiece, which allowed both the rotation of the spindle and the feedmotion to be executed during measurement.

With this new development in mind, it was decided to investigate the effect of feedmotion on the theoretical prediction of instability of a lathe, using both sinusoidal and random excitation methods.

EXPERIMENTAL SETUP

Figure 1(a) shows the experimental arrangement used in this investigation. A two-directional (60° apart) electromagnetic exciter was clamped in the toolpost of an EMAG type lathe and embracing a 3.3 in diameter, 28 in long mild steel workpiece supported by both the headstock and tailstock and machined in position to minimise the eccentricity. Four measuring positions along the workpiece were chosen, 6 in apart and identified by four microswitches fixed to the machinebed. These switches were triggered by a lever bolted to the carriage to ensure accurate positioning of the carriage and consequently the excitation points along the workpiece. The directions of excitations and measurement are shown in figure 1(b).

Figure 2 shows the block diagram of the procedure used in measuring and processing the data in case of sinusoidal (figure 2(a)) and random (figure 2(b)) force excitation. This procedure is explained in detail in[6].

EXPERIMENTAL RESULTS

Both the sinusoidal and random excitation results to be discussed were obtained under the same conditions

* Mechanical Engineering Department, University of Birmingham.

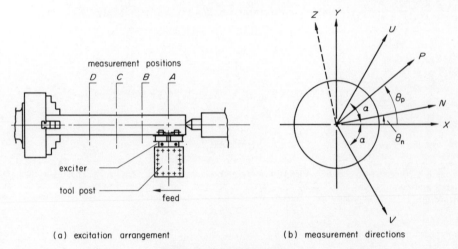

(a) excitation arrangement   (b) measurement directions

Figure 1. Experimental setup.

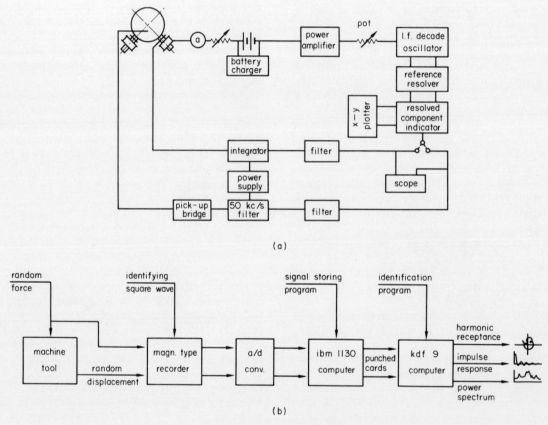

Figure 2. Identification procedure.

of spindle speed (280 rpm) and exciter preload (100 lb). The sinusoidal force was 7 lb amplitude and the random was 5.5 lb rms. The digital computation of the random force receptances was carried out in the manner described in[6]. The sampling rate was 800, the number of samples 1600, the number of delays 80 and the frequency resolution 2 Hz.

Two sets of receptances, at 0 and 2.7 in/min feed, were determined for each of the four positions A, B, C and D (figure 1), using both methods. For the case of the sinusoidal force method with feedmotion each point on the receptance was determined as follows. The carriage was placed at a position a few inches before the microswitch defining the required

measurement configuration. The sinusoidal force, at a particular frequency, was applied and the feed engaged. The receptance point was recorded only when the microswitch was triggered by the lever on the carriage. The procedure was repeated for a different frequency until the receptance was completed. For the random force, however, the complete receptance was determined during a single stroke of the carriage. Obviously, this occupied far less time than the tedious, point-by-point sinusoidal determination, and it was therefore decided to obtain another set of receptances at 10.8 in/min feed using only the random force method.

The number of individual direct and cross-receptances thus obtained was 80 (each position is identified by 4 receptances, 2 direct and 2 cross). Only two direct receptances for position A and in direction V, determined by the random force method under 0 and 2.7 in/min feed conditions, are shown in figure 3 for comparison. The operative receptances

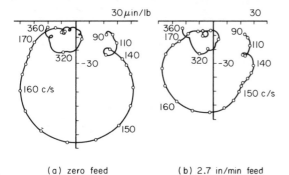

(a) zero feed   (b) 2.7 in/min feed

Figure 3. Direct receptance R_{VV}.

for the four positions and different feed conditions were calculated using the analytical method described in Appendix I (the operative receptance being between the normal to the workpiece surface and an assumed cutting force direction of $66°$ with this normal).

The results for 0 and 2.7 in/min feed are shown in figures 4 and 5.

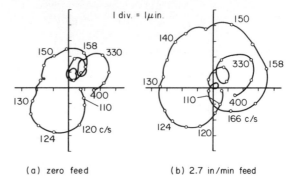

(a) zero feed   (b) 2.7 in/min feed

Figure 6. Vertical response of carriage relative to bed.

The relative motion between the lathe bed and its carriage at position A was also measured for 0 and 2.7 in/min feed: the results are given in figure 6.

PREDICTION OF STABILITY

Assuming force dependence on the chip thickness only, the stability charts for both stationary and moving carriage conditions were calculated using the

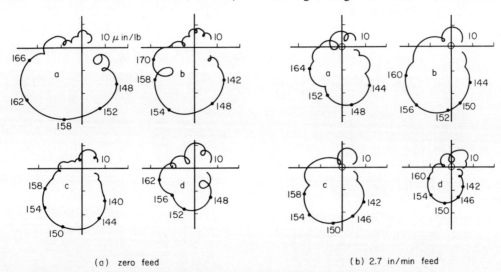

(a) zero feed   (b) 2.7 in/min feed

Figure 4. Operative receptances by random force method.

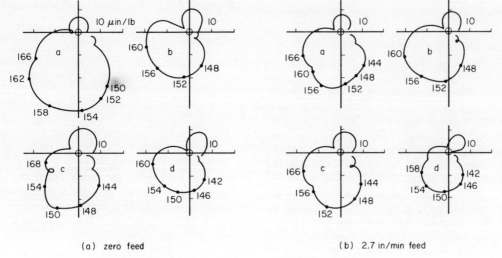

(a) zero feed   (b) 2.7 in/min feed

Figure 5. Operative receptances by sinusoidal force method.

method due to Sweeney and Tobias[7]. A computer programme was designed to compute these charts from the corresponding operative receptances and to plot the results directly on an *x-y* plotter linked to an 1130 IBM computer. The calculated charts for only the extreme positions *A,* close to the tailstock and *D,* close to the headstock, are shown in figures 7 and 8 (in the charts, the *com* value is given by the horizontal asymptote to the envelope of the unstable lobes, which in this case coincide). Figure 7 shows

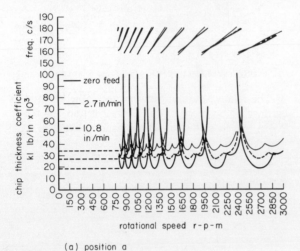

(a) position a

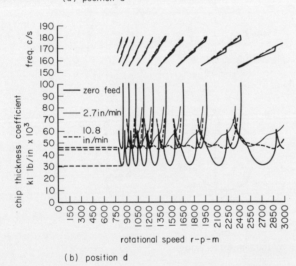

(b) position d

Figure 7. Stability charts (case of random excitation).

the results for the random method and figure 8 for the sinusoidal force method.

The variation of the *com* with feed velocity for the four positions is shown in figure 9. Taking the *com* values based on stationary carriage operative receptance as a reference, the percentage variation of those calculated from the 2.7 and 10.8 in/min feedrate receptances were obtained: these are given in table 1. The positive sign indicates an increase above the zero feed value of the *com*: the negative sign indicates the opposite.

The maximum negative inphase component of the operative receptance of position *A* was determined for different cutting force directions, from which the corresponding *com* values were calculated. The results for various feed conditions are shown in figure 10.

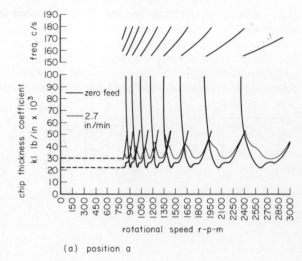

(a) position a

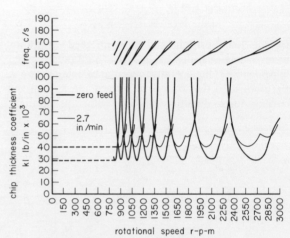

(b) position d

Figure 8. Stability charts (case of sinusoidal excitation).

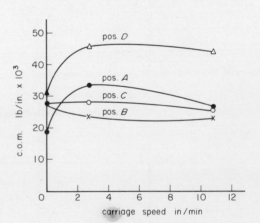

Figure 9. Variation of *com* with feed velocity.

TABLE 1
Relative variation of com

Pos.	% age variation in *com* (stationary basis)	
	2.7 in/min	10.8 in/min
A	+82	+44.5
B	−13.9	−15.8
C	+0.7	−7.9
D	+50	+44.1

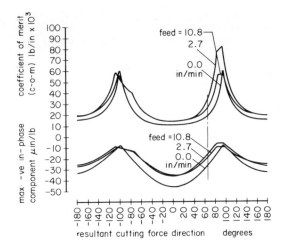

Figure 10. Variation of the max. negative inphase component and the *com* with force direction and carriage feed.

DISCUSSION OF THE RESULTS AND CONCLUSIONS

It is pertinent to point out that the results obtained by the random force method were sufficient to demonstrate the effect of the feed. In order to establish whether the results are subject to some peculiarities of either method used, it was decided to carry out the sinusoidal experiments as well. Any discrepancies in the corresponding receptances obtained by either method are mainly due to the relevant quantity measured and are discussed in other publications[5,6]

It is obvious from figures 3 to 5 that feedmotion alters the dynamic response and that the variation depends on the carriage position. Whereas this variation is large for positions A and D, it is marginal (but nevertheless significant) for positions B and C. The maximum negative inphase component decreases with feed in the case of the former positions, and slightly increases in the case of the latter.

Figure 9 shows that, in the investigated range of the feedrate, the *com* for positions A and D increases with feed to a maximum and then decreases, but to a value still above that with zero feed. For positions B and C, however, it drops, for higher feed, below the zero feed value. Regardless of the sign, the variation of the *com* with the feed velocity resembles the well-known frictional characteristics of lubricated surfaces, where the friction forces drop once relative motion takes place, to reach a minimum at some relatively-low velocity and then increase at higher velocity. The similarity in behaviour suggests that the variation of the damping (caused by friction) with feedmotion may be responsible for the observed variation of the dynamic response and hence the variation of the *com*.

Table 1 shows that the variation of the predicted *com* could be as high as 80% for position A. Because the values given in this table are based on the results of the random method, they may represent instantaneous values with some error due to noise and digital calculations. The corresponding value based on the sinusoidal results (which represent the mean

receptances[5]), is about 37% (see figure 8). Although these values are by no means identical, it was found that when one is large, the other is large too and vice versa. For instance for position B, the variation is -13.9% in the case of the random and -11.3% in the case of the sinusoidal method: the values for position D were 50% and 41% respectively. This confirms, at least qualitatively, that the observed differences between receptances measured under zero and finite feed conditions, are neither accidental nor due to the peculiarity of the method used.

It is relevant to mention that, as indicated in figure 10, the amount of variation of the *com* with different feeds depends on the cutting force direction used in the calculation of the operative receptances. For this investigation the force direction was assumed as $66°$ with the normal and the condition is represented by the vertical cross-section in figure 10. This assumption was based on experimental data of orthogonal cutting of mild steel with a $5°$ rake-angle tool.

The question which arises is that, while the relative response between the carriage and the bed increases with feedmotion (figure 6), indicating a drop in damping, the stability of the machine for the same position increases, indicating the opposite. This apparent contradiction is explained as follows.

Owing to the complexity of the system, the number of modes involved and the lack of information about the variations that take place on the sliding surfaces, the problem does not lend itself to rigorous quantitative, theoretical analysis. However, it is known that the average damping forces decrease with motion especially if it creates an oil film between the sliding surfaces; It is therefore reasonable to assume that the coefficient of damping of the slides while the lathe carriage is moving is smaller than that while it is stationary.

Bearing this in mind and assuming it to be the only change in the system induced by the carriage movement, one might expect, on the basis of a reference to the behaviour of a single-degree of freedom system, an increase in the response of the machine, for a particular configuration, when the carriage is set in motion. Although this over-simplified conclusion agrees with the observed increase in the relative response between the carriage and the bed (figure 6), it contradicts the other experimental results presented earlier. However, to show qualitatively that the apparent contradiction is in fact unreal, a two-degree of freedom system, with arbitrarily-chosen parameters, is employed.

Consider the system shown in figure 11(a) in which λ_2 and c purport to represent the stiffness and damping of the slides respectively. The solution of the forced vibration of such system can be found, for instance in[8]. Keeping constant all the parameters, except the damping, the frequency responses of the masses m_1 and m_2 and their relative response were calculated for different values of the damping: the results are given in figures 11(b) and (c). In these figures the vertical axis represents the nondimensional ratio of the amplitude to the static deflection A_s and the horizontal axis represents the ratio of the forcing frequency to the natural frequency of the main system p_1.

Figure 11(b) shows that as the value of μ decreases from 0.4 to 0.1 the resonance amplitude of m_1 decreases, whereas the amplitude of m_2 first decreases

the case under consideration, and depending on the position of the carriage on the bed, the discrepancy may amount to 41%, even using mean receptances.

It remains to be seen, however, to what extent the reported effect is representative of other

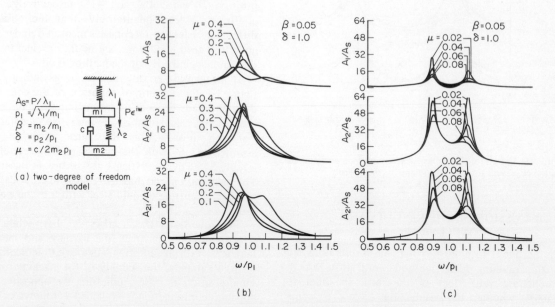

Figure 11. Variation of vibration amplitude with damping.

and then increases again and the relative amplitude increases throughout. For a further decrease of μ from 0.08 to 0.02 the three responses increase as shown in figure 11(c).

The discussed case illustrates that the relation between the damping of one mode and the amplitude of the other is not a straightforward inverse one, that is, a decrease in the former does not necessarily cause an increase in the latter. Instead, it shows that the amplitude of vibration may increase, decrease or remain unchanged with a change in the damping, depending on the value of the damping in relation to the other modal parameters. Obviously, with multi-degree of freedom systems the situation becomes more complex, especially if the variation of damping is coupled with a variation in stiffness, but the same conclusion could be drawn.

The analysis does not prove, but only demonstrates, the possibility that a diminution of the damping on the slides is the cause of the observed behaviour.

The effect of feedmotion on stability manifests itself in figure 8 not only by altering the value of the *com* but also by shifting the unstable lobes and hence changing the conditionally stable area of the chart. However, this effect becomes significant only when comparing theoretical with experimental stability charts.

From the above discussion it can be concluded that chatter predictions based on zero feed determinations are liable to be far from accurate. In

machines. This can only be discovered by conducting comprehensive dynamic tests on a large number of different machine tools.

ACKNOWLEDGEMENTS

This investigation has been carried out under a contract from the SRC. The authors wish to express their gratitude to Professor S. A. Tobias for his support and encouragement throughout the work.

REFERENCES

1. S. A. TOBIAS. Dynamic acceptance tests for machine tools, *Int. J. Machine Tool Des. Res.* Vol. 2. p. 267 (1962).
2. C. ANDREW. Chatter in horizontal milling machines, *Proc. Instn. Mech. Engrs.* 179, p. 877 (1964).
3. N. H. HANNA and S. A. TOBIAS. The nonlinear dynamic behaviour of a machine structure, *Int. J. Mach. Tool Des. Res.* Vol. 9, pp. 293–307 (1969).
4. I. CHOWDHURY, M. M. SADEK and S. A. TOBIAS. Determination of the dynamic characteristics of machine tool structures, *Proc. Instn. Mech. Engrs.* Vol. 184, Part 1 (1969–70).
5. N. H. HANNA and A. KWIATKOWSKI. Effect of feedmotion on the dynamic response of a lathe (to be published).
6. N. H. HANNA and A. KWIATKOWSKI. Identification of machine tool receptances by random force excitation, *Int. J. Mach. Tool Des. Res.* Vol. 11 (1971).
7. G. SWEENEY and S. A. TOBIAS. An algebraic method for the determination of the dynamic stability of machine tools, *Proc. Int. Res. in Production Engineering*, 495 (1963).
8. S. TIMOSHENKO. Vibration problems in engineering, van Nostrand, New York.

APPENDIX

DETERMINATION OF GENERAL CROSS-RECEPTANCE FROM MEASUREMENT ALONG OBLIQUE AXES

Referring to the configuration shown in figure 1(b), U and V are the directions of receptance measurement, enclosing an angle 2α which is bisected by the X-axis. P is the applied force the inclination of which with the X-axis is θ_p. N is the direction of the required cross-response and makes an angle θ_n with the X-axis.

Let the measured receptances in directions U and V due to a force in direction U be

$$R_{uu} = (a_{uu} + ib_{uu}) \text{ and } R_{vu} = (a_{vu} + ib_{vu})$$

Similarly for a force in direction V

$$R_{vv} = (a_{vv} + ib_{vv}) \text{ and } R_{uv} = (a_{uv} + ib_{uv})$$

where a and b are the inphase and outofphase components of the corresponding receptance respectively and $i = \sqrt{-1}$.

For a unit force in the P direction, its components in directions U and V are

$$P_u = \frac{\sin(\alpha + \theta_p)}{\sin 2\alpha} \text{ and } P_v = \frac{\sin(\alpha - \theta_p)}{\sin 2\alpha}$$

Therefore the responses due to these components

in the U and V directions become

$$R_{up} = R_{uu} \frac{\sin(\alpha + \theta_p)}{\sin 2\alpha} + R_{uv} \frac{\sin(\alpha - \theta_p)}{\sin 2\alpha}$$

$$R_{vp} = R_{vv} \frac{\sin(\alpha - \theta_p)}{\sin 2\alpha} + R_{vu} \frac{\sin(\alpha + \theta_p)}{\sin 2\alpha}$$

Assuming that the receptances in directions N and Z (perpendicular to N) are R_{np} and R_{zp} respectively

$$R_{up} = R_{np} \cos(\alpha - \theta_n) + R_{zp} \sin(\alpha - \theta_n)$$

$$R_{vp} = R_{np} \cos(\alpha + \theta_n) - R_{zp} \sin(\alpha + \theta_n)$$

from which

$$R_{np} = R_{up} \frac{\sin(\alpha + \theta_n)}{\sin 2\alpha} + R_{vp} \frac{\sin(\alpha - \theta_n)}{\sin 2\alpha}$$

Substituting for R_{up} and R_{vp} and rearranging

$$R_{np} = R_{uu} \frac{\sin(\alpha + \theta_p) \sin(\alpha + \theta_n)}{(\sin 2\alpha)^2}$$

$$+ R_{uv} \frac{\sin(\alpha - \theta_p) \sin(\alpha + \theta_n)}{(\sin 2\alpha)^2}$$

$$+ R_{vv} \frac{\sin(\alpha - \theta_p) \sin(\alpha - \theta_n)}{(\sin 2\alpha)^2}$$

$$+ R_{vu} \frac{\sin(\alpha + \theta_p) \sin(\alpha - \theta_n)}{(\sin 2\alpha)^2}$$

THE IMPACT DAMPER BORING BAR AND ITS PERFORMANCE WHEN CUTTING

by

M. D. THOMAS,* W. A. KNIGHT* and M. M. SADEK*

SUMMARY

The dynamic performance of an impact damper boring bar relative to that of a conventional solid bar of the same diameter is assessed by cutting tests. The device is optimised for a specific overhang by altering the mass ratio and the free mass stroke. The effect of the overhang variation on the chatter performance is also investigated. It is concluded that the impact damper boring bar allows cutting at greatly increased chatter-free widths of cut, particularly at larger values of the length to diameter ratio.

INTRODUCTION

In recently published work[1] different techniques used for increasing the stability of cantilever boring bars were assessed by means of forced vibration tests. It was pointed out that tuned absorbers need retuning each time they are used at a different overhang; Lanchester dampers[2] give only marginal improvement, as does machining flats on to the bar[3] which reduces its stiffness and serves only to define the principal axes. However, the advantage to be gained from these devices is, more often than not, totally eliminated by the reduction in stability due to the decreased stiffness of the boring bar.

An impact damper proved to be by far the most promising of the devices applied to boring bars[1].† With this, the predicted maximum stable width of cut was increased by 75%, despite the fact that the prototype bar lacked sufficiently fine adjustment of the gap and the inherent stiffness of the bar was low. A modified design of impact damper boring bar has now been produced, in which the gap can be more accurately optimised. This incorporates an improved configuration of the damper, which requires only a small cavity to be made in the bar, so that its inherent stiffness is not significantly reduced.

The parameters of this device have been optimised and its performance during cutting compared with that of a round solid boring bar of the same diameter. The effect of varying both the overhang of the bars and the cutting speed have been investigated.

DESIGN OF THE BORING BAR

The impact damper is essentially a simple device. It consists of a mass free to vibrate within fixed limiting stops against which it impacts at the end of each stroke. The impulses generated between this mass and the end stops restrain the main system whenever it starts to vibrate. The device should be mounted as near to the point of maximum amplitude of the main system as possible, as it is more efficient at larger amplitudes. The size of the free mass and the gap between it and the end stops are the two principal factors determining the performance of the damper.

Before proceeding to the design of the bar, it is necessarv to know what range of values the impact

mass and the gap should be allowed to take. It has been shown[4] that the optimum ratio between the free mass and the mass of the main system is 0.08, and so it is necessary to estimate the equivalent mass of the bar in order to determine the weight of the free mass, and hence the size of the cavity in the bar. For a $1\frac{7}{8}$ in diameter steel bar, used with a length to diameter ratio of 5:1, an equivalent weight of 3 lb was estimated. This value is an approximation, but it is sufficiently accurate for the weight of the impacting mass to be fixed in the range 0.1–0.4 lb, and for the mass to be designed so that it could be varied within these limits.

Previous work on the impact damper[5] has shown that gravity severely detracts from its performance, but that this can be overcome if the impact mass is supported by a light spring. Design charts, normally used when choosing impact damper parameters, are based on theory[5] which does not cover the case of a spring supported impact mass. However, experimental work[6] has shown that these design charts may be used, though with some reduction in accuracy, provided that the spring is light, so that the natural frequency of the impact damper system is much less than that of the main system.

The design chart[6] in figure 1 may thus be used to predict the range of values in which the optimum gap will lie. The stiffness of the bar and the exciting force

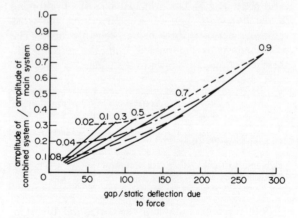

Figure 1. Design chart for the impact damper[6]. Solid lines are constant coefficient of restitution; broken lines are constant mass ratio.

* Department of Mechanical Engineering, University of Birmingham
† Patent No. 43421/70.

due to cutting were both estimated, and the optimum gap was calculated to be of the order of 0.010 in. As the gap may need to be adjusted to an accuracy of ±5%, this means that the inaccuracy in the setting of the gap must not be greater than ±0.0005 in.

The design of the impact damper boring bar is shown in figure 2. The impact mass A, which can be

As the theoretical optimum mass ratio[5] is 0.08, the weight of the free mass should be 0.16 lb. Using the design chart in figure 1[6], the optimum ratio of gap to the static deflection of the bar is 45, for a coefficient of restitution of 0.6 which is a typical value for steel on steel. Knowing the level of the mean cutting force at the stability boundary of a

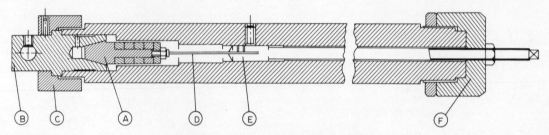

Figure 2. The impact damper boring bar.

varied in size, fits into the conical anvil combined with the tool holder B, and this may be fixed in position by the locking nut C. The impacting mass is supported by the spring steel rod D, which is fixed to the steel rod E. The gap, which is in this case the distance between the mass and its conical anvil, needs to be varied for optimisation, so it was necessary to be able to move the mass along the axis of the bar. In this prototype design, the gap could be varied by means of the rod joining the spring support E to the nut F. The whole rod could be screwed in and out, and as one complete turn of the nut F increased the gap by 0.025 in, it was simple to set the gap to an accuracy of ±0.0005 in. Despite the hole down the centre of the bar the stiffness was not seriously reduced when compared with a solid bar and the dynamic stiffness remained unaltered as the reduction in the mass of the bar, owing to the damper cavity, effectively compensates for the loss of stiffness.

SELECTION OF IMPACT DAMPER PARAMETERS

The optimum mass ratio was determined from forced vibration tests on the bar, firstly with the impact damper removed, and with the length to diameter ratio fixed at 5:1. The principal axes of vibration are in the horizontal and vertical directions, the cross responses between these being negligibly small. The modal constants determined from the responses in these directions, are shown in the table.

solid bar, it is possible to estimate that the oscillatory component of the cutting force will be of the order of 10 lb. This, in conjunction with the horizontal static stiffness given in the table, gives the optimum gap as 0.015 in.

These theoretical values are compared with experimental readings in figure 3, which shows the

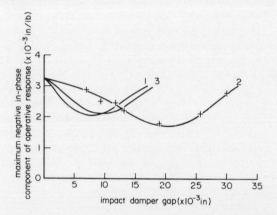

Figure 3. Theoretical stability boundary with tool horizontal, for three mass ratios. 1:0.045, 2:0.078, 3:0.095.

change in chatter performance of the bar with gap setting, for three different mass ratios. The operative responses[7] were calculated, assuming a cutting force angle of 60°, and the maximum negative in-phase component of this, which is inversely proportional to the maximum stable depth of cut, is shown plotted in figure 3. It can be seen that the mass ratio 0.08 gives the lowest value and is consequently most stable. This value is exactly the same as that expected from theory[4], and the gap at which it occurs is 0.020 in, which is of similar order to that estimated in designing the bar.

It should be noted that this optimum mass ratio is the result of only three different tests and cannot be considered exact, although adequate for the purpose of this work. As algebraic resolution to give the operative response is questionable when a non-linear device such as the impact damper is concerned, all further optimisation must be by cutting tests.

	Vertical mode	Horizontal mode
Natural frequency	421 Hz	388 Hz
Dynamic amplification factor	38	41
Dynamic stiffness	54 500 lb/in	31 300 lb/in
Equivalent weight	3.0 lb	1.9 lb

With the tool mounted in the horizontal direction, the normal to the cut surface coincides with the principal direction of the horizontal mode. The bar will be expected to chatter in this mode and the equivalent weight of the main system is therefore 1.9 lb.

The principal advantage of these forced vibration tests is that they produced an approximate solution in about one-tenth of the time of cutting tests.

CUTTING PERFORMANCE AT VARIOUS GAP SETTINGS

The only parameter of the impact damper, which is generally varied for optimisation, is the gap and this would normally be fixed at that value obtained from design charts.

However, it is of interest to investigate the effectiveness of the device for the case where the gap differs from the optimum. This is effected by means of cutting tests, with oblique cuts taken down the outside of a cylindrical workpiece. This method was chosen for ease of measurement and setting up, there being no significant difference between cutting the inside or outside diameter of the workpiece. The bar was clamped at a length to diameter ratio of 5:1 the feedrate throughout the tests being 0.008 in/rev. The cutting tools were tungsten carbide with a nose radius of $\frac{1}{64}$ in and an approach angle of 45°.

In figure 4, the variation in stability boundary is plotted against gap, with the radius of the circles

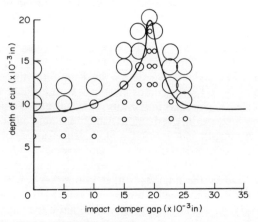

Figure 4. Experimental stability boundary at different gaps. Feed rate: 0.008 in/rev, cutting speed: 400 rev/min, mass ratio: 0.078, length to diameter ratio: 5:1.

corresponding to the amplitude of vibration of the bar, on the same scale as the ordinate. This curve is for the mass ratio determined from the forced vibration tests above, and for a cutting speed of 400 rev/min. The limit of stability was determined at each gap setting by increasing the depth of cut in steps and recording the amplitude of vibration of the bar, the stability boundary being that depth of cut at which chatter started to occur. The chatter frequency in this case associated with the mode of vibration whose principal axis is in the horizontal direction.

Figure 4 shows that the stability boundary slowly increases until the optimum is reached at a gap of 0.019 in, with the resultant value twice as high as that at zero gap. This gap setting corresponds almost exactly with that obtained from forced vibration tests, and shows that the resolution of harmonic responses to calculate the limit of stability was not greatly influenced by the nonlinearities of the device. If the gap is further increased, the stability boundary

rapidly decreases towards the same value as that at zero gap, but never goes below it. Thus, the performance of the impact damper boring bar is always better than that of the bar at zero gap, which in this case is equal to the performance of a solid bar of the same diameter. This is because, at large gaps, stable multi-impact motion of the damper is impossible. The optimum gap produces an improvement of 100% in the performance of the boring bar. However, slight deviations from this optimum will still produce significant improvement, but it is preferable to underestimate rather than to overestimate the gap as the curve falls very rapidly above the optimum condition. An error of 15% on the low side will still produce an improvement of about 50%.

COMPARISON BETWEEN THE IMPACT DAMPER BORING BAR AND A SOLID BAR

Figure 5 shows the stability boundary within the speed range of 0 to 500 rev/min for the impact damper boring bar with a gap setting of 0.019 in (curve 1) and for a similar solid bar (curve 2). It can

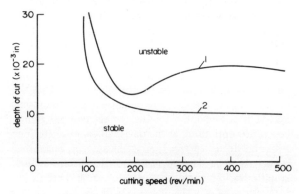

Figure 5. Experimental stability boundary at different speeds. Feed rate: 0.008 in/rev, gap: 0.019 in, length to diameter ratio: 5:1. Curve 1, bar with impact damper; curve 2, solid bar.

be seen that, at high cutting speeds, the limiting depth of cut has been increased by 100% owing to the inclusion of the impact damper, but at lower cutting speeds this improvement is less. The reason for this is that, as the impact damper mass is lying in the base of a cylindrical hole it will require a certain degree of excitation before it can achieve a motion with impacts on both end stops, this being when the damper is most effective. If the vibration is vertical, gravity must be overcome and, if the motion is horizontal, then some friction must also be overcome. At low speeds the exciting impulses, from inhomogeneities in the workpiece, are too widely spaced so that the impact damper takes an appreciable time to achieve an effective motion. If the speed of the workpiece is increased, then these pulses become more frequent and efficient operation of the damper can be achieved more quickly.

It should be emphasised that the impact damper does not need the continuous excitation resulting from chatter before it will operate. The sort of instantaneous impulses from inhomogeneities in the workpiece material occurring during cutting, and which would normally be the source of excitation causing chatter to build up, instead excites the damper sufficiently for it to start operating, and so damp out the vibration quickly before significant regeneration of the motion can occur.

Despite this reduction in performance at low cutting speeds, the overall improvement in the performance of the boring bar due to the impact damper, is very marked. Although there is no experimental work comparing impact dampers with tuned damped vibration absorbers, theoretical analysis[8] has shown that such an absorber, when favourably tuned, can match the performance of the impact damper. This is possible, however, only when there is very high damping between the absorber mass and the main mass (a damping ratio of 0.2), and this is difficult to achieve in practice, in particular in the range of frequencies commonly associated with boring bars.

The main complication in applying the tuned vibration absorber to a boring bar lies in the fact that the overhang is often changed during normal workshop use. The mass ratio and the ratio between the resonant frequencies of the main system and the absorber will thus alter and the device will no longer be tuned. In these circumstances, the performance of the bar with absorber during cutting may well be worse than that of a solid bar. An impact damper's performance is not affected by the frequency ratio, and only marginally by the mass ratio, so that the only result of altering the overhang would be that the optimum gap would change slightly. Figure 4 shows that it is not essential for the impact damper gap to be exactly optimised, as the device does not introduce a second degree of freedom to the system, but rather it increases the damping of the main system at high amplitudes. This will happen to some extent, no matter how poorly the device is optimised.

Figure 6 shows the effect on the stability boundary at 400 rev/min, varying the overhang of the boring

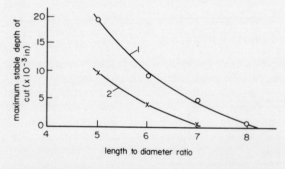

Figure 6. Experimental stability boundary at different length to diameter ratios. Feed rate: 0.008 in/rev, cutting speed: 400 rev/min, gap: 0.019 in. Curve 1: impact damper boring bar, curve 2: solid bar.

bar with the impact damper gap kept constant (curve 1). A similar curve for the solid boring bar is also shown (curve 2). The damped bar shows a marked improvement over the solid bar for all the overhangs tested, even though the static stiffness has altered, thereby changing the theoretical optimum gap. Under similar test conditions the performance of a tuned vibration absorber would be expected to deteriorate to a much greater extent.

Although the impact damper boring bar has a better performance than the solid bar, even when the overhang is changed, the improvement can be further increased by optimising the gap for the particular overhangs. As it is difficult to obtain an exact value for the change in stiffness and equivalent mass, it was assumed that the optimum gap was proportional to the overhang. The effect of the damper on stability, when the gap is altered according to this assumed relation, can be seen in figure 7. Comparison with

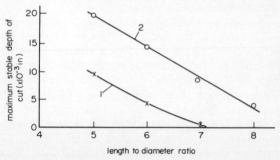

Figure 7. Experimental stability boundary at different length to diameter ratios. Feed rate: 0.008 in/rev, cutting speed: 400 rev/min. Curve 1: solid bar, curve 2: impact damper boring bar at optimum gap.

figure 6 shows that the stability of the boring bar is further increased when a better approximation to the optimum gap is chosen, although a considerable improvement is achieved even if this refinement is not introduced. Where a solid bar is effectively always unstable, as is the case at the length to diameter ratio of 7:1, it becomes meaningless to estimate the percentage increase in stability. It is better to consider the increase in overhang at which a stable cut of a given depth can be achieved. In this case a stable depth of cut of 0.005 in was only possible at a length to diameter ratio of 5.9:1 when a solid bar was used, but, when the impact damper boring bar was used, the maximum length to diameter ratio was increased to 7.6:1. There is no reason to suppose that further improvement could not be achieved if the true optimum gap was used at each overhang.

In a production version of this device, it would be very easy to calibrate the optimum gap for each overhang, and the gap could then be set mechanically in a few seconds, whenever the bar was set up, as small errors are insignificant, or in fact the gap need not be altered.

CONCLUSIONS

Use of the impact damper boring bar significantly increases the maximum stable depth of cut possible, compared with an equivalent undamped bar. A stable cut of 0.005 in can be taken at a length to diameter ratio of 7.6:1, where the solid bar could only be used at a ratio of 5.9:1. This improvement is not the maximum possible because the device could be accurately optimised for each overhang, and this calibration could then be used repeatedly. Alternatively, considerable improvement in stability can be achieved if for simplicity the device is optimised at one overhang and then locked at that setting for all other overhangs.

ACKNOWLEDGEMENTS

The authors would like to thank Professor S. A. Tobias, Head of the Department of Mechanical Engineering, University of Birmingham, for his guidance and advice during this research. Thanks are also due to Mr. E. J. Lamb for his assistance in the design of the bar, and to Birmingham University for providing the Carlyle Scholarship for the first author.

REFERENCES

1. M. D. THOMAS, W. A. KNIGHT and M. M. SADEK, 1970. 'Comparative dynamic performance of boring bars', *Proc. 11th Intern. Machine Tool Design and Res. Conf.*, Vol. A, p. 159.

2. R. S. HAHN, 1952. 'Design of the Lanchester damper for elimination of metal cutting chatter', *Trans. A.S.M.E.*, **73**, 331.

3. L. K. KUCHMA, 1957. 'Boring bars with improved resistance to vibration', *Engineer's Digest*, **2**, 68.

4. M. M. SADEK, 1966. 'Behaviour of the impact damper', *Proc. Inst. Mech. Engrs.*, **180**, Pt. 1, 38.

5. M. M. SADEK and B. A. MILLS, 1970. 'Effect of gravity on the performance of an impact damper: Part 1. Steady-state motion', *J. Mech. Engng. Sci.*, **12**, No. 4, 268.

6. P. C. PINOTTI and M. M. SADEK, 1970. 'Design procedure and charts for the impact damper', *Proc. 11th Intern. Machine Tool Design and Res. Conf.*, Vol. A, p. 181.

7. S. A. TOBIAS, 1965. *Machine Tool Vibrations*, Blackie, Edinburgh.

8. M. M. SADEK, 1972. 'The impact damper – an effective insert for controlling vibrations in machine tools', *Machinery*, 2 Feb., 152.

DESIGN OF MACHINE TOOL STRUCTURES:
JOINTS, SLIDES AND SPINDLES

A FINITE ELEMENT PROGRAM SYSTEM AND ITS APPLICATION FOR MACHINE TOOL STRUCTURAL ANALYSIS

by

H. OPITZ* and R. NOPPEN*

SUMMARY

A brief introduction to the finite element method is given, and a program system is described which permits structures to be analysed with respect to their static characteristics. The first part of the system is applicable to the deformation computation of structures approximated by flat elements and beams. The second part of the program system is used for the analysis of deformations and stresses in continuous three-dimensional components.

Some convergence tests, presented in the paper, demonstrate the accuracy to be obtained by the application of the program system. Further, various examples for a finite element analysis of problems arising in machine tool design are explained.

INTRODUCTION

The general intention is to design a structural member in such a way that it fulfils all important requirements at minimum costs. However, owing to the complexity of most design problems it is normally very difficult, if not impossible, to determine optimal solutions in the mathematical sense. It is therefore necessary to analyse various alternative design concepts with respect to their main characteristics and to base one's choice on those results.

In principle, this comparison can be carried out experimentally but, generally speaking, this method is rather time consuming and expensive. If the possibility is given one should therefore analyse the variants by computational methods. From the user's point of view it is important in this case that efficient computer programs, which are applicable to a wide range of different problems, are available.

All this holds for design in general, and for structural analysis in particular. In this field the finite element method has been widely used for the solution of problems arising in civil and mechanical engineering. In the design of machine tools finite element routines have already been used to determine the deformations of thin-walled columns[1,2].

In this paper an extended program system is described which is based on the finite element technique and permits a statically loaded structure to be analysed with respect to its elastic behaviour.

Some organisational foundations of this program system have recently been developed in a collaboration between the Manchester and the Aachen machine tool engineering divisions[3].

At first — without going into too much detail — the principal idea of the finite element method will be outlined. If required, further specific information on this section should be obtained from references 4, 5. After this some basic features of the program system are described, together with a few convergence tests. For these relatively simple examples exact solutions can be derived theoretically, so that the accuracy to be obtained by using computer procedures can be estimated. Another main part of the paper deals with the application of these finite element programs to various problems, arising in the design of machine tools.

THE FINITE ELEMENT METHOD

Very simple structures can be analysed exactly by using standard computation techniques. But, as the problems become more complicated with respect to either geometry or boundary and loading conditions, approximation methods have to be applied. In this field one has to distinguish whether the approximate solution describes the behaviour of the structure continuously in a closed form or only digitally at discrete points. At first sight the finite element method can be considered as belonging to the second group.

To apply this method of computation the structure to be analysed is—for means of calculation—subdivided into a number of finite elements which are of relatively simple geometrical shape and finite in dimensions. The elements are connected to each other at nodal points. For example, the column of figure 1 is represented by rectangular, triangular and

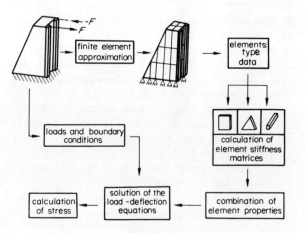

Figure 1. Finite element procedure.

beam-like elements. In this case the nodal points of the finite element are identical with the points of intersection of the grid. Except for simple structures this mathematical approximation to the original problem has to be set up manually and it is quite obvious that this procedure can become somewhat tedious for complicated geometries. It should be mentioned that, seen from a practical point of view,

* Laboratorium für Werkzeugmaschinen und Betriebslehre der TH Aachen

this fact is one of the most severe drawbacks of the finite element technique.

By assuming a certain displacement pattern, which describes the deflection under load for any point within the element as a function of the corresponding nodal values, one is able to derive an approximate solution for the elastic properties of this element. This is carried out by applying the principle of virtual work[4]. The elastic behaviour is expressed by a stiffness matrix, which relates nodal forces and thereby resulting nodal displacements to each other (figure 1). To a large extent the accuracy that is obtained by using the finite element technique depends on the choice of the displacement functions.

In reference 1 various types of these functions are discussed. According to figure 1 the following step in the analysis is given by superposing the load—deflection relationships for all elements of the model, i.e. all element stiffness matrices. This procedure is executed in such a way that, at every node, internal and external loads are in the state of equilibrium. Taking into account the boundary conditions (here in the form of prescribed displacements at certain nodes) and the loads which are externally acting upon the structure, a set of simultaneous equations is generated by the superposition process. In the case of a linear problem these equations are linear as well, which, of course, is of great advantage for the next operation, where the equations are solved for the unknown displacements. From this point it is then possible to derive all stresses acting within the structure. They are mainly given by certain differences of the first calculated displacements multiplied by geometrical and elastic constants.

From these general remarks about the prediction of the static behaviour of a structure by means of the finite element method it is evident that this technique is applicable to a very wide range of different structural analysis problems. This is because nearly all categories of structures can easily be represented by finite elements as long as the elastic properties of a few basic one-, two- and three-dimensional element types are known. And further it becomes clear that, in the field of machine tool design, well-known methods that only use beams as structural elements are just special cases of the general finite element procedure. It has been pointed out that the application of these techniques is seriously limited owing to the kind of approximation[6].

FINEL – A FINITE ELEMENT PROGRAM SYSTEM

The computer program system FINEL permits a structure to be analysed on the basis of the finite element technique with respect to its elastic behaviour. The problems which can presently be treated are assumed to be static and linear.

The program system consists of two parts, the first of which is used for the analysis of structures that can be represented by an assembly of flat and beam-like elements. Typical components of this kind are columns, headstocks, overarms, spindles, etc. The second part of the program system is applied to determine the characteristics of three-dimensional continuous structures such as gears, etc.

Subsequently the basic features of these finite element programs shall be outlined. In using part 1 of the computation system rectangular, triangular and beam-like elements are available for the idealisation of the structure (figure 2). The thickness of the flat

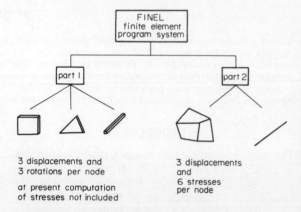

Figure 2. FINEL: finite element program system.

elements is supposed to be small compared with the side lengths so that the equations governing the two-dimensional state of stress can be applied with good accuracy.

In the case where the elements are not coplanar they incorporate two independent properties, namely membrane stiffness and plate-bending stiffness. Normally, these are automatically combined by the program, but in order to save computing time and storage requirements the 'unnecessary' parts can be suppressed by certain data identifiers, if just simple plates or membranes are analysed.

The derivation of the stiffness properties for the beam element is based upon the standard beam theory but provision has been made in the formulation of the stiffness matrix for the effects of shear forces to be included as well.

When using the FINEL 1 program without suppressing certain parts in the manner described above, the shape of the deformed structure is defined by six values for the displacements at each node. These are three transverse deflections, related to the Cartesian coordinate system in which the topology of the investigated system has been described and three rotations about the axes. By back-substituting these nodal values into the formerly explained displacement function one is able to determine the deformations at any point of the structure, but normally this step is not executed.

FINEL 1 was primarily developed for the analysis of components of metal cutting machine tools.

Of course, for high machining accuracy these structures are designed in such a way that the stiffness achieved is as large as possible. Because of this fact the resulting stresses are normally much below those that could be realised by the material. Consequently it is usually not necessary to check stresses in order to prevent a breakdown. Therefore the calculation of stresses has been left out of the FINEL 1 program. However, it should be mentioned that this limitation is not a serious one because the corresponding additional routines can easily be fitted in.

The important element of FINEL 2 is a general hexahedron, the nodal points of which are placed at the eight corners (figure 2). Owing to the fact that the derivation of the stiffness matrix for this element is performed in so-called natural coordinates[5], the relative positions of the nodes are arbitrary except for a few special cases of minor importance. This particular property makes it very easy to subdivide even geometrically complicated structures into finite elements.

When using FINEL 2, three displacements per node are computed which again are related to the coordinate axes of the entire structure. In contrast to FINEL 1 no rotations are determined in this case. In addition to the deformation analysis FINEL 2 carries out a stress analysis as well. This results in six stresses, which are calculated for every node. It is further possible to transform this general state of stress into a mean one by solving a 3 x 3 eigenvalue problem per node. The mean stresses determined in that way can then be superposed in order to achieve an effective stress that can be used for comparison with characteristic material properties.

The other element that is available in FINEL 2 is a pin-jointed bar. By definition, this type of structural element can only be subjected to axial forces, that means it could be used, for instance, in the analysis of trusses. So far, the bar element of FINEL 2 has only been used in connection with the general hexahedron for the representation of certain boundary or joint conditions.

Both parts of the FINEL program system are organised similarly to each other. The superposition of the element matrices yields a stiffness matrix for the complete structure, which is symmetric and of the banded type. Of course, both properties are taken advantage of in order to save core storage. The solution of the load—deflection equations follows the Gaussian procedure. One of the main advantages of the program system is that even large problems can be treated on relatively small computers as long as external storage devices can be used. This is achieved by automatically partitioning the whole problem at the beginning of the computation into sections.

The assembly of the main stiffness matrix is then performed successively by completing all sections, and also the solution procedure follows this technique.

It has already been mentioned in the introduction that the preparation of the data for finite element programs imposes a great problem. For certain structures with a geometry that is completely defined by only a few parameters an automatic data generation is most efficient. Typical problems of this kind are simple plate or box structures or gears. But if the topology is more general the preparation has mainly to be carried out manually, even though various aids such as digitisers are available. As the computing costs for a finite element analysis can be large, a reliable control of data must be given great attention. There are two kinds of checks which are of major importance, namely logical and visual tests. For both

groups programs have been written that are now attached to the FINEL system as service routines.

In the logical control the imposed boundary conditions are first checked whether they are sufficient for the structure to be stable. Then various tests on the element data are performed; for example, it is made sure that the four angles of a rectangular element are 90 degrees each within a certain accuracy. Further it is checked that at least one load is non-zero. If an error is detected, a comprehensive message is printed out that makes a correction easy.

The program for the visual control internally transforms the input data for the FINEL system and gives out a picture of the structure to be analysed. This picture is either drawn on a plotter or shown on the screen of a display system. By choosing certain identifiers, this program permits the structure to be looked at from different perspectives or to be turned round continuously, or permits particular cut-outs to be enlarged.

A further very important application of the program for visual control is the graphical presentation of computed results. For example, the deformations calculated for a structure can be multiplied by a scale factor and added to the coordinates of the corresponding node. By this it is possible to get an instantaneous visual impression of the behaviour of the structure because a lot of information is given in a concentrated form.

EXAMPLES FOR THE APPLICATION OF THE PROGRAM SYSTEM

(A) Convergence tests

In accepting a particular displacement function for an element type – normally a polynomial – the assumption is made that the deformation under load within the element follows this pattern, corresponding to the degree of the function, for instance a quadratic or cubic shape. Obviously for an element of given size and loading conditions the extent to which this assumption does hold depends on the choice of the displacement function.

In order to compare the behaviour of different element types with respect to the achieved accuracy, convergence tests are performed. For these rather simple problems theoretical solutions can be derived by standard computation methods and it is investigated how close the finite element solution for a particular mesh approaches the theoretical value and, further, how rapidly the error decreases as the mesh is refined.

An example for a convergence test is shown in figure 3(a). The square plate is simply supported and subjected to bending by a concentrated force, which is acting in the middle of the structure. From the results presented in the figure it is obvious that for a given number of nodes (vertical lines) the subdivision into reactangles (here squares) yields a higher accuracy for the calculated midpoint deflection than the approximation by triangles. Therefore, when analysing problems similar to this one, triangles should only be used if this is necessary from the geometry of the structure. As far as the second aspect is concerned, a fast convergence is obtained for these element types when refining the grid; and further it is seen that the rectangular element gives good results even for a coarse mesh. This fact is very important in order to reduce the amount of input data preparation.

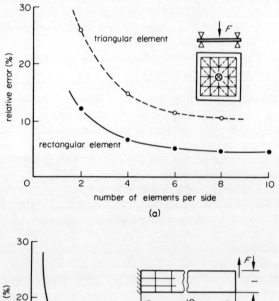

(a)

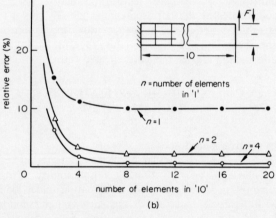

(b)

Figure 3(a). Convergence test: supported plate. (b) Convergence test: slender beam.

Another example is shown in figure 3(b), where the tip deflection of a slender cantilever has been calculated. For this the numbers of elements along length and along depth have been varied independently. The curves show that the error of the analysis rapidly decreases in particular as the number of elements along depth increases from 1 to 4.

These two problems have, of course, been investigated by using FINEL 1. An example for the application of FINEL 2 for three-dimensional continuous structures is given in figures 4(a) and 4(b).

According to the classical beam theory the stress along the parabolically shaped boundary of the beam should be constant. In the test this stress — due to bending — has been analysed with three-dimensional elements, using two different types of grids. The error for mesh 1 is less than 15% except for the region near the loading points, where the subdivision is more coarse. In mesh type 2 this part of the structure is approximated by more elements and it is evident,

from the corresponding curve, that a significant improvement has been achieved. When interpreting the presented results one has to be aware that several different types of errors influence the accuracy. First, the geometry of the structure has been slightly changed owing to the fact that the nodal points are interconnected by straight lines.

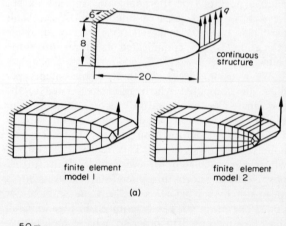

(a)

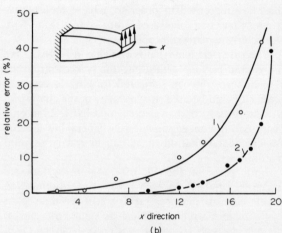

(b)

Figure 4(a). Parabolical beam: continuous structure and finite element models. (b) Relative errors of boundary stress.

Further, the finite element results have been compared with a value calculated by applying the beam theory. It is clear that this reference value is only an approximate one, which is particularly doubtful near the loading points because several specific assumptions become invalid there. Another component of the error is due to the fact that the loading cases are not identical for the original structure and the finite element model. Next, an error, as has been analysed in the previous examples, due to the actual element representation influences the calculated accuracy.

(B) Technical examples

In the left-hand part of figure 5 a hydrostatic slideway is shown. In order to determine the deformations of the outer component of the system this was approximated by rectangular finite elements and analysed with FINEL 1. In doing so a two-dimensional state of stress was assumed, which means that the structure shown represents a cut-out of, say, unit thickness out of the middle of a long slideway, where influences from both its ends are negligible. The system was subjected to a distributed load resulting from oil pressure being constant anywhere except for the regions near the two open ends, where it was assumed to decrease linearly.

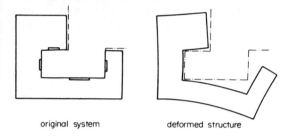

original system   deformed structure

Figure 5. Deformation analysis of a hydrostatic slideway.

In the right-hand part of figure 5 the shape of the deformed structure is presented in a qualitative form.

Having these calculated results, the designer is able to study the elastic behaviour of the structure. In particular, to give an example, he knows how the geometry of the oil slits varies as a load is applied and from there he can estimate what repercussions with respect to flow and pressure distribution have to be considered.

This investigation took a total time of about 1 hour in which data preparation and presentation of the results are included. The actual computation time was 12 seconds on a CDC 6400.

The following example deals with the analysis of two different types of flanges as they are used for screwed joints (figure 6). Together with the elastic

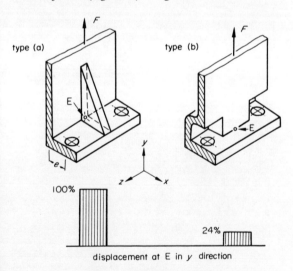

type (a)   type (b)

100%

24%

displacement at E in y direction

Figure 6. Deformation analysis of flanges.

properties of the screw and of the contact area the stiffness of the flange greatly influences the overall behaviour of the complete joint assembly. Because of the eccentricity e between the external load F and the forces from the screws, a relatively large deformation in vertical direction at point E occurs in structure (a). In flange type (b) the effect is significantly reduced. For the given system (b) the stiffness with respect to the deflection at point E is about 4.2 times larger than that achieved with flange (a). As points being situated above E in a positive y direction are considered, this value decreases, which

is due to a much higher stress and corresponding strain in structure (b) than in (a). Of course, this particular value calculated for point E depends on the dimensions chosen, etc., but it does stress a result, which has already been found experimentally[2], namely that load F and screw forces should act within the same plane. The complete analysis took about 3 hours (41 seconds on the computer) and from this it is obvious that the corresponding experimental investigation would require much more time, effort and costs.

Figure 7 shows a headstock, the static deformations of which have been calculated with the program. For this, an external force F was applied at the loading

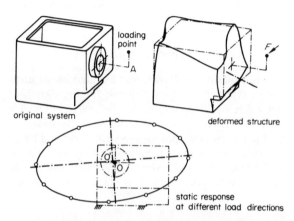

loading point

A

original system   deformed structure

static response at different load directions

Figure 7. Deformation analysis of a headstock.

point, changing in direction for each loading case. Altogether, twelve different loading cases have been considered simultaneously. In addition to this, force reactions from the bearings have been taken into account. The results are presented in figure 7 in the form of a static response at different load orientations. It is seen that, owing to the forces at the bearings, a constant displacement O–O' occurs at point A, on which the deformations resulting from load F are to be added symmetrically. In addition to these results, the shape of the deformed structure is given in figure 7 for the case when the external force F is acting horizontally.

A last example of an application of FINEL 2 will be given. At present the distribution of loads and

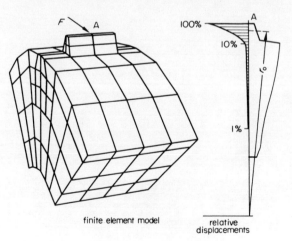

F   A   100%

10%

1%

finite element model   relative displacements

Figure 8. Deformation analysis of a gear.

stresses in helical gears is being investigated. To obtain a first idea of how to build up a finite element model sufficient to achieve good accuracy and yet not so detailed that it wastes computation costs, a preliminary test was carried out for a special type of gear. The particular aspect of this analysis was to find out to what extent the actual body of the gear would have to be considered for the final computation. Figure 8 shows the very coarse grid used and the deformations that have been calculated for nodal points on the centre line denoted by A at one end, when a tangential load F is applied. It is seen that the deflection decreases rather slowly, which for further computations means that in radial directions the entire body of the gear has to be subdivided into finite elements.

CONCLUSIONS

In this paper it was shown that by the idea of using various types of structural elements the finite element method is ideally applicable for the analysis of a wide range of different problems occurring in the design of machine tools. The program system FINEL is based upon the finite element technique and by means of its organisation permits even large structures to be analysed with respect to their static characteristics.

For the mathematical representation of the problems hexahedron, rectangular, triangular, beam-like and bar elements are available; most structures can be dealt with by these means. From the convergence tests that have been described in the paper it is obvious that a good accuracy is obtained by using the program system. The calculation of four technical examples has demonstrated that these finite element programs are a very efficient aid for the designer to analyse various structures and to predict the improvements that can be achieved by particular modifications.

REFERENCES

1. A. COWLEY and S. HINDUJA, 1971. 'The finite element method for machine tool structural analysis'. *Ann. C.I.R.P.*, 19.
2. H. OPITZ, 1971. *Moderner Werkzeugmaschinenbau*, Girardet, Essen.
3. R. NOPPEN and A. COWLEY. 'A finite element program system', to be published.
4. J. S. PRZEMIENIECKI, 1968. *Theory Of Matrix Structural Analysis*, McGraw-Hill, New York.
5. O. C. ZIENKIEWICZ, 1971. *The Finite Element Method in Engineering Science*, McGraw-Hill, London.
6. S. HINDUJA and A. COWLEY, 1971. 'The finite element method applied to the deformation analysis of thin-walled columns', *12th M.T.D.R. Conf.*, Macmillan, p. 455.

EXPERIMENTAL STUDY OF THE NORMAL STATIC STIFFNESS OF METALLIC CONTACT SURFACES OF JOINTS

by
C. DEKONINCK*

SUMMARY
An experimental method is developed for measuring the static plastic and elastic normal deformation properties of contact surfaces of fixed metal joints. The study mainly considers two-dimensional prismatic roughness-peaks. The influence on the elastic stiffness, of the application of a preload to the joint faces was investigated. A relation was found between the normal elastic stiffness and the height of the roughness-peaks.

NOTATION

b width of the top of a prismatic roughness-peak (mm)

$k_{z\,el}$ normal elastic stiffness of a joint (kg/mm)

l length of a prismatic roughness-peak (mm)

n number of prismatic roughness-peaks of a contact surface

$P_z{'}$ normal load applied to the testspecimen (kg)

s feed (mm)

S_w real contact surface $= nbl$ (mm^2)

$2\beta_0$ initial top-angle of a prismatic roughness-peak ($^\circ$)

$\delta_{z\,pl}$ normal plastic deformation of a roughness-peak (mm)

σ_d yield-strength of a material (kg/mm^2)

σ_s real limit tension of a roughness-peak (kg/mm^2).

INTRODUCTION

Investigations of the normal static stiffness of metallic contact surfaces of joints have been made by several authors[1-5].

This paper is concerned with plastic and elastic normal deformation components of contact surfaces and more distinctly the deformations of the roughness-peaks.

To determine accurately the deformation properties it is necessary to know exactly all parameters influencing these deformations. Two important parameters are:

(i) the roughness geometry of the contact surfaces,
(ii) the material properties of the surfaces.

The roughness geometry, obtained by some conventional machining operation like turning, milling, grinding or shaping, is very difficult to reproduce exactly, due mainly to the wear of tools and the granular structure of metals and the statistically random effects influencing the cutting process.

Moreover, the machining of large surfaces can possibly introduce a supplementary inaccuracy of the surface, namely *waviness*. We note, however, that waviness can also be introduced in initially-flat joined contact surfaces due to uneven contact pressures over the contact surfaces.

Considering these influencing factors, our experimental work for determining the plastic and elastic normal deformation properties of metallic contact surfaces has been carried out using mainly contact surfaces with accurately-machined two-dimensional prismatic roughness peaks, parallel to one side of the surface (figure 1).

The V-grooves are obtained by accurately milling. Moreover, no waviness of the surface could be

measured. We used top angles $2\beta_0$ of 45, 60, 90, 120 and 160° and feeds $s = 0.5$, 1 and 2 mm.

To obtain good material homogeneity, the test specimens are made from mild steel of good quality. The roughness-peaks are pressed against ground ($R_t \approx 2\mu$) mating surfaces, on the one side made of the same mild steel as the roughness peaks, and on

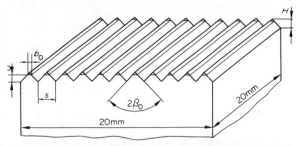

Figure 1. Contact surface with prismatic roughness-peaks.

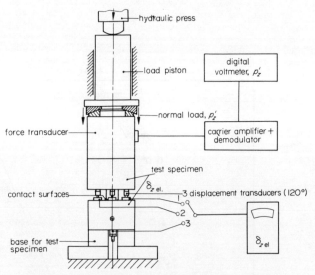

Figure 2. Testrig for applying normal loads to test specimens.

the other side against polished surfaces of hard steel ($H_{Vickers} \approx 1000$). During the initial normal compression a plastic and elastic deformation of the roughness-peaks occurs, as well as of the mating surface when using mild steel; when using hard steel for the mating surface, it only deforms elastically. Besides test specimens with these artificial geometries (figure 1) some others were used, for comparison, with contact surfaces shaped and ground in a conventional way.

Figure 2 is a schematic arrangement of the testrig for applying normal static loads to the contact

*Senior Lecturer, Laboratorium voor Machines en Machinenbouw, University of Ghent, Belgium.

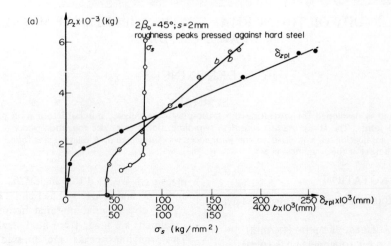

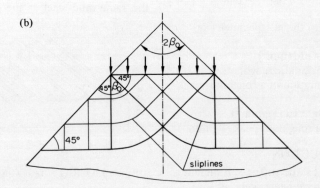

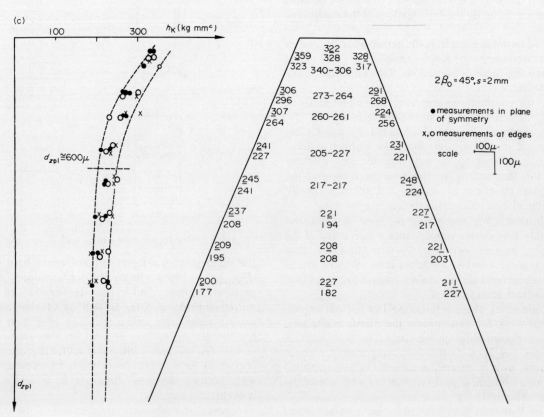

Figure 3. (a) Relationship between P_z' and δ_z pl, b, σ_s. (b) Plastic deformation model.
(c) Microhardness pattern of prismatic roughness-peak after plastic deformation.

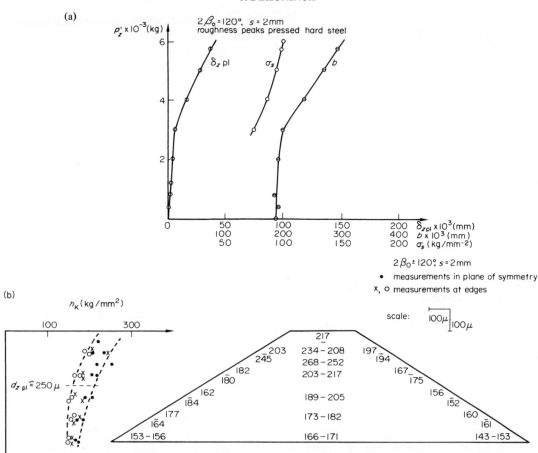

Figure 4. (a) Relationship between P_z' and $\delta_{z\,pl}$, b, σ_s.
(b) Microhardness pattern of prismatic roughness-peak after plastic deformation.

surfaces and measuring the corresponding normal deformations.

PLASTIC DEFORMATION PROPERTIES

Figure 3(a) shows the normal plastic deformation $\delta_{z\,pl}$ of roughness-peaks $2\beta_0 = 45°$, $s = 2$mm which were pressed against a mating surface from hard steel. The top of the roughness-peaks was ground to give a width $b_0 \approx 85\mu$, before the experiment was carried out.

Figure 3(b) represents the corresponding theoretical deformation pattern, according to, for example, Prandtl[6], when the plastic deformation starts.

Figure 3(a) also represents the measured values of the width b at the top of the roughness-peaks, corresponding to different normal loads P_z'. The real limit tension σ_s, in the contact zone was calculated by the formula

$$\sigma_s = \frac{P_z'}{nbl} \qquad (1)$$

with n = number of prismatic roughness-peaks on the contact surfaces

l = length of prismatic roughness-peaks.

Prandtl suggested a formula to relate the limit tension σ_{so}, at the initial plastic deformation, to the

geometry and material properties of a prismatic peak

$$\sigma_{so} = \sigma_{do}(1 + \beta_0) \qquad (2)$$

with σ_{do} = initial yield-strength of material
β_0 = initial semi-top-angle.

Figure 3(c) shows a polished perpendicular section of a plastically-deformed prismatic roughness-peak after a maximum normal load $P_z' = 5700$ kg; and the Knoop microhardness measurements (weight 300 g) at different locations of the section. At the left-side the hardness values are represented graphically as a function of depth. If we suppose that plastic deformations occur in the region where workhardening can be measured, figure 3(c) shows a plastic deformation to a depth $d_{z\,pl} \approx 600\mu$; the deformations in the plane of symmetry and at the edges are quite similar. Figure 4(a) represents analogous plastic deformation measurements for roughness-peaks $2\beta_0 = 120°$, $s = 2$mm and $b_0 \approx 190\mu$.

Figure 4(b) shows the corresponding Knoop microhardness measurements in a perpendicular section. This figure shows considerable workhardening in the plane-of-symmetry of the peak and relatively small hardening at the edges. This effect might be explained, considering the influence of elastic stresses, during the plastic deformations. The influence of the

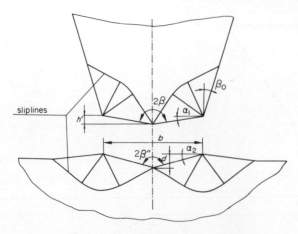

Figure 5. Plastic deformation model.

elastic strains during plastic deformation. could be detected experimentally by means of interferometric photographs of the roughness-peaks (elastic recovery). A convex curvature of the top plane of the peaks with width b could be detected. The actual value of the radius of curvature depends on several parameters:

(i) the initial width b_0,
(ii) the top angle $2\beta_0$,
(iii) the feed s,
(iv) the maximum load applied to the peak.

The compression of roughness-peaks and a mating surface of mild steel, yields analogous results as mentioned before; however, a supplementary plastic deformation component, according to figure 5, is superposed. Figure 6 represents the b values, obtained from several experiments, after a maximum load $P_z' = 5700$ kg. For $s = 2$mm a minimum in the $b - 2\beta_0$ curves can be detected. This effect can be explained considering:

(i) the Prandtl formula (2). From this we find a decrease of b with increasing $2\beta_0$ values, if we consider constant σ_{do} values;
(ii) the fact that for peaks with large $2\beta_0$ values we experimentally found small δz_{pl} values at $P_z' = 5700$ kg, compared to peaks with small $2\beta_0$ values. As a consequence, the corresponding work hardening

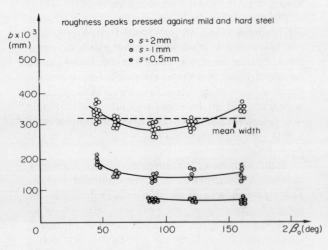

Figure 6. Relationship between b and $2\beta_0$.

and the corresponding σ_d values were much smaller in the first case;

(iii) the presence of elastic deformation components during plastic compression. These components introduce for example an extra widening of the contact zone during compression, especially for roughness-peaks with large $2\beta_0$ values.

For $s = 0.5$ mm the minimum in the $b - 2\beta_0$ curve almost disappears. This effect can be explained by the fact that the b-values have approximately the same value as the size of the steel grains; in this case the deformation mechanism according to figures 3(b) or 5 is no more effective and the limit tension of the roughness peaks is almost independent of geometry.

ELASTIC DEFORMATION PROPERTIES

Figure 7 shows a normal deformation curve for prismatic roughness-peaks $s = 2$mm, $2\beta_0 = 45°$, $b_0 \approx 90\mu$ pressed against a mating surface of mild steel. The initial loading curve OA includes:

(i) normal plastic deformation component as described before

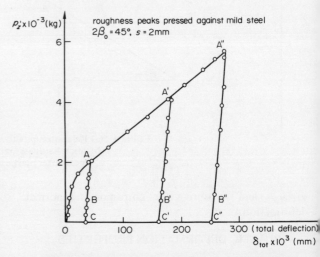

Figure 7. General deformation characteristic of a joint.

(ii) a normal elastic deformation component due to the body of the roughness-peak
(iii) a normal elastic deformation component of the base of the roughness-peak in an 'elastic foundation'
(iv) a normal elastic deformation component of the top of the roughness-peak in an 'elastic mating surface'
(v) a sideways elastic deformation of the body of the roughness-peak.

Fundamentally the elastic deformation can be calculated by means of the elastic contact theories.

When the normal load is decreased from A ($P_z' = 2000$ kg), the corresponding deformations are elastic (curve ABC). OC represents the plastic deformation component of the joint, due to the preload $P_z' = 2000$ kg. We experimentally found the elastic deformation curve ABC to contain a straight part AB corresponding to rather high mean contact pressures and a nonlinear part BC corresponding to small mean

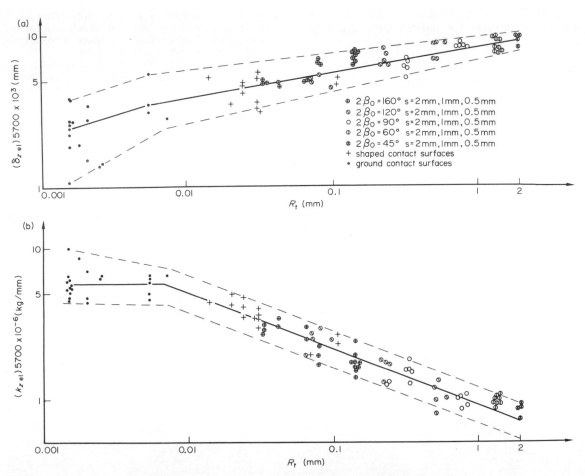

Figure 8. (a) Relationship between $k_{z\,el}$ and R_t after a preload of $P_z' = 5700$ kg.
(b) Relationship between $\delta_{z\,el}$ and R_t after a preload of $P_z' = 5700$ kg.

contact pressures. If, again, from C the load is increased, the elastic deformation curve is CBA. No hysteresis effect could be detected in this case. If the normal load is consequently increased beyond A ($P_z' = 2000$ kg), a new plastic deformation component is introduced and so on.

The reasons for the nonlinearity of the part BC, B′C′, B″C″ are many and very complex and will not be considered here. Experimentally we found an increasing slope of the linear parts AB, A′B′, A″B″ with increasing preloads. This means an increasing elastic stiffness of the joint with increasing preloads.

If we make a comparative study of the linear parts of the elastic deformation curves of metal joints, it is sensible to use the same values of the maximum preloads. We choose for this value $P_z' = 5700$ kg.

Figure 8(a) shows the normal elastic stiffness $k_{z\,el}$ of contact surfaces, measured, in the linear parts of the deformation curves, as a function of the height R_t of the roughness-peaks, measured after application of the preload $P_z' = 5700$ kg. Figure 8(b) shows the corresponding elastic deformation of the joints. These figures are related to prismatic roughness geometries and also to ground and shaped surfaces.

Some important conclusions can be drawn from figures 8(a) and (b)[7]:

(i) The elastic stiffness decreases much less than proportionally with increasing surface roughness. This result can be explained by the fact that, for contact surfaces with a high finishing degree such as ground surfaces, the plastic deformation of roughness-peaks, due to a preload, is relatively small and so is the corresponding increase of the elastic stiffness.

(ii) For the same s-values, the elastic stiffness of prismatic roughness-peaks increases with increasing values of $2\beta_0$.

This effect can be explained by the fact that b-values after a load $P_z' = 5700$ kg are very similar for the different roughness geometries considered (figure 6). This means that after the preload the height of peaks with large top-angles is smaller than that of peaks with small top-angles.

CONCLUSION

The normal plastic deformation properties of prismatic roughness-peaks can in general be predicted quite accurately by the Prandtl formula.

But, if the dimensions of the roughness-peaks are approaching the size of the steel grains, the Prandtl plastic deformation mechanism can no longer be detected, and the influence of the geometry of roughness-peaks on the limit tension decreases considerably.

The plastic deformation of roughness-peaks, due to preload of a joint, increases the normal elastic stiffness.

Experimentally this effect was found to be of increasing importance for surfaces with increasing roughness.

REFERENCES

1. R. H. THORNLEY et al., 1965. Int. J. Mach. Tool Des. Res. 5.2.
2. Z. M. LEVINA, 1967. Research on the static stiffness of joints in machine tools, 8th Int. Mach. Tool Des. Res. Conf., University of Manchester.
3. J. BUC and B. NOWICKI, 1967. The measurement of the real area of contact between two metal surfaces, 8th IMTDRC, University of Manchester.
4. R. E. SCHOFIELD and R. H. THORNLEY, 1972. Calculating the elastic components of deflections of plane joints formed from machined surfaces. 12th IMTDRC, University of Manchester. Macmillan, 89.
5. Y. ITO and M. MASUKO, 1972. Experimental study of the optimum interface pressure on a bolted joint considering the damping capacity. 12th IMTDRC, University of Manchester. Macmillan, 97.
6. PRANDTL, 1920. Uber die Eindringungsfestigkeit plastischer Baustoffe und die Festigkeit von Schneiden. Zeitschrift für Angewandte Mathematik und Mechanik. Band 1.
7. C. DEKONINCK, 1970. Mechanical deformation properties of contact surfaces of fixed mechanical joints. Ph.D. Thesis, Univ. Ghent (Belgium).

CALCULATING THE ELASTIC AND PLASTIC COMPONENTS OF DEFLECTION OF JOINTS FORMED FROM MACHINED SURFACES WITH FLATNESS ERRORS

by

R. E. SCHOFIELD* and R. H. THORNLEY†

SUMMARY

Contributions to successive MTDR Conferences have stressed the need for knowledge of the behaviour of joints in machine tool structures. At the twelfth conference, the authors presented theories for calculating the elastic and plastic components of deflection of plane joints formed from machined surfaces[1]. In this paper these theories are extended to the more general case of joints formed from surfaces with flatness errors. Again, a mathematical expression of surface finish is applied to the calculation of elastic and plastic deflections and from these, values of initial deflection and elastic recovery can be made. Experimental results show that the method is valid within the limits of accuracy of the surface finish expression.

NOTATION

P mean interface pressure
Y_c yield pressure
λ_p plastic component of deflection
λ_e elastic component of deflection
λ_t total deflection
H total peak-to-valley height
§   deflection at a point due to uniform pressure over a given area on the surface of a semi-infinite body of elastic material
μ Poisson's ratio
f feed or distance between cusps
θ half-angle of lay orientation
r half-length of diagonals of contact spots
h peak-to-valley height of one asperity or cusp
E Young's modulus of elasticity

INTRODUCTION

The complete and accurate analysis of machine tool structures is still impossible because of the inability to take the static and dynamic effects of joints fully into account. The authors have shown[1] that the factors affecting joint stiffness are

(a) flange deformation
(b) elastic deflection of the body material
(c) the compliance due to surface asperities and that (a) and (b) depend on (c).

Recognising the compliance due to the surface asperities as being the fundamental property in the analysis of joint stiffness the authors have proceeded to establish a method for its calculation[1]. It is suggested that this method provides a basis therefore for the calculation of overall joint stiffness. However, the method[1] is applied only to joints formed from surfaces free from flatness errors, that is, subject only to roughness error. The purpose of this paper is to extend the theory to the more general case of joints formed from surfaces subject to both roughness and flatness errors.

MECHANISM OF THE APPROACH OF SURFACES

As is discussed more comprehensively[1], when two surfaces first make contact, it is limited to very few points. The number and area of these contact spots increase as the approach continues. It is established in[1] that the elastic recovery, which manifests itself when the mean interface pressure on the joint is reduced, is due to the elastic deflection of the body material behind the contact spots. Under the initial loading condition the approach of surfaces depends on their bearing area properties and the initial approach can be calculated from the joint when the bearing area becomes large enough to support the load. The elastic recovery can be calculated by assuming it to be that of the contact spots if they are taken as an area of uniformly-distributed pressure on the surface of a semiinfinite body. This means that the recovery of the joint is that at the centre of the largest contact spot. It is also established[1] that the parameters governing the stiffness of the surface region are the material properties of the machined surfaces and the surface characteristics.

MATERIAL PROPERTIES

The material properties that govern joint deflection are the modulus of elasticity, Poisson's ratio and the flow or yield-pressure of the material under the contact conditions. The last is the only one of these to present any problem because of the variation of yield pressure with the angle of the contacting asperities. It has been shown however[2] that the Vickers hardness test is a suitable simulation of asperity contact and that the value of yield-pressure calculated from this test is sufficiently accurate for the calculation of the plastic component of joint deflection.

SURFACE ERROR CHARACTERISTICS

Surface error may be considered to fall into two categories, namely flatness and roughness. The flatness error is one in the nominal plane of the surface: the roughness error is one of much higher frequency superimposed on the former.

Roughness errors
Roughness errors are due to tool geometry and feed rate and such random effects as

* Department of Mechanical Engineering, Queen's University of Belfast.
† Department of Production Engineering, University of Aston in Birmingham.

- asperities thrown up by the machining process
- pieces of surface torn out by the tool causing a reduction in bearing area
- the built-up edge effect leaving particles of material on the surface.

These errors can be detected by a stylus-type instrument such as the Talyrond but this gives only an assessment of a very short sample length and cannot detect flatness errors. This method alone gives only an assessment of roughness such as CLA or peak-to-valley height, and even on a surface without flatness errors does not give an indication of its bearing properties. It has been suggested[3] that the bearing area curve and highspot frequency are far more valuable methods of assessment. The authors[4] have developed this hypothesis into a mathematical assessment and have shown it to be valid[1]. It consists of representing the surface as a number of asperities having the shape of the bearing area curve. The number of asperities is determined from the highspot frequency count.

Flatness errors

Flatness errors are due to corresponding errors in the slideways of the machine tools on which they are produced or, for example, in the profile of a slab milling cutter. A method of assessment of the flatness error[5] has been developed using the Talyrond instrument and so, by superimposing the bearing area and highspot frequency values described above, complete assessment of the surface is possible to a realistic degree of accuracy.

ANALYSIS OF JOINT DEFLECTION

Idealised surfaces

Consider the surfaces shown in figure 1(a). The bearing area curve of such a surface is shown in figure 1(b) whilst figure 1(c) shows the shape of the cusp derived by dividing the bearing area curve by the number of cusps on the surface. A surface of such cusps has the same bearing area as the original surface.

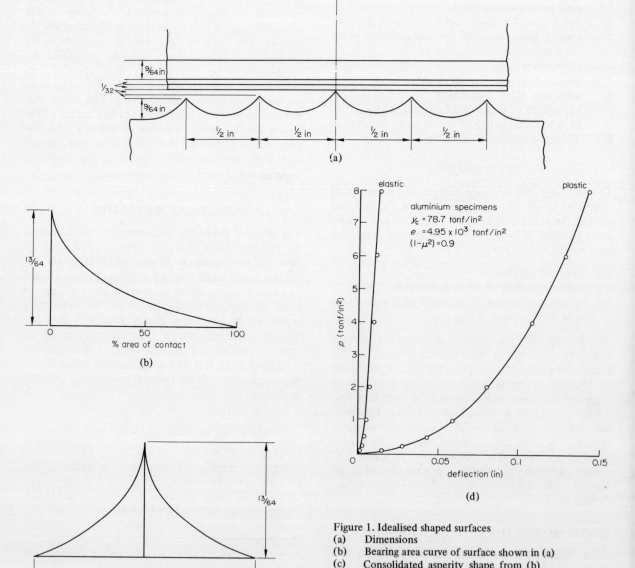

Figure 1. Idealised shaped surfaces
(a)   Dimensions
(b)   Bearing area curve of surface shown in (a)
(c)   Consolidated asperity shape from (b)
(d)   Relationship between P, λ_p and λ_t for a joint composed of the surfaces shown in (a). Theoretical curves; experimental points.

The geometry of these cusps can now be defined by a parabola of height H, where H is the maximum peak-to-valley height.

The plastic component of deflection may now be calculated in the same way as for flat surfaces using equation (1), that is,

$$\left(1 - \sqrt{1 - \frac{\lambda_p}{H_1}}\right)\left(1 - \sqrt{1 - \frac{\lambda_p}{H_2}}\right) = \frac{2P}{Y_c} \quad (1)$$

or if the surfaces are similar

$$\lambda_p = H\left[1 - \left(1 - \sqrt{\frac{2P}{Y_c}}\right)^2\right] \quad (2)$$

The elastic recovery of such a surface may now be calculated, as follows. Equation (18) from[1] gives an expression for the elastic recovery of the centre-point of a contact spot on a joint

$$\S_0 = \frac{1.12(1 - \mu^2)\sqrt{(PY_c)}f}{E} \quad (3)$$

This shows that the elastic recovery of the centre-point of a contact spot is directly proportional to its area.

Therefore the elastic recovery of a surface will be that of the largest contact spot.

It can be assumed that the highest cusp will provide the largest contact spot. Thus, if the geometry of the highest cusp is known, together with the plastic deflection, the largest area of contact can be calculated and from it the elastic recovery of the surface. λ_p can be calculated from equations (1) or (2). Cusps of this shape will produce a contact area in the form of a parallelogram which can be defined by the lengths of the diagonals $2r_1$ and $2r_2$. These values are defined in equations (3) and (4) in[1]

$$\frac{2r_1 \sin 2\theta}{f_1} = 1 - \left(1 - \frac{\lambda_p}{h_1}\right)^{\frac{1}{2}} \quad (4)$$

and

$$\frac{2r_2 \sin 2\theta}{f_1} = 1 - \left(1 - \frac{\lambda_p}{h_2}\right)^{\frac{1}{2}} \quad (5)$$

λ_p may be substituted in equations (4) and (5) to obtain values of r_1 and r_2.

The elastic deflection can then be obtained by following the same procedure as in equation (1).

Tests were carried out to verify this method of calculating the deflection of such a surface. The results are shown in figure 1(d), together with the theoretical predictions.

Conventionally shaped surfaces

If the distribution of asperity heights is assumed to be parabolic, equations (1) or (2) may be used to determine the total plastic deflection of the joint λ_p.

The above method was used to determine the properties of such surfaces. Figures 2(a), (b) and (c) and the results of the compression test and comparison with the theoretical predictions are shown in figure 2(d). Again the loading results are compared with the theoretical value λ_p from equation (1).

From equation (1) the elastic recovery of the joint will be equal to that at the largest contact point. It becomes necessary therefore to calculate the area of the contact spot consistent with a deflection of λ_p on the highest cusp. If the total deflection is less than the nominal asperity heights h, this is possible using the procedure set out above.

Sometimes, however, the presence of a flatness error greatly increases the actual contact pressure over the general areas of contact and the total deflection is greater than the nominal asperity height (see figure 3(d)). This greatly complicates the calculation of the contact spot shape and area and further work is needed before this can be achieved.

If the total deflection is less than h and if the value of λ is substituted in equations (4) and (5), with h_1 and h_2 as the nominal cusp heights, values of r_1 and r_2 can be obtained. This defines the shape of the contact area and the elastic deflection of the surface may be evaluated by following the procedure in equation (1).

If the surfaces are similar, the total deflection λ_p may be calculated from equation (2).

$$\lambda_p = H\left[1 - \left(1 - \sqrt{\frac{2P}{Y_c}}\right)^2\right]$$

and if the machine lays are orientated at $90°$ a square contact area may be assumed.

From equations (4) and (5)

$$r = \frac{f}{2}\left[1 - \sqrt{1 - \frac{\lambda_p}{h}}\right] \quad (6)$$

Substituting equation (2) in equation (6) yields

$$r = \frac{f}{2}\left[1 - \sqrt{1 - \frac{H}{h}\left\{1 - \left(1 - \sqrt{\frac{2P}{Y_c}}\right)^2\right\}}\right] \quad (7)$$

Equation (15)[1] gives

$$\S = \frac{3.526(1 - \mu^2)Y_c r_1}{2\sqrt{2\pi}E} \quad (8)$$

Substituting for r from equation (7) in equation (8)

$$\S = \frac{3.526(1 - \mu^2)Y_c f}{4\sqrt{2\pi}E}$$
$$\times \left[1 - \sqrt{1 - \frac{H}{h}\left\{1 - \left(1 - \sqrt{\frac{2P}{Y_c}}\right)^2\right\}}\right] \quad (9)$$

and

$$\S_0 = \frac{3.526(1 - \mu^2)Y_c f}{\sqrt{2\pi}E}$$
$$\times \left[1 - \sqrt{1 - \frac{H}{h}\left\{1 - \left(1 - \sqrt{\frac{2P}{Y_c}}\right)^2\right\}}\right] \quad (10)$$

$$\therefore \lambda_\theta = \frac{1.58Y_c f}{E}$$
$$\times \left[1 - \sqrt{1 - \frac{H}{h}\left\{1 - \left(1 - \sqrt{\frac{2P}{Y_c}}\right)^2\right\}}\right](1 - \mu^2) \quad (11)$$

This relationship is compared with experimental values in figure 3(d), taking $H = 1075$ μin.

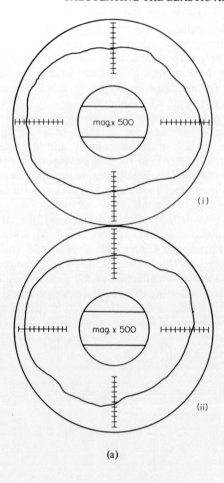

mag. x 500

(i)

mag. x 500

(ii)

(a)

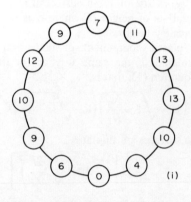

units x 0.0001 in

(i)

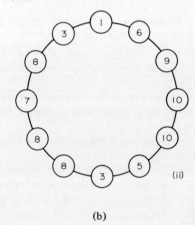

(ii)

(b)

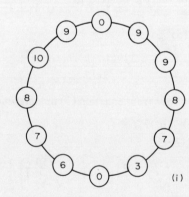

units x 0.0001 in

(i)

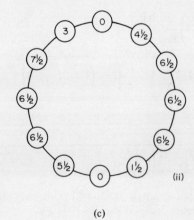

(ii)

(c)

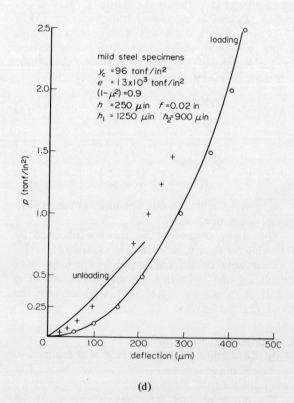

mild steel specimens
y_c =96 tonf/in²
e = 13x10³ tonf/in²
$(1-\mu^2)$ =0.9
h =250 μin f =0.02 in
h_1 = 1250 μin h_2=900 μin

loading

unloading

p (tonf/in²)

deflection (μm)

(d)

Figure 2. Conventionally shaped surfaces
(a)　Talyrond flatness traces
(b)　Analysis of Talyrond traces from (a)
(c)　Adjusted flatness errors from (b)
(d)　Relationship between P, λ_t and λ_e for a joint composed of the surfaces analysed in (a), (b) and (c). Theoretical curves: experimental points.

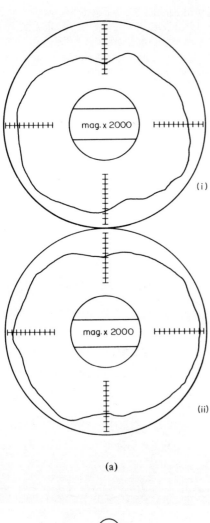

(i)

(ii)

(a)

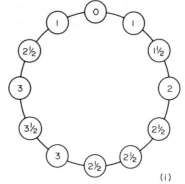

units x 0.0001 in

(i)

(ii)

(b)

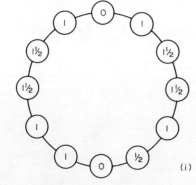

units x 0.0001 in

(i)

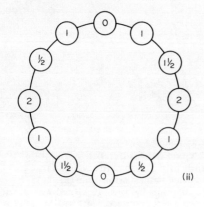

(ii)

(c)

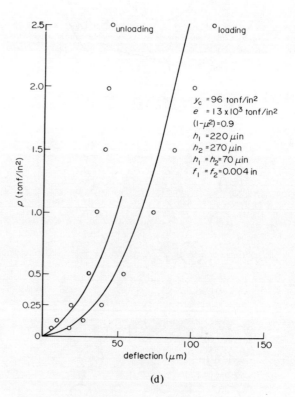

y_c = 96 tonf/in²
e = 13 x 10³ tonf/in²
$(1-\mu^2)$ = 0.9
h_1 = 220 µin
h_2 = 270 µin
h_1 = h_2 = 70 µin
f_1 = f_2 = 0.004 in

(d)

Figure 3. Ground surfaces
(a)   Talyrond flatness traces
(b)   Analysis of Talyrond traces from (a)
(c)   Adjusted flatness errors from (b)
(d)   Relationship between P, λ_t and λ_e for a joint composed of the surfaces analysed in (a), (b) and (c). Theoretical curves: experimental points.

Joints formed from surfaces machined by other methods

The deflection of joints formed from surfaces machined by any method may be calculated using this method. The only difference is in the shape of the asperities in each case. The bearing area curve is sufficient to calculate the initial deflections but a knowledge of the size and shape of the asperities is necessary to calculate the elastic recovery.

This information would be available from the Talyrond test[5] and a standard Talysurf test. Tests were carried out on ground and slab-milled surfaces. The flatness results and compression test results, and comparisons with theoretical values, are given in figures 3 and 4.

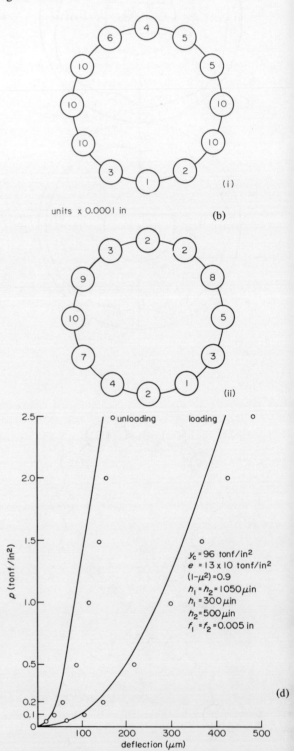

Figure 4. Slab-milled surfaces
(a)   Talyrond flatness traces
(b)   Analysis of Talyrond traces from (a)
(c)   Adjusted flatness errors from (b)
(d)   Relationship between P, λ_t and λ_e for a joint composed of the surfaces analysed in (a), (b) and (c). Theoretical curves: experimental points.

EXPERIMENTAL PROCEDURE

The experimental procedure is fully documented in [2]. The mean interface pressure is varied and the deflection measured while all other parameters are

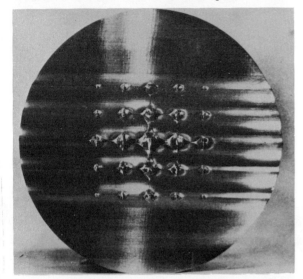

Figure 5. Specimen used in the test on a joint formed from surfaces of idealised shape.

kept constant. Tests were carried out on the idealised shaped surface shown in figure 1, on the conventionally shaped surface represented in figure 2 and on ground and slab-milled surfaces represented in figures 3 and 4 respectively.

DISCUSSION OF RESULTS

From the results of the compression tests it can be seen that the theoretical predictions are in relatively close proximity to the experimental values. This would appear to testify to the validity of the theory. The idealised shaped and conventionally shaped surfaces appear to allow prediction with a greater accuracy than the slab-milled and ground surfaces. This follows the trend reported in 1 where surfaces presenting the greatest difficulty of assessment tend to produce the greatest errors in prediction of deflection. In this case the last two surfaces have more random features than the first two. Even in the case of conventionally shaped surfaces, there is a geometric pattern to the roughness errors caused by the feed marks. This is only marginally so in the case of ground surfaces and not at all in the case of slab-milled surfaces.

It can be seen by comparing figures 2, 3 and 4 that greater values of deflection both under initial loading condition and elastic, on removal of the load, occur with greater flatness errors.

In the case of surfaces with significant flatness errors such as this the roughness errors are only of secondary importance, as is shown by comparing figures 2 and 4. The shaped surface represented in figure 2 has lower values of roughness peak-to-valley height h than the slab-milled surface but nevertheless shows greater values of deflection, both total and elastic. The sum of the total peak-to-valley height values H is slightly greater in the case of the

shaped than the slab-milled surfaces. The reason for the primary importance of the flatness errors is that they reduce the potential real contact area and so increase the effective mean interface pressure in these real contact areas. This will obviously increase the total deflection and by increasing the contact spot size will therefore also increase the elastic recovery and cause a more compliant joint[6].

This confirms conclusion (4) in 1 which states that the stiffest possible joint under elastic conditions is one formed from surfaces that have sufficiently large roughness errors to overcome the flatness errors, in such a way as to ensure that the contact spots are distributed over the complete surface.

CONCLUSIONS

(1)   The analysis made above, to calculate the initial deflection and elastic recovery of joints formed from machined surfaces with flatness errors, is valid.

(2)   The method of assessment of surface characteristics is valid but suffers from the inability accurately to assess the roughness characteristics of surfaces where the asperities do not have some basic geometric form.

(3)   On surfaces where flatness errors are significant, the roughness errors tend to be of secondary importance.

(4)   Flatness errors in excess of the roughness peak-to-valley height result in excessively compliant joints.

(5)   As stated in 1, the stiffest joint under elastic conditions is one where the roughness peak-to-valley height is sufficiently great to cause contact spots to be distributed over the complete joint surface.

ACKNOWLEDGEMENTS

The authors would like to thank Professor F. Koenigsberger for his encouragement during the project and for allowing the work to be carried out in the Machine Tool Division at UMIST. They would also like to thank the Science Research Council for their financial support.

REFERENCES

1.   R. E. SCHOFIELD and R. H. THORNLEY. Calculating the elastic and plastic components of deflection of plane joints formed from machined surfaces. *12th. Int. MTDR Conference, Manchester, 1971*, Macmillan, p. 89.
2.   R. CONNOLLY, R. E. SCHOFIELD and R. H. THORNLEY. The approach of machined surfaces with particular reference to their hardness. *8th Int. MTDR Conference, Manchester, 1967*, Pergamon.
3.   R. E. REASON. The bearing parameters of surface topography. *Rank-Taylor-Hobson, 1964*.
4.   R. E. SCHOFIELD and R. H. THORNLEY. Mathematical expression of surface finish characteristics. *Proc. Conference on Properties and Metrology of Surfaces I. Mech. E. Oxford, 1968*.
5.   R. H. THORNLEY, R. CONNOLLY and F. KOENIGS-BERGER. The effect of flatness of joint faces upon the static stiffness of machine tool joints. *Proc. I. Mech. E. Vol. 182 Part 1. No. 18 1967–68*.
6.   R. E. SCHOFIELD. Contribution to 5. *Proc. I. Mech. E. Vol. 182 Part I No. 18 1967–68*.

THE DAMPING EFFECT OF JOINTS FORMED FROM MACHINED SURFACES — THE STATE OF THE ART

by

R. E. SCHOFIELD*

SUMMARY

It is well established that one of the great obstacles to a complete understanding of the static and dynamic behaviour of machine tool structures is the inability to take the effects of the joints fully into account. The relevant properties of the joints are the static stiffness and the damping capacity. In both cases the aim must be not only to understand fully the parameters which affect these properties so that they may be optimised but also to be able to calculate and predict actual values. This latter requirement is necessary before any complete computer simulation of a machine tool structure can be made.

Many attempts have been made to investigate the damping effects but to a large extent have lacked direction and very little other than empirical observation has been achieved. However, the combination of results from these various investigations does lead to an indication of the true mechanism. This paper summarises the results of these investigations and draws the conclusion that the mechanism of damping is not merely due to friction as is often assumed but is a complex one of micro and macro friction and cyclic plastic deformation of the asperities.

INTRODUCTION

In a basic machine tool structure, the damping capacity is considerably greater than the sum of the damping capacities of the individual elements. The excess can only be a damping capacity which exists because of the presence in the structure of joints between the elements. These joints are either formed from machined surfaces which are bolted together or otherwise permanently fixed or sliding joints where the two surfaces are constrained to remain in some form of contact but slide relative to each other. In this latter category there is in almost every case a film of oil between the two metal surfaces which provides a variety of contact conditions from hydro-dynamic to complete metal to metal contact with the oil only present in the valleys of the surfaces.

LUBRICATED JOINTS

The general assumption when considering the energy dissipation in lubricated joints is that this is due to viscous damping. This appears to be a valid assumption since it has been found[1] that damping increases with an increase in excitation frequency Reshetov and Levina[2] state that, additionally, damping increases with an increase in viscosity of the lubricant. They have also found that an increase in mean interface pressure on the joint results in a reduction in damping. This is probably because an increase in mean interface pressure will result in a mutual interpenetration of the asperities thus causing a more 'definite' joint and reducing the relative motion between the surfaces. This effect would tend to increase with an increase in surface roughness[3]. It follows therefore that the conditions in a sliding joint leading to a maximum dissipation of energy would be the following: low mean interface pressure, smooth surface finish and a high-viscosity lubricant. Interfacial forces are difficult to control since they depend largely on the cutting forces generated in the machine tool. Generally an increase in metal removal rate will result in an increase in cutting force and a consequent increase in mean interfacial forces. A reduction in mean interfacial forces is therefore unobtainable. This means that mean interface pressure can only be reduced by increasing the area of contact in the slideways. This would, however, tend to reduce the static stiffness of the slideways as it is well established that this increases with mean interface pressure[4]. This is an example of an apparently consistent conflict of requirements between static stiffness and damping capabilities.

UNLUBRICATED JOINTS

The explanation of energy dissipation in lubricated joints is readily acceptable as being due to viscous damping from the presence of a fluid. Considerable energy dissipation is, however, still achieved in unlubricated joints and it is in the explanation of this mechanism that the greatest controversy arises.

Andrew, Cockburn and Waring[5] have shown that, when joints are oscillated in a direction normal to the plane of the joint, the energy dissipation is negligible. Many workers have shown that energy dissipation is significant whenever oscillation takes place parallel to the plane of the joint.

The immediate conclusion which is drawn from this is that this energy dissipation can be attributed to some form of friction mechanism. Brown assumes that the contact asperities are welded together under a normal pressure. Thus energy would be dissipated in the shearing of these localised welds. Andrew, Cockburn and Waring dispute this assumption and suggest that such welding of the contacting asperities is unlikely in a normal atmosphere at the magnitude of mean interface pressures likely in bolted joints.

Earles and Beards[7] observed the formation of an oxide layer on fretted scars at the contact points in the joint. This finding would tend to support the view of Andrew Cockburn and Waring but it must be remembered that the tests of Earles and Beards were carried out with very low interfacial pressures, i.e. in the order of 3.75 lbf/in^2 compared with say 0.1 ton/in^2 in a typical machine tool joint.

* The Queen's University of Belfast

Contact conditions

In any analysis of the energy dissipation mechanism it is essential to understand the contact conditions. Classical analyses of the friction mechanism have invariably assumed low interface pressures compared with those present in bolted joints in machine tools because a relative sliding between the surfaces is implicit in the analysis. The other common assumption is that the contacting surfaces are of the form of elliptical, spherical or cylindrical asperities. This is probably valid as a representation of contact conditions at low interface pressures and when sliding readily occurs. It is invalid in case of bolted joints because in this context the contact mechanism is completely different. It has been repeatedly established[3] that on initial contact between surfaces there will always be an element of plastic deformation and consequent mutual embedding of the surfaces.

Analyses which are based on the solely elastic deflections experienced when initial contact takes place between surfaces of the spherical asperity type are therefore completely invalid in this context.

Coefficient of friction

Goodman and Brown[8] indicate in their analysis that an increase in coefficient of friction results in a decrease in energy dissipation. Earles and Beards[7] show that energy dissipation is independent of the coefficient of friction and Kedrov[9] states that his damping coefficient is proportional to the friction force which means that he is taking energy dissipation to be directly proportional to the coefficient of friction, the exact opposite to Goodman's and Brown's findings.

The wide divergence in these views can probably be explained by the fact that different conditions have been examined in each case. As is outlined above, the friction mechanism depends on the surface texture and on the interface pressure. Another cause of confusion is the variation in the actual values of the factors affecting the coefficient of friction taken. In the case of macro-slip the mean interface pressure would be taken as the normal pressure value. On the other hand, in the case of micro-slip the normal pressure would be the yield pressure of the asperities which would be at least 50 times greater. Obviously a variation in choice of this value would lead to staggeringly different results.

Suggested mechanisms

Ito and Masuko[10] have shown that an optimum mean interface pressure exists between the joint surfaces for maximum energy dissipation. This indicates immediately that the mechanism of damping in a typical bolted joint cannot be treated in as simple a manner as has been attempted hitherto. That an optimum interface pressure exists implies either that a change in mechanism occurs or that the mechanism is always composed of different components and that at various values of interface pressure their significance varies.

In the case of Ito's and Masuko's work it is suggested that, at low pressures, sliding occurs between the surfaces; as the pressure increases, the work done and consequently energy dissipated increases until at some point mutual embedding of the surfaces begins to take place. At this point, sliding on a macro scale is reduced and another mechanism becomes predominant. It is suggested that this could be the cyclic plastic deformation at the contact spots. This would reduce with increased interface pressure and thus result in a decrease in energy dissipation. This would explain the optimum value of interface pressure.

Again, it must be stressed that the pressures employed by Ito and Masuko are an order of magnitude lower than typically experienced in a machine tool joint.

It follows from this that cyclic plastic deformation is likely to be the predominant mechanism in the case of machine tool joints. Thus the mechanisms to which energy dissipation could be attributed may be summarised as follows.

(a) Macro-slip involving friction damping is perhaps its simplest form.

(b) Micro-slip involving very small displacements of an asperity with respect to the other surface particularly if the asperities are embedded. Thus energy dissipation by this mechanism would also be in the form of friction damping. As mentioned above the normal pressure in this case would be very much greater than in the case of macro-slip.

(c) A cyclic plastic deformation of the contacting asperities. When two surfaces are compressed together by a normal force, plastic deformation of the asperities takes place. This deformation will continue until the real area of contact becomes great enough to support the load. The pressure at these contact spots will be the yield pressure of the softer contacting material. Therefore, if any tangential force is applied to the joint, an increase in pressure must occur at these contact spots which will obviously result in a further plastic deformation. This is particularly likely because the plane of the contact spots will not be parallel to the plane of the joint[4]. If the asperities are fully interlocked as in the case of shaped surfaces with the machine lays orientated at an angle of 90°, it is possible to achieve a cyclic plastic deformation without necessarily any micro-slip. This explanation would be consistent with a reduction in energy dissipation for an increase in normal pressure because such an increase would increase the real contact area and thereby reduce the tangential stress on the contact spots.

The most likely explanation of energy dissipation is that in the general case it may be attributed to all three mechanisms. Their relative significance would depend on the joint conditions. Obviously a low normal interface pressure would tend to increase the significance of the slip mechanisms as would an improvement in the quality of the surfaces in contact. Conversely, high normal interface pressures and relatively rough surfaces would tend to increase the significance of the plastic deformation mechanism. Because of the values of interface pressure and surface texture involved in machine tool joints it would seem that the latter is therefore the predominant mechanism of energy dissipation in their case. Evidence of this is contained in reference 11 where considerable energy dissipation was experienced but no signs of fretting due even to micro-slip could be detected.

RECOMMENDATIONS

Obviously the whole question of energy dissipation in dry bolted joints is far more complex than is implied by the methods employed in the majority of the investigations into it. Many empirical relationships have been established and various analytical relationships have been proposed. Unfortunately, the tendency in the latter is to base them on experimental data derived from tests where the conditions are less than realistic particularly in a machine tool context. For example, it has been common practice to employ normal interface pressures which are an order of magnitude below the pressures likely to be encountered in a typical machine tool joint. As discussed above, this is likely to give a grossly misleading impression of the significance of the various mechanisms proposed.

It is important that the significance of these mechanisms is established in relation to the variation in joint characteristics.

Additionally and more importantly in the machine tool context, the relationship between these mechanisms at the normal interface pressures encountered in machine tool joints must be investigated. It is only in this way that any sensible predictions of energy dissipation can be made.

CONCLUSIONS

(1) In a lubricated sliding joint, maximum energy dissipation may be achieved by using oil with as high a viscosity as possible and by reducing the normal interface pressure as much as possible consistent with the other possibly overriding design requirements. The surface finish of the sliding faces should be as smooth as possible.

(2) In unlubricated bolted joints, it would appear that a combination of mechanisms is involved in the dissipation of energy. These are frictional losses due to macro- and micro-slip and a cyclic plastic deformation of the asperities.

(3) In the case of bolted joints in machine tools, cyclic plastic deformation of the asperities would appear to be the most significant when the magnitude of the interface pressures involved are considered.

REFERENCES

1. P.E.R.A., 'Machine tool joints', *Rept.* No. 180.
2. D. N. RESHETOV and Z. M. LEVINA, 1956. 'Damping of oscillations in the couplings of components of machines', *Vestn. Mashinostr.*, No. 12, 3–13.
3. R. E. SCHOFIELD, 1968. 'Bolted joints in machine tool structures', *Proc. 6th Machine Tool Congress, Budapest.*
4. R. E. SCHOFIELD and R. H. THORNLEY, 1971. 'Calculating the elastic and plastic components of deflection of plane joints formed from machined surfaces', *12th Intern. M.T.D.R. Conf.*, Macmillan.
5. C. ANDREW, J. A. COCKBURN and A. E. WARING, 1967–8. 'Metal surfaces in contact under normal forces: some dynamic stiffness damping characteristics', *Proc. Inst. Mech. Engrs. (London)*, **182**, Part 3K, 92–100.
6. C. B. BROWN, 1968. 'Factors affecting the damping in a lap joint', *Proc. Am. Soc. Civil. Engrs.*, pp. 1197–217.
7. S. W. E. EARLES and C. F. BEARDS, 1970. 'Some aspects of frictional damping as applied to vibrating beams', *Intern. J. Mach. Tool Design Res.*, **10**, 123–31.
8. L. E. GOODMAN and C. B. BROWN, 1962. 'Energy dissipation in contact friction, constant normal and cyclic tangential loading', *J. Appl. Mech.*, **29**, 17–22.
9. S. S. KEDROV. 'Establishing damping factors in the joints of machine components', *Russian Engng. J.*, **46**, No. 9.
10. Y. ITO and M. MASUKO, 1971. 'Experimental study on the optimum interface pressure on a bolted joint considering the damping capacity', *12th Intern. M.T.D.R. Conf.*, Macmillan.
11. R. E. SCHOFIELD, to be published. 'Energy dissipation in bolted joints with fully interlocking asperities'.

SOME STATIC AND DYNAMIC CHARACTERISTICS OF BONDED, MACHINED JOINT FACES

by

R. H. THORNLEY* and K. LEES*

SUMMARY

This paper reports on investigations of the dynamic characteristics of joints with loads applied normal to the joint faces, in which an epoxy resin adhesive has been used as the interface bonding material. Joint face surface topography and a range of adhesives with varying properties were examined.

Bonding joint surfaces with this type of adhesive increases both the static and the dynamic stiffness. The stiffness is dependent on the type of machined surface and quality of surface roughness. The range of adhesives available for bonding steel-to-steel all have similar stiffness characteristics. A small increase in damping is achieved when using the epoxy resin adhesives compared with that obtained with dry joints.

INTRODUCTION

Earlier studies[1-4] have shown that by lubricating the metal-to-metal interface in a fixed joint the stiffness and damping characteristics of the joint are improved. To maintain optimum damping conditions with such films in service is often difficult. In machine-tool structures, it is desirable that the design of a fixed joint should be optimised to produce high stiffness and damping. If an interface material is to be used to improve these properties, adhesives will be more likely to give longer term stability than fluid films. Before introducing such interface materials as a general practice, both the static and dynamic behaviour of these films should be investigated. Work has already been carried out in this field, and this paper describes an investigation which examined the application to machine-tool joints of a number of high-strength epoxy resin adhesives. Some experimental results from reference 1 showed that this type of adhesive increased the stiffness of a joint and to a lesser extent its damping capacity. The influence of the type of machined surface on bonded joints is also investigated.

Joints in machine-tool structures are subjected to loads applied in both normal and tangential directions to the interface. In the present study only normal loading conditions are considered. Under this type of loading the damping mechanism in the adhesive layer would be to distort in shear and thus dissipate energy by a lateral shear action, without any significant sliding of the joint faces.[5]

TEST SPECIMENS – DESIGN AND MANUFACTURE

The joint specimens were basically composed of two mild steel square plates giving an interface area of 36 in² (232 cm²). To permit comparisons of joint dynamic displacements, without the effects of material stiffness, an equivalent solid specimen was used. Joint displacements alone could be determined by subtracting the results of the solid system from the jointed system.

A typical joint and equivalent solid specimen is shown in figure 1. To study the effect of bonding on the dynamic characteristics of a joint using various

Figure 1. Typical joint and equivalent solid specimen.

*The University of Aston in Birmingham

machined finishes, shaped, scraped and peripheral ground finishes were examined. To compare a number of adhesives, four specimens were used with the same surface finish, which was produced by shaping. The angle of lay orientation remained constant throughout the tests at 90° for both shaped and ground specimens.

During manufacture of the specimen surfaces, measures were taken to minimise any errors in parallelism on the specimen which may be generated by the slides of the machine. The scraped specimens were produced by bedding the surface to a master surface plate and scraping to achieve a minimum of 21 contact points per in² (6.45 cm²).

The surfaces were then measured and defined in terms of surface roughness and flatness. The surface roughness measurements, in terms of CLA values, are summarised in table 1.

TABLE 1

Specimen description	Surface roughness values CLA μm		Average of both surfaces
	Top surface	Bottom surface	
(1) shaped	87 (2.21)	99 (2.52)	93 (2.36)
(2) scraped	25 (0.63)	15 (0.38)	20 (0.51)
(3) ground	10 (0.25)	32 (0.81)	21 (0.53)
(4) shaped	89 (2.26)	92 (2.34)	90 (2.30)
(5) shaped	89 (2.26)	88 (2.24)	89 (2.25)
(6) shaped	107 (2.72)	93 (2.36)	100 (2.54)

Flatness deviations were measured by the technique given in reference 6 which enabled contour charts to be drawn for each surface. These were valuable aids in determining predominant peaks and valleys on the individual surfaces, and in comparing regions of these

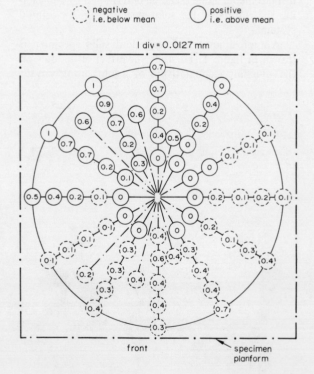

negative i.e. below mean   positive i.e. above mean

1 div = 0.0127 mm

front   specimen planform

flatness deviations of top plate shaped specimen (1) 0.0232 m²

Figure 2. Typical flatness deviation chart.

for each pair of surfaces. For the specimens, the peak-to-valley depth of the long-term surface waviness varied from 0.0004 in (0.0102 mm) to 0.0025 in (0.0635 mm) over the range of surfaces used. These were the maximum peak-to-valley depths measured over the 6 in (15.25 cm) surface length of each joint in the direction across the displacement measuring probes. Similar ranges of values existed on the scraped and the ground surfaces; the actual distribution of waviness over the surface, however, varied considerably for each specimen. A typical flatness deviation chart for one of the shaped specimens is shown in figure 2.

INTERFACE BONDING MATERIALS

A range of epoxy resin adhesives suitable for bonding metal-to-metal components is commercially available. Four types were eventually selected from this range, chosen on the basis that they would cure at room temperature and give a range of elastic moduli and shear strengths.

Type A

This type was selected to compare the effect of bonding on the various types of machined surfaces.

It was a solvent-free, cold-setting adhesive of medium viscosity (100–150 P at 21 °C) (10–15 N s/m²) producing joints of high mechanical strength, resistant to chemical attack, and providing full electrical insulation. The resin and hardener were mixed in the ratio of 100 to 4 by weight respectively. The mixture gave good flow properties and was stated to be unaffected when exposed to damp air. At room temperature (20 °C) it had a pot life of between 3 and 4 hours depending on the quantity mixed. When curing the adhesive at room temperature a minimum time of 24 hours was recommended, but curing could be accelerated by raising the curing temperature. However, for the application to machine-tool fixed joints, room-temperature curing would be essential. The modulus of elasticity, E, was given as 0.2–0.3 x 10⁶ lbf/in² (1.38–2.07 GN/m²).

Type B

This was also a solvent-free medium-viscosity adhesive (16–20 P at 21 °C) (1.6–2.0 N s/m²) and was capable of developing excellent bond strength when cured at room temperature (20 °C). The resin and hardener were mixed in the ratio of 100 to 8–10 by weight, respectively. As the hardener was hygroscopic, difficulties would be experienced with the mixture due to moisture pick-up. When applying a mixture in moist air, the adverse effects could be made less severe by allowing the mixture to stand in a covered container for about 15–20 min before applying it to the surfaces. In this way, the resin/hardener reaction was well under way before use. The mixture had a good pot life and a curing at 20 °C for a minimum of 24 hours was recommended. The modulus of elasticity was quoted as 0.45–0.50 x 10⁶ lbf/in² (3.1–3.45 GN/m²).

Type C

This was recommended for the bonding of low-strength materials to metals and had the lowest shear strength of the four types used. It was a cold-setting, medium-viscosity (200 P at 21 °C) (20 N s/m²) adhesive giving good bond strength after curing at 20 °C for at least 24 hours. The mixing of resin and hardener was 1 to 1 parts by weight and a satisfactory mix was indicated by a uniform brown colouration appearing after mixing. The pot life of one hour at 20 °C was shorter than for the other adhesives. The value of E was given as 0.13×10^6 lbf/in² (0.89 GN/m²).

Type D

This adhesive was an epoxy resin paste adhesive, being a high-viscosity mineral-filled paste, which when mixed with the low-viscosity liquid hardener became a thick cold-setting adhesive with excellent gap-filling properties. Bonds of good strength could be obtained by curing at 20 °C for at least 24 hours. The resin and hardener were mixed in the ratio of 100 to $4-4\frac{1}{2}$ by weight, respectively. The pot life at 20 °C was between one and two hours. The modulus of elasticity was given as 0.85×10^6 lbf/in² (5.85 GN/m²).

The common feature of curing at room temperature made all four types suitable for this application. Types B, C and D were used only on shaped surfaces of common planform shape and surface finish to allow direct comparison to be made of their static and dynamic characteristics.

METHOD OF APPLYING ADHESIVES

If two metal surfaces are separated by a low modulus of elasticity material, the effective spring characteristic of the joint takes on the value of the weakest spring in the system. During these tests it was intended that the adhesive should act as a filling element in all the valleys created by flatness deviations, leaving definite regions of metal-to-metal contact, the result desired being an increase in the actual area of contact between the surface and hence the effective stiffness. The influence of this on the damping characteristics could then be studied.

In order to make fair comparisons between joints and adhesives, it was necessary to use a common layer thickness on all joints to reduce the number of experimental variables. Owing to flatness deviations, it was almost impossible to apply a constant layer thickness, coupled with the fact that a minimum layer thickness was desired to maintain anything like the desired stiffness values. A compromise was therefore needed between the minimum layer thickness to fill the valleys, maintaining maximum metal-to-metal contact, and a constant layer thickness for each joint. On this basis it was first necessary to determine a nominal layer thickness suitable to all the joint surfaces and secondly to establish a technique for applying a thin, level layer of adhesive on each surface. By careful examination of the flatness deviation charts it was decided that a nominal adhesive layer thickness per surface of 0.001 in (25.4 μm) would be used. This thickness was applied on the highest peak of each surface, which ensured that each joint had ample adhesive but relied on the viscous properties and the weight of the top plate of the joint assembly to cause flow during curing. The joint faces thus came to rest on the high points of the surface.

The problem of applying this thin level layer was overcome by using a simple levelling arrangement consisting of a surface table and an adjustable levelling steel wiper blade. Having the blade with slip gauges, a volume of adhesive was applied to the surface; spreading and levelling to the desired thickness was then achieved by passing the specimen under the blade. This method was successful in removing surplus adhesive and applying a thin level layer. The only compressive load applied to the joint during the curing process was that due to the weight of the top plate: 13.0 lbf (57.6 N). During settling of the joint faces, amounts of adhesive did flow out from the joint boundary. For types B, C and D the curing time allowed was at least 24 hours at 23 °C. Type A was cured for 7 days to attain maximum bond strength, average temperature over this period being 24 °C.

A final check was carried out to determine the actual layer thickness attained. From the results of the measurements on the shaped surfaces, the maximum peak heights dry were the same after bonding, except for the specimen using type D adhesive. However, at what were the low points, the adhesive layer thicknesses indicated ranged from 0.001 in to 0.002 in (25.4 μm to 50.8 μm) for the four shaped specimens. The paste adhesive type D gave a layer thickness of 0.001 in (25.4 μm) at the higher points and 0.002 in (50.8 μm) at the lower points. The scraped and ground joints indicated from the periphery measurements, a surface layer thickness of 0.001 in (25.4 μm) all round. From these results it can be seen that the layer thickness varies in proportion to the flatness deviations for each type of surface.

ADHESIVE MECHANICAL PROPERTIES

From each batch of adhesives a sample was taken to provide test pieces to determine the modulus of elasticity E, the shear modulus G and the hardness values for each material.

The E value of each adhesive was determined by loading the test piece in compression up to a maximum of (5320 N) 1.120 lbf; the E values were calculated from the load–deflection curves. The values are compared in table 2 with those specified by the manufacturers.

The shear modulus was determined by loading the specimens in a tensile test machine in double shear. The calculated values of G from these tests are also given in table 2.

To complete this series of experimental tests and establish some comparative data on the mechanical properties of the adhesives, a series of hardness checks were carried out on the adhesive samples. For these tests a syringometer hardness comparator was used. The mild-steel specimens were tested by this method and by the conventional Vickers Hardness test. Comparative values were then obtained for mild

TABLE 2

Compression tests on adhesive samples adhesive type
A, B, C & D

Adhesive type	Modulus of elasticity E (10^9 N/m^2)	
	Expt. values	Specified values
A	1.47	1.38/2.07
B	2.48	3.1/3.45
C	1.47	0.89
D	1.62	5.85

Shear test on adhesive samples

Adhesive type	Shear modulus (10^9 N/m^2)
A	0.51
B	0.88
C	0.49
D	0.34

Surface hardness values of adhesive and joint samples

Adhesive type	Hardness, syringometer readings
A	59.3
B	70.8
C	50.0
D	36.4
Mild steel	34.0
	Brinell No. M.S. 141

steel on the syringometer scale and DPN hardness number. The reading from this instrument indicated the height of rebound of a free falling slide weight off the test surface. The slide weight was encased in a calibrated glass tube. Therefore, the higher the instrument reading the greater the resilience of a surface to impact loads.

Although this method is somewhat elementary it was effective in establishing hardness comparisons of the interface layer with that of the joint surface itself. The results are given in table 2.

TEST APPARATUS

The rig was capable of applying a uniform joint static preload through a hydrostatic oil thrust bearing and then superimposing a dynamic force from an electro-dynamic exciter. Joint interface displacements normal to the surface were measured with three capacitance-type pick-ups situated around the periphery of the joint top plate. Displacement signals from the probes after being filtered and amplified were recorded on an ultraviolet recorder or by oscilloscope after first being filtered and amplified.

Test procedure

Prior to any bonding, the joints were assembled dry, the surfaces having been thoroughly cleaned with carbon tetrachloride to remove all grease and contamination. Two of the shaped specimens were selected along with the scraped and ground joints to be tested in the clean, dry, unbonded condition. The following tests were performed.

Static tests

Static deflection tests were carried out on each joint and the equivalent solid, deflections being recorded

during loading and unloading for a joint preload range 0–413 lbf/in^2 (2840 kN/m^2). The results of these static tests are given in figure 3.

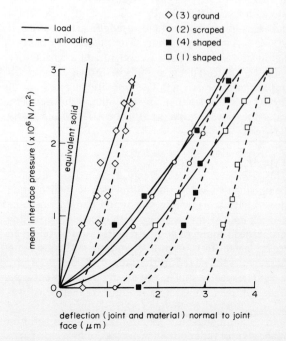

Figure 3. Joint characteristics: static loading, dry surfaces.

Dynamic tests

These tests were performed as a continuation of the static tests on each specimen and the equivalent solid without removal from the rig. For a range of static preloads, an excitation force of ±140 lbf (621 N) was applied over the frequency range 30–360 Hz. Figures 4 and 5 give the results, showing the variation of stiffness corrected for material stiffness, with preload and frequency.

Damping measurements were also recorded in terms of logarithmic decrement, from which the damping ratio ζ was derived for each of the above dynamic conditions.

Bonded joint tests

After completion of the above test programme each joint was prepared and bonded by the method described previously. The shaped specimen No. 1, scraped specimen No. 2 and ground specimen No. 3 were bonded with adhesive type A. The remaining shaped specimens, 4, 5 and 6, were bonded with adhesive types B, C and D, respectively. After curing, the static and dynamic tests described above were repeated on the complete range of bonded specimens. The results of the static tests are given in figure 6.

DISCUSSION OF RESULTS

Adhesive mechanical properties

Comparing the experimental E values with values given in the adhesive specification shows some discrepancy in actual values. The general trend of the values were in agreement except that obtained for type D. On examination, the type D test piece used in the

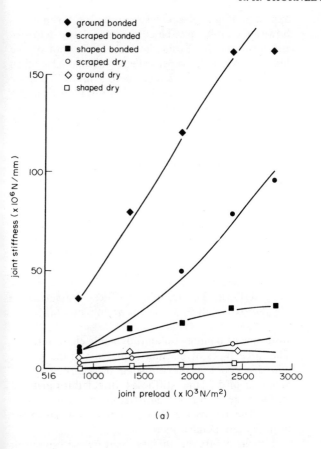

(a)

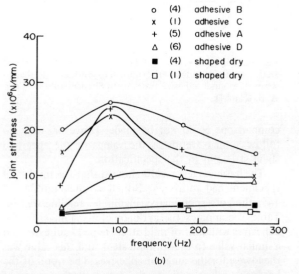

(a)

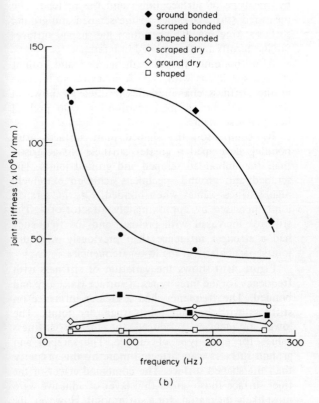

(b)

(b)

Figure 5(a) Stiffness versus preload corrected for material stiffness; 30 Hz; shaped, dry and bonded; adhesives types A, B, C and D.
(b) Stiffness versus frequency corrected for material stiffness; preload 1890×10^3 N/m²; shaped, dry and bonded; adhesives type A, B, C and D.

Figure 4(a) Stiffness versus preload corrected for material stiffness; 90 Hz; dry and bonded type A, shaped, scraped, ground;
(b) Stiffness versus frequency corrected for material stiffness, preload 1890×10^3 N/m²; dry and bonded; shaped, scraped, ground.

compression test was seen to contain a large number of microscopic air bubles. Although not as pronounced on the other test pieces, some small air bubble pockets were present. These air pockets could result in the material becoming more flexible, hence giving lower E values. If these air pockets prove to be a possible cause of increasing the flexibility of the interface layer, a more precise method of mixing perhaps under atmospheric controlled conditions would have to be adopted. Apart from this the adhesives in general were found to have easy handling characteristics.

From the shear tests, the shear modulus for each material was deduced, and are presented in table 2. As no values were given in the adhesive specification,

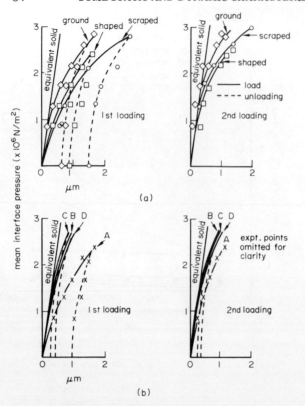

Figure 6. Joint characteristics: static loading, bonded surfaces. Deflection (joint and material) normal to joint face. (a) shaped, scraped, ground, adhesive type A. (b) Shaped, adhesives type A, B, C and D.

comparisons could not be made. However, they do follow similar trends as the values of the specified shear strengths in the specifications.

The hardness test results showed that the mild steel specimens had an average Brinell hardness number of 141, the corresponding syringometer reading was 34. Table 2 compares the syringometer readings for the adhesives with that of mild steel. Type D adhesive had a similar value to that of mild steel, and this value was the lowest for the range of adhesives. The trend of the hardness values reflects the trend in the order of magnitude of the shear moduli, i.e., the surface with the greatest rebound height possesses the highest shear modulus.

Influence of bonding on joint static characteristics

The variation of the static displacement with the normal interface pressure for dry joint conditions, shown in figure 3, produced similar relationships to those obtained by previous investigators[4, 6, 7-11]. During loading the shaped surfaces deformed greater (plastic plus elastic) than the scraped and ground surfaces. The comparisons change, however, after bonding the three types of machined surfaces with the same adhesive and the same mean layer thickness. Figure 6 shows that for each surface the characteristic of the deflection curve changes to one of a more linear relationship. The shaped surfaces approach a stiffness very close to that of the scraped and ground specimens. The relative increase in stiffness of the scraped and ground was much less than that for the shaped surfaces. After the first loading it would

appear that some plastic deformation had taken place, however, on subsequent loadings only elastic deformation was present. From the tests carried out on the bonding materials no permanent set could have taken place under normal circumstances in the adhesive layer at preloads of 413 lbf/in² (2840 kN/m²) as the test specimens remained elastic up to an equivalent compressive stress of 2520 lbf/in² (174 MN/m²). A possible explanation is that, if regions on contacting asperities formed during the curing process, a small amount of plastic deformation could take place during the initial loading. At these small points of contact extremely high pressures would develop and if thin layers of adhesive were present they could also plastically deform. The majority of the adhesive layer, however, would behave in an elastic manner under the magnitude of the preloads considered. The variation of adhesive type appeared to have little or no influence on the stiffness when used on similar joint surfaces. The resulting stiffness characteristics are approximately equal, as shown in figure 6(b).

Influence of bonding on joint dynamic characteristics

The three types of machined surface and the various adhesive types were compared by considering the variation of dynamic stiffness (corrected for material stiffness) with preload and frequency of excitation force. The results are shown in figures 4 and 5 for both dry and bonded joints.

For dry joints the stiffness increases with preload, the magnitude of the stiffness being strongly dictated by the type of surface finish and the preload. The measured stiffness values of the scraped and ground surfaces were much higher than the shaped surfaces Similar results are also quoted in reference 1.

After bonding, the shaped, scraped and ground surface specimens with type A adhesive, an increase in the stiffness characteristics resulted as shown in figure 4(a), the stiffness relationship with preload being almost linear.

By comparison the shaped joint surfaces after bonding developed a greater stiffness characteristic than the unbonded scraped and ground joints. The scraped and ground specimens achieved extremely high stiffness values when bonded, e.g. the scraped joint increased by approximately a factor of 4. The stiffness increased with preload, and the frequency had a stronger influence than previously with dry joints, particularly at the lower frequencies.

Figure 4(b) shows the variation of stiffness with frequency for the three types of surface finish dry and bonded. The frequency had a greater influence on stiffness with bonded joints than dry joints. The lower frequencies produced the highest stiffness values for each type of surface. The scraped and ground surfaces were affected more by this property than the shaped surface. The combined effect of the finer surface finish and a thin layer of adhesive were most likely the reasons for a stiffer joint. However, the adhesive appeared to offer high resistance to deformations at low frequencies; at high frequencies the internal molecular structure appeared to become less capable, indicating a more flexible nature. If greater internal motions in the layer do not take place at high frequencies this result would, therefore, influence the damping mechanism in the layer. Figure 5 compares the four types of adhesive, these having been applied on similar surfaces. The stiffness characteristic was similar to the previous result, in that it generally increased with preload. Each adhesive gave similar stiffness results; no outstanding differences were indicated. Over the preload and frequency range considered, adhesive type B produced higher values of stiffness than the other types. Type B had the highest modulus of elasticity. In order of decreasing magnitude types A and C gave similar results and type D the lowest value of stiffness. The shear moduli and the hardness values decreased in the same order as the stiffnesses. Each type of adhesive appeared to offer a greater resistance to oscillatory disturbances at the lower frequencies between 30 Hz and 100 Hz. As the frequency increased, this property diminished as discussed previously. Larger values of stiffness, however, prevailed at these higher frequencies for bonded joints than for the unbonded. Adhesive type D was the least sensitive to variations in frequency. This could be significant as type D had the lowest shear modulus and hardness value.

Influence of bonding on joint damping characteristics

The damping characteristics of the joint are allied to the stiffness characteristics in that decrease in stiffness would tend to indicate greater interface movement and hence an associated damping mechanism. The deduced values of damping ratio from the tests are given in table 3. These values were the average taken

TABLE 3

Damping ratios for joints dry and bonded

Specimen joint	Dry	Bonded with type 'A' adhesive
solid (0)	0.009	–
shaped (1)	0.013	0.014
scraped(2)	0.01	0.013
ground (3)	0.012	0.011

Specimen joint	Dry	Bonded with adhesive type	
solid (0)	0.009	–	
shaped (1)	0.013	0.014	A
shaped (4)	0.011	0.01	B
shaped (5)	–	0.011	C
shaped (6)	–	0.011	D

over five preloads. In each case the damping ratio was relatively constant with preload.

For dry joints the damping ratio was almost the same as the equivalent solid indicating little or no damping in the dry joints due to loads applied normal to the interface. When these joints were bonded with the same adhesive a slight increase in damping was produced from the shaped and scraped specimens, little difference was indicated by the ground specimen. Comparing the damping capacity of the four types of adhesive only small differences in damping ratio were indicated between the types of adhesive. The damping ratio again remained constant with increasing preload.

In general the damping capacity in terms of the

damping ratio remained the same when using epoxy resin adhesives, but the stiffness of the joint was shown to be substantially increased. The adhesive layer in the joint forms an equivalent solid coupling. Any damping which does take place must do so within the internal molecular structure of the resin.

The cured resins have been shown to be extremely stiff and resilient to impact loads; this does not make them the ideal damping medium but extremely useful to produce a high joint stiffness.

A theoretical study[12] examines the cyclic damping produced in thin layers of dissipative material placed between rigid surfaces. The interface was subjected to small cyclic normal displacements. The frequency of the motion and the magnitude of the displacement were assumed, such as to allow inertia terms in setting up the equations of motion for the dissipative material to be neglected. The general case was intended to study energy dissipation in thin layers of viscoelastic adhesives. Initially to simplify the problem, two- and three-dimensional cases were studied using a viscous fluid as the dissipative material and rectangular planform configurations. The interface material was then taken as an incompressible viscoelastic solid and the results compared with the fluid cases. Finally the three-dimensional cases of a circular and annular planform with a viscous fluid as the interface layer were considered.

The analysis demonstrates with respect to the damping capacity most of the effects of the geometry describing the dissipative layer. The geometry dependency was shown to involve simply the ratio of the maximum transverse amplitude of the displacement in the damping material, to the vertical displacement of the interface layer. In each case the energy dissipation was shown to be inversely proportional to the adhesive layer thickness, h.

The analysis provides some information relating the damping mechanism in adhesive layers of varying geometric configurations. However, it does not include the effect of such parameters as, surface topography of the bonded faces, preload and stiffness of the resulting assembly. The latter points being important when considering machine tool joints.

CONCLUSIONS

(1) If the layer thickness of an adhesive film is limited to the same order of magnitude as the flatness deviations of the surface involved, the static stiffness of the joint can be increased to a value which is in close proximity of that of the equivalent solid. A general rule for the layer thickness could be $0 < 2h < 0.002$ in (50.8 μm), where $2h$ is the layer thickness.

(2) The increase in static stiffness by bonding is dependent upon the type and quality of the surface finish. The greater the surface stiffness before bonding the less significant the increase in stiffness after bonding.

(3) For the range of epoxy resin adhesives tested the increase in static stiffness was generally the same, where the modulus of elasticity was restricted to $0.1 < (E \times 10^6 \text{ lbf/in}^2) < 0.9$. [$0.689 < E < 6.2$ (GN/m^2)].

(4) Bonding joint surfaces with epoxy resin adhesives increases the dynamic stiffness, the stiffness of the bonded surface is dependent upon the type of machine surface and quality of surface roughness. Generally, the lower the CLA values of the surface, the higher the stiffness. The increase in dynamic stiffness is only achieved by maintaining a minimum adhesive layer thickness sufficient to cover the surface but retaining some regions of metal contact. The relationship between the dynamic stiffness and pre-load has been shown to be almost linear for bonded joints.

(5) The type of epoxy resin adhesive used must be suitable for bonding metal-to-metal components. The range of epoxy resins available in this category all have a similar influence on the dynamic stiffness characteristics. In general, the higher the shear modulus the stiffer is the joint.

(6) For joints with dry surfaces the damping ratio tends to remain the same independent of type of machine finish and decreases slightly with increasing preload. With a bonded joint a small increase in damping can be achieved; however, in most cases the damping remained unchanged after bonding. This appears to be independent of type of machine surface, and, for the range tested, the type of adhesive.

ACKNOWLEDGEMENTS

The work reported here formed part of a programme of investigations carried out in the Machine Tool Division of Professor F. Koenigsberger at the University of Manchester Institute of Science and Technology, and was supported by a research grant from the Science Research Council.

REFERENCES

1. R. H. THORNLEY, 1969. 'A study of some parameters which influence the dynamic stiffness of joints in machine tool structures', *Ph.D. Thesis*, University of Manchester.
2. C. ANDREW, J. A. COCKBURN and A. E. WARING, 1967–8. 'Metallic surfaces in contact under normal forces: some dynamic stiffness and damping characteristics', *Proc. J. Mech. E*, 182, PT. 3K.
3. A. E. WARING, 1969. 'The damping in fluid-filled metal-to-metal joints undergoing normal vibration', *Ph.D. Thesis*, University of Bristol.
4. P.E.R.A., 1969. 'Machine tool joints. Part 2, effect of intermediate viscous films on stiffness and damping of cast iron joints', *Rept.*, No. 13, March.
5. T. J. MENTEL, 1967. 'Joint interface layer damping', *Trans. A.S.M.E.J. Engng. Ind.*, November, 757 – 804
6. R. H. THORNLEY, R. CONNOLLY, M. M. BARASH and F. KOENIGSBERGER, 1965. 'The effect of surface topography upon the static stiffness of machine tool joints'. *Intern. J. Mach. Tool Design Res.*, 5, 57–74.
7. R. CONNOLLY and R. H. THORNLEY, 1966. 'The static stiffness of joints between machined surfaces', *M.T.I.R.A. Rept.*, No 13, March.
8. R. CONNOLLY and R. H. THORNLEY, 1965. 'The significance of joints of the overall deflection of machine tool structures', *Proc. 6th Intern. M.T.D.R. Conf.*, Manchester.
9. R. H. THORNLEY, R. CONNOLLY and F. KOENIGSBERGER, 1967–8. 'The effect of the flatness of joint faces upon the static stiffness of machine tool joints', *Proc. Inst. Mech. Engrs.*, 182, Part 1, No. 18.
10. R. CONNOLLY and R. H. THORNLEY, 1967. 'Determining the normal stiffness of joint faces', *Trans. A.S.M.E. J. Engng. Ind. Prod., Engng. Div.*, Paper 67, No. 6.
11. D. N. RESHETOV and Z. M. LEVINA, 1956. 'Damping of oscillations in the couplings of components of machines', *Vestnik Machinostr.*, No. 12, 3–13.
12. C. GONG, 1963. 'Energy dissipation in thin layers of damping material subject to cyclic normal loading', *M.Sc. Thesis*, University of Minnesota.

REVIEW OF THE RESEARCH ON FIXED AND SLIDING JOINTS

by

N. BACK*, M. BURDEKIN* and A. COWLEY*·

SUMMARY

The paper summarises the results of research into the stiffness, damping, friction and wear characteristics of fixed and sliding joints. An attempt has been made to extract from the results of research conducted over the past decade or so, that information which is considered to be of direct interest to the machine tool designer.

NOTATION

a = coefficient relating normal pressure and deflection
A = constant
A_a = apparent area of contact
A_i = contact area of individual asperities
A_r = total area of contact
b = coefficient relating normal stiffness and pressure
C = coefficient relating normal deflection and pressure
C_L = surface roughness in CLA
C_d = damping coefficient normal to surfaces
h = surface finish (peak to valley)
K_τ = coefficient relating shear deflection and shear stress
K = coefficient relating normal stiffness and contact
L = length of cantilever
m = power law coefficient relating normal deflection and pressure
n = number of contacting points per unit area
N = number of asperities in contact
P_Δ = pressure at which full area of contact occurs with non-flat surfaces
P = normal contact pressure
P_{sy} = shear stress at elastic limit in shear
r = surface deflection ratio with and without flatness deviation
S = normal stiffness of surfaces
t = time
V = sliding velocity
W = normal load
Z = normal direction to surfaces
Z' = number of 'spots' per square inch on scraped surface
$\phi(Z)$ = distribution of surface asperities
δ = shear deflection of interface
δ' = deflection at the end of cantilever element
δ_b = deflection of cantilever with rigid surface connection
δ_j = deflection of cantilever resulting from interface deflection
Δ = flatness deviation
η = lubricant viscosity
λ = deflection normal to surfaces
μ = coefficient of kinetic friction
μ_s = coefficient of static friction
μ_y = coefficient of friction at the elastic limit
τ = shear stress applied to surfaces
ω = frequency of vibration

INTRODUCTION

Machine tools are not generally manufactured as a continuous casting or fabrication and the reasons for this are the difficulties in manufacture and transportation and for functional reasons such as the necessity to incorporate guideways. Most practical designs for machine tool structures, therefore, incorporate some form of connection between the basic elements. Such connections can be classified as bolted or fixed joints, and sliding joints; typical examples of the application can be seen in reference 1.

In both fixed and sliding connections, forces are transmitted across the joint interfaces and, therefore, one could expect that the overall static and dynamic characteristics of the machine tool are influenced by the compliance at these individual connections. In the authors' experience with testing and analysis of machines, the influence of compliance at sliding connections has been such that this feature limited the geometric accuracy in several types of machines. Other authors[2] have shown that damping in sliding elements can influence the overall dynamic response of the structure and, therefore, the cutting performance.

The principal design criteria of joints are, therefore, stiffness and damping but, for sliding connections, the friction and wear properties must also be given consideration by the designer since these can influence both the short- and long-term accuracy of the machine.

There are many publications of both fundamental and empirical nature which are concerned with the factors affecting the characteristics of both fixed and sliding joints. Consequently, a review of this work, showing the current state of the art, would not only be useful to the designer but also to prospective researchers in the field. It is hoped that this paper will particularly assist the designer.

In the following sections the work has been considered under the categories of stiffness, damping, friction and wear.

* The University of Manchester Institute of Science and Technology

THE STATIC STIFFNESS OF JOINTS

Most of the work carried out on the static stiffness of joints has assumed that the components surrounding the surfaces in contact are rigid and, therefore, only the approach of the surfaces is considered. Recent work carried out by the authors has shown that the total deflection at a jointed connection is more dependent upon the elasticity of the components surrounding the surfaces in contact, rather than upon the compliance of the contacting surfaces. This recent work does not, of course, make all the work on the surface contact compliance irrelevant because it is necessary to know such surface characteristics when formulating the solution of the complete joints.

The work on the contact compliance of interfaces is, therefore, the starting point for the work on jointed connections in machine tools. Factors influencing the surface or interface contact are considered in the following sections, but the designer should bear in mind the above comments when assessing the significance of the results to the overall characteristics of the connections.

The approach of machined surfaces

When a load is applied across the interface of small contacting metallic surfaces, then the approach of these surfaces is a function of the surface pressure. The resulting deflection at the interface is explained in terms of the deformation of the asperities, which are always present even on the finest machined surfaces.

Several different models have been proposed for the purpose of explaining the observed deflection—pressure characteristics and these show qualitative agreement over certain ranges of pressures and surface finishes. The so-called power law distribution of the asperity heights will be considered here.

For all the models it is widely accepted that the real area of contact between the surfaces is proportional to the applied normal load and given by the relationship

$$W = K \times A_r \qquad (1)$$

The real area of contact A_r is assumed to be the product of the number of contact points N, and the area of the individual contacts A_i

$$A_r = N \times A_i \qquad (2)$$

The areas A_i are assumed to be constant as the deflection increases. Suppose also that A_a is the apparent area of contact, n is the number of contacting points per unit area, $\phi(Z)$ is the probability that a contact is made at height or deflection λ; then the number of contacts is given by

$$N = n A_a \int_0^\lambda \phi(Z)\, dz \qquad (3)$$

With the relationships given in equations (2) and (3), equation (1) can be reduced to

$$P = \frac{W}{A_a} = a \int_0^\lambda \phi(Z)\, dz \qquad (4)$$

where $a = K A_i n$

P = the mean interface pressure

If the distribution of the contact points is assumed to be of the form (3)

$$\phi(Z) = b(Z)^{(1-m)/m} \qquad (5)$$

then equation (4) can be reduced to

$$P = a b m \lambda^{1/m}$$

or

$$\lambda = C P^m \qquad (6)$$

This is the most frequently used equation to describe the surface approach and the interface pressure. Research carried out by Levina[4,10,11,18], Ostrovskii[5], Tenner[17] and Dolbey[3] has shown that for the range of interface pressures and surface finishes used in the joint faces of machine tools, the above form of equation fits the experimental results. Usually, this equation is presented with P in kgf/cm^2, λ in μm and the corresponding coefficients C and m being constant for a particular pair of materials and surface finishes.

The work of Connolly[6,7] used an empirical relationship between λ and P. His equation which was valid over a pressure range of 8 to 500 kg/cm^2 was of the form

$$P = a e^{b\lambda} \qquad (7)$$

where a and b are coefficients depending upon the surface finish and material combinations.

Results from Dolbey[3] are shown in figure 1(a) and are for pressures up to 5 kgf/cm^2 which covers the range used for plain guideways. With the plastic surface combinations, hysteresis is present between the loading and unloading curves but plastic deformation was not observed. Similar experimental results were observed by Levina[4] and Ostrovskii[5] as shown in figure 1(b) and 1(c) respectively.

For fixed or bolted connections, the range of interface pressures used in practice is higher than that used for sliding connections. The work of Connolly[6,9], Schofield[12,13] and Thornley[7,8] covers this pressure range. Most of this work was carried out on mild steel specimens and the loading and unloading characteristics were of the form shown in figure 2. These loading and unloading curves are basically different from those at lower contact pressures. It was observed that, upon first loading, the total interface deflection was the sum of the plastic and elastic deformation. When unloaded without surface separation then the deformation was repeatable for subsequent loadings.

The plastic deformation of the interface is important for the initial relative position of the elements of the structure, but for the operating stiffness of the interface only the elastic deformation is relevant.

The operating stiffness of the interface depends upon the slope of the pressure–deflection characteristics at the working interface pressure.

Using the power law relationship in equation (6) this gives stiffness (S) per unit area as

$$S = \frac{dP}{d\lambda} = \frac{P^{(1-m)}}{Cm} \qquad (8)$$

when equation (7) is used then

$$S = bP \qquad (9)$$

For problems involving only normal loading of the interface either equation (8) or (9) can be used. However, the most common practical problems include the application of moments to the interface and for quick approximate calculations in these cases Levina[4] recommends a linearised approximation of the form

$$\lambda = kP \qquad (10)$$

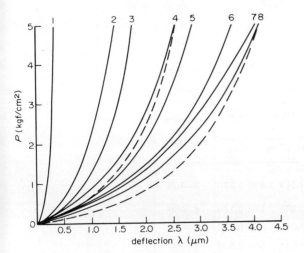

curve 1, cast iron lapped /cast iron lapped
curve 2, cast iron ground /cast iron ground
curve 3, cast iron ground /cast iron scraped
curve 4, cast iron ground /Ferobestos ground
curve 5, cast iron ground /glacier DX–plain ground
curve 6, cast iron ground/glacier DX–dimpled ground
curve 7, cast iron ground /glacier DU–as received
curve 8, cast iron ground/Tufnol ground

(a)

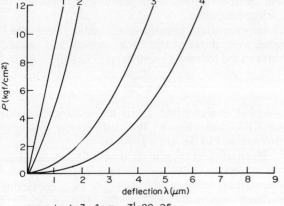

curve 1, $h=3-4\mu m$, $Z^1=20-25$
curve 2, $h=7-8\mu m$, $Z^1=20-25$
curve 3, $h=7-8\mu m$, $Z^1=12-15$
curve 4, $h=15-20\mu m$, $Z^1=12-15$

(b)

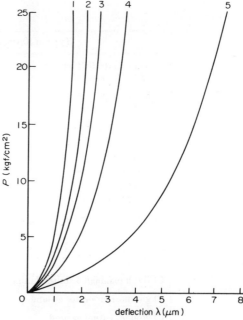

curve 1, periphery ground
curve 2, fine scraping, $Z^1$24–36
curve 3, finish planing
curve 4, conventional scraping
curve 5, coarse scraping

(c)

Figure 1(a), (b), (c). Influence of material and surface texture on the load deflection characteristics of contacting surfaces (references 3–5).

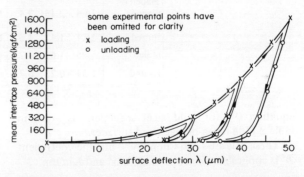

Figure 2. Load–deflection characteristics at high interface pressures (reference 6).

FACTORS AFFECTING THE NORMAL STIFFNESS OF MACHINED SURFACES

From equation (6) it can be seen that the magnitudes of the coefficients m and C influence the normal static stiffness. These constants depend upon the materials in contact; machining process, height of the asperities, relative orientation of the surface lays, hardness, flatness deviation and size of contact area.

The coefficients m and C are usually based upon data which have been obtained by loading relatively small specimens and hence the influence of waviness is small. Table 1 shows the range of the coefficients for different surface finishing processes which would be used in practice for plain guideways.

It is suggested by Levina[18] that these coefficients are applicable up to a contact pressure of 50 kgf/cm². For the higher pressure range, the coefficient 'b' of

TABLE 1

Type of Surfaces in Contact		C	m
CI hand-scraped/CI hand-scraped, $h = 3-5\ \mu m$ $Z = 20-25$ spots/in^2	references 4, 11, 18	0.3	0.5
CI hand-scraped/CI hand-scraped, $h = 6-8\ \mu m$ $Z = 20-25$ spots/in^2	references 4, 11, 18	0.5	0.5
CI hand-scraped/CI hand-scraped, $h = 6-8\ \mu m$ $Z = 15-18$ spots/in^2	references 5, 11	0.8-1.0	0.5
CI hand-scraped/CI hand-scraped, $h = 6-8\ \mu m$ $Z = 10-12$ spots/in^2	references 4, 11, 18	1.3-1.5	0.5
CI hand-scraped/CI hand-scraped, $h = 15-20\ \mu m$ $Z = 5-12$ spots/in^2		1.5-2.0	0.5
CI hand-scraped, $h = 6-8\ \mu m$, $Z = 15-18$/CI ground, $h = 1.0\ \mu m$ CLA	references 4, 18	0.8-1.0	0.4
CI peripheral ground/CI peripheral ground, $h = 1.0\ \mu m$ CLA	references 3, 5, 18	0.6-0.7	0.4-0.5
CI finish planing/CI finish planing	references 4, 18	0.6	0.5
CI ground/Ferobestos ground	reference 3	3.98	0.32
CI ground/Tufnol ground	reference 3	2.36	0.39

TABLE 2

Machined finish	Surface roughness (CLA μm)			Range of b value (μm^{-1})			Valid range of p (kg/cm^2)
	Max.	Min.	Av.	Max.	Min.	Av.	
shaped or planed	14.4	2.15	7.4	0.319	3.480	1.580	8-400
turned	10.4	1.42	4.6	0.880	5.170	2.120	8-480
milled	5.25	1.65	3.2	0.245	1.620	0.730	8-480
ground	1.54	0.25	0.73	0.835	5.860	2.820	8-160

equation (9) for mild steel specimens is given in table 2 for a range of machined surfaces.

These data are from Connolly[6,9] but the coefficient 'b' is applicable when p is in kgf/cm^2 and λ in μm.

Surface material

Irrespective of whether the mode of deformation of the surface is elastic or plastic it is evident that the surface rigidity will depend upon the material properties. It has been shown that the elastic properties of the surfaces are of interest for machine tool joints and, therefore, for a given surface finish, the surface stiffness is related to the modulus of elasticity of the materials.

In general, when different materials are used, the coefficient C of equation (6) varies inversely with the modulus of elasticity of the material, whilst the coefficient m remains at approximately 0.5. The ranges of C and m for different combinations are given in table 1.

Surface finish and lay orientation

The roughness of the surface of a given material depends upon the machining process and cutting conditions. Analyses of the resulting surface profile

has been carried out by Buc[16], Kragelstii[15] and Greenwood[14] who showed that the distribution of asperity heights can be described by normal, exponential, linear and power law distribution, depending upon the machining process used to generate the surfaces. Theoretical models of contact conditions have been considered and the results show the surface stiffness is increased with reduction in surface roughness.

Experimental investigations[6] using mild steel specimens showed that for shaped and turned surfaces the following empirical relationship is valid:

$$bC_L = A \qquad (11)$$

where A is a constant, C_L is the surface roughness in CLA and, hence, the stiffness is inversely proportional to the surface roughness.

When surfaces are generated with single point tools the surface lay will have a definite direction and the orientation between the lays on two contacting surfaces marginally effects the surface stiffness characteristics.

Schlosser[19] has also investigated the influence of surface finish on circular steel bolted flanges. For lapped, hand-scraped and ground surfaces with asperities less than 2 μm, then the surface stiffness characteristics were similar.

Influence of hardness

The influence of hardness on surface stiffness has been investigated[3] by using cast iron specimens at relatively low interface pressures. The results indicated that hardness had no effect on surface stiffness.

A similar investigation was carried out[9] which included different surface finished and contact pressures up to 1600 kg/cm^2. For the initial loading where the mode of deformation is the sum of plastic

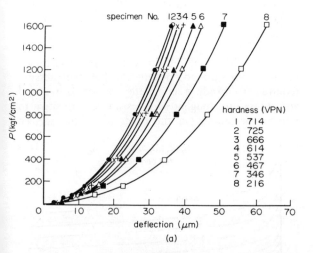

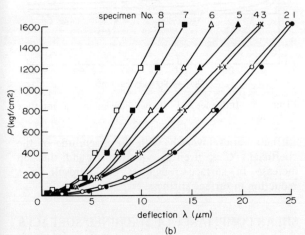

Figure 3(a), (b). Influence of material hardness on the loading (figure 3(a)) and unloading (figure 3(b)) characteristics (reference 9).

and elastic deformation then it was shown that the surface hardness had considerable influence on the approach of the surfaces. This characteristic is explained in terms of the plastic component of the deflection, where an increase in hardness results in a corresponding increase in the yield pressure and hence to a reduction in the plastic component of deformation. When the surfaces are unloaded, the mechanism is dominated by the elastic properties of the material which are not, of course, influenced by hardening. However, the surface stiffness on unloading is shown to decrease with increase in hardness. This is understandable since the area of

contact will decrease with increased hardness and, therefore, on unloading the elastic deformation will be greater. Typical experimental results showing the loading and unloading characteristics are given in figure 3(a) and 3(b). It can be concluded that for surfaces loaded to high surface pressures an increase in the hardness of the material will have a negative effect on the surface stiffness.

Flatness deviation and surface size

It is apparent from the basic models described previously that the surface stiffness would be proportional to the apparent area of the surfaces in contact, provided that the surfaces were flat. However, in practice the flatness deviation of a surface will increase with the size of the surface even though the machining quality is the same. This flatness deviation affects the distribution of the peaks or bearing area and results in the surface stiffness not being directly proportional to the apparent area of contact.

An investigation[6] on mild steel specimens of constant area in the form of an annulus, but, of different outside diameter, showed that the flatness deviation did increase with the size of the specimen.

The resulting surface stiffness as assessed by the coefficient 'b' in equation (9) varied with the size of the joint, as shown in figure 4(a). For shaped surfaces where the roughness was relatively high, the stiffness was less sensitive to flatness deviations than for ground surfaces.

To overcome the problem of flatness deviation on fixed joints Connolly[6] also considered the introduction of an epoxy resin as a bonding material between the surfaces. The resultant load—deflection characteristics of the bonded combination is given in figure 4(b). Also shown for comparison purposes are the equivalent unbonded specimens. It is quite conclusive from this comparison that the stiffness of the bonded combination is dominated by the properties of the bonding material. When flatness errors are not significant in the small ground specimens, then the stiffness is reduced by the bonding medium, but for larger surfaces subjected to flatness deviation this bonding considerably improves the stiffness properties.

Flatness errors between guideway surfaces can cause inconsistencies in the overall stiffness of identical types of machines. Tenner[17] has carried out static stiffness measurements on seven single column jig borers. The scatter in the results were in the order of 250% and this was explained in terms of the errors in manufacture of the mating surfaces.

Stiffness equations for surfaces which incorporate regular forms of flatness deviations have been formulated by Levina[11,18] and Tenner[17]. The combination of convex and sinusoidal surfaces against flat surfaces were analysed both theoretically and experimentally.

For the case of surfaces having a convex flatness deviation of maximum value Δ, then the following equations applied:

$$\lambda = (35P^2 C^4 \Delta)^{1/5} \quad \text{for} \quad \lambda < \Delta \qquad (12)$$

and

$$\lambda = \frac{\Delta}{3} + \left(PC^2 - \frac{8}{90}\Delta^2 \right)^{1/2} \qquad \text{for} \quad \lambda > \Delta \quad (13)$$

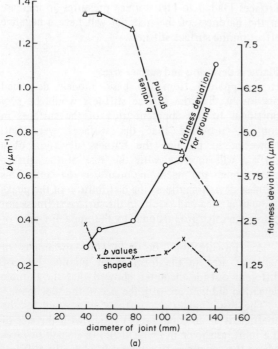

(a)

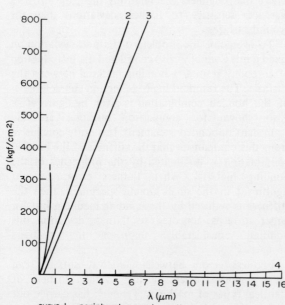

curve 1, peripheral ground small specimen
curve 2, peripheral ground small specimen, bonded with Araldite
curve 3, slab milled, large specimen, bonded with Araldite
curve 4, slab milled, large specimen

(b)

Figure 4(a), (b). Effect of flatness errors and bonding on interface stiffness (references 6, 8).

For flat surfaces when $\Delta = 0$, the deflection (λ_0) is calculated from equation (6). Thus, the deflection ratio r is defined as λ/λ_0, then the variation 'r' with flatness and pressure will give a qualitative measure of the effect of flatness errors. The results of these equations are shown graphically in figure 5 for the

case of convex surfaces. For low interface pressure, the error caused by flatness deviation is relatively large and results in a reduction of the overall contact stiffness.

In order to obtain contact over the whole area when the flatness deviation is Δ, then the minimum pressure P_Δ which must be applied is given by

$$P_\Delta = \frac{8\Delta^2}{15C^2} \qquad (14)$$

Recent work by Schofield[13] suggested that the stiffness of large surfaces which were, therefore, subjected to flatness deviation, could be increased by the use of rougher surfaces. Although the rougher surface will be less sensitive to flatness deviation as shown by reference 6, it does not necessarily follow that the surface will be stiffer. It can be seen from equation (14) that by increasing the roughness, i.e. increasing the coefficient 'C', then the pressure required to produce complete contact is reduced and, therefore, rougher surfaces are less sensitive to flatness deviation. On the other hand, the surface

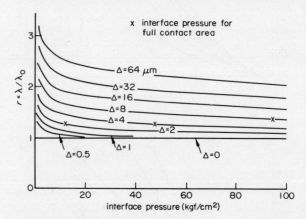

Figure 5. Influence of flatness deviation on interface deflection.

stiffness varies inversely with the magnitude of the coefficient 'C' as given by equation (8) and, thus, an increase in surface roughness could result in a reduction in surface stiffness.

SHEAR COMPLIANCE OF MACHINED SURFACES

In many practical configurations joints are also subjected to shear forces. Recent work on the characteristics of surface loaded in shear has been reported by Kirsanova[20] and Masuko[21]. It was found that, when forces are applied along the plane of the joint, elastic displacements take place within a specific range of loads. For a further increase in the shear force above this limit, the resulting deflections are basically plastic followed by slip. In general, the elastic and plastic ranges for shear deflections are analogous to those for normally loaded surfaces.

Failure of the joint can generally be determined on the basis of the coefficient of limiting friction. In practice this is unsatisfactory because irreversible displacements will occur before the joint fails. It is necessary to design such a joint so that the shear deformations remain below the elastic limit.

The results of measurements made[20] using grey cast iron specimens having an area of 225 cm^2 are shown in figure 6. It was found that for repeated loads which did not exceed the first loading limit

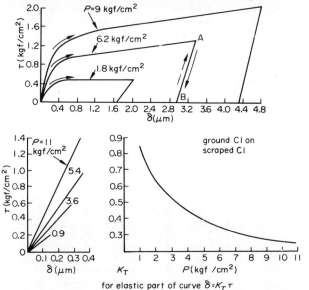

Figure 6. Shear deflection characteristics of interfaces (reference 20).

then the displacements were elastic only. It was also observed that at the elastic limit the ratio of the shear stress (P_{sy}) to the normal pressure (P) was constant over the measured pressure range of 1 to 15 kg/cm^2 and that this ratio was approximately half of the coefficient of static friction (μ_s).

Hence,

$$\frac{P_{sy}}{P} = \mu_y \approx \tfrac{1}{2}\mu_s \qquad (15)$$

The variation in μ_y and μ_s for different surface finishes is given in table 3.

TABLE 3

Type of machining	Dry		Lubricated	
	μ_y	μ_s	μ_y	μ_s
fine turning, $h = 1.6-6\ \mu m$	0.13	0.25	0.13	0.25
rough grinding, $h = 4-6\ \mu m$	0.12	0.18	0.12	0.18
grinding and lapping, $h = 1\ \mu m$	0.17	1.35	0.14	0.3
scraping, $h = 8-10\ \mu m$	0.12	0.22	0.12	0.22
fine scraping, $h = 1-2\ \mu m$	0.14	0.28	0.12	0.24

The elastic shear deflection δ is related to the shear stress by the coefficient K_τ. Thus,

$$\delta = K_\tau \tau \qquad (16)$$

The coefficient K_τ is also a function of the normal pressure P and the surface finish, and the variation of ground and scraped surfaces with normal pressure is shown in figure 6.

Masuko[25] has also investigated the shear stiffness of joints at interface pressures of 100 and 200 kg/cm^2. It is shown that the stiffness increases with normal pressure but the general trend is not given.

INFLUENCE OF JOINTS ON THE OVERALL STIFFNESS OF STRUCTURES

Connolly[22] attempted to incorporate the results of the basic work on surface compliance described

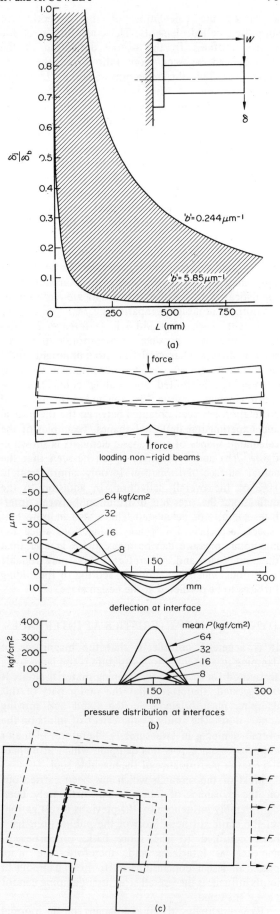

Figure 7(a), (b), (c). Examples of the calculation of deflections in jointed elements (references 22, 23).

above to the calculation of the deflection of simplified bolted elements. The contribution of the surface stiffness to the overall deflection of the element can be seen from ratio δ_j/δ_b shown in figure 7(a). δ_j is the deflection at length 'L' due to the deflection of the interface and δ_b is the corresponding bending deflection of the structure. Thus, as the length 'L' is increased, the joint becomes less and less significant. The influence of the interface stiffness is also shown for 'b' coefficients which are the extremes of the surface finish found in practice. Although this analysis is a good application of the basic work, the weakness lies in the assumption that the base of the column does not distort and, therefore, rotates as a rigid body on the contacting surface.

In reality, joints cannot be considered to be simple rigid elements which are connected by elastic surfaces and, therefore, it is necessary to consider more sophisticated models if accurate data are required for the purpose of design comparisons.

Recent work at U.M.I.S.T. (reference 23) has shown that it is possible to incorporate the basic surface stiffness characteristics into a non-rigid model of different jointed configurations similar to those encountered in bolted and sliding connections. A 'finite element' analysis[24] is used and extremely good correlation has been obtained between the theoretical and experimental deflected shapes. Examples of the calculated shapes of some basic elements is shown in figure 7(b) and 7(c) where it can be seen that the actual surface deformation is only approximately 10% of the overall deflection. In addition to the deflection, the pressure distribution between interfaces can also be calculated. This is also an important aspect, especially for sliding connections since the maximum pressure can be many times greater than the nominal and, therefore, this may enable conditions which are prone to scoring of the sliding surfaces to be predicted at the design stage.

DYNAMIC CHARACTERISTICS AT INTERFACES

It is generally considered that the magnitude of damping in a machine tool structure is too large to be accounted for in terms of damping material alone. It is suggested, therefore, that the major part of this damping must originate in the fixed and moving connections. The study of the effect of joints on the overall damping in structures is, therefore, of major importance since it is this damping which influences the chatter performance of the machine tool.

Most of the research which has been carried out on the dynamics of joints has unfortunately been considerably influenced by the previous investigations on the static stiffness. Most of the publications have been confined to the more basic properties of damping at interfaces and, therefore, the more practical design conclusions as to how damping in joints influences the overall structural damping cannot yet be answered.

Experimentally, the measurement of the normal dynamic characteristics of interfaces is much more difficult than for the previously determined static

characteristics. Several researchers[25-29] have attempted to measure the normal dynamic properties between interfaces but because of the basic inherent defects in experimentation[26] it is felt that the results from Andrew[25] are the most reliable for the purpose of formulating some general conclusions. The basic data from this work have been written in a more usable form and presented in figure 8.

We can consider that under normal dynamic loaded conditions the interface can be represented to a first approximation by a spring stiffness S and damping coefficient C_d. If this system is now excited by a normal harmonic force of $F \sin \omega t$, then the resulting relative displacement across the surface will be given by

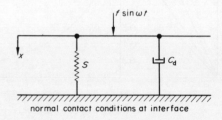

normal contact conditions at interface

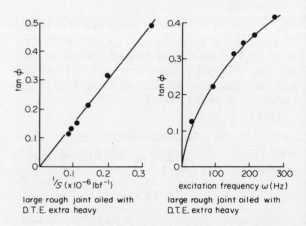

large rough joint oiled with D.T.E. extra heavy

large rough joint oiled with D.T.E. extra heavy

Figure 8. Influence of normal preload and excitation frequency on the normal damping of lubricated surfaces.

$$x = \frac{F \sin(\omega t - \phi)}{(S^2 + C_d^2 \omega^2)^{1/2}} \qquad (17)$$

where

$$\phi = \tan^{-1} \frac{\omega C_d}{S}$$

From the results given in reference 25 and also based upon the above model, the following qualitative conclusions can be reached.

(1) For dry surfaces in contact, the dynamic stiffness is similar to the static behaviour and the system has negligible damping.

(2) For a given surface combination, i.e. material and surface texture, the introduction of a lubricant film adds damping to the system. This damping coefficient is independent of the normal load as seen by plotting $\tan \phi$ against $1/S$, as given in figure 8. The slope of the curve gives the magnitude of the damping coefficient.

(3) The mechanism of the normal damping is analogous to squeeze film damping of a lubricant film.

Because of the lack of available data it is not possible to give quantitative values of the normal dynamic coefficient C_d. However, there is some evidence to suggest that this can be extremely high with high-viscosity lubricants. Unfortunately, however, because of the high value of S at the interface, it is probably not possible to use this damping potential to improve the overall damping of the structure. Better utilisation of this damping will only be possible by finding ways of reducing the interface stiffness S. It is possible that this may only be achieved at the expense of the static stiffness of the structure.

FRICTION AND WEAR IN SLIDING JOINTS

The specification of a sliding joint is more demanding than that for a fixed joint in that it is necessary to consider not only normal stiffness but also friction, lubrication and wear performance. Space will not permit us to deal in detail with the mechanisms of friction and wear of guideways but recent work[30-32] outlines that state of the art in this field. However, many of the factors discussed above also influence the friction and wear and these will be summarised below.

The friction characteristics of the sliding combination are influenced by the viscosity of the lubricant; the type of lubricant (boundary additive) and the apparent contact pressure. Typical forms of the friction characteristics are shown in figure 9(a). The main influence of the boundary additive is that it reduces the magnitude of the static coefficient of friction to about 0.1. This value increases slightly with the increase in sliding velocity and thus imparts a damping action. Stick-slip motion can be encountered with insufficient drive stiffness when plain mineral oils are used[31].

The increase in lubricant viscosity reduces the magnitude of the transition velocity between mixed and hydrodynamic lubrication. A similar effect can result from reducing the nominal contact pressure P at constant lubricant velocity η. The parameter $\eta V/P$ at the onset of hydrodynamic lubrication can be used to assess the influence of surface finish on the lubricant characteristics. Variations in this parameter for different surface combinations are shown in figure 9(b), where it can be seen that a cup ground surface sliding on a periphery ground surface gives the best surface combination based upon a criteria of hydrodynamic lubrication.

The basic data for the specimens numbered 1 to 9 were obtained from the data published by P.E.R.A[33]. These results were obtained from relatively small specimens where flatness errors could be assumed to be minimal. In contrast, the results 10 and 11 were obtained from larger specimens (100 in²)[29,30]. These two specimens were carefully manufactured and even so this still resulted in 100% difference in performance and even greater variations when compared with the small scale specimens.

This point leads us to conclude that joints in machine tools are an important reason for observed difference in performance between machine tools of identical design.

When the velocity increases, there is also a resulting separation of the sliding surfaces as shown in figure 9(c). The magnitude of this separation also depends upon the lubricant viscosity and the

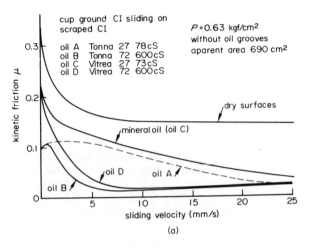

(a)

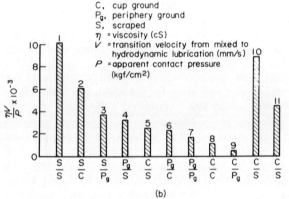

(b)

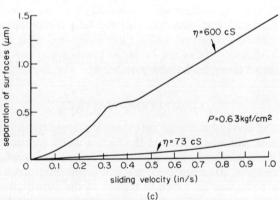

(c)

Figure 9(a), (b), (c). Effect of lubrication and surface texture on sliding performance of plain surfaces (reference 30).

apparent contact pressure. This phenomenon is especially significant when designing sliding joints on grinding machines.

On plane sliding joints the lubrication is usually assisted by means of oil grooves. Various configurations of these grooves are possible and the recommended forms are given in reference 34. These recommendations are based upon the minimum interruption of the hydrodynamic film, thus maximising the load carrying capacity. This criterion is significant for cutting speed mechanisms but for joints operating at feed speeds then this is not very significant.

Wear is also a factor to be considered when choosing the material and finishing process for a sliding joint. The surface finish mainly affects the running in process after which the rate of wear is generally independent of the original surface finish although the total wear is proportional to the original surface roughness. Many of the factors which influence the wear of plain guideways have been studied in detail by Domros[35].

The scuffing resistance of a sliding pair is of interest since the localised surface contact pressure may sometimes exceed the design pressure due to distortion of the joint. The pressure at which scuffing occurs for different surfaces and finishes is given in reference 35. For the plastic—cast iron combination no scuffing was observed, whereas for the cast iron—cast iron combination the scuffing resistance was greatly influenced by the method of surface finishing.

The type of cast iron micro-structure is also important from the aspect of wear and scuffing. Even with the same chemical composition of cast iron the structure can vary with effective cooling rate and hence with the distance from a chilled or densened surface. The most desirable surface structure for a cast iron combination is a pearlitic structure, the undesirable one being ferritic which can easily result from insufficient machining allowance due to distortion of the casting.

Hardened guideway surfaces are frequently used on beds of lathes, etc., because this surface is less prone to accident damage than unhardened cast iron. However, an industrial survey by Lapidus[36] on medium-sized lathes and milling machines concluded that a combination of unhardened cast iron against hardened cast iron was also the best combination from the aspect of machine life. Compared with unhardened surfaces the hardened combination gave a 2 to 3 times smaller wear rate.

CONCLUDING REMARKS

The survey has shown that there are ample empirical data and appreciation of the factors which influence the compliance, friction and wear of contacting surfaces. Although there are still many fundamental problems to be solved in respect of the quantitative explanation of many of the experimental observations, it would appear that the most effective way in which previous research could be exploited is in the application of the available basic data to the actual design of fixed and sliding structural elements.

With reference to the friction and wear of jointed connections, it is felt that the main limitation of the work is that it has been carried out on relatively small scale laboratory rigs. The interpretation of these results in realistic designs has not in general been given adequate attention. This is particularly true with wear investigation where it is doubtful

whether the available basic data will be relevant for quantitative analysis of slideway wear.

REFERENCES

1. F. KOENIGSBERGER, 1964. *Design Principles of Metal Cutting Machine Tools.* Pergamon Press, Oxford.
2. H. OPITZ and M. WECK, 1969. 'Determination of the transfer function by its application to the dynamic investigation of machine tools under machining conditions', *Proc. 10th M.T.D.R. Conf.*
3. M. P. DOLBEY and R. BELL, 1970. 'The contact stiffness of joints at low apparent interface pressures', *Ann. C.I.R.P.*
4. Z. M. LEVINA, 1965. 'Calculation of contact deformations in slideways', *Machines and Tooling, 36,* Nos. 1 and 8.
5. V. I. OSTROVSKII, 1965. 'The influence of machining methods on slideway contact stiffness', *Machines and Tooling, 36,* Nos. 1 and 17.
6. R. CONNOLLY and R. H. THORNLEY, 1967. 'Determining the normal stiffness of joint faces', *A.S.M.E. Paper No. 67,* Prod. 6.
7. R. H. THORNLEY, R. CONNOLLY, M. M. BARASH and F. KOENIGSBERGER, 1965. 'The effect of surface topography upon the static stiffness of machine tool joints', *Intern. J. Machine Tool Design Res., 5,* 57–74.
8. R. H. THORNLEY, R. CONNOLLY and F. KOENIGSBERGER, 1967. 'The effect of flatness of joint faces upon the static stiffness of machine tool joints', *Proc. Inst. Mech. Engrs.*
9. R. CONNOLLY, R. E. SCHOFIELD and R. H. THORNLEY, 1967. 'The approach of machined surfaces with particular reference to their hardness', *Proc. 8th M.T.D.R. Conf.*
10. Z. M. LEVINA, 1967. 'Research on the static stiffness of joints in machine tools', *Proc. 8th M.T.D.R. Conf.*
11. Z. M. LEVINA and D. N. RESHETOV, 1965. 'Machine design for contact stiffness', *Machines and Tooling, 36,* No. 12.
12. R. E. SCHOFIELD and R. H. THORNLEY, 1968. 'Mathematical expressions of surface-finish characteristics', *Conf. on Properties and Metrology of Surfaces,* Oxford, Institute of Mechanical Engineers.
13. R. E. SCHOFIELD and R. H. THORNLEY, 1971. 'Calculating the elastic and plastic components of deflection of plane joints formed from machined surfaces', *Proc. 12th M.T.D.R. Conf.*
14. J. A. GREENWOOD, 1967. 'The area of contact between rough surfaces and flats', *J. Lubrication Tech.,* January.
15. I. V. KRAGELSKII, 1965. *Friction and Wear,* Butterworths, London, Chap. 2.
16. F. BUC and B. NOWICHI, 1967. 'The measurement of the real area of contact between two metal surfaces', *Proc. 8th M.T.D.R. Conf.*
17. D. G. TENNER, 1968. 'Contact stiffness of friction slideways', *Machines and Tooling, 39,* No. 3.
18. Z. M. LEVINA, 1967. 'Research on the static stiffness of joints in machine tools', *Proc. 8th M.T.D.R. Conf.*
19. E. SCHLOSSER, 1957. 'Der Einfluss Ebiner Verschraubter Fugen auf das Statische Verhalten von Werkzeugmaschinengestellen', *Werkstatttechnik Maschinenbau, 1.*
20. V. N. KIRSANOVA, 1967. 'The shear compliance of flat joints', *Machines and Tooling, 38,* No. 7.
21. M. MASUKO, Y. ITO and C. FUJIMOTO. 'Behaviour of horizontal stiffness and the micro-sliding on the bolted joint under the normal pre-load', *Proc. 12th M.T.D.R. Conf.*
22. R. CONNOLLY and R. H. THORNLEY, 1965. 'The significance of joints on the overall deflection of machine tool structure', *Proc. 6th M.T.D.R. Conf.*

23. N. BACK, M. BURDEKIN and A. COWLEY. 'Calculating local deformations in machine tool connections', to be published.

24. A. COWLEY and S. HINDUJA, 1971. 'The finite element method for machine tool structural analysis', *Ann. C.I.R.P.,* 171–81.

25. C. ANDREW, J. A. COCKBURN and A. E. WARING, 'Metal surfaces in contact under normal forces — some dynamic stiffness and damping characteristics', *Proc. Inst. Mech. Engrs.,* **182,** pt. 3K, 92–160.

26. R. H. THORNLEY and F. KOENIGSBERGER, 1970. 'Dynamic characteristics of machine tool joints loaded and excited normal to the joint face', *Ann. C.I.R.P.,* **38.**

27. K. CORBACH. 'The dynamic stiffness of fixed and moving joints in machine tools', *M.T.I.R.A. Translation* T348.

28. P.E.R.A., 1968. 'The effect of intermediate viscous films on stiffness and damping', *P.E.R.A. Rept.,* No. 180.

29. M. P. DOLBEY, 1968. 'The normal dynamic characteristics of machine tool plain slideways', *Ph.D. Thesis,* U.M.I.S.T.

30. R. BELL and M. BURDEKIN. 'An investigation into the steady-state characteristics of plain slideways', *Proc. Inst. Mech. Engrs.,* **184,** pt. 1.

31. R. BELL and M. BURDEKIN. 'A study of the stick-slip motion of machine tool feed drives', *Proc. Inst. Mech. Engrs.,* **184,** pt. 1.

32. M. BURDEKIN, C. P. HEMINGRAY and A. COWLEY, 1971. 'Wear of slideways', *Tribology,* Feb.

33. P.E.R.A., 1958. 'Influence of type of surface, lubricant viscosity and load on the friction characteristics', *P.E.R.A. Rept.,* No. 59.

34. G. A. LEVITT and B. G. LURE, 1961. 'Improved lubricating methods for slideways of feed mechanisms', *Machines and Tooling,* **32,** No. 11, 19–25.

35. D. DOMROSS, 1964. 'Wear and friction of machine tool slideways', *Doctrate Thesis,* Aachen.

36. A. S. LAPIDUS, 1964. 'Wear of plastics for machine tool slideways', *Machines and Tooling,* **35,** No. 12, 20–6.

THE FRICTION AND WEAR OF PLASTICS, WITH SPECIAL REFERENCE TO MACHINE TOOL SLIDEWAYS

by

C. P. HEMINGRAY*

SUMMARY

A range of plastics have been tested under conditions simulating those commonly encountered on machine tool slideways. The resultant information is assessed, together with published data, to enable specific design recommendations to be promulgated. Aspects considered include the effect of velocity, load, and sliding distance on friction and wear.

It is suggested that a bronze-filled PTFE composite is likely to give good compromise between friction and wear when employed on machine tool slideways under dry conditions. In the lubricated state, the main function of such material would be to provide a backup in case of lubricant failure.

INTRODUCTION

Of recent years, there has been an increasing tendency to specify types of slideway other than the historically normal plain type, that is, cast iron on lubricated cast iron. The reasons are traceable to the stringent requirements placed on feed drive characteristics by numerical control and also to the demand for movement in very small increments. Three main alternative types have been widely employed: hydrostatics, rolling element guideways and nonmetallic facings. Although undoubtedly the cheapest, the last alternative is subject to some uncertainties in the areas of life and friction characteristics, and so attempts were made to elucidate the characteristics of a range of materials.

It is well known[1] that the properties of many plastics depend markedly on the precise conditions under which they are tested. Hence, although data is available for a range of suitable materials under certain conditions, such information is very sparse for the state of affairs encountered on machine tool slideways. This paper will therefore present the results of tests on a range of materials, largely based on PTFE (polytetrafluorethylene), under conditions intended to reproduce closely normal machine tool practice. PTFE derivatives were chosen not only as presenting the most promise (they have the lowest dry friction of any known material) but because of the great lack of useable practical data. It is hoped that the information included here will enable the design of slideways incorporating plastic materials to be approached with more confidence, and the performance of machines improved with minimal increase in cost.

REQUIREMENTS

There are in common use three alternative slideway types: plain, rolling-element (antifriction) and hydrostatic. A brief outline of the reasons leading to the choice of one or the other is outside the scope of this paper (see, for example[2]). We shall consider that plain slideways have been chosen, and can therefore list the

more important properties desirable in any material (or combination of materials) as follows:

(i) a low coefficient of friction, preferably even under dry friction conditions.

(ii) a coefficient of friction that does not decrease with increasing velocity.

(iii) a low rate of wear and a high resistance to abrasion.

(iv) good dimensional stability and adequate stiffness.

(v) reasonable cost, simplicity of application and good machining properties.

(vi) good chemical stability and good compatibility with lubricating oils and cutting fluids.

It must be said that, although it is very easy to list qualitative requirements, quantitative values are very much more difficult to state and often even more difficult to predict. A realistic judgement of the practical necessities can only be made on the basis of far more data than is possible in a single paper. Friction and wear are the only two parameters here considered in detail; the others are either easy to get from manufacturers' information, or are relatively unimportant, and will be discussed on the basis of available data.

AVAILABLE MATERIALS

Apart from cast iron, several materials have been used on machine tools: these include the following:

(1) Devametal. This is a bronze-graphite sintered material on a steel backing, and, although not a plastic, has plastic-like frictional properties. Use has largely been confined to the sliding faces of gib strips and keep plates.

(2) Ferobestos L.A.3. This material consists of an asbestos cloth bonded with thermosetting phenolic resin and impregnated with graphite filler to improve its frictional characteristics. It is widely employed at present, primarily to ameliorate the scoring that tends to occur with

* Staveley Engineering and Research Centre, Worcester.

TABLE 1
Properties of some slideway materials

property	material	PTFE-based								
		plain	VB 60	bronze	graphite	MoS_2	Turcite 'B'	Du	Sprelaflon	VX 2
property	filler	none	60% bronze	60% bronze	40% graphite	15% glass 3% MoS_2	50% bronze	lead/ bronze	lead/ bronze	65% bronze/ graphite
	source of data									
coefficient of friction	reciprocating testrig 0.79 MN m^{-2} 0 m s^{-1}	0.09	0.17	0.18	0.13	0.13	0.19	0.13	0.10	0.10
	0.79 MN m^{-2} 1.2 m s^{-1}	0.14	0.20	0.20	0.17		0.22			
	rotating-disc testrig 0 rev min^{-1} 100 rev min^{-1}	0.13 0.25	0.18 0.25	0.13 0.23	0.18 0.25	0.19 0.24	0.30 0.34	0.12 0.16		0.18 0.21
	initial friction reciprocating testrig rotating-disc testrig	0.075 0.080	0.09 0.09	0.06 0.11	0.075 0.110	0.08 0.10	0.065 0.090	0.06 0.05	0.065	0.08 0.05
	various sources 0.23 MN m^{-2} 0.76 ms^{-1}	0.10 0.23		0.14 0.28	0.11 0.26	0.15 0.29				
	data from ICI 0.89 MN m^{-2} 0.01 ms^{-2}	0.06	0.13		0.07	0.11				
wear rate* K (10^{-16} m^2 N^{-1})	reciprocating testrig	4 000	34	124	30	–	104			
	rotating-disc testrig	4 000	50	90	24	20	160	16		24
	data from ICI	1 430	3.4	3.2	46	1 120				
	various sources	12 000		5	26	7		2		
abrasion resistance	–	very poor	medium	medium	poor	poor	medium	medium	medium	medium
thermal expansion	–	← - - - - - - - - - - - - - -about 4 x that of steel (i.e. 43 x 10^{-6}°C^{-1})- - - - - - - - - - - - - - - - - →								
creep	–	about ½% change in length in one year under 2 MN m^{-2} (300 lbf in^{-2}) loading								

* Convenient practical units for K would be μm/km per N/mm^2; 10^{-16} m^2 N^{-1} = 0.1 μm/km per N/mm^2
In British units 10^{-16} m^2 N^{-1} = 10^{-10} in h^{-1} per (lbf in^{-2} x ft min^{-1})

exposed slideways, although whether it does so is by no means certain.

(3) Formica (urea formaldehyde) in various grades. At one time it was frequently used on large machines, but its present application is limited.

(4) PTFE-based materials. These plastic composites are based on PTFE which has the lowest dry coefficient of friction of any known material, but has a very poor wear resistance in its pure state. To overcome this major defect, a range of filler materials are commonly added: many different types of these composites are commercially available.

(5) Tufnol, based on cotton fabric bonded with a thermosetting phenolic resin, was employed extensively, primarily in rebuilding work, but its use in the original form for this application has diminished considerably in recent years.

These materials have all been considered in this investigation. If the study has concentrated on PTFE and its derivatives, it is because this material not only holds out the most promise, but is subject to the greatest uncertainties about its behaviour under use. Table 1 explains the nature of the materials just referred to.

FRICTION

Both the friction and wear tests were carried out on two small-scale rigs, one of which was designed to simulate slideway conditions as closely as possible, and reciprocated a loaded circle of the material over a steel slideway. An additional device, designed essentially to act as a cross-checking facility, utilised a small disc of the sample rubbing against a disc: wear measurement was by measuring the scar width. This latter rig used PV (pressure x velocity) values 2 to 5 times those associated with the linear device, and so the results cannot be expected to be very similar for both rigs.

TABLE 1 (*cont.*)
Properties of some slideway materials

			others			comments
Tufnol	Devametal	Ferobestos	Formica 'N'	sprayed Mo	cast iron	
•tton fibre/ phenolic resin	graphite/ tin/copper	asbestos/ graphite/ phenolic resin	urea formaldehyde	molybdenum		
0.26 0.18	0.2 0.2				0.2 0.15	Values determined after sliding upon hardened steel over 15 000 m. Velocity not quite zero, but 0.003 m s⁻¹. Values obtained by extrapolation from figure 1. Materials slid over cast iron.
						Values refer to materials sliding over steel.
						Values for samples sliding over cleaned hardened steel. Impractical but are included to show possibility of low coefficients of friction.
		0.2 (lub) 0.12	0.2 (lub) 0.12	0.19 (lub) 0.08	0.23 (lub) 0.12	Values for samples sliding over cast iron and carbon steel[9, 10, 11].
						Values for samples sliding over steel[6].
20	20					Sliding over hardened steel.
						Sliding over steel.
						Sliding over steel[6].
—		80 (lub)	80 (lub)	32 (lub)	2 (lub)	Sliding over cast iron and carbon steel[5, 10, 12].
medium	very good	good	good	very good	excellent	Very approximate.
	16×10^{-6} deg C⁻¹	10.8×10^{-6} deg C⁻¹				
------------------------------negligible------------------------------→						Creep limits thickness to 1.6 mm; 2.1 MN m⁻² is common point loading; 0.5 per cent change in 1.6 mm is 8 μm per year.

The frictional properties of PTFE-based materials cannot be expressed in simple terms: as a consequence efforts have been confined to investigating the effect of sliding velocity, contact pressures between slideways and traverse distances.

For a number of plastic materials the dependence of dry friction on sliding velocity was investigated, with conditions approaching those normally encountered on machine tool slideways, namely a contact pressure of about 0.79 MN m⁻² (115 lbf in⁻²). From the results shown on figure 1, it can be seen that for the PTFE-based materials tested, the coefficient of friction increased from about 0.07 under stationary conditions to between 0.09 and 0.12 as soon as breakaway occurred, and then showed a very gradual increase with velocity up to 0.16 at 0.5 ms⁻¹ (20 in s⁻¹). An increase of friction with sliding velocity was also found for the second rig, as shown in table 1, despite the use in this case of steel, compared to cast iron on the first device. The other materials examined

exhibit different characteristics: the friction of Devametal appears to be independent of velocity, and that of Tufnol, in common with most 'hard' plastics, drops rapidly with increasing velocity at low speeds, subsequently increasing slowly as the velocity is further raised.

The results shown in figure 1 are outlined in figure 2 which includes data also from other investigators. It is apparent that reasonable agreement is obtained, bearing in mind that the errors involved are not only experimental but also due to the differing conditions. The data obtained by PERA[3] are inconsistent with all available information.

Figure 3 shows the effect of sliding distance on the coefficient of friction of the various PTFE-based plastics studied: no other material studied exhibited (or could be expected to exhibit) any such variation. The main conclusion is that the coefficient of friction of most PTFE-based plastics sliding against steel approximately doubles in the initial phase, and then increases very slowly; the initial coefficient of friction is about 0.07, and this increases to about 0.13 after running-in. This increase of friction with sliding behaviour was confirmed by the second rig.

Information is available[4] on the effect of various periods of rest and also high velocities on the sliding behaviour of PTFE. However, although such

material is contemplated, and will be referred to later.

The results of a series of tests on the dynamic friction of Turcite 'B', a bronze-filled PTFE, as a function of sliding distance is shown in figure 4: little

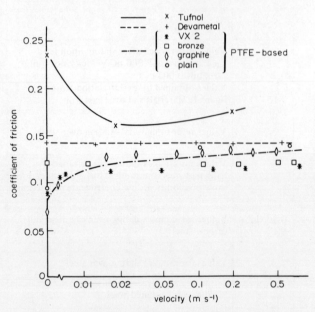

Figure 1. The effect of sliding velocity on coefficients of friction of various materials.

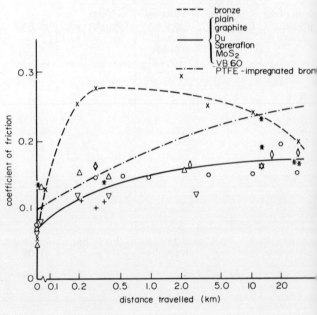

Figure 3. The effect of sliding distance on the frictional properties of PTFE-based material.

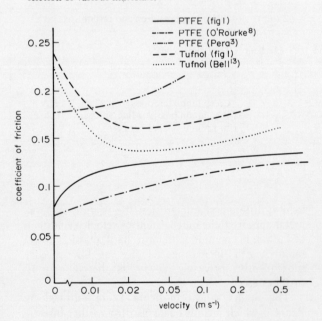

Figure 2. The effect of sliding velocity on coefficients of friction of PTFE and Tufnol.

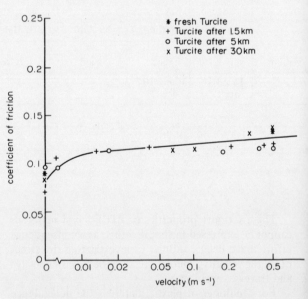

Figure 4. The effect of sliding velocity and distance on the coefficient of friction of Turcite 'B'.

conditions do have an impact on subsequent friction, the magnitudes observed are only about ±10% and hence relatively unimportant for machine tools.

The behaviour of the bronze-filled material shown in figure 3 was unusual; a similar material (VB60) is shown to exhibit 'normal' behaviour, and subsequent tests on this material also revealed much more representative data. This anomalous information is included to illustrate a possible danger if the dry use of PTFE

variation is apparent, as might be expected (the 'fresh' Turcite had to be run in for a few minutes, due to initial erratic frictional behaviour).

The influence of slideway pressure on coefficient of friction was not experimentally studied in the course of the present investigations; however, other sources gave information of value in confirming the practical results. An approximate empirical relationship from[5] between the coefficient of friction of PTFE, velocity and pressure is

$$f = C_v p^{-n}$$

where f is the coefficient of friction,
$\quad C_v$ is a constant depending on sliding speed,
$\quad p$ is the surface pressure, and
$\quad n$ is a constant for the material.

This is illustrated in figure 5.

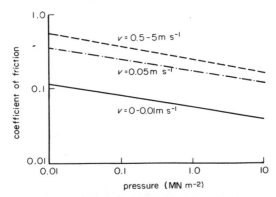

Figure 5. The effects of pressure and velocity on the coefficient of friction of unfilled PTFE (after Lewis[5]).

For unfilled PTFE values of $n = 0.15$ are quoted, with C_v increasing with velocity from 0.45 at 0.5 mms^{-1} (1.2 in min^{-1}) to 1.5 at 50 mms^{-1} (120 in min^{-1}) and 2.1 at high speeds (0.5–5 ms^{-1}, 1200–12 000 in min^{-1}).

Experimental results showed, with the pressure of 0.8 MN m^{-2} (115 lbf in^{-2}) used, a coefficient of friction of 0.1 at low speed was obtained rising to 0.15 at 50 mms^{-1} (120 in min^{-1}). From equation (1) the corresponding values are 0.06 and 0.19. Thus, although the agreement is not perfect, the equation does seem to indicate the correct trends.

Not all filled PTFE materials have been observed to exhibit a decrease of coefficient of friction with load, as shown on figure 6. The behaviour of the PTFE

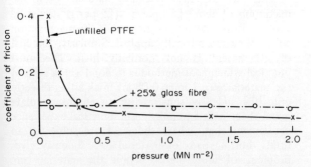

Figure 6. The effect of pressure on coefficient of friction (after O'Rourke[8]).

filled with 25% glass fibre as shown appears to constitute the exception rather than the rule, however.

Published data on coefficients of friction are given on table 1 for the purposes of comparison with the experimentally derived data. It should however be noted that because the physical properties of PTFE can vary markedly with time, temperature and test method, any table or graph expressing the physical properties of filled PTFE (including such tables and graphs as appear in this paper) must be regarded as simple and approximate statements of a complex subject. Similar comments hold for wear.

WEAR

The influence of sliding distance on wear for various materials under dry conditions was investigated on both rigs; some results from the reciprocating testrig are shown on figure 7. For all the PTFE-based materials tested, two wear regimes are visible: a running-in

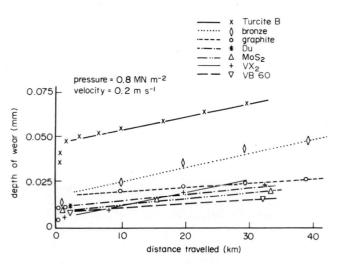

Figure 7. The effect of sliding distance on the wear of PTFE-based materials.

period of a few hundred metres characterised by rapid wear, followed by a steady state where a constant, low rate of wear is observed. The former behaviour is (for short durations) of relatively minor significance, as any inaccuracies caused by such wear can, by suitable design, be eliminated by adjustment at the commissioning stage.

The slope of the wear–distance relationship once the running-in phase is over can be used to evaluate a wear factor K which is used on table 1 to compare the materials tested.

$$K = \frac{L}{p \times v \times t} \ \text{m}^2 \, \text{N}^{-1} \quad \left[\frac{\text{in}^3 \ \text{min}}{\text{lbf ft h}} \right]$$

where L is the depth of wear (m: in)
$\quad p$ is the material pressure (Nm^{-2}: lbf in^{-2})
$\quad v$ is the sliding velocity (ms^{-1}: ft/min)
$\quad t$ is the time of wear (s: hours).

(It should be mentioned that imperial units have been chosen to ensure compatibility with the results quoted in the literature.)

The sliding speed was approximately constant in the tests so that K can be expressed as a function of velocity and time, that is, distance.

The factor K can only be regarded as a constant where similar conditions are being used; for such materials as PTFE-based plastics it will vary with both pressure and velocity. For this reason, there is no attempt in table 1 to collect all available wear factors, but only those for which the conditions of test approximate to those common on machine tool slideways. Even so, it is evident that published data shows greater variation than the results derived from the present tests. This is not unexpected; exactly the same materials (and often the same testing methods) were used experimentally, whereas different techniques and different samples were associated with the published data.

From table 1, for the PTFE-based materials running dry, it is evident that VB60, graphite-filled and molybdenum disulphide filled PTFE have good wear properties; the wear factor K associated with each is about 30×10^{-16} m^2 N^{-1} (3μm/km per N/mm^2). Tufnol and Devametal also have good wear characteristics ($\sim 20 \times 10^{-16}$ m^2 N^{-1}), whilst, even lubricated, Ferobestos and Formica have been observed to have somewhat poorer wear characteristics, both having wear factors of about 80×10^{-16} m^2 N^{-1}. Good wear properties were also exhibited by VX2 but the material was not readily available commercially.

The wide range of results for wear rates obtained on different occasions and by different observers on nominally-similar materials emphasises the need for caution in interpreting friction and wear data. For example, subsequent tests on the same rig but by different observers, although confirming the wear factors obtained for VB60, obtained somewhat lower rates for Turcite 'B' than previously obtained. It is apparent that the factors affecting wear rate and its measurement are rarely under sufficiently close control for reliable life predictions to be made; and the conditions extant on machine tool slideways are very difficult to reproduce on any rig, even if the conditions were assumed to be invariant with the machine.

OTHER FACTORS

Factors other than friction and wear, some not as obvious as those mentioned in the introduction, are also relevant to the selection of slideway material.

The resistance of slideway materials to scoring and abrasion is important and must be considered. However, even if a material eliminates scoring initially (that is, foreign particles preferentially embed themselves in the plastic), loading the material with these particles can readily turn it from a slideway material to an abrasive lap. The only real protection is either total enclosure of the slideway or adequate sealing. A small-scale trial, using contaminated oil, showed:

(i)   A 'soft' material such as filled PTFE prevents scoring of the mating slideway, as no scoring on it was observable after sliding a distance of 3 km (10 000 ft) using as a lubricant heavily-contaminated oil: a cast iron sample scratched the slideway quite badly under the same conditions.

(ii)   The resistance to abrasion of the PTFE-based material was quite low, for quite bad scratch marks were to be seen on the material surface.

For most slideway materials, lubricant, in however small a quantity, is beneficial. Available information would indicate that this is not necessarily true for PTFE-based materials: intermittent lubrication has

been observed to give high rates of wear. This was to some extent confirmed in the present series of tests, during which various samples of VB60 were tested under nominally identical conditions. Normally, great care was taken to ensure that dry conditions prevailed, but in one test run a very small quantity of grease was deliberately introduced. Compared to the behaviour observed during the dry tests, slideway wear rate was doubled and the friction level was increased by 60%. It is considered that the somewhat anomalous frictional characteristics of the bronze-based PTFE shown in figure 3 are due to this cause, and the associated high wear rates recorded in table 1. Lubricated, the friction of PTFE-based materials has been observed as about 0.1 at low velocities.

For PTFE-based materials, a coefficient of thermal expansion about four times that of steel can be assumed.[6] Other materials have the coefficients of thermal expansion indicated in table 1.

PTFE-based materials, alone among slideway materials, exhibit some degree of flow under load: about 0.5% a year under a pressure of 2 MNm^{-2} (300 lbf in^{-2}) can be assumed. Hence the thickness of a sheet about 2.5 mm (0.1 in) thick can be expected to change by about 12.5 μm (0.0005 in) per year under a peak load not uncommon on machine tool slideways. For this reason the thickness of such materials should be kept as low as possible.

PTFE-based materials do not normally present any machining problems, but where the composition is non-uniform (as with 'Du'), problems of availability of precisely flat materials arise, although such aspects are capable of being overcome in practice.[7]

No problems of chemical compatibility have been reported or can be expected with any of the slideway materials studied.

The cost of any slideway material, even the maximum of about £50 per m^2 (£5 per ft^2) must be very small compared with the cost of the machines on which they are normally applied. Similarly, the cost of attachment is not normally high. The most practical attachment method is to bond a thin strip of the material, say 2 mm (0.08 in) thick either directly onto the slideway substrate or to an intermediate metal sheet with an epoxy resin. Practical experience has shown that, where slideways are subjected to fluids, bolting pads of material to a slideway gives problems of liquid ingress behind the material: care must be exercised in such areas.

The problems associated with the lower contact stiffnesses associated with plastic materials can largely be ignored in practice, as joint stiffness is dominated by macro- and not micro-effects.

Finally, there is some information available on the effect of surface finish and material of the mating surface on the wear rate of plastic materials. Surface finishes[8] above 0.8 μm CLA (32 μin CLA) cause a serious increase in wear rate, with reduced performance below 0.1 μm CLA (4 μin CLA). Increase of hardness reduces the wear of the plastic, but actual material has little effect. For this reason, hardened mild steel was used in the wear tests, finished to 0.2 μm CLA (8 μin CLA). Peripheral grinding is the preferred finishing process.

CONCLUSIONS

It is apparent that, under dry sliding conditions, all PTFE-based composites have low coefficients of friction, a realistic figure for machine tool conditions being 0.15. Furthermore, only PTFE-based plastics have coefficients of friction that increase with sliding velocity under nonlubricated conditions. Hence these materials are recommended for dry use.

Although all PTFE-based plastics should prevent scoring of machine tool slideways to some extent, VB60, because of its high percentage of bronze filler, should have a higher abrasion resistance than most of the alternative materials, and is recommended.

If a harder material is required, to give better abrasion resistance under contaminated conditions, Devametal, which has a reasonably low and constant coefficient of friction, should be considered.

Under lubricated conditions, coefficients of friction down to 0.1 can be expected from cast iron on cast iron slideways with the correct choice of lubricant, which PTFE-composites will improve on only marginally (if at all) when lubricated and do not attain when dry. Hence, PTFE materials operating under dry conditions do not offer significant frictional advantages over plain cast iron when lubricated. Nevertheless, PTFE materials offer advantages in the following areas:

(1) For contouring machine tools, where stick-slip can be a serious problem especially on vertical slideways, PTFE-based materials when lubricated will remove all such effects and offer a very valuable backup feature in case of lubricant failure. However, it should be remembered that under trace lubrication PTFE-based materials may exhibit higher (although not catastrophic) wear rates than when dry.

(2) When lubrication is difficult, a filled PTFE material can be used dry and still give good frictional characteristics and adequate life. Ram slideways on vertical turning lathes and plano-millers, particularly when numerical control is used, are examples of situations of this kind.

(3) Because it is possible to run PTFE-composites totally dry, it is feasible to equip numerically controlled machines with slideways based on this material and eliminate all slideway lubrication, thus cutting overall machine cost, whilst still retaining adequate life.

For continuous dry use, the nominal pressure on slideways where PTFE materials are used should be kept to below 0.35 MNm^{-2} (50 lbf/in^2). However, when loads and/or movements are intermittent, for example on the ram of a large vertical turning machine, pressures of up to 1 MNm^{-2} (150 lbf/in^{-2}) may be acceptable, with transient peaks to 2 $MN\,m^{-2}$ (300 lbf/in^{-2}).

ACKNOWLEDGEMENTS

The author would like to thank Messrs B. J. Davies and H. G. Harris of Staveley Machine Tools for their encouragement and direction in the work leading to the production of this paper, and the Director of the Machine Tool Industry Research Association for permission to publish it.

REFERENCES

1. P. P. BOWDEN and D. D. TABOR. The friction and lubrication of solids. Part 1 (Oxford, 1950) and Part 2 (Oxford, 1964).
2. M. BURDEKIN, A. COWLEY and P. HEMINGRAY, 1971. Wear on slideways. *Tribology*, February, p. 15.
3. Frictional characteristics of PTFE—impregnated bronze sliding on cast iron. *PERA Report No. 80.*
4. R. P. STEIJN, 1966. The effect of time, temperature and environment on the sliding behaviour of poly-tetrafluorethylene. *ASLE Trans.*, Vol. 9, p. 149.
5. B. B. LEWIS, 1967. Predicting bearing performance of filled teflon PTFE resins. *Trans. ASME* (Jnl. of Eng. for Ind.), p. 182.
6. Filled PTFE: properties and application design data. *Fluon Technical Service Note F.13.* Vinyls Group, I.C.I. Limited, Welwyn Garden City.
7. D. GRINDROD and G. FARNWORTH, 1971. A modular range of NC machine tools, *Proc. 12th Int. MTDR Conference,* Macmillan, 31.
8. J. T. O'ROURKE, 1965. Advances in fluorocarbons. *Modern Plastics.* V43 No. 1, p. 161.
9. A. S. LAPIDUS. Wear of plastic machine tool slideways-*Machines and Tooling.* Vol. 32, No. 12, p. 20.
10. J. P. GALTROW and J. K. LANCASTER, 1967. Carbon-fibre reinforced polymers as self-lubricating materials. *RAE Technical Report 67378.*
11. Friction and wear of filled-teflon compounds. *Journal of Teflon*, Vol. 3. No. 4, p. 7.
12. S. RIDGWAY, 1966. Plastic laminates as slideway inserts. *The Engineer*, No. 11, p. 715.
13. R. BELL and M. BURDEKIN, 1966/7. Dynamic behaviour of plain slideways. *Proc. I. Mech. E.* Vol. 181, Part 1, No. 8, p. 169.

THE DYNAMIC STIFFNESS OF ANTIFRICTION ROLLER GUIDEWAYS

by

J. G. M. HALLOWES* and R. BELL†

SUMMARY

The stiffness and friction characteristics of roller guideways are removed. It is shown that, whilst the static stiffness of roller guideways is adequate, the absence of any inherent dynamic stiffness magnification can induce undesirable resonances. The effectiveness of additional squeeze film devices to provide increased dynamic stiffness is analysed and shown to be a very effective and economic extension of the design of antifriction guideways.

NOTATION

B_{SF} = width of rectangular squeeze film pad
C_{SF} = damping coefficient of squeeze film pad
C_o = specific roller load
d = diameter of roller
f_R = resonant frequency of moving mass and roller elements
F = external normal force
h = squeeze film gap
k = Palmgren stiffness for roller element
K_R = mean stiffness for roller element unit
K_{RM} = measured stiffness of experimental guideway system
L_{SF} = length of rectangular squeeze film pad
l = length of roller
M = sliding mass
M_{SF} = effective mass of the oil in squeeze film
Q = load per roller
S = Laplace operator
W_{SF} = load bearing capacity of squeeze film
β_S = shape factor for rectangular pad squeeze film
δ = deflection of roller
η = absolute viscosity of oil in squeeze film
ρ = density of oil in squeeze film

INTRODUCTION

The use of rolling element guideways to provide an effective antifriction guideway system is now well proved; a range of devices are now available that provide a number of possibilities for machine tool applications. The alternatives, i.e. plain guideways or hydrostatic guideways, both offer an advantage in competition with the rolling element systems; both of these systems rely on the presence of thin fluid films and thus have very attractive dynamic stiffness characteristics. The rolling element guideway has no equivalent inherent mechanism to provide high dynamic stiffness. In consequence, it is possible to find occasionally that machines employing rolling element guideways exhibit resonant modes of the moving elements on the rolling elements that limit the metal cutting performance of the machine tool. A simple design solution was alluded to by Tlusty[1] in a paper concerned with machine tool vibrations; an example was cited of the use of added squeeze film

devices to increase the limit depth of cut for a machine that employed a rotary slide guided by rolling element guideways. This paper gives the results of a study of the effectiveness of added squeeze film elements to enhance the dynamic stiffness of rolling element guideways. Results are included that give guidance for the machine tool designer on the use of this technique.

The literature on the design and application of rolling element slideways is dominated by a valuable group of Russian papers[2-4]. The stiffness, friction and load-carrying capacity of a range of guideway configurations have been discussed at length. A number of publications are also available[5,6] on proprietary devices that are available in this country.

In this paper the role of the squeeze film units is described as that of increasing the dynamic stiffness of the guideway. The term damping is carefully avoided as the results quoted in the paper do not conform to those that could be attributed to a simple increase in a viscous damping coefficient.

THE BASIC CHARACTERISTICS OF ROLLING ELEMENT ANTIFRICTION GUIDEWAYS

The basic parameters of importance in the assessment of the dynamic stiffness of a rolling element slideway are the normal stiffness and the frictional resistance to sliding. Both of these terms are influenced by the preload placed on the guideway elements. It is necessary, therefore, to assess the influence of preload on the stiffness and friction force characteristics of a particular antifriction guideway element before analysing the need to enhance its dynamic stiffness.

Consider the load deflection characteristic of a roller slideway element; a number of factors have to be included, i.e. the number and dimensions of the roller elements, hardness of the slideway, mean load on each roller, the guideway errors and the maximum permissible mean load on each roller.

The Palmgren[7] expression for the load–deflection characteristic for a steel roller between hard surfaces is

$$\delta = 0.6 \frac{Q^{0.9}}{l^{0.8}} \mu m \tag{1}$$

and the stiffness $k = 1.85\, l^{0.8}\, Q^{0.1}$ kg/μm (2)

* Staveley-Asquith Division, Halifax.
† Machine Tool Engineering Division, University of Manchester Institute of Science and Technology

the maximum permissible load, for non-plastic deformation, is

$$Q_{max} = C_o dl \qquad (3)$$

(C_o = 3.6 kg/mm² for $l = d$, C_o = 2.7 kg/mm² for $l = 2d$).

The expression for stiffness given in equation (2) contains no reference to the influence of errors in roller form and guideway manufacture. It is possible[8] to derive a simplified expression for the mean stiffness of an individual roller element which, because it does not contain a complete allowance for error terms, is rather optimistic. A plot of this expression is shown

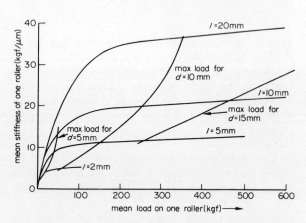

Figure 1. The mean stiffness of a roller element.

in figure 1 for the case of a total guideway assembly with an assumed error of 10 μm.

The curves in figure 1 emphasise the influence of preload on stiffness. Excessive preload merely reduces the life rating of roller elements. The selection of roller length and diameter offer, in turn, a more direct influence on roller stiffness and life than does the careful selection of initial preload. It should be noted that the conditions for maximum load are based on an assumed surface hardness value of RC60. The roller elements used in this investigation had a hardness of approximately RC60.

The friction characteristics of antifriction roller slideways are difficult to attempt to analyse. The overall effect is a much smaller coefficient of static friction than can be achieved with the use of plain slideways. The influence of preload can be most effectively assessed by the consideration of some experimental data published by Levina and Kalymykov[4]. These results are shown in figure 2. The friction force is shown to be proportional to preload and to diminish with the diameter of the roller element. The other important feature is that the friction force—velocity characteristic is substantially a Coulomb friction characteristic.

The basic characteristics of antifriction roller element guideways can now be summarised. The static stiffness can be described by the number and dimensions of the roller elements with a properly chosen preload. Excessive preload values increase stiffness but undermine the life of the guideway and increase the friction force levels. The friction force levels are such that this class of guideway will provide

static coefficients of friction that are less than 0.005 (the lowest value published for a guideway is 0.04[9]).

The roller antifriction slideway will, therefore, display a lightly damped motion in the direction of guided motion. It has been shown by Levina[2] that the preload has no influence on damping at high feed speeds (>1000 mm/min) but has discernible influence about low feed speeds although the influence of force amplitude is noted. These results

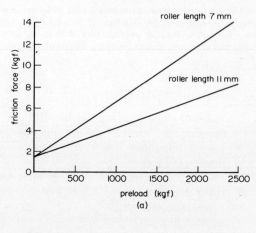

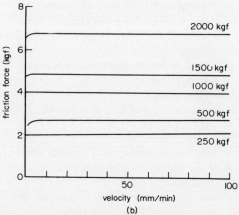

Figure 2. The influence of preload on the friction characteristics of a roller element antifriction guideway (after Levina and Kalymykov[4]): (a) The influence of preload on breakaway friction; (b) The friction force—velocity characteristics for the 11 mm long roller.

have been supported by the results of Polacek and Vavra[10].

The normal mode of motion when subjected to an external force has no effective source of damping and, therefore, the resonant mode of the sliding mass vibrating on the roller elements is lightly damped and constitutes a potentially hazardous mode of vibration (see the example quoted by Tlusty[1]). Levina[2] comments that the damping of roller slideways is similar in the tangential and normal modes for a lightly preloaded guideway and that normal mode damping is slightly increased with increased preload. The tangential mode can be damped, about zero velocity, by the use of a combined guideway. The use of an added squeeze film element will be shown in this paper to constitute a most effective and inexpensive technique for the enhancement of the dynamic stiffness of the normal mode.

THE DYNAMIC CHARACTERISTICS OF A SQUEEZE FILM

The squeeze film envisaged in this study is a rectangular pad with a working gap in the range of $25 \rightarrow 150\ \mu$m. It is, therefore, considered adequate to use the equation developed[11,12] for the zero end flow condition modified by the shape factor, β_S, quoted by Cameron[13] to allow for free (two-dimensional) flow, which is given below.

$$W_{SF} = \frac{\beta_S\ \eta\ (\mathrm{d}h/\mathrm{d}t)\ B_{SF}^3\ L_{SF}}{h^3} \qquad (4)$$

If the dynamic deflections under load are much smaller than the normal gap, h, then an expression for a viscous damping coefficient can be derived, i.e.

$$C_{SF} = \frac{\beta_S\ \eta\ B_{SF}\ L_{SF}}{h^3} \qquad (5)$$

It has also been shown by Van Herck[14] than an allowance for an equivalent mass term should be included for dynamic excitation. The term proposed is

$$M_{SF} = \frac{\beta_S\ \rho\ B_{SF}^3 L_{SF}}{12h} \qquad (6)$$

Van Herck also indicated a negative pressure limit that may be taken as a design guide.

THE DYNAMIC STIFFNESS OF A ROLLER GUIDEWAY WITH SQUEEZE FILM DEVICES

The lumped element diagram for the normal mode of the guideway is shown in figure 3. If it is assumed that there are four antifriction rolling element devices and four identical squeeze film blocks, then the dynamic stiffness of the system is given by equation (7).

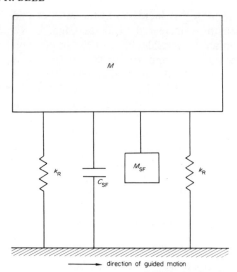

Figure 3. The lumped element representation of the normal mode of motion for a rolling element guideway with an added squeeze film.

$$\frac{F(s)}{x(s)} = 4k_R\ \left(1 + S\frac{C_{SF}}{k_R} + S^2\frac{M + 4M_{SF}}{4k_R}\right) \qquad (7)$$

The desired end product of this design procedure is to guarantee that dynamic stiffness is never less than the static stiffness $4k_R$. This is achieved if the damping factor contained in the quadratic term in the dynamic stiffness is not less than 1.

$$C_{SF} \geqslant \left[(M + 4M_{SF})k_R\right]^{1/2}$$

i.e.

$$\frac{\beta_S\ \eta\ B_{SF}\ L_{SF}}{h^3} \geqslant \left[\left(M + 4\frac{\beta_S\ \rho\ B_{SF}^3 L_{SF}}{12h}\right) k_R\right]^{1/2} \qquad (8)$$

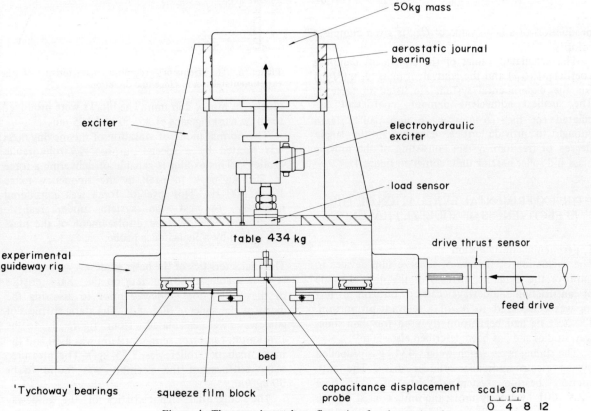

Figure 4. The experimental configuration for the study of the dynamic behaviour of antifriction guideways.

The term β_S[13] is increased as the ratio of pad width to length is increased. The damping term can be maximised by consideration of the interaction of the shape factor β_S and the width of the pad β_{SF}. In the experimental work described later, the pad design utilises the full width of the slideway surface. The choice of working gap, h, is chosen so as to cover a range of values that could be readily achieved in the manufacture of the guideway (i.e. $25-125$ μm). The shape factor is varied by the use of two squeeze film geometries, i.e. four square pads (76 mm x 76 mm) and two rectangular pads (76 mm x 286 mm). The influence of the absolute viscosity of the oil, η, is studied by the use of representative values that are available in a proprietary range of hydraulic system oils.

The equivalent mass term, M_{SF}, may require specific attention in some cases but for a fixed pad geometry there is then a clash between the values of C_{SF} and M_{SF} that are required. In this study the

TABLE 1
The calculated squeeze film dynamic parameters

Pad dimensions $L_{SF} \times B_{SF}$ (mm x mm)	Squeeze film gap h (μm)	Equivalent damping coefficient C_{SF} ($\times 10^6$ N$_S$ m^{-1})		Equivalent mass M_{SF} (kg)
		Oil viscosity 0.6 poise	Oil viscosity 5.5 poise	
76 x 76	25	55	480	42
	50	7	60	21
	125	0.44	3.8	8
286 x 76	50	51	440	153
	125	3.3	28	61

production of a large value of C_{SF} is given complete priority.

The calculated values of the equivalent damping coefficient (G_F) and the equivalent mass term (M_{SF}) for the experimental system are listed in table 1. The smallest equivalent damping coefficient calculated for the 76 mm x 76 mm pad is large enough to provide critical damping in the single degree of freedom system consisting of the moving mass and roller bearing units shown in figure 4.

THE EXPERIMENTAL EVALUATION OF THE EFFECTIVENESS OF SQUEEZE FILM UNITS

Experimental configuration
The evaluation of the use of squeeze film devices to improve the dynamic stiffness of the normal mode of antifriction guideways carried out on a test rig was considered to be of a relevant proportion. The test rig had been formerly used for plain slideway studies and has been described elsewhere[15].

The sliding mass which weighs 434 kg is propelled by an electro-hydraulic cylinder drive. The antifriction bearing system consists of four type RZA 1001 Tychoway units; the units consist of two roller tracks, 14 rollers are in contact at any time

(roller diameter 5 mm, roller length 5 mm). The contacting surface was a hardened steel insert. The squeeze film inserts (see figure 4) consisted of either four 76 mm^2 square blocks of two rectangular

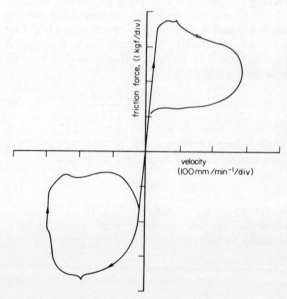

Figure 5. The friction force—velocity characteristic of the experimental guideway.

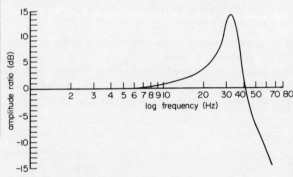

Figure 6. The frequency response characteristic of the experimental drive, in tangential direction.

blocks 76 mm x 286 mm. The blocks were produced to allow working gaps of 25, 50 and 125 μm.

The normal mode of vibration of the moving mass was excited by a special purpose electrohydraulic exciter[16]. This exciter is capable of delivering a force of 200 kg peak to peak in the frequency range 10 → 600 Hz. This level of force was considered necessary to test the system under realistic conditions. The dynamic displacement of the mass was sensed by a capacitance probe.

The characteristics of the basic guideway system
It is desirable to obtain data on the characteristics of the antifriction guideway prior to assessing the merit of the squeeze film units. The static stiffness of guideway was calculated, from figure 1, to be 196 kg/μm (an error term of 10 μm was used and the mean load per roller was 7.75 kgf). The measured static stiffness of the assembly was found to be 90 kgf/μm.

The tangential characteristics of the guideway were measured and found to conform to the

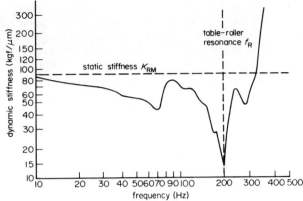

Figure 7. The normal dynamic stiffness of the experimental guideway.

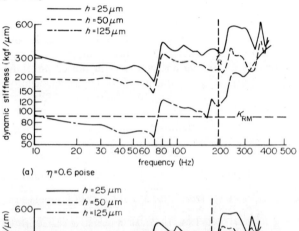

(a) $\eta = 0.6$ poise

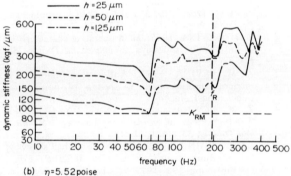

(b) $\eta = 5.52$ poise

Figure 8. The increase in normal dynamic stiffness due to the inclusion of four square (76 mm × 76 mm) squeeze film devices.

magnitudes reported by Levina[2] and Polacek and Vavra[10]. The frequency response characteristic, given in figure 6, displays the characteristically lightly damped response. The friction velocity characteristic, given in figure 5, has a breakaway friction level that is equivalent to a static coefficient of friction of less than 0.01.

The dynamic stiffness of the normal mode is depicted in figure 7. The principal resonance, occurring at 195 Hz, is the mode due to the mass vibrating on the stiffness of the roller bearing elements. The other resonant frequencies are transmitted from the bed structure. The inherently low damping of this normal mode is clearly demonstrated; the dynamic stiffness at this frequency is reduced to some 20% of the static stiffness.

The influence of the squeeze film devices
The squeeze films built into the guideways utilised the full width of each guideway (76 mm). They were initially used in the form of four square pads

(76 × 76 mm) and a second set of two rectangular pads (76 × 286 mm) were used to use the maximum available area. The squeeze film thickness was made variable in the range 25, 50 and 125 μm. The latter two gap values were considered to be typical of values that the designer might wish to employ. The selection

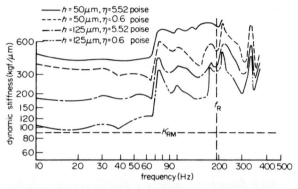

Figure 9. The increase in normal dynamic stiffness due to the inclusion of two rectangular (76 mm × 286 mm) squeeze film devices.

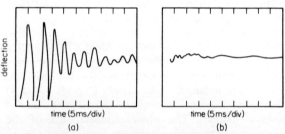

Figure 10. The relative magnitudes of the normal deflections of the experimental guideway with and without squeeze film compensation, when subjected to an impulse load: (a) uncompensated; (b) 4 pads (76 mm × 76 mm) $h = 25$ μm, $\eta = 5.52$ poise.

of the oil was made in the same practical manner; two viscosity values were chosen from a proprietary range of hydraulic oils. It was thought that this would be the procedure favoured by the designer rather than to employ very high viscosity fluids. The values of absolute viscosity, η, used in these tests were 0.64 and 5.5 poise.

The results obtained with the four pad configuration are shown in figure 8. In each case, the influence of the principal resonance which induces a minimum value of dynamic stiffness is nullified. The use of reduced squeeze film gap and the higher viscosity lubricant results in a general increase, across the measured band of frequencies, of dynamic stiffness.

The results shown in figure 8 achieve the desired removal of lightly damped mode of vibrations. It was decided, as this is essentially a low cost extension of the basic design, to evaluate the effect of using two squeeze film units of the maximum area that could be accommodated on each slideway track. The results are shown in figure 9. The trend described in figure 8 is continued, i.e. the use of the small gap and more viscous oil, produces the maximum mean level of dynamic stiffness across the measured frequency band. In this case, the low-frequency dynamic stiffness was observed to be some five times stiffer than the uncompensated antifriction guideway. The deflection across the rollers at the resonant frequency of the system (f_R) was reduced by a factor of fifty.

The results are further illustrated by the comparison of the two impact load responses shown in figure 10. The response of the antifriction guideway in figure 10(a) is typical of a lightly damped system. The trace shown in figure 10(b) is almost devoid of an oscillatory component.

CONCLUSIONS

The aim of the investigation was to assess the value of employing squeeze film devices to remove potentially undesirable lightly damped modes of vibrations from kinematic configurations that employ rolling element bearings. It has been shown that this simple addition to the guideway design is most effective and can be employed in those situations where large moving masses supported on antifriction rolling element bearings may exhibit resonant modes that impair the dynamic stability of a machine tool.

The one unsatisfactory point that has emerged from this investigation is the lack of a precise design procedure for the calculation of the squeeze film parameters and it is concluded that further attention should be directed to this problem. The existing design aids point to the concept of a combined damping and mass representation for the oil film. The results obtained in this study (figures 8 and 9) show that a general increase in dynamic stiffness is always attained. The phase characteristics of the compensated guideway, however, do not conform to the lumped representation for the squeeze film given in figure 3. A sample phase characteristic is shown in figure 11.

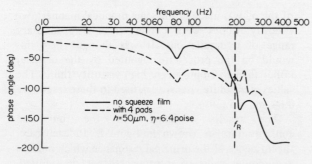

Figure 11. The phase shift frequency characteristic of the normal deflection of the experimental system with and without squeeze film compensation.

It must be emphasised that the errors involved in this procedure are not a major obstacle as it would appear that in general the use of the lowest gap and the most viscous lubricant give the best overall magnification of dynamic stiffness. The improvement of dynamic performance stiffness over a wide band of frequencies for applications involving large fluctuating forces may prove to be as valuable as the magnification of dynamic stiffness at the natural frequency of the mass and rolling element bearings (f_R). In this context, it is interesting to note that the stiffest configuration required the maximum thrust from the exciter to induce coherent deflection signals. The effectiveness of the squeeze film is dominated by

two factors: (i) the provision of an adequate supply of lubricant to avoid the collapse of the squeeze film, and (ii) the range of pressures occurring in the squeeze film. Experience gained during this investigation suggests that the use of the lower viscosity fluid offers an advantage of the use of the higher viscosity in that without a positive supply system the more viscous oil appeared to be expelled from the squeeze film block during prolonged dynamic excitation. This factor is thought to explain some variations in the overall trends in the stiffness characteristics given in this paper.

ACKNOWLEDGEMENTS

The authors wish to thank Professor F. Koenigsberger for the facilities to carry out this research. The work formed an extension to the programme of research of machine tool slideways supported by the Science Research Council. The project formed a design orientated dissertation topic and, in this context, the contributions of the Machine Tool Trades Association and the Staveley Asquith Division, Halifax, are gratefully acknowledged.

REFERENCES

1. J. TLUSTY and M. POLACEK, 1968. 'Experience with analysing stability of machine tools against chatter', *Proc. 9th M.T.D.R. Conf.*, Birmingham, Pergamon Press, Oxford, pp. 521–70.

2. Z. M. LEVINA, 1965. 'Main operating characteristics of anti-friction slideways', *Machines and Tooling*, **36**, No. 7, 10–16.

3. Z. M. LEVINA, 1963. 'Anti-friction slideways in modern machine tools (A review of non-Russian publications), *Machines and Tooling*, **34**, No. 3, 41–6.

4. Z. M. LEVINA and N. I. KALYMYKOV, 1962. 'Friction losses in anti-friction slideways', *Machines and Tooling*, **33**, No. 1, 11–18.

5. J. DE FRAINE, 1972. 'The choice of ball slideways for machine tools', *Proc. 12th M.T.D.R. Conf.*, Macmillan, pp. 193–197.

6. J. BAUER, 1969. 'Schlittenrollen lager in Werkzeugmaschinen', *Die Maschine*, **23**, No. 9, 59–63.

7. A. PALMGREN, 1945. *Ball and roller bearing engineering*, S.K.F. Philadelphia.

8. J. G. M. HALLOWES, 1972. 'The dynamic stiffness of antifriction guideways', *M.Sc. Dissertation*, U.M.I.S.T.

9. KLUBER LUBRICATION MUNCHEN GmbH, *Pamphlet*, 498.3.71, 2/2000/100.

10. H. POLACEK and Z. VAVRA, 1967. 'The influence of different types of guideways on the static and dynamic behaviour of feed drives', *Proc. 8th M.T.D.R. Conf.*, Manchester, Pergamon Press, Oxford, part 2, p. 1127.

11. D. D. FULLER, 1966. *Theory and Practice of Lubrication for Engineers*, Wiley, p. 134.

12. D. F. HAYS, 1963. 'Squeeze films for rectangular plates', *Trans. A.S.M.E.*, **85**, 243–6.

13. A. CAMERON, 1966. *Principles of Lubrication*, Longmans, p. 389.

14. P. VANHERCK, 1969. 'Dimensioning of fluid film dampers', *Ann. C.I.R.P.*, **17**, 65–72.

15. R. BELL and M. BURDEKIN, 1969–70. 'An investigation into the mechanism of the frictional damping of machine tool slideways', *Proc. 1st Mech. Engrs.*, **184**, part I.

16. M. DOLBY and R. BELL, 1970. 'An electrohydraulic exciter for machine tool testing', *Ann. C.I.R.P.*, **18**, 207–12.

OPTIMISATION OF HYDROSTATIC SLIDEWAYS INCLUDING STRUCTURE ELASTICITY

by

M. S. GIORGI,* S. G. POLLINI* and M. M. FAVARETO*

SUMMARY

The total stiffness of a hydrostatic slideway (especially with counteracting pads) depends on both the bearing and the structure stiffness; if we have a given space for the system, an increase of structure stiffness obtained by increasing its size involves a reduction of the available bearing areas and consequently of the bearing stiffness itself; an optimum configuration giving maximum total stiffness must then exist.

A computer program has been developed in order to calculate the total stiffness of each configuration under test; the procedure starts by optimising the bearing stiffness for rigid bodies, by computing structure deflections caused only by fluid pressures with no external load and by changing restrictor values in order to maintain the predetermined optimum restrictor ratios.

An external load being applied, the new pressure distributions and the associated structure deflections are computed by iterative methods and the final attitude is determined.

The computer results obtained have been analysed and design criteria have been drawn up which may be used by technicians in order to achieve nearly optimum solutions, with no need of large computing facilities.

The same program may be used for computer-aided design. These methods have been developed in the design of a hydrostatic slideway to be used for fine boring operations on transfer lines.

1. INTRODUCTION

The major characteristics of hydrostatic bearings are low friction, no wear, high stiffness, good inherent damping, controllability and adjustability of the system, low maintenance requirements on the hydrostatic components. These features can overcome some disadvantages such as investment and operating costs of the supply system, oil drain and protection problems, possible design difficulties.

Hydrostatic slideways have found many applications on machine tools: from very large boring and plano-milling machines to grinding machines and blanking presses[1].

Some problems can arise in the static design of these slideways when using counteracting pads of different areas: because of the many variables involved (maximum and minimum supported load, minimum acceptable film thickness, etc.) it is difficult to obtain general, relatively simple, criteria for choosing restrictor to pad resistance ratios giving the maximum achievable stiffness; the solution must be reached by numerical methods or by trial and error.

Care must also be given to the mechanical structure because, if it is not properly designed, it can impair considerably the actual stiffness of the slideway.

If we have a given space for the guideway bearing system, an increase of structure stiffness obtained by increasing its size involves a reduction of the available bearing areas and consequently of the bearing stiffness itself. The problem becomes then to find the optimum configuration featuring the maximum stiffness.

The present paper deals with this subject and illustrates the design method, developed by the authors, of optimising the total stiffness of such guideway systems with the aid of a digital computer.

The method has been used to design a general-purpose hydrostatic slideway unit particularly for use in fine boring operations on transfer lines.

Some computations have also been carried out on a series of geometrical configurations, obtaining interesting results from which some general design criteria can be derived.

2. DESIGN METHOD

2.1. The problem
Let us now consider how structure elasticity affects the hydrostatic bearing stiffness: starting from a bearing designed for rigid bodies, oil pressures deflect the slide structure and change the film thicknesses on the sills; pad resistances and restrictor ratios are now altered from the optimum design values. The first effect can be compensated by changing the restrictors in order to maintain the desired ratio. Because of the non-uniformity of the oil films, the bearings are less sensitive to displacements and their stiffness is reduced. Moreover, when an external load is applied (e.g. against a larger pad) the slide displaces but the gib, loaded by reduced pressures, recovers part of the displacement itself.

When starting to design a slideway unit of given overall dimensions, one finds some limitations to the space available for the guideway system; also because the inner part of the slide is required for the location of the feed mechanism (hydraulic cylinder, screw and nut, etc.); within this available space the designer has to obtain the most rigid configuration.

The problem of choosing a good balance between bearing areas and structure stiffness now arises.

The design method presented in this paper has been studied to solve this problem and to design quickly and more reliably such bearings.

2.2. Geometry and basic assumptions
The basic geometry analysed is shown in figure 1: the whole structure, loaded by pad pressures at a, b, c is considered to be jointed to an infinitely rigid body in the section where the slide increases its thickness and bears the surface of the lateral pad (not shown in figure 1), counteracting pad b.

* R.T.M. Institute, Vico Canavese, Italy

This structure is more complex than that studied by others[2] who considered only the gib and lateral beam deflections; actually in many instances the slide thickness t_a in that zone is of the same order of magnitude as the others.

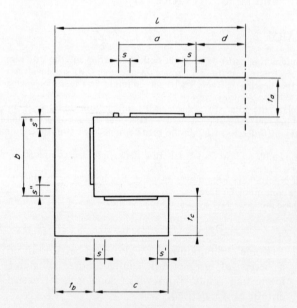

Figure 1. Basic structure geometry.

Some assumptions have been made in order to simplify the deflection computations.

(1) A two-dimensional model is considered and a slide of unit axial length analysed.

(2) It is assumed that uniform mean pressures act on pad width a, b, c, including their sills.

(3) The connections between beams a and b, b and c are considered to be infinitely rigid.

(4) Film thicknesses are measured at the centre of the sills.

The deformations are computed taking into account bending and shear stresses in the beams.

Different shear and elastic moduli can be used to simulate additional compliances due to bolted joints.

The basic boundary parameter to be kept constant is the guideway bearing width l; any other dimension can be varied independently.

2.3. Design procedure

The design procedure consists of two main parts: the first one is concerned only with the bearing stiffness optimisation for rigid bodies; the second deals with structure deflections under various loads and computes the total stiffness.

A synthetic flow chart of the method is shown in figure 2.

The required input data are as follows,

> Maximum and minimum permissible loads W_{max}, W_{min}.
> Normal load W_0 (corresponding to no working load condition).
> Normal working load ΔW.
> Supply pressure P_s.
> Clearance g.
> Minimum acceptable film thickness h_{min}.

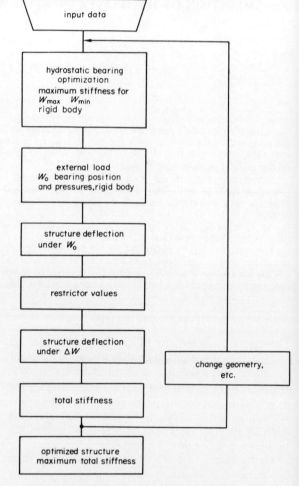

Figure 2. Design method flow chart.

> Minimum film thickness $h_{a\,min}$ under W_{max}.
> Available system size l.
> Other structure dimensions, pad and sill sizes, etc.
> Bending and shear moduli of the three beams.
> Oil viscosity.

The first series of computations is carried out automatically in order to find the restrictor ratios X_a, X_c that provide the minimum displacement between the maximum and the minimum load for the given $h_{a\,min}$.

With these data the bearing position under normal load W_0 is determined together with pad pressures and restrictors.

At this stage structure deflections caused by these pressures are computed and the position of the deflected slide is imposed by the condition that the smaller film thicknesses on pads a and c are in the same ratio as they were previously.

The deflected position now being defined, the new pad resistances are computed and the new restrictors determined in order to maintain the previous optimum ratios.

The external load ΔW is now applied to the slide; the final position where the new pressures and deflections balance the load is computed.

The total stiffness of the analysed geometry is determined.

Automatic preset changes in the geometry, and/or other factors are produced and the new configurations analysed.

The program has been provided with safety controls that show any irregular condition (minimum film thicknesses lower than acceptable ones, too compliant structures, etc.).

The high flexibility of the program provides a very rapid and economic means of optimising slideway design.

3. GENERAL RESULTS

In order to obtain some general ideas about the behaviour of hydrostatic elastic slideways the design method has been used to determine the stiffness performances of a series of simplified slide geometries, which are shown in figure 3.

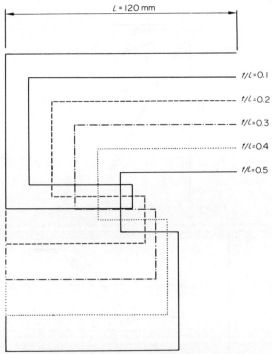

Figure 3. Analysed simplified geometries.

Their characteristics are as follows,

System size	$l = 120$ mm
Sills	$s = s' = s'' = 7$ mm
Beam lengths	$c = b = 0.5\,(1-t_b)$ variable
Beam thicknesses	$t_a = t_b = t_c$ variable
Pad a width	$a = 1-t_b$ or $= 0.5\,(1-t_b)$ variable
Supply pressure	$P_s = 50$ and 30 kg/cm^2
Clearance	$g = 100;\ 80;\ 60\ \mu$m
Minimum film thickness	
under W_{max}	$h_{a\,min} = 35;\ 40;\ 45\ \mu$m
Minimum acceptable	
film thickness	$h_{min} = 10\ \mu$m
Maximum load	$W_{max} = 5000$ kg/m
Minimum load	$W_{min} = -5000$ kg/m
Normal load	$W_0 = 0$ kg/m
Working load	$\Delta W = 2000$ kg/m
Bending modulus	$E = 21\,000$ kg/mm^2
Shear modulus	$G = 8\,000$ kg/mm^2

The geometry series has been chosen in such a way that each configuration is identified by varying the single parameter t/l and that the guideway geometry is kept of reasonable form.

Some of the obtained results have been collected in figures 4 to 8.

Figure 4 shows the total stiffness compared with the hydrostatic one, for rigid bodies, as a function of t/l. In the case $c/a = 0.5$ (full pad on the a surface) and

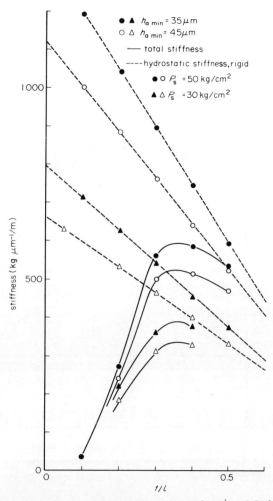

Figure 4. Total and hydrostatic stiffness for $c/a = 0.5$ (full pad on a).

clearance $g = 100\ \mu$m, as can easily be seen, the maximum stiffness is achieved for t/l values around 0.35–0.4; this stiffness is about 75% of the corresponding hydrostatic one and about 45% of the maximum hydrostatic one for a very thin, rigid body ($t/l = 0$).

From figure 4 the effect of the choice of minimum film thickness under W_{max}, $h_{a\,min}$ is seen to be rather appreciable (15% increase from 45 to 35 μm).

For good structures an increase of supply pressure increases, in nearly the same proportion, the total stiffness; while for relatively weak structures ($t/l = 0.2$) a 66% increase in pressure gives only a 25% gain.

In figure 5 the effect of clearance and pad a dimensions are shown: a reduced clearance can improve stiffness only in a narrow region around the optimum; for $t/l = 0.3$ no advantage has been found also for 60 μm clearance. The use of a reduced pad a size ($c/l = 1$, $d/a = 1$) seems to be disadvantageous because of the strong reduction of the total stiffness.

The effect of reduced pad a position ($c/a = 1$, d/a variable) is shown in figure 6: an appreciable increase can be noticed when displacing it towards the inside of the slideway.

An idea of the order of magnitude of deflections occurring on the slide is given by figure 7 for $P_s = 50$ kg/cm^2.

For $t/l = 0.2$ the slideway deformations and displacements under ΔW are also given.

Especially for rather weak structures the effect of

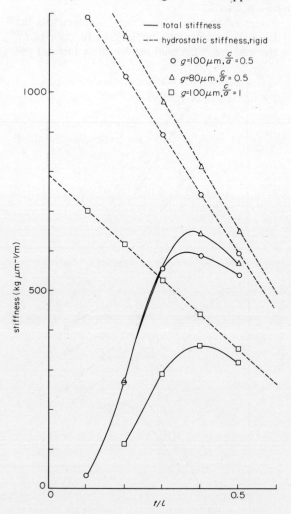

Figure 5. Total and hydrostatic stiffness for various clearances and reduced pad a.

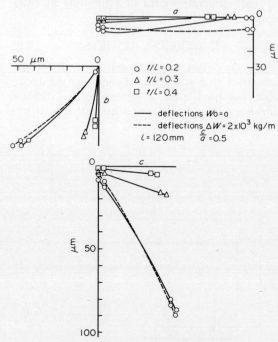

Figure 7. Typical structure deflections.

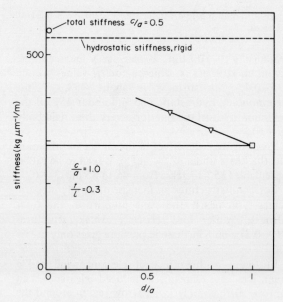

Figure 6. Effect of reduced pad a position.

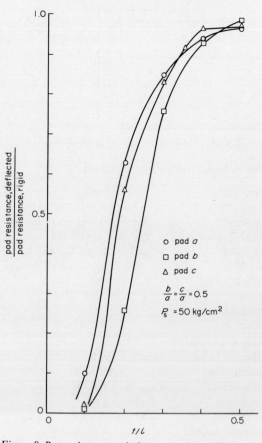

Figure 8. Pad resistance variations caused by deflections.

beam a is very important. In fact in both cases t/l = 0.2 and 0.4, a strongly reduced gib thickness (t_c/l = 0.1 and 0.2 respectively) has been studied showing that it reduces total stiffness by less than 1%.

Figure 8 shows that, using strong structures, there is practically no need for compensating restrictors for pad resistance variations caused by deflections; for nearly optimum structures (t/l = 0.3–0.35) the compensation is already necessary.

4. DISCUSSION

The results obtained from the study of the simplified structure show that optimum configurations can be obtained using t/l ratios from 0.35 to 0.4 and higher values if bolted joints are used in the structure.

Bolted joints should be used as near as possible to the gib in order to limit the effects of their compliance.

For optimum structures a total stiffness of about 70% of the hydrostatic one may be expected.

There is practically no advantage in reducing the upper pad a.

Optimum t/l ratios are not affected by the supply pressure (at least in the range 30–50 kg/cm²). For optimum configurations increasing supply pressure means increasing practically in the same proportion the total stiffness.

Small advantages (in stiffness) can be obtained by reducing the clearance below 100 μm, for the studied case.

Small restrictor compensation for deflections is required by optimum structures.

5. CONCLUSIONS

A method for computer-aided design of hydrostatic slideways for maximum stiffness, and taking into account structure elasticity, has been presented.

The method has been used for the design of a general-purpose slideway unit for transfer lines.

A series of simplified structures has been also analysed and simple design criteria leading to good solutions have been derived.

REFERENCES

1. M. FAVARETO and G. B. RAZELLI, 1970. 'Design of hydrostatic slideways under impulsive loads', *C.I.R.P., Ann.,* XVIV.
2. G. PORSCH, 1969. 'Uber die Steifigkeit hydrostatischer Führungen unter besonderer Berücksichtigung eines Umgriffes', *Doktor Ingenieurs Dissertation,* T. H. Aachen.
3. M. FAVARETO and G. B. RAZELLI, 1970. 'Design of hydrostatic journal bearings under impulsive loads', *Proc. 11th M.T.D.R. Conf.*
4. M. FAVARETO, 1970. 'Come non fare una guida idrostatica', *La Meccanica Italiana,* Apr.

DESIGN OF HYDROSTATIC BEARINGS FOR EXACTING APPLICATIONS

by

W. B. ROWE* and K. J. STOUT†

SUMMARY

While many bearings can be designed from very simple rules, more severe problems arise where the operating requirements involve widely varying speeds and temperature conditions. Exacting applications where hydrostatic bearings may prove of advantage are to be found in machines for primary processes, such as rolling mills and centrifugal casting machines, and in secondary processes involving high rotational speeds, such as ball mills and high-speed lathes. This paper discusses some effects and gives consideration to the consequences of various design alternatives.

NOTATION

A_f	effective friction area
$\overline{A_f}$	A_f/D^2
$\overline{B}$	flow factor $\pi D/(6A)$
C_D	diametral clearance
C_V	specific heat
D	journal diameter
H_f	friction power
H_p	pumping power
H_t	total power
J	mechanical equivalent of heat
K	power ratio H_p/H_f
K_c	capillary resistance factor $\dfrac{128L}{\pi D^4}$
L	bearing length
N	rotational speed
P_s	supply pressure
S	orifice constant
S_H	speed parameter
S_{H0}	optimised speed parameter without viscosity variations
ΔT	adiabatic temperature rise in lubricant
h	radial clearance
h_0	concentric clearance
m	rotating mass
n	number of recesses
q	actual flow rate
q_0	flow rate without viscosity variations
β	concentric pressure ratio
β_0	pressure ratio without viscosity variations
ϵ	eccentricity ratio
η	dynamic viscosity
η_B	effective viscosity in the bearing
η_0	optimised value of viscosity
η_R	effective viscosity in the restrictor
λ_0	bearing stiffness
ρ	mass density
ω_0	natural frequency

INTRODUCTION

In many applications, such as slideways for machine tools, a satisfactory design can often be achieved by the simplest of methods. More care must be taken however in designing for, say, a grinding spindle which operates at speed. Under these circumstances, it is desirable to optimise the design for minimum power for the following two reasons:

(1) to avoid excessive flow rate
(2) to avoid excessive temperature rise.

Design for minimum power has been explained by Fuller[1] and Opitz[2]. The principles are not difficult to apply in the design of bearings which operate at ambient or near-ambient temperatures.

It is, however, for bearings that operate at speed and over a wide range of temperatures that the practical design problems become severe. Exacting applications where hydrostatic bearings may prove an advantage are to be found in machines for primary processes, such as rolling mills and centrifugal casting machines, and in secondary processes involving high rotational speeds such as ball mills and high-speed lathes. Alternative types of bearing are likely to be susceptible to wear, thus giving rise to a requirement for expensive overhauls. The designer may therefore decide that the advantage of long operational life obtainable with hydrostatic bearings outweighs the disadvantage of high flow rate, which is likely to be found necessary. Discussion is limited to journal bearings which are likely to prove the most common configuration; the basic geometry is illustrated in figure 1.

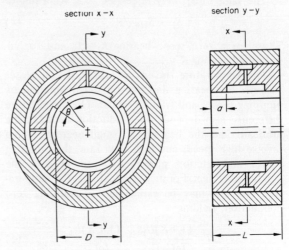

Figure 1. The geometry of a hydrostatic journal bearing.

*Principal Lecturer ⎱ Department of Production Engineering, Lanchester Polytechnic, Coventry.
†Lecturer ⎰

Three parameters found critical in the design of high-speed bearings and bearings involving thermal problems are viscosity, clearance and flow rate. It is possible to determine relationships between possible values of viscosity, clearance and flow rate. The relationships are defined by the process of optimisation for minimum power.

LOAD CAPACITY AND OPTIMISATION FOR MINIMUM POWER IN NON-EXACTING APPLICATIONS

Previous papers[3] have shown that minimum power is obtained when friction power H_f equals 1 to 3 times pumping power H_p

i.e. $$H_f = KH_p \qquad (1)$$

and $$1 < K < 3$$

Equation (1) leads to the following form which has been found most convenient for analysing optimised bearings.

$$S_H = \frac{\eta N}{P_s}\left(\frac{D}{C_D}\right)^2 = \frac{1}{4\pi}\sqrt{\frac{K\beta B}{\overline{A}_f}} \qquad (2)$$

The value of the parameter S_H for a particular design therefore determines the value of K which is the ratio H_f/H_p. Hence it may be deduced that the value of S_H determines the relative proportions of hydrodynamic and hydrostatic lift, a conclusion that can be rigorously proved by rearrangement of Reynolds' equation. The value of S_H is useful as a measure of speed; the value of K provides an indication of the type of behaviour to be expected at speed.

The optimum value when viscosity is the variable parameter is termed S_{H0} and is given for the condition $K = 1$, so that

$$S_{H0} = \frac{1}{4\pi}\sqrt{\frac{\beta\overline{B}}{\overline{A}_f}} \qquad (3)$$

Bearing performance has been computed from thin-land theory for various values of S_H, as indicated in figure 2 for a 4-recess journal.

The value of S_H may be related to K according to

$$S_H = S_{H0}\sqrt{K} \qquad (4)$$

Thus, for a zero speed bearing $K = 0$, and for an optimised bearing $K = 1$.

The temperature rise in the lubricant as it passes through the bearing depends on the dissipation of pumping power and friction power. The temperature difference also depends on conduction between the lubricant and the bearing. In small bearings which involve high speeds and high flow rate, the pumping power and friction power predominate over the conduction effect. For such applications, the temperature rise ΔT may be estimated by assuming zero conduction, yielding

$$\Delta T = \frac{(1 + K)H_p}{J\rho C_v q} = \frac{(1 + K)P_s}{J\rho C_v} \qquad (5)$$

From equations (2), (4) and (5) it is deduced that there is a direct relationship between the speed

parameter S_H, the power ratio K and the adiabatic temperature rise ΔT.

RELATIONSHIP BETWEEN VISCOSITY, CLEARANCE AND FLOW RATE FOR $K = 1$

When designing high-speed bearings, the problem that arises is how to achieve a design that does not involve excessive flow rate. Some possible design alternatives which may be considered are:

(i) High values of power ratio $K = H_f/H_p$
(ii) Low values of design pressure ratio β
(iii) Thick bearing land widths.

None of these alternatives is altogether advisable. High values of K increase the adiabatic temperature rise ΔT and lead to variable viscosity effects as discussed in more detail later. It is also found that designing for a high value of K limits the permissible range of operating temperature and also increases the power to be dissipated as heat.

Low values of design pressure ratio β reduce flow rate according to the relationship

$$q_0 = P_s\beta\frac{\overline{B}h_0^{3}}{\eta} \qquad (6)$$

However, reducing β below a value of 0.25 severely reduces stiffness as illustrated in figure 2, and a higher

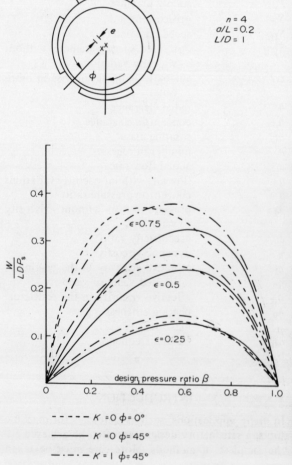

$$\begin{aligned}n &= 4\\a/L &= 0.2\\L/D &= 1\end{aligned}$$

----- $K = 0 \ \phi = 0°$

——— $K = 0 \ \phi = 45°$

—·—·— $K = 1 \ \phi = 45°$

Figure 2. Load capacities of a 4-recess bearing, non-rotating $K = 0$, and rotating $K = 1$.

supply pressure is required for the same operating eccentricity, so that any benefit is lost. The optimum value of design pressure ratio is $\beta = 0.5$.

Bearing land width should not be greater than one-quarter the bearing length as argued previously[3] since a higher supply pressure will be required and the bearing will be more susceptible to thermal collapse problems when operated over a wide temperature range[4]. It is found that flow rate is not reduced for thick land widths if the bearing is always optimised for $K = 1$.

In designing for minimum flow rate the problem therefore reduces to selecting a suitable combination of viscosity and clearance, while maintaining $K = 1$. From equation (2) it follows that viscosity and clearance may be varied according to the rule

$$\frac{\eta}{C_D^2} = \text{constant} \qquad (7)$$

The minimum acceptable value of C_D will depend either on manufacturing tolerances or on the minimum acceptable viscosity dictated by equation (2). The relationships between viscosity η, clearance C_D, flow rate q, and total power H_t are illustrated in figure 3.

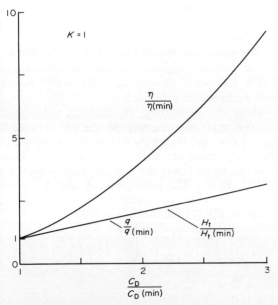

Figure 3. Relationships between viscosity, flow rate and power dissipation for optimised bearings when clearance is increased from its minimum value.

EFFECT OF TEMPERATURE VARIATIONS IN THE LUBRICANT

It is necessary to distinguish the various situations in which temperature variations occur, so that it may be possible to distinguish in turn the correct design philosophy for each situation. Two principal situations may be encountered in practice.

(1) Slow-speed bearings

The temperature of the lubricant increases as it passes through the bearing either due to the high value of supply pressure or alternatively due to conduction from the bearing where there is a nearby heat source. If the bearing operates at very low speed, it may be assumed that $K = H_f/H_p$ is much less than 1, in which case the effect of the temperature rise is to reduce the effective viscosity in the bearing η_B to a value lower than the viscosity in the restrictor η_R. For large values of the viscosity ratio η_R/η_B the concentric pressure ratio is reduced. The operating value of concentric pressure ratio may be derived by equating flow through the restrictor to flow through the restrictor to flow through the bearing. For capillary control

$$q_0 = \frac{P_s(1-\beta)}{K_c \eta_R} = \frac{P_s \beta \bar{B} h_0^3}{\eta_B}$$

and hence

$$\beta = \frac{1}{1 + K_c \bar{B} h_0^3 \eta_R/\eta_B} \qquad (8)$$

If the design pressure ratio has been calculated on the basis of $\eta_R/\eta_B = 1$ there may be a considerable reduction in the bearing load capacity. Some computations for a non-rotating journal bearing with 4-recesses are shown in figure 4. The load capacity at an eccentricity

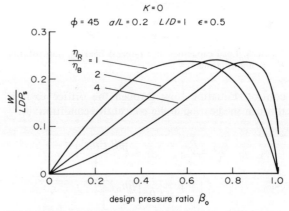

Figure 4. Effect of viscosity variation through a non-rotating 4-recess journal bearing.

ratio of $\epsilon = 0.5$ is reduced for all values of intended design pressure ratio up to $\beta_0 = 0.65$ when $\eta_R/\eta_B = 2$.

In designing bearings for variable viscosity the designer should consider the range of viscosity ratio likely to occur and select a value of design pressure ratio accordingly. Operating a given bearing over its widest possible viscosity range will involve low values of concentric pressure ratio on occasions, and high values at other times. From figure 2, it is possible to examine the influence of the direction of loading at low values of β. It is found that the load capacity is greatly reduced when the line of eccentricity is towards an inter-recess land.

The directional effect on load capacity may be pronounced in 4-recess journal bearings but is negligible in 5-recess and 6-recess bearings, which also yield improved load capacities. It is therefore suggested that 5- or 6-recess bearings should be employed where viscosity ratio effects are likely to occur.

The load capacities of 6-recess journal bearings are illustrated in figure 5, where $K = 0$ for the nonrotating bearing and $K = 1$ for an optimised bearing with hydrodynamics. The improvement with 6-recess bearings is evident by comparing figure 5 with figure 2.

Equation (8) was derived for capillary control. It is possible to compare the effect of viscosity variation for capillary control with the effect for orifice

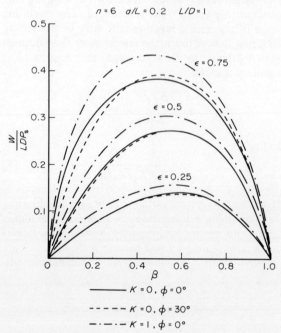

$n = 6$ $a/L = 0.2$ $L/D = 1$

Figure 5. Load capacities of a 6-recess bearing, non-rotating $K = 0$, and rotating $K = 1$.

control. Equating flow through the orifice to flow through the bearing in the concentric condition

$$q^2 = \frac{P_s}{S}(1 - \beta) = \left(\frac{\beta P_s \bar{B} h_0^3}{\eta}\right)^2$$

$$\frac{1 - \beta}{\beta^2} = SP_s \left[\frac{\bar{B} h_0^3}{\eta}\right]^2$$

If β_0 is the pressure ratio when $\eta = \eta_0 = \eta_R$ then β may be expressed in terms of η_R/η_B

$$\frac{1 - \beta}{\beta^2} = \frac{1 - \beta_0}{\beta_0^2} \left[\frac{\eta_R}{\eta_B}\right]^2 \qquad (9)$$

With capillary control, it is found that the pressure ratio is reduced from 0.5 to 0.33 when the effective viscosity in the bearing is halved. However, from equation (9) it is found that the pressure ratio is reduced from 0.5 to 0.3 in an orifice controlled bearing.

It may be concluded therefore that capillary control is preferable to orifice control in applications where viscosity variations are involved. It is also apparent from equation (8) that a change in ambient temperature variation produces no change in the ratio η_R/η_B and hence no variation in pressure ratio with capillary control. However, the same viscosity variation $\eta_R/\eta_B = 2$ due to an ambient temperature change with an orifice controlled bearing produces the same large change in pressure ratio as caused by a viscosity variation through the bearing.

(2) High-speed bearings

A design problem that arises in high-speed bearings which normally operate at an elevated temperature is caused by cold starting. If the bearing is operated at speed before the system has warmed up, the friction losses will be higher than normal. While the friction losses are increased, the flow rate and pumping power are reduced so that $K = H_f/H_p$ will be much higher than normal.

The high values of K result in a large temperature variation in the lubricant passing through the bearing as shown by equation (5). The temperature rise leads to a reduction in viscosity in the bearing, which tends to reduce the increase in K but also reduces the pressure ratio β as previously demonstrated by equation (8) for capillary controlled bearings and by equation (9) for orifice controlled bearings.

Both these effects, the reduction in pressure ratio β and an excessive increase in the power ratio K, are undesirable. The reduction in β reduces the hydrostatic load capacity. High values of K cause severe cavitation which detracts from the hydrodynamic load support for a given value of eccentricity. High values of K also increase the tendency to thermal collapse[5] although this effect is not included in the following simplified analysis, which is intended to be used by designers as a guide for order of magnitude calculations.

A typical example of the cold starting problem is a bearing which has been designed for operation at $120°F$ and for a viscosity $\eta_0 = 2.4 \times 10^{-6}$ Reyns, with a pressure ratio $\beta_0 = 0.5$.

However, on starting, the ambient temperature is $70°F$ and the viscosity of the oil at inlet is $\eta_R = 9.6 \times 10^{-6}$ Reyns. As a result of the high value, the temperature of the oil at the bearing lands increases to $90°F$ and the viscosity falls to $\eta_B = 4.8 \times 10^{-6}$ Reyns. The variation of K and β for this example may be calculated by rearrangement of equations (2), (3) and (8).

$$K = \left[\frac{\eta_B}{\eta_0}\right]^2 \left[\frac{\beta_0}{\beta}\right] = \left[\frac{\eta_B}{\eta_0}\right]^2 \left[\beta_0 + (1 - \beta)\frac{\eta_R}{\eta_B}\right] = 6$$

$$\frac{S_H}{S_{H0}} = \frac{\eta_B}{\eta_0} = \sqrt{\frac{K\beta}{\beta_0}} = 2$$

$$\beta = \cfrac{1}{1 + \cfrac{(1 - \beta_0)}{\beta_0} \cdot \cfrac{\eta_R}{\eta_B}} = 0.33$$

From equation (5), the approximate supply pressure at which the condition in the above example arises is given by

$$P_s = \frac{J\rho C_v \Delta T}{1 + K} = \frac{20}{7 \times 0.0075} = 400 \text{ lbf/in}^2 \text{ (approx)}$$

The values of S_H and β derived from the variations in viscosity may be used to compute the bearing load characteristics. Figure 5 illustrates the increased load capacity due to the hydrodynamic effect at an eccentricity ratio $\epsilon = 0.5$, and for values of β not less than 0.5. Values of β less than 0.5 lead however to severe cavitation. Thus it appears that the bearing quoted will operate satisfactorily, but the higher values of K may prove unsatisfactory.

The results shown in figure 6 indicate that where a wide temperature range is involved, high values of design pressure are to be preferred to low values, since the hydrodynamic effect may be better utilised before

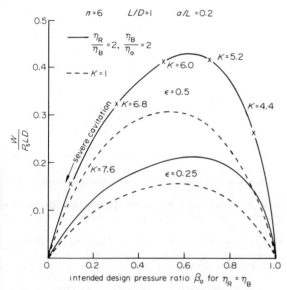

Figure 6. Effect of viscosity variations through bearing at the cold starting condition. Bearing optimised for hot temperature $\eta R = \eta B = \eta_0$.

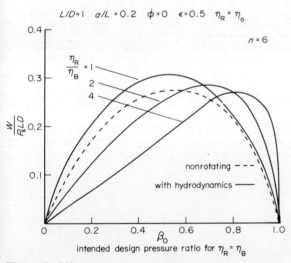

Figure 7. Effect of viscosity variations through bearing. Bearing optimised for cold temperature $\eta R = \eta B = \eta_0$.

severe cavitation is incurred. An alternative approach to the design of a bearing which suffers a substantial temperature rise is to optimise for the lowest starting temperature. This approach increases the flow rate requirement but tends to prevent the temperature rising so high as in bearings optimised for high temperature operation, since the power rate will never be greater than $K = 1$, But, if temperature variations occur as the lubricant passes through the bearing, the load capacity will be reduced, as illustrated in figure 7. The loss of load capacity is explained by the reduction in viscosity in the bearing which reduces both the

concentric pressure ratio and the hydrodynamic contribution.

A possible compromise between the two previous methods is to optimise the bearing for a viscosity which is half the value at its coldest condition.

A further alternative (which removes the problem) is to heat the oil up to its normal level before commencing operation.

EFFECT OF TEMPERATURE VARIATIONS ON THE WHIRL ONSET SPEED

It may be shown that the onset of whirl tends to be prevented by the hydrostatic load contribution in a hybrid hydrostatic bearing. However, it is still possible for whirl to occur and the circumstances favourable for the whirl condition are bearings which involve large mass, low hydrostatic stiffness and a large hydrodynamic contribution.

The frequency of whirl and the rotational speed for onset of whirl may be calculated from the following conditions[6,7].

(1) The first whirl onset frequency is equal to the first undamped natural frequency of the rotating mass calculated from the effective structural stiffness and the hydrostatic bearing stiffness. Where the structural stiffness is high in comparison with the hydrostatic stiffness it is sufficiently accurate to write

$$\omega_0 = \sqrt{\frac{m}{\lambda_0}}$$

(2) The rotational speed equals twice the whirl frequency. At the onset of whirl this is the condition

$$N = \frac{\omega_0}{2\pi}$$

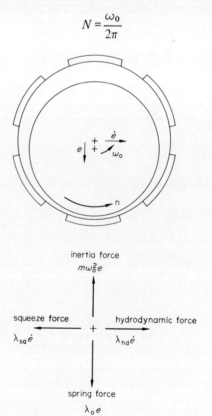

Figure 8. Vector diagram of forces at the onset of whirl (after Leonard and Rowe).

The whirl condition arises because the squeeze forces when whirling at half the shaft speed are equal and opposite to the hydrodynamic forces. Also when the whirl speed equals the undamped natural frequency, the inertia forces balance the stiffness contribution which is in phase with the bearing eccentricity. It may be shown that the vector diagram of forces at the onset of whirl is as shown in figure 8 for small amplitudes.

From figures 4 and 7 it is seen that there may be a reduction in the hydrostatic load contribution with viscosity variation through the bearing and hence a reduction in the whirl onset frequency. Even if the bearing is designed for the condition described with reference to figure 6, the whirl onset frequency may be reduced by the same degree, since a large proportion of the load support is hydrodynamic in nature. Figure 6 does not reveal the extent to which the hydrostatic contribution may be reduced.

CONCLUSION

The effects of viscosity variations have been described for thin land bearings and a guide has been given which should help in assessing whether a particular design is likely to prove acceptable. It has been shown that the design pressure ratio should be increased if it is desired to make some allowance for viscosity variations. Orifice controlled bearings are particularly susceptible to a variable viscosity distribution through the bearing as well as to bulk viscosity change. Capillary control is therefore preferable to orifice control.

It has also been shown that viscosity variations may increase the susceptibility to whirl.

REFERENCES

1. D. D. FULLER, 1947. Hydrostatic lubrication, *Machine Design*, parts 1–10.
2. H. OPITZ, 1967. Pressure pad bearings, *Proc. I. Mech. E.* Conference on lubrication and wear, London, September, pp. 67–82.
3. W. B. ROWE, J. P. O'DONOGHUE and A. CAMERON, 1970. Optimisation of externally pressurised bearings for minimum power and temperature rise, *Tribology*, 3, p. 153.
4. W. B. ROWE and J. P. O'DONOGHUE, 1971. A review of hydrostatic bearings design, *Proc. I. Mech. E.* Conference on externally pressurised bearings, London, November, pp. 157–187.
5. W. B. ROWE and J. P. O'DONOGHUE, 1970. Design procedures for hydrostatic bearings, *Machinery Publishing Company*.
6. R. LEONARD and W. B. ROWE. Damping in hydrostatic bearings and the mechanism of instability. To be published.
7. F. KOENIGSBERGER and A. COWLEY, 1971. Dynamics of hydrostatic bearing systems, *Proc. I. Mech. E.* Conference on externally pressurised bearings, London, November, pp. 284–308.

A MEASURING SYSTEM FOR THE EVALUATION OF SPINDLE ROTATION ACCURACY

by

E. J. GODDARD*, A. COWLEY* and M. BURDEKIN*

SUMMARY

The paper discusses currently available procedures and instrumentation systems for the measurement of spindle rotation accuracy. After describing the operating principles of measuring systems which have been previously proposed and discussing their merits and limitations, a new system is described which offers advantages in terms of simplicity of use and of economy of components.

1. INTRODUCTION

The rotational performance of a machine tool spindle has a direct influence upon both the surface finish and the geometric shape of the finished work, the significant motions being radial, angular and axial. Whilst roundness is an important property of the resulting workpiece produced in boring and turning, there is a basic difference between the two operations. In boring the workpiece is stationary and the radial errors of the rotating tool affect the roundness of the hole produced as it moves around a complete revolution, whereas in turning, where the tool is stationary and the work rotates, the radial errors between the instantaneous centre of rotation and the tool tip are transferred to the workpiece. In determining the roundness errors on a lathe, therefore, only the distance between the tool tip and the instantaneous centre of rotation of the workpiece is of interest, and equipment is required to measure and record variations of this dimension throughout the complete speed range of the machine.

The standard machine tool testing procedure prepared by Schlesinger[1] in 1932 have become the most widely accepted International Standard, the seventh edition being printed in 1966. Lathe spindle rotational accuracy is assessed with a dial indicator clocking on to a precision mandrel fitted into the taper in the spindle nose, the clock being positioned close to the nose. The limitations of this method are that any eccentricity in the location of the mandrel is included in the reading, no recording facility is available and the method is limited to low speeds of spindle rotation.

Alternative methods of measurement which have been proposed recently utilise electrical displacement transducers with associated recording facilities. A feature common to these methods is the display of of the measurements in the form of a polar diagram, the significance of which is the same as that of a standard Talyrond diagram, the shape of the diagram being intended to represent the shape of the machined workpiece.

The important features of these measuring systems are described in this paper and a new system is proposed and described in detail.

2. EXISTING MEASURING SYSTEMS

A method of measurement which appears to have originated at the Czechoslovakian Machine Tool Research Institute (V.U.O.S.O.) some years ago[4] makes use of two electrical displacement transducers mounted at 90° to one another, each acting upon an accurately round test sphere set in the spindle with approximately 0.005 in eccentricity. The electrical signals from the two transducers are taken to the X and Y plates of an oscilloscope. When the spindle is rotated a basic circle, the radius of which is proportional to the eccentricity of the test sphere, is traced out on the oscilloscope and any out of roundness errors due to spindle rotation appear superimposed upon the basic circle. This method is perfectly satisfactory for machining operations in which the cutting tool rotates (i.e. boring). However, the system is not suitable for operations in which the spindle under examination supports a rotating workpiece (i.e. turning or cylindrical grinding).

It was pointed out by J. W. Pearson of the Lawrence Radiation Laboratory (L.R.L.) at the University of California that the two-axis method does not reflect the errors machined on to a workpiece with a single-point stationary tool because the nominal depth of cut is affected by the radial spindle motion towards and away from the tool and almost unaffected by transverse motion. He concluded that there is a 'sensitive direction' of spindle motion defined as 'workpiece axis motion' that occurs in a direction which is directly towards or away from a cutting tool on a line joining the tip of the cutting tool, and the instantaneous centre of rotation of the workpiece.

2.1. The L.R.L. system

In order to obtain a polar display from a two-axis system in the case of a stationary tool rotating workpiece machine, a separate means of generating a base circle is required together with a signal from a single transducer fixed in place of the cutting tool, so providing a signal of radial error which can be superimposed upon the basic circle.

A method fulfilling this requirement was proposed by Bryan, Clouser and Holland[2] in 1967 of the L.R.L.

---

* University of Manchester Institute of Science and Technology.

The system referred to here as the L.R.L. system is shown in figure 1.

Two identical circle generating cams are used, each being circular with equal eccentricities of 0.005 in and set with their maximum displacement positions relatively at 90°. Two gauge heads fixed in the same angular position measure the displacement of the cams. When connected to the X and Y axes of an oscilloscope a basic circular trace is created on the oscilloscope screen.

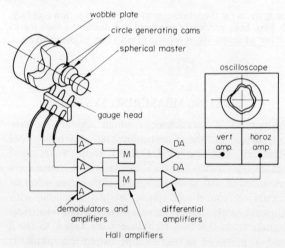

Figure 1. L.R.L. method – sketch of rig and electrical set-up.

A spherical master is centred as accurately as possible using a wobble plate device, and a signal is obtained from the surface of the master sphere through a relative displacement transducer fixed in the same plane as the tip of the cutting tool and the instantaneous centre of rotation.

The signals from the two gauge heads measuring from the two eccentrics are each electrically multiplied by the signal from the gauge head on the spherical master using Hall multipliers, and added to the original before being displayed as X and Y components on the oscilloscope screen.

2.1.1. System theory
Let

  E = eccentricity of test sphere

and

  C = eccentricity of circle generating cams at any angle of spindle rotation, θ

Component due to eccentricity of test sphere = $E \cos \theta$

Components due to basic circle generating cams

and
$$= C \sin \theta \text{ (on } X \text{ plates)}$$
$$= C \cos \theta \text{ (on } Y \text{ plates)}$$

Total signal to X plates of scope

$$= C \sin \theta + E \cos C \sin \theta$$
$$= C \sin \theta + \frac{EC}{2} \sin \theta \qquad (1)$$

Total signal to Y plates of scope

$$= C \cos \theta + E \cos \theta\, C \cos \theta$$
$$= C \cos \theta + \frac{EC}{2} (1 + \cos 2\theta) \qquad (2)$$

2.2. The V.U.O.S.O. system
As an alternative, but basically similar in principal to the L.R.L. method described above, Vanek[3] of V.U.O.S.O. proposed a scheme in 1969 in which the basic circle is obtained electrically using a resolver, whereby two output signals, phase displaced by 90°, are obtained as a function of the angular rotation of the machine spindle, and are each modulated by a displacement signal taken from two linear differential transformer displacement transducers fixed horizontally and in the same plane as the tip of the cutting tool and the centre of the spindle. A diagram of the

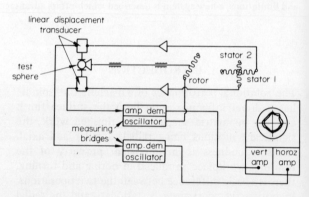

Figure 2. V.U.O.S.O. method – sketch of rig and electrical circuit.

system is given in figure 2. The carrier frequency voltage from one of the oscillators is first fed to the rotor of the resolver, and the voltage from each of its two stator windings is then amplitude modulated by the sine and cosine of the angular rotation of the spindle respectively. Each of these voltages is fed to the individual linear differential transformers, whose outputs are thereby amplitude modulated by a further signal proportional to displacement.

Thus multiplication of the spindle displacement signal by sine and cosine functions is carried out. The basic circle is obtained by displacing the linear differential transformers from their null positions, the circle radius being proportional to this displacement. The spindle rotation error also moves the transducers from their null position, causing the error signal to be superimposed upon the basic circle.

2.2.1. System theory

Let

  E = eccentricity of test sphere

and

  D = transducer offset from mid-position giving radius of basic circle at any angle of spindle rotation, θ

Component due to eccentricity of test sphere = $E \cos \theta$

Total displacement of transducer = $D + E \cos \theta$

Resolved signal to X plates of scope

$$= (D + E \cos \theta) \sin \theta$$
$$= D \sin \theta + \frac{E}{2} \sin 2\theta \qquad (3)$$

Resolved signal to Y plates of scope

$$= (D + E \cos \theta) \cos \theta$$

$$= D \cos \theta + \frac{E}{2}(1 + \cos 2\theta) \qquad (4)$$

2.3. Appraisal of L.R.L. and V.U.O.S.O. systems

Both the L.R.L. and V.U.O.S.O. systems are similar in principle but differ in mechanical and electrical detail. In the L.R.L. system, careful manufacture, matching and subsequent setting of the cylindrical eccentric cams and master sphere is required and the electrical system is of moderate complexity. Although in the V.U.O.S.O. system the test ball can be readily purchased from most bearing manufacturers, it demands a Grade I specification and special selection by the manufacturer, with subsequent selection to ensure the roundness is < 10 μin when the ball is fixed in its final position. Care is also required in the V.U.O.S.O. system in setting the relative positions of the two transducers, in order to ensure that they are both at their null positions at the same time that their bridge circuits are electrically balanced. If this is not done thoroughly and rechecked at intervals throughout a test series, the basic circularity of the oscilloscope trace will be upset.

In both systems the resulting output signal has three components: first, the radial motion of the spindle axis in the sensitive direction (i.e. in the plane of the tool tip and the instantaneous centre of rotation of the workpiece); second, the eccentricity in the setting of the master test sphere; and third, any out of roundness of the master test sphere itself. The latter can be minimised in manufacture or by selection of the sphere and the eccentricity of the master sphere reduced to a minimum by careful setting.

However, as pointed out by Bryan, Clouser and Holland[2], in the event of the sphere eccentricity being excessive, a difficulty can arise in that the basic circle of the polar display may become distorted. The distortion produced is both a distortion of circularity and an angular distortion, and increases as either test sphere eccentricity or magnification is increased, and also as the base circle decreases.

The relevant calculations show that a similar distortion is present in both the V.U.O.S.O. and the L.R.L. systems at a given setting of eccentricity of the test sphere.

The terms $\cos 2\theta$ and $\sin 2\theta$ are present in the expression for the total X plate and total Y plate signals in each system, and its significance in terms of distortion will be dependent upon the relative value of the constants C, D and E for a particular test rig set up.

In practice it is necessary, of course, to regulate the size of the basic circle to fit on to the screen of an oscilloscope and, given that the available space is 10 cm by 8 cm, a circle of 6 cm diameter is a practicable size. At an overall magnification of 1 cm = 100 μin this represents 300 μin radius. If we regard 0.0001 to 0.0002 in as the best practicable setting for the eccentricity of the best sphere, then the range of

ratios of D/E (V.U.O.S.O. system) with which we are concerned is of the order of 3:1 to 1.5:1.

The shape of the basic circle can be determined by calculations from expressions (1) and (2), and has also been determined experimentally using the V.U.O.S.O. type rig, as this was readily available. For convenience, and in order to obtain sufficient experimental accuracy on the particular lathe being used, the overall magnification was reduced by a factor of 10 from that which would normally be used for spindle rotation tests, and the test sphere eccentricity and transducer displacement from the null position correspondingly increased by a factor of 10. The test sphere eccentricity was set at 0.006 in and transducer displacements from the null position of 0.018 in, 0.012 in and 0.009 in giving D/E ratios of 3.0, 2.0 and 1.5 respectively, were obtained by movement of the cross slide.

The traces are shown in figures 3, 4 and 5. Each of these traces was taken at a lathe speed of 30 rev/min,

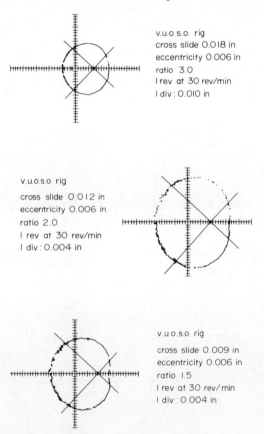

Figures 3, 4, 5. Typical traces obtained using the V.U.O.S.O. system.

and represents one revolution of the lathe spindle, obtained by switching the oscilloscope beam. On each trace can be seen two diagonal lines intersecting at 90°. These lines were obtained with the spindle held stationary in two positions 90° apart. The lines were produced by displacing the transducers either side of their null position by moving the cross slide with the handwheel. The point of intersection of the diagonals represents the origin (or centre) or the trace, and in each case is on the X axis. The point is to the right of the centre of a 'circle of best fit' and the amount by which it is displaced increases with decrease in the ratio D/E. The angular distortion of the traces can be seen by comparing the curved length of the trace between adjacent $90°$ lines, and clearly this also increases with decrease in the ratio D/E.

Measurements of the chordal lengths of the left- and right-hand quadrants give left-hand/right-hand ratios of 2.3, 2.6 and 3.1 to 1 respectively for D/E ratios of 3.0, 2.0 and 1.5 to 1. Clearly this amount of angular distortion presents a false impression of the true shape of the resulting workpiece to a degree which is not satisfactory for our purpose.

In any event, even though the particular values of the constants C, D and E were such that a nominally true and acceptable basic circle was developed, the shape displayed is not necessarily a faithful representation of the resulting shape of the workpiece, as has been well revealed by Reason[5] where he illustrates the effect of the magnification of a circular shape by taking the example of a nominally circular shape which has a small error in roundness.

This component has a small error of reduced radius in six regular places around its periphery, and he illustrates this error graphically by polar diagrams having a suppressed zero, to magnifications of ×2, ×4 and ×8. The actual shapes of his diagrams, which are reproduced in figure 6, are a convex hexagon, a regular

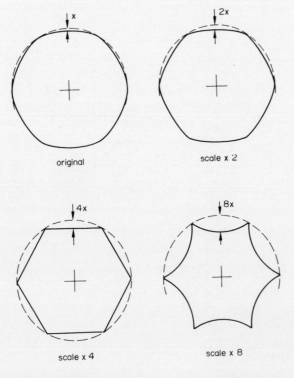

Figure 6. Effect of changing the radial magnification on apparent shape of specimen.

hexagon, and a concave hexagon respectively. Thus, the shape of the almost circular component appears as a quite different shape, particularly at the higher magnifications.

The U.M.I.S.T. report 'The specifications and tests of metal cutting machine tools' issued in 1970 proposes that spindle rotational accuracy be assessed over 10 revolutions, representative of a portion of the

workpiece produced by the tool being transversed in an axial direction, this being more representative than a single revolution.

The spindle rotational error is expressed by definition as a numerical value of error corresponding to the minimum radial zone width between two concentric circles containing the polar trace, and constructed from the minimum zone centre; see figure 7.

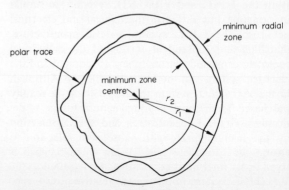

Figure 7. Minimum zone centre and minimum radial zone for a polar trace.

The U.M.I.S.T. proposal furthermore classifies spindle rotational error according to the following table.

Class	0	1	2	3	4	5	6	7	8
Error μm less than	0.25	0.5	1	2	5	10	16	20	40

In the method of assessment of the roundness of a specimen on a Talyrond machine the specimen is held stationary, and its circular profile swept by a stylus attached to a rotating shaft whose rotational error has been made negligible.

The measurement of lathe spindle rotational accuracy is in practice the reversal of the Talyrond machine assessment in that the circular specimen in the form of the test sphere has a known high degree of accuracy, and is rotated with the spindle against a stationary stylus (i.e. the transducer).

The considerations for or against a polar display have a close parallel with the considerations made during the development of the Talyrond instrument, when it was found necessary to investigate the basic properties of polar graphs which are employed for display, and fully described by Reason[5].

Clearly, there are several alternative features of spindle rotation that might be assessed, such as diametrical, radial and high frequency variations, and it is in these instances where a polar display is more desirable. Polar co-ordinates provide a realistic display that, with practice, can be immediately related to the shape of the specimen, but the departures from true roundness can àlso be revealed in Cartesian co-ordinates.

Thus, it is necessary to balance the provision of the extra instrumentation required to obtain a polar display, plus the possibility that errors in the basic circle shape may be present to a greater or lesser degree, against the less realistic but experimentally simplified

Cartesian co-ordinate method from which all the data required to determine the spindle rotation accuracy measurements to meet practical requirements can be obtained.

3. THE PROPOSED SYSTEM

The proposed system employs a Cartesian co-ordinate method of presentation and its choice has been made because of the relatively simplified experimental equipment required, and the complete avoidance of the possibility of radial errors, errors of apparent centre shift, and circularity errors, whilst at the same time providing the means whereby the minimum radial zone is determined, and a numerical value of spindle rotational accuracy obtained. The mechanical components required fit to the chuck (or alternatively to the taper socket in the spindle nose) the toolpost and the tailstock, no attachments being required at any other point on the machine. A diagram of the equipment is shown in figure 8, and a photograph showing the rig on a lathe is shown in figure 9.

A steel ball fixed to a centre-piece with Araldite serves as the test sphere. The centre-piece is adjusted for run-out by a wobble device. The steel ball is $\frac{1}{2}$ in diameter Grade I specially selected by the manufacturer, and was checked on a Talyrond machine as having a roundness error of less than 10 μin.

An adaptor forming part of the wobble device is either parallel shanked for fixing into a lathe chuck, or taper shanked for fitting into the taper socket of the machine spindle. The probe of a linear displacement transducer (Philips 9310) is acting on the test sphere. The transducer is connected as a differential transformer and fed from a Philips PR 9307 carrier wave measuring bridge. The probe is mounted in a bracket fixed in the tool post, the centre line of the probe being in the horizontal plane containing the centre

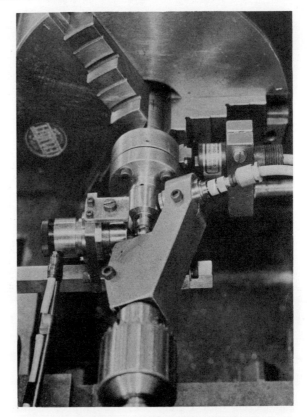

Figure 9. Proposed system mounted in lathe.

line of the machine. A leafspring, bearing against the test sphere, made of 0.004 in brass shim is held in a bracket and positioned between the end of the probe and the sphere in order to take the tangential frictional force arising owing to the rotation of the test sphere, rather than allow it to act upon the probe in a transverse direction. The mass of the Philips probe is specified as 1 gramme, and the force

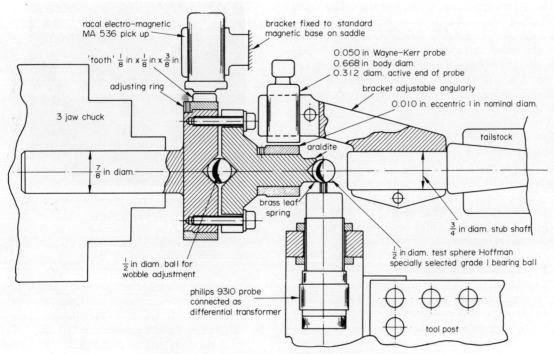

Figure 8. Diagrammatic arrangement of proposed system.

at the null position 20 grammes; thus the probe will follow accelerations of up to 20 grammes. The mass and spring force of the brass shim leafspring have been designed to have similar characteristics so as not to reduce the performance of the probe. The detail of this arrangement is shown in figure 10.

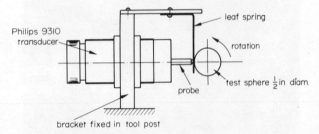

Figure 10. Displacement transducer and test sphere.

Although a wobble device is provided for setting the run-out of the test sphere, in practice the best that could be achieved would not entirely eliminate all the errors of run-out due to sphere eccentricity. For this reason, another relatively coarse eccentric is provided in the form of a cylindrical bush having 0.009 in eccentricity between its bore and outer cylindrical surface. This eccentric is situated on the centre-piece, and can be set to any angular position and locked with a grubscrew. The purpose of the angular setting is to enable the signal from the coarse eccentric to be set approximately in phase with the eccentricity of the test sphere, the final phase setting being carried out with the machine rotating by angularly moving the transducer acting upon the coarse eccentric by means of its bracket supported in the tailstock subshaft.

It was found preferable to minimise tailstock misalignment and other similar errors by having a coarse non-contacting displacement transducer to act upon the coarse eccentric, and for this purpose a 0.050 in Wayne–Kerr capacitance probe was used. The radial position of the probe is set so that the minimum clearance is approximately 0.020 in. In conjunction with the 0.008 in to 0.009 in coarse eccentric this gives rise to a voltage variation of approximately 0.4 to 0.6 V per revolution. A potentiometer is incorporated in order to achieve matching with the amplitude signal of the test sphere.

A twin-channel storage oscilloscope, having the facility whereby the signal in one channel can be inverted, is fed with the transducer signals, the signal from the coarse eccentric being taken to the invert channel and the other connected in a manner such that infeed of the cross slide will result in the trace being moved in an upward (positive) direction on the Y axis. A switchable filter unit is available for optional use at higher speeds of rotation.

3.1. Experimental technique

The experimental procedure for the measurement of lathe spindle rotational accuracy is as follows:

(i) *Calibration.* The calibration of the Philips probe is done by taking a clock indicator reading of the flat surface of the Philips probe bracket. By using a clock of 0.0001 in per division and the carrier frequency measuring bridge and oscilloscope set to give an overall approximate magnification of 1 cm = 0.001 in, the gain control on the bridge unit is adjusted to give 1, 2 and 3 cm displacement on the Y axis of the oscilloscope for relative clock indicator readings of 0.001 in, 0.002 in and 0.003 in respectively. Finally the gain on the oscilloscope is increased by a factor of 10 to given an overall magnification of 1 cm = 100 μin.

(ii) *Testing.*

(a) Within the coarse eccentric signal in normal mode, the oscilloscope X gain is set for 1 revolution to equal 8 cm on the X axis of the oscilloscope screen.

(b) Select 'CHOP' position, and obtain a good fit of the coarse and test sphere eccentric signals by using the coarse eccentric amplitude control and phase setting control by adjusting the angular position of the coarse eccentric bush, and the angular position of the Wayne–Kerr bracket in the tailstock for fine final setting. See figures 11 and 12 for typical traces.

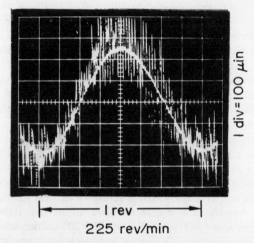

Figure 11. Signal from coarse eccentric matched with signal from test sphere.

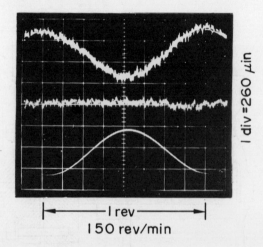

Figure 12. Signals from coarse eccentric and test sphere before and after adding.

(c) Invert coarse eccentric signal and add both signals.

(d) Allow 10 traces to accumulate with the oscilloscope in the storage position.

(e) Repeat at speeds throughout the complete speed range of the machine.

(iii) *Evaluation of test result.* A typical 10-revolution trace is shown in figure 13 in Cartesian co-ordinates together with the equivalent polar diagram. The method of obtaining the equivalent minimum radial zone (MRZ from the Cartesian co-ordinate graph is clearly shown. Ten single revolution traces are superimposed upon one another in the Cartesian co-ordinate diagram, and in arriving at the radii R_1 and R_2 the smooth curve drawn along the

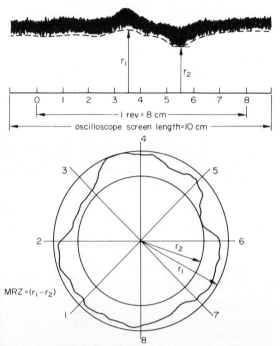

Figure 13. Typical 10 revolution Cartesian trace and equivalent polar trace.

underside of the 10 trace cloud has been used, thereby eliminating the effect of unwanted high frequency vibrations to an acceptable level. The difference between the radii R_1 and R_2 is the numerical value of spindle rotational error.

(iv) *Photographic recording.* In the event of a permanent record being required a polaroid photograph of the trace can be taken. In a test series this is only necessary occasionally since, with practice, the values of R_1 and R_2 can be read directly from the sorted traces on the oscilloscope screen.

4. CONCLUSIONS

The measuring system described is capable of evaluating the important components of spindle rotation accuracy. It may be concluded that the system of measurement yields the same information as either the L.R.L. or the V.U.O.S.O. polar display systems, but without the complexity of instrumentation which they involve and without the need for accurate and careful choice of components and skilful use.

REFERENCES

1. G. SCHLESINGER, 1966. *Testing Machine Tools,* 7th edn., Machinery Publishing Co. Ltd.
2. BRYAN, CLOUSER and HOLLAND, 1967. 'Spindle accuracy', *American Machinist, Spec. Rep.,* No. 612, 4th Dec.
3. J. VANEK, 1969. 'Measurement of accuracy of rotation of machine tool spindles' (in Czech), V.U.O.S.O., 28th July (English transl., U.M.I.S.T.).
4. *U.M.I.S.T. Rept.,* 1970. 'Specifications and tests for metal cutting machine tools', part I and II, 20th Feb.
5. R. E. REASON, 1966. *1966 Report on the Measurement of Roundness,* Rank–Taylor–Hobson, Leicester.

AUTOMATION:
GROUP TECHNOLOGY AND NC CONTROL

CONDITIONS FOR THE INTRODUCTION OF GROUP TECHNOLOGY

by

R. LEONARD* and F. KOENIGSBERGER*

SUMMARY

The merits of Group Technology for batch manufacturing situations have been amply demonstrated by previous authors. However, it is evident that certain industrial environments are more suitable for group production than others. The present work identifies a number of circumstances advantageous to group manufacture and indicates how some of these conditions may be obtained when initially absent.

The paper shows that the company environment must be considered in its entirety before group technology can be successfully established. Such diverse factors as labour rates, capital equipment, inspection requirements, labour flexibility and even component weight must be duly considered before a satisfactory production system can be finalised.

INTRODUCTION

There is a tendency to believe that we live in an age of mass production. It is thus surprising to learn that even in such a highly industrialised society as the United States of America, 75% of metalworking production takes place in the form of batches of less than 50 pieces (Merchant[1]) and because of such factors as the advent of NC this percentage is expected to increase. It is self-evident therefore that a procedure which improves the efficiency of batch manufacture has a widespread range of application. Group technology is a method of obtaining some of the advantages of mass production when only batch quantities are present. The gains ensuing from group manufacture are a greatly reduced overall production time, variety reduction and a corresponding reduction of work in progress. Increased standardisation results in a lower number of jigs, fixtures, set-ups and items in the finished stores. The production control function is greatly simplified and this yields a corresponding improvement in customer delivery dates and hence customer goodwill.

The basic concept of group technology is to think in terms of individual components and not their final function within a completed product. Thus, it is necessary to establish a classification system which embraces the whole product range of the factory and which avoids undesirable ambiguities, such as one designer calling a part a bush, because it forms part of a bearing, and an identical part a spacer, if it fulfils that function. After the components have been classified subsequent variety reduction will result in some form of standardisation and in increased quantities of similar parts being required, so larger batches will exist and these may justify the use of special fixtures or machines. The next stage concerns the production sequence for each part. If it is found that an adequate number of similar parts undergo the same operations, then the production facilities of the factory can be arranged so that parts may pass from one operation to the next without interruption. Hence, the factory is arranged in either group production lines (flow lines) or in 'cells' (figure 1) and not in a conventional functional layout.

The question naturally arises, if group technology is so advantageous, why has it not achieved universal application? One answer is that group technology requires a total management commitment, and the amount of preparatory effort necessary before successful operation is considerable (Edwards[2]). The whole range of company operations must be re-thought on group technology lines and a significant amount of the factory layout arranged on the cell concept. Varying levels of effort will be required by some companies before the introduction of group manufacture: equally some product ranges are more suitable for this method of production than others. The aim of this paper is to outline the conditions which would be considered most suitable for group technology to be implemented. It would be extremely fortuitous if all the conditions were found to prevail; the ease of implementation and subsequent operations increases, however, with the number of points satisfied.

OPTIMAL CONDITIONS

A large number of small batches
This is the prime condition for the introduction of group technology. If the conditions of production were such that the converse were true, a small number of large batches or a single component, then special-purpose machining and mass production techniques are indicated.

Accurate production information
A considerable amount of data used by management is inaccurate, either marginally or excessively. In the experience of the authors, job production times can be seriously in error, often to such a level that their use is counter-productive. It is also the experience of the authors that even the methods of manufacture which management think exist, and believe are controlled by them, can be erroneous. It is not uncommon for certain planned operations to be entirely omitted or vice versa. A simple example of incorrect data occurred when a marker-out used a

* Machine Tool Engineering Division, UMIST.

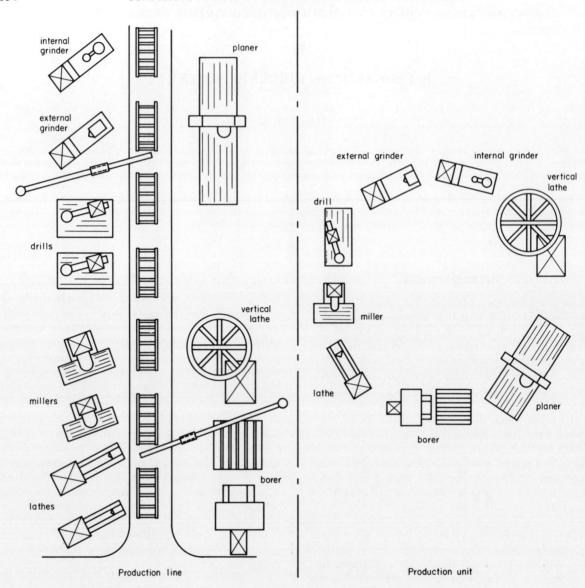

Figure 1. Group layouts.

special template which he had the initiative to make. This reduced the time for marking-out a control panel from 30 min to 5. A second error of information for the same job occurred later when a radial arm drill operator 'stacked' the control panels and drilled through 10 instead of 1, thus reducing the production time to $\frac{1}{10}$ of that stated. For a group production line to operate successfully, it is imperative to use the correct data. It is self-evident that the line cannot be balanced if erroneous information has been used. Thus, when group production is being considered, it is necessary to ascertain the accuracy of production data. The optimum condition occurs when the data available is found to be adequate. If this is not so, a costly programme is required to generate correct information.

Customer delivery requirements
A marked advantage of group manufacture is a greatly reduced throughput time, achieved mainly through the absence of queuing time at each operation. Thus, it is possible to meet customer demands for urgent deliveries, and delivery dates can be given with increased confidence. A senior manager of a firm which has successfully operated on group production principles for a number of years, stated that the greatest benefit to his company is the knowledge that two days after a job enters the line it emerges completed (Connolly[3]). Before the introduction of group manufacture, production control has been a major problem, the average production time being three months, many jobs actually becoming lost en route, with the need for replacement batches to be initiated. Thus, group technology is particularly advantageous when it is important to meet delivery dates, which it is hoped is normal business practice.

Company control of raw material
When attempting to meet urgent customer demands, it is of little value to have a group line with a production time of two days, if a six-week delay in

the delivery of raw materials exists. Thus, for optimum conditions to be established for meeting delivery dates, considerable control must be secured over the supply of raw materials. The problems of stocking raw material stores are significantly reduced when group manufacture is adopted, mainly because of variety reduction and raw material standardisation.

Light components
Although there are many successful applications of group technology, even with components heavy enough to require the use of an overhead crane, there are distinct advantages when the articles for manufacture are light. Conveyor belts or rollers, placed between the machines, greatly aid the automatic flow of work along the line, and avoid the possibility of jobs being sidetracked. If the services of a crane are required to lift a component from a machine, it is unlikely that the next machine in the line will be ready to accept the job and so it must be placed on the floor. The amount of work handling can become significant under these conditions and the possibility of work being sidetracked emerges. One company, making components of considerable weight, state that the advantages produced by group manufacture are so significant that it is worthwhile

Simple jobs and flexible labour
In a group production cell, flexibility of labour between operations is desirable. On the other hand, the level of flexibility decreases as the degree of skill required for successful operation increases. Consequently, labour movement between unskilled jobs can be more easily achieved than between skilled operations. For instance, a lathe operator may not be able to operate an internal grinding machine or a horizontal boring machine with the same efficiency as on his own work. Even if there existed complete union cooperation, the decrease in efficiency which results from moving skilled personnel from one task to another can be considerable. It must be noted, however, that when a high level of operator skill is necessary or costly equipment is used, labour does not in general move between stations, and techniques of cell loading are preferable (El-Essawy[7]).

Inexpensive plant
If group technology is applied in relatively low-cost plant, it is often advantageous to establish an excess of work stations over personnel (figure 2). In this case the manager must be able to move the labour at his disposal to the necessary stations so that a smooth flow of work is achieved. This procedure must result

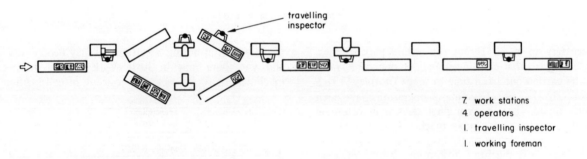

travelling inspector

7. work stations
4. operators
1. travelling inspector
1. working foreman

Figure 2. Movement of labour between workstations for relatively simple operations.

investing in complex work handling machinery between stations (Schaffran[4]). When component size is considered the results obtained by Opitz[5] may be recalled. A survey of workpiece statistics indicated that even a large final product incorporated many small articles fastened to a few large pieces.

Minimum inspection
The normal engineering practice of inspecting components after each operation is highly undesirable in group manufacture if it requires the movement of work to an inspection table. The automatic flow of work is interrupted and handling may become excessive. It is particularly advantageous if the component requires minimum inspection between operations. However, if quality control necessitates intermediate inspection the services of a 'travelling inspector' and special inspection aids are advocated (Marklew[6]). This method utilises one inspector for each cell, although other duties may be fulfilled as well. The inspector checks the work between operations without interrupting the flow or moving the components from the line.

in an under-utilisation of plant. The cost incurred by machinery not being fully used depends primarily on the value of the individual machines. If the machines are relatively cheap, say about £1000, the maximum production efficiency of the cell is achieved if the labour is always working at the correct work stations; little account need then be taken of machine utilisation. However, if the hourly cost of the investment value approaches the hourly cost to a company of a man's wage, the problems of optimising the cost of production become more complex. If the cost of plant significantly exceeds the labour cost, under-utilisation of machinery cannot be sanctioned. When considering the purchase of plant for a group technology cell, a detailed description of the facilities required should be determined and this, together with a comprehensive survey of existing machines, will make it possible to decide on the purchase of plant which only fulfils specific functions, resulting in a relatively inexpensive solution. The most publicised example of a machine tool limited to machining requirements is the short-bed lathe (figure 3) proposed by Moll[8]. An analysis of the sequence of operations

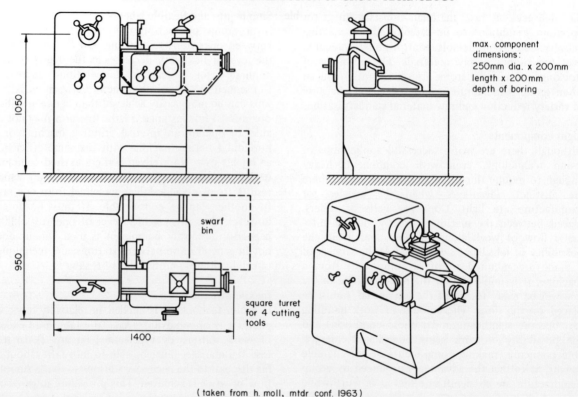

(taken from h. moll, mtdr conf. 1963)

Figure 3. The Moll design of short-bed lathe.

necessary in a cell layout for certain part families, often reveals that certain very simple operations need to be carried out from time to time (Thornley[9]). In a group flowline, after the major operations have been performed, other less complex ones may often be carried out on very simple machines. As the capital costs of these simple machines are not excessive, the problems of machine utilisation are somewhat reduced.

Similarity of components and production operations

It is unlikely that the whole factory output can be arranged on group technology principles. Some components, such as large crankshafts for marine engines, do not fall easily within a parts family and so these would still be produced conventionally. The amount of work which falls naturally into a cell grouping increases with component similarity and decreases with the complexity of manufacture. Figure 4 shows a typical routing card, for which the operation sequence is not likely to be repeated by a large number of components.

A balanced machine utilisation

It has been stated previously that when the capital cost of plant is considerable, a high level of machine utilisation is essential. Regrettably, it is not possible to rely on an acceptable level being achieved by pure chance, even if considerable attention has been given to the construction of a cell. The cell layout must of necessity be based on an average workload, calculated over a long time. The actual workload over a short period, however, can only approximate to the capabilities of the cell. Thus it will occur, for example,

that horizontal boring requirements may be excessive and yet there may be little work for the drilling machines; in such cases, the concept of cell loading

quantity required 30	description sprayer body									
material specification and size m/c from stamping b 1057										
material specification sprayer body ms en. 3a.										
size stamping										
routing sequence	23	51	23	23	23	23	23	51	23	83
dept.	op.no.	operation								
23	1	part off surplus metal, face length								
51	2	carburise								
23	3	face flange and boss at rear								
23	4	turn dias, face and radius flange groove screw, turn $\frac{13}{16}$ in. dia. bore $\frac{7}{16}$ in. x 3in. dp.								
23	5	face flange and length bore C/bore $\frac{21}{32}$ in. cone tap $\frac{3}{8}$ in. bsp complete								
23	6	drill tap and s/face complete dj 3788								
23	7	mill flats								
51	8	harden end face								
23	9	grind end face lap. face.								
83	10	hardness test								
25	11	grind and glaze								

Figure 4. Job sequence card for an engineering component.

becomes important. Thus, when the work schedule for a given period indicates a naturally-occurring deficiency of work of a given class, the production planner injects certain jobs into the work programme and restores the balance of the cell.

Maximum advantage of a classification system

The cost of establishing a satisfactory classification system is considerable (Middle[10]). It is a major task to code many thousands of drawings and so it is desirable that the company should obtain the maximum benefit from this investment. One of the areas where a well-designed coding system has obvious advantages is in the field of data retrieval for determining and using standard parts at the design stage, the degree of advantage depending on the amount of design work undertaken. If the products made by the company have a relatively short life and new designs are frequent, the use and advantages of such standard parts can be quickly established. If, however, the product life is long and spare parts are required by past customers for some considerable time, the change to standard components will occupy a number of years and the undoubted advantage of variety reduction will take time to emerge. It is self-evident, however, that advantages can never appear unless some pioneering work has been carried out.

CONCLUSIONS

The advantages to be gained by firms engaged in batch production who adopt group technology are considerable, but some products and industrial environments are more amenable to this system of manufacture than others. Thus, it is imperative for a company, contemplating the adoption of group technology, fully to investigate any problems that may be peculiar to its product range. The points described in this paper should be given adequate consideration before a new production system is finalised.

REFERENCES

1. M. E. MERCHANT, 1969. Trends in manufacturing systems concepts, *10th MTDR Conf.,* Pergamon.
2. G. A. B. EDWARDS, 1969. The management problems of introducing Group Technology. Paper to *1st International Conference on Group Technology,* ILO, September.
3. R. CONNOLLY, G. MIDDLE and R. H. THORNLEY, 1970. Organising the manufacturing facilities in order to obtain a short and reliable manufacturing time, *11th MTDR Conf.,* Pergamon.
4. J. SCHAFFRAN, 1969. The application of Group Technology in the Societe Stephanoise de Constructions Mecaniques. *Group Technology Seminar,* Turin.
5. H. OPITZ, 1964. Workpiece statistics and manufacture of parts families. *VDI Zeitschrift,* **106,** No. 26, September. M.T.I.R.A. Translation T146.
6. J. J. MARKLEW, 1970. An example of the cell system of manufacture at Ferranti. *Machinery and Production Engineering,* August.
7. I. F. K. EL-ESSAWY, 1971. Component flow analysis. Ph.D. Thesis, Department of Management Sciences, U.M.I.S.T.
8. H. MOLL, 1963. A user's viewpoint of the machine tool of the future. *Proceedings of the 4th MTDR Conf.,* Pergamon.
9. R. H. THORNLEY, 1972. Group Technology: a complete manufacturing system. *The Chartered Mechanical Engineer,* January.
10. G. H. MIDDLE, R. CONNOLLY and R. H. THORNLEY, 1971. Organisation problems and the relevant manufacturing system. *Int. Prod. Res..* 1971, 9, No. 2, pp. 297–309.

COMPUTER CONTROL OF MACHINE TOOLS

by

D. FRENCH* and J. KNIGHT*

INTRODUCTION

In recent years there has been an increasing interest in control of machine tools by digital computer (DNC). To determine the problems of DNC a research program was initiated to control three machine tools by digital computer. The system of control was different for the three machines, and was composed of a machining centre with GE control, a milling machine with ICON control and a lathe — retrofitted with electrohydraulic pulse motors. The computer used in this research was an IBM 360/44 situated approximately 1000 ft from the machine tool laboratory.

The paper discusses the software problems relating to DNC and the control of three different types of machine tools utilising two distinct modes of control.

SOFTWARE REQUIREMENTS FOR DNC

Computer programming for ON-LINE control of machine tools requires the following.

(1) There must be communication between the operator and the computer. For this purpose the interface unit is the operator control panel, and from which machining routines can be selected, computer operation can be cancelled, and management functions recorded.

(2) For a system in which the DNC replaces the reading head the code format must be transmitted to the control unit in a series of pulses. In a system using 8-channel tape 9 'bits' of information must be generated, 8 for the code (EIA or ASCII) and the ninth to supply the timing pulse required to trigger the control unit logic circuit, which is normally actuated by the sprocket holes on the tape when used with photo-electric reading heads. If a control unit utilises a mechanical reading head, the tape sprocket holes are not used to trigger the logic circuitry of the control unit; consequently the program for this type of machine control must transmit 8 parallel 'bits' of data at the exact speed of the control unit internal clock.

In addition, dependent on the control system being used for DNC, the type of pulse trains can differ; for example, on the general electric control system, a complete pulse represents one 'bit' of data, whereas the ICON control system the same data train would consist of high and low voltage levels.

(3) When the computer is used to drive pulse motors directly, a program must be written to propagate the pulses required for performing the machining operation at the desired feedrate.

(4) The DNC data, comprising the machining routines, are stored on disc. The data are stored in two forms; EIA coding for tape reader substitution, and numerical data for the propagation of digital pulse trains.

Each EIA character is a binary number which can be represented by a decimal counterpart, the decimal character being stored on the disc. During the transmission of the EIA code, the decimal character is converted to binary on the computer output terminals.

The segments of disc storage are termed 'blocks'. One disc block contains the capacity for storing 180 EIA characters. The maximum length of an EIA data block is 170 characters. To facilitate programming and to decrease the time to read EIA data into core, one EIA data block is stored on one disc block. When the EIA data are ready for transmission, the entire disc block is read into core. During transmission, each character is checked, and transmission is terminated when an EOB character is recognised. Surplus, unrelated information remaining on the disc is disregarded.

Nine decimal numbers represent the data required for one machining motion by digital pulse train propagation. One block of disc storage contains 17 digital pulse train blocks; the entire disc block is read into core. One of the nine numbers of each digital pulse machine tool motion is a pulse count. Termination of a machine tool motion occurs when the count is obtained. Machining restarts on the information provided by the next nine numbers. After 17 machine tool motions are completed, the next disc block is read into core.

Each EIA and digital pulse machining routine is separated and catalogued according to the machine tool it controls. Listed on the catalogue are the following.

(a) The number of routines available for transmission.
(b) The name of each routine, i.e. BASEBALL BAT.
(c) The disc starting block for each routine (BSTA).
(d) The disc finishing block for each routine (BEND).
(e) The number of blocks required for storage (NBLKS).
(f) The code that signifies to which machine tool the routine has been adapted (AORB).

The catalogue is used in the control program for reading the routine blocks into core and for equating program variables to the catalogue variables.

* Department of Mechanical Engineering, University of Waterloo, Waterloo, Ontario, Canada

(5) The control program must respond to the input data from three machine tools.

Control data are transmitted from the computer to three separate machines; the Icon control unit, the GE control unit, and the digital pulse motors of the lathe. The control program must have facilities for reading the operator control panel to satisfy the three different machining conditions.

(6) Restarting a new routine on any machine tool.

(7) Cancellation of the control program.

The facility for entire control program cancellation is available to the operator during any period in which communication with the computer has been established.

(8) Control program response to external interrupt signals.

Interrupts are used for two operations. The first interrupt originates from the control units and/or from a switch at the digital pulse motors. This interrupt starts the transmission of EIA and/or digital pulse train data. The transmission of each EIA character requires a second source of interrupt signals. The spacing between each character is exact, and an external frequency generator producing 50 measured pulses per second provides the interrupt signals.

(9) Priorities for each of the three machining systems.

Maximum control of the machine tools is required for efficient machining operations. One of the main objectives of DNC is to provide extensive software control of the cutting routines. The control program contains the following secondary features which provide the operator with maximum machine tool control.

(1) The control program operates a series of lights indicating the operator's position in routine selection and indicating any error in routine selection. The lights are situated on the operator control panel.

(2) The control program senses errors in routine selection. Each machining routine has an assigned value, termed AORB. The value of AORB determines to which machine tool a machining routine is applicable. There are four values of AORB for each machining possibility; the Icon, the GE, the digital pulse motors, and an AORB value for transmitting a routine to both the Icon and the GE control units. For example, if the operator was selecting a machining routine for the Icon system, the value of AORB for that particular routine selected is checked against the value accepted for an Icon routine. If the AORB value of the selected routine does not match the Icon AORB value, an error message is generated.

(3) A machining routine can be started and stopped at any block of data in the routine. This option is available during routine selection.

(4) The operator has the facility for selecting a new routine when a machining routine is completed on one of the three machine tools. If either of the other two machine tools is in operation, they are excluded from the selection. The operator can select a program for the idle machine tools only. This facility safeguards the machining operations in progress.

(5) An emergency stop facility is available to the operator. The emergency switch is located on the operator control panel, and, when closed, transmission of data is automatically stopped. The lights located on the operator control panel will indicate the routine block number where transmission was terminated.

COMPUTER CONTROL PROGRAMMING

The control program supervises the operations of three independent machine tools. The two milling machines are controlled via tape reader substitution, and the lathe via both tape reader substitution, and Control Unit substitution.

MAIN initiates the control program and executes the following operations.

(1) Allocates core storage for the program variables.

(2) Sets up request control blocks (RCB). The RCB is an area in core where both the processor and the channels have access to vital information concerning various I/0 operations.

(3) Reads the catalogue into core for use in the machining routine selection.

(4) Attaches priority interrupt levels (PIL) to the subroutines that will receive external interrupt signals.

(5) Queues the READ subroutines. CALL MQUEUE is used to enter a request for execution of a foreground task (FGT). FGT's are executed by task priority.

Two modes of data transmission are employed, which requires two modes of transferring control from the START subroutines. The transmission of EIA data is dependent upon the external frequency generator. The pulses from the generator are utilised as interrupt signals to transmit the EIA characters. Therefore, control must be passed from STARTA and STARTB to RTT subroutines which execute on a priority interrupt level basis. These subroutines are SENDA and SENDB. The transmission of digital pulse trains is based on the internal clock of the computer. External signals are not required to trigger the digital pulse train transmission. Therefore, STARTP transfers control by 'queueing' subroutine PULSES into a low priority task level.

The transmission of EIA data to the Icon and G.E. Control Units is executed by an interaction between subroutines (SEND and OUT). Subroutine OUT prepares each EIA character for transmission, then activates the SEND PIL. The interrupt signal from the frequency generator realises the EIA character value on the output terminals. To prepare the next character for transmission, SEND queues subroutine OUT back into a priority task level. The two subroutines continue this cycle until the EOB character has been transmitted. Subroutine OUT then activates the START PIL to transmit the next block of data when the appropriate starting interrupt signal is received. The priority task level of OUT is higher than the priority task level of PULSES.

The digital pulse trains for the lathe's Fujitsu digital pulse motors are propagated from the PULSES subroutine. Subroutine PULSES is on the lowest priority task level. Execution is temporarily terminated when any of the other transmitting operations are required.

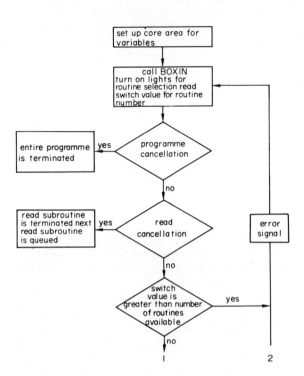

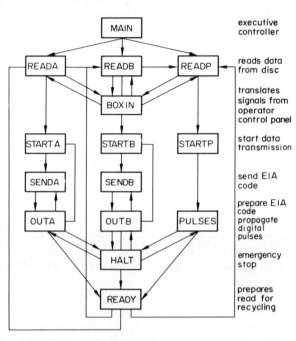

flow chart for MC control programme

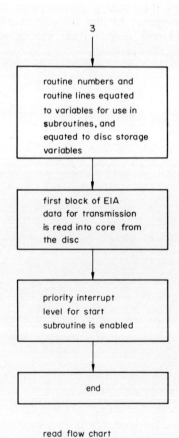

read flow chart

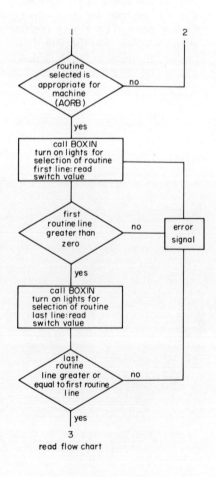

read flow chart

PULSES produces linear pulse trains on two axes and in four directions: X,Y, $X,-Y$, $-X,Y$ and $-X,-Y$ requiring four output terminals.

The length of a single pulse dictates the feedrate and is controlled by the internal clock of the computer. All calculations in the iterations are performed by addition and subtraction to minimize the time required for each iteration.

The data for pulse propagation is stored on disc as in the EIA coding. Nine numbers are stored for each machining segment, these are as follows.

INCX: The increment for X-axis motion and the number used in the overflow calculations.

INCY: The increment used for Y-axis motion.

INCZ: The increment for Z-axis motion.

MS: The time in milliseconds for each pulse length.

MTD: The direction on the coordinate plane that the tool is to move. MTD (machine tool direction) consists of four numbers; 1 for X,Y, 2 for $X,-Y$, 3 for $-X,Y$ and 4 for $-X,-Y$.

NP: The number of iterations to be made by the computer.

X: The operator is free to select any starting position in the machining routine. X is the distance on the X axis from the original datum point to the new starting point. A rapid traverse to the new starting point occurs over the distance designated by X.

Y: Y is the distance on the Y axis to the new starting point. The Y tool feedrate is considerably slower than the X feedrate to allow a slower approach to the workpiece.

EOB: EOB signifies the end of that particular machining segment.

During execution to PULSES, the emergency stop facility is called regularly. PULSES is entirely self-sufficient requiring no other control than a starting interrupt to carry it to completion.

The priority interrupt system provides a further facility that cannot be programmed. Once the priority interrupt levels for the start subroutines have been activated, the operator has the facility to delay the start of the program. This facility provides the operator with the opportunity to affect any adjustments or inspection required by the machine tool.

The rapid response time of the computer permits the program to control simultaneously the three machine tools. The operation of transmitting EIA code to both control units is on the same priority task level. Therefore, with a transmission rate of only 50 Hz, both control units can be receiving the EIA code at the same time. The propagation of digital pulse trains is on the lowest priority task level. When EIA data are being transmitted to the control units, the digital pulse trains are temporarily terminated. The length of time that the pulses are stopped depends on the number of characters in the EIA block. Pulse train termination is usually less than a second. While the control units are executing the EIA data, the digital pulse propagation continues.

The control program, once initiated, can only be cancelled by the operator control panel, and in the case of emergency by the computer console typewriter, and will recycle itself continuously until cancelled by the operator.

A LOW-COST HARDWARE INTERPOLATION SYSTEM FOR DNC

by

M. SIMPSON*, D. FRENCH* and W. LITTLE*

INTRODUCTION

The diagram in figure 1 shows a configuration for controlling a number of machine tools from an on-line computer. The operator panels (OPC1–OPC4) communicate with their own interpolation units (INTERP1–INTERP4) through the on-line computer.

Interface units (INT1–INT8) are used to route information to and from the computer to the control panels and interpolation units. A teletype unit is used to communicate with the system software for generating management reports. The disc is used as a random

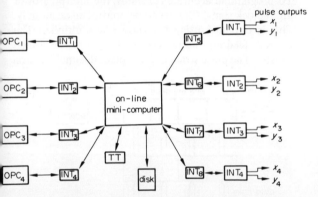

Figure 1. System configuration for on-line computer total of machine tools.

access bulk storage device for storing parts programmes.

All interpolation units operate asynchronously from the on-line computer. Once data is transferred from the computer to an interpolation unit no further attention is required until more data is required to execute a new 'cut' instruction. Whenever an interpolation is in a 'busy' state a 'flag' is set in the computer interface. The computer scans the interpolation flags to determine when new data are required by a particular unit. In addition the computer services the control panels, loads new blocks of data (from tape or disc) and prepares these data for transmission to the interpolators. The control panel communicates with the computer on an interrupt basis rather than by flag scanning. This is a necessary condition because certain operator actions require instant action by the computer. The system has the facility of allowing the operator panel instructions to override other control actions, as necessary, and then to return control to the original function.

THE INTERPOLATION UNIT

The interpolation unit comprises six sections or blocks.

(1) A computer interface.
(2) An instruction decoder/data distributor.

(3) A specialised DDA.
(4) A logical axis switch.
(5) A motion distance detector.
(6) A feedrate controller.

The computer interface (figure 2)
Instructions from the computer are transmitted (via the computer interface) in parallel 12 bit words. The

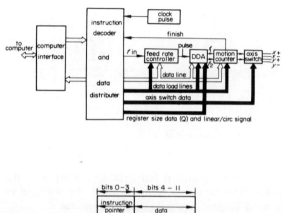

Figure 2. Functional block diagram of input unit.

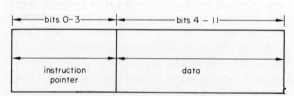

Figure 3.

format of the word is shown in figure 3. The first 4 bits constitute an instruction pointer, the instruction decoder determines from these 4 bits what action should be taken, these actions include; hold interpolation, continue interpolation, clear registers, and load data.

An instruction decoder/data distributor
When a load data instruction is detected, bits 4 to 11 are outputted onto the data lines. Bits 0 to 3 determine into which register the data are to be loaded. These bits generate the appropriate load signals on the data load line. Whenever both axes of the motion counter go to zero, the finish line is activated. This is interpreted by the instruction decoder which then clears the flag on the interface unit Axis switch and register size data are loaded into a storage register prior to interpolation; these data are static throughout the entire interpolation. The interpolator is driven from a pulse train generated by a clock in the computer. Initially gating within the

*Faculty of Engineering, University of Waterloo, Waterloo, Ontario, Canada instruction decoder inhibits the clock pulses from entering the interpolator; when all data have been loaded a start interpolation instruction is transmitted to the instruction decoder from the computer; clock pulses are then allowed to enter the system (f_{in} is the controlled clock input). When the operator or program, calls for a hold, the computer transmits a hold interpolation instruction to the instruction decoder to inhibit the clock pulses.

Specialised DDA (figure 4)

The 5 decade rate multiplier/decade counter constitutes a digital integrator. With $L/C = 1$, the output

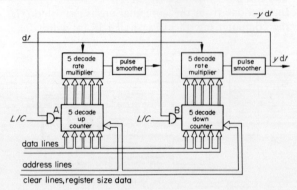

Figure 4. DDA.

from one integrator is fed back to the input of the other. This closes the loop to form the solution to $Y = -Y$ (digital approximation). Therefore

$$\text{signal A} = (L/C) \cdot (Y\mathrm{d}t)$$
and $$\text{signal B} = (L/C) \cdot (-Y\mathrm{d}t) \text{ if } L/C = 1$$

Initial conditions $(-Y(0)); (Y(0))$ are loaded into the decade counters prior to enabling the pulse input. The number of decades used in the integrator is determined by the register size data. The offset values are entered initially into the UP and DOWN counters dependant on the cutter path requirements. Linear interpolation is obtained with $L/C = 0$ which causes the feedback from the integrators to be inhibited. Data loaded into the $-Y$ and Y counters will remain static throughout the linear interpolation process. The pulse rates $-Y\mathrm{d}t$ and $Y\mathrm{d}t$ will be directly proportional to their corresponding initial conditions. For a linear segment with slope $\pm\Delta Y/\Delta X$, ΔY would be loaded as $-Y(0)$ and ΔX as $+Y(0)$, negative slopes being handled by a 'sign' command that switches the output train to either a forward or reverse pulse output line.

Pulse smoothing

A problem which exists when using rate multipliers is that the distribution of the output pulse train is uneven with respect to time. Consider a rate multiplier receiving an input pulse f_1; the rate multiplier is set by BCD inputs. For every 10 input pulses the rate multiplier cycles through its internal counting all states once. If the BCD input is set at 4, the input f_1 and output f_0 would appear as illustrated.

For every $10 f_i$ pulses one would obtain $4 f_0$ pulses, therefore the average frequency over 1 frame is $f_0 \times 4/10$. However, the pulse spacing is uneven, and

this would be worse if several stages of rate multipliers were cascaded. This unevenness in the pulse train would result in poor cutting conditions along the cutter path being generated. A method of reducing this problem is by pulse smoothing. The technique consists in

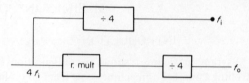

Figure 5. Pulse smoother circuit.

dividing the output frequency of the rate multiplier f_0 by a '÷ N divider' circuit and increasing the input frequency by a similar factor (figure 5).

Feedrate controller (figure 6)

To programme accurate feedrates, the interpolator or input clock signal ($\mathrm{d}t$) has to be controlled accurately. This is obtained using a 4-stage rate multiplier, The desired feedrate number is entered into the 4-digit register. The pulse strain $\mathrm{d}t$ is a 4-place decimal fraction

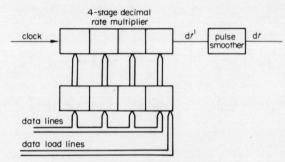

Figure 6. Feedrate control.

of the clock frequency. A pulse smoother is incorporated to obtain a better pulse distribution. Data in the register can be changed whilst the clock is active enabling manual feedrate control from the operator panel.

Motion counter (figure 7)

When data have been loaded into the $-\dot{Y}$ and Y counters and the pulse input is activated, output pulses will be generated by the interpolator; a count of these pulses must be obtained in order to control the cutter path length.

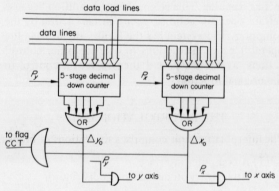

Figure 7. Motion counter.

The motion counter counts the pulses sent to each axis and, when the correct pulse count is obtained for a particular movement, the motion counter signals the axis switch to inhibit pulse output. At the same time the interpolator busy flag is cleared and the system returns to a HOLD state.

CONSTRUCTION

The complete interpolation unit with the exception of the data loader/instruction decoder was built on a standard DEC double size board (approximately 100 IC's). MSI technology was used throughout to minimise size, cost and power consumption.

CONCLUSIONS

The hardware interpolation unit described above was designed and built to plug into a mini computer. It was a low-cost method for DNC using pulse motor drives, although with minor modifications it could be used for other types of drive systems. The system built and being used in the research laboratories of the University of Waterloo illustrates a low-cost method of DNC for up to 4 machine tools.

REFERENCES

1. R. PHILLIPS, 1968. 'Many digital function generated with rate multipliers', *Electron. Design,* February 1.
2. W. ARNSTEIN, H. W. MERGLER and B. SINGER, 1964. 'Digital linear interpolation', *Control Engng.,* June.
3. R. L. OWEN, M. F. PARTRIDGE and T. R. H. SIZER, 1960. 'A digital deferential analyzer', *Electron. Engng.,* **32,** 740.
4. 'Real time digital differential analyser', 1954. *Proc. Western Computer Conf.,* pp. 134–9.
5. SCHMID, 1963. 'Analog computers with digital elements', *I.E.E.E. Trans. Computers,* December.

THE DESIGN OF AN OPERATOR PANEL FOR ON-LINE COMPUTER CONTROL OF MACHINE TOOLS

by

M. SIMPSON*, D. FRENCH* and W. LITTLE*

INTRODUCTION

The design of the operator panel was based on certain criteria.

(1) The operator control panel (OCP) should communicate directly with the computer.

(2) The OCP should have no direct link with the machine tool.

(3) It must be capable of being located at a distance from the computer, possibly up to 100 ft.

(4) To minimise the possibility of induced noise affecting the system then, the following hold.

(a) The number of cable connections between the OCP and the computer should be minimal.

(b) All interconnecting cables should be shielded.

(c) A 'twisted-pair', bi-directional transmission system should be used.

(d) Select a logic family that is reasonably immune from noise, but has reasonable power consumption and well developed logic complexity.

(e) The OCP must be shielded.

(f) All logic power supplies should be filtered.

(g) Circuiting should be used to protect the logic against power supply transients induced in the supply lines after regulation.

If a 'single-wire' communication system is adopted, instructions/data to and from the OCP must be serialised prior to transmission. This system would use a minimum of cabling but would be slower than parallel data transmission.

If a single control panel is serviced by the control computer, then the speed difference between serial and parallel transmission would not be noticeable by the control panel operator as events at the panel activated by the operator will be spaced, in time, by seconds. This time interval is very much longer than that required to transmit a typical message in serial form. However, if a number of control panels were connected to a single control computer, data transmission speed would be much more important. For these reasons it was decided to use parallel datal transmission, whilst encoding data and instructions so as to realise the minimum cable criteria.

The panel was designed to be used in conjunction with a lathe, with two-axis control, or a machining centre, with three-axis control. The position of the push buttons, lamps and digital displays are based on this application. However, it is possible to use the panel for other ON-LINE applications as the software programmes determine the function of the panel in a particular application.

* Faculty of Engineering, University of Waterloo, Waterloo, Ontario, Canada.

THE FRONT PANEL (figure 1)

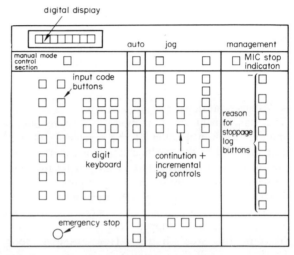

Figure 1. Front panel layout.

The front panel comprises 66 lighted push buttons and a nine-character digital display. Lamps behind the push buttons are activated only by the control computer, as are the digital displays. Registers within the panel logic system store lamp and display data, all registers can be read or written by the main computer. The layout of the panel has the control functions grouped in four groups or modes: these are manual, automatic, jog and management modes. These modes are software realisation only. The control panel logic itself cannot discriminate between one mode or another.

1. Manual mode

The machine operator can use the complete system as a conventional N/C controller in this mode, by entering the instruction block by the appropriate push buttons. In addition a facility is provided for the operator to display the information he has just entered as a check prior to depressing the execute (end-of-block) push button. Absolute or incremental positions can also be displayed on the panel when in this mode.

2. Jog mode

This mode enables the operator to pre-position the cutter prior to running a program. The jog push button can cause the machine to move continuously or in conjunction with selector push buttons, in a series of incremental steps.

3. Auto mode

If a system utilises a disc, drum, or magnetic tape storage media, then a part program could be stored; each part program would have a unique program number. In the auto-mode, the operator would enter a program number via the nine-digit keyboard array and initiate a 'program search' for that particular program by depressing the 'search' button; once the program was located, the control computer would signal the control panel to activate the 'found' lamp. The operator could then initiate the program execution by depressing the appropriate panel button. If the program was required to run a second time, the 're-load' button would restart the program. This condition can be compared with a tape rewind condition in a conventional numerically controlled machine tool, but would be very much faster.

4. Management mode

The management function mode is used to obtain more efficient utilisation of the machine tool and computer. When the machine stops, or is stopped by the operator, the panel enters a management function mode (MFM). Software in the computer will log from which panel the stop originated and the length of time the stop occurred. The operator must, after the machine has stopped, depress one of the 'reasons for stop' buttons (for example, tool breakage) to tell the computer under which reason the time was to be allocated. Failure to depress one of the 'reasons for stop' buttons would cause the down time to be allocated to operator absence. At the end of the shift or on request a management report is produced by the computer.

System configuration (figure 2)

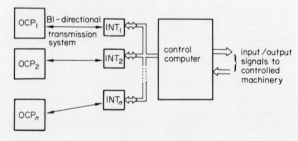

Figure 2. System configuration.

Operator control panels (OCP1–OCP) each communicate with the control computer through an interface (INT_1–INT_N) module. The computer program determines the type of panel/machine interactions that are required.

CONTROL PANEL LOGIC STRUCTURE (figure 3)

The control panel logic can be divided into a number of blocks or sections. These actions are the following.

(1) Switch encoder,
(2) Control section,
(3) Digital display registers, decoders and read selectors,

(4) Lamp registers, decoder, read selector,
(5) Lamp drives,
(6) Bi-directional data transmission system.

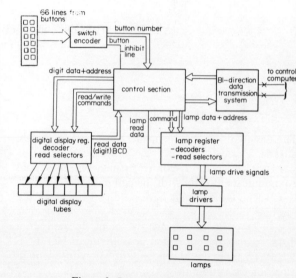

Figure 3. Control panel logic structure.

The control section controls all communications as well as decoding instructions from the control computer. It routes data to and from the lamps and digital display modules, sets flags and interrupt lines to the control computer.

When a push button is depressed the switch encoder encodes into pure binary an assigned button number; switch bounce illumination circuit is incorporated at this point. The control section sets a button inhibit line to the encoder to disable the buttons from the control section as a precaution against two or more buttons being operated together. The encoder button number is outputted to the transmission lines (via the bi-directional system) together with flag and interrupt signals. The control computer interrupt system is activated and scans the peripheral interface boards to determine the appropriate panel. The software causes the necessary action to take place according to the activated button.

When an instruction is issued by the control computer, the control section responds by 'freezing' any input from the switch encoder until the instruction has been executed. It also places the panel into a 'busy mode' to ensure that there will be no more instructions issued from it until the present instruction has been executed.

The digital display sections when activated by the control section, decode the data and address lines and either read or write one digit of the digital display.

The similar condition applies to the lamp register section, except that one lamp is read or written per instruction. The ability to read the status of any digit or lamp is incorporated to allow checking of instructions issued by the computer.

Lamp drive circuitry is used to supply the necessary current drive to the lamps. It is important in a machine shop environment to use lamp bulbs that emit an adequate amount of light to be read from a distance of several feet.

The bi-directional transmission system comprises of integrated circuit line driver/receiver devices. The outputs of these drivers are bi-polar, and have a high common mode voltage rejection. This is due to the fact that they operate essentially as differential amplifiers. A similar line drive/receiver module is used at the computer.

CONCLUSION

This panel is used in conjunction with a mini computer and hardware interpolators, enabling computer control of a number of machine tools or other systems. The basic configuration can be used, with some panel modification, as an on-line data acquisition system for production planning or like operators. The panel would be modified to reduce its complexity and thus reduce its cost.

THE MODULAR CONCEPT IN COMPUTER NUMERICAL CONTROL

by

R. S. MACLEAN,* J. W. BRUCE* and B. DAVIES*

SUMMARY

In the design of numerical control systems for machine tools several advantages may be gained by basing the control system on a small, general-purpose, digital computer rather than on special-purpose electronics, particularly on more complex machines. At the National Engineering Laboratory such a CNC system is being developed in modular form so that it may be applied to a wide range of machines. Initial development has taken place on a Herbert Co-ordatrol drill. This paper describes the system with particular emphasis on the software which, it is believed, has some valuable new features.

1. INTRODUCTION

When computer numerical control (CNC) is applied to a multi-axis machine tool with contouring and positioning capability, the task of translating control tape data is carried out in a small general-purpose dedicated digital computer. By using a computer program to replace the logical circuitry of the conventional controller a greater degree of flexibility than is possible with conventional NC becomes available permitting extensions, modifications and by updates coping with obsolescence.

International progress in recent years in the development of CNC systems has been rapid. Most of these are described in recent survey articles.[1,2] Of particular interest in Europe is the CNC system of Kongsberg Våpenfabrikk A/S[3] and Electronic Control Systems[4] as well as that developed at the Cranfield Institute of Technology.[5]

Progress, however, has not been without setbacks as Rosenberg has pointed out[6] and these demonstrate the continuing need for vigilance over costs and performance. The key to such improvement must lie in exploiting to the full one of the great advantages of CNC — its versatility.

The benefits of decreased development time and cost, lower production costs because of standardization, and easier maintenance may be expected if the same basic control system modules can be adapted to a wide range of machines quickly and cheaply. This modular approach requires a range of standard modules to be available from which a system can be built up for a particular machine tool. A wide range of options can be offered if all are made compatible with the basic software architecture. Good software organisation and efficient use of the computer is vital to the success of a CNC system.

This paper describes the rules of the system software for the NEL CNC system modules and discusses the possibilities of various drive systems before describing the NEL CNC prototype and the interface electronics modules which are needed between computer and machine servodrive.

One important objective in all the R & D into machine tool control systems at NEL is to produce innovative devices which have considerable commercial potential for U.K. industry. It is therefore necessary at the outset to recognise and attempt to evaluate the constraints placed on the R & D as a consequence. In the case of CNC it was clear that the advent of the low-cost, highly versatile, small computer should make a very significant reduction in the relatively high cost of conventional NC controllers and should therefore be of particular interest to machine tool manufacturers. Fortunately the UK small computer industry took advantage of the growing expertise in software for small computers in the U.K. and have produced machines that are particularly suitable to CNC and readily usable.

2. CONTROL SYSTEM ELEMENTS

The conventional machine controller is made up of the following elements.

(1) Input elements.
   (a) A tape reader.
   (b) Switches and push buttons used by the operator.
   (c) Position transducers fitted to the machine tool.
   (d) Limit switches and contacts on the machine.

(2) Computational elements.
   (a) A decoder for processing incoming information.
   (b) An executive controller to handle this information.
   (c) A controller for handling positional information.
   (d) An interpolator.
   (e) A controller of sequential operations.
   (f) Servo loop compensation elements.

(3) Output elements.
   (a) Servo power amplifiers.
   (b) Relay drive units or equivalent solid-state devices.
   (c) Output elements to provide information for the operator.

In a CNC system the role of the general-purpose computer is to take over the work of the computational elements. Some additional electronics are required to act as an interface between the computer and input and output elements. It may be convenient to include counters and rate-multipliers in this interface so that the computer is relieved of the chore of issuing and receiving strings of pulses at high speeds.

*National Engineering Laboratory, East Kilbride, Glasgow

The input and output elements mentioned above are still needed in a CNC system, but others may be added, e.g. to couple the control to a DNC system.

2.1. Input

It is assumed that the input to the system is in the form of control tape data, such as post-processor output from programs such as 2CL run on the UNIVAC 1108 at NEL. This will normally be read in from a tape reader. Alternatively CNC may be one element of a DNC system when the post-processor output will be fed directly and not through the intermediary of punched paper tape.

A CNC system can be adapted easily to accept control tape data in any standard code such as EIA or ISO.

3.   SOFTWARE

The basis of the NEL CNC system software is a small highly efficient modular program which at present provides up to $2\frac{1}{2}$ axis motion in about 4k bytes. The system is designed to implement real-time data acquisition, and control, and to resolve the problems of supervising parallel processes which may subsequently arise in its interpretation.

These parallel processes comprise separate sequential sub-programs which may have to run simultaneously and asynchronously. It is necessary therefore, to adopt a method of organising, sharing and protecting common data and any interrupted or suspended operations associated with it. One way of realising this is to use the principles developed by Dijkstra to indicate, by a semaphore, when an operation may, or may not, be performed on relevant buffers. A binary flag which can take only the values 1 or 0, indicating 'active' or 'inactive', will accomplish this.

Program logic is organised in two broad classes of activity: foreground and background operations. These two levels of control may be distinguished by regarding all control aspects as foreground, or high-priority tasks, all else being background. It is then convenient to consider the executive routine as a background processor which has as its basic decision rule, after all initialisation has been completed, that the overall sequence of operations will be the following.

(1) UPDATE system status.
(2) INTERPRET next control block.
(3) SEQUENCE computed control fuctions.

This order of operation also resolves the problem of system initialisation.

The significance of the basic rule is that the system will always have a 'current' and 'next block' in process. Furthermore, the interlocking character of the semaphores, status words of the system and interrupt structure can ensure fast reliable operation on more than one machine tool at the same time.

The foreground activities of the system comprise all the control and associated operations required by the system function handlers. The most important of these are the axis control routines, which may be configured on a point-to-point basis or as a multi-axis

contouring system. The latter will require linear and circular interpolation between pairs of axes. From the system module point of view the two modes described are rather similar. In both cases the background operations will convert control tape data into appropriate control values and store these in the control buffers. The process of setting up control indicators is therefore independent of the manner in which they will be interpreted. In this way it is possible to locate all the rules for control interpretation in the relevant control routine which simplifies the task of modular construction.

Control is implemented, for the current block, by a secondary level of indicators, or flags, forming the 'demand' and 'inhibit' status words of the system. A sequencing operation may then have regard to current demands and attempt to satisfy them. Moreover, it will be possible to maintain a dynamic machine state table from the inter-relationship of status flags and to service more than one machine tool with only one common control program.

Inhibit or 'go' flags determine when a particular system function may be initiated. As the response-time ratio of computer to machine tool may be of the order of 1000:1 or more, it is sensible that demands could be satisfied by decrementing current demand in a number of discrete steps. This continues until the demands in the current block are satisfied. For a

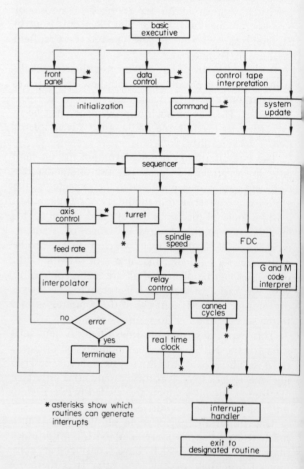

Figure 1. System organisation (software)

system with a servo resolution of 0.0001 in operating at 180 in/min rapid traverse, one of these steps will be decremented in 8.5 ms. The 'next block' is then shunted in and the process continues.

All background processing is supervised by the system Executive which consists of six major parts as illustrated in figure 1. The sequencer will be seen to be nodal in controlling foreground operations.

The Basic Executive is involved in the following classes of activity.

3.1. Initialisation

This comprises setting up parameter tables, initialising various control values, especially modal settings, e.g. assuming input data to be absolute coordinates until reset for incremental mode by the appropriate (G91) preparatory code. Initialising also involves flushing all system and control buffers and generally advancing the system to await start-up.

The parameters used are listed in table 1 and contain critical values relating to the functional specification of the machine tool being controlled. It is then

TABLE 1
Parameter table

no.	parameter	use
1	maximum x-axis travel	limit calculation
2	maximum y-axis travel	limit calculation
3	maximum z-axis travel	limit calculation
4	maximum spindle speed	over-ride limit
5	maximum feedrate	over-ride and ramp up/down
6	ramp slope	ramp-up/down
7	system increment in 10^{-4} inches	increment step output
8	number of tools in turret	turret control

possible to generalise the control program for any type of machine or machine tool to which computer numerical control is applicable. The control program may thus be easily reconfigured for any acceptable type of machine tool having regard to the type of control for which the basic system is configured. It is intended that this will be the basis of a mixed point-to-point and contouring capability where each machine tool would have its own parameter table.

3.2 Initialising and Priming the INTERRUPT Handler

The interrupt response code is set to an initial state which changes dynamically as the program executes. Interrupt code must be quite brief because the system interrupt response time must be taken into account when reckoning total response time. In the present system (February 1972) the time to commencement of interrupt procedure is minimally $22\mu s$.

The interrupt handler has to ascertain the entry point of interrupt response code from a hardware generated pointer which is also subject to hardware defined priority levels. It is therefore possible, and even likely, that an interrupt may itself be interrupted by a call of higher priority. In the present system up to 8 primary levels and 64 secondary levels are available.

3.3. Control tape interpretation

Control tape interpretation and the maintenance of a circular buffer of control data is a central feature of system operation; see figure 2.

The system accepts punched paper tape post-processor output coded in ISO standard format. Interpretation is carried out by a character search routine which, on recognising one of the conventional characters comprising blocks of control data, jumps via a

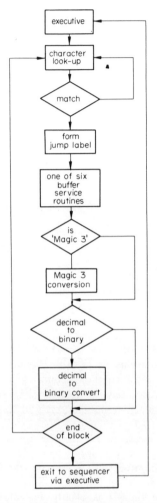

Figure 2. Control tape interpretation.

parallel word-address dispatch table to the appropriate service routine. The multi-digit number associated with a given character (table 2) is converted from 'Magic 3' form if necessary and then from decimal to binary. The relevant demand flag is then set and the control value transferred to the appropriate control buffer. The difficulties of zeroing control values is therefore avoided as only flagged buffers are accessed.

The circular buffer is replenished character-wise, thus reducing the intermittent load on processing time when tape reader input is in units of blocks.

3.4. UPDATE

UPDATE occurs at the end of 'current block' operations at which point 'next block' is updated to 'current block' status. Control then passes to the Sequencer which attempts to satisfy current demands as reflected in the demand status word. A scanning process, tied to another parallel dispatch table, automatically directs operations to the appropriate control routine.

TABLE 2
Word addresses in a full block

word header character	followed by	meaning
N	three digits	block sequence number
*		TAB character
G	two digits	preparatory function code
*		TAB
X	sign and five digits	absolute X coordinate
*		TAB
Y	sign and five digits	absolute Y coordinate
*		TAB
Z	sign and four digits	absolute Z coordinate
*		TAB
R	sign and four digits	absolute Z coordinate for CYCLE rapid move
*		TAB
F	three digits	feedrate in 'Magic 3'
*		TAB
S	three digits	spindle, speed in 'Magic 3'
*		TAB
T	two digits	turret selection 01 to 08
*		TAB
M	two digits	miscellaneous function code

The Sequencer is capable of detecting the end-of-block operations at which time control is returned to the Executive where the cycle of updating and interpreting a new next block continues. The Sequencer also has the capacity to generate 'factory data' if it cannot hand over to a control routine because these either are not required or are busy and therefore cannot return to the executive. When this occurs, and no other operations are possible, the system can output the present state of the machine on punched paper tape. A factory data collection (FDC) block will contain the machine number, elapsed time since last FDC block, current sequence block number, whether it is cutting metal, if it is stopped and why, and the job number from the control tape. The accumulated output can be processed off-line to produce a utilisation report or to maintain a workshop scheduling system data file. It is clearly feasible to output, directly to an off-line storage device, such as a disc or drum, where real-time scheduling can be in operation.

3.5 System displays

The maintenance of system displays is also delegated to the Executive. The conventional 'block sequence number' and the coordinate of a selected axis are displayed on numerical indicators. Other front panel options are also processed before 'update' so that

TABLE 3
Command language vocabulary

ASSIGN, n	Loads control tape on to disc and assigns to machine 'n'
START, n	Starts operations on machine 'n'
STOP, n	Immediately stops machine 'n'
SUSPEND, n	Stops machine at end of current block as when operator wishes to intervene Operation commences on START
TEST, n	Generates status report on machine 'n' and prints-out on the teletype
ABORT, n	Causes designated machine and program to terminate completely

front panel actions, with the exception of STOP, are honoured when the current block is completed.

3.6. Multi-machine working

When more than one machine tool is being controlled it is necessary to store control tapes or access control data from a supervisory computer. It is also desirable that the status of individual programs be interrogated and error conditions indicated to the operator. For this purpose a simple COMMAND language and teletype facility, with a vocabulary of six words will suffice; see table 3.

4. FOREGROUND PROCESSES

The main aspect of foreground processing is axis control. Each control routine is entered from the Sequencer if the appropriate demand flag is set. A control routine will accept a call in one of several ways.

If the axis is active the call will be reflected, otherwise it will accept and sense whether the call is an initial one, or whether the axis is in ramp-up/down mode when it will output a proportional increment. If the axis is operating at programmed feedrate it will output a full increment or step, the magnitude of which will depend on the system increment, for example 0.025 in for a system resolution of 0.0001 in. On decrementing the step to less than 10^{-3} in, the control program resolution (for 0.0001 in transducer resolution), an interrupt is generated and the increment buffer is replenished at high speed.

Control routines are responsible for deleting their own demand flags on completion of an axis demand, and for setting inhibiting conditions on other parts of the system. In a point-to-point system, for example, x- and y-axis motions will not be permitted when z-axis motion is taking place. There is also a considerable problem of ensuring sufficient time for relatively slow mechanical operations to take place. A real-time clock facility is primed to generate an interrupt at the end of the elapsed programmed time so that these can be taken into account.

Feedrate is computed in terms of 64 steps where '0' equals 'stop' and '63' equals 'rapid traverse'. The ramp-up/down function is also carried out in the feedrate subroutine. A test is carried out in ramp-up to ascertain whether programmed feedrate can be reached. If it cannot, feedrate is automatically ramped down. The slope of the ramp function is determined by a parameter in the parameter table. Feedrate over-ride (0–120%) is also provided for in the conventional manner.

Spindle speed and turret control are serviced by their own routines and are more or less conventional. When spindle speed over-ride is programmed it will only be implemented if the proportion permits escalation to the next speed step. In all cases the maximum spindle speed and feedrate are limited to maximum values in the parameter table.

5.   CONTROL SYSTEM HARDWARE

In a CNC system it is possible to provide a wide range of options for power drive units and position-measuring devices without calling for any basic change in software.

The simplest option is that shown in figure 3 where the output of the computer is a velocity demand

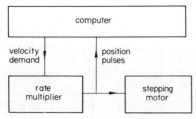

Figure 3. Stepping motor control.

which, by means of a rate multiplier, controls the rate of pulses supplied to a stepping motor. While feedback of pulses into the computer provides positional information, these pulses do not come from a measuring transducer, and so the control is in effect 'open-loop'.

Another relatively simple system is that shown in figure 4 where the position servo loop is closed

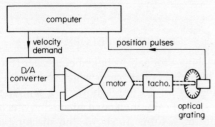

Figure 4. Servo loop closed directly through the computer.

through the computer, in this case by the use of position pulses from an optical transducer. Such a system is quite adequate where the frequency of the pulses is such that the computer can deal with them with reasonable time to spare. The system shown has a rotary encoder fitted to the machine lead screw, alternatively a linear optical grating could be used. The pulse frequency depends on the maximum speed of movement and on the size of the transducer increment. For example, for a movement along one

axis at 180 in/min with a position increment of 0.0001 in a position pulse will arrive every 33 μs. As several axes may have to be serviced simultaneously, it will be seen that the fast operating time of the computer is essential and even then buffer electronics will be needed.

One arrangement of such buffer electronics is shown in figure 5. This is the system used at NEL in the developments on a Herbert Co-ordatrol drill described below. In this the velocity demand from the computer is converted to a train of pulses by a rate multiplier. These are added to an up/down

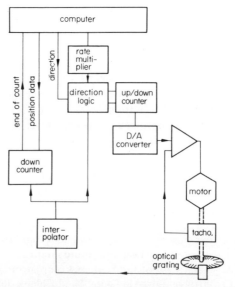

Figure 5. Block diagram of control of one axis of Co-ordatrol drill.

counter whose count is decremented by pulses from an optical transducer fitted to the machine lead screw. The output of the up/down counter controls the servo motor via a digital-to-analogue converter. However, the transducer pulses are also fed to a down counter which is set initially from the computer with the required positional step. When this counter reaches zero a pulse is sent to the computer requesting the output of a new positional step and, simultaneously, the rate at which this step is to be made.

In a point-to-point control system a change of position called for by a new block of information on the tape will be broken down into steps, each con-

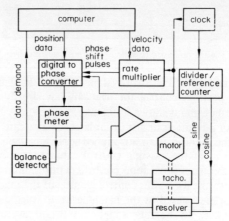

Figure 6. Block diagram of control system using resolver position feedback.

sisting of several increments of the position measuring transducer. Considering only these steps, the system provides closed loop control because a new step is not initiated until the transducer on the output has signalled that the previous step is complete. However, within one of these steps the system is open loop.

A rather similar arrangement is shown in figure 6 but in this case the position feedback is from a resolver used in time phase mode. The output of the rate multiplier goes to a digital-to-phase converter. Each input pulse causes a change of time phase corresponding to a position increment. The converter can be arranged to act effectively as a counter, making a separate counter unnecessary.

It is assumed in these systems (figure 3–6) that the error signal to the servo amplifier will be compared with a signal from a tachogenerator providing velocity feedback. This is suitable where a lead screw is used. One possible alternative is to use a hydraulic ram, controlled by an electrohydraulic valve. In this case ram velocity is approximately proportional to current input, and a velocity feedback signal is not needed.

6.   THE NEL CNC PROTOTYPE

A prototype CNC system developed at NEL to test the software and some of the hardware options which have been described is based on a Herbert Co-ordatrol drill. It is possible to use it to check out the basic principles of the software/hardware module.

A general view of the Herbert Co-ordatrol and its CNC controls are given in figure 7 and a block diagram

Figure 7. Co-ordatrol drill with NEL system fitted.

of the modified control system is shown in figure 8. The position control for one axis has already been described and is shown in figure 5.

An advantage of the Co-ordatrol as a research tool for the modifications introduced at NEL is the ease of removal of the existing control cabinet, tape reader and the main servo loop controls without interfering with the heavy power control circuits of the machine, which are housed in the machine itself. Such aspects are likely to be increasingly important as retrofit action increases. The control cabinet was replaced by one containing the minicomputer and its interface

logic. A new high-speed tape reader has been provided to make it easier to develop contouring facilities.

The principal alterations to the machine were to the servo motors and position feedback transducers. On the x and y axes the synchros used for position measurement were replaced by radial optical gratings fitted to the lead screws. The a.c. servo motors were replaced by d.c. servo motors. Low-power servo motors are adequate on these axes because the

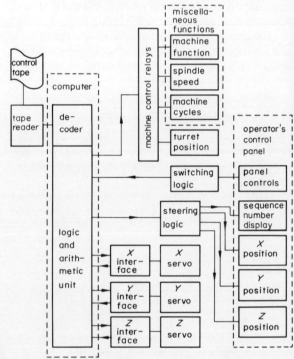

Figure 8. Block diagram of modified control system for Co-ordatrol drill.

Co-ordatrol table has air-bearing slides which are clamped while drilling takes place.

NEL's expertise built up over many years in the fields of optical grating transducers and NC proved invaluable.

7.   FUTURE DEVELOPMENTS

The machine tool and its linked controller form only a part of a chain of elements which make up a manufacturing system. Figure 9 shows in a simplified way a type of vertically integrated hierarchy which these elements may form. It shows the machine and its controller acting as a satellite in a very simple DNC system.

The feedback of information about machine performance to a supervisory computer is one of the more obvious developments of a CNC system. Another is the creation of adaptive control by using monitoring devices to provide data on cutter performance for the control of feeds and speeds, to keep to a minimum the time lost when not cutting metal and to optimise the time of cutting according to dynamic scheduling rules.[7] Interaction between inspection equipment and

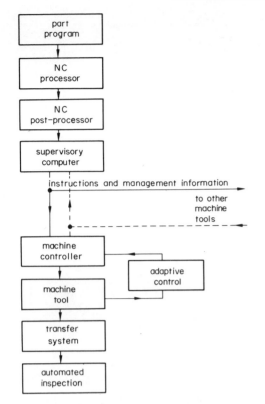

Figure 9. The NC machine tool in an NC manufacturing system.

the machine controller can provide in-process and post-process control. Much work remains to be done not only on the hardware of the monitoring and inspection devices but, in addition, on the computer software needed to coordinate and optimise the performance of the various elements of the system. Once again a modular approach is needed so that the hardware and software can be readily tailored to fit the exact requirements of the machine tool system to be designed and specified.

It can be argued that a far greater increase in productivity may be expected by improving the management of a machine shop than by increasing the efficiency of the machine controls themselves; this argument does not invalidate the need for CNC. The workshop of the future is likely to rely heavily on computers as an aid to efficient management and there must be close interaction between the supervising computer and its dependent machine tools. The presence of a small computer in the machine controls, if not obligatory, will prove a great asset in making this close interaction a reality.

ACKNOWLEDGEMENTS

This paper is presented by permission of the Director, National Engineering Laboratory, Department of Trade and Industry. It is Crown copyright and reproduced by permission of the Controller, H. M. Stationery Office.

REFERENCES

1. C. BARNETT and B. DAVIES, 1971. 'A review of developments', *Symp. on Comput. Control of Machine Tools, Glasgow*, Paper 1.
2. P. G. MESNIAEFF, 1971. 'The technical ins and outs of computerised numerical control', *Control Eng.,* **18**, No. 3, 65–84.
3. J. E. TORJUSEN, 1971. 'Kongsberg computer numerical control', *Symp. on Comput. Control of Machine Tools, Glasgow*, Paper 9.
4. C. DELGRASSO and P. ZATTORI, 1971. 'A numerical control designed around an internal minicomputer', *1st World Meeting on Machine Tools, Milan.*
5. C. J. CHARNLEY and S. SRINIVASAN, 1971. 'The design and development of a numerical control system using a small digital computer', *12th M.T.D.R. Conf.,* Macmillan, pp. 433–438.
6. J. ROSENBERG, 1971. 'The trouble with DNC', in: *NC: 1971. The Opening Door to Productivity and Profit* (Proc. 8th Ann. Conf., 1971) (Ed. Mary A. De Vries). Numerical Control Society, Princeton, N.J.
7. E. H. FROST-SMITH, 1971. 'Optimisation of the machine process and overall system concepts', *Ann. C.I.R.P.,* **19**, 385–94.

A CONVERSATIONAL MODE FOR DIRECT NUMERICAL MACHINE TOOL CONTROL

by

J. FRISCH*

SUMMARY

A direct numerical control metal processing facility using a 12-K minicomputer and peripheral equipment is described. The conversational computer language FOCAL has been expanded to include various machine tool performance commands which are applied to manufacturing equipment from a standard input terminal. Examples are presented of two- and three-dimensional contours to illustrate the simplicity of programming for a conversational mode.

NOTATION

Program command abbreviations

A   ASK: user gives values to define each variable.

C   COMMENT: information line only.

D   DO: execute and return to line following DO command.

F   FOR: command following, is executed at each new value.

G   GO: starts indirect programme at lowest numbered line number.

I   IF: standard programming command for control transfer.

P   PERFORM: generates machine motions, cutting routines, CRT displays.

Q   QUIT: returns control to the user.

R   RETURN: terminates DO subroutines, returning to the original sequence.

S   SET: defines identifiers in the symbol table.

T   TYPE: evaluates expressions and types out = and result in current output format.

General symbols

F linear feedrate (in/min).

R rotation (deg $\times 10^{-2}$).

X,Y,Z coordinates of positive cutter motion (in $\times 10^{-3}$)

INTRODUCTION

During the past few years the most important advances in the numerical control of machine tools have been the change from the use of NC tapes and controllers to direct numerical control (DNC) or direct computer control. Designs and equipment configurations to achieve this method of computer-aided manufacturing have been described by Spur and Wentz[1], G. S. Nann[2], Charnley and Trinivasan[3]. The particular details of various commercially available systems have been recently tabulated and compared by Mesniaeff[4]. The major feature is essentially the on-line connection of NC machine tools to a process computer, which permits the sending of control data directly, thereby avoiding the now conventional mode of using tape. Two advantages thus gained have

also been described by Parsons[5]; they are real-time programme editing and handling of information necessary for management control. Both aspects which are essential to optimising process performances are further enhanced by the use of closed-loop adaptive control systems. This permits wider latitude in required accuracy levels since adaptive control can be used for corrective actions, for example, during the cutting process.

Most of the DNC systems cited in the literature are large machine tools connected to computers with largescale core storage. With the rapid reduction in the cost of minicomputers, it was of interest to the author to establish a direct numerically controlled machine tool facility, using a small computer to control the manufacturing process. A numerically controlled milling machine and lathe, without controllers, were selected as most representative of general-purpose machine tools. Since the actual cutting-time is only a small fraction of the total part time in manufacturing, the use of a small computer for other than DNC purposes is an important consideration in the economy of small plants.

Having multiuser capability for varied digital computations within the minicomputer system was therefore a factor in selecting for our purposes a 12K-Core PDP-8e computer (Digital Equipment Corp.). The computer language[6] FOCAL provided by the manufacturer was found to be easily adapted to changes for a conversational mode necessary to give direct numerical machine tool control. A P-command with additional coding for performance of various machine tool functions was added to the software of FOCAL and for separate identification the name was changed to UC/DNC/L (University of California/ Direct Numerical Control/Language). User programmes, which are shown later, can be more easily prepared with the diagnostic message ability of FOCAL. Programme changes and modification can be made on-line when either the milling machine or the lathe motors are connected through their power supplies to the computer.

The user programmes presented here are examples of the brevity and simplicity of software requirements for the cutting of relatively complex shapes. With minor modifications these programmes can be used to

* Department of Mechanical Engineering, University of California at Berkeley

provide CRT-display or automatic drafting of the work-piece on available peripheral equipment. Through the use of analog/digital equipment it is possible to prepare programmes which permit adaptive control and on-line process modifications by means of the interactive conversational language available in direct machine tool control.

EQUIPMENT

The main components of the equipment used for the preparation of DNC programmes and manufacturing are given in figures 1 and 2. As shown in figure 1, the

Figure 1. DNC computer console with tape readers and CRT-display.

computer console contains the 12 K-Core PDP-8e computer with extended arithmetic and a real-time fixed interval clock. The latter is being replaced by a programmable clock which should make it possible to obtain further improvements in smooth cutting operations. The preparation and loading of pro-grammes, in addition to a standard teletype, is also accomplished through the high-speed tape reader/punch or the magnetic tape deck. When the DNC-system is not in use additional teletypes are available for time share FOCAL users.

Programme text or configurations of parts before DNC-machining can be viewed on the Textronix 611 Display scope shown in figure 1. A Gerber 600 automatic drafting machine, available in the laboratory, is used for making part drawings from tape generated with the DNC programming language. For feedback control of metal processing variables a 10-bit analog/digital converter with 8-channel multiplexer is incorporated in the computer system.

The components shown in figures 4 to 7 were cut on the DNC-vertical turret milling machine (Bridge-port Model B12-2J) shown in figure 2. The 9 x 42 in table, cross-slide and knee are moved by SLOSYN stepping motors in 0.001 in steps. A 12-in rotating table with step motor, is also available for DNC operations. The positive X, Y, Z coordinate axes of cutter motion relative to the workpiece, according to

accepted practice[5], are shown in figure 3. The stepping motors are actuated by either one of two performance commands, P MOV(X,Y,Z,R) to move the table (and workpiece) into a desired position and P CUT (F,X,Y,Z,R) which is the cutting routine. Both these commands as well as others are explained later. The quill with a 10-station turret is located in the drilling and boring head of the machine. A command P QIL is used for quill and tool motion as well as turret rotation. However, the control box shown in figure 2 provides manual override for these machine operations.

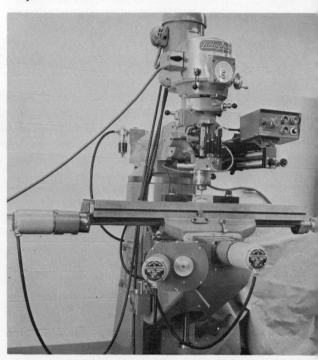

Figure 2. NC-milling machine showing DNC cutting of three-dimensional surface.

Also available for metal cutting and research is an engine lathe retrofitted for direct numerical control operation. Such retrofitting considering the use of the existing DNC-language is relatively easy, since the same command routines are used for pulsing of stepmotors, regardless of the machine tool configuration.

COMPUTER PROGRAMMES

The interpretative direct numerical control UC/DNC/L language which is used to programme and operate the machine tools is based on the FOCAL language extended for use on the PDP-8e computer. In the present configuration 8 K-FOCAL is located in fields 0 and 1, and the additional functions are placed in the second field.

The essential feature of the modification was the replacement of the FOCAL Library command L with a new command P for Perform, which effects the linking to the appropriate routine in the second field. Since the system has both on-line and off-line simulation capability, the information generated by the user programme may be applied either directly to manufacture a part, or may be put on tape for verification on a NC-drafting machine or a display scope. In the latter case the programme can, subsequently, be read in by the computer for operating the machine tools.

For use of the programming language to operate the DNC-milling machine, the following performance routines are available as P-addressable functions:

(1) P MOV(X,Y,Z,R). This is a utility routine for moving the milling machine table to an initial position before processing routines are utilised. The X, Y and Z variables are in mils and the rotation command R in hundredths of a degree for the step-motor-operated rotating table. For example, a command P MOV (1000, 1000, 1000, 100) will move the machine table 1 in in the positive X,Y,Z directions defined by coordinates shown in figure 3 and, give a one degree motion to the rotating

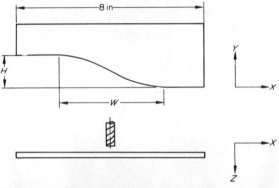

Figure 3. Template with sinusoidal contour showing positive X-Y-Z coordinates of cutter motion.

table. All values of X,Y,Z,R represent the desired increment of motion; this routine will not, however, necessarily move the table in a straight line. If straight line motion of the milling machine table is essential, the following cutting routine should be used.

(2) P CUT(F,X,Y,Z,R). This is the routine used to control machine tool motion while cutting takes place. In the case of the milling machine, the feedrate F can be set between 0.125 and 12 in/min with the vertical knee motion Z at 1/10 of the set value. X, Y, Z and R are absolute coordinate values relative to the specified origin when this routine is initialised, and are given in mils for X, Y, Z and 1/100 degree for R.

(3) P QIL (argument) is the routine which controls the motion of the tool mounted in the milling machine quill as shown in figure 3. There are three arguments used in the routine. D puts the quill down; if it is already down it remains there. U puts the quill up and R which rotates the quill turret to any of the 10 possible positions. If the quill is down, it is raised before the turret rotates and left in the up-position. A delay of 892 ms has been programmed so that the quill control mechanism will have enough time to complete the desired motion.

(4) P DWL (time) is a routine which causes a delay or dwell of time specified in milliseconds. Computation is suspended during the specified time period.

(5) P PQS (argument) is a dual routine to QIL.

Instead of controlling tool and quill motion, it generates a tape which can be verified on a Gerber 600 automatic drafting machine and may also be read back by another programme to operate in a simulation of NC machining. There are four arguments in this routine. OD puts the drafting pen or quill down; OU puts the drafting pen or quill up; OR which in the quill routine rotates the turret is ignored by the automatic drafting machine; OS is the stop command.

(6) P PNC (F,X,Y,Z,R) is a dual routine to CUT to generate a tape which can be verified on a Gerber 600 automatic drafting machine. To initialise the routine as in CUT, the feed F is set to 0, rather than the 0.125 to 12 in/min feedrate. If $F = -1$, the resulting tape is used for plotting purposes and feedrate information is not given on the tape. With regard to the X-Y-Z-R increments, a zero value is not punched out and R is ignored by the drafting machine.

(7) P VUE (argument) is a routine which generates a display of the programmed part. The oscilloscope or Textronix 611 display scope shown in figure 1 are used respectively for either point-to-point or complete figure display. The argument in the routine uses the increment values of the CUT routine divided by a scale factor, to show the part at the desired size.

The above routines are basic for most standard machine tool processes and verification methods. However, depending on available computer storage, additional P-routines can be developed. For example, a routine for engraving alphanumeric characters and a drilling routine in connection with P QIL are currently being completed.

The milling machine and automatic drafting machine operate on incremental position information instead of absolute coordinates. The CUT and PNC programmes, described above, have been designed for absolute coordinate information input and since $n+1$ coordinate values are required to obtain n increments, both routines must be initialised at least once during programme execution.

While using the above programmes it must be realised that this interactive language is an open loop-incremental control scheme. If there is no feedback from the machine to the computer, the programmes are designed to recognise completion of the routines. Commands to the stepping motors are timed by an internal clock in the computer so that enough time is available for the motors to complete one step before the next stepping pulse arrives. However, closed-loop control can be established with peripheral equipment to monitor machine tool performance. Occurrences such as excessive cutting forces or jamming, for which the computer would continue to send stepping pulses, can then be successfully avoided.

The following user programme generates, on the DNC milling machine, the template with a sinusoidal contour as shown in figure 3. The required input information is the machining accuracy (step), the cutter radius and the width and height of the sinusoidal portion of the profile. Virtual and actual cutter radii are given to the cutting routine for rough and final machining of parts.

```
Ø1.Ø5  C  MACHINE A SINUSOIDAL PROFILE ON AN 8 IN LONG PLATE.
Ø1.1Ø  A  "STEP"STEP,"CUTTER RADIUS"CR,"WIDTH"W,"HEIGHT"H
Ø1.15  S  X=Ø.Ø; S Y=Ø.Ø; P CUT(Ø,X,Y,Ø,Ø)
Ø1.2Ø  P  CUT(2,−42ØØ,H−CR,Ø,Ø); P QIL(ØD);A "QUILL DOWN"DN
Ø1.25  S  X=−2ØØØ; P CUT(4,X,H−CR,Ø,Ø)
Ø1.3Ø  F  X=−2ØØØ, STEP,−2ØØØ+W; D 2.Ø
Ø1.35  F  X=−2ØØØ+W, STEP, 42ØØ; P CUT(2,X,−CR,Ø,Ø)
Ø1.4Ø  P  QIL(ØU); A "QUILL UP"DN
Ø1.45  Q

Ø2.Ø5  S  XA=X+2ØØØ; S Y=(H/2)∗FCOS((3.1416/3ØØØ)∗XA)+(H/2−CR)
Ø2.1Ø  P  CUT(4,X,Y,Ø,Ø); R
```

The P-CUT commands generate the straight part followed by the sinusoidal contour and another straight portion to complete the 8 in long overall contour.

The sprockets shown in figure 4 are produced with the user programme shown below. The information required to change to different sprocket sizes are requested by the programme through the ask routines in lines 1.05, 1.10 and 1.15 for pitch, accuracy (number of segments) and number of teeth, respectively.

by giving a new slot number as requested in line 1.04. The inside and outside radii of the disc can be changed in line 1.06 and rough or finish cutting is accomplished by typing virtual or actual cutter diameters requested in line 1.08.

Figure 5. Slotted disc produced by DNC-cutting. Required I/O information is cutter radius, number of slots and incremental steps.

Figure 4. Sprockets produced by DNC-cutting. Required I/O information is number of teeth, pitch and incremental steps.

The user programme opposite is for slotted discs, such as the one shown in figure 5, used in optical equipment. Production with different numbers of slots using the same programme can easily be achieved

```
Ø1.Ø1  C  MACHINE A SPROCKET.
Ø1.Ø5  T  !,"PITCH IN INCHES";A P;S P=P∗1ØØØ
Ø1.1Ø  T  "NUMBER OF SEGMENTS";A NO
Ø1.15  T  "NUMBER OF TEETH";A NU
Ø1.17  S  PI=3.14159265
Ø1.2Ø  S  TH=2∗PI/NU;S YØ=P∗FCOS(TH/2)/2/FSIN(TH/2)
Ø1.25  F  I=1,NO+1;D 2
Ø1.35  P  CUT(Ø,Ø,Ø,Ø,Ø);P CUT(6,X(1),Y(1),Ø,Ø);A KO
Ø1.36  C  ABOVE INITIALIZES AND MOVES TO 1'ST PT
Ø1.4Ø  F  I=1,NU;D 3
Ø1.45  A  KO;P CUT(6,Ø,Ø,Ø,Ø);C MOVE TO ORIGIN & STOP
Ø1.47  T  "END"!;Q

Ø2.Ø5  S  OM=(I−1)∗PI/3/NO;C AT IMAX, OMMAX=6Ø DEG
Ø2.1Ø  S  X(I)=P∗FCOS(OM)−P/2
Ø2.15  S  Y(I)=YØ+P∗FSIN(OM)

Ø3.Ø5  S  PH=(I−1)∗TH;S X1=FCOS(PH);S X2=−FSIN(PH)
Ø3.1Ø  S  X3=−X2;S X4=X1;COMPUTER ROTATION MATRIX −− X1,X2,X3,X4
Ø3.15  F  J=2,NO+1;D 3.97;CUT FROM BOTTOM TO TOP OF TOOTH
Ø3.2Ø  F  J=1,NO;S K2=NO−J+1;D 3.9; CUT FROM TOP TO BOTTOM OF TOOTH
Ø3.25  R
Ø3.9Ø  S  XX=X2∗Y(K2)−−X1∗X(K2);S YY=X4∗Y(K2)−X3∗X(K2);D 3.99
Ø3.97  S  XX=X1∗X(J)+X2∗Y(J);S YY=X3∗X(J)+X4∗Y(J);D 3.99
Ø3.99  P  CUT(6,XX,YY,Ø,Ø);CUT TO XX,YY
```

```
∅1.∅1  C  MACHINE A SLOTTED DISK.
∅1.∅2  S  X=∅; S Y=∅; S Z=∅;P CUT(∅,X,Y,Z,∅)
∅1.∅4  A  "NUMBER OF SLOTS?" NUM
∅1.∅6  S  RO=656∅/2; S RI=4∅∅∅/2
∅1.∅8  A  "DIAMETER OF CUTTER IN THOUSANDS?" CD; S CR=CD/2
∅1.1∅  S  R=RO+CR;P CUT(8,R,∅,∅,∅)
∅1.12  P  QIL(∅D)
∅1.14  F  TH=∅,5,36∅; D 1∅; CUT OUTSIDE DIAMETER
∅1.16  P  QIL(∅U)
∅1.18  S  DTH=18∅/NUM; S TH=∅; S ANG=∅
∅1.2∅  S  R=2∅∅∅+CR;D 9;P QIL(∅D);S R=3∅∅∅−CR;D 9
∅1.22  S  A=TH+DTH; S T=TH+DTH−(CD/R).∅174533; F TH=TH,T; D 9
∅1.24  S  TH=T; D 9; S TH=A;S R=2∅∅∅+CR;D 8
∅1.26  S  A=TH; F B=∅,2,C; S TH=A−B; D 8
∅1.28  P  QIL(∅U);S ANG=ANG+2∗DTH; S TH=ANG
∅1.3∅  IF (ANG−36∅) 1.2∅,1.2∅, 6.1∅

∅6.1∅  Q

∅8.1∅  S  X=R∗FCOS(TH∗.∅174533)+CR∗FSIN(TH∗.∅174533)
∅8.2∅  S  Y=R∗FSIN(TH∗.∅174533)−CR∗FCOS(TH∗.∅174533)
∅8.3∅  P  CUT(6,X,Y,∅,∅)
∅8.9∅  R

∅9.1∅  S  X=R∗FCOS(TH∗.∅174533)−CR∗FSIN(TH∗.∅174533)
∅9.2∅  S  Y=R∗FSIN(TH∗.∅174533)+CR∗FCOS(TH∗.∅174533)
∅9.3∅  P  CUT(8,X,Y,∅,∅)
∅9.9∅  R

1∅.1∅  S  X=R∗FCOS(TH∗.∅174533); S Y=R∗FSIN(TH∗.∅174533)
1∅.2∅  P  CUT(6,X,Y,∅,∅)
1∅.9∅  R
```

The contour of the fluidic circuit component shown in figure 6 consists of a series of straight lines and circular arcs. Such contours are generated by means of a user programme called OFFSET which, although relatively brief, is too long to present here. However, the essential feature of this programme is the requirement of input data to describe the desired finished shape and to calculate the path of the tool centre to generate the lines and arcs. The input data values are cutting speed, as well as X-Y-Z increments and segment length for each arc or line. Although it is apparent that this particular routine parallels NC machining, the ability to make on-line programme changes and modifications, or more easily to request viewing and plotting routines demonstrates a distinct advantage.

The hyperbolic paraboloid shown in figure 7 was machined using the user programme shown below for the surface equation

$$Z = \frac{Y^2}{4000} - \frac{X^2}{6000}$$

The surfaces were generated from a $3\frac{3}{4} \times 2\frac{1}{4} \times 1\frac{1}{4}$ block with the machining accuracy requested in line 1.09 of the programme.

```
∅1.∅5  C  MACHINE A SADDLE SURFACE.
∅1.∅7  C  BLANK SIZE: X=3.75,Y=2.25,Z=1.25 MINIMUM
∅1.∅8  C  CLEAR CLAMPS.ORIGIN IS AT CENTER OF SURFACE.
∅1.∅9  A  "STEP SIZE" STEP
∅1.1∅  S  X=∅;S Y=∅; S Z=∅;S R=∅;P CUT(∅,X,Y,Z,R)
∅1.15  P  CUT(6,−15∅∅,−13∅∅,∅,∅)
∅1.2∅  P  QIL(∅D); A "QUILL DOWN"DN
∅1.3∅  F  X=−15∅∅,STEP,15∅∅;D 2
∅1.35  P  QIL(∅U);A "QUILL UP"DN;P CUT(7,∅,∅,∅,∅)
∅1.4∅  Q

∅2.1∅  F  Y=−13∅∅,1∅∅,13∅∅; D 3
∅2.2∅  S  X=X+STEP;S Y=Y−1∅∅;S Z=(Y↑2/4∅∅∅−X↑2/6∅∅∅);P CUT(7,X,Y,Z,∅)
∅2.3∅  F  I=−13∅∅,1∅∅,13∅∅;S Y=−I;D 3

∅3.1∅  S  Z=(Y↑2/4∅∅∅−X↑2/6∅∅∅)
∅3.2∅  P  CUT(7,X,Y,Z,∅)
```

The inversion of the programme to machine the bottom surface was accomplished by the statement in line 2.30. However, if this line is omitted, a minus Z in the P CUT commands of line 2.20 and 3.20 would accomplish the same task.

Figure 6. Fluidic circuit component produced by DNC-cutting, with OFFSET programme.

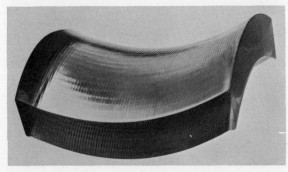

Figure 7. Hyperbolic paraboloid surfaces produced by DNC-cutting. Required I/O information is part size and incremental steps.

Since editing of machining programmes can be accomplished with a modifying M-command while the machine tool is stopped, the choice of simplified programme writing, as in this case, is one of experience.

CONCLUSION

It has been shown that the development of an interactive computer language in conjunction with a minicomputer provides considerable optimisation of the manufacturing processes with DNC machine tool systems.

The computer software has been designed in such a way that minor changes in user programmes result in relatively brief routines for automatic drafting and CRT displays, thereby permitting rapid error detection and programme verification.

The addition of analog—digital equipment to provide closed-loop control will permit on-line process control by using the conversational mode to generate parametric changes requested during machine tool operation.

Comparing the versatility and relative cost of minicomputers with conventional NC machine tool control, it is our experience that a DNC system, with the interactive language presented here, provides economically and technically a more favourable solution to computer-aided manufacturing.

ACKNOWLEDGEMENTS

The author is grateful to Dr. N. Akgerman of Batelle Institute and to Messrs. D. Davis, J. Kohli and F. Melsheimer of the University of California, for their assistance and help during the development of the computer programmes and laboratory facility.

REFERENCES

1. G. SPUR and W. WENTZ, 1972. A system for direct numerical control of machine tools. *12th MTDR Conf.,* Macmillan, p. 421.
2. G. STUTE and R. NANN, 1972. Report on a special DNC system. *12th MTDR Conf.,* Macmillan, p. 429.
3. C. J. CHARNLEY and S. SRINIVASAN, 1972. The design and development of a NC system using a small computer. *12th MTDR Conf.,* Macmillan, p. 433.
4. P. G. MESNIAEFF, 1971. Computerised numerical control. *Control Engineering,* March, pp. 65—84.
5. N. R. PARSONS, 1971. NC machinability data systems. Soc. of Manufacturing Engineers, Int. St'd. Book No. 0-87263-029-3.
6. Programming languages. PDP-8 handbook series, Digital Equipment Corp., 1970.
7. A. D. ROBERTS and R. C. PRENTICE, 1968. Programming for NC machines, McGraw-Hill.

SMALL AND MINILANGUAGES IN NUMERICAL CONTROL

by

R. WEILL*, J. C. MORENS* and M. STRADY*

INTRODUCTION:
THE PROBLEM OF THE LARGE SYSTEMS FOR PROGRAMMING NUMERICAL CONTROL

It is a matter of fact that the first numerically controlled machine tools were designed for contouring work applied to complex shaped workpieces. The preparation of the movements of the tool had therefore to be computed by special programs which were developed for this particular purpose. One of the most comprehensive languages was APT developed at M.I.T. at the beginning of the 60s and supported later by a special organisation: the APT Long Range Program (A.L.R.P.). Other languages were developed by private firms for special applications, but their development could not generally be supported after a certain interval of time[1]. For this reason, the large systems which exist actually belong to one of the following classes,

(1) Languages from the IBM Company, big computer companies, like ADAPT, AUTOSPOT, etc.

(2) Languages supported by industrial or public organisations, similar to A.L.R.P., which were developed especially in Europe as subsets to the APT system: for example, IFAPT, EXAPT and 2 C,L.

At the beginning, this second class of systems was developed as a family of APT-like languages offered to smaller users who could not afford the expenses of a large computer which is needed by the APT language. But the needs of the users are so various that these subsets became gradually larger and larger and incorporated many of the features of the APT language. Although they do not ask for core sizes as big as those for APT, they have to be implemented on large- or medium-sized computers and their processing environment, hardware and personnel became important. All of them have added special features which do not exist in APT, like turning modules, technology subprograms, etc., which are of great value for the user, but increase the size of the system. The trend is to give to these languages more and more capabilities and this effort has its justification in the future implementation of the fully automated manufacturing system. The integration and optimisation of all the functions in a manufacturing system requires a high-level language for the control of the production hardware.

On one hand, therefore, the large systems have a natural tendency to become more and more comprehensive, but, on the other hand, this tendency has many drawbacks which can be related essentially to the following:

(1) The service of a big computing system is generally not available in small companies and the computer cost is high for such users.

(2) The personnel for an efficient use of such systems have to be specially trained and of a high standard.

(3) The maintenance of the system and its reimplementation on new computers are expensive.

For these reasons, other solutions have been investigated in industry. A fast development of small and minilanguages has taken place in recent years and the importance of these new technologies as programming aids for industry has increased continuously in this field. As a first step, the 'small' languages have been developed, after three or four years, on the basis of the use of conventional but smaller computing systems, like IBM 1130. MINIFAPT, PICNIC, MITURN and UNIAPT will be given as examples of languages of this generation.

More recently, since about two years ago, a new class of languages, the 'minilanguages', have appeared on the industrial scene and are making rapid progress in performance and implementation. MINIPROG will be mentioned and evaluated as an example of this class.

This last development is due not only to a reduction of the capabilities of the programming system, but also to the introduction of the minicomputer for some years in industry. New computing systems with memory sizes of 4 K to 32 K have been developed and offered to the customers for reasonable prices. The development of new microcircuit technologies offers higher performances for the computing system and lower costs for the user. Another development has taken place in the field of memory technologies: lithium ferrite memories, dead memories or read only memories, etc. Also, in the software field, new microprogramming technologies have been developed and offer the possibility of implementing big systems on small computers.

But the most interesting advantage of the minicomputer development is probably the possibility of having direct access to information-processing equipment in the workshop itself. Minicomputers can be purchased by a small-size company and their maintenance is straightforward. They are available 100% of the working time for manufacturing personnel who were obliged before to wait for program processing in computing centres and who lost time for error corrections for each program. Even if the installation has a lower performance than big computing systems, an overall gain in time is given by the elimination of waiting and error correction times. The availability of the computer in the workshop is also an excellent incentive for manufacturing personnel who are growing used to these technologies and will be prepared for a more computerised system in the future, especially for the introduction of direct numerical control.

* CIRP, Paris, France.

To illustrate this evolution, examples will be given in the field of the small and minilanguages. An appraisal of the future of these developments will be presented in the conclusion.

2. SOME TYPICAL SMALL LANGUAGES

PICNIC[2]

This small language was developed by P.E.R.A. (Production Engineering Research Association) of Melton Mowbray (Great Britain) for point-to-point programming. About a hundred societies are using PICNIC and twenty of them are members of the PICNIC Bureau Club which processes 20 to 30 programs for workpieces in a month. Programming time is reduced by a ratio of 4 to 40, costs by a ratio of 50%. PICNIC is implemented on computers like IBM 360/25, IBM 1130 (16 K), ICL 1901. It is also used in time-sharing on the Honeywell computer GE 615. Adaptations exist for about thirty machine and control systems.

The processor of PICNIC has modules for input, decoding, geometry, machining and motions. Modules for milling and area clearance can be added. The computer should be of 24–28 K bytes core size with a card or tape reader, a line printer and an on-line tape punching machine. The input language is a subset of 40 APT words. By a concentration technique, PICNIC is implemented with its post-processor on small computers which are of common use in production departments in industry.

The instructions for tooling are identical with those for APT. Special cycles with technological data, for deep drilling especially, are incorporated. For pockets, there are two options, based on a spiral movement and two directions of rotation (CLIMB or NORMAL). It is possible to leave a boss of material in the pocket and to define the thickness of stock for finishing. There are also programming aids like MATRIX and TRACUT, repetitions of figures and selection of points in a PATERN. Calculations can be performed on most arithmetic and trigonometric operations.

MINIFAPT[3]

MINIFAPT was developed by the University of Lyon (INSA) for point-to-point problems and simple milling straight-cut problems. It is implemented on a IBM 1130 computer of 16 K bytes memory size with a disc. The input language is an excerpt of IFAPT and therefore compatible with APT. It includes MACRO for predetermined cycles. The instruction COPY is in the language. The processor is written in FORTRAN and can be implemented on any computer with a FORTRAN compiler. It is written in assembler language for the IBM computer, allowing shorter processing times. The syntactic part is conversational. The CLTAPE is not compatible with that of APT, but standardisation of post-processors is possible for different machines and systems. The structure of the processor is modular, instruction by instruction, and is very easy to adapt to modifications and corrections of errors (card by card). MINIFAPT is a small language which is offered as a complete package

including the processor and the post-processor for a machine tool control system at a reasonable cost for small industries.

MITURN[4]

The system MITURN was developed at T.N.O. in Delft (Netherlands) on the basis of statistical data concerning families of workpieces (group technology). It is well known that industrial components belong to families defined by dimensions, ratios of diameter to length, configurations, etc. It is therefore possible to derive an automatic preparation for the working sequence, the selection of tools, machining conditions, tolerances, machining time and punching of the control tape. The MITURN system is intended for turning operations in numerical control and is implemented for time-sharing use in a computing system.

The programming language uses technological words defining diameters, faces, etc., and their numerical value. The processor derives automatically the machining sequence from the set-up, the final and the initial contours, etc. A part of the program is optional and the programmer is free to choose technological data. Owing to the concise language, the input program is shorter than the corresponding program for APT or EXAPT.

The system includes 10 types of operations with a tool file defined by the programmer. Technological data are chosen in the following order: depth of cut, feed, speed. The depth of cut results from the geometry and rigidity of the workpiece and the tool. The feed is defined by the shape of the chip and the speed by the formulae of Taylor. A check is made to ensure that power is sufficient for machining. Other algorithms apply to the finish in relation to surface roughness. Fixed cycles are defined for machining chamfers at an angle of $45°$, grooves, etc.

MITURN appears as a specific language based on group technology and with a good efficiency in turning operations belonging to predetermined cycles of operations.

UNIAPT[5]

UNIAPT has been developed in the United States by the United Computing Corporation. It is implemented on computers like PDP8 with 12 K memory size and disc units of 64 K to 262 K words. The APT processor has been rewritten using techniques like overlay, condensed intermediate files and disc units as core extension. The post-processor is generated at low cost by a generalised post-processor. Although processing time is longer than for normal computers, the overall time is shorter because time is gained on correction and waiting times.

UNIAPT is a new tool bringing the capabilities of the APT system to smaller users on a minicomputer for a reasonable cost.

3. THE DESIGN OF MINILANGUAGES: AN EXAMPLE, MINIPROG

When the programmer has to prepare the program, he has essentially to execute two functions: (1) to write the formal program, (2) to check the program on the machine tool. The first function has to do with the observation of the machining sequence and machining data. He has to foresee checking operations and the corresponding points on the workpiece. He has also to divide the program into logic homogeneous phases which are easy to check on the machine.

The second function consists of the checking of the machining parameters on the machine: cutting conditions, tolerances, etc., and of the optimisation of the working cycle, i.e. the maximisation of the ratio

$$\frac{\text{cutting time}}{\text{total working time}}$$

This is more important for mass production than for single workpieces.

The efficiency of programming is therefore a direct function of the time needed by the information to return to the operator for correction and varies according to the complexity of the workpiece.

In the case of point-to-point and straight-cut programming, an attempt was made to automate the programming work. There are no difficult or complex geometrical problems, but there are long calculations which are a source of multiple errors. It is possible to use a specific software and to process the data in a computing centre or it is possible to use an automatic programming system, but this latter solution is expensive. In both cases, the time for correcting the errors is too long.

The other solution which is becoming more and more popular consists of a minilanguage which is processed in the workshop on a minicomputer. This system is available at any time for programming and correcting the tape. It is also called an aid for manual programming. The minicomputers used for processing have good capabilities and are equivalent to the conventional office calculators used in the design office.

Minilanguages are normally designed for point-to-point and straight-cut problems. The geometrical calculations are applicable to X and Y coordinates in a horizontal plane. The instructions can be derived directly from the blueprint without passing through a symbolic language.

The minilanguage is a complete facility for tape preparations. It has a basic processor and a service library for the tape preparation. This library rapidly processes the corrections, editing and punching of the tape. A simple vocabulary and a logical syntax are sufficient for point-to-point problems. This simplicity allows the programmer to skip the phase of intermediate programming. For such a system the programmer must be very used to the machine language. It is possible for him to use facilities for duplication, for correction, for listing of the tape to gain time and for automatic operation.

MINIPROG, an aid to manual programming

MINIPROG is a complete system including a minicomputer, a teletype and a processing program. Calculations are executed automatically, but the programmer has to indicate the machine functions. He has also to indicate at the beginning of the program the format of the output blocks. No post-processor is necessary and the system allows programming of different control systems. The MINIPROG system delivers a punched tape which can be used directly on the machine tool.

MINIPROG is specially designed for geometrical calculations in point-to-point work. Geometrical figures, like aligned points, can be defined and transformed geometrically by translation or rotation. The geometrical figures are memorised. In a simple instruction, the programmer defines the parameter for a machining sequence: for example, a drilling cycle for aligned points in a translated system.

The language has a format with addresses and three zones: geometrical description, geometrical transformation and operation. They are defined by symbols which can be retrieved in the program: P for point, L for line, etc. The addresses are X, Y, Z for coordinates, R for radius, etc. A point can be defined by different ways, for example, by the coordinates

$$P1\ X20\ Y30$$

A pattern is defined by the initial point, the angle of the line with OX, the distance between two points, the number of points

$$LIN\ X10\ Y40\ A30\ D15\ N4$$

Geometrical transformations are defined as translations (TRL), rotations (RØT), symetries (SYM). Operations are defined by ØP and EØP, end of cycle. The machine functions, like preparatory functions, speed, feed, etc., are directly punched on the tape.

There are two versions: one with a 4 K core size and another with an 8 K core size for conversational syntaxic programming. They are implemented on minicomputers like PDP 8.

4.   CONCLUSIONS

Large programming systems have been in use for several years and have helped programmers for programming of complex shaped workpieces. But their implementation in industry is difficult because of their high costs of processing time.

Therefore, attempts to develop smaller languages have been made in recent years and more economic systems exist on the market. Nevertheless, smaller companies cannot afford the expenses of a computing centre and the training of specialised personnel. For this reason, and also because new technologies have been developed in the field of microcircuits and microprogramming, a new generation of minilanguages has been designed and is to be implemented in industry. The language and the minicomputer are delivered as a complete package which can be used at any time in manufacturing and allows easy corrective interventions. The processing time is longer, but the whole turn-around time is shorter and the service is cheaper.

This new technology is not only a better way to use numerical control in manufacturing, but also a way to introduce the computer in the workshop and to come nearer to the integrated manufacturing system.

REFERENCES

1. R. WEILL, 1972. 'Current status of N/C languages in Europe', *Intern. Engng. Conf.*, Chicago, Ill., U.S.A.

2. P.E.R.A., 1969. 'PICNIC, NC programming using a small computer', Rep. No. 201, Automation, October.

3. P. MAY and J. RENAULT, 1971. 'Le système MINIFAPT', *Machine Outil,* No. 268, Aug.–Sept.

4. J. KOLOC, 1971. 'MITURN, a computer-aided production planning system for numerically controlled lathes', *TNO, Centrum voor Metaalbewerking,* Delft, Netherlands.

5. J. H. WRIGHT, 1970. 'UNIAPT is APT on the PDP 8 computer', *Proc. 7th Ann. Meet. Tech. Conf. Numerical Control Soc.,* Boston, Mass., U.S.A.

PROGRAMMING A POINT-TO-POINT NC MACHINE FOR CONTOURING OPERATIONS

by

C. HUSEMEYER,* M. S. WONG,* J. L. DUNCAN* and M. C. deMALHERBE*

SUMMARY

This paper reviews some aspects pertaining to the cost justification of point-to-point NC machine tool utilisation and describes two simple methods of extending their use for a particular milling machine to contouring with an acceptable accuracy and surface finish. In both methods, tapes are produced using a time-sharing terminal with access to a general-purpose computer or a 4K word address minicomputer with a teleprinter and tape punch-reader. Examples of components produced by these methods are shown.

INTRODUCTION

Numerical control (NC) is a relatively new form of automation which is generally applied in metalcutting machine tools, although forming, drafting, assembling, wiredrawing, welding and inspection machines are also making use of numerical control systems nowadays.

It was introduced in the late 1940's and has gained great momentum during the past twenty years. The popularity of NC machines does not lie in the fact that they are novel but almost solely in their ability to permit the manufacture of superior products more economically than with manual control or other conventional means of manufacture.

A widely-used numerically controlled machine tool is basically a vertical milling machine and these machines are available with three or more degrees of freedom. The control systems are divided into three general classifications: positioning-control, point-to-point control and continuous-path control. A positioning-control system is only capable of performing operations such as drilling: point-to-point control can perform drilling and straight-line milling operations, generally along mutually perpendicular axes: continuous-path control is used for complex contouring operations and requires some form of interpolation as will be discussed below.

Most NC machine tools used in industry today are of the point-to-point type. The difference in cost between point-to-point and continuous-path contouring machines is considerable so, unless a contouring machine can be well utilised, production and jobbing facilities may have to use point-to-point machines and some other method of machining contours.

There are, however, a number of instances where a limited contouring ability in the control system of a point-to-point machine is useful. Point-to-point machines can be programmed to follow a simple contour, but a vast amount of information is required on the control tape. This paper describes a computerised method for producing a control tape to extend the use of a typical point-to-point machine to contouring without encountering large tape-preparation costs. This does not imply that point-to-point machines should be extensively employed for contouring, but rather that it is possible to perform a limited number of contouring operations of the type illustrated, such as toolmaking.

* McMaster University, Hamilton, Ontario, Canada

FUNDAMENTALS FOR CONTOURING

Continuous-path contouring requires continuous control of spindle movements with respect to the machine table, not only of the start and end point of a movement but of all the intermediate points as well. Careful programming of point-to-point sequences or cutter path motions allows point-to-point, straight-line milling machines to follow almost any curve. The method of filling in data about curved surfaces from a set of point data describing some points on a curve or a function representing a curve is aptly called interpolation. Continuous-path machines have built-in interpolators that apply linear, circular, parabolic or other forms of interpolation for tracing curves. There are two methods of linear interpolation that can be used for contouring with a point-to-point machine: stepwise-interpolation and chordal-interpolation. The choice of method depends on the characteristics of the particular control system. Figure 1 shows some of the differences between normal point-to-point and straight-line continuous-path control of the type that can be used with computer languages such as APT. The four examples are concerned with moving from point A to B for a total movement of say 3 units in the Y direction and 4 units in the X direction (the directions refer to the machine table axes). The first method refers to a machine that has a controller capable of driving the table in only one direction at a time. The tool centre moves from A to B in a stepwise fashion, so that each step is within the required error band. The method would normally require one block of tape information for each short line. The second method refers to a controller that can drive the table in any of the principal axes independently, or simultaneously trace a 45° line (in machines in which the feedrates are not independent). This method would require fewer blocks of tape information than the first mentioned. The third method refers to a control system in which independent (but discrete) feedrates can be selected so that an angle for the tool path can be selected which is close to the required path. This method requires still fewer tape information blocks. The fourth method refers to a continuous-path machine with linear interpolation (for comparison) where the tool centre can trace the straight-line joining any points A and B within the accuracy of the control system: such a system requires only one block of tape information for each line. Any curve could be traced by programming the tool tip to generate a series of straight-line cuts whose lengths depend on the required tolerance.

Since NC machines fall into two basic categories, namely, those that have independent control of the

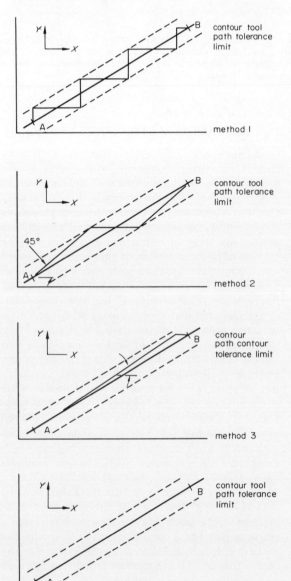

Figure 1. Four methods of defining contour tool path tolerance limits.

principal drive axes and those that do not, there are two methods of generating contours with point-to-point NC systems: chordal, as in the second and third methods, and stepwise-interpolation as in the first method. The basic approach to tape preparation is similar for both methods so only chordal interpolation will be described in detail.

AN ALGORITHM FOR CONTOURING

The method of presenting information to a numerical control system emphasising a block format control

tape is well known. Each block of information contains data pertaining to such parameters as feedrates, coordinates to which the machine table and spindle are required to move, as well as auxiliary data such as coolant, tool changing and the type of operation to be performed, that is drilling, tapping, milling and so forth. The method of preparing the tape is to write a programme that calculates a series of coordinates to be punched into a control tape in accordance with some algorithm, together with the necessary auxiliary data, all in accordance with the block format the machine control system is designed to read. The resulting tape will allow the control system to trace the required contour for the centre of the cutting tool to produce a machined contour within the required tolerance. The computer hardware required is a general-purpose high-speed computer with a high-speed tape punch output: or alternatively an 8K word address minicomputer with a high-speed punch output, a type of computer that is less convenient for it invariably requires a relatively large amount of time to perform multiplications.

Numerical control languages such as APT generate control tapes in two parts: the basic APT compiler calculates all the necessary data such as coordinates, feedrates, spindle speeds and so on, and a 'post processor' assembles the data into block format compatible with a particular NC system. This approach can be used when generating a contouring control tape for a point-to-point machine. A general-purpose computer with a FORTRAN compiler can be used to calculate the necessary data and a suitably programmed minicomputer can be used as a post processor. Transfer of data between computers can be by cards, paper tape or magnetic tape. Two computers are suggested because a general-purpose high-speed computer can be accessed from a time-sharing terminal and also minicomputers are relatively cheap. A two-pass method of producing a control tape is also useful for translating tape codes. The punched tape output of a general-purpose computer is usually in ASCII standard code while the EIA standard code has been adopted for NC machine tools. Alternatively, this problem could be overcome by adding a short machine language routine to the programme for the general-purpose computer to change codes before the tape is punched, but it is far simpler to change codes and assemble the data into a suitable block format with a minicomputer.

The most important part of the procedure for generating a contouring control tape for a point-to-point NC machine is the algorithm for generating the string of coordinate sets to guide the tool centre along the contour within the required tolerance band.

For chordal interpolation, the method involves combining X and Y feedrates so as to approximate the required contour by an oblique chordal line whose slope is chosen to make a relatively-long cut without deviating from the required contour by more than a specified amount. A typical NC system has a positioning accuracy of ±0.001 in and a repeatability of ±0.0003 in and is therefore quite suitable for this type of work when accuracies of ±0.005 in or better are required. An application for this type of chordal interpolation would be in some classes of toolmaking where the final accuracy and surface finish is achieved by hand finishing. The chordal line traced by the tool is allowed to deviate from the required contour by the specified accuracy on either side. The algorithm described below illustrates the principles for producing the basic coordinate and feedrate data for contouring.

The tool path is the path traced by the centre of the tool (end mill) and the slope of the tool path line is determined by the relative X and Y-feedrates. The numerical control system sets the machine feedrates according to the 'tool function code' (USA industry standards) and reads X-feedrate or Y-feedrate depending on the 'miscellaneous function code'. Since the feedrates will change from about 1 in/min to about 44 in/min for a range of values of 'tool function code' (say 1 to 18 arbitrarily set by the control system manufacturer), it is possible to relate the feedrate to the tool function code by means of a polynomial of degree 2 (or higher if greater accuracy is required) using the method of least squares as follows:

$$F = at^2 + bt + c \qquad (1)$$

$$t = \frac{-b \pm \sqrt{b^2 - 4a(c - F)}}{2a} \qquad (2)$$

where

F = feedrate in X, Y or Z axis (in/min)
T = tool function code (integer 1 through 18)
a, b and c = leastsquares coefficients of the polynomial.

Using the relationships of equations (1) and (2) and a suitable programmed computer, it is possible to match the tool path to the first derivative of the equation representing the desired contour in chordal steps and to interrupt each chord at the point where the shortest distance between the tool path and the required contour is greater than the specified accuracy. The principles involved will be clarified by the following description of a test where the tool centre was programmed to follow an elliptical path to machine an elliptical hole in a plate.

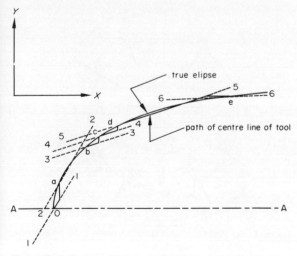

Figure 2. Path traced by tool centre in relation to the true ellipse.

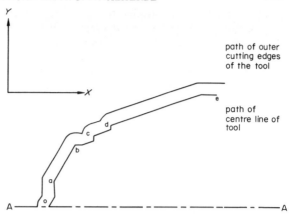

Figure 3. Contour cut by the tool as it moves along the tool centre o, a, b, c, d, e.

The tool employed for cutting an elliptical contour was an end mill having a diameter of 0.50 in, and was used to cut a 0.20 in deep slot in a 0.50 in thick aluminium plate. The resulting slot had an outside oval wall with major and minor axes greater than those of the ellipse by the diameter of the tool. The path traced by the tool in relation to the true ellipse is shown in figure 2, and the actual contour cut by the tool is shown in figure 3. These illustrations are exaggerated by allowing the tool path to wander approximately 0.10 in from the true contour, thus leaving large cusps in the machined slot. When the allowable error is reduced to ±0.005 in, the cusps are less noticeable.

The method by which the contour was traced follows the pattern given in figure 2 where:

(1) Chordal movements of the cutting tool followed the curve

$$\frac{(X - P)^2}{A^2} + \frac{(Y - Q)^2}{B^2} = 1 \qquad (3)$$

where: P, Q are the coordinates of the centre of the ellipse
A is the length of the major axis (in)
B is the length of the minor axis (in).

(2) A value was chosen for the feedrate in the X-direction. This was actually a tool function code number. This X-feedrate remained constant throughout the contouring operation.
(3) The polynomial that was fitted to the range of feedrates and tool function codes was then used to determine the X-feedrate (in/min) and corresponding Y-feedrate, subject to the following conditions, for, although it was desirable to have the slope of the tool path equal to the first derivative of equation (3), there were some limitations.

(a) The maximum allowable slope of the tool path was approximately 1:4. This was because an increase in slope meant an increase in the Y-feedrate (X-feedrate was fixed) and the possibility of exceeding the maximum allowable feedrate along the hypotenuse per cut for the tool and material being used.
(b) The number of times the Y-feedrate could be changed had practical limitations.

(c) Although it was possible for the derivative of equation (3) to be zero at some point on the contour, a zero Y-feedrate would cause the tool to stop and was therefore disallowed. Whenever zero was calculated for the Y-feedrate code by the computer, the lowest feedrate other than zero was submitted.

(4) The tool entered the workpiece at the minimum X value on the axis parallel to the X-axis and sank to the final depth of the slot, after which the centre of the tool was at coordinate o on the elliptical path.

(5) The next tape instruction was for the tool to move to point a situated obliquely with respect to o and therefore requiring simultaneous engaging of both the X and Y-feed mechanisms. The values of tool function code were previously read into the control system and thus the tool moved along line 1,1 which has a slope proportional to the respective feedrates. The NC system will obviously stop one of the feed mechanisms as soon as the tool has reached the final position in that axis. Thus the X-feed was disengaged when the X-coordinate of a was satisfied and the tool continued in the Y-direction only, until both coordinates of a were satisfied.

(6) The next block of information on the tape contained the coordinates of point b, and since no feedrate change has been made, the tool moved along a path parallel to 1,1 (viz. 2,2), except that this time the Y-coordinate was satisfied first, the corresponding feed mechanism was stopped and the tool moved horizontally until both coordinates were satisfied.

(7) The Y-feedrate was then changed (by appropriate codes in the tape) resulting in a tool path parallel to 3,3 and the coordinates of point c were punched into the following information block. Similar processes applied to points d and e and so on to complete the ellipse.

The computer logic for generating a complete tape obviously includes a number of checks and subroutines that are beyond the scope of this paper.

APPLICATIONS OF CONTOURING

Tests conducted on a circular groove (see figure 4) using a Talyrond surface finish testing machine showed that when a tolerance of ±0.005 in was called for in the tape generating programme, it was possible to machine a circular contour (ellipse with equal major and minor axes) with a minimum circumscribing circle that deviated 0.0040 in from the leastsquares circle and a maximum inscribed circle that deviated 0.0047 in from the leastsquares circle. Figure 5 shows two polar plots produced by the Talyrond: each division is 0.001 in, and the leastsquares circle is superimposed on the plot.

Although the contouring method was developed to machine a generalised ellipse or circle by a method of approximate linear interpolations, using an NC system that was not originally intended for contouring, it could easily be expanded to machine any contour. The methods used could also be expanded to make use of data from an APT type language, suitably simplified to make use of point-to-point hydraulic machines for contouring. Numerical

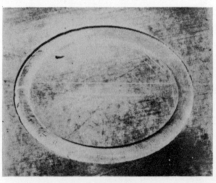

oval groove

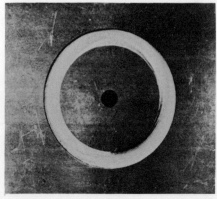

circular groove

Figure 4. Two grooves machined in an aluminium plate using a linear approximation contouring method.

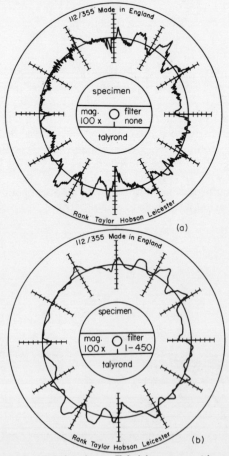

Figure 5. (a) Polar plot on Teledeltos paper with no damping and 100 x magnification.
(b) Plot with damping to reduce the number of undulations per revolution to 450.

	no.	method of calculating coordinates		
		approximate linear interpolations	linear interpolation (APT)	stepwise contouring
optimum no. of coordinates	1	±150	31	668
actual no. of coordinates	2	210	31	668
maximum feedrate	3	16 in/min	16 in/min	16 in/min
minimum feedrate	4	5.5 in/min	16 in/min	16 in/min
tolerance called for	5	±0.005 in	±0.005 in	±0.005 in
maximum outside error	6	+0.0040 in	±0.006 in	+0.0017 in
maximum inside error	7	−0.0047 in	±0.006 in	−0.0075 in
floor-to-floor time for 2	8	±11 min	±1.8 min	±30 min

Figure 6. Comparison of three methods of NC contouring.

methods could be used to fit a series of curves to the coordinates generated by the APT type language. The point-to-point machine could then follow the curves in chordal steps, derivatives having been found by finite difference methods. The fact that a hydraulic system requires five times more time to generate a given contour than a more sophisticated continuous-path machine (see figure 6) does not preclude its use

Figure 7. Typical die machined by a point-to-point machine using 2-dimensional approximate linear interpolations.

for contouring. Figure 6 compares three methods of contouring from which it can be seen that linear interpolations with a hydraulic machine are comparable in all respects other than time. Conventional methods of machining a contour would require considerably more time than even this relatively-primitive type of computerised numerical control and in the case of some complex contours, conventional methods would be unfeasible.

Methods could also be developed to machine three-dimensional surfaces such as are required in diemaking. Figure 7 shows a typical die machined by a point-to-point hydraulic machine using 2-dimensional approximate linear interpolations. Figure 8 shows a

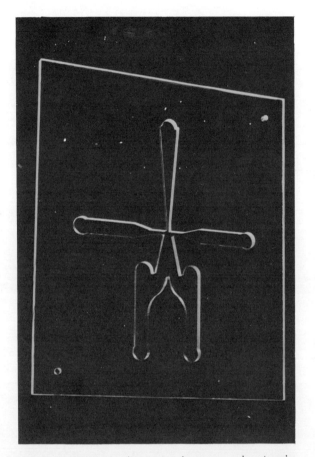

Figure 8. Fluid logic element cut in perspex using stepwise interpolation.

fluid logic element cut from a perspex sheet. This fairly complex shape was generated using stepwise interpolation, which has the advantage over approximate linear interpolations of making better use of the available range of feedrates. In fact, any 2-dimensional contour can be machined by combining the appropriate curves in the form of mathematical functions in the data generating part of the programme (processor).

CONCLUSION

The paper shows that an ordinary point-to-point NC machine tool can be used for a limited amount of contouring. The above contours, all of which were machined using an NC system designed specifically for point-to-point work, required a minimum amount of effort for both the tape preparation and the machining processes once the necessary computer software had been set up. Generation of the contour can be achieved by chordal or stepwise-interpolation and the control tape can be prepared using a computer programmed with a suitable algorithm.

Reasonable accuracy can be obtained, but the cost of computing and machining increases rather rapidly with increased accuracy. Accuracies of around ±0.005 in can readily be obtained.

The work presented aims at the extension of point-to-point machines for cases where no contouring facility exists.

REFERENCES

1. E. H. BOWMAN and R. B. FETTER, 1967. Analysis for production and operations management. Richard D. Irwin, Inc.
2. P. W. MILLYARD and R. C. BREWER, 1960. Some economic aspects of numerically controlled machines. *I.P.E. Journal*.
3. A. H. LOW, 1970. An introduction to metalcutting technology in NC languages. Numerical control programming languages, North Holland.
4. J. J. CHILDS, 1969. Principles of numerical control. Industrial Press Inc., New York.
5. G. G. ERTELL, 1961. Numerical control. Wiley–Interscience, USA.
6. EXAPT 1, 1967. Part-programmer reference manual. NEL report No. 293, National Engineering Laboratory, East Kilbride, Glasgow.
7. PERA, 1963. Numerical control and economic survey. *Report No. 19*, UK.
8. R. CHRISTIE, 1968. Why Canada trails in continuous path NC. *Canadian Machinery and Metal Working*, pp. 82–83.
9. Digital Equipment Corp. Introduction to programming. Small computer handbook series, prepared by Digital Equipment Software Writing Group.
10. Digital Equipment Corp., 1970. Small computer handbook, prepared by Digital Equipment Software Writing Group.
11. Digital Equipment Corp., 1968. PDP-8/L users handbook. Digital Equipment Corp.
12. J. D. McCARROLL, 1969. Computer-aided part programming for numerical control: an industrial study. Institute of Science and Technology, the University of Michigan.
13. L. MILLER, 1962. Engineering dimensional methodology. Edward Arnold Ltd., London.
14. A. W. JUDGE, 1957. Engineering precision measurements. Chapman & Hall Ltd.
15. I.B.M., 1952. Precision measurements in the metal working industry. Syracuse University Press, Syracuse, N.Y.

THE OPTOSYN NUMERICAL CONTROL SYSTEM

by

A. RUSSELL*

SUMMARY

The principal features and operation are described of an Optosyn numerical control system developed at the National Engineering Laboratory and being produced commercially for machine tool operation. The system is based on a refined AC moiré reader, which accurately translates optical grating angle into temporal phase angle. Measurement feedback is constituted in absolute form by the inclusion of resolvers for coarse measurement. The system resolution and accuracy is 0.0001 in from a coarse grating of some 78 lines/in and the chosen operating frequency of the system permits traverse rates up to 24/in s. A reflection version of the moiré reader can be used with an unsealed grating to form a low cost digital readout (DRO) for the digital control of machine tools.

INTRODUCTION

1. Absolute positioning

An absolute positioning system defines the position of the moving member of a machine tool with reference to an absolute zero on the machine tool. An incremental positioning system defines the position of the member by its coordinate distance from the last position occupied. The system described in this paper is absolute but the optical grating transducer used can equally form the feedback element for an incremental system with appropriate modifications to the input data translation.

Absolute programming of cutter tool position defines the desired position of the member with reference to a programme zero; this should not be confused with absolute positioning as the absolute data are often compared with present position and translated into an incremental difference which is used to drive an incremental positioning system. The reverse is also valid in that incremental programming can be made to drive an absolute positioning system. The specification for the Optosyn system called for both absolute programming and absolute feedback of position; variants could be manufactured to customer requirements.

2. Workpiece positioning

The method for setting the workpiece to an absolute datum is illustrated in figure 1. The polygon workpiece is to be positioned so that side ae is parallel to the y-axis and point p is at absolute coordinate p_x, p_y. For purposes of explanation let $p_x = 2$ and $p_y = 3$.

The workpiece should be placed on the table to the nearest inch in reference to the table datum. Data is now fed into the control to bring the tool head (or gauge) to a point on the surface ae near e. Data is now fed manually to control so as to bring the head to the proximity of a. The workpiece is then moved to bring the head accurately to surface ae repeating the machine movement of e to a several times.

With alignment completed point p requires to be set up without moving the workpiece. Input data are

set at $x = 2$ and $y = 3$ and the head will move to within ±1 in of point p. Tool offset switches are provided so that data at these switches is varied until accurate registration is achieved at p (input data $x = 2$, $y = 3$). The machine is then ready for control tape operation.

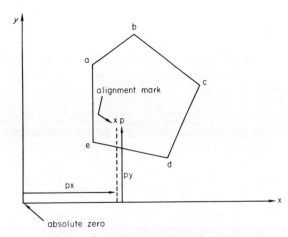

Figure 1. Absolute positioning of workpiece.

Further switches labelled 'datum' are provided so that bipolar data (centreline zero) can be converted into positive values.

3. Position movement, resolution and accuracy

The system accuracy depends on the fine transducer accuracy and the closed-loop control. The fine transducer, which is described later, has an accuracy of 0.0001 in and finer resolution, but the system 'bit size' is standardized at 0.0001 in.

The system is excellent for high-speed point-to-point operation because, in addition to the fine transducer, two resolvers are used for coarse and intermediate positioning: therefore in high speed traverse only very low frequencies need be considered.

To utilise the accuracy performance of the fine transducer, direct measurement is made, that is, linear gratings are used for linear measurement and radial gratings for angular measurement.

*National Engineering Laboratory, East Kilbride, Glasgow

GRATING TRANSDUCERS: STATE OF THE ART

1. General

The optical grating is unsurpassed as a measurement element. Its application in the NC machine tool field has been relatively limited mainly because the current demand is for systems of lower cost and lower accuracy. Even in the high accuracy field, however, its use is outweighed by other inductive and capacitive direct systems. The reasons for this cannot be quantified but it would appear that the main objections are:

(a) variations due to contamination
(b) fitting difficulties
(c) cost.

Factor (a) necessitates adequate sealing of the transducer from the environment, an action that makes a large demand on space, while factor (b) poses the problem of holding the Fresnel gap (space between grating and index grating) sensibly constant over the machine stroke (± 0.002 in for a grating of 1000 lines/in density).

Sealed systems operated by a pushrod are in use but tend to be expensive as the index transport is required to be very accurate if it is not to detract from the machine accuracy. The accurate fitting of such devices tends to be expensive.

It would appear that an acceptable system must have an exposed grating, be easily fitted and be relatively cheap.

2. DC systems

There are many successful DC grating systems[1] in current use but a host of possible variants that can affect performance. As these systems are amplitude-conscious, variations in DC level, which are potential error sources, can be caused by changes in

(a) lamp brightness
(b) grating transmission ratio
(c) line–space ratio
(d) line–space contrast
(e) electronic translation
(f) power supplies.

It will be seen that great care in grating manufacture and in electronic design is necessary to realise the inherent grating accuracy.

A common means for interpolating the basic grating is the resistor phase shift network. The sine and cosine waveforms are applied to invertors which produce the $\overline{\text{sine}}$ and $\overline{\text{cos}}$ waveforms. These four signals are fed to a phase-shift network and resultant slave waveforms are available at $36°$ intervals. By this means suitable level detectors which sense the zero level switchpoints of the ten waveforms can be utilised to produce a divided-by-10 action: thus, a grating with a pitch (spatial wavelength) of 0.001 in can be resolved into steps of 0.0001 in.

It will be appreciated that any DC variation will cause an error θ_e at the transitions (for triangular waves)

$$\theta_e = \pi(V_e/V) \text{ radians} \qquad (1)$$

where V_e is the voltage variation and V the peak-to-peak basic signal amplitude.

Interpolation factors of more than ten are not normally found in a DC system; this is not the extent of waveform accuracy but usually the limit set by cost of interpolation. It is therefore an advantage if a system possesses means for lowcost interpolation.

3. AC systems: spatial phase

There are certain advantages to be had if an AC system of measurement is achieved simply by modulating the light source[2]. If this is done to represent $a + b \sin\omega t$ where ω is the angular frequency of the modulation, a is the standing DC level and b the depth of modulation, it can be shown that two resolver-type outputs can be achived by suitable filtering to give outputs of the form

$$V_0 = 2\,bd\,\sin\omega t\,\sin\theta$$

$$V_1 = 2\,bd\,\sin\omega t\,\cos\theta$$

where $d\sin\theta$ and $d\cos\theta$ represent the spatial modulation at the grating/index junction. Here, the advantage is that both AC amplifiers and commonly used two-phase servos can be utilised. Such a system is not as prone to the errors mentioned in section 2.2 as the DC system because the sin/cos ratios remain more constant. Spatial phase errors tend to be of a cosine nature, but severe grating imperfections still cause error as the imperfection may only be 'seen' by one of four indices.

Improvements can be obtained if the grating is observed only at one point and not at several.

4. AC systems: temporal phase

The most useful feature of optical gratings is the accurate average displacement of the line-pairs (mark and space). The only technique that can accurately measure the line-pair displacement and remain unaffected by the imperfections listed in section 2.2 is the technique of temporal (or time) phase measurement. This is possible because at least one spatial cycle of the grating image is scanned at any one time and the resultant information is phase-compared with a reference phase signal. Waveshape imperfections which would result from the imperfections listed in section 2.2 introduce only second-order errors and a judicious amount of waveform filtering can spread these over the complete cycle of movement.

Interpolation or subdivision of the temporal phase signal is conveniently achieved by introducing a large number of synchronised clock pulses per cycle of the reference phase and using the clocks to measure the difference between the reference phase ($\sin\omega t$) and the unknown phase, $\sin(\omega t \pm \theta)$. In practice the clock pulses are obtained from a crystal oscillator and are divided electronically to obtain the phase of the reference signal; this ensures the necessary synchronisation.

The obvious advantages of temporal phase have led to transducer developments such as the Ferranti[3] spiral index transducer and the recent Philips[4] rotating polygon transducer. The basis of operation of both is similar in that the reference $\sin\omega t$ signal is achieved by scanning the optical grating, in the first instance, by a spiral rotating index piece of the same linear pitch as the grating and, in the second instance, by an image of the main grating reflected from a polygon comprised of many small mirrors. Both the spiral and polygon are driven from synchronous motors running at the reference frequency or part thereof.

THE OPTOSYN CONCEPT

The Optosyn transducer[5] utilises an integrated circuit consisting of fifty photodiodes in linear array and self-scanned from a shift register and gates all contained on a single chip. Moiré fringes from a grating and index are made to fall on the diode array so that fifty points on a fringe are viewed. The electrical output derived from this gives fifty ordinates per cycle of scan.

The optical system for the production of moiré fringes is shown in figure 2 and the resultant output in figure 3. In figure 2 a cross-section is shown through a

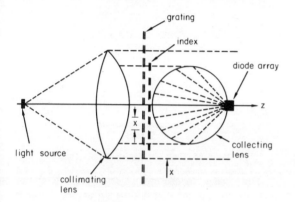

Figure 2. Optosyn optical system.

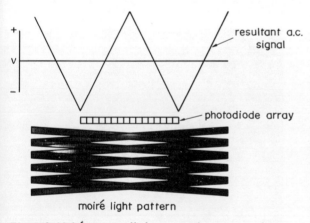

Figure 3. Moiré pattern, diode array and output waveform.

plane XZ at right angles to the grating lines, from the light source through to the diode array. Only a small number of grating lines are shown. A filament light source is used which produces the cone of illumination shown; this is collimated at the lens so that parallel light passes the grating/index junction, and a pattern of light created at this junction is collected by a cylindrical lens and focused on to the diode array. The array is fitted in the Y-plane so that the index can be 'tilted' to form one complete moiré pattern over

the fifty cells. The photodiode array sees the average intensities and for every element the intensity is the result of its position along the Y-axis. The result for a coarse grating is the serrisoidal or triangular distribution shown in figure 3.

As the photodiodes are scanned serially in time and the electrical output AC coupled, the transducer gives alternately positive and negative signals. Its crossover point, say negative to positive, is well defined and moves in time linearly with the displacement of the fringe system. The clock rate used is 2MHz, which is divided by the interpolation factor 256 to produce a reference frequency of 7.8125 kHz. As the chosen grating pitch or spatial wavelength is 0.0128 in, the phase shift increments in terms of clock pulses are 0.0001 in (divided by 128).

The Optosyn grating transducer forms a stable, accurate and wide-tolerance measuring element which operates in temporal phase *without moving parts*. The device is reasonably cheap and easily adjusted for correct operation; the concept is an improvement on currently-available moiré readers.

Improvements are currently being introduced whereby the diode charging currents can be viewed instead of the diode voltages. This removes any standing pattern from the waveform due to diode differences.

Owing to the high sensitivity of the diode array the lamp can be underrun: a typical value is 1.3V on a 6V lamp.

System description

The development of a suitable temporal phase transducer for direct machine measurement led to the adoption of temporal phase-operated resolvers for the coarser measurements. All transducers are operated from a common clock pulse generator and phase scaler.

The input data from thumbwheels or paper tape reader are fed to a conversion unit where a single shift register and associated gates form the metric-inch conversion where necessary, datum and tool shift conversion, and the resultant BCD (binary coded decimal) is converted into the more manageable pure binary (figure 4).

A crystal-controlled clock pulse generator and 8-bit binary counter is used as a reference phase scaler. The 8-bit counter generates 256 unique 8-bit patterns during each complete reference cycle so that the binary data values can be compared with these and an accurate digits-to-phase pulse generated when the two patterns are in agreement.

The complete binary position demand value consists of a 19-digit binary number, which is 1 part in 524288 or, in our case, 52.4288 in resolved to 0.001 in. This number is split into three parts: $2^0 - 2^6$ fine measurement and $2^5 - 2^{12}$ and $2^{11} - 2^{18}$ for intermediate and coarse measurements respectively. It will be observed that there is an overlap of two binary places in each case, to ensure accurate positioning in one state before passing control to the next finer state.

The coarse part of the input data is first converted into time phase at the digits-to-phase converter: this signal is then selected and fed to the phase comparator where the disagreement between the demand signal and the coarse resolver signal generates an error signal at the phase comparator which is proportional to the sign and magnitude of the difference. The motor and

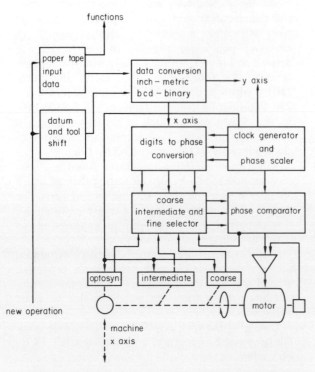

Figure 4. Block schematic of Optosyn control system.

machine axis is driven until the position feedback from the coarse resolver is in synchronism with the demand phase. Secondly, at balance the intermediate value is set up as before and a phase comparison made. Finally, the Optosyn is balanced to match the fine position of the binary position demand and the system achieves balance at the desired position to the accuracy of the Optosyn transducer. At the permissible fast traverse rates the three-stage balancing is achieved in a very short time.

System performance

The system positioning accuracy depends only on the accuracy of the Optosyn and the servo system. However a certain degree of accuracy is required from the resolvers if control is not to be passed from one stage to the other in the wrong cycle.

The coarse resolver is required to be accurate to ±0.4 in (intermediate resolver 'stroke' is 0.8192 in) and it has been found that the total error including leadscrew, gears and servo is ±0.1 in. The intermediate resolver total error is ±0.003 in, which is within the necessary ±0.006 in to match the Optosyn 'stroke' of 0.0128 in.

The Optosyn accuracy (and thus the overall accuracy) is within ±0.0001 in: this performance has been upheld even when using badly scored and contaminated gratings.

Most of the circuits in the system are digital and therefore immune from temperature variations. The analogue circuits used (filter and phase comparator)

were tested over the range 20–70°C and no phase shift observed. However as these are DC coupled to a squaring circuit volts drift was examined and found to be about 1 part in 5000 (2.5 μin) and insignificant. Phase comparator drift over the same range was found to be 20mV, which represents a spatial variation of 0.00003 in and this was attributed to the zener diodes used for clipping the comparator squarewave output.

HARDWARE

1. Optosyn control system

Figure 5 shows the chip containing the 50 photodiodes and associated shift-register and gates, used in the system for converting the spatial moiré light pattern into high-frequency temporal phase.

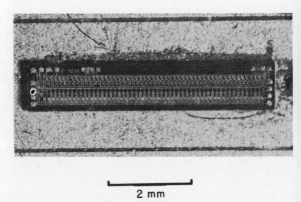

2 mm

Figure 5. Chip containing 50 photodiodes.

The coarse measurement resolvers are shown in in figure 6 and are attached to a gearbox on the leadscrew of the machine tool. A tachogenerator is fitted directly to the leadscrew for velocity-feedback information.

Figure 6. Coarse measurement resolvers.

Position controls are the manual position control panel, the tape reader and operator's pendant switchbox. The manual position control panel can be removed from the console to the machine table for ease in setting up the workpiece as described previously in section 1.2. The pendant switchbox is also detachable. All relevant switch actions are interlocked and the machine state is clearly indicated by illuminated legends on the switch buttons.

2. Reflection moiré readers

Several types of reflection head using the Optosyn principle have been designed mainly for retro-fitting DRO applications. They are designed for use with a coarse reflection grating mounted to a metal spar; the spar can be attached to the machine tool by two bolts thus eliminating the need for accurately-machined mounting faces on the machine tool.

The type B reader is shown in figure 7 and a sketch of it in figure 8. Light is collimated as previously

described and passes through a semicoated mirror, lying at 45° to the light plane, the index-piece and on to the grating. Reflected light passes to the mirror and a portion reaches the cylindrical collecting lens and photodiode array. The index is 'tilted' as before to provide one moiré pattern over the complete diode array.

The reflection readers have been operated with gaps up to $\frac{1}{2}$ in from the grating and variations in gap do not introduce error, provided the movement is purely at right angles to the grating. Low cost digital readout (DRO) can be affected from the reflection readers by a novel phase-digitiser also developed at NEL[6].

CONCLUSIONS

The Optosyn system provides a means for very reliable and accurate measurements from coarse optical gratings in temporal phase without recourse to moving parts. The AC techniques used enable economic interpolation to provide very fine resolution from a coarse grating and the system remains accurate under light variations which would normally cause error in a DC system.

ACKNOWLEDGEMENTS

The work reported in this paper was undertaken for Newall Engineering Co., Ltd. who have agreed to publication. The paper is published by permission of the Director, National Engineering Laboratory, Department of Trade and Industry. It is Crown copyright and reproduced with the permission of the Controller HMSO.

Figure 7. Moiré reflecting head type B.

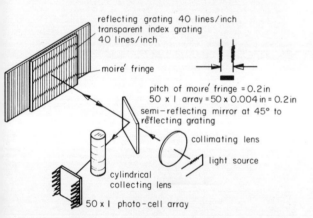

Figure 8. Schematic arrangement of moiré reflecting head type B.

REFERENCES

1. A. RUSSELL. 'A method for extracting absolute digital displacement information from optical gratings', *Proc. 9th IMTDR Conference.* Birmingham University, 1968. Oxford: Pergamon Press, 1969.
2. A. RUSSELL. British Patent Application 60481/70.
3. D. G. PETTIE. 'An introduction to the Ferrant Multiax continuous path control system', *Ferranti Report No. NCD/GEN/102,* June 1968.
4. PHILIPS. 'Linear Measuring System, PE2271', Publ. 77.082 BF. Eindhoven: Philips Co. Ltd.
5. A. RUSSELL. British Patent Application 31794/70.
6. A. R. ROBERTSON. British Patent Application 55532/70.

A STEPPING MOTOR DRIVE ASSEMBLY ESPECIALLY DESIGNED FOR CNC SYSTEMS

by

J. PLAS* and J. BLOMMAERT*

SUMMARY

This paper deals with the problem of sampled data computerised numerical control. The technique implies the partition of the tool path into successive segments of equal travel duration, the main problem being the choice of the sampling period and its influence on the contouring accuracy. The authors first introduce the most important theoretical considerations, and afterwards describe more thoroughly a particular drive system, built by the CRIF, using electrohydraulic stepping motors.

NOTATION

C	= friction coefficient
f	= frequency
$G(i)$	= lower limit of frequency interval
$H(i)$	= higher limit of frequency interval
I_t	= rotational inertia
i	= reduction factor
K	= scaling factor
L	= segment length
M_A	= available torque
M_L	= load torque
M_S	= static torque
p	= pitch of leadscrew
R	= radius
r	= rate
S	= number of steps per revolution
T_s	= sampling period
t	= time
V	= travel speed
$X(i)$	= slope of torque-speed line (i)
x	= longitudinal travel component
δ	= positioning resolution
$Z(i)$	= origin ordinate of torque speed line (i)
y	= transverse travel component
ϵ	= geometrical error during circular interpolation
φ	= segment angle
τ	= time constant
ω	= angular velocity

INTRODUCTION

Computer-controlled machine tools have appeared quite recently in many countries. Although many devices are already on the market, it must be admitted that 'softwired' controllers are still in the development stage, so it is difficult to classify them clearly and foresee which among them will expand the most.

Nevertheless, one can retain the following comparison criterions:[3, 6, 9, 10]

— size of the computer(s)
— number of axes under control
— number and complexity of tasks accomplished by software
— degree of computer intervention within the positioning loops
— means of data input
— computer assistance for part programming
— management reporting.

At the request of the Belgian metalworking industry, the CRIF started two years ago a study on the future trends of numerical controls. This study led to the conclusion that any new development should be compatible with the concept of integrated manufacturing systems (IMS), but, on the other hand, for economical and practical reasons, should not differ too much from the existing versions. Consequently, the use of a cheap process computer appeared to be the best compromise between these two requirements. We shall not detail the capabilities of this principle, but will mention two essential ones: versatility due to software implementation of many functions and continuous dropping of the computer prices. In order to obtain a valuable practical experience in this field, the CRIF decided to design such a system and test it in actual workshop conditions. First of all this paper will shortly describe the whole system, according to the preceding criterions. Then the positioning loop (that is, the stepping motor drive assembly) will be studied more thoroughly.

GENERAL DESCRIPTION OF THE COMPUTER-CONTROLLED SYSTEM INVOLVED

Computer configuration

The computer used is a DEC-PDP8E type with a 4K words of 12-bit core memory.

The following hardware options are added:

- an extended arithmetic element (EAE)
- a power fail detect and restart circuit
- 24 digital input—output channels.

This configuration allows high-speed calculations in double precision: for example, the multiplication of two 12-bit numbers takes 7.4 μs and the division of a 24-bit number by a 12-bit the same time. The input—output channels are used for parallel transmission of data and commands to (and from) all the peripherals which are connected in parallel on the same lines: each peripheral has its own address and all the addresses are demultiplexed in the machine tool cabinet.

This method increases the hardware versatility, because the number of peripherals is not strictly limited. On the contrary, the addition of a new one only requires the generation of its address and its direct connection on the existing lines. Typical

*Research Centre of the Belgian Metal Working Industry, Section Mechanical Engineering (CRIF), Heverlee, Belgium

peripherals are: positioning servos, displays, operator's panel, push buttons, teletype, and so on.

Finally, the I/O channels are equipped with line drivers/receivers, in order to provide remote control of the peripherals. At the moment, these are located at about 100 m from the computer. Consequently, the computer can be installed in the most suitable place and eventually serve other purposes, such as programming.

Software implemented functions

Due to the capability of the process computer for data manipulation, it will perform the following tasks:

- reading, checking, decoding and storing the part's programme tape
- supervising and interpreting all limitswitches, pushbuttons from machine tools or operator's panel
- rough interpolation
- issuing orders to positioning servos, displays, lamps, and so on.

In other words, the software package consists of a set of specialised subroutines and one general background programme responsible for the proper sequencing of the whole system.

Although software problems are most important, they are not the object of this paper and so will not be discussed further.

Hardware implemented functions

As said before, the computer receives data and issues commands via I/O buffer registers and line drivers/ receivers. These signals are taken over by a common bus. Such a principle was chosen in order to conform to the computer's internal signal handling (DEC *Omnibus* trade mark). Each peripheral consists of

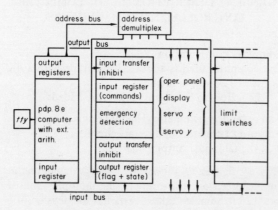

Figure 1. Hardware implemented functions.

of three supplementary functions besides its own functional circuits. The first is the address decoding: it allows, during proper address detection, the peripheral to make use of the common bus to communicate with the computer and isolates all the other peripherals. The second, the *flag*, tells the computer that the peripheral needs a communication: it means that data is to be sent or commands received. The advantage of such a function is the optimisation of computing time because the computer takes care only of those peripherals whose flag

is set. The third function is performed by buffer memories. They store the computer orders as soon as the address signal has disappeared. Figure 1 summarises the hardware implemented functions of the system. It can be seen that essentially they consist of slave elements, such as servo assemblies (see later), displays and interfaces for limitswitches and pushbuttons.

Controlled axes

The command generation function is shared between the process computer and the hardwired drive assemblies. The tool path can be linear or circular. The computer takes care of the rough interpolation: it always divides the tool path into linear segments. The resulting data (distance and velocity along the respective axes) are transmitted to hardwired linear interpolators at fixed time intervals. This is consequently a sample data system. The sampling period must be long enough to cover all the required calculations and delays relative to all the controlled peripherals. But on the other hand this sampling period may not exceed a given limit determined by the required path accuracy, for each axis will remain on the given straight line, whilst the computer takes care of each of the other peripherals. This will be detailed further on. In our case, the system could be extended to six axes of motion, although the present amount equals two axes. For each of them, the total travel length can reach + and −8 m, with a positioning resolution of 5 μm and a top positioning speed of 5 m/min.

The final drive is achieved by electrohydraulic stepping motors. This choice relies on economical considerations: indeed, the price of positioning servos using stepping motors is lower owing to the fact that these can perform the functions of both driving and measuring[8]. Furthermore, the application of such motors eliminates the need for any A/D or D/A converters. Nevertheless, our system is not restricted to the open-loop principle: the output positioning and speed commands are compatible with closed-loop servo systems.

Input of machining data

The part's programme, punched on paper tape in standard ISO code and word address format, is read by the computer teletype. Then, the operator of the machine tool can dial in any coordinate or function. The reason for doing so is essentially compatibility with the conventional numerical control.

Programming assistance

In its present state, the system does not offer any computer assistance for the automatic calculation of coordinates, feeds, speeds, and so forth. On the other hand, when reading the part's programme tape, the computer checks each character and in the case of format error, issues a message on the teletype printer. The correction can be immediately introduced by typing the right character on the keyboard.

Management reporting

Provision has been made to issue any message on the teletype printer without influencing the machining control. At the moment, the messages refer only to emergency situations like: lack of power, out of range, no tape.

THE NEED FOR AN ELECTROHYDRAULIC STEPPING MOTOR

As said before, we adopted the open-loop positioning principle for economic reasons[7]. Furthermore, we imposed the following features:

top positioning speed V_{max} = 5 m/min
positioning resolution δ = 5 μm

Finally, the test table to be driven has a travel surface of 600 mm x 400 mm and weighs about 1000 kg.

From all this data, it appears that hydraulic amplification is needed at the output shaft of the electric stepping motor. The following calculations clarify this assertion.

Positioning resolution[11, 12, 13, 14]

The motor used is a Pratt SM 10 EHSM, providing 480 steps/rev at the output shaft. The reduction factor i between the motor and the ballscrew equals 1.67 and the latter has a pitch p of 4 mm. Finally, the positioning resolution is given by

$$\delta = \frac{1}{S} \cdot \frac{1}{i} \cdot p = \frac{1}{480} \cdot \frac{1}{1.67} \cdot 4 = 0.005 \text{ mm}$$

Maximum positioning speed

This speed is a function of the maximum continuous running frequency of the motor.

In our case, this maximum equals 16 kHz and the top speed can be deduced as follows:

$$V_{max} = 60 \cdot f_{max} \cdot \delta = 60 \times 16.10^3 \times 5.10^{-6}$$

$$= 4.8 \text{ m/min}$$

Dynamic stability

The inertia load connected to the motor shaft is about $I_t = 0.75 \times 10^{-3}$ kgm^2. Such a value, combined with the imposed V_{max} and δ lies above the capabilities of most electric stepping motors, but well within the range of electrohydraulic types. Nevertheless, these are not unconditionally stable: the load inertia is strictly limited upwards. Here the manufacturer imposes $I_{tmax} = 1.4.10^{-3}$ kgm^2 and so the load we have yields enough stability margin.

Following error

The hydraulic amplifier introduces an inevitable error which is a function of the hydraulic amplification. Thus, for a supply pressure of 70 kg/cm^2, the resulting linear following error equals 32 μm at 0.6 m/min, 210 μm at 4.8 m/min. These values are not negligible and must certainly be taken into account by the computer, for circular interpolation.

INFLUENCE OF THE SAMPLING PERIOD ON THE WHOLE SYSTEM

As said before, our system belongs to the sample data type. In fact, many manufacturers have adopted this principle[2, 4, 5] (for example, Westinghouse, Allen-Bradley, Kongsberg and others). This choice is easily justified, for the computer cannot be asked to take care of every elementary step of the controlled elements, because the time between two successive commands would then be too short to leave enough time for background functions (such as speed 5 m/min, resolution 5 μm, command period 100 μs). The important point is the relationship between sampling period and software/hardware task distribution.

TWO LEVELS OF COMPUTER INVOLVEMENT CAN BE DISTINGUISHED

First solution

In the first level the computer accomplishes all interpolation by geometric calculations. It issues step results at fixed time intervals which are successively applied to the command input of the servo loop comparators. In this case of sample and hold, the sampling period is mainly correlated to the dynamic characteristics of the servo system. L. Evans mentioned this problem and found, in a particular case, a maximum period of 4 ms.

Practical systems exhibit about the same value: for example, 10 ms for the Westinghouse New World system. To conclude this first point: the period is relatively small and so the amount of computing work during it is somewhat restricted. But on the other hand the servo hardware is limited to a minimum.

Second solution

The second level of computer involvement makes use of hardwired linear interpolators and, as in the first case, issues the commands for these devices, at fixed time intervals. *As already mentioned, we have chosen this principle.*

In the case of linear interpolation, the sampling period could be as long as the travel time, because the linear interpolator itself, if properly fed, is capable of achieving the required path accuracy. The influence of servodynamics can be suppressed if the speed variations are properly shaped (see later). Circular motions are divided by the computer in successive linear segments. Here, besides dynamic errors, a geometrical error is added, due to the partition of the tool path. Figure 2 shows this geometrical error ϵ as a function of the circle radius r and segment length l.

The sampling period t_s being constant, it follows that the segment length l is a function of the tangential speed v, that is, $l = v \cdot t_s$. The error ϵ is given by $\epsilon = r(1 - \cos\varphi/2)$, and finally with $\varphi = \frac{360}{2\pi} \cdot \frac{l}{r}$ (deg), one sees that ϵ can be expressed as a function of v, t_s and r. This relationship is given in figure 2, for t_s = 100 ms (our value). It can be deduced that, even for such a relatively long sampling

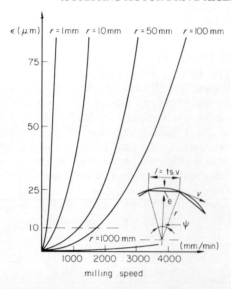

Figure 2. Circular interpolation – geometrical error.

period, the geometrical error remains acceptable for a wide range of speeds and circle radii.

In conclusion, the presence of a hardwired linear interpolator yields a fundamental change in the software configuration and capabilities, because the same path accuracy can be guaranteed with relatively a much longer sampling period.

The last point to be studied here is the influence of the servodynamic characteristics on the path accuracy. The full theoretical study falls outside the scope of this short paper and we shall now try to summarise the fundamental results.

— The complete system of servodrive and hydraulic amplifier responds to a sudden speed change command by an exponential evolution, whose time constant equals about 30 ms in our case. There is no constant speed error.

— This speed delay yields a following error: the actual path lags behind the theoretical one.

— In the case of linear interpolation, the following errors (x and y) have no influence on the path accuracy, provided the time constants relative to each axis are exactly equal.

— In the case of circular interpolation, in other words, a succession of linear interpolations of equal period T_s, there is a speed variation at each vertex, and consequently the actual tool path differs from the theoretical polygon.

From one segment to the following, the path deviation is quite small: 15 μm maximum, depending on the speed, the radius and the time constants. The fundamental problem is to avoid any accumulation of the successive following errors. Here again, the computer can help solve this problem. Indeed, it can be demonstrated that an accumulation of errors is avoided if the speeds are modified prior to each vertex, at a point which can be calculated on the basis of the speed, the radius, the period T_s and the servo time constant. The only condition to be fulfilled is that the speeds are to be stabilised when the tool

approaches the breakpoints. Therefore, it can be understood that the servo drives must fulfil the following requirements: ability to modify the speed reference independently from the position reference, when a predefined anticipation point is reached; very short time constant, much shorter than 100 ms; accurate and stable transfer function, to provide exactly the same speed evolutions in both axes, in function of time.

SPEED REQUIREMENTS TO BE FULFILLED BY THE DRIVE UNIT

A fundamental characteristic of a motor is its torque–speed curve.[12, 13] In the particular case of stepping motors working in open loop, this characteristic acts as an absolute limit; exceeding it (statically or dynamically) is sanctioned by loss of synchronism. In this section, it will be explained how acceleration and deceleration are to be shaped in order to avoid any overshoot of the torque-speed curve.

Torque-speed characteristic

Figure 3 shows the torque-speed characteristic of the electrical stage of the used EHSM (Pratt PM.6)[14].

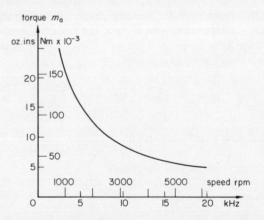

Figure 3. Electric stepping motor: torque/speed characteristic.

In fact, both stages have their own limitations, but the ones imposed by the electric motor are much more severe. Consequently, this simplifies the study a great deal, for the hydraulic amplifier can be assumed ideal and the study can be restricted to the electric motor, which is coupled to a constant load — the input shaft of the hydraulic amplifier.

This torque–speed characteristic denotes quite a fast torque decrease; essentially, this will restrict the acceleration capabilities in the high speed range.

Correlation between torque–speed and speed–time evolutions

For each frequency, the full available torque M_A is given by the curve of figure 3.

The load torque M_L consists of three terms:

the inertia torque $I_t \cdot \dfrac{d\omega}{dt}$

the friction torque $C \cdot \omega$

the static torque M_S

If the full available torque is constantly utilised

$$M_A = M_L = I_t \frac{d\omega}{dt} + C\omega + M_S$$

To solve this equation, it is necessary to express M_A in terms of ω. As this relationship is not linear, an exact mathematical treatment would be too complex for our purpose, but a good approximation is given by dividing the $M_A - \omega$ curve of figure 3 into a set of straight lines. Between the two frequencies $G(i)$ and $H(i)$, a line i is defined by:

$$M_A(i) = X(i) . \omega + Z(i)$$

so, for each line (i)

$$X(i) . \omega + Z(i) = I_t \frac{d\omega}{dt} + C\omega + M_S$$

Such an equation is very easily solved

$$\omega(i)_{max} = \left[G(i) - \frac{Z(i) - M_S}{C - X(i)} \right] \exp \left[- \frac{C - X(i)}{I_t} t \right]$$
$$+ \frac{Z(i) - M_S}{C - X(i)}$$

By plotting side-by-side the successive $\omega(i)$ functions gives the total evolution of ω_{max} versus time. Practically, we divided the $M_A - \omega$ curve into 10 segments and treated the problem on a digital computer, in order to allow the introduction of many different parameter values, and also to change easily from acceleration to deceleration.

One typical result, where the parameter values are the closest to reality, is given in figure 4 where:

curve a shows the limit of the allowed acceleration area,
curve b shows the limit of the allowed deceleration area from a frequency of 16 kHz downwards.

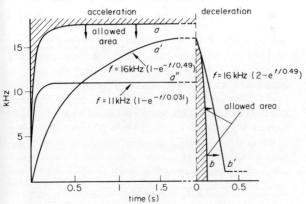

Figure 4. Stepping motor: practical acceleration and deceleration curve.

$$I_t = 7.10^{-6} \text{ kgm}^2$$
$$C = 5.10^{-6} \text{ Nm s}$$
$$M_S = 36.10^{-3} \text{ Nm}$$

So, in all cases, the practical speed evolution must remain within the allowed area and at each frequency the acceleration must be smaller than that found on the limiting line.

In conclusion, the chosen speed evolution is a compromise between the two above requirements, the time to reach a top frequency and the ease of practical realisation.

In figure 4, the curves a', a'' and b' illustrate our solution:

a' is followed when the top frequency f_{max} to be reached is 16 kHz
a'' is followed when the top frequency f_{max} to be reached is 11 kHz
b' is followed when decelerating from 16 kHz downwards.

Note that the deceleration curve b'' corresponding to 11 kHz is not drawn, because it coincides with a vertical line.

a' and a'' are exponential curves of the type $f = f_{max}(1 - e^{-t/\tau})$
b' is also an exponential curve, but this time of another type $f = f_{max}(2 - e^{t/\tau})$

Such a shape, contrasting with most other systems, preserves the motor from any shock.

a' and b' have a time constant τ equal to 0.49 s and are used for positioning over long distances,
a'' and b'' have a time constant τ equal to 0.031 s and are used for positioning over short distances and for milling where such a small value is needed any way (see above).

The choice between a long and a short distance and the setting of the corresponding f_{max} and τ are done by the computer, which always aims at minimising the positioning time.

Optimisation of positioning time
As mentioned above, the positioning function can be accomplished in two ways:

– acceleration and deceleration with a time constant of 0.49 s and a top frequency of 16 kHz.
– acceleration and deceleration with a time constant of 0.031 s and a top frequency of 11 kHz.

We shall now demonstrate how to decide between these two solutions. The positioning time t_p is divided into:

– the acceleration time t_a
– the top speed travel time t_t
– the deceleration time t_d
– the time needed to cover the safety distance t_s

$$t_p = t_a + t_t + t_d + t_s$$

t_a and t_d are obtained graphically from the distance–time relationship plotted by integration of the two equations of speed (see above)

t_t is simply given by the $\dfrac{\text{distance}}{\text{speed}}$ ratio

t_s is a constant related to the top speed: 1 s for 16 kHz and 0.1 s for 11 kHz in our case.

The sum t_p is plotted in figure 5 versus total distance for the two ways of positioning (t_{p11} and t_{p16}). It can be deduced quite easily that the break-even point lies around 260 mm; below this point the

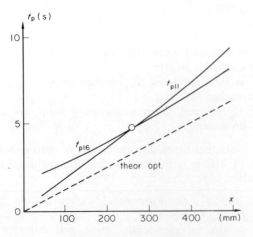

Figure 5. Total travel time.

computer imposes 11 kHz and 0.031 s, and above it imposes the other solution. The positioning time is always minimised. If this distance is too short ($<$ 10 mm for instance), the computer imposes a step speed command equal to the maximum starting frequency.

Finally, the dotted line indicates the theoretical value of t_p when the speed command is a square signal of 16 kHz amplitude. It can be seen that, for distances smaller than 260 mm, the efficiency of the

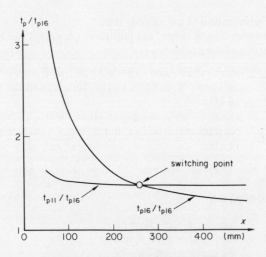

Figure 6. Minimisation of total travel time.

lowest line is quite high. This fact is clearly demonstrated in figure 6, giving the normalised values of $t_{p\,11}$ and $t_{p\,16}$ as a function of the distance.

From figure 3, it could be concluded that the largest time constant is going to impoverish the whole system: but, with regard to figure 6, this is false. Conclusion: the range of 50 . . . 260 mm (the most important one) is covered with the shortest time constant, which yields an increase in t_p of only 50% above the optimal value. Over 260 mm, the time constant of 0.49 s loses its relative importance with respect to the total positioning time; here the increase above the optimal time equals roughly 30%.

FUNCTIONAL DESCRIPTION OF THE STEPPING MOTOR DRIVE ASSEMBLY

This last section deals with a more detailed description of one particular hardwired element mentioned in figure 1 – the servo assembly. We shall limit our

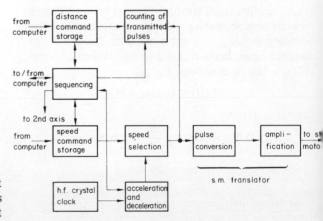

Figure 7. Servo assembly: functional block diagram.

study to the functions directly connected with the positioning itself, so functions such as *flag* and *address* will not be described (see figure 7).

Speed command storage
This function is performed by two 12-bit registers. The first one is directly loaded by the computer during axis motion, at its sampling instant, but only if its former content has already been transmitted to the second one. The second register, the active one, provides the input for the speed selection circuit. On reception of a command from the sequencing circuitry, it imposes instantly the new input value that is already available in the first register. From that moment on, the first register sets its flag signal, telling the computer a new speed command is needed. This information will be received at the next sampling instant.

The function of the two registers can be summarised as follows: the second one looks after the motion itself; the first links the computer to the system.

Another point to be justified is the length of these registers: 12 bits. This length provides a speed resolution of $V_{max}/2^{12} = 1.2$ mm/min = 20 μm/s. Assuming a linear interpolation at the maximum speed, the largest deviation that can affect each elementary segment of the linear path is equal to

$$20 \times T_s = 20 \times 0.1 = 2\,\mu m$$

It is clear that such an error, although smaller than the resolution, may not be cumulative. The computer takes care of this requirement.

Distance command storage
For the same reasons as above, the separation of computer linkage from motion control, this function is also performed by two registers. Let us mention that the transfer from the first register to the second is not in synchronism with the transfer occurring between the two speed registers; this point was already discussed when considering the circular interpolation.

The length of these registers also equals 12 bits; this covers a distance of $2^{12} \times 0.005 \cong 20$ mm. This amount, although much smaller than the total travel length, is more than sufficient, because each distance command refers to one sampling period. The maximum that could be reached is $V_{max} \times T_s = 80 \times 0.1 = 8$ mm.

Counting the transmitted pulses

This function is of prime importance. Indeed, as in all open-loop servos, the counting of the pulses transmitted to the motor is the only way of measuring the distance accurately.

The counter used is also 12 bits long; it is preset by the second distance register and counted down by the pulses sent to the motor. During motion, it requires a new preset immediately after passing zero. Finally, it plays another important role: it tells the sequencing circuit when to initiate the transfer of new data in the second speed command register. During positioning or linear interpolation, this is done when its content passes zero, and during circular interpolation, when its content reaches the 'anticipation distance' calculated by the computer.

Acceleration and deceleration

This function is performed by a circuit fed by a constant high frequency which generates the required acceleration and deceleration exponentials, with one of the two imposed time constants (either 0.49 s or 0.013 s). A simplified block diagram is given in figure 8, made up of the following essential elements:

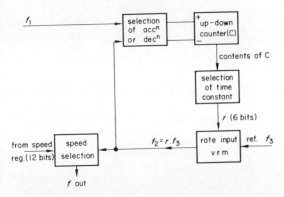

Figure 8. Stepping motor: digital input frequency generation.

– a 20-bit up-down counter C, the contents of which serve as data input for the VRM circuit.
– a 6-bit variable rate multiplier (VRM); this is a special purpose counter, the capacity of which

ranges from $\dfrac{0}{64}$ to $\dfrac{63}{64}$ in steps of $\dfrac{1}{64}$, according to

the data r presented at the rate input. So if the VRM receives a frequency f_3 at the reference counting input, it provides an output frequency

f_2 equal to $r \cdot \dfrac{f_3}{64}$.

– a time constant selector: its task is to transfer to the VRM rate input those output bits of C that are required for the generation of the desired time constant.

– a mode selector which provides the sign of the power of the exponential and thus distinguishes acceleration from deceleration. This circuit is fed by a constant frequency f_1 slightly higher than f_3.

It is easily justified that such a configuration is able to generate exponentially varying frequencies. Indeed, consider the acceleration first: in this case the mode selector connects f_2 (output of VRM) to the minus input of the counter C, and f_1 to the plus input of C. Assuming that at $t = 0$, C is reset (so $f_{2,0} = 0$) and that f_1 and f_3 are present

$$f_2(t) = r(t) \cdot f_3/64$$

with
$$r(t) = K \int_0^t (f_1 - f_2) \, dt$$

The $\int$ notation is used instead of Σ, to simplify the demonstration.

So
$$f_2(t) = \frac{Kf_3}{64} \int_0^t (f_1 - f_2) \cdot dt$$

Solving this equation leads to

$$f_2(t) = f_1 \left[1 - e^{-Kf_3 t/64} \right].$$

a relation that coincides with the one given before, and is also drawn in figure 3.

The time constant τ is $64/Kf_3$ and to obtain the imposed values of 0.49 s and 0.031 s, Kf_3 must equal 130 and 2065 respectively. In our case $f_3 = 1.01 . 10^6$ Hz and then

$$K = 2^{-13} \quad \text{for } \tau = 0.49 \text{ s}$$
$$K = 2^{-9} \quad \text{for } \tau = 0.031 \text{ s}$$

When decelerating, the mode selector inverts the two inputs of C: f_1 is applied to the minus input and f_2 to the plus input. Assuming that at $t = 0$, C is filled with certain contents that yield $f_2 = f_1$

$$f_2(t) = \frac{Kf_3}{64} \int_0^t (f_2 - f_1) \cdot dt + f_1$$

Solving this equation

$$f_2(t) = f_1 \left[2 - e^{Kf_3 t/64} \right]$$

This relation also coincides with the one given before.

Two remarks come to the fore:

The time constant is independent of the acceleration or deceleration mode;

the deceleration has to be interrupted before the frequency reaches zero.

Practically, a logic circuit detects the 1 kHz level, stops the counting in C and so maintains f_2 constant, until the target is reached. It has been shown that $f_2(t)$ is a sole function of f_1, f_3 and K. The two frequencies f_1 and f_3 are supplied by a crystal clock and so remain fairly constant. The factor K is a pure digital datum and cannot shift, by principle.

In conclusion, the shape of $f_2(t)$ is perfectly constant and forseeable in each axis of motion; it follows that since all axes are synchronised by the same references f_1 and f_3, the linear interpolation function can be assumed to be perfect and independent of any systematic disturbances, even when the speeds are not stationary.

Speed selection

The acceleration and deceleration circuitry issues the frequency f_2 which always varies between 0 and 16 kHz. The role of the speed selector is to scale f_2 in order to obtain a final frequency f_{out} corresponding to the programmed positioning speed. Of course, this scaling does not affect the time constant. The circuitry contains two variable rate multipliers connected in series, and so f_{out} can vary from 0 to $\left(\frac{63}{64} \cdot \frac{63}{64}\right) f_2$ with an increment of $\left(\frac{f_2}{64.64}\right) = \left(\frac{f_2}{4096}\right)$.

Let us denote here that this extremely wide range of speeds is provided by two integrated circuits only, and that each of the elementary steps is perfectly defined and stable. Consequently, the conclusions of the preceding section are still valid: the stationary speeds have the same properties as the non-stationary ones.

Pulse motor translator

It has been stated above that the drive assembly could be applied to either a closed-loop or an open-loop servo. In other words, the frequency f_{out} (from the speed selector) may be used as a speed command for the two servo types. As we chose the open-loop principle (stepping motors), f_{out} is fed to their translator. Such a translator has two main sections: the pulse converter and the power stage. The pulse converter supplies the required sequence to the stator phases. In the case of the Pratt PM6 motor, three out of six phases $(A, B \ldots F)$ are always energised at a time, so the sequence is as follows: $ABC, BCD, CDE, DEF, EFA, FAB, ABC \ldots$. Evidently, this sequence suggests the adoption of a shift register: the complete sequencing is then generated by two integrated circuits only. Furthermore, the inversion of sign is obtained without any difficulty, by changing from right-shift to left-shift.

The function of the power stage is to provide the stator coils with the required current pulses. This is best accomplished by using a current source with a high output impedance.

Another factor can help the current switching: careful study of the above sequence shows that the phases A and D, B and E, C and F are respectively complementary to each other. This means that the overvoltage generated by the switching off/on of coil A can always be used to help the switching on/off of coil D, and the same for couples B—E and C—F. This implies the use of fast switching diodes, but on the other hand, increases the performances.

CONCLUSION

As stated in the introduction, our study has been towards a concept in which the computer plays an irreplacable role. Thus, the servo drive assembly has been designed as a true computer peripheral. The presence of the necessary interfaces increases the system's complexity, but on the other hand, allows

ON LINE dynamic calculations, such as optimal acceleration, deceleration, velocity control (for the shortest positioning) and anticipation points (for the highest contouring accuracy).

The exclusion of any analogue components complicates a certain number of functions, such as acceleration and deceleration, but the qualities of a purely digital circuitry compensate quite easily for these drawbacks. Remember the speed resolution, and acceleration or deceleration accuracy.

Although the open-loop concept is not the only possible alternative, it does respond very easily to what the computer implicates.

REFERENCES

1. L. EVANS, 1971. Optimal contouring control system for digital applications. Present and future techniques. Paper presented at the *CIRP third international seminar on optimisation of manufacturing systems*, Pisa (Italy), 24—25 June.

2. Ø. BJØRKE, 1971. On-line numerical control systems. Paper presented at the *CIRP third international seminar on optimisation of manufacturing systems*, Pisa (Italy), 24—25 June.

3. G. SPUR and W. ADAM, 1971. Automatisierung des datenflusses und optimierung des zerspanungsprozesses durch den einsatz von prozessrechnern. Paper presented at the congress held prior to the *XXI Gen. Assembly of the CIRP*, Warsaw, August.

4. W. C. CARTER, 1970. Criteria for computerised contouring control. Paper presented at the *34th Annual Machine Tool Electrification Forum*, sponsored by the Westinghouse Electric Corp., Pittsburgh, Pa., May 26—27.

5. JOHN L. PATRICK, 1970. Expandable contouring: tomorrow's NC today. Paper presented at the *34th Annual Machine Tool Electrification Forum* sponsored by the Westinghouse Electric Corp., Pittsburgh, Pa., May 26—27.

6. G. STUTE and R. NANN, 1971. DNC-rechnerdirekt-steuerung von werkzeugmaschinen, *Zeitschrift für industrielle Fertigung 61*, p. 69.

7. S. INABA, 1968. Contouring NC featuring cutter radius full size offset and direct digital electrohydraulic pulse motors. Technical paper of the *American Society of Tool and Manufacturing Engineers*, MS 68-218, Philadelphia, April.

8. GORDON B. BATY, 1968. Recent advances in open-loop numerical control. Technical paper of the *American Society of Tool and Manufacturing Engineers*, MS 58-166, Philadelphia, April.

9. C. BARNETT and B. DOWIES, 1971. Computer NC: a review of developments. Paper presented in the *NEL Birniehill Symposium*, September 21.

10. DOUGLAS A. CASSELL, 1971. Direct computer control. *NC Scene*, January p. 6.

11. R. BELL, A. C. LOWTH and R. B. SHELLEY, 1970. The application of stepping motors to machine tools. The Machinery Publishing Co. Ltd.

12. J. F. JACKSON and R. BELL, 1971. An electro-hydraulic stepping motor for numerically controlled machine tools. *Proc. 11th MTDR Conf.*, Pergamon.

13. J. PLAS and J. BLOMMAERT, 1970. Les moteurs pas à pas sur machines-outils à commande numérique: introduction à leur constitution et à leur emploi. *CRIF Report MC 33*, April.

14. Pratt hydraulics. Stepping motor drive system. Pratt Engineering Corp., Bankfield Works, Haley Hill, Halifax.

QUALITY CONTROL BY USING AUTOMATIC INSPECTION PROCEDURE

by

C. WATKINS*

SUMMARY

The use of a computer with multiple dimensional jigs coupled to it in the audit inspection area of the auto-machine shop at Raleigh Industries Ltd was planned by the author in conjunction with Elliott Process Automation in 1966/67. It was installed in August 1968 and continued to be developed from that date under his supervision.

INTRODUCTION

In the original write-up of the project two further developments were envisaged:

(i) locating of the test-jigs near the production machines instead of in the central inspection room,

(ii) increasing the number of inspection points.

Although the main course was easily planned and easy to follow, a number of half-expected difficulties were met and had to be overcome to achieve success. The final solutions to these problems came in 1971 as the direct result of work by Mr L. Cordingley, the engineer-in-charge, aided by Mr G. Pike on the programming side.

Progress up to this breakthrough has been described in a paper already published but the present developments have not been described elsewhere. They may be divided into the following sections:

(i) transmission of signals,
(ii) revision of programme,
(iii) simplification of the generation of signals,
(vi) method of machine control,
(v) effect on personnel and general procedures.

TRANSMISSION OF SIGNALS

The first gauging fixture was moved onto the shop floor in 1968 and a second in 1969. The first of these merely signalled that the component being gauged was either correct or defective. To carry out real machine control it was necessary as a minimum to transmit to each machine being controlled the actual measurement of every dimension being gauged.

Experiments were carried out using the signals generated by the computer for display on the Creed printer. These were made to operate transistor-controlled lights mounted in a control box at the machines in question. It was found that, although spurious signals generated by local electrical apparatus did not interfere with the transmission to the computer, the same noise signals sometimes caused a shift in the dimension indicated at the machine. To employ a normal solid-state high-level line driver and receiver on each lamp circuit was regarded as highly uneconomical, in view of the large number of transmission lines throughout the installation. The transmission consisted of a coded lamp turn-on or off and comparatively long and idle periods between information. It was decided that the most economical method to output these test results (and other

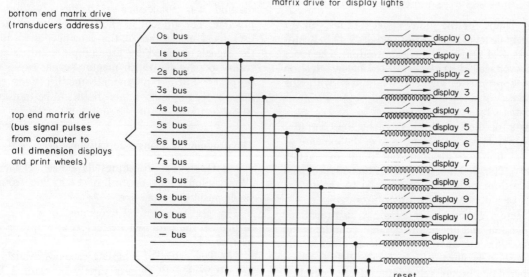

Figure 1. Electrical circuit for one dimension.

* Raleigh Industries Ltd, Nottingham, England.

operating signals to be disclosed later) was for the computer to generate 24 V pulses which could be used to drive latching reed relay matrices located at shop floor distribution points. The technique was pursued in the second shop floor jig installation and found to be successful. This jig installation used

Figure 2. Cabinet containing equipment for control of eight machines.

filament lamps for the dimensional readout, but for subsequent installations a set of two decimal indicator tubes was used for each dimension readout. The tubes were low voltage (24 V) bar segmented and by selection of the appropriate segments, numbers with the appropriate sign were displayed to ±19 (this represented ±2 x ½ total tolerance).

Signals for the lamps generated by the computer were output in decimal form to save the expense of additional shop floor decoding and to improve system reliability. This decision also meant that a less expensive shop floor printer could be used to record the mean of each dimension and other information.

In figure 1 the electrical circuit is illustrated for one dimension, and one of the cabinets containing the equipment for the control of eight machines in one machine shop is illustrated in figure 2. It must be remembered that progress along other lines of development described below was going on at the same time and in fact all went on hand-in-hand. Indeed each was complementary and necessary to the others. One of these developments was the complete modification of

the programme to enable the computer to deal with the new setup but, as will be outlined later, local scanning logic has been introduced at the measuring interface and this has enabled the programme to be simplified. At the same time it had to be changed to operate on quality control principles instead of audit inspection.

Actually two out of the eight inputs to the computer have been modified to this new procedure: the other six remain unmodified and still operate in the audit inspection area. This brings us to a detailed consideration of the modification to the programme.

REVISION OF PROGRAMME

The original system necessitated a general programme controlling the overall operation of the installation, a subsidiary programme giving the details of each component tested and a third programme controlling the application of the required sampling plan. After correct calibration of a line, production pieces were compared with the readings of a setting piece whose mid-limits and tolerances on each dimension had previously been programmed into the computer. If all dimensions on the production piece were within tolerance, an accept signal (green light) was transmitted. If one or more dimensions were out-of-tolerance, the nonacceptable (red light) signal was sent out. If requested, any information appertaining to an individual piece tested was printed out on the Creed teleprinter in the computer room. Sample test results were displayed in numerical histogram form on a typewriter, again in the computer room.

In order to increase the capacity of the system, that is, provide space in the computer store for more jigs to be added and their inputs to be processed on-line, programming a permanent set of tolerances into the computer was considered, which would cover most of the piece parts checked. The results would then be interpreted externally by physically drawing in the required tolerances on the histogram. This method was abandoned as being very clumsy in practice. The present method, to be described later, uses local electronics at the gauging jigs to convert the signals received by the computer into standard form, so releasing much memory capacity for other purposes. Further programme modifications enable individual component sample results to be transmitted to the operator out on the shop floor.

The programme changes actually made

(1) Operating procedures have been changed to allow independent control of each line from one single pushbutton per line.

The previous method demanded that a code was dialled on the data switches before any options could be entered.

(2) The range of sampling schemes available has been increased to allow a variety on both AQL 1% and 1.5%. A provision for double sampling schemes has also been made.

(3) No separate tapes are needed to i/p ASC nos. The number of the scheme is entered on the data

switches, the correct option set up, and the push-button referring to the line on which the scheme is to be operated is then pressed.

(4) The option codes have been shortened and simplified to allow more personnel to operate the machine.

(5) Once information appertaining to a particular line, that is, number of dimensions to be checked, and the tolerance on each dimension has been pro-grammed into the computer, it is now unnecessary to repeat this every time the line is recalibrated. A special option enables the line to be recalibrated at the push of the line button.

(6) More checks to ensure that correct options are entered into the computer have been implemented. Audible and visual warnings are now given for any incorrect procedure.

(7) Programme modifications allow further analyses of the histograms. On option request, a printout of the sample mean and standard deviation of each dimension can be obtained.

As the convention +TU (+ve side of histogram) ≡+metal, knowledge of the process standard deviation, and hence the theoretical range, can be used to calculate an optimum setting point on the tolerance band, to enable a maximum number of components to be produced within tolerance before the tool needs resetting. Coupled with this facility is out-putted a number termed the *coefficient of incapability*. To facilitate easy transmission to the shop floor, a coefficient of 10 is made to represent a range of 20 TU, that is, the total tolerance band. A co-efficient of 10 indicates that the process is only just capable of producing pieces to the required tolerance on the dimension referred to by the coefficient of incapability.

Hence a theoretical range of 0 gives a $CI = 0$, that is, *zero incapability* implies total *capability*.

A CI value of $1 \rightarrow 10$ indicates that the process is under control. CI values greater than 10 imply increasing *incapability*.

The process standard deviation for each dimension is estimated by choosing samples greater than 200. Once a determination has been made, the setting points and upper control limits are calculated. These limits are then used as lower and upper action limits respectively for process control charts for the mean of each dimension checked in the sample. In calculating these limits, the computer is programmed to take into consideration the error in determining the mean of a small sample.

Process control chart for mean of a typical dimension

The setter is required to set his machine so that the mean of a small sample taken from the machine is as close as possible to the top side of the setting point line.

When the mean drifts to the upper control limit the tool is reset.

The knowledge of the process standard deviation, and the ability speedily to compute the standard deviation of a sample of each dimension checked, enables the range or spread of each dimension to be controlled more easily than by conventional methods of estimating the standard deviation of the sample from several ranges.

The theoretical range calculated from the standard deviation of a small sample of continuous pieces from the machine will considerably underestimate the pro-cess standard deviation. However, this discrepancy can be remedied by statistically reducing the process standard deviation by a factor depending on the size of the small sample and on the degree of certainty required.

This reduced process standard deviation is known as the CI zone and is used as an upper action limit for the range control chart. As the CI is directly propor-tional to the range, the CI can be plotted on the range chart with the upper action limit and the plots scaled accordingly.

It is emphasised that the CI for a small sample will be much less than the CI for a sample greater than 200.

To avoid anomalies, the CI for a small sample is referred to as the *control index* for the sample. The CI for a large sample (greater than 200) is known as the *coefficient of incapability*.

(8) A provision has been made in the programme to compensate for false readings during the testing of a component, due to swarf wedging between the probe of the transducer and the surface being contacted. Inputs are ignored if any dimension shows a reading of greater than +2T. The results are neither included in the sample results, nor in the analysis.

Note: convention determines that +metal ≡ +TU, so the presence of any swarf constitutes +metal.

(9) Conversely, input results are monitored for short feeds, that is, excessive absence of metal. It is assumed that if the process is in control and the machine capability does not exceed the tolerance allowed, no pieces having any dimension less than $-2T$ will be produced under normal conditions. It can be safely assumed that any dimension giving a reading of less than $-2T$ constitutes a short feed. This piece is counted as a legitimate reject to comply with the sampling plan but is not included in the analysis results.

PROGRAMME CHANGES TO ACCOMMODATE FURTHER MULTIPLEXING OF THE LINES

To make maximum use of the existing hardware facilities and to save superfluous and uneconomical additions of hardware, the transmission of the data to the multiplexed lines is simplified by coding the data by programme and then transferring this information from the central processor to external decode logic. An additional advantage of using this method is that only one sixty-cored cable is required to carry all the data to and from each of the twelve jigs connected to each of the multiplexed lines. A twenty-four bit code is then decoded out on the shop floor.

As a greater amount of data is now being transmitted from the computer, it was considered necessary to increase the rate of output by giving more frequent priority to the programme controlling the output. This was achieved by increasing the frequency from 64 pulses/s to 128 pulses/s. It is now possible to complete a component test in less than $\frac{1}{4}$ s.

A completely new programme block was written to encode the data for transmission.

The codes used are as follows.

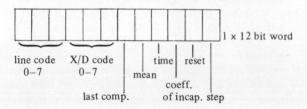

line code X/D code   time reset
0–7   0–7   mean
                              coeff.
      last comp.   of incap. step

One of the disadvantages of multiplexing a line is that a problem arises when the number of dimensions being checked varies from jig-to-jig connected to that line. On a standard nonmultiplexed line, the number of dimensions undergoing inspection is programmed into the computer during the calibration procedure, and this number remains constant. However, on a

multiplexed line, the situation alters when a component having, for example, four dimensions under inspection is sample tested immediately after a component having eight dimensions under inspection. The line will be programmed to accommodate eight dimensions, the maximum number allowed on that line. Hence the first four dimensions will read correctly, but the next four will give false readings. The problem is overcome by making the last four dimensions give a reading that the computer will be able to differentiate from all the others, and hence recognise and ignore the redundant dimensions. This prevents unnecessary processing of invalid inputs and so avoids outputting meaningless results.

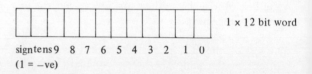

sign tens 9 8 7 6 5 4 3 2 1 0
(1 = –ve)

This programme also controls the amount of time that one of the additionally multiplexed jigs is connected to the line. When the test is completed (or the allocated time has elapsed) a step pulse is sent out to the line multiplexer so that a continuous scan can be made for any other jig requesting access to the computer.

In figure 3 an illustration of the results of an actual sample is given.

SAMPLE 1 200 pieces machine drifting out of tolerance

5   9.25   25.01.72

system accept	36	2908	9	1	200	200	0								
1.455 0.0060							2	55	106	34	2	1			
0.292 0.0030							2	24	42	53	45	30	3	1	
0.306 0.0030							1	4	**14**	46	56	37	27	13	2
0.220 0.0055					2	8	18	44	38	49	27	11	2	1	

histogram analysis	total	mean	std. dev.	CI	setting points	CI zone
1.455 0.0060	200	+3	0.7	2	7	2
0.292 0.0030	200	+4	1.3	4	4	3
0.306 0.0030	200	+6	1.5	5	4	4
0.220 0.0055	200	+3	1.6	5	3	4

SAMPLE 2 6 pieces machine setting corrected

5   11.00   25.01.72

system gonogo	37	2908	10	1	6	6	0
1.455 0.0060	4	2					
0.292 0.0030			1	1	3	1	
0.306 0.0030				1	4	1	
0.220 0.0055				1	3	2	

histogram analysis	total	mean	std. dev.	CI
1.455 0.0060	6	−7	0.5	2
0.292 0.0030	6	−4	1.0	3
0.306 0.0030	6	−4	0.6	2
0.220 0.0055	6	−3	1.1	3

Figure 3. Results of two samples.

One further point must be made here: the computer retains control of the time allowed for sampling. An operator of a jig places a piece part in his jig and presses the operating button. The computer

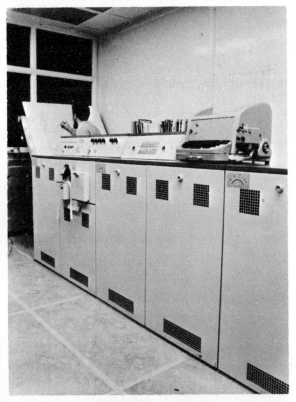

Figure 4. General view of computer.

continuously monitors the jigs and once it receives a signal will lock onto that jig for an allocated time period dependent on the sample size.

If the sampling is completed before this period has elapsed the computer immediately on completion of the operation will continue its monitoring of all jigs: it will do this automatically even if the operation is not completed at the end of this interval. The reason for this procedure is to prevent one operator taking up too much of the computer's time for any one sample or indeed monopolising it altogether.

MODIFICATIONS TO ELECTRONICS AT GAUGING POINTS

In the original system and as now in operation at the inspection audit area, the transducers are of the DC/DC type transmitting a voltage of 80 mV per 0.001 in linear movement to a maximum of 0.050 in.

The computer is programmed to take this analogue signal, to digitalise the reading into twentieth of whatever tolerance has been read in for the given dimension and to record in these units the deviation right and left of the initial setting point. The transducer is set to the centre of the tolerance either by using a mid-limit ground setting piece or some measuring device to give the same results.

The accuracy of measurement in the above method is about two tenths of one thousandth of an

inch. The read out is accurate to one twentieth of the tolerance. Both these limits have proved acceptable for audit inspection and for process control.

The above methods and equipment had several defects when a wider and more sophisticated method of process control was sought. Firstly, they were wasteful of computer space when more than one jig had to be switched into the same computer gauging point. Reprogramming was necessary for each piece part to be checked and this was true even where a standard set of tolerances was programmed, so that only the address of a jig and the subaddresses of the dimensions had to be included to ensure correct

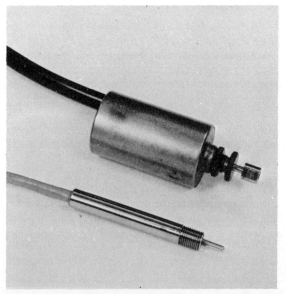

Figure 5. Comparison of transducers with and without built-in electronic system.

Figure 6. Multidimensional measuring jig by probe and gauge.

location of signals in the two-way communication system. Secondly, the transducers, themselves, carried the electronics which produced the DC/AC exciting voltage for the LDVT and the AC/DC demodulator and phase-sensitive detector. They were bulky and often difficult to install even where the movement of the anvil contacting the piece part was transmitted by levers.

A solution to the above problems was eventually found by separating the electronics from the probe at the gauging station and using conventional AC probes. The local electronics do a number of operations. Firstly, an adjustment is provided so that a standard output measured in volts is obtained when the transducer is set at the centre of the tolerance. Secondly, the gain adjustment is provided so that the computer always receives an analogue signal which it can digitalise directly into twentieths of the tolerance without being told what the tolerance is. Thirdly, AC probes are much slimmer and so more easily accommodated in a jig. Fourthly, by adjusting the gain locally, there is no restriction on buying one type of transducer, though having settled on one it is as well to stick to it all the way through one subsystem to prevent confusion arising and the mixing of transducers in jigs.

The operator controlboxes have had to be enlarged to contain the signal lamps for indicating the size of each dimension as against merely three control lamps in the audit inspection boxes and it has been found convenient to house the extra local electronics at the back of each box. Normally, provision is made for eight dimensions to be measured on any one jig and their electronics are mounted on eight boards sliding into a card frame.

Audit inspection is normally carried out on an attribute basis so that the inspector does not wish to know the dimension of a given part, but only whether it is defective or nondefective. In reporting back to the shop floor if a histogram is found to extend beyond the limits of the dimension, and more detailed information is required, a wider tolerance can be programmed and this information obtained. However, where direct process control is required it is useful to be able to see directly and immediately what the histogram of the sample is like when it extends in both directions (or either direction) a little beyond the limits. With the release of computer space as indicated above this has now become possible and the display can go up to nineteen twentieths of the tolerance in either direction, that is, a total width of 38 twentieths of the tolerances.

Thus the present process control set-up for one shop is summarised as follows. The equipment consists of:

(1) Several gauging jigs dealing with up to eight dimensions using AC transducers.

(2) Each of the above has an electronic control and display box. The electronics set the mid-point voltage and gain to give a standard signal transmitted by cables to the computer. There is on this box a control button which the operator presses after he has loaded a piece to the gauge. The result of any such test is displayed in steps of one twentieth of each tolerance concerned by means of two 'number lamps'.

(3) A relay cabinet which takes the output pulse signals from the computer to switch through the reed relay matrix. A local power source operates the above number lamps.

(4) At the central point of the group of gauging heads a printer is located which provides a printout of the mean results per dimension of one gauging operation or the statistics arising from the taking of a sample.

The actual quality control procedures used and how they are adapted to this particular system are described in the next paragraphs.

QUALITY CONTROL PROCEDURES

The basic concept underlying the whole of the above extension of the computerised system (as quoted from the original project) is 'to enable sources of error to be readily identified and often corrected before the error exceeds the permitted tolerance': to do this statistics of samples replace the normal control charts. These statistics are:

(1) The spread of variability of the process at any given moment. This is measured in units which would give a zero reading if all measurements of a dimension were in one computer column (that is, if virtually within 1/20 of the tolerance of that dimension); 10 if exactly equal to the tolerance (that is, exactly within 20 computer columns); and exceeding 10 if greater than the tolerances.

(2) The location of the mean of the sample, that is, in which computer column it lies.

(3) The ideal setting point, that is, the expected mean values of the samples to give the maximum life between resetting for tool wear, and the ideal point at which the reset should take place. Actually where the variability is less than the tolerance the setting point is taken as the bottom or top limit plus or minus twice the sum of estimated standard deviation of the mean of the sample. Normally these are based on an initial sample of 200 items and recalculated at appropriate intervals. In other words, the result is the equivalent of using normal control charts.

In order to carry out such apparently complex procedures not only must the operator (and more-to-the-point, the setter) be trained in them but must have an appreciation and be fully converted to their usefulness. This implies a close association and liaison between shop floor and the computer room. The method by which this was achieved was to appoint a quality technician, who had been with the project since its inception, to back up this area. This proved to be an essential post and well carried out by the first one appointed. In practice the shop floor system has greatly simplified the machine setting by indicating precise mean settings, on the standardised tolerance scale using ±10% tolerance increments. It was possible for the machine to be set with a precision not previously possible in an economical time.

There was also a need foreseen from the beginning

of this development to provide printed records under certain circumstances. These could arise mainly in two ways. Firstly, where because an item is safety critical permanent records of process inspection during production were required to be stored batch-by-batch; secondly, when there is permanent inability for an operation to meet requirements or where a comparison is required with a proposed new machine part. In the second case, the higher production supervision, production engineers and even design staff are likely to be involved so that it is essential that the quality technician understands fully the implication and limitation of the data with which he is dealing.

The introduction of the system has affected two considerable areas of production at Raleigh Industries as will be appreciated from the above outline and in doing so affected the attitude of a number of the higher supervisors and executives of the company. The account cannot close therefore without some reference to these factors, the full effects of which have yet to be felt.

EFFECT ON PERSONNEL

The initial installation had little effect on anyone outside the quality control organisation and even in this area it was mainly limited to the audit inspection room where, after being a seven days' wonder, it grew to be merely part of the background. Exceptions were the supervisor setter and operator of one section where, due to the rapid feedback of information arising from the sample inspection, the resultant sorting was reduced from 40 h to 2 h and rectification to zero.

Here the progress of events forced attention upon it. After the first months following installation of the system, progress was slow and continued so until L. A. Cordingley became the engineer-in-charge when development was speeded up. It was further aided by the appointment of G. Pike to look after the software side.

However, despite the initial slowness in developments the accountants checked and approved the level of savings being made in the summer of 1970. This together with the increased pressure for greater production at higher quality levels caused production and executive management to show an increasing level of interest in proposals for process control based on the computer system. Already tentative steps had been taken by the installation of lines to separate areas of the factory, one 400 yd from the computer room. These developments have been discussed above, but they lead to a great change in the attitude of higher management and when two projects for the extension and enlargement of the system were proposed in 1971, both were accepted.

Pressure from above and also pressure for greater conformance to tighter limits forced designers and production engineers to take into account the new technique and its effects; reluctantly, because computers in general have a bad name even with the more progressive engineers. Also there is always resistance to new procedures, methods and equip-ment all of which tend to have their teething troubles. This reluctance is not by any means reduced when the system is not only new to the firm but the first in the entire field. However, once this typical reluctance was overcome and the possibilities ahead appreciated the pressure towards the development of new techniques and methods in quality control by the computer become, if anything, too great.

The shop floor supervisors and operators were at first both suspicious and interested. They were much closer to the day-to-day working and so intimately connected with progress. Perhaps the accurate predictions of the results of some of their actions made by the computer section were the most convincing arguments for the usefulness of the system.

It is not too much to say that the computer process control is bringing about a revolution in the attitude of all personnel affected, whether in design, production engineering or quality control, towards process control for automatic machines and that of the present computer system in particular. However, to take the full advantage of the rapidity and completeness of data arising a whole new series of procedures has had to be worked. This will be true in any system and it will be essential to intensify the cooperation between all functions. The effects have not yet been fully worked out here at Raleigh Industries Limited but progress is being made.

CONCLUSION

It is difficult to summarise properly the above paper which in a way is a description of the latest step in realising a system under consideration in the early sixties but whose final goal was only partially foreseen. The ground covered is the progress from general thought in the early sixties to an actual project in 1966/67; its installation in 1968 followed by a period of stagnation in development, but much hard thought and experimental work to this latest stage of the introduction of process control. This has not only brought advances in techniques but also changes in the attitude of personnel from the shop floor to higher management.

To look into the future, it is considered by the computer process control team at Raleigh Industries Limited that this last step is only one more on the way to a completely computerised control for an automatic machine shop. This control will eventually include:

   (i)   process control and audit when necessary,
  (ii)   machine loading,
 (iii)   progress of work,
 (iv)   machine efficiency.

It would appear that a small local computer is the best approach with possibly even more localised electronic units keyed in. There seems no reason why this computer should not be part of a larger computer system providing control data for all activities of any large firm.

The team at Raleigh consider that it is a natural development following on the work here and elsewhere.

REFERENCES

1. C. WATKINS, 1969. Computerised inspection of components. *SIRA Conference*, May.

2. C. WATKINS, 1972. Quality assurance by computer. *PERA Conference* on The small low-cost computer: its application in manufacturing. 23rd and 24th February.

3. C. WATKINS, 1970. Quality data through inspection of variables. *Lausanne Conference*, June.

4. C. WATKINS, 1971. Electronic gauging and small-lot production. *International Conference*, Pozen. 10th to 12th March.

A METHOD OF ANALYSING THE LOGIC DESIGN OF PNEUMATIC SEQUENTIAL CIRCUITS

by

R. M..H. CHENG*

SUMMARY

In the absence of systematic analysis, a proposed 'logic' design of pneumatic sequential circuits may be tested by using a model, or some form of simulation process. These often incur unnecessary expense of time, effort and equipment. A simple time-chart method of analysis is described in this paper. It is based on the principle of 'reduction-to-absurdity' and is to be distinguished from a simulator.

NOTATION

Cylinder movements

$A+, A-$ = extension stroke and retraction stroke of pneumatic actuator A.

Bistables

Y_i = bistable i

Y_{is}, Y_{ir} = set and reset input signals of Y_i respectively.

y_i, y_i' = set and reset states of Y_i respectively.

Logic operators

$a.b.c. \ldots$ = a AND b AND c . . .; sometimes represented simply as abc . . . . . .

$a+b+c. \ldots$ = a OR b OR c . . .

a' = NOT a

INTRODUCTION

In recent years, pneumatic control has been replacing electrical control in many industrial applications. These vary from the simple light-duty assembly machines of electronic components, to a fully automatic investment-casting plant. Other uses are found in numerous machine tools, materials handling, packaging machines, tube benders and so on. The essential feature of pneumatic control lies in an arrangement of air cylinders which perform the various operations by reciprocating in a controlled manner, with respect to speed as well as the sequence of reciprocations. These pneumatic sequential machines, as they are often called, contain in the control circuit various types and quantities of directional control equipment. In an all-pneumatic circuit, spool valves or poppet valves are commonly used to carry out the logic. Limit sensors are also fitted to the cylinders to provide feedback information to the control logic, so that one operation follows the other in perfect order. There is also an increasing tendency to use other logic devices for directional control, because of faster operating speeds, size and weight, easy integration with other control equipment and sometimes costs. They include electromagnetic relays, solidstate switches, fluidic devices and so forth.

Usually, though not always, two feedback signals are provided for each cylinder, one for each extreme position. Because of the constant interaction of these feedback signals on the remainder of the logic circuit, a circuit designer has to be competent and ingenious enough to see that the circuit contains no erratic cylinder movements. Moreover, having come up with a circuit (or being confronted with an alternative design) he must be able to judge whether the circuit can carry out what he wants. In other words, he should be able to analyse a circuit, systematically and efficiently.

Unlike the aspect of design of pneumatic sequential circuits, circuit analysis is very rarely mentioned in the literature. By introducing a simple analysis method in this paper it is hoped to initiate some discussion and further work in this area. The method of analysis will be introduced by a number of simple examples.

A SIMPLE EXAMPLE

Consider the following sequence of movements of two air cylinders A and B:

$$A+, B+, B-, A-.$$

The + refers to the outward stroke of the cylinder (extension), and the − to the return stroke (retraction), according to the usual practice[1]. Two logic circuits are proposed for this sequence and are shown in figures 1(a) and (b). If both circuits are logically operable, figure 1(a) is certainly more desirable since it contains less equipment. A logic analysis is therefore necessary before accepting any circuit.

One way to find out if the circuit is operable is to construct a working model. This is not always practicable, as it involves time, effort and equipment.

An alternative is simulation on paper. By taking note of all flow patterns at all times, the cylinder movements are worked out and compared with the desired sequence. Applying this technique to figure 1(a), it may be shown that the circuit breaks down at the end of the movement $B+$, when the valve Y_B is subject to control signals at both ends, one coming from the limit sensor $a2$, and the other from $b2$.

* Mechanical Engineering Department, University of Birmingham.

Since pneumatic spoolvalves are not generally designed to cater for such ambiguous situations, the circuit is to be rejected. On the other hand, there is no breakdown of any sort with the circuit of figure 1(b). It is therefore an acceptable design.

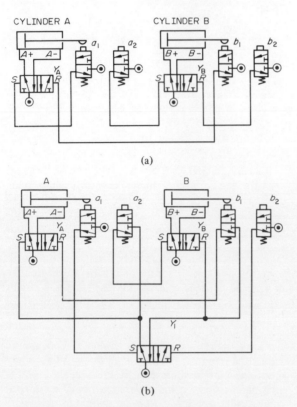

(a)

(b)

Figure 1. Two proposed logic circuits for $A+, B+, B-, A-$.

When a circuit contains a large number of control valves and consequently there are many interconnections, any serious attempt to apply the simulation technique just described will be a formidable task. It would require a good layout of the circuit diagram, a systematic recording of the current flow patterns at every instant and a good deal of time and concentration. An error made at any one point will nullify all subsequent effort, leading to an erroneous conclusion. Although it is true that the digital computer **may** profitably be employed, such simulation programmes are not yet available to the public. In any case, access to a suitable computer could be a problem in itself.

Because of these difficulties, it is the purpose of this paper to report a simple method which is useful principally as a means of testing the acceptability of any given pneumatic sequential circuit. The prerequisite to applying this method of analysis is that the circuit is described in the form of a set of switching functions in terms of the basic logic operators *AND, OR, NOT* and *BISTABLE*[3]. If the circuit consisted of such devices as relays, solidstate switches and fluidics elements, such a description is straightforward and may readily be obtained from the circuit diagram. On the other hand, for a circuit that contains a pneumatic spool-type valve for directional control, such a description is less direct. An account of the logic operators and the logic

equivalence of pneumatic valves is given in the Appendix for readers who are not familiar with the subject.

Irrespective of the type of logic devices involved, a pneumatic sequential circuit may generally be specified by a list of bistables with their set and reset switching functions. For the convenience of subsequent discussion, the following definitions will be useful. If a bistable device is to be used directly to govern the cylinder movement (as Y_A and Y_B in figure 1), it is called a primary bistable: otherwise it is an auxiliary bistable. Y_1 in figure 1(b) is an example of an auxiliary bistable.

It may be shown that figure 1(a) has the following logic specifications:
Primary bistables: Y_A, Y_B
Auxiliary bistables: nil.

$$Y_{As} = \text{setting function of bistable } Y_A$$
$$= a1 \qquad \qquad \dots(1.1)$$

$$Y_{Ar} = \text{resetting function of } Y_A$$
$$= b1 \qquad \qquad \dots(1.2)$$

$$Y_{Bs} = a2 \qquad \qquad \dots(1.3)$$

$$Y_{Br} = b2 \qquad \qquad \dots(1.4)$$

where $a1$, $a2$, $b1$, $b2$ are the outputs from the corresponding limit valves in figure 1(a).

Similarly for figure 1(b):
Primary bistables: Y_A, Y_B
Auxiliary bistable: Y_1

$$Y_{1s} = a1 \qquad \qquad \dots(2.1)$$

$$Y_{1r} = b2 \qquad \qquad \dots(2.2)$$

$$Y_{As} = y_1 \qquad \qquad \dots(2.3)$$

$$Y_{Ar} = b1 . y_1' \qquad \qquad \dots(2.4)$$

$$Y_{Bs} = a2 . y_1 \qquad \qquad \dots(2.5)$$

$$Y_{Br} = y_1' \qquad \qquad \dots(2.6)$$

where y_1 and y_1' refer to the set and reset states of the auxiliary bistable Y_1 (see Appendix).

Using the specifications in this form, we shall proceed to introduce the method of analysis.

PRINCIPLE OF THE METHOD OF ANALYSIS

The principle of reduction-to-absurdity is employed in this method. As such the method is to be distinguished from a simulation.

Basically, one begins by making an assumption. If all subsequent development agrees with the assumption, then it is correct. Any contradiction and incongruity will nullify the assumption.

Presently, the assumption to be made is that the circuit functions properly, or specifically that the cylinders will actually operate according to the required sequence of movements.

Consequently, one can determine the output signals from all limitswitches associated with the cylinders. Based on this information, and on the set of switching functions that describe the circuit, one can test whether the bistables will set and reset in such a way that the sequence of cylinders will actually take place as assumed. The test is very simply, based on the following two conditions:

Condition (a): no bistable should be subject to set and reset signals simultaneously.

Condition (b): when the bistable is supposed to be in the set state, the reset signal must be off, and viceversa. (The set state of a primary bistable corresponds to + of the cylinder it governs, and viceversa.)

To carry out such an analysis efficiently, a timing chart may be used.

A TIMING CHART FOR ANALYSIS

The timing chart contains as many columns as there are stages of cylinder movements. Taking figure 1(a) as a working example, the timing chart has 4 columns as shown in figure 2. The first row contains the stage numbers, and the cylinder movements are

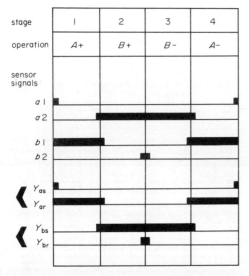

Figure 2. Time chart for analysing figure 1(a).

entered in the second row. The vertical dividing lines indicate the instant at which the next movement will commence. By following the movements of cylinder A for example, deduce the waveform of the signals from the limit valves $a1$ and $a2$; that is, the durations for which the cam acts on the valves' action. As A begins to move forwards the cam carried by its piston loses contact with $a1$. The waveform $a1$ in the timing chart drops to 0-value at a point slightly to the right of the first dividing line. Note that for the purpose of this analysis, it is not important to adhere to any strict time scale as long as the **sequence** of events is correctly represented, and that an appropriate time delay is allowed for the switching response of the logic devices. Towards the end of this $A+$ stroke, the cam engages with sensor $a2$, and the waveform $a2$ rises to a 1-value at a point slightly to the left of the dividing line between stage 1 and stage 2. The remainder of the waveforms for $a1$ and $a2$ may be similarly worked out by considering respectively cam engagement. The same applies to

waveforms $b1$ and $b2$ for cylinder B. These waveforms constitute the basic constraints as a result of the assumptions that have been made.

The waveforms of switching functions on the same bistable are then determined and grouped adjacent to one another. Since $Y_{As} = a1$ and $Y_{Ar} = b1$ according to equations (1.1) and (1.2), their signal waveforms are simply those of $a1$ and $b1$ respectively. Similarly the waveforms of Y_{2s} and Y_{2r} are those of $a2$ and $b2$, as shown in figure 2.

The next and final step of analysis is to study the set and reset waveforms of the same bistable one-by-one. If it is found that conditions (a) and (b) are satisfied by the waveforms of every bistable, the circuit is acceptable.

Referring to figure 2, it is seen that Y_{As} is on towards the end of stage 4. But then Y_{Ar} is also on. This violates condition (a). Without proceeding any further, it may be concluded that this circuit (figure 1(a)) is not operable, and the analysis is complete. The analysis also demonstrates where the fault lies.

Where the circuit contains an auxiliary bistable, the analysis is slightly more complicated unless information is also available on exactly how the auxiliary bistable is supposed to behave in the sequence. When this information is known, the assumption to be made also includes the correct switching of the auxiliary bistable. Corresponding to the latter (there being no limit valve associated with it), the waveforms of the set and reset states of the bistable also form the basic constraints. Otherwise the waveforms for the set and reset signals of the auxiliary bistable have to be worked out first (according to equations (2.1) and (2.2)), so as to establish the waveforms for the output states. The analysis for figure 1(b) is shown in figure 3, assuming

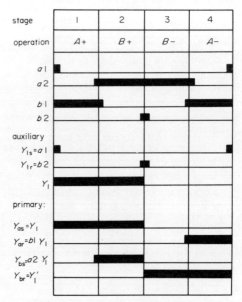

Figure 3. Time chart for analysing figure 1(b).

that it is not known exactly when Y_1 is in the set or reset state.

Since Y_{1s} and Y_{1r} are seen to satisfy conditions (a) and (b), one can proceed with the analysis. Based on the waveforms of state of the auxiliary bistable Y_1 as well as those of the mechanical control action of the limitvalves, the switching functions of the primary bistables can then be worked out.

From equation (2.3), Y_{As} is shown to have the same waveform as y_1. However, from equation (2.4), $Y_{Ar} = b1 \cdot y_1'$. The waveform is *ON* whenever both waveforms for $b1$ and y_1' are *ON*. Examination of Y_{As} and Y_{Ar} shows that they satisfy both conditions (a) and (b). Similarly, Y_{Bs} and Y_{Br} may be shown to satisfy these conditions too. Thus the circuit is found acceptable.

TWO OR MORE AUXILIARY BISTABLES

The following example is chosen to illustrate the analysis of a rather unusual circuit. It has two auxiliary bistables. The circuit was first discussed by Bouteille[4].

Two cylinders are arranged as shown in figure 4 for handling components coming through a chute. In

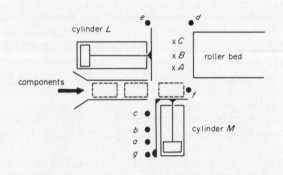

● normally-off limit valve

Figure 4. A materials handling arrangement.

the following description of the work sequence the original notations are retained as far as possible:—

Components come in position and act on valve f.

Cylinder M pushes the first component to position C, acting on limitvalves a, b and c, and then retracts.

The next component is pushed to position B by cylinder M, acting on valves a and b.

The third component is pushed to position A by cylinder M, acting on valve a only.

Cylinder L pushes the three components to the roller bed, and retracts after acting on valve d.

The sequence to be considered is thus

$$M+,\quad M-,\quad M+,\quad M-,\quad M+,\quad M-,\quad L+,\quad L-.$$

According to the design given by Bouteille, the switching functions are

Primary bistables: Y_M and Y_L

$$Y_{Ms} = efg(y' + x) \qquad \ldots\ldots (3.1)$$

$$Y_{Mr} = c + bx + ay \qquad \ldots\ldots (3.2)$$

$$Y_{Ls} = egx'y \qquad \ldots\ldots (3.3)$$

$$Y_{Lr} = d \qquad \ldots\ldots (3.4)$$

Auxiliary bistables: X and Y

$$X_s = c \qquad \ldots\ldots (3.5)$$

$$X_r = ay \qquad \ldots\ldots (3.6)$$

$$Y_s = bx \qquad \ldots\ldots (3.7)$$

$$Y_r = d \qquad \ldots\ldots (3.8)$$

The basic constraints a to g are first determined and are shown as the first seven rows of waveforms in figure 5. Since there is no information available concerning the states of the auxiliaries X and Y at any instant of time, they have to be determined from the switching functions (3.5) to (3.8). The procedure is shown in the next six rows of figure 5, where the sequence of setting and resetting X and Y follows the labelled arrows 1 to 8. Thus, starting with both X and Y in the reset state, one proceeds along arrow 1 to set X according to (3.5), making $x = 1$ (arrow 2). Then Y is reset (arrow 3), according to (3.7), making $y = 1$ (arrow 4). Shortly afterwards, X is reset (arrow 5) according to (3.6), making $x = 0$ again (arrow 6). Eventually Y is reset (arrow 7) according to (3.8), making $y = 0$ (arrow 8). The set and reset waveforms for X satisfy the two conditions, and so do the waveforms for Y. Thus it is reasonable to carry on.

The switching signal waveforms for Y_M and Y_L are then determined according to (3.1) to (3.4) in the usual manner. These are shown as the last four rows of figure 5.

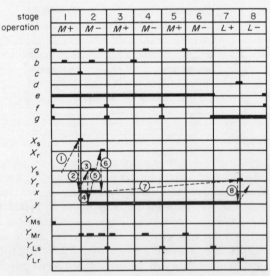

Figure 5. Time chart for analysing Bouteille's design.

It is seen that Y_{Ms} occurs only once, instead of three times, whereas Y_{Mr} occurs six times. Similarly, Y_{Ls} occurs three times, instead of only once. Thus although condition (a) is satisfied in all cases, it is not so for condition (b). The waveforms of the primary switching signals contradict the assumed sequence of operations, and the circuit is not operable.

CONCLUSION

Through the use of a number of examples, a method has been developed which enables a logic circuit to be analysed systematically and efficiently. The method may be summarised in the following steps:

(1)   from the circuit diagram, a set of switching functions is first obtained.
(2)   assume that the cylinders operate in the correct sequence.
(3)   construct a timing chart based on the correct sequence.
(4)   waveforms of mechanical control action of the limit sensors are determined as the basic constraints.
(5)   for those auxiliary bistables (if any) whose switching duration are exactly known, establish the waveforms of their output states as part of the basic constraints.
(6)   for the remaining auxiliary bistables (if any), determine the waveforms for the set and reset switching signals.
(7)   determine the set and reset waveforms for all primary flipflops.
(8)   apply conditions (a) and (b).

Where there are time delays and/or other forms of control signal also used in the directional circuit, the method may be suitably modified.

Subsequent development of this method of analysis has led to a powerful design method, which is beyond the scope of this paper. The reader is referred to reference 2.

ACKNOWLEDGEMENT

The author is grateful to Professor S. A. Tobias and Dr. K. Foster of the University of Birmingham for their interest and support, and for providing facilities for this work.

REFERENCES

1. F. X. KAY, Pneumatic circuit design, Machinery Publishing Ltd. (1966).
2. R. M. H. CHENG and K. FOSTER. On a systematic method of designing fluidic-pneumatic control circuits, *Proc. I. Mech. E., 186* (1972).
3. L. LEWIN. Logical design of switching circuits. Nelson & Sons Ltd (1968).
4. D. BOUTEILLE. Recent developments in piston fluid logic for general automation. *Hydraulic Pneumatic Power,* (Dec 1967) p. 708.
5. CETOP (European Oil Hydraulic and Pneumatic Committee). Symbols for hydraulic and pneumatic equipment and accessories for fluid power transmission. Provisional Recommendations, 1965.

APPENDIX

Logic operators

The following are the definitions of a set of fundamental logic operators[3]:

AND An *AND* operator provides an output only when all the input signals are present simultaneously.
 OR An *OR* operator provides an output when at least one input is present.

NOT Also known as *INVERTOR* or *NEGATOR*, produces an output when the input is absent.

These three operators are combinational operators, providing an output depending on the present input condition only.

BISTABLE: Better-known as the set-reset bistable or *FLIPFLOP*. Such a device Y, say, has two input lines designated Y_s (set input) and Y_r (reset input). The corresponding set and reset outputs are designated y and y' such that when one is on the other is off. Once Y is set it will remain in that state (that is, y on) even after the set signal has been removed. The *BISTABLE* is thus said to possess memory. Similarly once Y is reset, it remains so (that is, y' on) even after the reset signal has been removed.

Control actions of pneumatic valves

A number of standard control actions can be employed to operate a pneumatic directional control valve so that its spool takes one position or the other. A list of these is given in BS 2917: 1957. For the purpose of this paper, these control actions may be classified as (a) designed control actions and (b) restorative actions:

(a)   designed control actions:
       manual
       mechanical (cam acting on roller, plunger and so on)
       solenoid
       pilot (air pressure)

       air motor
       electric motor
       hydraulic motor
       cylinder
       centrifugal
       compensation
       thermal.

(b)   restorative control actions.

All spring-return actions are restorative. They are arranged to act on the valve permanently, to be overcome by a designed control action at specific instances, in order to displace the valve. Such control actions are commonly provided by a mechanical tension or compression spring, or by an airspring (that is, a constant-bias air pressure).

Logic equivalence of pneumatic valves

A pneumatic control valve is often more complicated than a simple logic gate. Each of its outputs may be expressed as a switching function of its inputs. In the present context, input signals include all pressure signals connected to the input ports, as well as the designed control action required to operate the valve. The spring-return action is a bias, and is not considered as a separate input signal.

Two-position pneumatic valves may be classified as follows:

A valve having one restorative spring control and one designed control action is a combinational device.

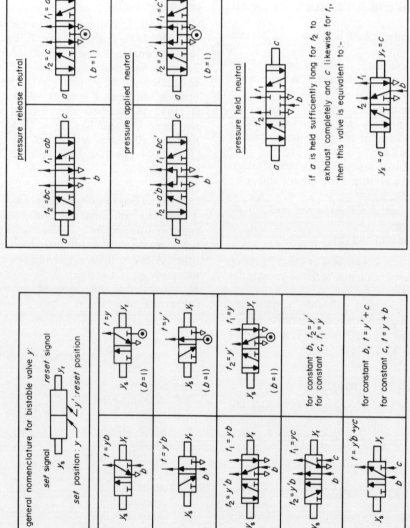

Figure 8. Logic equivalence of pneumatic 3-position valves.

Figure 7. Logic equivalence of pneumatic bistable valves.

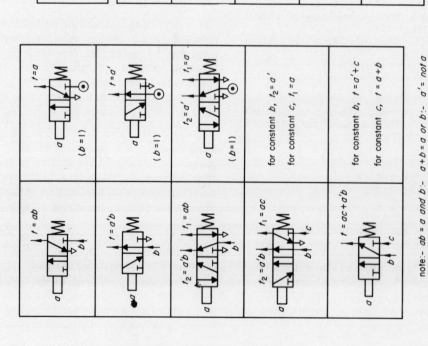

Figure 6. Logic equivalence of pneumatic combinational valves.

It has one stable position, to which it is restored on the removal of the designed control signal.

A valve that is operated by two designed control actions has two stable positions. It is therefore a *BISTABLE*.

Figure 6 shows a number of common combinational pneumatic valves. The symbols are according to the recommendation of CETOP[5]. For every output t_1 or t_2, there is a corresponding switching function describing the logic property of the valve. An exhaust port is not considered as a separate output unless the exhaust air is purposely directed to switch another valve.

Figure 7 shows some common bistable valves. Because of the bistable characteristic of such valves, the outputs t_1, t_2 are more appropriately expressed in terms of the current state or position of the valve and not of the designed control actions.

Figure 8 shows some of the 3-position valves. They are less common in automatic sequential applications.

METAL CUTTING
AND TOOL WEAR

EFFECT OF STRAIN-RATE SENSITIVITY ON SCALE PHENOMENA IN CHIP FORMATION

by

J. LARSEN-BASSE* and P. L. B. OXLEY†

SUMMARY

The results of machining experiments have shown that the strain-rate in the shear zone in which the chip is formed is inversely proportional to the depth-of-cut. In this paper it is shown how the increase in strain-rate with decrease in depth-of-cut and the resulting increase in work material flow stress can explain the increase in specific cutting pressure experienced on moving into the range of small depths-of-cut.

INTRODUCTION

In recent studies of orthogonal machining with sharp cutting tools, Stevenson and Oxley[1, 2, 3, 4] were able to determine the stresses, strains and strain-rates in the primary zone, that is, in the shear zone where chip formation takes place. They showed that the magnitude and direction of the resultant cutting force are determined primarily by the flow stress properties in the shear zone, adjusted to the strain, strain-rate and effective temperature of this zone and that, conversely, measurements of cutting forces and shear angle can be used to determine a material's stress—strain:strain-rate properties. The flow stress—strain-rate data obtained in this way from their experimental work on a resulphurised low carbon steel correlate well with the limited amount of data available for similar material obtained by other means and at considerably lower strains.

different study at a cutting speed of 0.04 ft/min and a depth-of-cut of 0.011 in. It was found that the maximum value of the maximum shear strain-rate in the shear zone $\dot{\gamma}_{max}$, which occurs approximately at the 'shear plane' AB (figure 1), can be represented by the empirical expression

$$\dot{\gamma}_{max} = C\frac{V_S}{t} \qquad (1)$$

where V_S (figure 1) is the shear velocity, t the depth-of-cut and C is a constant depending on the strain-hardening properties of the material. For the steel in question the value of C is 2.59 for V_S (in/s) and t (in) The shear velocity is given by

$$V_S = \frac{U \cos \alpha}{\cos(\phi - \alpha)} \qquad (2)$$

where U (figure 1) is the cutting-speed, α the tool rake angle and ϕ the shear angle. Equation (1) shows that the depth-of-cut, which determines the scale of the process, has a major effect on the strain-rate at which the material in the shear zone is deforming. The purpose of this paper is to clarify the extent to which changes in flow stress with increasing strain-rate might possibly contribute to some commonly known scale phenomena in material removal processes.

The strain-rates in question are expected to fall in the range $10^3 - 10^6$ s^{-1}. From experiments with dynamic punching[5] and with various types of modified

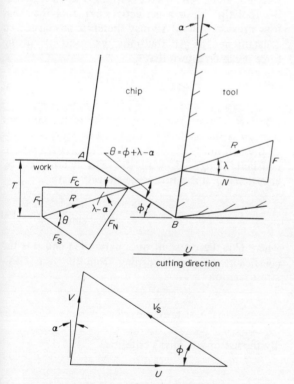

Figure 1. Force relations in chip formation.

Their experimental work covered a range of cutting-speeds from 17 to 817 ft/min and depths-of-cut between 0.005 and 0.0108 in. The relations developed were shown to cover also data obtained in a

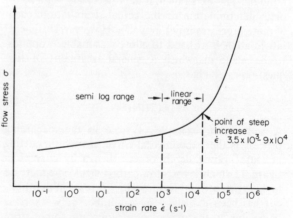

Figure 2. Sketch of the generalised flow stress—strain-rate relationship in semilogarithmic coordinates.

Hopkinson pressure bar equipment[6, 7] it is known that the flow stress of a metal behaves approximately as sketched in figure 2 and it can be seen that a pronounced change in behaviour is observed just within

* Department of Mechanical Engineering, University of Hawaii, USA.
† School of Mechanical and Industrial Engineering, University of New South Wales, New South Wales, Australia.

this range. For values of the true strain rate $\dot{\epsilon}$ between 10^{-1} and 2.2×10^3 the flow stress at a given strain is approximately proportional to $\log \dot{\epsilon}$ which is consistent with a mechanism of thermal activation of dislocations over short-range barriers. Between strain rates of 10^3 and 1.5×10^4 the flow stress is approximately proportional to $\dot{\epsilon}$. This corresponds to the viscous damping region and the slope is a measure of the damping constant. At still higher strain-rate values the curve turns steeply. The turning point moves towards higher strain-rates at increased strain levels, presumably because the number of moving dislocations increases with the total number of dislocations present, that is, with the strain level[6]. Different experiments, using different techniques, report somewhat different values of the strain-rate at the turning point. At a strain of 25%, Dowling et al.[5], using a dynamic punching technique, reported that the pronounced turn in the curves occurred at $\dot{\epsilon} \simeq 3.5 \times 10^3$ for aluminium, copper and mild steel, and at a slightly higher value for brass. Dharan and Hauser[6], who used a modified split Hopkinson bar technique, found the turn at $\dot{\epsilon} \simeq 9 \times 10^4$ for aluminium at 20% strain.

For the strain levels common in machining processes, it would appear that the pronounced increase in flow stress with strain-rate would occur at strain-rate values between 5×10^3 and 10^5 s^{-1}. If strain-rate sensitivity is an important factor in scale phenomena in machining operations, it can be expected that, when the depth-of-cut is decreased, the corresponding increase of the strain-rate in the shear zone will result in an increased flow stress in this zone and thus an increase in specific cutting pressure; that is, the cutting force in the cutting direction/area of cut. The effect should be noticeable at all strain-rates and become pronounced as the strain-rate exceed $10^4 - 10^5 \text{ s}^{-1}$.

The validity of the correlation outlined above will now be examined for machining, sliding abrasion and grinding. The experimental data necessary for a complete determination of the actual strain-rate in each of these processes is not available. However, sufficient information is at hand to allow reasonable estimates to be made on which a general treatment of the question can be based.

MACHINING

A well-known example of scale in the machining process is the experimental curve of specific cutting pressure against depth of cut shown by curve I in figure 3, which is for a low carbon steel. The increase in specific cutting pressure with decrease in depth-of-cut has been explained by a number of workers as being due to the influence of ploughing forces from the tool tip. In practice the tip does always have a certain radius of curvature, the effect of which, in terms of ploughing and frictional forces, increases in relative proportion as the depth-of-cut is reduced[9,10].

In order to study the possible effect of strain-rate sensitivity on this relationship we will proceed as follows. Based on the experimentally-determined equation (1) and theoretical and experimental analyses

of deformation and forces in machining, Stevenson and Oxley[1,2,3,4] have developed a set of general expressions which relate cutting forces and angles to the flow properties in the shear zone. These expressions have been applied to machining data obtained for a low carbon steel to develop relations between stress, strain and strain-rate for this material for the range of general interest in machining[11]. These relations were found to correlate closely with data obtained by other means. We shall use these stress–strain:strain-rate relations together with the general expressions to extrapolate into the range of small depths-of-cut. The calculations are for a perfectly sharp tool and possible ploughing forces are thereby ignored. It is assumed that this extrapolation is permissible, that is that the shape of the material flow lines does not change with the depth-of-cut and that equation (1) applies even down to very small depths where the number of grains in the cut becomes small. Since no applicable experimental determination of the shear angle is available at very small depths it will be necessary to estimate a reasonable value for this angle.

Stevenson and Oxley[1,2,3,4] found the shear line AB (figure 1) to be a direction of maximum shear strain-rate and maximum shear stress and, apart from a small region near the tool nose, both strain and strain rate were found to be approximately constant along AB. There is also very little temperature variation along AB, again excepting a small region at the tool tip[12]. For a perfectly sharp tool the shear flow stress k_{AB} along AB may therefore be considered constant and as AB transmits the resultant cutting force it can be shown that

$$k_{AB} = \frac{R \cos \theta \sin \phi}{tw} \tag{3}$$

where R (figure 1) is the magnitude of the resultant force, θ is the angle this force makes with AB and w is the width-of-cut. When the shear strain-rate is constant along AB, with its value determined by equation (1), it can further be shown[2] that

$$\tan \theta = 1 + 2 \left(\frac{\pi}{4} - \phi \right) - \frac{Cn}{\sin \phi} \tag{4}$$

where C is the constant of equation (1) and n is the coefficient of strainhardening from the usual flow stress relation

$$\sigma = \sigma_1 \epsilon^n \tag{5}$$

By the use of von Mise's criterion

$$\sigma = \sqrt{3} \, k \quad \text{and} \quad \epsilon = \frac{1}{\sqrt{3}} \gamma \tag{6}$$

and the experimental finding[2] that the strain γ_{AB} at AB is half the total strain in the shear zone, it can further be shown that

$$\epsilon_{AB} = \frac{1}{2\sqrt{3}} \frac{\cos \alpha}{\sin \phi \cos(\phi - \alpha)} \tag{7}$$

and from equation (4) that

$$n = \frac{\sin\phi}{C} \left[1 + 2\left(\frac{\pi}{4} - \phi\right) - \tan\theta \right]. \qquad (8)$$

The flow stress σ_{AB} at **AB** is obtained from equations (3) and (6) and it is now possible to determine the constant σ_1 from the relation

$$\sigma_1 = \frac{\sigma_{AB}}{(\epsilon_{AB})^n} \qquad (9)$$

The usually measured components F_C and F_T of the resultant force (figure 1) are obtained from

$$F_C = twk_{AB}\cos(\lambda - \alpha)/\sin\phi\cos(\phi + \lambda - \alpha) \qquad (10)$$

where λ is given by

$$\lambda = (\theta + \alpha - \phi) \qquad (11)$$

and

$$F_T = \sqrt{\left(\frac{F_S}{\cos\theta}\right)^2 - F_C{}^2} \qquad (12)$$

where

$$F_S = k_{AB}tw/\sin\phi \qquad (13)$$

This set of equations can be used in two ways. If F_C, F_T and ϕ are measured for given values of α and U, the equations make it possible to determine the stress–strain:strain-rate properties of the work material at rather high levels of strain and strain-rate. Conversely, one can use existing flow stress relations to predict cutting forces, if ϕ is known. In the present case, a computer programme was used to calculate the horizontal force component F_C for a number of values of the shear angle ϕ and depth-of-cut t. The cutting speed U was kept constant at 20 ft/min and the rake angle α at $30°$ which were the conditions

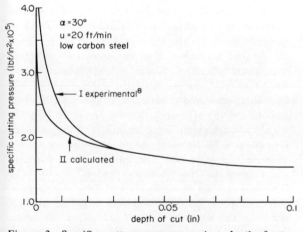

Figure 3. Specific cutting pressure against depth-of-cut. Curve I: experimental data[8]. Curve II: calculated data, this study. Both curves are for low carbon steel, rake angle $30°$, cutting velocity 20 ft/min.

used in obtaining the experimental curve (curve I in figure 3). The material properties used in the calculations were those determined by Hastings $et\ al.$[11] for a low carbon steel, namely,

$$n = 0.246 - 0.041 \log_{10}\dot{\epsilon}_{AB} \qquad (14)$$

and

$$\sigma_1 = 61.31 + 21.91 \log_{10}\dot{\epsilon}_{AB} \qquad (15)$$

where σ_1 is in 1000 lb/in². These expressions were obtained from machining data and cover the intermediate strain-rate region where the flowstress is approximately proportional to the logarithm of the strain-rate. It is realised that the use of these expressions at strain-rates above $1-5 \times 10^3$ s⁻¹, will result in an underestimation of the flowstress. The deviation should become especially noticeable above the turning point which is expected to fall between strain-rates of 10^4 and 10^5 s⁻¹. However it is felt that for this preliminary study the use of equations (14) and (15) is sufficiently accurate, particularly since this will only lead to an underestimation of the strain-rate effect.

For constant values of U, w and α and selected values of t and ϕ equations (1)–(4), (6)–(7), (9)–(11) and (14)–(15) are now used to calculate σ_{AB} and F_C. A value of 2.59 was used for the constant C.

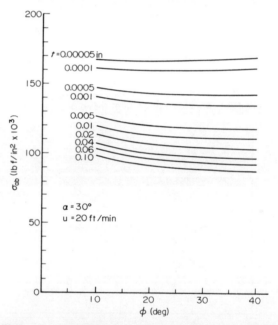

Figure 4. Calculated values of the flow stress in the shear zone for various depths-of-cut as a function of the shear angle.

Figure 4 shows the values obtained for σ_{AB}, the flow stress along AB in the shear zone, and figure 5 shows the calculated values of the specific cutting pressure (F_C/tw). It remains to select a reasonable value for the shear angle ϕ. Values of this angle generally fall in the range $20-35°$ for $\alpha = 30°$. For a steel similar to that now considered, Fenton and Oxley[13] found that ϕ varies only slightly with depth-of-cut between 0.002 and 0.01 in, while Hastings $et\ al.$[11] found some decrease in ϕ as the depth-of-cut is decreased. In order to compare the calculated values with the experimentally obtained curve I of figure 3, it was decided to select a value of ϕ which would give identical values of the specific cutting pressure for the experimental and the calculated data at the largest depth-of-cut. This ϕ value was then used in the calculations for all depths-of-cut. This procedure would be expected to result in an underestimation of the cutting-force at small depths as it takes no account of the expected decrease in ϕ at lower depths. The value of ϕ used was 23.5°. The calculated values of the specific

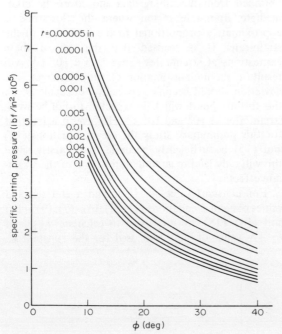

Figure 5. Calculated values of the specific cutting pressure for various depths-of-cut as a function of the shear angle.

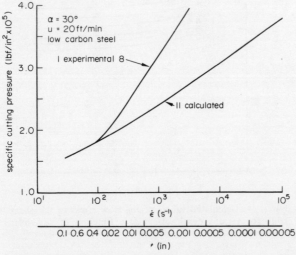

Figure 6. Specific cutting pressure. Data from figure 3 plotted against the logarithm of strain rate in the shear zone.

cutting pressure obtained for this shear angle from figure 5 are plotted as curve II in figure 3.

The similarity between the two curves of figure 3 is rather remarkable and it is clear that strain-rate sensitivity could explain much of the increase in specific cutting pressure with decrease in depth. At small depths-of-cut there is some deviation between the two curves, with the experimental curve rising somewhat faster than the calculated one. This is seen more clearly in figure 6 which shows semi-logarithmic plots of the two curves. The deviation could be due to several factors. As mentioned above, the calculated values are conservative at the higher values of strain-rate (smaller depths-of-cut) because

no account was taken of the possible decrease in the shear angle ϕ as the depth decreases, and because a semilogarithmic relation between flow stress and strain-rate ($\sigma \propto \log \dot{\epsilon}$) was used with ranges of strain-rate where a more rapidly increasing function ($\sigma \propto \dot{\epsilon}$) might be expected. Furthermore, the calculated values of the flow stress were not adjusted for the effects of the temperature in the shear zone. The velocity modified temperature of this zone was actually calculated according to the method outlined by Stevenson and Oxley[2]. However the temperature variations with depth-of-cut were relatively minor ($T = 134°C$ for $t = 0.00005$ in and $T = 194°C$ for $t = 0.1$ in at $\phi = 23.5°$), and were therefore not incorporated in the final calculations. If account had been taken of temperature variations the difference between the two curves would decrease somewhat. It is also possible that the two steels have somewhat different strain-rate sensitivities (data for the steel used in the experimental work from which curve I is taken are not available). Finally, the ploughing component of the tool force, which is felt increasingly in the experimental values as the depth-of-cut decreases, but which is excluded in the calculated data, should contribute to a deviation of the kind found. It will be noticed that all these factors tend to bring the two curves closer together. It can therefore be concluded that in machining the strain-rate sensitivity of the work material is one of the major factors (besides ploughing) responsible for the rapid increase in the specific cutting pressure as the depth-of-cut is decreased.

SLIDING ABRASION

A well-known scale phenomenon in the process of sliding abrasion is the effect of abrasive grain size on the rate of material removed per unit distance of sliding (wear rate). Below a 'critical' grain size the wear rate increases rapidly with the size of the abrasive grain D. Above this value the wear rate is constant or increases very slowly with D. The critical grit size is usually found to fall in the range 40–80 μm. A number of explanations for the effect have been proposed. Goddard and Wilman[14] and Avient et al.[15] suggest that the space between the fine abrasives rapidly becomes clogged with debris and the cutting efficiency thereby decreases. Rabinowicz[16] has calculated a minimum size of the particles formed in adhesive wear, based on a surface energy criterion. He proposed that adhesive wear particles are formed in the abrasion process, due to the frictional contact between the metal and the abrasive and that these particles interfere with the cutting action of the abrasives at small grain sizes. Mulhearn and Samuels[17] made a microscopic study of abrasive papers. They found that the finer abrasive grains had somewhat more cracks and surface flaws than the coarser abrasives and proposed that this accounts for the effect. Larsen-Basse[18, 19] found that the location of the 'knee' in the curve depends on the applied load and the length of the specimen in the direction of sliding and proposed that the effect was, to a large extent, a geometrical effect due to the relatively faster wear and dulling of the finer abrasive grains than when sample size and load are kept constant, as is generally the case.

In the following we will evaluate some of the published wear data in the light of the strain-rate effects outlined above. It is well-established that the material removal process in abrasion is one of micro-chip formation[15, 17, 20]. Owing to the flow of material around the abrasive grains into ridges along the grooves, only approximately 15% of the groove volume is actually removed as a microchip[18]. While little is known about the details of the material removal process it has been proposed[17, 21] that the abrasion process may be treated as a process of micromachining. We shall follow that general assumption here.

In order to estimate the strain-rates it is first necessary to arrive at an estimate of the depth-of-cut. This is so because in abrasion testing the specimen is simply pulled over an abrasive medium under a constant applied load and the weight loss (and sometimes the cutting force) determined. Little information is available on the actual average depth-of-cut in abrasion. For $\frac{1}{2} \times \frac{1}{2}$ in copper specimens abraded by 120 μm SC abrasive papers under 1500 g load, Du[22], in a rather limited series of experiments, found that the average depth of the grooves formed is approximately 0.12 D where D is the nominal grit size. When this result is combined with the findings that the average groove width is essentially independent of the applied load (the number of grooves is proportional to the load) and approximately directly proportional to the grit size, and that grits of different sizes appear to be geometrically similar and to cut in geometrically similar ways[20], it seems that for the purpose of this study one can assume an 'average depth-of-cut' $\bar{t}$ such that

$$\bar{t} \simeq 0.12\,D \qquad (16)$$

It has been proposed that the maximum size of the grooves formed in low-speed sliding abrasion is determined primarily by the balance between the flowstress of the metal forced out of the groove and the strength of the abrasive grit and its bonding medium. One could thus expect the maximum groove size to decrease somewhat with increasing hardness of the base metal and possibly with increasing cutting-speed (increasing strain-rate). In the latter case, the strength of the abrasive and of its bonding medium will also increase and the direction of change in groove size is not clearly predictable. Variations in metal hardness and in cutting velocity could result in some variation in the value of $\bar{t}$ used above. The magnitude of the effect is not known and may be minor since the largest grooves will be most affected. A difficulty in evaluating abrasion data is that very few investigators have measured the cutting-force. This limits the present preliminary study to a few investigations between which there is little difference in the speed of sliding. No data for the rake and shear angles of the microchip formation process in abrasion are available. To a first approximation one can, however, set the shear velocity V_S equal to the speed of sliding, U, that is, $V_S \simeq U$. By substituting this ex-

pression and $t = \bar{t}$ of equation (16), equation (1) can be rewritten

$$\dot{\gamma}_{\max} = C\,\frac{U}{0.12D} \qquad (17)$$

where D is the abrasive grain size. The constant C may have a value somewhat different from the value found in machining because the strain level of abrasion is considerably higher.

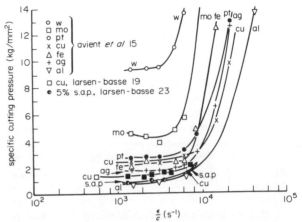

Figure 7. Calculated specific cutting pressure in sliding abrasion as a function of the estimated strain-rate. Data by Avient *et al.*[15] for commercially pure metals, by Larsen-Basse[19] for OFHC copper and[23] for an experimental Al–5% Al$_2$O$_3$ alloy.

Figure 7 shows plots of the specific cutting pressure in abrasion obtained from

$$p = \frac{\text{cutting force (kg)}}{\text{wear rate (mm}^3/\text{mm})} \quad (\text{kg/mm}^2) \qquad (18)$$

where the wear rate is the volume lost per unit distance of sliding. The pressure is plotted against $U(\sqrt{3} \times 0.12D)$, that is, against the estimated value of $\dot{\epsilon}/C$. The majority of the data are from Avient *et al.*[15] who worked with commercially-pure metals of 1–2.5 in diameter, abraded against emery papers of 5–150 μm grit size under 1 kg load at a velocity of 5 cm/s. The results for the sintered aluminium powder (SAP) with 5% oxide were obtained by Larsen-Basse[23] for 3 mm diameter samples worn at 1 kg load and a velocity of 3.5 cm/s against silicon carbide abrasive papers of between 20 and 120 μm grit size.

Larsen Basse's curve for the SAP alloy follows the data by Avient *et al.* for pure aluminium extremely well. This supports a conclusion of a previous paper[23] that the deformation of SAP-type alloys at room temperature and at the strain levels of interest in abrasion, takes place in the metallic matrix and not in the oxide particles. All the curves of figure 7 become very steep in the general strain-rate region where this would be expected, if strain-rate sensitivity played an important role in determining the shape of the curves. The constant C of equation (1) is not known for most of the metals in question. If the value of 2.59 found for steels is used for the pure iron and values of 1.94 for copper and 3.64 for aluminium, as indicated by recent machining studies at the University of New South Wales, the strain-rate values at the turning point become 1.7×10^4 for iron, 1.8×10^4 for copper and 2.2×10^4 for aluminium. These values are clearly in the range where one would expect a pronounced change in the flow stress:strain-rate curve at the high strains in question. A strain value of $\epsilon = 4.8$ has been estimated for abrasion of a medium carbon steel[24].

It can be concluded that strain-rate effects have a large influence on the shape of the wear-rate against grit size curve in abrasion. Some of the other factors outlined above such as fracture, wear and dislodgement of grits and clogging, will undoubtedly contribute to the overall effect in amounts depending largely on the experimental conditions (specimen size and applied load). These factors would primarily influence the slope of the curve above the turning point and it would become steeper with increasing clogging, grit wear and so on, that is, with increased specimen size and applied load and with decreased hardness of the specimen material. It would be desirable to double-check the effect of strain-rate sensitivity by considering the influence of sliding velocity on the cutting force in abrasion although in this case account would have to be taken of temperature. As shown by equation (1) the strain rate should be approximately proportional to the speed of sliding if the depth-of-cut remains constant. Unfortunately it appears that no studies of the influence of cutting speed on both cutting force and wear rate have been published. Nathan and Jones[25] measured the influence of velocity of sliding on the wear rate of a number of metals and found a slight increase in the wear rate with sliding velocity. Since the cutting force was not measured it is not possible to evaluate these results in terms of strain-rate sensitivity. It seems possible that in this case the increased sliding speed caused heating effects, which caused a decrease in the flow stress which, in turn, outweighed the increase due to increased strain-rate. These authors do, in fact, mention noticing appreciable heating of the specimens and explain their results as due to temperature effects.

GRINDING

Shaw[26] has reported a curve for grinding with constant size of the grinding grain but varying depth-of-cut, which has a general shape similar to curve I of figure 3. The work material was a hardened ballbearing steel (AISI 52100), surface ground by a 60 grain size wheel. Shaw represents the abrasive grains as roughly spherical and explains the effect qualitatively as due to the increasingly high negative value of the effective rake angle as the depth-of-cut is decreased.

We shall here estimate the extent to which the strain-rate sensitivity of the work material could be responsible for the effect. Since the detailed data necessary for a complete study of this case are lacking the estimate must remain rather a rough one. It is assumed that the material removal process in grinding is sufficiently similar to machining to allow the strain rate to be estimated from equation (1). The constant C of that equation is expected to vary somewhat with

the work material and with the rake angle (that is, with the total strain value). However, since minor variations in the value of C make little difference in the general trend of the curves, and since no experimental C values exist, either for grinding or for the steel in question, the value of $C = 2.59$ was again used. Because values of α and ϕ are not available, the substitution of $V_S \simeq U$ was applied as for the case of abrasion.

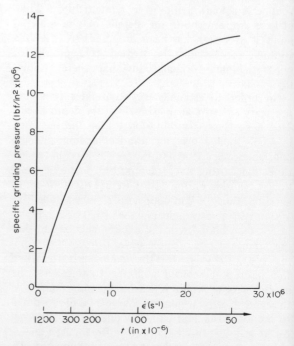

Figure 8. Specific grinding pressure against strain-rate for grinding a hardened steel (AISI 52100). Data from Shaw[26].

The results are shown in figure 8. The strain-rate values are here in the range where the flow stress is expected to rise linearly, or even more steeply, with the strain-rate. A direct plot of specific cutting pressure against strain-rate has therefore been used. It is seen that the curve in general rises quite steeply, as would be expected if strain-rate effects were of major importance. Above a strain-rate of about 5×10^6 s^{-1}, the curve levels out somewhat rather than becoming even steeper, as initially expected from the flow stress: strain-rate relationship of figure 2. It should be pointed out that the data are not compensated for the rather high temperatures encountered in grinding and it is proposed that differences in frictional heating with depth-of-cut could well be responsible for the levelling out. As mentioned above, Shaw[26] has proposed that as the depth-of-cut decreases for constant size of the abrasive grain, the rake angle becomes increasingly negative. We shall accept this explanation for the purpose of the present discussion. The effect of increasing negative values of α on V_S (equation (2)) is not immediately clear. If ϕ is very small, as would be expected for large negative values of α, little change in V_S will result. The changing value of α will thus, in itself, have little or no effect on the strain rate or the flow stress. However, as the depth-of-cut is decreased for abrasive grains of a rounded shape the rubbing between the grains and the metal will increase as a proportion of the total work. The quantity of resulting frictional heat will then decrease less rapidly than the volume of the chip. The temperature of the shear zone will therefore increase with decreasing depth-of-cut. The resulting softening could be responsible for the upper part of the curve in figure 8 being less steep than initially expected.

While the results discussed here do not prove conclusively that strain-rate sensitivity is an important part of a material's resistance to grinding, they do certainly tend to support such a conclusion. This is seen more clearly in figure 9 which is a composite of figures 3 and 8 and the data for pure iron of figure 7 adjusted for $C = 2.59$. The curves plotted are in log-log coordinates for easier overview. It will be noticed that the curves for machining and grinding appear to join one another. This becomes even more apparent when it is recalled that the calculated machining curve is on the conservative side, that both machining curves should become rather steep in the strain-rate range $10^4 - 10^5$ s^{-1}, and that the grinding curve is not temperature-compensated. If that data were reduced to the same temperature for which the

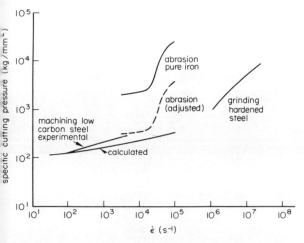

Figure 9. Composite of figures 3, 8 and the data for pure iron of figure 7. The adjusted abrasion curve represents the cutting pressure necessary to form the grooves.

machining data was obtained, one would expect the grinding curve to move upwards and become steeper. The curve for abrasion of pure iron parallels the curves for machining of a low carbon steel and grinding of a hardened steel. As mentioned above, only about 15% of the groove volume is removed in abrasion[18]. Since the specific cutting pressure was calculated on the basis of the amount of material actually removed (equation (18)) the calculated values will be approximately 6.6 times higher than the pressure actually needed to form the grooves. When this is taken into account the adjusted curve of figure 9 is obtained. This curve is very close to the machining data, a fact that tends to justify an assumption made above, that the abrasion process to a first approximation can be considered as a case of micromachining. The steepness of the abrasion curve above the turning point may be somewhat higher than dictated by strain-rate effects alone, because rather large specimens were used to obtain the data in question. This would mean that wear and deteriora-

tion of the cutting abrasive grains and clogging of the space between the grains could be considerable, which would result in decreased material removal for the fine grain sizes, that is, a calculated curve rising more steeply towards the higher values of strain-rate than if strain-rate sensitivity alone controlled the shape of the curve.

CONCLUSIONS

Some well-known scale phenomena in material removal processes involving chip formation can be explained in large measure in terms of the strain-rate sensitivity of the work material. As the scale of the cut is decreased the strain-rate increases. This results in an increased flow stress of the work material and thus an increased specific cutting pressure. Under practical conditions other factors, such as ploughing from rounded tools in machining or wear and deterioration of abrasive grains in abrasion, will work in conjunction with the strain-rate effect and accentuate the phenomena.

ACKNOWLEDGEMENT

This work was carried out at the University of New South Wales. One of the authors, J. Larsen-Basse, is grateful to the University of New South Wales, and to Professor R. A. Bryant in particular, for making facilities available and for a stipend that made the work possible.

REFERENCES

1. M. G. STEVENSON and P. L. B. OXLEY, 1969/70. An experimental investigation of the influence of speed and scale on the strain-rate in a zone of intense plastic deformation, *Proc. Inst. Mech. Engrs.*, **184**, p. 561.
2. M. G. STEVENSON and P. L. B. OXLEY, 1970/71. An experimental investigation of the influence of strain-rate and temperature on the flowstress properties of a low carbon steel using a machining test, *Proc. Inst. Mech. Engrs.*, **185**, p. 741.
3. M. G. STEVENSON and P. L. B. OXLEY, 1971. The influence of the high strain rate, high temperature stress-strain behaviour of a material on its machining characteristics, presented at *21st General Assembly of CIRP, Warsaw*, September.
4. P. L. B. OXLEY and M. G. STEVENSON, 1967. Measuring stress—strain properties at very high strain rates using a machining test, *J. Inst. Metals*, **95**, p. 308.
5. A. R. DOWLING, J. HARDING and J. D. CAMPBELL, 1970. The dynamic punching of metals, *J. Inst. Metals*, **98**, p. 215.
6. C. K. H. DHARAN and F. E. HAUSER, 1970. Determination of stress—strain characteristics at very high strain rates, *Experimental Mechanics*, **10**, p. 370.
7. S. K. SAMANTA, 1971. Dynamic deformation of aluminium and copper at elevated temperatures, *J. Mech. Phys. Solids*, **19**, p. 117.
8. A. W. J. CHISHOLM. The theory of cutting tools, in H. W. Baker (Editor), *Modern Workshop Technology, Part II*, Macmillan, p. 83.
9. G. BOOTHROYD, 1965. Fundamentals of metal machining, Edward Arnolds, London, p. 19.
10. P. ALBRECHT, 1960. New developments in the theory of the metal cutting process: Part I, The ploughing process in metal cutting, *Trans. ASME*, p. 348.
11. W. F. HASTINGS, M. G. STEVENSON and P. L. B. OXLEY, 1972. Assessing machinability from fundamental work material properties, *12th Int. MTDR Conf.*, Macmillan, p. 507.

12. M. G. STEVENSON, 1971. Influence of the stress–strain properties of the work material at high strain rates and temperatures in the machining process. *Harold Armstrong Conference on Production Science in Industry, Melbourne,* The Institution of Engineers, Australia, p. 17.

13. R. F. FENTON and P. L. B. OXLEY, 1969/70. Mechanics of orthogonal machining: predicting chip geometry and cutting forces from work material properties and cutting conditions, *Proc. Inst. Mech. Engrs,* **184,** Part 1, p. 927.

14. J. GODDARD and H. WILMAN, 1962. A theory of friction and wear during the abrasion of metals, *Wear,* **5,** p. 114.

15. B. W. E. AVIENT, J. GODDARD and H. WILMAN, 1960. An experimental study of friction and wear during abrasion of metals, *Proc. Roy. Soc.* (London), **258,** p. 159.

16. E. RABINOWICZ, 1964. Practical uses of the surface energy criterion, *Wear,* **7,** p. 9.

17. T. O. MULHEARN and L. E. SAMUELS, 1962. The abrasion of metals: a model of the process, *Wear,* **5,** p. 478.

18. J. LARSEN-BASSE, 1968. Influence of grit diameter and specimen size on wear during sliding abrasion, *Wear,* **12,** p. 35.

19. J. LARSEN-BASSE, 1972. Some effects of specimen size on abrasive wear, *Wear,* **19,** p. 27.

20. J. LARSEN-BASSE, 1968. Influence of grit size on the groove formation during sliding abrasion, *Wear,* **11,** p. 213.

21. L. E. SAMUELS, 1971. Abrasive surface finishing processes: mechanisms, *Harold Armstrong Conference on Production Science in Industry, Melbourne,* The Institution of Engineers, Australia, p. 247.

22. F. O. DU, 1969. Surface deformation and strain due to abrasion, *M.S. Thesis, University of Hawaii.*

23. J. LARSEN-BASSE, 1968. Abrasion resistance of some SAP-type alloys at room temperature, *Wear,* **12,** p. 357.

24. J. LARSEN-BASSE and K. G. MATHEW, 1969. Influence of structure on the abrasion resistance of a 1040 steel, *Wear,* **14,** p. 199.

25. G. K. NATHAN and W. J. D. JONES, 1966. The empirical relationship between abrasive wear and the applied conditions, *Wear,* **9,** p. 300.

26. M. C. SHAW, 1971. A new theory of grinding, *Harold Armstrong Conference on Production Science in Industry, Melbourne,* The Institution of Engineers, Australia, p. 1.

DESIGN AND PRELIMINARY RESULTS FROM AN EXPERIMENTAL MACHINE TOOL CUTTING METALS AT UP TO 8,000 FEET PER SECOND

by

G. ARNDT* and R. H. BROWN*

SUMMARY

The design and operation of an explosively actuated experimental system suitable for both machining and deformation studies at speeds within the range 500 to 8000 ft/s is described. The results of the first 92 test shots, in which aluminium workpieces were fired past fixed high-speed steel tools, are presented and analysed. For the cutting conditions investigated (cutting velocity 1770 to 5200 ft/s, depth-of-cut 0.005 to 0.088 in, width-of-cut 0.375 in, rake-angle $-20°$ to $+10°$) the chip thickness was found to be less than the depth-of-cut, with a cutting ratio varying from 1 to 3, as an exponential function of cutting speed. Shear angles up to 75° were calculated. Variations in chip−tool interface length, metallurgical and photographic examinations indicate chip melting. This supports the logical hypothesis that the chip material progressively approaches the 'fluid' state with increasing cutting speed.

INTRODUCTION

The history of metal cutting has, to a large extent, been a story of increasing cutting speed[1,2]. It is logical to speculate about what will happen if the speed is increased well above the currently used values. This will have the obvious practical advantage that components will be machined more rapidly. It has also been suggested[3] that favourable changes in the cutting mechanism may occur under the conditions of higher strain-rate in the shear zone. On the other hand there may be some undesirable features resulting from high speeds: tool wear may be excessive, surface finish unacceptable and inertia effects could lead to very high cutting forces.

There have been several preliminary studies into high and ultra-high-speed machining[3−7], the term ultra-high-speed machining (UHSM) generally being used for cutting speeds above 500 ft/s. In most of these, small firearms were used to shoot workpieces past fixed cutting tools. Vaughn[3], whose investigation was one of the most comprehensive of these studies, employed cutting speeds from 250 to 4000 ft/s. He chiefly used high-speed steel tools cutting steel and aluminium. He concluded that the cutting mechanism was the same as that at conventional speeds, except that the chip thickness was usually less than the depth-of-cut. The tool temperature remained essentially constant over the speed range tested. When machining AISI 4340 steel at 2000 ft/s, cutting forces were found to decrease by approximately 40% from the values at conventional speeds. Surface finish was improved by high speeds. A review paper by one of the authors[2] discusses some other papers that have considered the effects of cutting at high speeds[8−14]. With increasing cutting speed it is generally agreed that the cutting ratio r_c and shear angle ϕ increase, while the coefficient of friction at the rake face μ and the length of the chip−tool interface l_i decrease.

While the previous studies have demonstrated the feasibility of UHSM, they have not provided a clear understanding of the basic cutting mechanics. This involves such interrelated topics as inertia (momentum), temperature, adiabatic shear and stress wave considerations. It is for this reason that the present study was undertaken. In the long term, it is expected that an understanding of the mechanics of UHSM will lead to significant increases in cutting speed for at least some industrial machining operations.

The present paper describes the experimental techniques and results from a pilot study of cutting at ultra-high-speed. The complete details of the project are reported in a thesis by Arndt[15] and some theoretical aspects are considered elsewhere[16].

EXPERIMENTAL SYSTEM

It was decided that the system should be capable of exceeding the elastic wave propagation speed of at least some common materials. Lead, which has a relatively low propagation speed ($c_\varrho = 7100$ ft/s) was chosen as a benchmark, and the system designed to achieve 8000 ft/s. A maximum cutting length of 8 ft was selected. Considerations of cost and safety indicated the best driving mechanism to be a proven artillery gun. A computer programme was written to study ballistic performance. This indicated that a modified 40 mm Bofors gun would be suitable, and it established optimum firing conditions[17]. The selected conditions were

barrel length	180 in
bore diameter (smooth)	1.625 in (nominal)
chamber volume	42 in^3
propellant	cordite, type W/T 120−040
charge weights	up to 1.1 lb
shot start pressures	up to 6 tonf/in^2

The ballistic analysis indicated that speeds from approximately 1500 to 8000 ft/s should be attained for projectile weights of 1.1−0.2 lb, without exceeding the safe maximum barrel pressure of 21 tonf/in^2.

The general arrangement of the experimental system is shown in figure 1. The individual components are described below.

(1) Gun drive

For economy, safety, noise containment and convenience, the gun was mounted in an underground bunker on campus. It was bolted to a concrete floor

* Department of Mechanical Engineering, Monash University, Victoria, Australia.

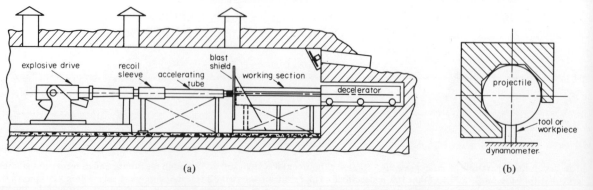

(a)

(b)

Figure 1. General arrangement of the ultra-high-speed test system at Monash University. (a) Bunker cross-section. (b) Cross-section through the working section.

(a)

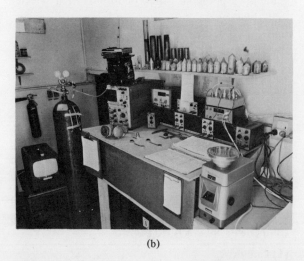

(b)

Figure 2. Experimental arrangement. (a) Bunker interior and general experimental arrangement. (b) Control hut interior, showing firing control panel, electronic counters, balance, CRO for pressure recording, gas supply for flash photography, and closed-circuit TV monitor.

and fired mechanically (percussion primer) by remote control from an adjacent control hut, figure 2.

(2) Gun barrel

The standard rifling grooves in a 40 mm barrel were machined out to their root diameter, 1.6302 in. This relatively large size allows firing of either a full-bore projectile or a sabot-type composite projectile, whose 'carrier' is discarded at the muzzle. The sabot may be used, for example, to carry a rectangular cutting tool. A standard cylindrical projectile has been used for the tests up to the present. The projectile has shown no tendency to rotate, so that alignment grooves or keys are unnecessary.

(3) Cartridge case – projectile assembly

Standard 40 mm cartridge cases were fitted with re-usable steel extension pieces, which located the projectile and shear die ring inside the explosive chamber. The various types of projectile used are shown in figure 3. Shot-start pressure was controlled by varying the driving band thickness, t_{db}, shown.

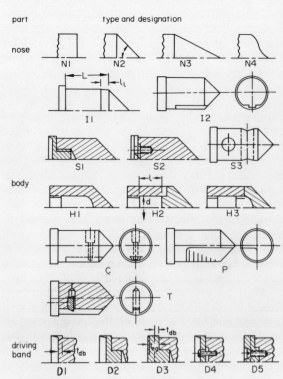

Figure 3. Projectile configurations and designation.

(4) Recoil section

The standard barrel length of the gun has to be extended to achieve the very high velocities. To permit the extension, a dog-clutch type sleeve was constructed: one end of this was screwed to the recoiling barrel and the other to the fixed extension tube. A cover tube prevented gas escape. Measurements of barrel recoil (up to 8 in) indicated the explosive energy in a shot.

(5) Acceleration tube

The additional, fixed, extension tube has not yet been tried in the system, but it will be installed as stage 2 of this project. The alloy steel tube has been honed-out to size and will be rigidly mounted to the concrete floor.

(6) Gas and blast deflection

For photographing the cut, obscuration by smoke and burning propellant must be kept to a minimum. In addition, instrumentation must be protected from the blast pressure. Three radial gas escape slots near the muzzle, followed by a ¼ in plate and two baffle plates close to the working section, provided suitable protection.

(7) Working section

As indicated in figure 3, either the tool or the workpiece may be located on the projectile. For cuts up to 8 in long, the latter system was used and the tool was rigidly mounted in a holder, as shown in figure 4. Chips were caught by a 12 ft long tube

Figure 4. General view of working section.

leading tangentially to a drum where the chips 'spun' to rest. In the case of the tool mounted on the projectile, chip collection is more difficult. At the present stage of testing, space has been provided in the projectile, but this has limited the maximum cutting length.

An octagonal shape was chosen for the projectile guide-track, as in figure 1(b). This, while accurately locating the projectile, allows access to its lower surface. Three track inserts, giving a total length of 8 ft, are clamped to the track assembly structure which is bolted to the floor. Ample space for workpieces, tools and instruments is provided.

(8) Workpieces

Most cuts have been made on an aluminium alloy (percentage composition Cu 5.5, Pb 0.51, Bi 0.44, Fe 0.15, Si 0.08). Some tests have also been performed on lead and low carbon steel.

Of the projectile shapes shown in figure 3, body type I1 with a lead-in of $l_\varrho = 0.38$ in proved most convenient and was used in the majority of tests. Body type I2 gave true orthogonal cutting, while type C was used to carry the heavier work materials. Several longer cuts were taken with the tool-carrying projectile, type T. In this case the workpiece, a ¼-in plate, was rigidly clamped to the track.

(9) Cutting tools

High-speed steel tools have been used for nearly all tests. An orthogonal cutting edge with $3°$ clearance angle has been used, and rake angles between $-20°$ and $+30°$. The toolholder consisted of a 2 in steel cube with a retaining plate and set screws to hold and tool bit. The depth-of-cut was set by adjusting the entire holder assembly vertically, using a dial indicator.

(10) Projectile decelerator

The projectiles were stopped in a 6 ft long steel box with ¼ in wall thickness. The box was three-quarter filled with wet beach sand and was closed with a lid bolted on top. Apart from sand resistance, additional energy was absorbed by this box sliding freely over a wooden table. The distance the box moved (up to 15 in for a box weight of 1500 lb) indicated the deceleration energy.

(11) Control and measurement

Cutting (that is, projectile) velocity is controlled by charge weight, projectile weight and shot-start pressure. Depth-of-cut is set by dial indicator. The gun is fired from the control hut by an electric circuit which activates the firing mechanism only after 5 safety microswitches are closed. A rigid countdown sequence is adhered to, and the bunker is completely closed for every shot.

Both ballistic and machining variables were recorded, the main ballistic data being: the barrel pressure-time characteristic, blast pressure, recoil, decelerator and projectile-penetration distances. The machining data included: precut and postcut projectile velocities, chip thickness, chip–tool contact length, surface finish, and chip and workpiece microstructure. Techniques for cutting force measurement and for photographing the cut are still being developed.

SYSTEM PERFORMANCE

Cutting velocity was measured by electronic counters recording a high-frequency signal which was gated by two impact probes mounted on the working section. The probes were set an accurately known distance apart (±0.001 in) and they operated when fractured by the projectile. Comparison readings with several probes and counters in series indicated a velocity measurement accuracy of ±5%. This relatively poor performance is caused by the explosive environment surrounding the probes (for example, shock, debris, vibration). Figure 5 shows the theoretical and the measured velocity characteristics of the system. The theoretical characteristics apply for a barrel length of 180 in, while the experimental ones are for the 101 in barrel. There are some differences in the shape of the two sets of characteristics, but the velocities are of similar order of magnitude. Since muzzle velocity increases with barrel length, it is to be expected that when the actual barrel is extended to 180 in, the cutting velocities will exceed the theoretical predictions. In particular this indicates that the system should be capable of velocities even higher than the 8000 ft/s

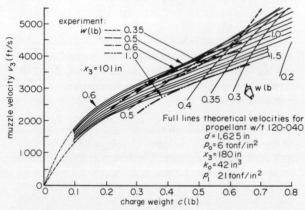

Figure 5. Comparison of predicted ($x_3 = 180$ in) with actual ($x_3 = 101$ in) muzzle velocities.

selected as maximum. A piezoelectric pressure transducer mounted inside the recoiling barrel-cartridge assembly recorded pressure distributions. This enables shot-start pressure to be determined. The recorded maximum pressure was checked against the value indicated by a standard copper cylinder crusher gauge placed in the cartridge. In all 92 shots performed, the maximum pressure has been less than half the 21 tonf/in² allowable in this gun, that is, a large factor of safety prevails.

The cartridge cases are used again (after appropriate hardness tests) and propellant quantities are small, so that the operating costs of the system are quite low. The only operational difficulties arise from time-consuming adjustments to instrumentation.

PRELIMINARY MACHINING RESULTS

The purpose of the study to the present time has been to prove the feasibility of the system, rather than to collect sufficient data for a conclusive analysis of ultra-high-speed machining: that is the aim of future work. In the following sections some experimental results are presented for one basic tool—workpiece combination: stationary high-speed steel tools machining moving aluminium projectiles.

(1) Collection of experimental data

From the measured machining variables listed earlier under the section 'Control and measurement', the cutting ratio, the shear angle and the momentum force were derived.

The measurement of the mean depth-of-cut and the mean chip thickness requires some explanation. The nominal depth-of-cut was set by dial gauge before cutting, but examination of the projectiles after cutting indicated that appreciable deflection of the tool or workpiece occurred during cutting. This deflection caused the tool to dig in, so that, for a nominal depth-of-cut variation of 0.005 to 0.040 in, actual values up to 4 or 5 times these values pertained during a cut. The deflection arose from additive elastic and plastic movements in the projectile and tool holder under the action of high momentum forces (calculated below). It was found that the projectiles were quite well preserved, and so the actual depth-of-cut could be determined by a micrometer (across the projectile diameter). This was measured at three points along the length of the projectile. In most cases, the projectiles showed some increase in size, or mushrooming, from the effect of impact with the sand. This effect could be measured and appropriate corrections made to find the depth-of-cut.

Several methods for arresting the chips were tried. The spinning drum method appeared to give least damage to the chips. At the lower speeds tested (2000 ft/s) the chips were in the form of quite large fragments. It is probable that the chip was produced as a continuous ribbon and inevitably suffered some damage in the arresting process. As speed was increased the chip became more discontinuous.

Because of the variable depth-of-cut, the circular segment shape of the projectile and the damage on arrest of the chips, special attention was needed in measuring chip thickness and cutting ratio. Only chip fragments were measured, that clearly had not been severely damaged at arrest. The thickness variation of these was measured by a micrometer with pointed anvils. It was found that the chips had the same general circular-segment cross-section of the cut. As the only reliable basis for comparison, the maximum chip thickness for each shot was determined. The probability of collecting the section of chip right at the end of the cut (when the cutting depth is maximum) is quite low. Hence, as a reasonable point for comparison of the measured maximum chip thickness and the depth-of-cut, the point midway between the end and the centre of the projectile was selected. The depth-of-cut at this point $d_{\varrho=75\%}$, was divided by the maximum measured chip thickness t_{cm} to give the cutting ratio r_c for each shot. The shear angle was determined from the orthogonal cutting relationship

$$\phi = \arctan\left[\frac{\cos\alpha}{1/r_c - \sin\alpha}\right] \qquad (1)$$

where α is the rake-angle.

Knowing the shear angle, it is possible to calculate the momentum force F_m and its components parallel and normal to the cutting direction. The change in momentum caused by the transformation of uncut chip material into the chip gives rise to the momentum force, which may be written

$$F_m = \frac{btV^2\rho}{\cos\phi}\left[\frac{1}{1 + \tan\phi\,\tan\alpha}\right] \qquad (2)$$

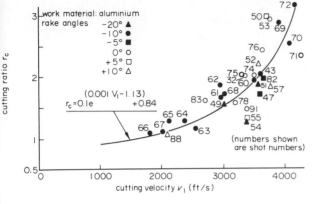

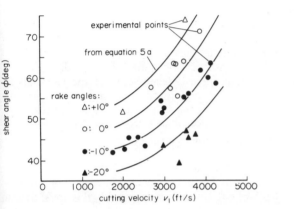

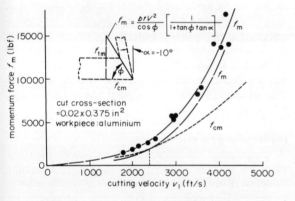

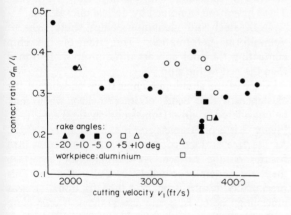

Figure 6. Experimental variation of (a) cutting ratio, (b) shear angle, (c) momentum force, and (d) contact ratio with cutting velocity.

where b, t are width and depth-of-cut, respectively, V is the cutting velocity, and ρ is the workpiece density. The measured density of the aluminium used was $\rho = 0.101$ lb/in^2. The cross-sectional area of cut was taken at its mean depth, that is $\ell = 50\%$. An investigation of the variation of F_m with V is only meaningful if b, t, ρ and α are kept constant (equation (2)). For this reason actual values of bt, for which ϕ was deduced as above, were corrected to a constant nominal cross-sectional area of $bt = 0.02 \times 0.375$ in^2. This was used in plotting F_m.

Observation of discoloration and build-up on the tool rake-face after cutting enabled the chip contact length to be determined. Two regions could be distinguished: one dark region close to the cutting edge, followed by a lighter region in which there was a greater build-up of chip material. The combined mean length was taken as that of chip–tool contact. Values of the contact length obtained at different cutting conditions cannot be mutually compared. The ratio of depth-of-cut to contact length is more meaningful. This ratio was calculated using the mean depth (that is, $\ell = 50\%$).

Values of the cutting variables, determined as described above, are shown plotted against speed in figure 6. The significant trends in these variables are discussed below.

(1) Cutting ratio

For the orthogonal cutting system used, cutting ratio is the basic variable defining the geometry of the cut. Values determined as described above are plotted in figure 6(a). It is seen that r_c increases with speed with values in the range 1 to 3, that is, at the highest speeds tested, chip thickness is three times less than the depth-of-cut. Vaughn[3] reported similar findings. For AISI 4340 steel and cutting velocities from 417 to 2893 ft/s, his chip velocity measurements yield cutting ratios from 1.07 to 2.62. He obtained similar results for Iconel X. It thus appears that the phenomenon applies irrespective of material, and in fact it represents the fundamental difference between ultra-high-speed machining and conventional-speed machining, where r_c is always less than unity.

Because of the unusual values of r_c obtained, the validity of the testing method might be questioned. There is a possibility that at high speeds, appreciable sideflow of the chip may occur, analogous to a liquid jet impinging on a flat plate. If this were the case, values of r_c found from chip thickness measurements or for that matter, chip length, would not be meaningful. Although most of the chips collected were broken in the stopping process, they were still sufficiently intact to be able to measure chip width. This indicated no significant sideflow, certainly none of such magnitude as to account for a 2:1 to 3:1 thickness reduction. This evidence, together with the fact that Vaughn[3] derived his cutting ratios from measurements of chip velocity (which could be lower, but certainly not higher, if sideflow existed) and obtained similar results, points to the conclusion that sideflow is negligible at speeds at least up to 4000 ft/s. Thus the phenomenon of chip thickness less than depth-of-cut, and hence shear angle ϕ greater than 45°, appears to be a characteristic of ultra-high-speed machining.

This phenomenon may be intuitively expected by analogy with fluid flow. In two-dimensional supersonic flow, the shock set up at the point of a deflecting plate is inclined at an angle β to the original flow direction (here, shock angle $\beta = 180 - \phi$). With increasing Mach number, β decreases: the shock wave approaches the deflecting plate. In the present case, the situation of $t_c < d_m$ can be explained by considering the theoretical shear plane to be a plastic wavefront emanating from the cutting edge. The plastic wave propagation speed will be a constant for the material (2800 ft/s for aluminium[2]). Thus, with increasing cutting speed, the ratio of cutting speed to plastic wave propagation speed will increase, so that the plastic wavefront will be progressively displaced towards the toolface, thereby resulting in an increasing shear-angle ϕ. When the two speeds are equal, a cutting ratio of $r_c = 1$ would be expected for a zero rake-angle and an ideally plastic material. It is seen from figure 6(a) that the cutting ratio is close to unity ($r_c = 1.35$) at a cutting speed of 2800 ft/s. The difference may be attributed to experimental scatter, the non-ideal work material, rake-angle effects or approximation in calculating the plastic wave propagation speed. The agreement seems reasonably close to show the feasibility of this theory.

Takeyama *et al*[14], considering lower speeds, empirically derived an expression for cutting ratio in terms of rake-angle and speed.

$$\frac{1}{r_c} = e^{K_1(90-\alpha)}(K_2 e^{-K_3 V} + 1) \tag{3}$$

It is noted that in the higher-speed condition represented by figure 6(a), the cutting ratio is apparently independent of rake-angle. This indicates that K_1 in equation (3) approaches zero at very high speeds. Or, rewriting the equation, that

$$r_c = Ae^{(BV-C)} + D \tag{4}$$

The curve superimposed on the experimental points in figure 6(a) is of this type. Its equation is

$$r_c = 0.1\ e^{(0.001V-1.13)} + 0.84 \tag{5}$$
$$\text{(applicable for } 1700 < V < 4000)$$

(2) Shear angle

While equation (5) is seen to be a reasonable fit to the experimental points in figure 6(a), on closer examination the effect of rake-angle α is still detectable. Values of the shear angle ϕ were calculated from equation (1) for each point, and are plotted in figure 6(b). The lines in the figure were derived by using equation (5) and substituting in equation (1) for each of the four rake-angles ($+10°$, $0°$, $-10°$, $-20°$).

The very high values of shear angle obtained are of interest. These point towards a high cutting efficiency. The results given by both Tanaka *et al*[14] and Fenton and Oxley[13] are in agreement with the measured values. Their work indicated that a shear angle of $45°$ should result when cutting steel at

2000 ft/s with a $\alpha = 10°$. Under these conditions the shear angle measured here was $43°$.

(3) Momentum force

In the absence of experimental cutting force and/or yield shear stress data, no conclusions are possible concerning the mechanical force components. However, the force F_m arising from momentum change can be calculated from equation (2). This force and its components in the direction of cutting F_{cm} and perpendicular to it F_{tm} are shown in figure 6(c). The three curves are lines-of-best-fit through the experimental points. As expected, the F_{cm} and F_{tm} curves intersect at $V_1 = 2400$ ft/s, that is, where ϕ is $45°$ (see figure 6(b) at $\alpha = -10°$). The order of magnitude of these forces is extremely high compared to cutting forces for aluminium at conventional speeds. The forces would be even higher for a higher density material such as steel.

From a practical viewpoint, the very high values of momentum force indicate some of the problems in applying ultra-high-speed machining. The energy required for cutting will be very high, and there are difficulties in designing a tool sufficiently strong to withstand the forces involved.

(4) Contact ratio

Experimentally determined values of the ratio of depth-of-cut to contact length (d_m/ℓ_i) are shown in figure 6(d). The general trend is for d_m/ℓ_i to decrease with increasing cutting velocity, that is, for the chip to adhere longer to the tool at higher speeds. Interface lengths up to 6.5 times the depth-of-cut were of obtained at the higher velocities.

(5) Surface finish

Surface finish in UHSM did not vary greatly from that at conventional speeds, but the surface traces taken did exhibit more rounded-off features.

(6) Metallurgical examination

A preliminary metallurgical examination of some chips has been performed, with the aim of obtaining evidence of microstructural changes and deformation modes. Microsections have been prepared by mounting, polishing and etching chip fragments. These have been examined by optical microscope.

Both steel and aluminium chips show signs of nonuniform deformation. The side of the chip contacting the tool appears more severely deformed than the rest of the chip.

Some of the aluminium chips show evidence of the existence of thin bands of intense shear which may be considered to arise from an adiabatic shear effect. Figure 7 is a photomicrograph exhibiting this effect. At the free surface of the chip, this band runs into fissures on the surface. At the chip–tool interface they are generally parallel to the tool face. This presumably results from the high frictional stresses prevailing there. In some cases there is complete separation of a layer of chip material which is deposited on the toolface.

In some of the aluminium chips, relatively wide white-etching bands were observed. These appeared to be heat-affected zones in which the aluminium had recrystallised. Microhardness tests supported this conclusion. The bands were present on both sides and within the chip, suggesting that although the whole chip is subjected to intense heating, only preferred

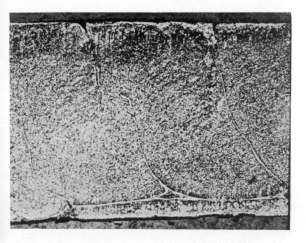

Figure 7. Longitudinal centre-section through chip showing shear bands (aluminium, V = 2600 ft/s × 45).

regions have time to melt and/or recrystallise. Such adiabatic shear bands may give rise to a net cutting-force decrease at very high cutting speeds[16].

Hardness measurement on the cut workpiece surface indicated a decrease in hardness with increasing velocity. This also suggests a softening of the aluminium due to melting and/or recrystallisation.

The metallographic evidence of melting is supported by visual observation of the region surrounding the cutting zone (tool holder and track faces). This revealed 'splashes' of aluminium and the presence of small globular aluminium particles.

CONCLUSIONS

(1) A survey has revealed that very little scientific work on UHSM has been published. That by Vaughn[3] is the most significant.

(2) An explosively driven UHSM test facility has been designed and constructed and has been shown to operate as required for a detailed study of the process. This development has constituted the main part of the present investigation.

(3) The test facility is capable of achieving strain-rates of 10^9 s^{-1}, a value higher than those attainable by any existing mechanical test system.

(4) It has been found that the cutting ratio is greater than unity at high cutting velocities. This may be attributed to a standing plastic wave, corresponding to the shear plane, which is pushed towards the toolface with velocity increase.

(5) Shear-angles may be as high as 75° at a cutting speed of 4000 ft/s, when cutting aluminium.

(6) The force arising from momentum change in the chip becomes extremely high as velocity in-

creases. This tends to cause the tool to dig into the workpiece.

(7) The chip–tool interface length increases with velocity.

(8) A metallurgical examination of the chip and workpiece indicates that melting may occur in UHSM, both at the chip–tool interface and along preferred layers within the chip. Metallurgical examination of these layers suggests that they are similar to adiabatic shear bands, which have been documented in the literature.

ACKNOWLEDGEMENTS

The authors express their thanks to Mr. R. R. Peach for assistance in building the test facility and during performance tests, and wish to acknowledge that the project was supported by the Harold Armstrong Memorial Fund.

REFERENCES

1. M. E. MERCHANT, 1969. Technological forecasting and production engineering research. Presidential address, CIRP General Assembly.

2. G. ARNDT, 1970. The development of higher machining speeds. *The Production Engineer*, **49**, pp. 470 and 517.

3. R. L. VAUGHN, 1960. Ultra-high-speed machining. *AMC Technical Report* 60–7–635 (1).

4. V. D. KUZNETSOV, 1945. Super-high-speed cutting of metals. *The Iron Age*, **155**, pp. 66 and 142.

5. R. F. RECHT, 1960. The feasibility of ultra-high-speed machining. *M.Sc. Thesis,* Denver University, Colorado.

6. V. D. KUZNETSOV, G. D. POLOSATKIN and M. P. KALASHNIKOVA, 1960. The study of cutting processes at very high speeds. *Fizika metallov i metallovedenie*, **10**, p. 425.

7. G. D. POLOSATKIN *et al.*, 1967. Cutting and grinding at ultra-high-speeds (in Russian), *Izvestiya vyschikh uchebnykh zavedenii, Fizika*, **5**, pp. 93–101.

8. R. L. VAUGHN, 1958. A theoretical approach to the solution of machining problems. *ASTME Technical Paper No. 164.*

9. R. L. VAUGHN, 1960. Recent developments in ultra-high-speed machining, *ASTME Technical Paper No. 255.*

10. G. D. POLOSATKIN, 1948. Rezaniye metallov so skorostyami ot 100–700 m/sek, *Sibirsk phys, tekhn. inst., Scientific Report.*

11. H. TAKEYAMA, T. MURAI and E. USUI, 1955. Speed effect on metal machining. *Jnl. Mechanical Laboratory of Japan 1*, pp. 59–61.

12. W. N. FINDLEY and R. M. REED, 1963. The influence of extreme speeds and rake angles in metal cutting, *Trans. ASME Series B*, **85**, pp. 49–67.

13. R. G. FENTON and P. L. B. OXLEY, 1967. Predicting cutting forces at super-high cutting speeds from work material properties and cutting conditions. *Proc. 8th Int. M.T.D.R., Conf.*, Pergamon Press, pp. 247–258.

14. Y. TANAKA, H. TSUWA and M. KITANO, 1967. Cutting mechanism in ultra-high-speed machining, *ASME Paper No. 67*-PROD-14.

15. G. ARNDT, 1971. Ballistically induced ultra-high-speed machining, *Ph.D. Thesis,* Monash University, Melbourne, Australia.

16. G. ARNDT, 1972. Ultra-high-speed machining. Submitted for presentation at *CIRP General Assembly.*

17. G. ARNDT and J. T. McHENRY, 1970. A computerised internal ballistic analysis of conventional gun systems with muzzle velocities of up to 8000 ft/s, *EXPLOSIVSTOFFE*, **18**, p. 253.

AN INVESTIGATION OF THE PERFORMANCE OF A QUICK-STOP DEVICE FOR METAL CUTTING STUDIES

by

R. H. BROWN* and R. KOMANDURI*

SUMMARY

The specimen obtained by a tool quick-stop device should not be grossly influenced by the stopping conditions. Some previous investigators have reported measurements of extremely high deceleration rates for explosive quick-stop devices. However, these measurements are open to doubt since nearly all assume a uniform acceleration for the first few thousandths of an inch of tool displacement, and also, it can be shown that the reported accelerations would require explosive 'shot start' pressures and energies higher than seems possible.

The instantaneous chamber pressure and the displacement—time behaviour of an explosive quick-stop device have been investigated using a piezoelectric pressure transducer and a photo-optic comparator. The initial acceleration is shown to be highly non-uniform, with acceleration rates of around 10^5 and 10^6 in/s² in the first few thousandths of an inch of displacement (the stage of tool disengagement). The influence of variation in explosive charge weight, chamber volume and shearpin size is reported.

INTRODUCTION

Much useful information on the mechanics of cutting can result from a study of the chip root after rapidly stopping a cut. For a valid representation of the cutting conditions, it is important that there should be very rapid deceleration of the tool relative to the work. Most 'quick-stop' devices have employed some form of shearpin which is fractured by the application of energy from an external source. Early devices simply used a hand-held hammer or energy from a spring. The deceleration rates were relatively low and the chip roots obtained were influenced by the stopping conditions. In some systems the tool was picked up by the workpiece, but in most it was accelerated away from the cut.

Recently a variety of explosively-powered quick-stop devices have been developed. In all of these the tool is accelerated away from the workpiece after fracture of a shearpin. The devices may be categorised into two basic types:

(i) those in which the exploding gas pressure acts directly on a piston in contact with the tool holder.

(ii) those in which an explosively-driven bolt impacts the top of the tool holder.

Many papers describing explosive quick-stop devices have quoted performance figures for the acceleration, stopping time or stopping distance of the tool. Most of these have indicated that the devices are suitable to study machining at quite high cutting speeds, up to 1000 ft/min and above. The present authors wanted to use a quick-stop for an investigation of segmental cutting, but on reviewing the literature it became clear that some of the measurements of performance involved considerable error and most of the quoted values of performance seemed higher than could be achieved with the type of explosive drives employed. This led to the present study.

REVIEW OF PREVIOUS QUICK-STOP PERFORMANCE

Quick-stop performance is probably best measured by the time (t_s) or distance (d_s) for the tool to reach the cutting velocity. Many authors have used 1000 ft/min as a standard cutting speed for this comparison. This will be used here. The acceleration of the quick-stop has also been used as a performance measure.

Table 1 summarises some of the values from recent literature. When a particular reference did not quote all the performance figures, the missing values have been calculated assuming a constant acceleration during the time the tool accelerates up to 1000 ft/min. Values calculated in this way are shown in brackets. Where the original paper used metric units these have been converted to inch values. The second column of the table briefly indicates the type of device: the the third column indicates the method used for performance measurement. All these methods basically involve measurement of the displacement—time characteristic of the toolholder. It is significant that many investigators have measured this characteristic at time or distance intervals that are relatively large compared to the time and distances they claim for stopping the cut. All the authors, with the exception of Ellis, Kirk and Barrow[4], have indicated that they consider the acceleration to be constant for at least a time (or distance) greater than that of their basic measuring interval. When one considers the nature of the force producing the acceleration this assumption seems rather unreasonable and may lead to performance measurements unrepresentative of the initial movement of the toolholder. The movement in the initial few thousandths-of-an-inch controls the nature of the chip root specimen. Thus, this is the stage at which accurate performance figures are required.

* Department of Mechanical Engineering, Monash University, Victoria, Australia

TABLE 1

Reference	Type of quick-stop	Method of measuring performance	d_s Displacement to reach velocity of 1000 ft/min (in)	t_s Time to reach velocity of 1000 ft/min (μs)	Acceleration (in/s^2)
Hastings[1]	Explosive charge driving piston in contact with toolpost	(a) High-speed film (acceleration estimated for first ½ in of movement)	(0.0003)	(3.3)	6×10^7
		(b) Capacitance plates (acceleration in first 0.010 in)	0.00005	(0.5)	4×10^8
Stevenson and Oxley[2]	As above	Breaking strain gauge (acceleration during first 0.002 in movement)	(0.002)	(20)	1×10^7
Philip[3]	As above	(a) High-speed film (3000 frames/s, i.e. 330 μs time intervals between measurements (b) Breaking wires (wires approx. 0.4 in apart) (c) Forces on shank holding quick-stop device (natural frequency 850 Hz)	(0.006)	(57)	3.5×10^6
Ellis, Kirk and Barrow[4]	Bolt from a gun striking toolholder in a slide mounting (¾ in free-flight)	Capacitance (Wayne and Kerr distance meter)	0.0025	31	(5.2×10^6) (6.5×10^6) Non-uniform
Williams, Smart and Milner[5]	Bolt from a gun striking pivoted toolholder (2 in free-flight)	High-speed film (measured time for first 0.1 in of movement)	0.0015	(15.2)	1.3×10^7
Spaans[6]	As above (free-flight distance not stated, apparently about 1 in)	Very-high-speed film (40,000 frames/s i.e. measurements at 25 μs intervals)	0.0007	7	2.8×10^7

FEASIBILITY OF QUOTED QUICK-STOP PERFORMANCE

Many of the performance figures quoted in the literature seem greater than would be expected by standard ballistic devices. This feature is indicated by approximate calculations of the required explosive pressures and energy outputs to give the stated performance figures.

We consider first devices of the type where an exploding gas pressure acts directly on a piston contacting the toolholder[1,2,3,7]. It is known from pressure measurements in the chamber of guns that the pressure reaches a certain value before the projectile strats to move (the 'shot-start' pressure). Subsequently the pressure rises as the projectile moves along the barrel, reaches a maximum and then falls slowly. The shot-start pressure is considerably less than the maximum pressure (usually about 1/8th to 1/5th of the maximum). By introducing a shearpin or other projectile-containing device, the shot-start pressure may be raised slightly, but it will still be considerably less than the maximum pressure reached after the initial projectile movement. The shot-start pressure will influence the initial acceleration. As the projectile starts to move, the acceleration increases with increase in pressure, to the maximum. The value of the maximum will be limited by the bursting condition of the chamber.

For the quick-stop used by Hastings[1] and by Stevenson and Oxley[2] we can make an approximate calculation of the shot-start pressure required to give the initial acceleration of 1×10^7 in/s^2 indicated by Stevenson and Oxley. The dimensions of the device are not known exactly, but from photographs and drawings the weight of the moving parts seems about 1 lb and the chamber piston about ¾ in diameter. Considering the explosive pressure acting as a force on the piston and taking force as mass times acceleration, the required shot-start pressure can be calculated as 58 000 lb/in². When it is realised that the maximum pressure may be about five times greater than the shot-start pressure, this is seen to be an extremely high pressure, considerably greater than the probable loading density (the ratio of charge weight to chamber volume) of propellant could produce and greater than the expected bursting pressure for the type of explosive chamber used.

A similar calculation can be made for the device used by Philip[3]. From the drawing in figure 1(a) of his paper, his piston seems about 1 in diameter and the moving toolholder might weigh as little as ¾ lb (it is probably slightly heavier than this). With these values, to achieve an initial acceleration of 3.5×10^6 in/s², a shot-start pressure of at least 5820 psi would be required. This is within the bounds of possibility, but it would require a very high loading density and a massive combustion chamber. Philip used a relatively-long time interval for his initial displacement measurements (330 μs). Thus his quoted acceleration figure is not the initial value, but a value influenced by the later rise in chamber pressure.

An approximate estimate of the possible accelerations, for devices in which an explosively-powered bolt strikes the toolholder, is more difficult than for the direct explosive drive systems. The kinetic energy of a relatively low-mass bolt is transferred to the toolholder causing shearpin fracture and acceleration of the toolholder mass. Impact will involve some energy loss. For the three devices of this type listed in Table 1 the free flight is large, so it is unlikely that gas pressure will still be accelerating the bolt at impact, that is while the impacting surfaces undergo plastic and elastic deformation. This is the time of interest, so far as quick-stop performance is concerned.

The estimation of impact accelerations depends critically on the shape and nature of the impacting surfaces, and will not be attempted here. We can, alternatively, make a rough estimate of the energy required to achieve the accelerations for the impacting devices listed in Table 1.

Williams, Smart and Milner[5] assume a uniform acceleration of 1.3×10^7 in/s² for the first 0.1 in of toolpost movement. From figure 3 of their paper the weights of the impacting bolt and the toolholder may be estimated as about 0.2 lb and 2 lb, with a 7 in radius of toolholder rotation. On these figures, neglecting impact energy loss and shearpin fracture energy, a driving energy of 3500 ft lb would be required. Since captive bolt guns of the type used have typical energy outputs of about 300 ft lb, the measured acceleration appears excessively high. This may be partly explained by the very large value assumed for tool block displacement during impact (0.1 in).

Ellis, Kirk and Barrow[4] have not indicated the total distance for which their tool block is accelerating. From their displacement—time records it would seem to be about 0.025 in. Using this and estimating bolt and toolholder weights as 0.3 lb and 2 lb, respectively; for an acceleration of 5.2×10^6 in/s² the driving energy of the gun should be 778 ft lb. When allowance is made for energy losses, this seems too high a value

for a gun of the type used. It must be recognised that the present calculation is based on assumed data, but there does seem some reason to question the measured values of time and distance quoted for stopping the cut.

Spaans[6] gives no indication of the dimensions of his quick-stop. Hence even a rough estimation of required driving energy is not possible. However, by comparison with the above two devices, it is seen that a high energy is needed to obtain 2.8×10^7 in/s² acceleration. Even though Spaans used a very high framing speed to record displacement, the time interval between measurements is still relatively large (25 μs). This may have led to error in establishing the zero-point for displacement, which is critical for a valid determination of the initial accelerations.

PERFORMANCE MEASUREMENT OF A QUICK-STOP DEVICE

A quick-stop device, almost identical to that developed by Gradman and Bramall[7] was constructed; see figure 1. The critical operating dimensions were taken from Gladman and Bramall's design, although some minor changes were made. The toolholder and its slide were tilted 5°, to ensure that the tool cleared the workpiece when cutting the end of a tube. The shank was increased to 2½ in square and strain gauges cemented to this to measure two components of force. The device was mounted in a massive toolpost. A safety helmet was added to guard the operator in case of a chamber burst. At a later stage the plunger size was varied to alter chamber volume and the cylinder head modified to incorporate a piezoelectric transducer. The main working characteristics of the device are listed in Table 2.

TABLE 2

Working dimensions and characteristics of the quick-stop device used by the authors.

Plunger	: top diameter (exposed to gas pressure) 0.75 in mass 0.146 lb wt.
Tool block and tool	: mass 0.678 lb wt.
Tool size	: ½ in square.
Shearpin	: material, cast iron (Meehanite) diameter 0.25 in 30° notches with root-diameter of 0.1875 in were machined in some shearpins. The notches were located directly under the sides of the toolblock.
Propellant	: type, mononitro cellulose (also known as T-powder) charge weights: 0.3 g and 0.4 g charges were used in different tests.
Fuse	: LMNR fuse head (electrically ignited).
Chamber volume	: varied by altering length of piston. Volumes of 2.724 and 3.176 cm³ were used.

The performance of the device was studied with various chamber volumes, charge weights and with notched and unnotched shearpins. The displacement—time characteristic was measured using an MTI

Fotonic Sensor, type KD-45A. This sensor consists essentially of a tube of randomly-oriented fine glass fibreoptics. Half the fibres transmit light from a source in the transducer; the other half collect light reflected from any surface close to the end of the tube of fibreoptics. The intensity of light reflected from a given surface is a function of the separation distance. The collected light is transmitted by the

instrument is particularly suitable for the present study because it has a frequency response flat from zero to 80 kHz. In the performance tests on the quick-stop, a flat reflecting surface was cemented to the top of a 5° negative rake tool, and the fibreoptic tube mounted normal to the tool surface and slightly above it. As the tool accelerated, the transducer recorded its instantaneous displacement from the initial position.

During the tests, the explosive chamber pressure was measured by a Kistler quartz pressure transducer,

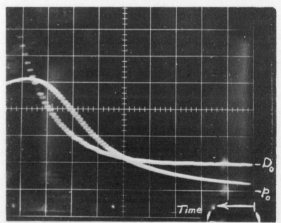

Figure 2. Oscilloscope trace obtained in a performance test. Pressure scale, 4000 lb/in² per division; displacement scale 0.0065 in per division; time scale, 50 μs per division. P_0 and D_0 indicate zero levels for these scales. For this test, chamber volume = 2.724 cm²; charge weight = 0.4 g; ¼ in shearpin notched to $\frac{3}{16}$ in diameter.

model 607. This has a natural frequency of 240 kHz and was used with a sensitivity of 1000 lb in⁻² V⁻¹. The pressure signal was recorded on a second channel of the oscilloscope used to record displacement. The pressure signal provided the trigger source for the oscilloscope sweep. A photograph of the pressure and displacement recorded in this way is shown in figure 2.

PERFORMANCE TEST RESULTS

For each test, the pressure and displacement data were plotted against time, taking the time origin as the instant at which toolholder movement started. Figure 3 shows such a plot from the data in figure 2.

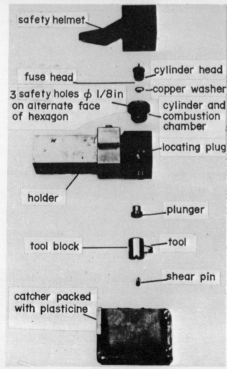

Figure 1. The quick-stop device used by the authors: (a) mounted on the cross-slide of a lathe; (b) exploded view of components.

fibreoptic tube to a photocell and converted to an electric signal which may be displayed on an oscilloscope. The signal can be accurately calibrated in terms of separation distance. For the particular transducer used, the calibration was 65 μin/mV. The

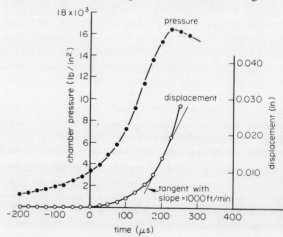

Figure 3. Pressure and displacement data from figure 2 plotted against time. Time origin taken as start of tool movement.

The pressure curve indicates a relatively-slow rise prior to tool movement, reaching a shot-start pressure of 3400 lb/in^2. Thereafter the pressure increases more rapidly and flattens-off to a maximum of about 16 400 lb/in^2, followed by a slow decline. The gradient of the displacement—time curve gives the velocity at any instant. Using the method suggested by Ellis, Kirk and Barrow[4], a tangent with slope equal to a velocity of 1000 ft/min (i.e. 200 in/s) has been drawn. The point of tangency gives the time (t_s) and displacement (d_s) at which the cutting tool velocity would match a cutting speed of 1000 ft/min or, in other words, the stopping time and distance at this cutting speed. From figure 3, t_s = 200 μs and d_s = 0.013 in. These are considerably greater than all the corresponding values quoted in Table 1 for equivalent quick-stop devices.

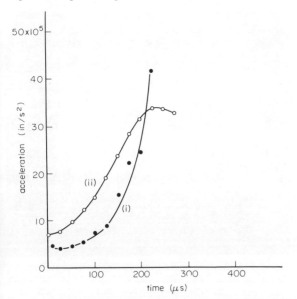

Figure 4. Acceleration—time characteristics for conditions of the test shown in figure 2. Curve (i) is the acceleration calculated from displacement measurements, curve (ii) is that from pressure measurements.

The acceleration of the toolholder can be found by double-differentiating the displacement—time data: this was done graphically. As a check on accuracy the data from some displacement—time records was fed into a computer using an x—y digitiser to read the oscilloscope photograph. A computer programme fitted a high order polynomial to the data and then double-differentiated this polynomial numerically to give acceleration values at 5 μs time intervals. The two methods were found to be in satisfactory agreement. The acceleration determined graphically from figure 2 is shown in figure 4. Initially the acceleration is about 5×10^5 in/s rising to about 2.8×10^6 in/s after 200 μs of tool movement. These are low compared to values in table 1.

To check that the measured acceleration is of about the right magnitude, accelerations to be expected from the measured pressures were calculated. This was a prime reason for recording the pressure. Knowing that the piston area is 0.4418 in^2 and the mass of moving components 0.824 lb wt the acceleration is easily determined from the pressure, neglecting

the energy loss to fracture the shearpin. In figure 4, the acceleration found in this way is compared with that from displacement. The two are of similar magnitude, with the acceleration from pressure measurement initially exceeding that from displacement. The acceleration determined from pressure passes through a maximum value. The acceleration from displacement would be expected to follow the same trend and would not be expected to exceed that calculated from pressure since the force applied by gas pressure is producing the acceleration. In several other tests the acceleration curve from displacement was found to lie below that from the pressure. The discrepancy in figure 4 is thought to be due to in-accuracy in acceleration determined when the displacement—time curve becomes very steep. At this later stage of tool movement, the pressure measurements are thought to give a more reliable estimate of acceleration. This type of inaccuracy from displacement—time measurements may well apply in some of the acceleration determinations reported by previous investigators.

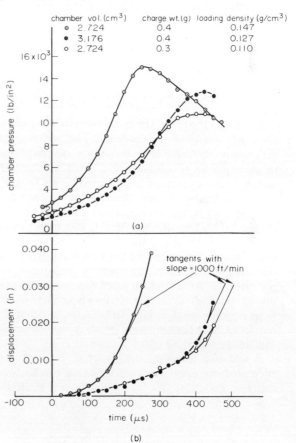

Figure 5. Pressure and displacement results from three tests with different chamber volumes and charge weights. In all cases, shearpin ¼ in diameter, unnotched.

The shape of the pressure—time curve in figure 3 and the corresponding acceleration—time curve in figure 4 are typical of ballistic behaviour. From this evidence, there is little justification for the assumption of uniform acceleration at the start of movement of the tool in a direct explosive quick-stop device. For devices with an impacting bolt, the assumption may be more reasonable, but this will depend on the nature of the impacting surfaces. To compare quick-stop performance it seems better to quote stopping time and distance at a standard cutting speed, say 1000 ft/min (or 300 m/min), rather than give a particular acceleration figure.

Loading density (the ratio of propellant mass to chamber volume) is a useful quantity to indicate ballistic behaviour. Rinehart and Pearson[8] quote experimental data showing that the maximum pressure increases, and the burning time decreases, as loading density is increased. The test data in figure 5 shows these effects for the authors' quick-stop device. The loading density was changed by altering both the mass of propellant and the chamber volume. For all three tests shown in the figure an unnotched shearpin was used. With a lower loading density, the maximum pressure is reduced and the time to reach the maximum is increased, indicating a slower burning rate. The shot-start pressures are also significantly reduced with low loading density. This probably results from the slower burning rate. These changes are reflected in the displacement–time behaviour: the time (t_s) to reach a velocity of 1000 ft/min, is greatly increased with low loading density. The displacement (d_s) to reach this velocity, is also increased, although not in the same proportion as the time increase.

Repeat tests were conducted with both notched and unnotched shearpins: table 3 summarises the results and gives an estimate of the expected variation in measured quantities. The latter estimate is based on three repetitions of one condition. The performance variation is reasonably large, but this is to be expected in an explosive process where a small variation in distribution of propellant grains can influence burning rate. Allowing for the variations, it can be concluded that there is a significant performance difference with change of loading density from 0.147 to 0.126 or 0.110 g/cm³. However, the difference between the two lower loading densities is not significant. Notching the shearpin does not significantly influence performance. At low strain rates this would be surprising, but high strain rate tests with notched plates have suggested that at high strain rates, notches do not act as stress-concentration points. This effect may explain the behaviour noted in table 3.

A test with a loading density of 0.18 g/cm³ was attempted, but resulted in bursting of the combustion chamber. Following this failure, the three ⅛ in safety

holes shown in figure 1(b) were provided above the combustion chamber. These allow the explosive gas to blow out the copper washer and exhaust to atmosphere at excessive pressures (above about 20 000 lb/in²). The system has been in satisfactory service since this modification, but loading densities above 0.147 g/cm³

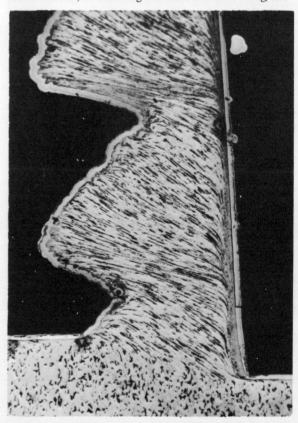

Figure 6. An optical micrograph of a chip root obtained at a cutting speed of 180 ft/min: depth of cut 0.0094, rake angle −5°, work material, low carbon steel. Segmental chip formation.

have not been used for fear of damaging the pressure transducer.

The quick-stop has been used to study segmental chip formation at cutting speeds of about 200 ft/min[9]. Figure 6 shows an optical micrograph of a specimen obtained from the device. The stopping time is considered sufficiently short for cutting speeds of this order, but it is doubtful if it would satisfactorily represent the cutting conditions at speeds approaching 1000 ft/min.

TABLE 3

shearpin		chamber volume (cm³)	charge weight (g)	loading density (g/cm³)	measured pressures		d_s displacement to reach 1000 ft/min (in)	t_s time to reach 1000 ft/min (µs)
diameter (in)	diameter under 30° notch (in)				shot-start (lb/in²)	maximum (lb/in²)		
0.25	0.1875	2.724	0.4	0.147	3000	16 000	0.014	220
0.25	no notch	2.724	0.4	0.147	2900	15 000	0.015	200
0.25	no notch	3.176	0.4	0.127	1400	12 800	0.017	420
0.25	no notch	2.724	0.3	0.110	1900	10 800	0.020	460
Estimated variation in measured quantities					±400	±600	±0.004	±50

METHODS FOR IMPROVING QUICK-STOP PERFORMANCE

For an explosive acting directly on a piston contacting the toolholder, initial acceleration depends primarily on shot-start pressure, piston area and the mass of moving components. The last-mentioned cannot be reduced significantly, so design should be directed to increasing shot-start pressure and piston diameter. Increasing the diameter yields an immediate improvement by increasing the accelerating force, but leads to higher stresses in the combustion chamber. Shot-start pressure may be increased, to some extent, by increasing the shearpin diameter. However, because some energy is lost in shearing the pin, this change also requires an increase in loading density which necessitates a massive combustion chamber to prevent failure.

At best, the shot-start pressure might be increased to about 6000 lb/in^2. If the piston diameter were increased to 1 in and the moving mass kept to 0.824 lb wt, the theoretical initial acceleration would be 2.2×10^6 in/s^2, which is still low compared with values claimed as satisfactory in the literature.

A better solution may be obtained by employing the high gas pressures that follow the shot-start condition. This could be done by using an accelerating bolt to strike the toolholder. To achieve a high pressure the bolt itself should be retained by a shearpin or shear diaphragm and it should impact the toolholder at the instant when the explosive pressure reaches its maximum value. With careful design it is thought possible to approach initial accelerations of about 10^7 in/s^2 with a device of this type.

CONCLUSIONS

(1) It has been shown that the acceleration from a direct explosively-driven quick-stop device is not uniform. This indicates that measurement of the tool displacement at relatively long time (or distance) intervals after firing, can give misleading values of initial acceleration.

(2) Many of the performance figures for quick-stop devices quoted in the literature appear excessively optimistic in relation to the explosive pressures and energy available from these devices.

(3) Some initial considerations for the design of a device to achieve very short stopping times are discussed.

ACKNOWLEDGEMENTS

The authors express their thanks to Mr. G. A. Gladman, formerly Senior Principal Research Scientist of CSIRO Applied Physics Division, for providing working drawings of a quick-stop device[7], and to Mr. R. R. Peach for assistance in the performance tests.

REFERENCES

1. W. F. HASTINGS, 1957. A new quick-stop device and grid technique for metal cutting research. *Annals CIRP*, **15**, p. 109.
2. M. G. STEVENSON and P. L. B. OXLEY, 1969/70. An experimental investigation of the influence of speed and scale on the strain-rate in a zone of intense plastic deformation. *Proc. Inst. Mech. Engrs.*, **184**, p. 561.
3. P. K. PHILIP, 1971. Study of the performance characteristics of an explosive quick-stop device for freezing cutting action. *Int. J. Mach. Tool Des. Res.*, **11**, p. 133.
4. J. ELLIS, R. KIRK and G. BARROW, 1969. The development of a quick-stop device for metal cutting research. *Int. J. Mach. Tool Des. Res.*, **9**, p. 321.
5. J. E. WILLIAMS, E. F. SMART and D. R. MILNER, 1970. The metallurgy of machining, Part 1: basic considerations and the cutting of pure metals. *Metallurgia*, **81**, p. 483.
6. C. SPAANS, 1971. The fundamentals of three-dimensional chip curl, chip breaking and chip control. Department of Mechanical Engineering, Delft University of Technology, Report WTHD No. 24, p. 15.
7. C. A. GLADMAN and F. J. BRAMALL. Discussion contributed to reference 2.
8. J. S. RINEHART and J. PEARSON, 1963. Explosive working of metals. Pergamon Press, Oxford.
9. R. KOMANDURI and R. H. BROWN, 1972. Microcrack formation in machining a low carbon steel. Paper submitted to *Annals CIRP*.

THE ACCURATE DETERMINATION OF CUTTING FORCES

by

J. TAYLOR* and G. C. I. LIN†

SUMMARY

The design and construction of a rigid three-component dynamometer is described. The calibrating procedure is presented as an application of the principles of designed experimentation enabling a very accurate calibration to be carried out. The dynamometer was used in conjunction with a carefully planned experimental procedure to detect the influence of clearance angle on cutting-forces.

INTRODUCTION

The most interesting development in metalcutting theory in recent years has been the discovery of the true nature of the connection between cutting mechanics and the basic work material properties. Fenton and Oxley[1] successfully predicted chip geometry and cutting-forces from work material properties and cutting conditions, while Stevenson and Oxley[2] derived the basic flowstress properties of a low carbon steel at high strain rates using an orthogonal machining test. Lin and Oxley[3] have successfully predicted forces and so on in oblique machining from material properties.

It is now possible to measure the flowstress properties at high strain rates from orthogonal machining tests and to use the results to predict cutting mechanics for a much wider range of conditions than that used in the original test. It is likely also that the derived results of material properties could be used to predict forces arising in other processes of rapid deformation. Future work along these lines will cover the investigation of a wide range of work in materials by the orthogonal machining test.

To carry out this work successfully it has been necessary to develop an accurate technique for measuring cutting-forces. The technique includes not only the instrumentation, but the organisation of the experiment and the compilation and analysis of the results. The technique is described in this paper together with an account of its application to the detection of a known minute effect in cutting mechanics.

THE DYNAMOMETER

Operating principles

Figure 1 illustrates the main operating principles of the three-component dynamometer. The cutting tool tip is mounted on a platform which fits into a tubular member supported at its rear end by a flexible diaphragm. This locates the end of the tubular member on the central axis, but permits it to be supported by the load carrying element at the front. The load carrying element, machined from aluminium alloy bar, has an enlarged centre section through which the tube passes with two radial arms

having enlarged ends for fixing in the body. The working section of the radial arms, which are vertical, is $1\frac{1}{4}$ in long with a $1\frac{1}{4} \times 1\frac{1}{4}$ cross-section. Rosette type straingauges are cemented in the position shown, and sense the cutting-force components F_c, F_1 and F_2.

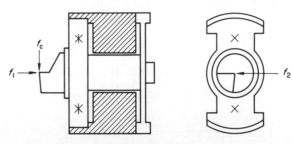

Figure 1. Design principles of dynamometer.

With the ends firmly located in the body, the upper and lower branches of the element are placed in tension and compression respectively, and the four vertical gauges can be connected to form a temperature-compensated circuit which theoretically should sense only the vertical component F_c. The output signal voltage V_s from this circuit is given by the equation

$$V_s = \frac{V(GF)F_c}{2AE} \qquad (1)$$

where V = supply voltage
$\quad GF$ = gauge factor
$\quad A$ = cross-sectional area of element
$\quad E$ = modulus of elasticity

The gauges placed at 45° to the axis respond only to the principal strains caused by the shearing actions of forces F_1 and F_2 on the element. The voltage signals are given by the equations

$$V_s = \frac{V(GF)F_1}{4AG} \qquad (2)$$

and

$$V_s = \frac{V(GF)F_2}{4AG} \qquad (3)$$

Here G is the shear modulus.

* School of Mechanical and Industrial Engineering, University of New South Wales, New South Wales, Australia.
† Warrnambool Institute of Advanced Education, formerly University of New South Wales, Australia.

Since E/G for aluminium is approximately 2.2, the sensitivity of the F_1 and F_2 circuits is slightly greater than that of the F_c circuit but as F_c is always greater than the other components, the load bearing member gives the same magnitude of response from all these forces. This arrangement, whereby the same member carries direct tension, compressive and shear stresses, is superior to other designs as it yields the maximum strain sensitivity in relation to the deflection of the element. The design stiffness of the element in the F_c direction is about 30×10^6 lb/in, and about 10×10^6 in the transverse direction. As the cross-section of the element is large in relation to the working length, the transverse deflection due to bending is only 3 per cent of that due to shear.

Construction

The load bearing element is locked into the front end of the body which is machined from a bar of medium carbon steel (figure 2). The outer rim of the flexible diaphragm is bolted to the rear end of the body. The tubular member passes through a close fitting central hole. It is intended eventually to fill the clearance

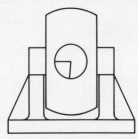

Figure 2. External view of dynamometer on mounting.

space between the tube and the hole with a suitable fluid to obtain viscous damping to enable dynamic measurements to be carried out. The body has been made massive to function as a seismic mass. When in use the dynamometer is mounted on a heavy castiron bracket which in turn is bolted directly to the lathe cross slide in place of the usual swivel slide. The monobloc type of construction facilitates rapid removal from the lathe to an identical mounting bracket on the calibrating fixture.

Calibration

For static calibration the dynamometer is transferred to a very rigid calibrating fixture. Three levers carrying deadweights are pivoted on needle roller bearings, and can load the dynamometer along the three specified directions. The loads are transmitted through three pushrods accurately aligned along these chosen directions. A special nosepiece is used in place of the normal cutting tool.

In order to measure the response of the dynamometer under all conditions of loading, it is desirable to organise the calibrating procedure as a three factor experiment in which the dynamometer is loaded along all three directions simultaneously, rather than check the response along one axis keeping the loads in the other two directions fixed. While the last method would detect cross-sensitivity, it would overlook the possible existence of interactions

between the three components. The force levels to be used are decided in advance and in the example shown (see appendix) are

$$F_c, \quad 0, \quad 300, \quad 600 \, \text{lb}$$
$$F_1, \quad 0, \quad 200, \quad 400 \, \text{lb}$$
$$F_2, \quad 0, \quad 200, \quad 400 \, \text{lb}$$

There are therefore twenty-seven combinations of load which are worked through in random order.

The dynamometer is used in conjunction with an SEL carrier amplifier type SE 4000 coupled to an RDK three-pen electronic recorder type B.34.

Table A1 (see appendix) shows the pen deflections given by the F_c net (the R_c response), which theoretically should not be influenced by F_1 and F_2. However, these components appear to influence the R_c response to some extent. Similar results are obtained for the R_1 and R_2 responses. The significance of the effects of F_c, F_1 and F_2 is tested in the analysis of variance of the R_c response. Bearing in mind that the pen deflections always fluctuate to some extent during actual measurements of cutting-forces, a reasonable level of discrimination between two pen traces appears to be 2 units which would indicate an error mean square of 1.

Table A2 shows that the F_1 and F_2 effects are significant and that there is a small deviation from linearity in the response from F_c. These effects must be considered in the calibration equations.

Equations connecting R_c, R_1 and R_2 with F_c, F_1 and F_2 can be derived from the coefficients of the significant effects. For example

$$R_c = 140.72\phi_1 - 1.62\phi_2 + 0.72\theta_1 \qquad (4)$$

ϕ and θ are orthogonal polynomials of $\dfrac{F_c}{300} - 1$ and $\dfrac{F_1}{200} - 1$ respectively.

For three levels

$$\phi_1 = \frac{F_c}{300} - 1, \qquad \phi_2 = \phi_1^2 - \frac{2}{3}$$

$$\theta_1 = \frac{F_1}{200} - 1$$

substituting for ϕ_1 and so on

$$R_c = -0.7 + 0.48F_c - 0.000018F_c^2 + 0.0036F_1 \quad (5)$$

Similar analyses are carried out for R_1 and R_2 giving

$$R_1 = -0.4 + 0.004F_c + 0.497F_1 \qquad (6)$$

$$R_2 = -1.43 + 0.011F_c + 0.0098F_1 + 0.502F_2$$
$$-0.000026F_2^2 \qquad (7)$$

Finally, expressed in terms of R_c and so on

$$F_c = 2.1R_c + 0.00016R_c^2 - 0.015R_1 + 0.7 \, \text{lb} \qquad (8)$$

$$F_1 = 2R_1 - 0.017R_c - 0.8 \qquad (9)$$

$$F_2 = 2R_2 + 0.0002R_2^2 - 0.063R_c - 0.41R + 3 \quad (10)$$

The small cross-sensitivity terms are considered to be due to small errors in fixing the straingauges. An error of $1°$ in gauge alignment can introduce a cross-sensitivity of 1.7 per cent if the perpendicular force is equal to the main force. Unfortunately, rosette gauges are more sensitive to errors in alignment since the effects in each element are likely to be additive.

THE EFFECT OF CLEARANCE ANGLE ON CUTTING FORCES

Introduction
The technique was first applied to the detection of a possible effect of clearance-angle on cutting-force. Kobayashi and Thomson[4,5] carried out a comprehensive series of force measurements in which some trials were repeated with a change in clearance from $1°$ to $3°$. A statistical examination of these results showed that the increase in clearance gave an average decrease of 6.6 per cent in cutting force. Undoubtedly, the low values of clearance used intensified the effect, and a proportionate decrease could not be expected at higher clearance angles. Therefore, the measurement of a possible effect at higher clearance angles was undertaken as a particularly searching test of the technique just described.

Design of experiment
The detection of any small effect in metalcutting calls for careful planning and execution of the experiments in view of the large variability of machining investigations. The large variability is a consequence of the presence of numerous factors affecting performance in any metalcutting operation, our inability to exercise complete control over these factors and our incomplete knowledge of the process.

The random fluctuations of each factor about its nominal expected level gives rise to a variance in the dependent variables. Because of the additive property many small variances will result finally in a large variance in the measured quantity. One of the authors[6] attempted to determine the average variance of tool-life determinations and arrived at a figure of 0.0043 for the residual mean square in log tool life, equivalent to a standard deviation of ±15 per cent of the mean level. Cutting force measurements are unlikely to suffer such large residuals but in a recent statistical determination of cutting force Braithwaite and Hague[7] found a standard deviation of ±8 per cent when the effects of work material variation were included. When such large errors are possible meaningful results can only be obtained from a correctly designed experiment susceptible to the appropriate statistical analysis[6,8].

In the current example cutting-force measurements were carried out over the following range of conditions:

work material, EN8
tool material, carboloy 350
rake-angle R, $6°$, $13°$
uncut chip thickness t, 0.00446, 0.00892 in
cutting speed S, 200, 300, 450, 675 ft/min
clearance angle C, $1°$, $2°$, $4°$, $8°$ and $16°$.

By testing all combinations of the 80 chosen levels, a four-factor experiment was obtained, the results of which could be assessed by the analysis of variance. This is a particularly effective and appropriate technique for metalcutting research in view of the large residuals, and possibility of strong interactions between the independent variables.

The residuals of many metalcutting variables tend to a log-normal distribution. Also equations of the type

$$y = Ax^b z^c \quad \text{and so on} \tag{11}$$

predominate. Therefore in order to satisfy essential conditions connected with the analysis of variance, the relationship between the dependent variable Y and the independent was assumed to take the following form

$$Y = \overline{Y} + \sum_1^4 a_K(\log C)^K + \sum_1^3 b_m(\log S)^m$$

$$+ dR + et_1 + \sum_1^4 \sum_1^3 f_{Km}(\log C)^K(\log S)^m \text{ etc.}$$

$$+ \sum_1^4 \sum_1^3 g_{Km} R(\log C)^K(\log S)^m \text{ etc} \tag{12}$$

$$Y = 100[(\log_{10} F - 1)]$$

$$F = \text{resultant cutting-force}$$

The significance of the various terms can be assessed by noting if the addition of each extra term results in a significant reduction in the residual mean square. This is very tedious, and a much simpler method is possible using orthogonal polynomials. If the levels of the independent variables are equally-spaced on the appropriate scale, tabulated values[9] of the polynomials can be used.

The equation for Y was expressed as follows

$$Y = \overline{Y} + \sum_1^4 a_K^1 \phi_K + \sum_1^3 b_m^1 \theta_m$$

$$+ d^1\gamma_1 + e^1\lambda_1 \sum_1^4 \sum_1^3 f_{Km}^1 \phi_K^1 \theta_m^1$$

$$+ \sum_1^4 \sum_1^3 g_{Km}^1 \gamma_1 \phi_K \theta_m + \sum_1^4 \sum_1^3 h_{Km}^1 \lambda_1 \phi_K \theta_m$$

$$\text{etc.} \tag{13}$$

ϕ_K is the orthogonal polynomial of order k of $f(\log C)$
θ_m is the orthogonal polynomial of order m of $f(\log S)$
and so on.

The mean square of each term and the coefficient a_K^1 and so on can be evaluated independently of each other in one analysis of variance. Substitution of the significant polynomials by the appropriate function will give an equation in terms of the original variables.

Experimental procedure

All the tests were carried out on one bar of 6 in diameter EN8 steel which was mounted in a Denham SS8 centre lathe. By means of an undercutting tool mounted in a rear toolpost, a short tubular specimen could be machined in the end of the bar before each test. The tube thickness was maintained at 0.2 ± 0.0003 in, and the length was approximately 0.25 in, sufficient to enable steady force readings to be recorded. The test tools which were mounted in the dynamometer were arranged to machine the end of the tube (figure 3).

Brazed sintered carbide tools (carboloy 350) were used with precision ground shanks fitting into the

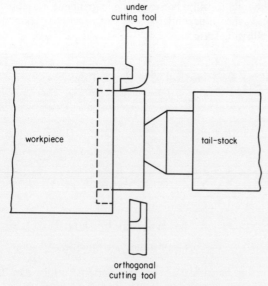

Figure 3. Experimental arrangement.

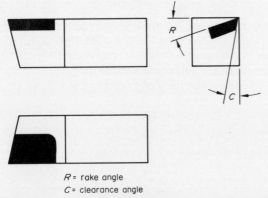

R = rake angle
C = clearance angle

Figure 4. Cutting tool.

dynamometer (figure 4). All tool faces were lightly ground with a 240 grit metal bonded diamond wheel before each test. By using shims the cutting-edge was maintained in line with the axis of the dynamometer and the workpiece to an accuracy of ± 0.0005 in. Tests were run at the conditions already described and measurements of F_c and F_1 taken. For this experiment the dynamometer operated as a two-component instrument. The resulting cutting-force F was calculated.

$$F^2 = F_c^2 + F_1^2 \qquad (14)$$

F_c = component parallel to cutting velocity
F_1 = feed component.

RESULTS AND ANALYSIS OF VARIANCE

Tables 1 and 2 show the resultant cutting force and Y values respectively and the qualitative analysis of variance is given in table 3. The highest order interaction yields a mean square of 0.51 representing a standard deviation of approximately ± 1.6 per cent of the average cutting force. Although this is significant when compared with the overall accuracy of the instrumentation, it is considered sufficiently

TABLE 1

Original values of resultant cutting force F (lb)

rake R	speed S ft/min	uct t_1 (in)	clearance angle C				
			1°	2°	4°	8°	16°
13°	200	0.00446	345.5	341.5	332.9	340.5	338
		0.00892	567.2	562.7	570.0	554.8	559
	300	0.00446	308.7	305.5	303.7	304.1	288
		0.00892	682.9	675.7	658.0	666.9	648.9
	450	0.00446	307.8	327.5	330.2	335.9	328
		0.00892	668.8	650.7	641.8	653.4	632.2
	675	0.00446	347.0	345.6	345.1	339.8	334.4
		0.00892	552.3	555.5	546.0	546.9	539.0
6°	200	0.00446	401.3	393.8	392.5	392.5	387.5
		0.00892	732.1	709.6	715.0	684.5	673.5
	300	0.00446	357.6	350.7	358.4	362.2	360.0
		0.00892	718.6	721.1	718.9	720.3	722.8
	450	0.00446	391.6	395.5	394.9	386.6	396.1
		0.00892	684.5	691.4	675.9	670.5	670.3
	675	0.00446	376.7	351.1	371.2	368.0	346.2
		0.00892	614.5	609.4	596.9	586.1	597.2

TABLE 2

Transformed values from resultant force

rake R	speed S (ft/min)	uct t_1 (in)	clearance angle C				
			1°	2°	4°	8°	16°
13°	200	0.00446	153.6	153.3	152.2	153.2	152.9
		0.00892	175.4	175.0	175.6	174.4	174.7
	300	0.00446	149.0	148.5	148.2	148.3	146.0
		0.00892	183.4	183.0	181.8	182.4	181.2
	450	0.00446	148.8	151.5	151.9	152.6	151.6
		0.00892	182.5	181.3	180.7	181.5	180.1
	675	0.00446	154.0	153.9	153.8	153.1	152.4
		0.00892	174.2	174.5	173.7	173.8	173.2
6°	200	0.00446	160.3	159.5	159.4	159.4	158.8
		0.00892	186.8	185.1	185.4	183.5	182.8
	300	0.00446	155.3	154.5	155.4	155.9	155.6
		0.00892	185.6	185.8	185.7	185.8	185.9
	450	0.00446	159.3	159.7	159.6	158.7	159.8
		0.00892	183.5	184.0	183.0	182.6	182.6
	675	0.00446	157.6	154.5	157.0	156.6	153.9
		0.00892	178.9	178.5	177.6	176.8	177.6

TABLE 3

Analysis of variance of log resultant force
(all effects considered qualitative)

Source of variance	Degrees of freedom	Sum of squares	Mean square	Variance ratio
Clearance, C	4	12.24	3.06	5.95
Speed, S	3	140.84	46.95	91
Rake, R	1	589.23	589.23	1146
U.C.T. t_1	1	13523.96	13523.96	26291
$C{\times}S$	12	6.18	0.52	1
$C{\times}R$	4	2.15	0.54	1.05
$C{\times}t_1$	4	4.28	1.67	2.08
$S{\times}R$	3	59.75	19.92	39
$S{\times}t_1$	3	352.52	117.5	228
$R{\times}t_1$	1	9.63	9.63	18.7
$C{\times}S{\times}R$	12	10.13	0.84	1.64
$C{\times}S{\times}t_1$	12	9.78	0.82	1.58
$C{\times}R{\times}t_1$	4	1.05	0.26	
$S{\times}R{\times}t_1$	3	74.96	24.99	.48
Residual				
$C{\times}S{\times}R{\times}t_1$	12	6.17	0.51	

TABLE 4

Quantitative analysis of variance of log resultant force

Source of variance	Degrees of freedom	Sum of squares	Mean square	Variance ratio
C Linear (C_1)	1	11.03	11.03	16.5
S Linear (S_1)	1	56.1	56.1	84
S Quadratic (S_2)	1	54.45	54.45	82
S Cubic (S_3)	1	30.30	30.30	45
R	1	589.23	589.23	900
t_1	1	13523.96	13523.96	20000
Interactions				
S_1R	1	55.06	55.06	83
S_3R	1	3.68	3.68	5.5
S_1t_1	1	41.22	41.22	62
S_2t_1	1	260.64	260.64	391
S_3t_1	1	50.55	50.55	76
Rt_1	1	9.63	9.63	144
S_1Rt_1	1	3.61	3.61	5.4
S_2Rt_1	1	76.31	76.31	106
Residual				
$CRSt$ + non-sig. effects	65	42.98	0.66	

small, in view of the numerous sources of variance, to be considered as a residual. A rigorous test of this assumption would have required a complete duplication of the experiment. The mean square estimate of all the other effects are given and the variance ratio related to the residual.

The effects of speed, chip thickness and rake are, as expected, highly significant. The clearance effect, although small, is significant at the 1 per cent level, that is, the probability that the observed effect arises from chance fluctuations is 1 per cent. It is therefore assumed that a real effect exists. Of the ten possible interactions, those involving clearance do not appear to be significant. However, the qualitative analysis is not a sufficient test for quantitative variables.

Where quantitative variables are involved, the sum of squares for each effect can be partitioned into linear, quadratic, cubic components and so forth, each carrying one-degree of freedom. The significance

of each component can be assessed separately increasing the sensitivity of the experiment and enabling the mathematical form of the effect to be found. When the levels are equally-spaced and tabulated values of orthogonal polynomials are used, little extra computation is required.

The quantitative analysis is given in table 4 which shows only the significant effects. Nonsignificant sums of squares with their degrees of freedom have been pooled with the old residual sums of squares to yield a new estimate of the residual.

The influence of the clearance angle comes solely from a linear relationship which is not highly significant (<0.1 per cent). Since Y is a logarithmic function, the relationship between cutting force and clearance can be expressed in the form

$$\log F = P + q \log C \tag{15}$$

$$\text{or } F = PC^q \tag{16}$$

From the analysis of variance the mean cutting force for all levels of S, R and t_1 is

$$F_m = 479C^{-0.00872} \qquad (17)$$

C = clearance angle (deg).

Since there is no interaction involving C the form of equation remains unchanged when the other variables (speed, rake and chip thickness) are altered. For each combination of the levels of these variables there will be a unique value for P but q is constant. The mean variation with C is shown in figure 5. The effect of C can be summarised by the statement that a 100 per cent increase in clearance will reduce the cutting force by 0.6 per cent. The fact that such a small effect can be detected in a metalcutting experiment is a convincing demonstration of the utility and power of statistical methods when applied to the design of experiments.

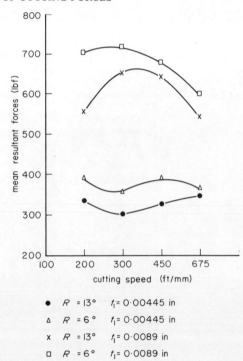

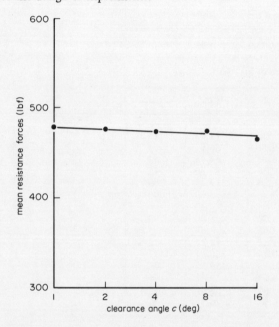

Figure 5. Variation of mean cutting force with clearance angle.

●	$R = 13°$	$t_1 = 0.00445$ in
△	$R = 6°$	$t_1 = 0.00445$ in
×	$R = 13°$	$t_1 = 0.0089$ in
□	$R = 6°$	$t_1 = 0.0089$ in

Figure 6. Variation of cutting force with speed, chip thickness and rake angle.

VARIATION OF FORCE WITH OTHER VARIABLES

Figure 6 shows the expected variation of cutting-force with speed, chip thickness and rake angle for a 4° clearance. The change in shape of the curves as the levels of rake and chip thickness are altered in the physical manifestation of the two-factor and three-factor interactions. For this example the cutting-force cannot be expressed as a simple empirical relationship similar to equation (16). It seems highly likely that the complicated response can be explained in terms of the blue brittleness of steel. This theme will be developed in later papers.

It is considered that the dynamometer used in conjunction with correctly designed experiments is capable of very accurate force measurements and can be used for measuring the properties of materials at high rates of strain.

REFERENCES

1. R. G. FENTON and P. L. B. OXLEY. Mechanics of orthogonal machining: predicting chip geometry and cutting forces from work material properties and cutting conditions, *Proc. Inst. Mech. E., 184, Part I* (1969–70) p. 927.

2. M. G. STEVENSON and P. L. B. OXLEY. An experimental investigation of the influence of strain rate and temperature on the flowstress properties of a low carbon steel using a machining test, *Proc. Inst. Mech. E., 185* (1970–71) p. 741.

3. G. C. I. LIN and P. L. B. OXLEY. To be published.

4. S. KOBAYASHI and E. G. THOMSON. A critical comparison of metalcutting theories with new experimental data, *J. Eng. Industry, Trans. ASME(B) 82* (1960) p. 333.

5. S. KOBAYASHI and E. G. THOMSON. The role of friction in metalcutting. *J. Eng. Industry, Trans. ASME(B) 82* (1960) p. 324.

6. J. TAYLOR. Carbide cutting tool variance and breakage: unknown factors in machining economics, *Proc. 8th MTDR Conf.* (1967) p.487.

7. G. R. BRAITHWAITE and A. G. HAGUE. A statistical analysis of machining variables. *The Production Engineer* (1970).

8. J. TAYLOR. The use of designed experiments in metalcutting research, *The Production Engineer* (1961) p. 654.

9. O. C. DAVIES. The design and analysis of industrial experiments. Oliver and Boyd (1960).

APPENDIX

CALIBRATION OF THREE-COMPONENT DYNAMOMETER

TABLE A1

Deflection of R_C trace

F_1 (lb)		0			200			400		
F_2 (lb)		0	200	400	0	200	400	0	200	400
F_C (lb)	0	0	0	0	0	2	4	0	0	0
	300	141	140	142	142	143	142	144	143	144
	600	281	280	282	281	280	281	284	281	283

1 Unit = 0.02 in

TABLE A2

Analysis of variance

source of variance	sum of squares	DF	mean square
F_C linear	356449.4	1	356449.4
F_C quadratic	15.6	1	15.6
F_1 linear	9.4	1	9.4
F_1 quadratic	.9	1	.9*
F_2 linear	3.2	1	3.2*
F_2 quadratic	0	1	0*
Interactions			
$F_C F_1$	7.2	4	1.8*
$F_C F_2$	3.0	4	.75*
$F_1 F_2$	2.4	4	.6*
$F_C F_1 F_2$	2.1	8	.26*
Total	356493.2	26	

* Not significant when compared with a residual mean square of 1.

Coefficients of significant polynomials

	R_C
F_C linear	140.72
F_C quadratic	−0.54
F_1 linear	0.72

	R_1
F_C linear	1.2
F_1 linear	99.3

	R_2
F_C linear	3.33
F_1 linear	1.95
F_2 linear	98.4
F_2 quadratic	−0.35

Using orthogonal polynomials

$$R_C = \overline{R}_C + 140.72\left(\frac{F_C}{300} - 1\right) - 3 \times 0.54\left[\left(\frac{F_C}{300} - 1\right)^2 - \frac{2}{3}\right]$$

$$+ 0.72\left(\frac{F_1}{200} - 1\right) \tag{i}$$

$$R_C = -0.7 + 0.48F_C - 0.000018F_C^2 + 0.0036F_1 \tag{ii}$$

Similarly

$$R_1 = -0.4 + 0.004F_C + 0.497F_1 \tag{iii}$$

$$R_2 = -1.43 + 0.011F_C + 0.0098F_1 + 0.502F_2$$

$$-0.000026F_2^2 \tag{iv}$$

Solving for F_C, F_1 and F_2

$$F_C = 2.1R_C + 0.00016R_C^2 - 0.015R_1 + 0.7 \text{ lb} \tag{v}$$

$$F_1 = 2R_1 - 0.017R_C - 0.8 \text{ lb} \tag{vi}$$

$$F_2 = 2R_2 + 0.0002R_2^2 - 0.063R_C - 0.041R_1 + 3 \text{ lb} \tag{vii}$$

SOME EFFECTS OF DRILL POINT SHAPE ON THE CHISEL EDGE CONTRIBUTION TO CUTTING FORCES

by

RAFFAELLO LEVI* and UWE KOCH†

SUMMARY

The contribution of the chisel edge to drilling forces was assessed by making tests with and without a pilot hole. The temperature distribution on the flank was also evaluated using temperature-sensitive paints. By altering the drill point shape according to a given pattern the effects of several variables were evaluated, and the results derived in the form of response surfaces.

INTRODUCTION

The chisel edge is known to contribute a major portion of the thrust force during drilling. The large negative rake angle and peculiar cutting conditions justify the magnitude of the cutting force per unit length exceeding by far that of the lips.

Both large thrust forces and large unit forces acting on the tool edge are undesirable. They lead to unwanted machine and workpiece deflections and premature tool failures unless the feed, i.e. penetration rate, is reduced.

Some basic investigations on the mechanics of the drilling process[1,2] identified the wedge action of the chisel edge, and provided the equations for thrust prediction. Improved models were proposed in the course of further researches, offering a more detailed analysis[3].

The results of these works have thrown a good deal of light on a rather complex phenomenon. However, they do not always offer the right kind of answer to some practical shop problems.

Several modifications of the drill point were proposed in order to reduce thrust[4-7]. They often entail a shortening of the chisel edge, which is broken into a short central portion and several straight or curved secondary cutting edges. Sometimes it may be impractical to alter radically the drill point shape, yet some reduction in thrust is needed. All one can do is to adjust the grinding parameters, and see what happens. The work described in this paper shows that sizeable results may be obtained with reasonably small departures from the conventional shape. Some observations on thermal effects and contribution to torque are also given, in order to throw some additional light on the chisel edge effects.

TEST METHOD

The chisel edge contribution was assessed using the technique developed by Kurrein about half a century ago[8,9], that is from drilling tests performed under nominally identical conditions but for the presence of pilot holes.

Two main factors were considered as independent variables, namely clearance angle α, and chisel edge angle ψ. The cutting feed and speed were added as

additional variables in order to broaden the range of validity of results. The experimental plan is a $3^2 \times 2^2$ factorial; the factors and levels are listed in table 1. This plan is seen to be a subset of an experiment described previously[10]. Three response variables were analysed, namely thrust T, torque M and a thermal load parameter d, measured by the distance between cutting edge and a given isotherm on the flank.

Laboratory equipment and experimental material are described in detail elsewhere[11,12], along with some results obtained in the course of this investigation. Standard twist drills of 25 mm diameter were used throughout the tests. Workpiece material was free machining leaded steel.

EXPERIMENTAL RESULTS

Average responses are given in table 2 which shows the main effects of each factor on the three dependent

TABLE 1
Factors and levels

Factors	Levels		
	0	1	2
A clearance angle (deg)	6	9	12
B chisel edge angle (deg)	50	55	60
C spindle speed (rev/min)	156	200	--
D feed a (mm/rev)	0.125	0.200	--

TABLE 2
Average effects of main factors

Factors	Levels	Response		
		T (kg)	M (kg cm)	d (mm)
A	0	217	32	3.0
	1	210	31	2.7
	2	195	26	2.4
B	0	183	24	2.1
	1	207	30	2.7
	2	233	35	3.3
C	0	211	30	2.2
	1	204	29	3.2
D	0	174	25	1.2
	1	241	34	4.2

* Istituto di Tecnologia Meccanica, Politecnico di Torino.
† Aspera Motors S.p.A., Torino.

variables. Previous investigations on drilling mechanics showed that the linear effects of single factors tell only a part of the story. Second-order effects must be taken into account in order to obtain a clear picture of the situation. This is because the main four angular parameters determining drill point shape, namely clearance angle α, chisel edge rake angle γ_w, point

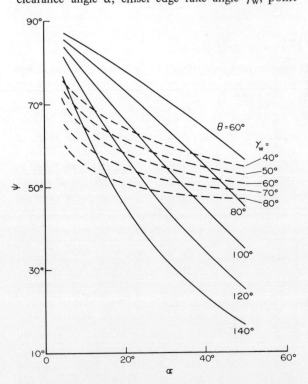

Figure 1. Relationship between clearance angle α, chisel edge angle ψ, point angle θ and chisel edge rake angle γ_w for a bevel-ground twist drill point.

angle θ and chisel edge angle ψ are interrelated, as shown in figure 1. It is clear from figure 1 why some interaction must be considered. On the other hand, second-order effects are hardly likely to concern speed.

The effect of feed on thrust was the object of exhaustive investigations – see, for example, reference 2 – whose results do not suggest the need of further replication.

ANALYSIS OF RESULTS

Table 3 shows the analysis of variance for thrust. The main effects are all significant, three at 0.1% level and one at 1%. The high level of significance pertaining to both A and B confirms the indications of table 2; a sizeable reduction in thrust at the chisel edge may be obtained by a few degrees alteration of the chisel edge angle, and the clearance angle. No comments are needed concerning the effect of feed. Speed does enter the picture because of its very low value in the region under consideration; the maximum cutting speed at the chisel edge corner was only 2 m/s. For these low values cutting forces are known to be influenced by cutting speed.

Feed is present in both significant interactions because it is one of the main parameters determining

the dynamic geometry of the drill point[13,14]. It is well known that, while contact between clearance face and workpiece can be avoided for the main cutting edges (lips), such a result can be obtained for only a portion of the chisel edge. The length of its cutting edge along which both clearance and rake face are in contact with the workpiece is a critical factor

TABLE 3
Analysis of variance for thrust T. Three- and four-factor interactions supply the estimate of error ($s^2 = 52.5$ with 20 DF)

	ΣQ	DF	$\Sigma Q/DF$	$F_{exp.}$
A	3085.39	2	1542.70	29.4†
B	14667.72	2	7333.86	139.6†
AB	392.78	4	98.2	1.9
C	568.03	1	568.03	10.8*
AC	242.38	2	121.19	2.3
BC	44.39	2	22.20	0.4
ABC	158.78	4	39.70	– –
D	39402.25	1	39402.25	749.9†
AD	339.50	2	169.75	3.2
BD	1944.17	2	972.08	18.5†
ABD	258.66	4	64.66	– –
CD	148.02	1	148.02	2.8
ACD	79.40	2	39.70	– –
BCD	53.06	2	26.53	– –
$ABCD$	368.44	4	92.11	– –
T	61572.97	35		

* Significant at 1% level.
† Significant at 0.1% level.

determining thrust, and is influenced directly by feed.

The joint effect of A and B may be further analysed by determining their linear and quadratic components, and their interaction[15]. From such an analysis the terms of a second-order equation may be derived easily.

Results are best shown under a graphical form for further considerations. Figure 2 shows the response surface of thrust T versus α and ψ. Linear components

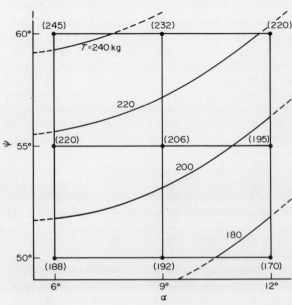

Figure 2. Response surface of thrust component T due to the chisel edge versus α and ψ. Treatment combinations and relevant experimental results are also shown.

are seen to predominate, the curvature of the response surface being on the verge of significance only. The range of thrust forces exceeds one-third of the minimum value, pertaining to the A_2B_0 treatment combination. Response surfaces for torque M and the distance shift of the reference isotherm d are shown

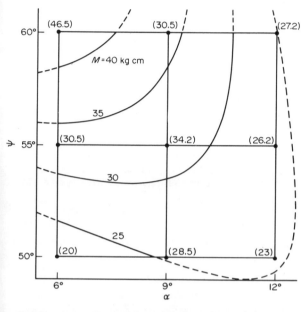

Figure 3. Response surface of torque component M due to the chisel edge versus α and ψ. Treatment combinations and relevant experimental results are also shown.

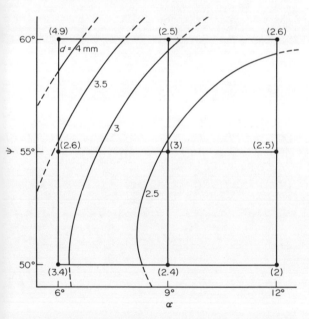

Figure 4. Response surface of distance shift d on the flank surface of the reference isotherm due to the chisel edge versus α and ψ. Treatment combinations and relevant experimental results are also shown.

in figures 3 and 4. The torque component at the chisel edge is seen to be even more affected percentage-wise by factors A and B than thrust. A look at figure 1 shows why — the rake angle, which affects directly main cutting force (i.e. torque) is strongly dependent from both α and ψ. Torque range equals the minimum value, pertaining again to the A_2B_0 combination.

Distance d follows much the same trend of T and M in the upper part of figure 4, but departs from that trend in the lower right-hand quarter, where some form of stagnation is seen to exist. Improved cutting conditions due to more favourable cutting angles entail a decrease in temperature only to a given extent; then the effect of the decrease in thickness of the tool in the cutting edge region prevails. This effect is even more marked when the temperature is not measured at the cutting edge but at some distance, as in this case, where temperature-sensitive paints are used. The observed shape of isotherms reproduced published results[16] (quoted by Saxena et al.[17]).

DISCUSSION

Experimental results show the improvements in cutting conditions to be expected by modifying the drill point shape acting on the chisel edge angle and the clearance angle only. No information is given concerning the effects of these parameters on tool life; it seems that there is no chance of obtaining information unless a rather large number of life tests are performed. This is because twist drills are known to exhibit an even more erratic behaviour than single point tools as far as their life is concerned.

Coefficients of variation as high as 0.6 are not unheard of, as against an average value of 0.3 and maxima of 0.4 for single point tools[18]. Furthermore, the peculiar shape of the twist drill life density of probability curve[19,20] makes it awkward to apply standard statistical techniques based on the normal distribution.

A large clearance angle and a small chisel edge angle mean a weak spot at the chisel edge corner, where it merges with the lip. It is likely that the presence of secondary cutting edges in modified drill point shapes helps to overcome this drawback.

The effects of several other parameters affecting drill point shape[21] were not considered in the course of this work. The curves shown in figure 1 are strictly valid only for bevel-ground drill points as they were derived under some simplified assumptions[22,23]. They do provide, however, a useful approximation also for other drill point shapes.

REFERENCES

1. C. J. OXFORD Jr., 1955. 'On the drilling of metals — basic mechanics of the process', *Trans. Am. Soc. Mech. Engrs.*, **77**, 103.

2. M. C. SHAW and C. J. OXFORD Jr., 1957. 'On the drilling of metals — the torque and thrust in drilling', *Trans. Am. Soc. Mech. Engrs.*, **79**, 139.

3. S. BERA and A. BHATTACHARYYA, 1967. 'On the determination of torque and thrust during drilling of ductile materials', *Proc. 8th M.T.D.R. Conf.*, **2**, p. 879.

4. M. KRONENBERG, 1963. *Grundzüge der Zerspanungslehre*, Vol. 2, Springer, Berlin, p. 211.

5. C. J. OXFORD Jr., 1967. 'A review of some recent developments in the design and application of twist drills', *Proc. 8th M.T.D.R. Conf.*, Vol. 2, p. 845.

6. I. VINNIKOV and M. FRENKEL, 1964. *Drilling Practice*, Peace, Moscow, p. 111.

7. A. BHATTACHARYYA, A. J. BHATTACHARYYA, A. B. CHATTERJEE and I. HAM, 1971. 'Modification

of drill point for reducing thrust', *Trans. Am. Soc. Mech. Engrs., Ser. B*, **93**, 1073.

8. M. KURREIN, 1927. 'Die Bearbeitbarkeit der Metalle in Zusammenhang mit der Festigkeitsprüfung', *Werkstattstechnik*, **21**, 613.

9. M. KURREIN and F. C. LEA, 1947. *Cutting Tools for Metal Machining,* Griffin, London, p. 106.

10. U. KOCH and R. LEVI, 1971. 'Some mechanical and thermal aspects of twist drill performance', *Ann, C.I.R.P.*, **19**, 247.

11. R. LEVI and U. KOCH, 1970. 'Analisi delle forze di taglio nella foratura in funzione della geometria dell'utensile', *Atti 2° Conv. Naz. Macch. Ut.*, FAST, p. 57.

12. R. LEVI, 1967. 'Drill press dynamometers', *Intern. J. Machine Tool Design Res.*, **7**, 269.

13. R. A. WILLIAMS, 1969. 'Dynamic geometry of a twist drill', *Intern. J. Prod. Res.*, **7**, 253.

14. R. A. WILLIAMS, 1970. 'A study of the basic mechanics of the chisel edge of a twist drill', *Intern. J. Prod. Res.*, **8**, 325.

15. O. L. DAVIES (Ed.), 1963. *The Design and Analysis of Industrial Experiments*, Oliver and Boyd, Edinburgh.

16. M. TSEUDA, Y. HASEGAWA and Y. NISINA, 1961. 'The study of the cutting temperature in drilling (1) on the measuring method of cutting temperatures', *Trans. Jap. Soc. Mech. Engrs.*, **27**, 1423.

17. U. K. SAXENA, M. F. DE VRIES and S. M. WU, 1971. 'Drill temperature distributions by numerical solutions', *Trans. Am. Soc. Mech. Engrs., Ser. B*, **93**, 1057.

18. J. G. WAGER and M. M. BARASH, 1971. 'Study of the distribution of the life of HSS tools', *Trans. Am. Soc. Mech. Engrs., Ser. B*, **93**, 1044.

19. N. D. SINGPURWALLA and A. A. KUEBLER, 1966. 'A quantitative evaluation of drill life', *Am. Soc. Mech. Engrs.*, Paper No. 66-WA/Prod-11.

20. R. LEVI, 1970. 'Extreme value analysis of twist drill performance', *Metal Processing and Machine Tools*, R.T.M., Vico Canavese.

21. S. FUJII, M. F. DE VRIES and S. M. WU, 1971. 'Analysis of the chisel edge and the effect of the *d*-theta relationship on drill point geometry', *Trans. Am. Soc. Mech. Engrs., Ser. B*, **93**, 1093.

22. G. F. MICHELETTI and R. LEVI, 1967. 'The effect of several parameters on twist drill performance', *Proc. 8th Intern. M.T.D.R. Conf.*, p. 863.

23. R. LEVI, 1968. 'Forze di taglio e usura dell'utensile nelle lavorazioni di foratura', *Atti 1° Conv. Naz. Macch. Ut.*, FAST.

THE PROFILE OF A HELICAL SLOT MACHINED BY A DISC-TYPE CUTTER WITH AN INFINITESIMAL WIDTH, CONSIDERING UNDERCUTTING

by

M. Y. FRIEDMAN*, M. BOLESLAVSKI†, and I. MEISTER*

SUMMARY

An exact analytic solution is presented for the profile of a helical slot machined by an elementary disc-type cutter with zero width, taking into account undercutting. The profile of the helical slot produced by a profiled cutter can be built by superposition of the solutions of the infinitesimal cutters.

INTRODUCTION

The classical use of the theory of envelopes for the problem of the profile of a helical slot machined by a profiled disc-type cutter, gives an exact solution without taking into account undercutting.

In order to take into account the phenomenon of undercutting, we view a disc-type cutter with a certain given profile as composed of many elementary cutters with infinitesimal width, which differ from one another by the radius and position on the axis of the cutter. One can build the profile of the helical slot by superposition of the solutions of the infinitesimal cutters.

The following is an exact analytic solution for an elementary disc-type cutter with zero width, which takes into consideration undercutting.

THEORETICAL

The geometry of the cutting set-up is described in figure 1. The coordinate system $(x_0 y_0 z_0)$ is attached to the machine (fixed in space). Coordinate system $(x_1 y_1 z_1)$ is attached to the workpiece, and coordinate system $(x_2 y_2 z_2)$ is attached to the elementary cutter.

At the beginning of cutting, $t=0$, system 1 coincides with system 0. The workpiece is given a constant velocity v in the direction of z_0, and a constant angular velocity Ω about z_0.

The tool is fixed in space relative to the machine. The position of system 2 (the cutter) relative to system 0 (the machine) is defined by the distances a and b, and angle α.

The transformation matrix from system 1 to system 2 through system 0 is:

$$[T_{12}] = [T_{02}][T_{10}] \tag{1}$$

After building the matrices and multiplying the transformation equations (2) result.

A disc-type cutter with zero width may be described by the two equations

$$\left.\begin{array}{r} z_2^2 + y_2^2 = R^2 \\ x_2 = 0 \end{array}\right\} \tag{3}$$

where R is the radius of the elementary cutter.

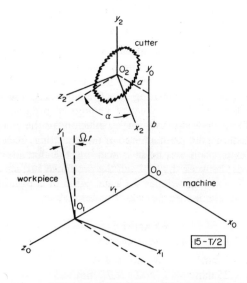

Figure 1. Geometry of milling set-up.

For the sake of simplicity an elementary cutter which passes through the origin was chosen. For a definite profile one has to solve for cutters that are displaced from the origin: in this case x_2 will not equal zero. Because of the helical movement of the workpiece, the lower half of the elementary cutter will trace a slot, the profile of which is given by the equations (3), transformed to system 1.

$$
\begin{bmatrix} 1 \\ x_2 \\ y_2 \\ z_2 \end{bmatrix}
=
\begin{bmatrix}
1 & 0 & 0 & 0 \\
a\sin\alpha + vt\cos\alpha & \sin\alpha\,\cos\Omega t & -\sin\alpha\,\sin\Omega t & \cos\alpha \\
-b & \sin\Omega t & \cos\Omega t & 0 \\
-a\cos\alpha + vt\sin\alpha & -\cos\alpha\,\cos\Omega t & \cos\alpha\,\sin\Omega t & \sin\alpha
\end{bmatrix}
\begin{bmatrix} 1 \\ x_1 \\ y_1 \\ z_1 \end{bmatrix}
\tag{2}
$$

* Department of Mechanical Engineering, Technion-Israel Institute of Technology, Haifa, Israel
† Department of Mathematics (Descriptive Geometry), Technion-Israel Institute of Technology, Haifa, Israel
NOTE: The paper forms part of the M.Sc. Thesis of I. Meister.

The transformation is accomplished by equations (2), and we get the two equations (4) which define the surface of the slot.

$$\left.\begin{aligned} [(-\cos\alpha\,\cos\Omega t)x_1 + (\cos\alpha\,\sin\Omega t)y_1 + (\sin\alpha)z_1 \\ + (-a\cos\alpha + vt\sin\alpha)]^2 \\ + [(\sin\Omega t)x_1 + (\cos\Omega t)y_1 - b]^2 \\ = R^2 \\ (\sin\alpha\,\cos\Omega t)x_1 - (\sin\alpha\,\sin\Omega t)y_1 + (\cos\alpha)z_1 \\ + (a\sin\alpha + vt\cos\alpha) = 0 \end{aligned}\right\} \quad (4)$$

Actually the number of parameters is one fewer, because one can define a new angular parameter θ

$$\Omega t = \theta \qquad (5)$$

and then

$$vt = (v/\Omega)\theta \qquad (6)$$

where the ratio (v/Ω) is given by the pitch of the helix.

In order to get the profile of the slot in a section perpendicular to the workpiece axis z_1, substitute for z_1 any constant value, for example

$$z_1 = 0 \qquad (7)$$

We get two equations which, after elimination of t or θ, give the relationship between x_1 and y_1 which describes the profile of the slot. In fact there is no possibility of eliminating t or θ analytically and the profile is calculated by use of a computer programme.

The proposed theory gives a solution for the profile considering the phenomenon of undercutting, because no assumption was made concerning the character of contact between the tool and the produced surface. In the theory of envelopes the assumption of continuous contact eliminates the case of undercutting.

EXAMPLE

The following data were chosen:

$a = 1$ mm	$\alpha = \pi/3$
$b = 25$ mm	$v/\Omega = 1.75$ mm/rad
$R = 20$ mm	

The calculated profile is described in figure 2. The origin is the rotation axis of the workpiece. The actual profile of the slot produced for a workpiece with a given radius R, will be that part of the described profile which is limited by an arc of radius R. If the

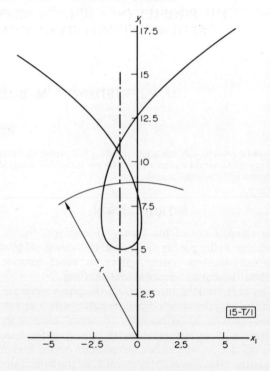

Figure 2. Profile of helical slot.

radius of the workpiece is large enough, there exists a certain point in which there is no undercutting. In this point the slope of the helix coincides with the direction of the cutter as defined by angle α.

In fact we have a cutter with a certain profile given by

$$R = f(x_2) \qquad (8)$$

This cutter is divided into a number of elementary cutters, spaced each from another a distance Δx_2. The distance of the nth cutter from the origin O_2 is

$$x_2 = (n-1)\Delta x_2 \qquad (9)$$

and its radius is

$$R = f[(n-1)\Delta x_2] \qquad (10)$$

Substitution of x_2 and R from the last two equations in (3) results in a series of solutions, the envelope of which is the profile of the slot considering undercutting.

STRESS ANALYSIS OF SEGMENTED CIRCULAR SAWBLADES

by

M. MAHOMED*, M. C. deMALHERBE*, M. A. DOKAINISH* and R. B. YOUNG†

SUMMARY

The problem of fatigue cracking at the slot bases of diamond-impregnated segmented circular saws is investigated. An improved slot shape is suggested minimising stress concentration and thus prolonging the fatigue life of saws. Plane stress conditions are assumed and the finite element method is used in order to find the limits of the principal stress cycles. The various slot shapes tried are compared for improved fatigue life by means of an octahedral shear stress fatigue criterion developed by G. Sines.[1]

NOTATION

F_R, F_T radial and tangential cutting forces (lb)

σ_1, σ_2 maximum and minimum principal stresses (lb/in^2)

P_1, P_2 principal stress amplitudes (lb/in^2)

S_1, S_2 static stresses (lb/in^2)

R_1, R_2 stress range (lb/in^2)

R'_1, R'_2 residual stresses (lb/in^2)

σ_{oct} octahedral shear stress (lb/in^2)

D, r_H, γ, ψ slot geometry variables.

INTRODUCTION

In the class of segmented or slotted circular saws, the type of primary interest here are diamond impregnated saws, used for cutting hard materials such as stone and concrete (represented in figure 1).

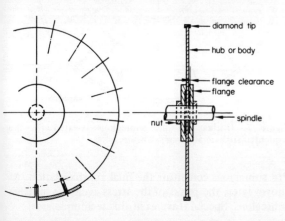

Figure 1. Plan and cross-sectional views of diamond-tipped circular saw.

Diamond circular saws have diameters ranging from 250 mm to about 3 m, the hub or body usually being made of a tool steel hardened to around 42–44 Rockwell. Radial slots are located on the periphery at equiangular intervals. Diamond-impregnated tips are brazed to the periphery of the saw on the segments between slots. The tips are made from a variety of materials ranging from bronze to tungsten carbide. On assembly a saw is held firmly by large flanges having diameters equal to one-third the saw diameter to prevent lateral vibration of the saw.

*McMaster University, Hamilton, Ontario, Canada.

†Boart Research Laboratory, Johannesburg, South Africa.

There is no basic difference between the cutting actions of a slotted saw and a continuous periphery saw. However, the slots serve other important purposes:

(i) they aid in the removal of chips during cutting and prevent loading of the diamond cutting edges;

(ii) they provide a larger cooling area and are responsible for introducing lubricant into the cutting region;

(iii) they prevent the distortion of the hub during the brazing of the diamond tips on the hub.

Diamond-tipped saws achieve their cutting by the abrasive action of the diamond grits on the work in a similar manner to a grinding wheel.

Diamond circular saws are used either singly or mounted in gangs. There are many stone-cutting machines on the market, ranging from small hand-operated machines to large multiple blade orthogonal stone cutting machines.

The slot shapes commonly used on segmented circular saws consist of parallel sides with semicircular bases, as shown in figure 1. The problem encountered with saws having this geometry is that they fail through fatigue cracking at the slot bases after prolonged usage. In this paper, a qualitative study that was made on the effect of varying slot shape on the fatigue characteristics of circular saws is discussed.

ANALYSIS

Formulation for stress analysis

When considered as a free body, a sawblade is essentially a rotating irregularly shaped plate loaded externally by cutting forces. By D'Alembert's principle the problem may be resolved as the sum of two static problems:

(i) a sawblade loaded by cutting forces.

(ii) a sawblade loaded by centrifugal forces due to rotation, if elastic conditions are assumed.

These two problems are shown diagrammatically in figure 2.

In the absence of lateral vibrations the blade is considered to be in a state of plane stress. Due to concentration of stress at the slot bases, each slot base experiences a cycle of complex stresses for each revolution of the sawblade.

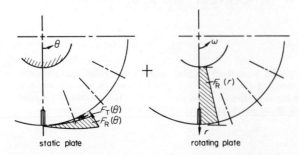

static plate   rotating plate

Figure 2. Resolution of sawblade problem to two static problems.

Sines[1] and Little[2] state that fatigue cracks are initiated by alternating shear stress and propagated by alternating normal stress. For the fatigue analysis a criterion of failure based on alternating octahedral shear stress (discussed later) was used and for this criterion the amplitudes of the alternating principal stresses were required. These were obtained by the finite element method of stress analysis.

For the stress analysis, a sawblade having the following dimensions was used as a model:

    outer diameter 30 cm (12 in)
    central-hole diameter 1 in
    flange diameter 10 cm
    blade thickness 0.0625 in
    number of slots 17.

Applied forces

For the stress analysis, it was necessary to determine the distribution of the cutting forces over the cutting region. At this time, there is no clear answer to the cutting mechanism for stone. Hence, an analogy was drawn between the cutting mechanism of a diamond saw and grinding, in order to obtain an approximate cutting-force distribution. It was assumed that side forces due to lateral contact of the saw with the work are insignificant. Based on work by Backer[3], Marshall[4] and Reichenbach[5] in the field of grinding, it was concluded that if depths-of-cut are kept below 1 in, the cutting forces in stone cutting obey a linear distribution as follows:

    radial cutting force $F_R = k_1\theta$
    tangential cutting force $F_T = k_2\theta$ (1)

where k_1 and k_2 are constants for one set of cutting parameters, and θ is the angle (deg) measured from the vertical. Mock distributions are shown in figure 2.

To equilibrate internal centrifugal forces, equivalent nodal forces were applied as required by the finite element method.

Formulation for the finite element method

Since the greatest concentration of stress exists at the slot bases in the immediate vicinity of the cutting region, it was necessary critically to analyse this region.

A technique used in order to 'home-in' on the critical regions by means of a number of successive approximations is described below.

In this technique, a first approximation to the nodal displacements and stresses is made by performing a coarse finite element analysis of the whole structure without particular attention to the regions of high concentration. A fair sized region surrounding the concentration is then isolated from the main structure and a refined finite element mesh applied to the isolated region, which is termed a subregion. The subregion is then treated as a separate problem, and displacements obtained from the previous approximation are prescribed on the boundaries sectioned off. A second subregion may be isolated from the first subregion and this sequence may be continued until the desired accuracy is obtained.

Outlines of the subregions that were isolated successively for the stress analysis of the sawblade by this technique are shown in figure 3. The 2nd and

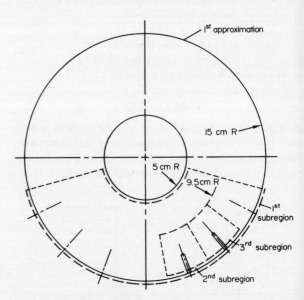

Figure 3. Outlines of the subregions isolated for the sequential stress analysis of sawblade.

3rd subregions constitute the final approximation. As shown later, the peaks of the stress cycles lie in these subregions. Special features of this technique are:

(i) it was possible to obtain accurate stresses in the critical regions without much consideration to the entire blade;

(ii) a refined mesh could be applied and still remain within computer memory size limitation;

(iii) by St. Venant's principle, slot shapes could be varied in the final approximation without retracing the whole procedure. This step was achieved by defining a 'generalised' slot shape in figure 4 by means of the following 'slot variables':

r_H = radius of slot base circle
d = slot depth
γ = slope of slot sides
ψ = angle defining intersection point of slot side and slot base circle.

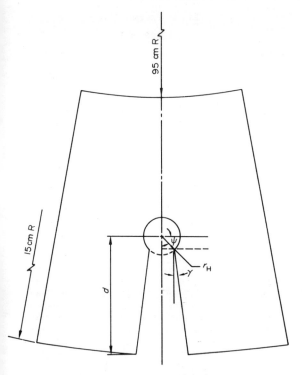

Figure 4. Generalised slot shape.

A number of slot shapes were tried by varying one slot variable at a time and keeping the others the same as the basic slot shape, which was parallel sided with a semicircular base having $r_H = 1/16$ in, $d = 1$ in; $\gamma = \phi$ deg., $\psi = \phi$ deg. These shapes are listed in table 1.

of the orthogonal normal static stresses, or, shown mathematically:

$$\tfrac{1}{3}\{(P_1 - P_2)^2 + (P_2 - P_3)^2 + (P_1 - P_3)^2\}^{1/2}$$
$$\leqslant A - a(S_1 + S_2 + S_3 + R_1' + R_2' + R_3') \quad (2)$$

where P_1, P_2 and P_3 are the amplitudes of the alternating principal stresses and S_1, S_2 and S_3 are the orthogonal static (or mean) stresses. The orthogonal axes $1'$, $2'$, $3'$ for the residual stresses R_1 and so on need not be in the same directions as those for the static normal stresses. A is a constant for the material, proportional to the reversed fatigue strength, and a gives the variation of the permissible range of stress with static stress.[1,2] Both A and a are given for the desired cyclic lifetime. The expression on the left must not exceed the righthand side or failure will occur before the desired lifetime.

For a biaxial state of stress and in the absence of residual stresses, equation (2) reduces to

$$\frac{\sqrt{2}}{3}\left\{(P_1^2 + P_2^2) - P_1 P_2\right\}^{1/2} + a(S_1 + S_2) \leqslant A \quad (3)$$

The square root quantity in the brackets, is the octahedral shear stress σ_{oct} and for the present purpose, the expression to the left of the inequality is termed a 'fatigue stress'. Clearly, the fatigue stress has to be minimised to improve fatigue life. This quantity was used to compare the slot shapes tried.

TABLE 1
List of slot shapes tried.

Set	Slot shape no.	D(in)	r_H (in)	γ(deg)	ψ(deg)
1	1	1.0	1/16	0.0	0.0
	2	1.0	3/32	0.0	0.0
	3	1.0	1/8	0.0	0.0
	*4	1.0	3/32	0.0	−48.0
2	5	1.0	1/16	2.5	0.0
	6	1.0	1/16	5.0	0.0
	7	1.0	1/16	7.5	0.0
	*8	1.0	1/8	−3.5	0.0
3	9	1.0	1/8	0.0	−60.0
	10	1.0	1/8	0.0	−48.0
	11	1.0	1/8	0.0	−24.0
4	+12	1.0	1/8	−7.0,0.0	0.0
	+13	1.0	1/16	−7.5,7.5	0.0

* These shapes do not fall completely in the classification as two variables were changed simultaneously.
+ These were hybrid shapes that were not symmetric about the centrelines.

Fatigue criterion
Based on diverse experimental data available, Sines[1] has developed a criterion of fatigue failure which includes the effect of different combinations of alternating stress with static stresses. It is the simple statement that the permissible alternation of the octahedral-shear stress is a linear function of the sum

RESULTS

Principal stress contours
Figures 5(a) and 5(b) show principal stress contour plots obtained when the effects of centrifugal forces and a sample cutting force distribution ($k_1 = 0.26$, $k_2 = 0.2\ k_1$) in equation (1) were superimposed.

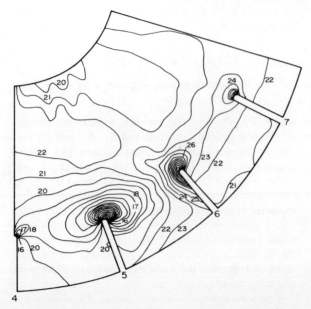

Figure 5(a). Maximum principal stress contours obtained due to combined centrifugal forces and cutting forces. Blade speed 1000 r.p.m., feed 300 mm/s, depth-of-cut 0.9 in.

Key for contour plots

Figure 5(a)

no.	stress	no.	stress
1	−1000.000	27	300.000
2	−950.000	28	350.000
3	−900.000	29	400.000
4	−850.000	30	450.000
5	−800.000	31	500.000
6	−750.000	32	550.000
7	−700.000	33	600.000
8	−650.000	34	650.000
9	−600.000	35	700.000
10	−550.000	36	750.000
11	−500.000	37	800.000
12	−450.000	38	850.000
13	−400.000	39	900.000
14	−350.000	40	950.000
15	−300.000	41	1000.000
16	−250.000	42	1050.000
17	−200.000	43	1100.000
18	−150.000	44	1150.000
19	−100.000	45	1200.000
20	−50.000	46	1250.000
21	10.000	47	1300.000
22	50.000	48	1350.000
23	100.000	49	1400.000
24	150.000	50	1450.000
25	200.000	51	1500.000
26	250.000		

Figure 5(b)

no.	stress	no.	stress
1	−3500.000	22	−1400.000
2	−3400.000	23	−1300.000
3	−3300.000	24	−1200.000
4	−3200.000	25	−1100.000
5	−3100.000	26	−1000.000
6	−3000.000	27	−900.000
7	−2900.000	28	−800.000
8	−2800.000	29	−700.000
9	−2700.000	30	−600.000
10	−2600.000	31	−500.000
11	−2500.000	32	−400.000
12	−2400.000	33	−300.000
13	−2300.000	34	−200.000
14	−2200.000	35	−100.000
15	−2100.000	36	10.000
16	−2000.000	37	100.000
17	−1900.000	38	200.000
18	−1800.000	39	300.000
19	−1700.000	40	400.000
20	−1600.000	41	500.000
21	−1500.000		

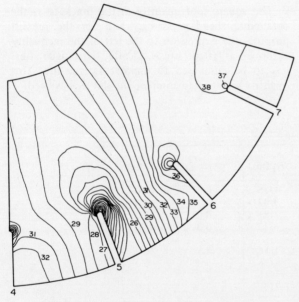

Figure 5(b). Minimum principal stress contours.

These diagrams represent an area sectioned off from the lower portion of the sawblade. The basic slot shape is shown.

The stress contours indicate that the most critically stressed regions are those surrounding slots 5 and 6. Also, the region surrounding slot 5 is basically in a compressive state while that surrounding slot 6 is in a tensile state.

Principal stress cycles

In figure 6, the principal stresses σ_1 and σ_2 at the critical angular position on the slot base circle are plotted for slots 1 to 7. Stresses for slots beyond the 7th tend toward zero.

If the problem is considered to be a purely static one, then figure 6 simply represents the variation of principal stress with angle. However, since the blade is rotating and the force is fixed with respect to 'ground', this curve can be interpreted as the variation of principal stresses with time. In other words, if t(min) is the time taken for a slot to rotate from position 4 to position 5, then slot position 5 is considered the 'new' position of 4 after time t. If T (min) is the time period for one revolution of the saw, $T = 1/N$, where N is the blade speed (rpm). Having a total of n slots, then $t = T/n$. An interval on the abscissa axis of figure 6 in time coordinates would then be of magnitude t as shown for interval 1−2. In this way, the problem is converted from a static problem to a dynamic one, and figure 6 represents the principal stress cycles at the point of critical stress. From this figure, it can be seen that the two principal stresses σ_1 and σ_2 are in phase.

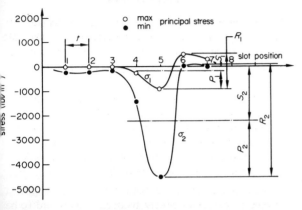

Figure 6. Variation of principal stresses with angular position of slot. The R's are stress ranges.

Waisman[6] and Miner[7] show how an actual stress cycle may be replaced by an equivalent one having constant amplitude. The principal stress cycles shown in figure 6 may similarly be replaced by cycles having constant amplitudes P_1 and P_2, superimposed on static stresses S_1 and S_2.

Table 2 gives a list of the critical principal stress amplitudes obtained respectively for the various shapes studied. The corresponding fatigue stress calculations are tabulated in table 3, the last column of which contains the reduction in fatigue stresses as compared to the basic slot shape (no. 1). In figure 7

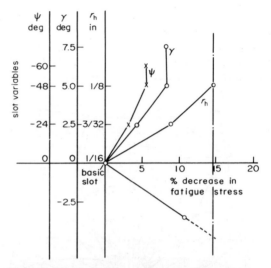

Figure 7. Effect of varying slot shape on fatigue stress.

this reduction is used as a basis to compare the effect of changing one slot variable at a time. This figure represents the first three sets of table 3. It can be seen that the greatest improvement is caused by increasing the slot base radius r_H. Improvement caused by increasing angle γ falls off after 5° and improvement caused by increasing angle ψ (in the negative direction) falls off after $-48°$. By making slot angle γ negative a substantial decrease in fatigue stress results, as shown by the line below the abscissa.

It is important to note that by increasing either γ or r_H, while keeping other variables constant, the

TABLE 2
Tabulation of peak principal stresses for various slot shapes.

slot shape number	position 6 (subregion 3)		position 5 (subregion 2)	
	σ_i (lb/in²)	σ_1 (lb/in²)	σ_1 (lb/in²)	σ_1 (lb/in²)
1	440	0	−460	−4810
2	380	0	−233	−4243
3	280	0	−147	−3880
4	264	0	−241	−4450
5	440	0	−400	−4570
6	440	0	−371	−4365
7	440	0	−363	−4352
8	270	0	−153	−4078
9	67	0	−200	−4367
10	102	0	−202	−4286
11	155	0	−151	−4025
12	270	0	−146	−3838
13	440	0	−367	−4036

TABLE 3
Tabulation of fatigue stress calculations and improvements in fatigue stress for various slot shapes

slot shape number	principal stress amplitudes		static stresses		octahedral shear stress	sum of static stress	fatigue stress	reduction
	P_1 (lb/in²) × 6	P_2 (lb/in²)	S_1 (lb/in²)	S_2 (lb/in²)	σ_{oct} (lb/in²)	$S_1 + S_2 = S$ (lb/in²)	$\sigma_{oct} + aS$	%
1	450	2405	−10	−2405	1044	−2415	802	0.0
2	307	2122	73	−2122	936	−2049	731	8.3
3	213	1940	67	−1940	870	−1873	683	14.8
4	253	2225	11	−2225	994	−2214	773	3.6
5	420	2285	20	−2285	993	−2265	776	4.4
6	405	2182	35	−2182	948	−2147	734	8.5
7	400	2176	40	−2176	946	−2136	733	8.5
8	212	2039	58	−2039	915	−1981	717	10.6
9	134	2184	−66	−2184	1000	−2250	775	3.4
10	152	2143	−50	−2143	976	−2193	757	5.6
11	153	2013	0	−2103	959	−2013	758	5.5
12	208	1919	62	−1919	859	−1857	673	16.0
13	404	2108	36	−2018	871	−1982	673	16.0

slot is widened and the available cutting area of the saw is consequently reduced. Though the fatigue life may increase, the cutting efficiency decreases. This fact was kept in mind when designing the hybrid shapes no. 12 and no. 13 as shown in figure 8.

In order to obtain on an improved shape, two hybrid slot shapes were designed by a combination of the variables r_H and γ. The vertical sectioning line drawn on figure 7 shows that the maximum decrease in stress was obtained by a large r_H and a negative γ. This fact was used in shape no. 12 of figure 8(a) in which the right hand slot side has $\gamma = -7\frac{1}{2}°$ and the slot base circle radius $r_H = 1$ in. The other shape is no. 13 shown in figure 8(b) in which an attempt was made to take advantage of the improvement caused by making γ positive and negative for the respective slot sides. The result was a parallel sided slot which was similar to the basic slot but skewed to the centre-line by $7\frac{1}{2}°$.

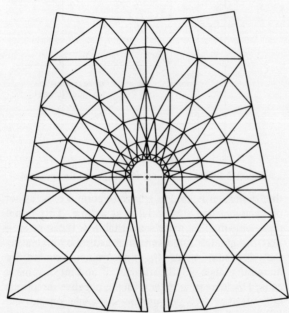

Figure 8(a). Hybrid slot shape (no. 12).

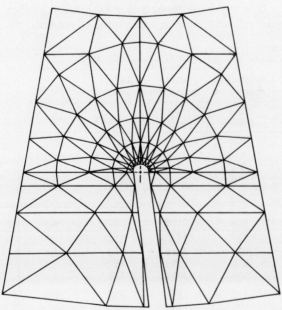

Figure 8(b). Hybrid slot shape (no. 13).

From the last set in table 3, it can be seen that the two hybrid shapes give the largest reduction in fatigue stress, and thus the greatest improvement in fatigue life is expected from them.

CONCLUSIONS

Of the shapes tried it was found that the hybrid slot shapes of figure 8 would theoretically behave the best under fatigue conditions. Other than reducing the fatigue stress, these two slot shapes have the following advantages:

(a) They do not reduce the cutting area of the blade, which is important economically.

(b) Because of the positive rake-angle, they are expected to improve chip removal and coolant inflow. In fact, shape no. 12 has a greater capacity for chip removal because it is recessed.

(2) The critical point on the slot base circle at which fatigue cracking was likely to occur was found to be at $30°$ to the horizontal, and not at the slot root. Most of the cracks found on a used circular saw blade supported this fact.

(3) In some stone cutting operations, diamond circular saw blades rotate at about 4000 cycles per min. For an 8 h day, this would mean that the saw has to endure about 2 million cycles of stress a day. Under these conditions, it can only be hoped to prolong the fatigue life of a saw and not to countermand fatigue failures completely.

(4) The stresses in table 3 can be seen to be generally low in magnitude. This is due to the fact that a small depth of cut and a small rotational speed were used in the model. In use, the saw speed is about 4000 surface ft/min and the largest possible depths of cut are taken. Under such conditions the cutting forces become very large and their distribution non-linear. The model chosen was for comparative purposes only.

(5) From the fatigue criterion, equation (2), it can be seen that compressive residual stresses improve fatigue life. Metallurgical methods such as shotpeening may be used to introduce compressive residual stresses.

(6) The propagation of fatigue cracks can be inhibited by removing the surface of the slot base from time-to-time: this could be done by drilling and is a practical way of prolonging the life of a saw.

REFERENCES

1. G. SINES, 1959. Behaviour of metals under complex static and alternating stresses, *Metal Fatigue,* edited by G. Sines and J. L. Waisman, McGraw-Hill.

2. R. E. LITTLE, 1965. A simplified method for determining fatigue stress using Mohr's circle, *Machine Design.*

3. W. R. BACKER, E. R. MARSHALL and M. C. SHAW, 1952. The size effect in metal cutting, *Trans. ASME.,* Vol. 74, pp. 61–72.

4. E. R. MARSHALL and M. C. SHAW, 1952. Forces in dry surface grinding, *Trans. ASME*, Vol. 74. pp. 51–59.

5. G. S. REICHENBACK, and others, 1956. The role of chip thickness in grinding, *Trans. ASME*, Vol. 78, pp. 847–59.

6. J. L. WAISMAN, 1959. Factors affecting fatigue strength, Metal Fatigue, edited by G. Sines, and J. L. Waisman, McGraw-Hill.

7. M. A. MINER, 1959. Estimation of fatigue life with particular emphasis on cumulative damage, *Metal Fatigue,* edited by G. Sines, and J. L. Waisman, McGraw-Hill.

A STUDY ON RECRYSTALLISATION IN METAL CUTTING BY FOURIER ANALYSIS

by

R. RAMASWAMI*

SUMMARY

The study of recrystallisation in metal cutting is of significant importance since it influences tool wear and surface finish. The formation of recrystallised crystallites in the secondary shear zone (flow layer) when machining mild steel with HSS tools and the determination of size of such crystallites by X-ray diffraction techniques were reported by the author in earlier papers. In this paper an attempt has been made to fit the X-ray intensity profiles by Fourier series and transforms. Computer programmes have been developed for solving Fourier series and the result is discussed. The similarity in structure of the negative wedge built-up-edge, and the secondary shear zone is also analysed. The changes in X-ray diffraction patterns give a vivid picture of metallurgical changes in the flow layer. Fourier analysis lends sophistication to the analytical approach to the X-ray diffraction technique.

NOTATION

B observed breadth of intensity profile for the secondary shear zone

b observed breadth of intensity profile for undeformed metal

d interplanar spacing

e microstrain in the flow layer

K constant in Scherrer equation

n integer

β true breadth of profile due to crystallites and plastic deformation

β_1 true breadth of profile for the heated specimen

β_2 true breadth of profile due to plastic deformation

ϵ crystallite size (angstrom)

λ wavelength of X-ray (angstrom)

θ Bragg angle

INTRODUCTION

The investigation of recrystallisation in metalcutting is of primary importance since it influences tool wear and surface finish. Extensive work has been carried out on recrystallisation and phase transformation in the secondary shear zone when machining steel with carbide tools, but little work has been done with HSS tools.

A detailed investigation of different types of built-up-edge was carried out by Heginbotham and Gogia[1]. The different types of built-up-edge, their structural details and their influence on tool wear and surface finish were discussed by Ramaswami[2,3]. Opitz and Gappisch[4] thought that the temperature was lying above transformation temperature in the case of machining steel with carbide tools.

Recrystallised grains were observed by Konig[5] in the secondary zone of chip obtained during high-speed machining with carbide tools. A similar structure was observed by Venkatesh and Ramaswami[6,7] during the machining of mild steel with HSS tools. The size of such crystallites was determined by making use of X-ray diffraction techniques. In this article the similarity in structure between the negative wedge built-up-edge and the flowlayer is discussed. An attempt has been made to fit the X-ray intensity profiles by Fourier series and transforms. Computer programmes have been developed for solving Fourier series to fit the profiles and to extract true profiles. The result is discussed in subsequent paragraphs.

EXPERIMENT

The study was carried for an orthogonal turning operation. The work material was seamless mild steel tube (o.2% C) of 200 mm diameter. Feed was kept constant at 0.314 mm/revolution and the speed was varied. A HSS tool with 5% cobalt was used. The chips were frozen by means of an explosive quick-stop device. The chips were mounted suitably for metallographic and X-ray diffraction studies.

STRUCTURE OF NEGATIVE WEDGE BUILT-UP-EDGE AND FLOWLAYER

As reported by Heginbotham and Gogia various types of built-up-edge (BUE) were observed. Though the loss of apex in the positive wedge built-up-edge resulting in rectangular built-up-edge is due to the effect of temperature, a significant change in structure was observed in the next type of built-up-edge, namely negative wedge BUE.

Figure 1 shows microphotographs of different parts of the negative wedge BUE obtained at 35 m/min. Finely dispersed crystallites, though not fully-resolved, are observed from the micrograph taken at a higher magnification. Figure 2 shows electron micrographs of (a) structure observed in the negative wedge BUE and (b) the structure in the flowlayer on the contact side of the chip. The

* Department of Production Engineering, Guindy Engineering College, Madras, India.

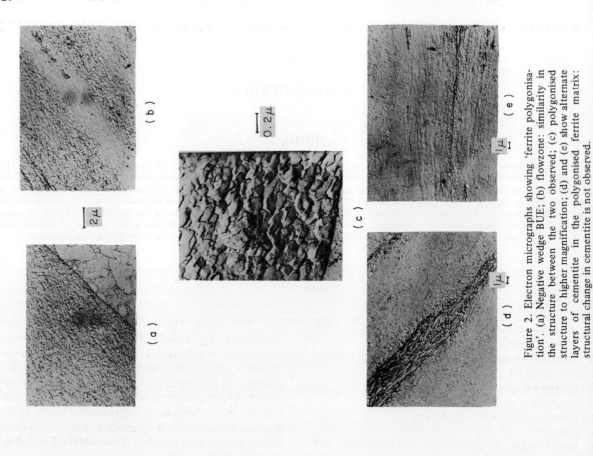

Figure 2. Electron micrographs showing 'ferrite polygonisation'. (a) Negative wedge BUE; (b) flowzone: similarity in the structure between the two observed; (c) polygonised structure to higher magnification; (d) and (e) show alternate layers of cementite in the polygonised ferrite matrix: structural change in cementite is not observed.

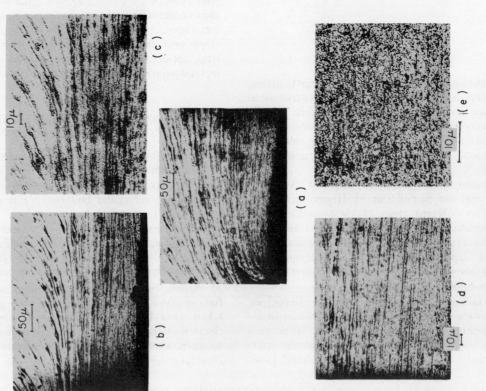

Figure 1. Negative wedge built-up-edge. (a) BUE near the tool tip; (b) central part of BUE; (c) and (d) finely-dispersed particles and streaks of cementite; (e) crystallites in the negative wedge BUE.

structure, which appears to be a minute polygonisation of ferrite, is not only evident in the entire negative wedge BUE but also on the contact side of the chip, thus indicating that the negative wedge and the secondary flow zone had been subjected to the same degree of deformation under identical conditions of temperature. Figure 2(c) shows the polygonised structure to higher magnification. Figures 2(e) and 2(f) show alternate layers of cementite in the polygonised ferrite matrix. From that it is clear that a structural change in cementite does not occur.

Figure 3 shows the morphology of flow layer in detail. The flow layer was obtained at a cutting speed of 40m/min. A striking similarity in structure of the negative wedge BUE and the flow layer can be observed from the micrographs. Recrystallised ferrite crystallites can be observed in the zone. They are however not clearly resolved as in electron micrographs. Temperature measurement by an inserted tool thermocouple method gave a value of 550 °C, indicating that the temperature at the interface was definitely above strain recrystallisation temperature, when the flow layer emerged.

The presence of recrystallised crystallites could be identified and an estimate of their size could also be obtained by X-ray diffraction methods, discussed in the ensuing paragraphs.

X-RAY DIFFRACTION TECHNIQUE

Specimen preparation and operation conditions
The specimen was mounted in araldite so that the bottom side of the chip was perpendicular to the X-ray beam. X-ray diffraction patterns were obtained by the back-reflection method and intensity profiles with the use of a microphotometer. The operating conditions were kept as follows:

camera film distance	2 cm
radiation	iron $k\alpha$
filter	manganese
exciting voltage	32 kV
tube current	17 mA
exposure time	30 min.

Diffraction patterns and intensity profiles
Figure 4(a) shows the diffraction pattern and intensity profile of undeformed metal. The small circle was indexed as (220) plane and the big circle as (211) plane. The Bragg angle for the small circle is $72°15'$ and for the big circle $55°24'$. The angle of spread at the bottom of the profile was $2°$. The X-ray diffraction pattern is spotty and this is characteristic of crystals of infinite size. In X-ray studies the expression 'crystals of infinite size' is used for crystals larger than one micron. Each large crystal gives rise

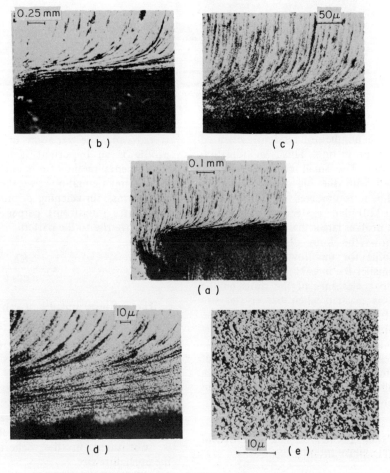

Figure 3. Morphology of flowzone. (a) and (b) curved zone of intense shear; (c) continuous flowlayer; (d) finely-dispersed particles in the flowlayer and streaks of cementite; (e) recrystallised crystallites in the flowlayer.

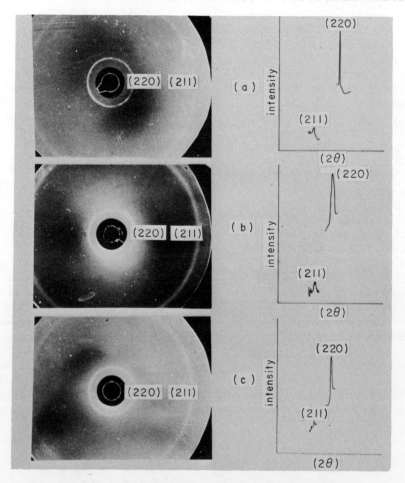

Figure 4. X-ray diffraction patterns and the respective intensity profiles for (a) undeformed metal; (b) flowlayer; (c) flowlayer heated to 300 °C. Iron $k\alpha$ radiation: exposure 30 min: film-camera distance 2 cm.

to a spot. Due to a number of randomly-oriented grains, the lines are spotty rings. This is an indication of large crystal size. The small circle is of high intensity compared with the big circle. From the intensity of profile it is observed that the profile corresponding to (220) plane is steep: a steep profile is characteristic of profiles larger than 5 microns.

Figure 4(b) shows the X-ray diffraction pattern and intensity profile for the flowzone. When the crystals become smaller the lines become smooth and continuous due to coalescence of the spots on the circle. The stage is reached when the crystal size becomes about 10^{-4} cm. The intensity profile is also broadened as evident from the profile. The line broadening becomes detectable when the particle size is reduced below one micron. Broadening of the intensity profile is an indication of the presence of minute crystallites which were clearly observed from electron micrographs.

DETERMINATION OF CRYSTALLITE SIZE FROM INTENSITY PROFILES

The broadening of intensity profiles by crystallites is due to lack of resolving power of such crystallites. This is analogous to the broadening of lines from a diffraction optical grating when the number of lines

is reduced. The variation of breadth of intensity profile is an important factor which provides an experimental method for determining crystallite size.

Scherrer's equation[8] gives the relationship between the intrinsic broadening β and the average size of crystallites (angstrom) perpendicular to the planes which give rise to the pattern

$$\epsilon = \frac{K\lambda}{\beta \cos \theta}$$

The intrinsic breadth β can be obtained by any one of the following methods:

(1) If the half-peak-breadth (b) of the intensity profile for crystals of infinite size is small it may be neglected. So the breadth B obtained from the intensity profile for the flowzone can be taken as β.
(2) If the broadening b of the undeformed metal is appreciable, pure diffraction breadth can be obtained by subtracting b from B. The relationship $\beta = (B - b)$ is substituted in the Scherrer equation to get the crystallite size.
(3) When the broadening is due to two factors, namely crystallites and microstrain introduced by plastic deformation, the specimen is given a recovery anneal to separate the breadth into two components.

The size is then determined from the breadth due to crystallites alone.

(4) Intensity profiles are derived analytically based on certain assumptions. Profiles could be obtained by making use of fourier series and transforms. The profiles thus obtained may be used for further calculation to determine the crystallite size.

The last two methods were adopted in the present investigation.

Generally it is accepted that the total width is a sum of effects by crystallite size and lattice distortion caused by plastic deformation. Referring to microphotographs of the flowzone at low magnification shows traces of residual plastic deformation from texture lines. At the same time well-defined polygonisation of ferrite in the flowzone has also been observed. This is an indication that the zone has not fully recrystallised due to lack of time.

The total broadening

$\beta = \beta_1$ (due to grain size) $+ \beta_2$ (due to lattice distortion). To distinguish the broadening due to the two effects, the sample was given a recovery anneal by heating it at 300 °C for one hour to effect crystal recovery without promoting recrystallisation. Figure 4(c) shows the diffraction pattern and intensity profile for the heated sample. This profile does not include the increase in breadth due to plastic deformation. It is also evident from the reduction in breadth of the intensity profile. The breadth β_2 is a measure of the microstrain in the zone.

For calculation the intensity profile of the small circle, that is, corresponding to (220) planes, was considered as the intensity was stronger compared to the intensity profile of the large circle. Figure 5

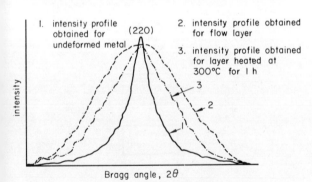

Figure 5. Comparison of intensity profiles obtained under various conditions.

shows the three profiles corresponding to (220) planes for undeformed metal, flow zone and the flow zone heated to 300 °C. The profiles were graphically recorded on stripchart, when the film was fed by an electric motor. Numerical values of intensity at various points were also noted. The intrinsic breadth β_1 due to crystallites was determined by subtracting the half-peak-width of the intensity profile for undeformed metal from the half-peak-width of the intensity profile for the flow zone heated at 300 °C. Making use of the Scherrer

equation and β_1 the crystallite size perpendicular to the (220) plane was found to be 1800 angstrom.

Microstrain

Microstrain is the result of lattice distortion introduced by plastic deformation. The broadening due to strain depends on, in a way, what may be taken by taking logarithmic deviation of Bragg's law

$$n\lambda = 2d \sin\theta \quad \text{which gives}$$

$$\frac{\delta d}{d} = \frac{\delta\theta}{\tan\theta}$$

The microstrain (e) was calculated and found to be 0.1%. This gives only the microstrain due to 'locked in' stresses and does not indicate the total plastic strain the flow zone had experienced.

FOURIER ANALYSIS

There are many methods of expressing the shape of intensity profiles. One of the methods is to express the intensity profiles by Fourier series. Under ideal conditions the intensity profile for undeformed metal should be a straight line. However, little broadening was ever observed due to instrumental and experimental factors. In the method earlier adopted this width was deducted from the width for the flowlayer to get the intrinsic or true profile without the effect of instrumental errors. Such a step is straightforward and simple. An improved method, to extract a true or intrinsic profile devoid of broadening due to instrumental factors, is by 'convolution' relation of Fourier series.

If $g(x)$ and $h(x)$ indicate the contours of intensity profiles of undeformed metal and the flowlayer, then the true profile $f(x)$ is given by the convolution relation[9]

$$h(x) = \int_{-\infty}^{+\infty} f(x)\, g\,(x-y)\, \mathrm{d}y \qquad (1)$$

The functions $h(x)$ and $g(x)$ are known in numerical form from microdensitometry. Function $f(x)$ is an unknown function which is to be derived from $g(x)$ and $h(x)$. The expression given in equation (1) for h in terms of f and g is known as the fold of f and g. Similarly f may be termed the unfold of h and g and the Fourier method giving $f(x)$ is termed 'unfolding'.

Using the more general approach devised by Stokes[10] a range of values of x is considered from $-a/2$ to $+a/2$ outside which $g(x)$ and $h(x)$ are zero, that is, outside which the intensity is considered to have fallen to its background value.

The contours of $f(x)$, $g(x)$ and $h(x)$ can be expressed by the following set of Fourier series.

$$f(x) = \sum_{-a/2}^{+a/2} F(t)\mathrm{e}^{-2\pi i x t/a} \qquad (2)$$

$$g(x) = \sum_{-a/2}^{+a/2} G(t)e^{-2\pi ixt/a} \qquad (3)$$

$$h(x) = \sum_{-a/2}^{+a/2} H(t)e^{-2\pi ixt/a} \qquad (4)$$

where $t = 0, \pm1, \pm2, \pm3, \pm4$ and so on. F, G and H are complex coefficients whose values are given by equations of the form

$$F(t) = \frac{1}{a} \int_{-a/2}^{+a/2} f(x)e^{2\pi ixt/a} dx \qquad (5)$$

By substituting the functions (2) and (3) in equation (1)

$$h(x) = \int \sum_{t} \sum_{t'} F(t)G(t')e^{-2\pi iyt/a}e^{-2\pi i(x-y)t'/a} dy \qquad (6)$$

$$= \sum_{t} \sum_{t'} F(t)G(t')\int e^{-2\pi iy(t-t')/a}e^{-2\pi iy(xt')/a} dy \qquad (7)$$

Since the integral of $e^{-2\pi iy(t-t')/a}$ is zero if $t \neq t'$ and equal to a if $t = t'$ the only terms left in the summation are those with $t = t'$. The series therefore reduces to

$$h(x) = a \sum_{t} F(t)G(t)e^{-2\pi ixt/a} \qquad (8)$$

and using equation (4)

$$H(t) = aF(t)G(t) \qquad (9)$$

or

$$F(t) = \frac{H(t)}{aF(t)} \qquad (10)$$

Substituting equation (10) in equation (2)

$$f(x) = \sum \frac{H(t)}{aG(t)} e^{-2\pi ixt/a} \qquad (11)$$

which is the required unfold of h and g.

The procedure for finding $f(x)$, the function describing the pure diffraction broadening, is as follows:

(1) The Fourier components of distributions of intensity $h(x)$ and $g(x)$ of the observed broad line and sharp line respectively are determined.
(2) Each Fourier component of $h(x)$ is divided by the corresponding component of $g(x)$.
(3) The resulting quotients are used in a Fourier synthesis to find $f(x)$. From a plot of $f(x)$, the breadth is determined immediately.

The Fourier components $H(t)$ and $G(t)$ of the functions $h(x)$ and $g(x)$ are complex and may be obtained by the use of Beevers–Lipson strips. The summation process is much facilitated by programming it on a digital computer. A computer programme was developed to solve the problem with the following assumptions.

- The profiles are symmetrical and therefore the sine series are omitted.
- The profile is not repetitive and therefore only half-range is considered.

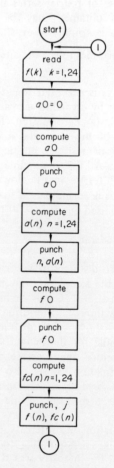

Figure 6. Flowchart for programme 1.

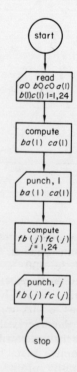

Figure 7. Flowchart for programme 2.

The computerisation is divided into two parts. The first part deals with the determination of the Fourier coefficients and the profiles for various conditions. Figure 6 shows the flow chart for the computation of the first part. The numerical values obtained from the microphotometer were used for this purpose. Figure 8 compares the profile obtained by plotting numerical values obtained from microdensitometry and also the values obtained by Fourier method for the flow layer heated to 300 °C. The profile obtained by Fourier analysis provides a close fit to the profile obtained by numerical values from microdensitometry.

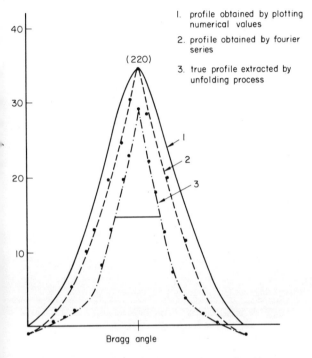

Figure 8. Comparison of profiles for specimen heated at 300 °C.

The second part of the computation deals with extracting profiles giving the intrinsic breadth as explained in the deviation. Figure 7 shows the flow-chart for the second part. Figure 8 also shows the extracted profile. This profile gives the true breadth excluding the effect of instrumental factors for the flow zone heated at 300 °C. The half-peak-width of the profile was determined, and making use of this breadth and the Scherrer equation the crystallite size was calculated to be 980Å.

CONCLUSIONS

The X-ray diffraction technique can be used as a powerful tool to investigate metallurgical changes at the work–tool interface. The presence of re-crystallised crystallites is clearly evident from electron micrographs. Optical micrographs also show the features, though not clearly resolved. The change of the X-ray diffraction pattern from spotty ring to continuous diffused line gives a qualitative picture of the presence of crystallites. The change in breadth is an index of crystallite size. The grain size obtained by X-ray technique can be taken as 0.1 μm. The value obtained by this technique gives the grain size perpendicular to the (220) plane and as such it cannot be compared directly with the size of crystallites obtained from micrographs. However, the presence of very small crystallites is an indication of the enormous strain and subsequent recrystallisation the zone had experienced. The changes in X-ray diffraction patterns give a vivid picture of the metallurgical changes in the flow layer. A quantitative estimate of grain size has been obtained by X-ray diffraction techniques. Fourier analysis lends sophistication to the X-ray diffraction technique.

ACKNOWLEDGEMENTS

The author would like to thank Dr. V. C. Venkatesh, Professor and Head of the Production Engineering and Machine Tool Section, Indian Institute of Technology, Madras, for his guidance. Thanks are also due to Metallurgy Department of IIT, Geology Department of University of Madras and the Computer Centre of Guindy Engineering College for their help in experimental and computational work.

REFERENCES

1. W. B. HEGINBOTHAM and S. L. GOGIA. Metal-cutting and builtup nose. *Proc. Inst. Mech. Engrs.*, **175** (1961), pp. 892–917.
2. R. RAMASWAMI. Investigation of the secondary shear zone in metalcutting. *Int. J. Mach. Tool Des. Res.*, **11** (1971), pp. 75–83.
3. R. RAMASWAMI. The effect of the built-up-edge BUE) on the wear of cutting tools. *Wear*, **18** (1971), pp. 1–10.
4. H. OPITZ and M. GAPPISCH. Some recent research on the wear behaviour of carbide cutting tools. *Int. J. Mach. Tool Des. Res.*, **2** (1962), p. 43.
5. W. KONIG, Dr. Ing. Diss., T. H. Aachen (1962).
6. V. C. VENKATESH and R. RAMASWAMI. The secondary shear zone and the negative wedge on HSS tools. *Annals of CIRP*, **18** (1970), pp. 513–519.
7. V. C. VENKATESH and R. RAMASWAMI. Investigation of the secondary shear zone by X-ray diffraction technique. Paper presented at the *11th International MTDR Conference*, (1970).
8. A. TAYLOR. X-ray metallurgy. John Wiley, New York (1961).
9. S. BARRET and T. B. MASSALSKI. Structure of metals. McGraw-Hill, New York (1966), p. 156.
10. A. R. STOKES. *Proc. Phys. Soc.*, **61** (1948), p. 382.

TOOL-LIFE TESTING BY RESPONSE SURFACE METHODOLOGY COUPLED WITH A RANDOM STRATEGY APPROACH

by

R. VILENCHICH*, K. STROBELE† and R. VENTER‡

SUMMARY

The paper presents a study of tool-life testing combining response surface methodology with Monte Carlo random strategy so as to obtain improved predictions with a 95% confidence level for the life of a tool used in machining operations. The analysis is based on a second-order model in which the tool-life is expressed as a function of the three not entirely independent variables, namely, speed, feed and depth-of-cut. The results and the reliability of the new model are discussed and compared with results obtained using the surface response methodology approach only.

NOTATION

T	tool-life (min)
V	cutting speed (ft/min)
f	feedrate (in/rev)
d	depth-of-cut (in)
K	constant
m, n, p	experimentally-determined constants
y'	true response of tool-life
x_1, x_2, x_3	logarithmic transformations of V, f and d
y	observed tool-life: $y = \ln T$
e	error in tool-life y
X	24 x 10 matrix of variables x_i
Y	24 x 1 vector of tool-life
X^T	transpose of X
Y_r	random values of tool-life.

INTRODUCTION

Tool-life testing requires a considerable number of tests if a reasonable functional relationship is to be established between the tool-life and cutting speed, feedrate and depth-of-cut. For each combination of cutting tool and workpiece material, a separate set of tests has to be conducted. This requires a large total number of tests and consequently can become an expensive undertaking.

These parameters, often thought of as having independent effects on the life of a tool, do show different degrees of interaction with one another, an effect overlooked by many researchers. In the present study interaction coefficients, as well as quadratic coefficients, are used in a model based on the three parameters cutting speed, feedrate, and depth-of-cut.

The one-variable-at-a-time approach used for the past decades is time-consuming and ignores any second-order effects of the variables. A better approach is the response surface methodology. This technique views the response of a dependent variable as a surface and was first applied to tool-life testing by Wu[1], following the successful application of this method in the study of optimisation problems in chemical processes.

Until recently the random nature of tool-life was ignored in almost all analytical work in metal cutting. This effect was investigated by slightly different approaches and with different aims by Leslie and Lorenz[7] and by Taylor[8]. The investigations were, however, limited to the conventional Taylor-equation for tool-life where only the speed of cut is varied to obtain a prediction of tool-life.

This paper analyses the second-order model presented by Wu[2]. The model is extended and it was possible to improve tool-life prediction. This is done by taking into consideration the indeterminate nature of tool-life in the model and combining the response surface methodology with the Monte Carlo random strategy.

THE MODEL

Studies in tool-life behaviour have enabled the following relationship to be defined:

$$T^n = \frac{K}{Vf^m d^p}$$

This is an extension to the basic Taylor equation, taking into account the effects of feedrate and depth-of-cut on the tool-life.

The relationship can be written as

$$y' = b_0' + b_1' x_1 + b_2' x_2 + b_3' x_3$$

The logarithm of both sides is taken and V, f and d are normalised according to:

$$x_1 = \frac{2(\ln V - \ln 700)}{\ln 700 - \ln 300} + 1$$

$$x_2 = \frac{2(\ln f - \ln 0.022)}{\ln 0.022 - \ln 0.010} + 1$$

$$x_3 = \frac{2(\ln d - \ln 0.100)}{\ln 0.100 - \ln 0.049} + 1$$

* Windsor University, Engineering, Windsor, Ontario, Canada
† McMaster University, Mechanical Engineering, Hamilton, Ontario, Canada.
‡ de Beers Industrial Diamond Division, Johannesburg, South Africa.

If the equation is written as:

$$y = b_0 + b_1 x_1 + b_2 x_2 + b_3 x_3 + e \qquad (1)$$

then y is the observed tool-life, b_0, b_1, b_2, b_3 are estimates of the parameters b_0', b_1', b_2', b_3' and e is the experimental error.

However, it has been found that over wide operating ranges this simple first-order model does not adequately describe the relationship between tool-life and the three variables, namely, speed-of-cut, feedrate and depth-of-cut.

In theoretical models the three variables were for long considered to have independent effects on the tool-life, which might be a reasonable approximation for a small range, but introduces large deviations from observed results over wider operating ranges. This was found to be the case with the simple model presented by equation (1).

This model, given by equation (1), has therefore been extended to a second-order model in which both the quadratic and the interactive effects of the three variables are considered. The new enlarged model is

$$y = b_0 + b_1 x_1 + b_2 x_2 + b_3 x_3 + b_4 x_1^2 + b_5 x_2^2 + b_6 x_3^2$$
$$+ b_7 x_1 x_2 + b_8 x_1 x_3 + b_9 x_2 x_3 + e$$

or

$$y = b_0 + b_1 x_1 + b_2 x_2 + b_3 x_3 + b_4 x_4 + b_5 x_5 + b_6 x_6$$
$$+ b_7 x_7 + b_8 x_8 + b_9 x_9 + e$$

This model reflects the degree of interaction in the respective coefficients and shows on analysing the results why it is important not to ignore the second-order effects.

ANALYSIS OF THE MODEL

A set of 24 experimentally-obtained cutting conditions were analysed: for convenience, these have been presented in table 1. These tests[1,2] were selected for the study to be able to compare the results of the two methods. Figure 1 illustrates the choice of cutting conditions for the 24 tests when plotted as a function of the transferred variables x_1, x_2, x_3 respectively, as given by Wu[1].

The ten parameters of the second-order model $b_0 - b_9$ are best found using the established method of least squares. Equation (2) can be written in matrix notation for the block of 24 tests as

$$Y = Xb$$

where Y is a vector of tool-life and b the vector of coefficients. The coefficients b are then determined from equation 3, where

$$b = (X^T X)^{-1} X^T Y \qquad (3)$$

It should be noted that a choice had to be made between the loss of orthogonality of the design or loss of accuracy when calculating the coefficients of the second-order equation. This is by no means an inherent problem of this method. In this case the choice of experimental data which enabled a comparison of the two methods made it necessary to use data that were not orthogonal.

TABLE 1

CUTTING CONDITIONS AND TOOL-LIFE[1]

test no.	speed V (ft/min)	feed f (in/rev)	depth-of-cut d (in)	recorded tool-life T (min)
1	330	0.010	0.049	160
2	700	0.010	0.049	37
3	330	0.022	0.049	165
4	700	0.022	0.049	27
5	330	0.010	0.100	172
6	700	0.010	0.100	35
7	330	0.022	0.100	120
8	700	0.022	0.100	18
9	480	0.015	0.070	66
10	480	0.015	0.070	83
11	480	0.015	0.070	71
12	480	0.015	0.070	82
13	226	0.015	0.070	240
14	1020	0.015	0.070	14
15	480	$0.007\frac{1}{4}$	0.070	110
16	480	0.034	0.070	54
17	480	0.015	0.034	99
18	480	0.015	0.143	70
19	226	0.015	0.070	222
20	1020	0.015	0.070	15
21	480	$0.007\frac{1}{4}$	0.070	93
22	480	0.034	0.070	42
23	480	0.015	0.034	106
24	480	0.015	0.143	65

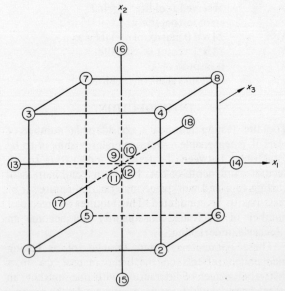

Figure 1 Selection of tests. Tests 19–24 are identical to tests 13–18.

Ideally, all normalised elements of the matrix X should be integers obeying a certain recurrence formula if an orthogonal polynomial is to be fitted to the data[4]. The actual data analysed by Wu[1,2] did not produce these integers or any values close to them, and rounding these quantities to the nearest integer would have resulted in a loss of accuracy in the tool-life prediction. The authors chose to sacrifice the orthogonal property for the sake of more accurate results of tool-life[4]. This problem could easily be overcome by using a variable-speed, variable-feed lathe for testing which would give infinite control of the parameters in any selected cutting range.

TABLE 2

COEFFICIENTS OF SECOND-ORDER EQUATION

no. of tests	b_0	b_1	b_2	b_3	b_4	b_5	b_6	b_7	b_8	b_9
12	4.3358	−0.8455	−0.1637	−0.0884	4.6562	−19.0491	−11.4991	−0.0814	−0.0269	−0.0926
12[1]	4.3200	−0.8463	−0.1635	−0.0886	−0.1338	0.0000	0.0000	−0.0813	−0.0263	−0.0938
14	4.3188	−0.7761	−0.1637	−0.0884	−0.0647	4.0008	−4.0697	−0.0794	−0.0269	−0.0926
16	4.3235	−0.7761	−0.1740	−0.0884	−0.0650	0.1050	−0.0854	−0.0793	−0.0269	−0.0926
18	4.2948	−0.7762	−0.1710	−0.0858	−0.0724	0.0021	0.0174	−0.0793	−0.0269	−0.0925
18[1]	4.2930	−0.7781	−0.1706	−0.0881	−0.0718	0.0005	−0.0195	−0.0813	−0.0263	−0.0938
20	4.2931	−0.7409	−0.1708	−0.0858	−0.0657	0.0015	0.0169	−0.0783	−0.0269	−0.0925
22	4.2998	−0.7409	−0.1783	−0.0858	−0.0660	−0.0171	0.0180	−0.0783	−0.0269	−0.0925
24	4.2978	−0.7409	−0.1782	−0.0967	−0.0660	−0.0171	0.0239	−0.0783	−0.0269	−0.0928
24	4.2954	−0.7436	−0.1796	−0.0996	−0.0650	−0.0200	0.0250	−0.0816	−0.0256	−0.0938

Table 2 shows the coefficients of equation (2) as obtained from different tests. Except for the coefficients b_5 and b_6 for the 12-test and 14-test model, all other coefficients are consistent and converge rapidly.* The reason for this inconsistency should be pointed out as the 10-variable, 12-test model was not analysed by Wu[2]: he chose to ignore the second-order effects of x_2 and x_3. If this effect is, however, not eliminated from the mathematical model in the 12-test case, the second-order effects of x_1, x_2, x_3 become so large that the prediction of tool-life is no longer accurate. The reason for this lies in the concentration of 4 test points at the origin of the three axes x_1, x_2, x_3 with only one test point at the corner of the cube (see figure 1). When more tests are added in the star arrangement this effect is counteracted and the coefficients converge rapidly. Table 2 also shows coefficients for the 18-test and 24-test model as obtained by Wu[2].

Comparing coefficients from the 18-test model upwards to the 24-test model, one sign change is observed and except for b_8, which remains constant, a convergence is observed. Both these observations indicate that 24 tests are required to produce a reliable tool-life prediction.

THE MONTE CARLO RANDOM STRATEGY

The life of a tool is determined by measuring previously-defined criteria of tool-failure and recording the time when this occurs as the tool-life. Depending on the use, the following criteria could apply:

1. complete failure of a tool.
2. flank wear reaches a certain level.
3. crater wear reaches a certain level.
4. change of surface finish or size of turned part.
5. increase of tangential force on tool.
6. increase in thrust force on tool.
7. any combination of the above.

Most frequently the criterion of flank-wear is chosen for tool-life since it increases linearly with time and is fairly easy to measure accurately (although not always). Often flank-wear is not uniform and the

width of wear measurement is subject to some judgement.

In the tests conducted by Wu[1] flank-wear was chosen as a tool-life criterion. Apart from the variation in tool-life due to measurement difficulties alone the amount of flank-wear, that is the tool-life, will show an inherent variation for identical cutting conditions.

The Monte Carlo random strategy[5] is used to allow for the random value of tool-life at a particular set of cutting conditions. Tests 9–12 from table 1 show a definite variation of tool-life, although the cutting conditions are identical; similarly, tests 19–24 are the same as tests 13–18. These test results were used to obtain an estimate of the variation of tool-life for any set of cutting conditions. A 30% variation of tool-life was found adequately to encompass all tests and this value was subsequently used in the random sampling approach, by fitting a normal distribution about the mean value of each of the 24 experimentally measured tool-life values.

Taylor[8] suggests that, in the absence of required data, 95% confidence limits of ±30% should be applied to individual determinations of tool-life. As it turned out this figure was suitable for the present study. As long as further data on the variability of tool-life under different conditions is not available it appears to be wide enough for this type of criterion of tool-life. This model is easily extended to determine confidence limits for any other criteria listed above. A suitably conducted test will establish the tool-life variation for different criteria expressed as a percentage of the life at various cutting conditions. If this percentage changes over the operating range, the change will be reflected in the confidence limits of tool-life predictions.

The 24-test model, which gave the best equation for tool-life, was then analysed using random values for tool-life within the specified range, in accordance with equation (3). The values for Y are no longer fixed at the recorded values, but instead are randomly selected about the mean value and denoted by Y_r. Solving equation (3) will result in the polynomial being fitted to an infinite population of 'experimental' tests. The advantage of this approach is clear when it is realised that 24 tests only give a coarse grid of

* This convergence must be interpreted in the context of the response surface methodology model only.

TABLE 3

VARIATIONS OF TOOL-LIFE AS OBTAINED FROM THE MONTE CARLO APPROACH

| | RESPONSE SURFACE METHODOLOGY WITH RANDOMLY SELECTED TOOL-LIFE | | | | | | RESPONSE SURFACE METHODOLOGY[2] | | |
| test no. | 30% variation in T | | | 50% variation in T | | | experimentally-measured life T | calculated life T | 95% confidence limits |
	mean life T	standard deviation	95% confidence limits	mean life T	standard deviation	95% confidence limits			
1	156.4	11.7	133–179	158.9	21.8	116–201	160	156	127 –195
2	43.7	3.2	37– 50	43.5	5.0	34– 53	37	44	35.5– 52
3	153.5	10.8	132–175	154.7	16.4	123–187	165	155	126 –192
4	31.6	2.4	27– 36	31.6	3.8	24– 39	27	31.5	25.5– 39
5	162.3	10.3	142–183	160.0	21.9	117–203	172	163	133 –202
6	41.2	3.1	35– 47	41.4	4.9	32– 51	35	41	33 – 51
7	111.6	8.1	96–127	111.5	12.9	86–137	120	111	90 –138
8	20.6	1.2	18– 23	20.8	2.6	16– 26	18	20	16.4– 25
9	73.2	3.5	66– 80	72.2	5.9	61– 84	66	73	64 – 84
10	72.8	4.0	65– 81	71.8	6.2	59– 84	83	73	64 – 84
11	72.7	3.4	66– 79	72.9	5.6	62– 84	71	73	64 – 84
12	73.8	3.9	66– 82	72.1	6.0	60– 84	82	73	64 – 84
13	249.9	14.4	221–278	249.0	25.1	200–299	240	250	210 –300
14	12.7	0.8	11– 14	12.7	1.2	10– 15	14	12.7	10.7– 15.3
15	96.3	6.5	83–109	93.7	9.9	74–113	110	97	81 –117
16	47.7	3.0	41– 54	46.8	5.0	37– 57	54	47	39 – 57
17	98.3	6.1	86–110	97.9	10.6	77–119	99	99	83 –118
18	65.5	4.1	57– 74	66.2	6.9	53– 80	70	66	55 –80
19	249.1	15.3	219–279	248.6	25.0	200–298	222	250	210 –300
20	12.9	0.8	11– 15	12.8	1.6	10– 16	15	12.7	10.7– 15.3
21	95.9	5.4	85–107	94.2	10.2	74–114	93	97	81 –117
22	46.9	3.0	41– 53	46.7	5.1	37– 57	42	47	39 – 57
23	98.8	6.0	87–111	98.3	9.9	79–118	106	99	83 –118
24	66.0	4.3	58– 75	65.7	6.9	52– 79	65	66	55 – 80

experimental points which can be improved by adding to these points a distribution as it would be found were more tests conducted. The solution will be expressed in terms of a predicted mean tool-life for every cutting condition associated with a corresponding standard deviation.

Results are summarised in table 3. A 50% random variation of tool-life is also presented, which gives a useful basis for comparing these results with those obtained by Wu[2], also included in the table.

The following points should be noted:

1. The mean life obtained when considering the random nature of tool wear for both 30% and 50% variation of tool-life agrees well with that found by considering the experimentally determined life alone[2]. This means that the extended model investigated did not produce any shift of the mean lives, or cause any irregular distortion of the results. The 30% and 50% variations in tool-life are illustrated in figure 2.

2. For all tests 95% confidence limits were obtained and are shown in table 3. It was possible by this approach to reduce these limits considerably and they may become more meaningful for practical application.

3. The 50% variation in tool-life produces confidence limits that correspond to those found by using the response surface methodology only. This certainly shows that the 95% confidence limits obtained by the response surface methodology are

pessimistic, namely, assuming at least a 50% variation in the experimental values of the tool-life.

4. Comparing the results of tests 13–18 with tests 19–24 gives an indication of the reproducibility

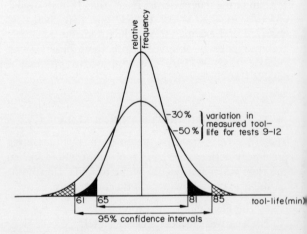

Figure 2 Variations in tool-life as produced by the Monte Carlo technique.

of the random strategy approach when used for similar tests.

5. The second-order model behaves well when subjected to the Monte Carlo technique. It was found that the 30% variation in tool-life corresponded to a 4–5% variation of the coefficients in the tool-life equation. This indicates that reasonable accuracy can be obtained with four significant figures.

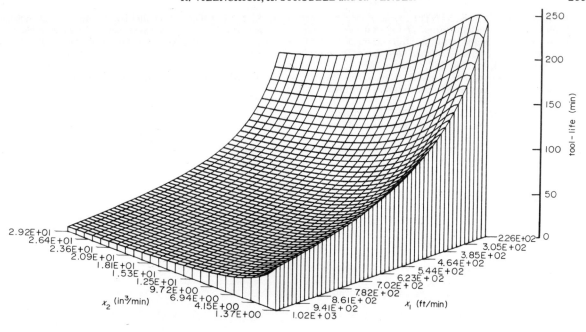

Figure 3 Tool-life prediction as a function of cutting speed (x_1) and volume of metal removed (x_2).

Throughout the analysis it has been assumed that there is no statistical variation or experimental error in the variables V, f and d. The accuracy of these quantities, however, has to be specified in experimental work to suit any particular model. Variations of speed, feed and depth-of-cut were investigated and permissible variations are listed below. Accuracies within the limits given will ensure that the tool-life prediction is accurate to ±1%.

The permissible variation in surface speed (V) = 2%

The permissible variation in feedrate (f) = 10%

The permissible variation in depth-of-cut (d) = 15%

Figure 3, a three-dimensional plot, shows tool-life as a function of speed-of-cut and volume of metal removed. The surface generated by a computer[9] is based on the second-order model that was studied. The plot shows the strong dependence of tool-life on the speed-of-cut and the lesser influence of volume of metal removed, which is a function of feedrate and depth-of-cut. This dependence is reflected also in the permissible variations, or required accuracies listed above.

The advantage of such a three-dimensional plot is obvious since it is automatically produced by the computer programme with all scales for any set of tests. If a particular tool-life is required, the variables V and d can be read-off from the graph, if f is chosen to give a particular surface finish. Since there are no calculations involved, such plots would be of great use to machine shops if supplied in a suitable form as a reference manual.

CONCLUSION

It was shown that tool-life testing by response surface methodology, coupled with a random strategy

approach, produces advantages over the response surface methodology alone.

The convergence of the coefficients in the tool-life equation was investigated for an increasing number of tests. It was shown that a minimum of 24 tests are required to produce a reliable equation for tool-life prediction.

The new approach produces a more realistic model by taking into consideration the random nature of tool-life. Though ignored in the past, this fact has more recently received attention by several people[7,8].

It was shown that the application of the random strategy approach did not distort the model or produce any shift of results. The irregular effect due to both experimental errors and inherent variation of tool-life was added to the model and the results are presented as a predicted mean life associated with 95% confidence levels. These proved to be narrower and thus more useful for practical applications.

The effect of experimental errors in speed-of-cut, feedrate and depth-of-cut on the model were determined. Limits of permissible errors are specified to ensure a prediction of tool-life within 1%.

A three-dimensional plot of tool-life against speed-of-cut and volume of metal removed is presented. This plot, if given for common combinations of workpiece and cutting tools, could provide quick and useful information about the tool-life as a function of all three cutting variables.

REFERENCES

1. S. M. WU. Tool-life testing by response surface methodology – Part 1. *Journal of Engineering for Industry*, May 1964.
2. S. M. WU. Tool-life testing by response surface methodology – Part 2. *Journal of Engineering for Industry*, May 1964.

3. R. VILENCHICH and R. VENTER. An analytical method applicable to the determination of individual concentrations of components, resulting from gamma induced iodination of methane. *Canadian Journal of Chemical Engineering,* Vol. 48; 584–587, October 1970.

4. N. L. JOHNSON and F. C. LEONE. Statistics and experimental design in engineering and physics. John Wiley & Sons 1964.

5. E. H. BOWMAN and R. B. FETTER. Analysis for production and operations management. Richard D. Irwin, Inc. 1967.

6. An evaluation of the present understanding of metal cutting. Battelle Memorial Institute, August 1958.

7. R. T. LESLIE and G. LORENZ. Comparison of multiple regression in machining experiments. *Advances in Machine Tool Design and Research*, vol. 8, Pergamon Press Ltd., 1967.

8. J. TAYLOR. Carbide cutting tool variance and breakage: unknown factors in machining economics. *Advances in Machine Tool Design and Research*, vol. 8, Pergamon Press Ltd., 1967.

9. P. ZACHAR. Applications of Computer Techniques to engineering problems. McMaster University, M.Eng. Thesis, 1972.

GRINDING AND
SURFACE PROPERTIES

RELATIONSHIP BETWEEN WHEEL CHARACTERISTICS AND OPERATING PROBLEMS IN HIGH-PRODUCTION PRECISION GRINDING

by

RICHARD P. LINDSAY* and ROBERT S. HAHN*

SUMMARY

The grinding behaviour of wheel—work pairs is discussed in terms of the *cutting metal removal parameter*, the *ploughing metal removal parameter*, the *ploughing—cutting transitional force intensity*, the *threshold force intensity* and the *wheel breakdown force intensity*. The sparkout *metal removal rate* and *interface normal force* are locked together as sparkout time progresses and their relationship can be graphically portrayed. Changes in finished workpiece size, taper and finish in production grinding are related to variations in sparkout behaviour.

INTRODUCTION

In production precision grinding operations it is often important to set the grinding machine parameters so as to produce parts of a given quality at minimum cost. In order to keep the cost down and the production rate up, the grinding machine must be set to operate in the shortest possible grind-cycle time. For example, in grinding a certain automotive part, a reduction of cycle time from 12 s to 11 s corresponds to a saving of over 1 million dollars a year. Consequently it is worth while to develop highly efficient grinding cycles. In order to do this it is necessary to know the quantitative relationships between the various important process variables. The workpieces arriving at the precision grinding operation generally have varying amounts of stock, with varying degrees of runout. These workpieces must be ground to a given surface finish, roundness, size and taper in spite of initial stock and out-of-roundness variations. The problem is to set the machine parameters (feedrates or grinding forces, sparkout times, diamond dressing lead and depth of cut, wheel and work speeds) so as to produce the desired surface finish, size tolerance, roundness, taper and surface integrity in the shortest possible cycle time, taking account of initial runout and stock variations. Before discussing grinding cycles it is necessary to develop working equations that adequately represent the grinding process.

BASIC RELATIONSHIPS IN PRECISION GRINDING

In grinding metals three distinct processes[1] take place at the interface of the abrasive grain and the workpiece.

(a) Rubbing, where the grain rubs on the work, causing elastic and/or plastic deformation in the work material with essentially no material removal.

(b) Ploughing, where the grain causes plastic flow of the work material in the direction of sliding, extruded material being thrown up and broken off along the sides of the groove, resulting in low rates of stock removal.

(c) Cutting, where a fracture takes place in the plastically stressed zone just ahead of the rubbing grain, causing the formation of a chip and resulting in fairly rapid stock removal rates.

Grinding parameter nomenclature

To illustrate this behaviour and the basic relationships between grinding parameters, consider a steady state grinding process. The cross-slide is advancing at a rate $\bar{v}_f$. Simultaneously the work radius is receding at a rate $\bar{v}_w$, while the wheel radius is wearing at a rate $\bar{v}_s$. Induced force intensities (force per unit width of contact) exist during the steady state: F'_n, normal to the wheel—work contact surface, and F'_t, tangential to the interface. The wheel peripheral speed is V_s and the work speed is V_w. The work support mechanism is represented by a spring K_w, while the wheel support system is represented by another spring K_s. The wheel—work contact surface also has some stiffness, K.

In explaining the grinding process it has been customary to report results in terms of downfeed (on surface grinders) or infeed rate (on cylindrical grinders). Under these conditions the grinding performance depends on the rigidity of the machine tool. For example, a downfeed of 0.002 in on a very stiff machine will induce a much larger wheel—work contact force than the same downfeed on a more compliant machine. Accordingly, grinding performance (surface finish, wheelwear, stock removal rate and surface integrity) will differ from one machine to another, even though the same wheel and work material are used. Consequently, in order to become independent of machine tool rigidity it is convenient, in plunge grinding operations, to relate grinding performance to the wheel—work interface contact force intensity, instead of to downfeed or feedrate.

* Cincinnati Milacron—Heald Division, Worcester, Massachusetts, USA.

Material removal equations

Consider the steady state grinding process previously described. Since both bodies are being 'mutually machined', in the steady state mode, the cross-slide infeed rate must equal the sum of the wheel and work radial removal rates:

$$\bar{v}_f = \bar{v}_w + \bar{v}_s \qquad (1)$$

The total material removal rates (in³/min in of width) are as follows,

$$\text{work: } Z'_w = \pi D_w \bar{v}_w \times 60 \text{ (in}^3\text{/min in)} \qquad (2a)$$

$$\text{wheel: } Z'_s = \pi D_s \bar{v}_s \times 60 \qquad (2b)$$

where D_w = diameter of work

D_s = diameter of wheel

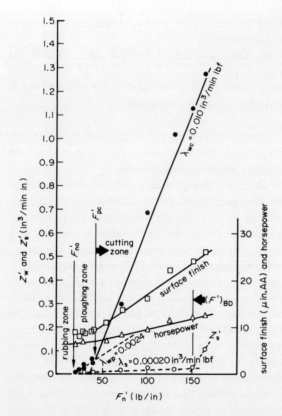

Figure 1 Wheel–work characteristic chart showing metal removal rate Z'_w, surface finish and horsepower versus interface normal force per unit width.

Wheel: A80K4V
Dress lead: 0.003 in/rev
Dress depth: 0.0004 in on diameter
Wheel speed: 12 500 ft/min
Work speed: 250 ft/min
Coolant: Cimcool 5 star
Material: AISI52100 at R_c60
Equivalent diameter (D_e): 2.0 in
External, plunge, climb grinding

Figure 1 is a plot of Z'_w and Z'_s (material removal rates) versus total applied force intensities; also shown are the surface finish and horsepower. From this graph it can be seen that for this work material, which is an easy-to-grind (ETG) material, there is a small force intensity threshold F'_{no} below which no metal removal takes place. Ploughing takes place from F'_{no} to F'_{pc}, the ploughing–cutting transition. Above F'_{pc}, the

cutting process occurs. The region below F'_{pc} is very important in grinding cycles using a sparkout.

The wheel wear rate Z'_s is also plotted in figure 1. The wheel wears at a linear rate up to some force intensity F'_{BD}. Above this breakdown force intensity, wheelwear becomes excessive and it is undesirable to operate in this region in precision grinding.

In the cutting zone, linear relationships exist for Z'_w, Z'_s and the force intensity. Thus we can write linear equations

$$\text{work: } Z'_w = \Lambda_{wc}(F'_n - F'_{pc}) \qquad (3a)$$

$$\text{wheel: } Z'_s = \Lambda_s F'_n \qquad (3b)$$

where the constant of proportionality in each case is Greek letter lambda (Λ) and has the following units: 'cubic inches of material being removed per minute, per pound of interface normal force intensity'. Λ_{wc} is the 'metal removal parameter' and Λ_s the 'wheel removal parameter'.

Using equation (2), the work removal of equation (3a) can be written in two ways:

$$\text{work: } \bar{v}_w = \frac{\Lambda_{wc}(F'_n - F'_{pc})}{\pi D_w 60} \qquad (4a)$$

$$\text{or } F'_n - F'_{pc} = \frac{\pi D_w 60 \bar{v}_w}{\Lambda_{wc}} \qquad (4b)$$

$$\text{wheel: } \bar{v}_s = \frac{\Lambda_s F'_n}{\pi D_s 60} \qquad (4c)$$

Equations (4a) and (4c) apply to controlled-force grinding where the force intensity[2] is prescribed while equation (4b) applies to feedrate grinding where the feedrate is prescribed. In practice, the wheelwear velocity, $\bar{v}_s$ is generally small compared to $\bar{v}_w$ so $\bar{v}_f$ can be substituted for $\bar{v}_w$ in equation (4b) giving

$$F'_n - F'_{pc} = \frac{\pi D_w 60 \bar{v}_f}{\Lambda_{wc}} \qquad (5a)$$

Equation 5a gives the induced force caused by applying a feedrate $\bar{v}_f$ onto the grinder. It can be seen that the induced force is directly proportional to the workpiece diameter and width and inversely proportional to Λ_{wc}. Thus, doubling Λ_{wc} causes the induced force to be halved, if the feedrate $\bar{v}_f$ is held constant.

From the standpoint of gross metal removal the cutting region and Λ_{wc} are significant. From the standpoint of precise sizing and producing smooth surface finishes the region around F'_{no} and F'_{pc} is very important.

The data shown in figure 1 presents in a graphical way the complete grinding action and is called a wheel–work characteristic chart. It is helpful in developing fast, efficient grinding cycles.

Production grinding cycles

Production grinding machines generally consist of a rotating work-holding device into which workpieces are automatically loaded and unloaded; a grinding wheelhead mounted on a cross-slide which is capable of moving the wheel radially into contact with the workpiece. Figure 2 shows a record of the radial cross-slide displacement (upper trace) during two consecutive grinding cycles of a production grinding machine. The lower trace is a record of the normal interface

waterpump brg.
wheel 97AI20IM4VFMD2

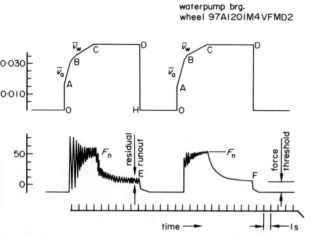

Figure 2 Typical grinding cycle showing cross-slide position and grinding force versus time.
Workpiece: automotive double-groove water-pump bearing
Operation: grind ball tracks

force existing at the wheel—work contact area during the grind cycle, as obtained by a force transducer. In a controlled force grinding machine the cross-slide is moved at a high velocity from 0 to A by the applied force F_a, bringing the wheel close to the unground running out workpiece. At point A the dashpot c is engaged, causing the slide to approach the workpiece at the approach velocity $\bar{v}_a$, which is simply

$$\bar{v}_a = \frac{F_a}{c} \qquad (6)$$

As the cross-slide moves from A to B it picks up the running out workpiece. Large instantaneous force pulsations occur on the wheel during the early stages of rounding-up, as shown by the first force trace. As rounding-up takes place these force pulsations reduce, and tend toward an average grinding force F_n according to

$$F_n = F_a - c\bar{v}_w \qquad (7)$$

where $\bar{v}_w$ is the plunge grinding velocity. Most of the stock is removed and further round-up accomplished from B to C. At C the cross-slide strikes a stop and comes to rest. However, the grinding wheel continues to grind owing to residual spindle deflection and sparks out during the interval CD. At D the slide is retracted and the workpiece hopefully is at size with the proper surface finish, roundness and microprofile. The work is unloaded and a new workpiece reloaded from H to 0 in preparation for the next cycle.

The engagement process
There are three critical processes during the grinding cycle that determine workpiece quality. During the

engagement process, when the wheel is just making contact with a non-running-out workpiece (AB in figure 2), high local force intensities may occur when a sharp edge of the work engages the wheel. If these local force intensities exceed the breakdown force intensity F'_{BD} in figure 1, excessive local wheelwear will take place and the profile of the wheel will be damaged, resulting in an unacceptable workpiece micro-profile. (Tolerances on micro-profile are sometimes as low as 30 μin.) The local force intensity can be found from equation (5a) in terms of the approach velocity $\bar{v}_a$ thus

$$F'_n = \frac{60\pi D_w \bar{v}_a}{\Lambda_{wc}} + F'_{pc} \qquad (8)$$

The maximum approach velocity for engaging a non-running-out workpiece such that local force intensities may equal (but not exceed) the breakdown force intensity is found from equation 8 by setting $F'_n = F'_{BD}$ and solving for $\bar{v}_a$.

$$v_{a\,max} = \frac{\Lambda_{wc}(F'_{BD} - F'_{pc})}{60\pi D_w} \qquad (9)$$

This approach velocity can be set by adjusting the applied force F_a and the dashpot c (equation (6)).

The rounding-up process
The second critical process in the grinding cycle is the rounding-up process. This has previously been studied[3] both analytically and empirically and the results combined into a digital computer programme which prints out, for each work revolution, the progressively diminishing workpiece runout, the peak force intensity existing on the grinding wheel at the wheel—work contact point, the peak force itself and the amount of wheelwear. Also given is the approach force F_a required to prevent wheel breakdown, the amount of stock required to true up the workpiece to a given accuracy, and the time required to round up the workpiece.

Experimental verification of the predicted behaviour has shown the simulation to be reasonable[3].

The sparkout process
The third critical process in a grinding cycle is the sparkout process (CD in figure 2). Figure 3 shows three recorder traces taken during a controlled-force, plunge, rough-grind and sparkout cycle. A force transducer measured the normal force existing at the wheel—work interface and is shown in the top trace. The radial progression of the workpiece surface is shown in the middle trace and was recorded from a work-riding probe—transducer system. The cross-slide motion is shown in the bottom trace and was recorded from a linear transducer. Points of interest during the grinding cycle are as follows.

(1) During the 4 s long rough grind, the force was constant at 75 lb and the radial grinding rate $\bar{v}_w$ was 0.001 075 in/s. The cross-slide infeed rate $\bar{v}_f$ was 0.001 150 in/s meaning, for this steady state, that the wheelwear rate, $\bar{v}_s$ was 0.000 075 in/s (0.001 150– 0.001 075).

(2) At the end of the rough grind, the cross-slide meets a positive stop and its motion effectively ceases.

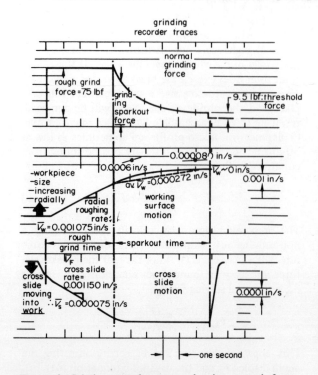

Figure 3 Grinder recorder traces showing normal force, workpiece and cross-slide motions.

Wheel:	97A80L6VFMB
Wheelspeed:	9100 ft/min
Workspeed:	250 ft/min
Wheel diameter:	1.0 in
D_e:	1.61 in
Dress lead:	0.003 in/rev
Dress depth:	0.0004 in
Coolant:	Cimperial 20
Material:	AISI52100 at R_c60

Internal, plunge, rough grind – sparkout cycle.

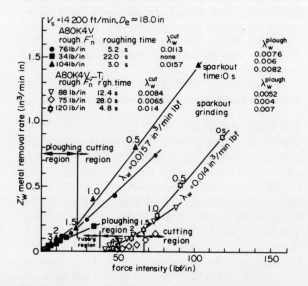

Figure 4 Metal removal rate versus force intensity during the sparkout process.

Wheel:	A80K4V and A80K4V T
Wheelspeed:	14 200 ft/min
Workspeed:	250 ft/min
Wheel diameter:	3.1 in
Work diameter:	3.7 in
Dress lead:	0.004 in/rev
Dress depth:	0.001 in
Coolant:	Cimperial 20
Work material:	AISI52100 at R_c60

Internal, plunge, rough grind–sparkout cycle.

Now, the 75 lb force which existed at the wheel–work interface during the rough grind, and has bent the wheel-holding quill, begins to 'grind itself out'. The workpiece continues to be ground as shown by the work-riding probe. One second after sparkout has begun, the grinding force has decreased to 40 lb and the grinding rate has diminished to 0.0006 in/s. After 5 s of sparkout, the grinding force has decayed to 11 lb, and the grinding rate to 0.000 080 in/s. After 6 s, the $\bar{v}_w$ rate is effectively zero, while the force still remains about 9.5 lb, a threshold force has been reached which will never decay further because no grinding action will occur.

Thus, in summary, during sparkout, we can measure a decreasing of the 'wound-up' force stored in the grinding quill accompanied by a similar diminution in the radial grinding rate, $\bar{v}_w$. Since both force and grinding rate are decreasing together, and because, as already seen, a powerful, linear relationship exists between the metal removal rate and force, we plot the the instantaneous Z'_w rates represented by the grinding velocities of the middle trace of figure 3 versus the grinding forces shown in the top trace.

Sparkout-cycle grinding tests were performed using two different wheels, one treated with a solid lubricant and one untreated, and the 14 200 ft/min wheel-speed information obtained during the sparkout is shown plotted in figure 4. Thus, for a rough force of 104 lb/in, using the untreated wheel, the (Z'_w, F'_n) position at the beginning of sparkout is noted as 'sparkout time: 0 s'. After 0.5 s of sparkout, both Z'_w and F'_n have diminished, and this instantaneous point is noted '0.5'; and so on, through 3+ seconds of sparkout. Connecting the points we see the exact relationship previously noted in figure 1, even though figure 1 was for individual, rough, plunge grinds, while figure 4 is the behaviour during a sparkout. For the untreated wheel, a steep, 'cutting' region exists down to 25 lb/in (for the 104 lb/in, data, this is reached in 1.5 s of sparkout), then a shallow, 'ploughing' region down to zero-force intensity.

Additionally, the treated wheel was sparkout-tested and similar behaviour resulted, except that now a third region exists. At forces above about 67 lb/in, cutting occurs accompanied by high Λ_w values. Between 67 and 40 lb/in, the Λ_w values are lower and are thus indicative of ploughing. Below 40 lb/in, no material removal occurs and additional sparkout time will be wasted in this, the rubbing region. Thus, the three classical regions of grinding behaviour: cutting, ploughing and rubbing are shown to exist during sparkout.

THE TIME CONSTANT

It has been shown that the sparkout process follows an exponential decay with a time constant τ given by[4]

$$\tau = \frac{w}{K_t[(\Lambda_w/60\pi D_w) + (\Lambda_w/60\pi D_s)]} \quad (10)$$

Equation (10) can be used to predict the time constant of a grinding system. It can be measured by plotting the normalised instantaneous grinding force

$F' - F'_{th}$ divided by the total grinding force at the beginning of sparkout $F'_o - F'_{th}$ versus the sparkout time. F'_{th} here is the threshold force intensity of the sparkout process, and F'_o is the initial grinding force intensity at time = 0, or the beginning of sparkout.

Figure 5 is a plot of normalised force intensity versus sparkout time for two wheelspeeds and various rough grinding forces. Obviously, these same grinds had a sparkout Λ_w and, since equation 10 predicts the time constant to be inversely proportional to Λ_w,

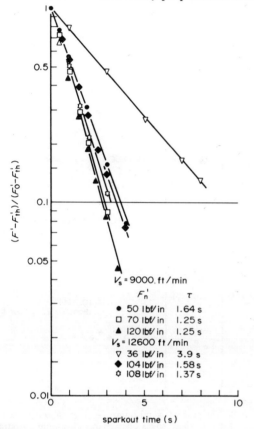

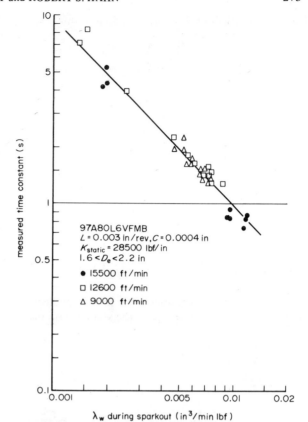

Figure 6 Measured time constant versus Λ_w during sparkout.

Wheel:	97A80L6VFMB
Wheelspeed:	9000, 12 600, 15 500 ft/min
Workspeed:	250 ft/min
Wheel diameter:	1.15 in
Work diameter:	2.3 in
Dress lead:	0.003 in/rev
Dress comp:	0.0004 in
Coolant:	Cimperial 20

Internal, plunge, rough grind–sparkout cycle.

Figure 5 Normalised sparkout force intensity versus sparkout time.

Wheel:	97A80L6VFMB
Wheelspeed:	9000 and 12 600 ft/min
Workspeed:	250 ft/min
Wheel diameter:	1.15 in
Work diameter:	2.3 in
Dress lead:	0.003 in/rev
Dress depth:	0.0004 in
Coolant:	Cimperial 20

Internal, plunge rough grind–sparkout cycle.

a graph of the measured time constant versus Λ_w would be interesting. Figure 6 shows this relationship for data obtained using various wheelspeeds. Note that the data from the three wheelspeeds, 9000, 12 600 and 15 500 ft/min, all fall on the same curve, which is the minus one slope. The effect of wheelspeed here is to increase Λ_w which in turn produces a smaller time constant. However, even at high wheelspeeds, when the wheel is purposely dulled by a long rough grind (and because of this produces a low Λ_w during sparkout), the resulting τ will be large. Therefore rough grinding time and wheelspeed will affect Λ_w and therefore τ. Additionally, any parameter that influences Λ_w will affect τ. Thus, large dress leads and depths of dress will increase Λ_w and reduce τ, while high conformities will reduce Λ_w and increase τ.

On grinding cycles where size, taper and surface finish are determined by a sparkout of fixed time, several factors affect the size, taper and finish variations. First of all, for cycles with a long sparkout time, the influence of the initial condition or force level from which sparkout began is completely lost. Size variations under these conditions are determined solely by variations in the threshold force F_{no}. For shorter sparkout times, errors in size may be caused by either variations in the initial force level, variations in Λ_{wc} and/or Λ_{wp} and variations in the threshold force.

For the case where the ploughing regime is negligible, a single sparkout time constant results and the quill deflection during sparkout can be plotted as in figure 7. The solid line represents a standard cycle or, for instance, grinding with a new wheel that has been freshly dressed with a sharp diamond. The dotted line represents the situation where (a) the initial force level has changed from F_n to $F_n + \Delta F_n$, (b) the time constant has changed from τ_0 to τ_1 due to changes in Λ_w in equation (10), and (c) the threshold force on corresponding quill deflection has changed by ΔF_{no}. Using equations (7) and (4a),

$$F_n = F_a - c\,\frac{\Lambda_{wc}}{60\pi D_w^w}\,(F_n - F_{no}) \quad (11a)$$

$$\text{or } F_n = \frac{F_a + (c\Lambda_{wc}/60\pi D_w^w)F_{no}}{1 + c\Lambda_{wc}/60\pi D_w^w} \quad (11b)$$

The system deflection X_q, corresponding to this force is

$$X_q = \frac{F_n}{K_t} = \frac{\dfrac{F_a}{K_t} + \dfrac{c\Lambda_{wc}}{60\pi D_w^w} \cdot \dfrac{F_{no}}{K_t}}{1 + \dfrac{c\Lambda_{wc}}{60\pi D_w^w}} \quad (12)$$

and the system deflection during sparkout will be

$$X_q(t) = \left[\frac{\dfrac{F_a}{K_t} + \dfrac{C\Lambda_{wc}}{60\pi D_w^w}\dfrac{F_{no}}{K_t}}{1 + \dfrac{C\Lambda_{wc}}{60\pi D_w^w}} - \frac{F_{no}}{K_t} \right]$$

$$\exp\left[\left(-K_t\left\{ \frac{\dfrac{\Lambda_{wc}}{60\pi D_w} + \dfrac{\Lambda_s}{60\pi D_s}}{w} \right\} \right) t \right] + \frac{F_{no}}{K_t} \quad (13)$$

The term in brackets represents the initial condition at the beginning of sparkout, and the second term the

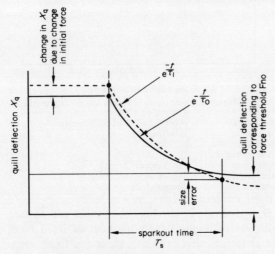

Figure 7 Quill deflection versus time.

threshold quill deflection after a long time. Equation (13) can be used to assess the size error at the end of a sparkout time T_s due to variations in Λ_{wc} and F_{no}. The constancy of the initial deflection term in brackets is held within close limits on controlled-force grinding machines. This is not the case on conventional feed-rate machines where the grinding force at the onset of sparkout varies strongly with stock variations.

The influence of system rigidity K_t on size-holding ability at the end of a 6 s sparkout (about 3 time constants) is illustrated in the table below in the grinding of the two parallel ball tracks (figure 8) in automotive water-pump bearings. In one setup the rigidity of the grinding quill and wheelhead was increased from 23 000 lb/in to 50 000 lb/in (4180 N/mm to 9100 N/mm) at the outer wheel, all other factors remaining the same.

System rigidity (lb/in)	Size-holding ability standard deviation σ (in)
23 000	110 × 10⁻⁶
50 000	32 × 10⁻⁶

It is seen that the more rigid system gives greater precision. This is due primarily to reduction of the variation in threshold quill deflection represented by the last term in equation (13). Variation of the threshold force F_{no} and the cutting and ploughing metal removal parameters Λ_{wc}, Λ_{wp} as the grinding wheel becomes smaller can cause a disruption in size at wheel change in production internal grinders. This is illustrated by the size plot of the bore of a double-groove automotive water-pump bearing shown in

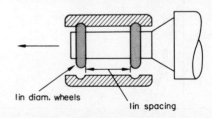

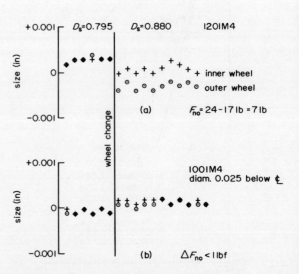

Figure 8 Size plot of water-pump bearings showing a wheel change for (a) wheel specs giving large threshold variation; (b) wheel specs giving small threshold variation.

figure 8(a): the bore size is plotted just before wheel change. A size drop of about 0.0006 in (0.015 mm) and the appearance of 0.0004 in (0.01 mm) size difference or taper between the two simultaneously ground balltracks indicates higher force levels at the end of the sparkout for the large wheel. This was confirmed by a measurement of the threshold force for the new and old wheels, 24 lb (107 N) for the new large wheel, 17 lb (75 N) for the used small wheel.

This illustrates the use of wheel–work characteristic charts as in figure 1, for explaining and improving grinding performance. By selecting wheels that exhibit a very small change in F_{no} as wheelsize changes, the size plot across a wheel change was improved to that shown in figure 8(b) where a positive jump of about 0.0001 in (0.0025 mm) occurred at wheel change.

CONCLUSION

The grinding behaviour of wheel–work pairs with a given coolant, dressing conditions and wheel and workspeeds can be measured and presented in the form of wheel–work characteristic charts as in figure 1. These charts apply equally to controlled force or feedrate grinding.

The critical quantities in determining the grinding behaviour are

- the cutting metal removal parameter Λ_{wc}
- the ploughing metal removal parameter Λ_{wp}
- the ploughing–cutting transitional force intensity F'_{pc}
- the threshold force intensity F'_{no}
- the wheel removal parameter Λ_s
- the wheel breakdown force intensity F'_{BD}

The selection and evaluation of grinding wheels, coolants and operating speeds can be made effectively by measuring these quantities.

The development of grinding cycles for optimising certain criteria can be accomplished in terms of the above, practically oriented quantities and equations.

These equations also form the basis for control of the grinding process.

REFERENCES

1. R. S. HAHN, 1962, 'On the nature of the grinding process', *Proc. 3rd M.T.D.R. Conf.*, Pergamon Press, Oxford, pp. 129–54.
2. R. S. HAHN, 1964. 'Controlled force grinding–a new technique for precision internal grinding'. *Trans. A.S.M.E., Ser. B: J. Eng. Ind.*, 68, 287–93.
3. R. S. HAHN and R. P. LINDSAY, 1971, 'On the rounding-up process in high production internal grinding machines by digital computer simulation', *Proc. 12th M.T.D.R. Conf.*, Macmillan, London, p. 235.
4. R. S. HAHN and R. P. LINDSAY, 1969, 'The influence of process variables on material removal, surface integrity, surface finish and vibration in grinding', *Proc. 10th M.T.D.R. Conf.*, Manchester, Pergamon Press, Oxford, pp. 95–117.

SURFACE GRINDING WITH HIGH WHEEL SPEEDS AND METAL REMOVAL RATES

by

W. KÖNIG* and M. DEDERICHS*

SUMMARY

Two methods of surface grinding can be distinguished: (a) grinding with high downfeed rate and low work speed, and (b) grinding with low downfeed rate and high work speed. This paper shows which of the two methods gives the better grinding results in terms of cutting forces, surface roughness and integrity, grinding wheel wear at high wheel speeds and high metal removal rates. The workpiece materials used were low and high alloyed steels, stainless steels and high temperature alloys. The grinding wheels differed in hardness, grain size and bond material.

INTRODUCTION

In recent years, research into the techniques of cylindrical plunge grinding, with an increase in the wheel speed of the grinding wheel to 90 m/s, has enabled a significantly greater grinding efficiency to be achieved without influence to workpiece quality; a much faster chip removal rate is used than has hitherto been reached in conventional grinding[1,2,4,7]. On the grounds of kinematics and the mechanics of chip removal, a similar improved efficiency may also be expected in surface grinding.

It is therefore necessary to investigate more closely the possible methods of improving the economics of surface grinding, especially as there are many examples of the use of surface grinding in practice. The economics of these methods are much more favourable when it is possible, with the help of higher wheel speeds, to grind the required profile directly from raw stock, so that the workpiece can be finished on one machine without preparation such as milling or planing.

The results and experience gained in using high chip removal rates and high wheel speeds in plunge grinding cannot be directly applied in surface grinding since different laws apply, particularly in achieving higher chip removal rates. In cylindrical plunge-grinding the chip removal rate over the one millimeter grinding width depends only on the average work-piece diameter and on the infeed speed. A change in work speed has no effect on the rate of chip removal.

In surface grinding the relationships are different. Here the specific chip removal rate is the product of the depth of cut and the work speed:

$$Z' = a \times v_w$$

Thus the same chip removal rate can be the result of a deep cut and a low work speed, or a high work speed and a shallow cut. Here it needs to be shown which of these two alternative approaches to surface grinding is the best in terms of cutting force, surface quality, surface integrity and grinding wheel wear rate.

Investigations have been carried out in the Laboratorium für Werkzeugmaschinen und Betriebslehre der RWTH Aachen, on a high-speed surface grinding machine. This machine can provide a driving power of 55 kW at the grinding wheel spindle up to 5000 rev/min.

The cooling lubricant used in the investigations was a grinding oil with a viscosity of 15° Engler at 20 °C. To ascertain the influence of the coolant on the working results an oil–water emulsion with a concentration of 7% was also employed. The high-pressure coolant was applied at a rate of 120 l/min

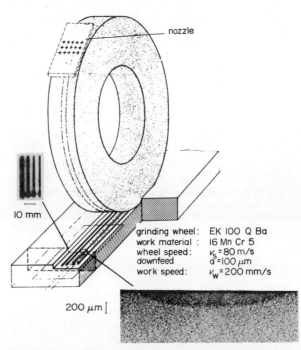

grinding wheel : EK 100 Q Ba
work material : 16 Mn Cr 5
wheel speed : v_s = 80 m/s
downfeed : a = 100 μm
work speed : v_w = 200 mm/s

Figure 1. Cleaning effect on grinding wheel surface.

under a pressure of 12 kp/cm². The coolant nozzles produced jets which struck the grinding wheel with high energy, cleaning and wetting it, so that on the one hand the wheel surface was freed from embedded particles, and on the other the grains were provided with friction-reducing lubricant before cutting.

Evidence for the effect of the high-pressure cooling is shown in figure 1. This shows the bottom surface of a slot produced by grinding. As may be seen in the photograph on the left, and as is shown again in the sketch, burn marks are visible at the base of the slot. These arose from local hot spots as a consequence of high friction between the loaded wheel and the ground surface. The distance between the dark-coloured stripes corresponds to that of the holes in the jet plate. On the other hand, the parts of the grinding wheel cleaned by the impact of the coolant produced

* Lehrstuhl für Technologie der Fertigungsverfahren am Laboratorium für Werkzeugmaschinen und Betriebslehre der RWTH Aachen, Germany.

a faultless bright cut. For all subsequent tests jet plates with staggered holes were used to ensure that each point on the wheel was cleaned.

Aluminium oxide was used for the grain of all the grinding wheels. For the bond, ceramic as well as resinoid bond was used; in addition the hardness and grain size were varied.

The following materials were investigated: Ni-basis alloy RGT 12, Co-basis alloy ATS 115, stainless steel Remanit 1880 SST (X10 Cr Ni Mo Ti 1810), high-speed steel DMo 5 (S 6-5-2) alloy tool steel Bora 12 (X 210 Cr 12), cast iron GGG 70, cementation steel C 15 and heat-treatable steel 50 Cr V4.

RESULTS OF THE INVESTIGATIONS

Cutting force

Knowledge of the cutting force in grinding enables conclusions to be drawn about the cutting process and statements to be made about the working results to be expected. Thus the total value of the cutting force largely determines the wear of the grinding wheel, while the relationship between tangential and normal force gives a measure of the cutting capacity of the wheel. Not least, the tangential force provides information about the driving power required for a given work process.

Since the condition of the outer surface of the grinding wheel has a not insignificant effect on the magnitude of the cutting force,[3] care must be taken to ensure that measurements of the cutting force are taken under constant working conditions. In the present investigations the cutting force was therefore always measured immediately after the wheel had

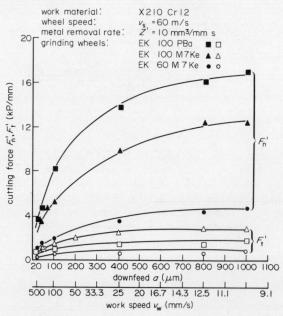

Figure 2. Cutting forces versus downfeed and work speed for grinding X 210 Cr 12 steel.

been dressed, so that the condition of the outer surface of the grinding wheel could be taken as more or less constant.

As mentioned earlier, the same chip removal rate

can be obtained by different combinations of depth of cut and work speed.

The extent to which the cutting force is influenced by the depth of cut and work speed, that is, the variations in these two quantities required to keep the chip removal rate constant, is shown in the following examples.

Figure 2 shows the specific cutting forces plotted against depth of cut a and work speed v_w for the steel Bora 12, the product of these two variables being held constant at the value of $Z' = 10$ mm³/mm s. The test was carried out first with an EK 100 P Ba resinoid wheel and then, for comparison, with ceramic bond wheels EK 100 M7Ke and EK 60 M7Ke.

From the diagram it may be seen that there is a clear rise in cutting force as the depth of cut is increased with correspondingly reduced work speed. The force normal to the workpiece surface reaches, for example, with the resinoid wheel a value of about 3.5 kp/mm when the work speed is 500 mm/s or 30 m/min and the depth of cut is 20 μm; with a workspeed of $v_w = 10$ mm/s or 0.6 m/min and a depth of cut of $a = 1$ mm and thus the same chip removal rate, this force increases about five times. This tendency was confirmed with all the wheels tried. From this it follows that in order to keep the cutting force low for a given chip removal rate the ratio of work speed to depth of cut should be kept as high as possible.

From figure 2 it may also be inferred that the normal cutting force using a grinding wheel with resinoid bond is significantly greater than with a ceramic-bonded wheel of hardness M.

Looking now at the tangential cutting force, it may be seen that in the case of the resinoid wheel this force is relatively small compared with the normal force. An explanation of this relationship is to be found in the greater compliance of the grains in the resinoid wheel[7,8] which are already set up in a different cutting geometry in the dressing of the wheel. Thus because the impact between dressing tool and grain is relatively elastic, there is a tendency for a few of the grains to be split in dressing, this leading to a strongly negative rake angle. This tendency for few of the grains to split, because they are supported in a relatively elastic medium, also leads to a strongly negative rake angle for cutting in the grinding operation itself.

An extremely negative rake angle, as may be supposed for a resinoid wheel, from Werner's model[6] requires very high normal cutting forces, while the tangential forces necessary for chip formation are smaller. The results of the tests with wheels having different bonds can be explained by this model; this theory must be confirmed by carrying out appropriate tests.

In figure 3 are shown the cutting forces for three test materials using the resinoid wheel EK 100 P Ba and the ceramic wheel EK 100 M7Ke, again with the depth of cut and work speed chosen so that their product is constant.

The result here is that the steel Bora 12 gives the highest values both for the resinoid and for the ceramic wheel. Smaller normal forces arise in grinding

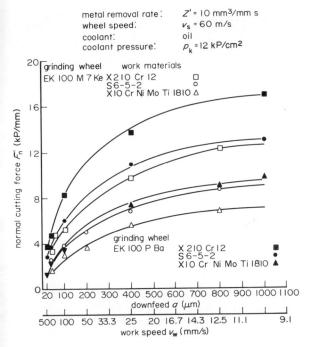

metal removal rate: $Z' = 10 \text{ mm}^3/\text{mm s}$
wheel speed: $v_s = 60 \text{ m/s}$
coolant: oil
coolant pressure: $p_k = 12 \text{ kP/cm}^2$

Figure 3. Normal cutting force versus downfeed and work speed for different grinding wheels and work materials.

DMo5, whilst Remanit gave the lowest values for the three materials.

The steels lie in the same order for tangential forces, with the lower values again attaching to the resinoid wheel.

The higher cutting forces in grinding the material Bora 12 may be explained by the high proportion of the alloy element chromium, together with 2% carbon content, thus producing heat- and wear-resisting chrome carbides which make mechanical removal of material very difficult.

While the tendency of the rust- and acid-resisting steel Remanit 1880 SST to toughen indeed leads to higher cutting forces than those obtaining in grinding structural steel[2,4], nevertheless these values lie slightly below those for vanadium carbide high-speed steel S 6-5-2 (DMo 5).

Grinding wheel wear

The wear resistance of the grinding wheel is one of the most important criteria in determining the economics of the process, especially when grinding at high chip removal rates. In the case of a wheel used in plunge grinding, the wear is the combination of that of the diameter of the grinding wheel and that of the edges. The loss of diameter, or the corresponding radius, of the grinding wheel during the production of a number of workpieces results in a dimensional error which in many cases can be compensated by a measurement control system, so that the accuracy of the product does not suffer.

However, the permanence of the grinding wheel edges is vital for the profile of the workpiece. If the grinding wheel edges go out of their predetermined tolerance, the necessary profile can only be restored by dressing the wheel.

Wheel edge wear and radius wear therefore cannot be regarded as independent, since they influence each other. That is to say, a constant rounding of the edges is exactly compensated by the increase of the radial wear.

On this basis, the following diagrams derive a criterion for the grinding process which takes account of the relationship between radial and edge wear of the grinding wheel.

Corresponding to the explanation of cutting forces already set out in figures 2 and 3, there is the expected additional wear as the depth of cut is increased. Here we may speak of the grinding ratio, defined as the ratio of the volume removed from the workpiece to that worn off the grinding wheel, decreasing for a constant chip removal rate (figure 4).

In grinding Bora 12 steel and Remanit 1880 SST this ratio has low similar values, but for high-speed steel DMo 5 it is appreciably higher. Also it is clear that in surface grinding a grinding ratio many times greater can be used than in deep grinding with a low work speed and large depth of cut.

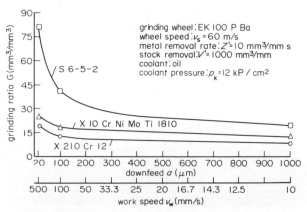

Figure 4. Grinding ratio versus downfeed and work speed for different materials.

The interdependence of wear rate and grinding ratio in the wheels used is set out briefly below.

Varying the grain size in the wheel for a constant chip removal rate of $Z' = 10 \text{ mm}^3/\text{mm s}$ using ceramic-bonded wheels of hardness M shows a big increase in the grinding ratio in going from grain size 46 to size 60, while grain size 100 leads to unfavourable wear rates. This tendency is the same for all the workpiece materials investigated, irrespective of the absolute quantities involved.

Harder wheels of similar grain size ought to offer a greater resistance to the breaking out of grains or of whole groups of grains. This is confirmed by comparing the grinding results obtained with a soft J wheel and with a hard M wheel, with a chip removal rate of $Z' = 10 \text{ mm}^3/\text{mm s}$. As expected, the harder bond of the M wheel results in less wear and thus a larger grinding ratio.

As is already known from other studies, a higher wheel speed of the grinding wheel results in less wear. This is true for practically all workpiece materials with the exception of nickel alloys.

In comparing the grinding ratio of resinoid wheels of various grain sizes, similar tendencies were found as in the case of ceramic bonded wheels. However, this did not hold for grain sizes greater than 100, since here the wheels became loaded under the previous working conditions. Confirmation of this was given by very high roughness when using wheels of grain size 220.

Surface roughness

The quality of a workpiece produced mainly by grinding will in most cases be defined by its surface roughness, which is greatly affected by the number of cutting edges presented in the contact zone between the grinding wheel and the workpiece. This in turn is primarily determined by the grinding wheel itself, in its grain size, structure and bond, and secondarily by the choice of grinding conditions. The latter includes sparking-out, that is, grinding with zero depth of cut as a final operation to achieve good surface quality.

The diagram in figure 5 shows the effect of sparking-out on the obtainable surface quality for different grinding parameters. As can be seen from the diagram, the centre line average (CLA) falls sharply with the first strokes of sparking-out, and then remains fairly constant.

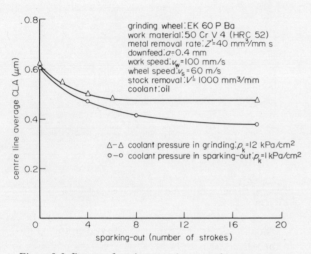

Figure 5. Influence of coolant supply on sparking-out process.

The relationship of the original roughness to the roughness after sparking-out is only slightly affected by the chip removal rate and is not, as might be expected, determined by the depth of roughness of the grinding wheel working surface. This is apparently for the following reason. In sparking-out, as in the grinding finish itself, coolant is delivered in large quantities at high pressure into the grinding zone. This results in a hydrodynamic wedge of lubricant in the gap between the wheel and the workpiece, generating forces in the grinding wheel–workpiece system and causing a deflection of the grinding wheel spindle, so that after a few sparking-out strokes no more material presents itself for removal. Consequently the original roughness can only be removed to a limited extent.

Trials with greatly reduced oil flow during sparking-out showed that the finished roughness was largely reduced.

This means that, in order to obtain the best surface quality, sparking-out must be done with reduced coolant flow.

From the diagram shown previously it can be seen that after about two sparking-out strokes no further improvement in surface quality is obtained so long as the high-pressure coolant flow is unaltered. Taking this finding for the number of useful sparking-out strokes, the investigations were carried further.

The workpiece surface is generated by the cutting edges which project furthest from the wheel and thus penetrate the workpiece most deeply. Kassen[5] showed that a change in the grinding wheel characteristics was always accompanied by a change in the density of cutting edges. The density of cutting edges, indicated by the static number of cutting edges, in turn determines how many edges are available for a given depth of profile. This means that the size of a chip, principally its length and thickness, is dependent on the average case where two cutting edges follow each other.

While the chips become thinner with finer grain, their length remains substantially constant. Meanwhile the number of successive edges becomes progressively greater.[5] The implication for the chip-formation process is that for the same grinding conditions the fine-grain wheel, which generally exhibits a higher density of edges, produces the same volume of workpiece

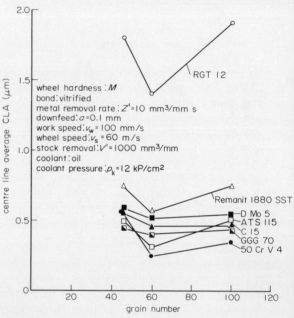

Figure 6. Surface roughness versus grit number for vitrified wheels.

material in a greater number of chips. Thus it is necessarily the case that the individual chip volumes must become smaller as the number of edges increases.

From this relationship it follows that in grinding with a low density of cutting edges only a poor surface roughness can be obtained.

The practical tests show this relationship only to a limited extent. In the case of the ceramic-bonded wheel with hardness M, it was valid only to a grain size of 60. With finer grinding wheels, particles of workpiece material appeared to become embedded in the wheel (figure 6). This embedding prevented free chip formation and created a greater roughness, which can be observed particularly with workpiece materials RGT 12 and Remanit 1880 SST, accelerating loading of the pores of the wheel. The harder and less tough materials such as DMo 5 and 50 Cr V4 show, on the other hand, only a slight increase in roughness with grain size 100.

It is well known that for given grinding conditions soft grinding wheels wear faster than hard. This behaviour leads to poor surface quality since the rapid wear brings forward new grains and, with them, new sharp cutting edges; these have only a small negative rake angle as Gühring has already ascertained[2]. Following new investigations into the kinematics and mechanics of grinding by Werner[6] these sharp edges are very unsuitable for a free-running chip formation process. This is especially apparent for the smearing, wheel loading and therefore hard-to-grind materials RGT 12 and Remanit 1880 SST.

The widely differing cutting forces produced by resinoid and ceramic-bonded wheels lead one to expect that the type of bond has an effect on surface quality. Also, the significantly more elastic resinoid bond could affect the cutting process.

A comparison of the results of working with the wheels EK 60 M Ke and EK 60 P Ba at the high chip removal rate of $Z' = 40$ mm^3/mm s showed that the surface quality achieved was slightly better with the resinoid than with the ceramic-bonded wheel. This may be traced back to the greater deflection of the outermost grains of the resinoid wheel under cutting load, on account of its lower E modulus. This leads to a reduction in the effective depth of cut, and so to reduced stress on the grains which, because of the same effect, have already in dressing obtained a favourable strongly negative rake angle, and thus show optimum chip formation capacity.

Furthermore it has been confirmed that improved surface quality can also be obtained in surface grinding by increasing the wheel speed with otherwise constant grinding conditions. In particular for the material RGT 12 an improvement of 100% in CLA can be obtained using a ceramic-bonded wheel with the circumferential speed raised to $v_s = 90$ m/s.

Using a resinoid wheel the improvement in surface quality for RGT 12 by raising the wheel speed is even more marked. Here a reduction of the CLA value from the previous 3.6 µm at $v_s = 60$ m/s to 0.85 µm at $v_s = 90$ m/s was noted. The high value of 3.6 µm indicated that at $v_s = 60$ m/s the chip volume was too great and loading of the wheel followed.

Cooling fluid

The characteristics of the coolant have a great influence on the working results, as earlier studies with structural steel have already shown.[1, 2]

For use in high-speed grinding the most suitable coolant, other than an oil—water emulsion, is a pure grinding oil of 3° to 4° Engler at 50°C. Emulsions, on account of their high proportion of water, have high thermal conductivity and heat of vaporisation; however, the better wetting and lubricating capacity of the oil leads to lower temperatures in the cutting zone.

At the same time, the introduction of the grinding oil results in lower cutting forces and thus, on account

of reduced wear, a higher grinding ratio. By using this grinding oil the grinding ratio was improved by up to three times that obtainable with the emulsion, according to workpiece material.

It has already been pointed out that a high wear rate of the grinding wheel brings a relative reduction in the surface quality. If one compares figure 7 and the subsequent figure 8 from this standpoint, one can establish that in those cases where a low grinding ratio prevails, the surface roughness takes on a high value. In general, it can be said that the grinding oil used here enabled significantly smaller surface roughnesses to be obtained than with the emulsion.

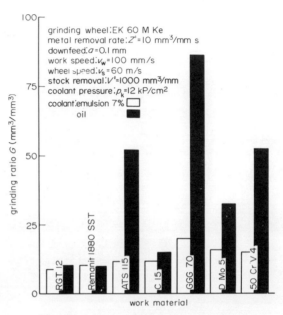

Figure 7. Influence of coolant on grinding ratio.

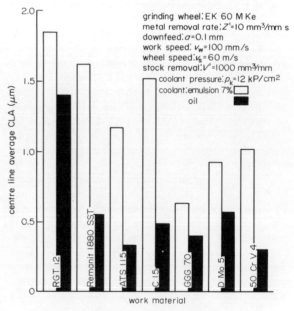

FIgure 8. Influence of coolant on surface roughness.

Thermal effects in the workpiece surface

To ascertain the structure of these effects samples were taken from the ground workpiece and examined under an optical microscope. The influence of the various grinding parameters on the thermal effects in the workpiece surface may be briefly set out as follows.

Increased work speeds have a beneficial influence on the thermal effects in the workpiece surface.[2] For

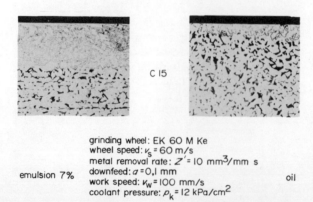

grinding wheel: EK 60 M Ke
wheel speed: v_s = 60 m/s
metal removal rate: Z' = 10 mm³/mm s
downfeed: a = 0,1 mm
work speed: v_w = 100 mm/s
coolant pressure: p_k = 12 kPa/cm²

emulsion 7%   oil

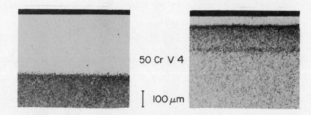

50 Cr V 4

$\quad$ 100 μm

Figure 9. Influence of coolant on thermal damage of the workpiece.

example, when grinding DMo 5 with a depth of cut of 4 mm and a work speed of 10 mm/s a structural change took place in the workpiece surface to a depth of about 2.2 mm while, by raising the work speed by 50 times and reducing the depth of cut by the same factor, with the high chip removal rate of Z' = 40 mm³/mm s, only a small hardening of the surface zone of about 50 μm depth could be found.

The influence of the coolant on damage to the workpiece surface is shown in figure 9. Using grinding oil there is a clear reduction in the zone affected by grinding. With the steel C 15 there was, for example, a reduction of this zone from a depth of about 300 μm to 100 μm.

Significantly greater structural effects were found with heat-treated work material 50 Cr V 4 whether emulsions or grinding oils were used; yet even with this steel the positive influence of grinding oil is apparent. The reduced thermal loading of the workpiece when using grinding oil is due to the higher wetting and lubricating capacity of the oil, and hence correspondingly lower temperatures in the cutting zone.

CONCLUSIONS

The following general remarks may be made about the results of the investigations.

In surface grinding, as in plunge grinding, higher wheel speeds of the grinding wheel have a positive influence on the working results obtained for cutting force, surface roughness and accuracy.

On the assumption that the product of depth of cut and work speed remains constant, cutting force, driving power and grinding wheel wear increase when greater depths of cut and lower work speeds are used, while the surface quality improves.

Thermal loading of the workpiece surface can be avoided with increasing wheel speed, assuming suitable lubricating coolant and constant chip removal rate. The danger of thermal damage arises when, with constant chip removal rate, a very low work speed and a very deep cut are chosen.

The exact determination of chip removal rate in surface grinding, as the product of depth of cut and work speed must be regarded as very important for the future so that the results of different studies may be compared.

ACKNOWLEDGEMENTS

The work reported here formed part of a programme of investigations carried out in the Laboratorium für Werkzeugmaschinen und Betriebslehre der TH-Aachen. The authors wish to thank the Arbeitsgemeinschaft Industrieller Forschungsvereingungen e.V. (AIF), the Deutsche Keramische Gesellschaft e.V. (DKG) and the Verband Deutscher Schleifmittelwerke e.V. (VDS) for their financial support.

REFERENCES

1. W. ERNST, 1965. 'Erhöhte Schnittgeschwindigkeit beim Außenrund-Einstechschleifen und ihr Einfluß auf das Schleifergebnis und die Wirtschaftlichkeit', *Dissertation*, TH Aachen.
2. K. GÜHRING, 1967. 'Hochleistungsschleifen – Eine Methode zur Leistungssteigerung der Schleifverfahren durch hohe Schnittgeschwindigkeiten', *Dissertation*, TH Aachen.
3. H. FRANK, 1953. 'Das Abrichten von Schleifscheiben mit Diamanten und der Einfluß auf das Schleifergebnis beim Außenrund-Einstechschleifen', *Dissertation*, TH Aachen.
4. F. SPERLING, 1970. 'Grundlagenuntersuchung beim Flachschleifen mit hohen Schleifscheibenumfangsgeschwindigkeiten und Zerspanleistungen', *Dissertation*, TH Aachen.
5. G. KASSEN, 1969. 'Beschreibung der elementaren Kinematik des Schleifvorganges', *Dissertation*, TH Aachen.
6. G. WERNER, 1971. 'Kinematik und Mechanik des Schleifprozesses', *Dissertation*, TH Aachen.
7. H. J. SCHREITMÜLLER, 1971. 'Kinematische Grundlagen für die praktische Anwendung des spitzenlosen Hochleistungsschleifens', *Dissertation*, TH Aachen.
8. M. A. YOUNIS, 1971. 'Zusammenhang zwischen der Kontaktsteife und den Schleifbedingungen', *Industrieanzeiger*, 93, No. 15.

GRINDING FORCE PREDICTIONS BASED ON WEAR THEORY

by

T. C. BUTTERY*

SUMMARY

The purpose of this paper is to explore the extent to which an abrasive wear approach gives a consistent and coherent picture of the grinding process. The parameters required to determine the normal grinding force will be described and the theoretical calculations compared with experiment. As a result of adopting a 'wear theory' of grinding two important parameters emerge; namely, the rate of metal removal per unit sliding distance and a 'K' factor which is an indication of the efficiency of the individual grits.

INTRODUCTION

There is increasing evidence to support the view that grinding should be regarded as an interaction between surfaces rather than a conventional cutting tool process. It would therefore seem more logical to apply wear theory rather than cutting tool theory to the grinding process.

Generally theoretical treatments of grinding have assumed it to be a cutting process; that is to say, each contact between the work and an abrasive grain produces a chip. Wear processes, on the other hand, involve contact between asperities, only a small proportion of such contacts producing a wear product. In grinding theory the rake angle of the grits has largely been disregarded but in abrasive wear theory the angle is critical. Indeed a negative rake angle is essential since it is assumed that the normal load on the abrasive grit is supported by its frontal facets. There is ample evidence in the literature[1] to support the view that most of the grits in a grinding wheel have negative rake angles.

The major inconsistences are observed with grinding when it is compared with conventional cutting tool processes, namely the ratio of normal to tangential force (grinding coefficient) and the observation reported by a number of workers and confirmed in the present investigation that the forces observed when grinding hard or soft materials are substantially the same.

The purpose of the present work was to develop an expression for the normal force in the grinding process in abrasive wear terms. Although many studies of the grinding process involving force measurements have been reported, the theoretical predictions could not be reliably checked against these as some of the parameters required had not been measured or even considered. It was therefore necessary to construct a simple grinding dynamometer in order to measure all the parameters for a complete analysis. Force measurements were made for surface grinding, experiments being confined to steels and in particular a 1% steel (DAK5) heat-treated to various conditions.

THEORY

It can be readily shown[2] that the volume wear rate per unit sliding distance V/L in terms of load (W), material hardness (H) and half-angle (θ) of the scratches formed on the abraded surface is given by an equation of the general form

$$\frac{V}{L} = \frac{C\,W\cot\theta}{H} \qquad (1)$$

where, for a process which is 100% efficient (whole of scratch volume removed), C is a function of the geometry assumed for the abrasive grits.

When allowance is made for the chip running up the front face of the grit it has been suggested that[3] a value of $C = 0.5$ is representative, giving a wear equation

$$\frac{V}{L} = \frac{0.5\,W\cot\theta}{H} \qquad (2)$$

Values of θ observed in fine grinding give width/depth ratios between 50/1 and 5/1; an average value is about 20/1.

Even when the width/depth ratio is taken into account, experiments with abrasive papers[4] have shown that predicted wear rates are greater than those observed experimentally. Two possible explanations for this discrepancy are (1) only a proportion of the grits are cutting, or alternatively (2) only a proportion of the scratch volume is actually removed. To allow for both possibilities, equation (2) must be modified, two new parameters being incorporated α the proportion of the grits actually cutting and β the proportion of the groove volume removed. Equation (2) now becomes

$$\frac{V}{L} = \frac{0.5\alpha\beta W\cot\theta}{H} \qquad (3)$$

Any discrepancies between the observed and predicted wear rates is accounted for in wear theory by a so-called K factor. The K factor has effectively been sub-divided, in abrasive wear theory, into three distinct parts.

$$K = \alpha\beta\cot\theta \qquad (4)$$

Each of the parameters has a definite physical significance and it should be possible to measure or derive a value for each in any particular process.

When applying the expressions developed from abrasive wear concepts to grinding it is assumed that the same conditions are applicable, therefore, the

*Department of Mechanical and Production Engineering, City of Leicester Polytechnic.

geometry of the groove cut is determined by the shape of the abrasive grits, the hardness of the material being ground and the applied normal force.

In order to use the wear equation (3) to predict normal grinding force the rate of metal removal per unit sliding distance must be expressed in terms of the parameters of the particular type of grinding being studied. The resulting expression is then substituted in the wear equation (3) to obtain normal force.

Consider surface grinding with a wheel depth of cut d (in), a table speed v_t(in/s) and a cross feed w (in per traverse); the volume rate of removal per second V is given by

$$V = v_t w d \qquad (5)$$

Since the surface speed of the wheel $v_w \gg v_t$ the volume wear rate per unit sliding distance is given by

$$\frac{V}{L} = \frac{v_t w d}{v_w} \qquad (6)$$

This may be combined with equation (3) to give the normal grinding force ($N = W$). Thus

$$N = \frac{2 v_t w d H}{v_w \alpha \beta \cot\theta} \qquad (7)$$

Evaluating the constant terms the normal force in lb for surface grinding is given by

$$N = \frac{10\ 843\ T d w p_m}{D r \alpha \beta \cot\theta} \qquad (8)$$

where

D = wheel diameter in in
r = wheel speed in rev/min
T = table speed in ft/min
d = wheel depth of cut in in
p_m = hardness of material being ground in kg/mm^2
w = work width in in (plunge grinding) or cross feed in in per traverse

Abrasive wear theory can be applied to other grinding processes by substituting the rate of metal removal per unit sliding distance in the wear equation. Thus the parameters required to interpret grinding in abrasive wear terms are as follows.

(i) The rate of metal removal per unit sliding distance.
(ii) The width/depth ratio of the scratches formed on the surface.
(iii) α the proportion of the grits cutting.
(iv) β the proportion of the groove volume removed.
(v) The hardness of the material being ground.

Grinding force predictions
Many studies of the grinding process involving dynamometer force measurements have been reported[5-8] and in the first instance this earlier published work was used to test theoretical predictions of normal grinding force. Unfortunately some of the parameters required for calculation were not measured or even considered by other workers and it has been necessary to assume representative values. The parameters in question

were those which make up the K-factor equation (4), namely α, β and $\cot\theta$. In the calculations a value of $\cot\theta$ typical fine of grinding was adopted ($\cot\theta = 0.1$) and α and β were assumed to be unity.

Force predictions based on Marshall and Shaw's[5] work and using equation 8 are shown in table 1. For hardened steel the force predictions are within about

TABLE 1
Normal force predictions for materials of different hardness

Hardness of specimen (VPN)	Predicted force (lb)	Observed force* (lb)
830	4.76	5.1
545	3.12	5.1
230	1.32	5.1

*Marshall and Shaw[5]

10% of the observed value. However, for the same material in the annealed condition the calculated values are only about 25% of the experimental value. It would appear that hardened steel is a unique case; the force calculations suggest that both the whole of the scratch volume is removed and every grit is cutting. A similar pattern of behaviour was shown when wear theory was applied to other experimental results reported in the literature (assuming α and $\beta = 1$). Good agreement with experimental results when fully hardened steels were being ground but considerable underestimates for forces produced when grinding soft steels. A similar effect has been observed with abrasive papers[4] experimental results showing that hardened steel is more readily abraded than would be expected relative to soft steel.

GRINDING EXPERIMENTS

Surface grinder
All the grinding experiments were carried out on a Jones and Shipman 540 Surface Grinder. The only modification to the machine was to fit a tachometer to the grinding wheel shaft to monitor wheel speed. The output from the tachometer was displayed on a Honeywell Ultraviolet Recorder.

Grinding dynamometer
General layouts of both the dynamometer and measuring circuits are shown in figures 1 and 2.

The dynamometer consists of a simple cantilever A with an integral vice B which holds the specimen. When the specimen is in place the surface to be ground is at the neutral axis of the cantilever; this arrangement was adopted to reduce twisting of the beam. The vertical and horizontal displacements of the beam under load were observed by two transducers C and D positioned at right angles to one another and locating against two glass flats. The transducers were connected to two Pendeford Multimeters whose outputs were in turn fed into a Honeywell Ultraviolet Recorder (figure 2). The dynamometer was calibrated against dead loads when in position on the surface grinder. Interference between the horizontal and vertical force measurements was

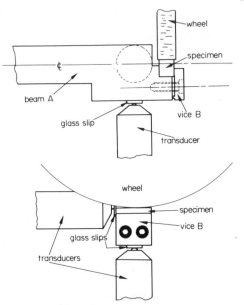

Figure 1. Dynamometer.

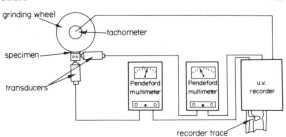

Figure 2. Dynamometer measuring circuit.

negligible. Calibration was checked at the beginning and end of each series of experiments.

Table speed was obtained by measuring the time taken to make a cut from the ultraviolet recorder trace and, since specimens of known length were used, the average table speed could be calculated.

The specimens for surface grinding were 0.375 in square and one inch long; they were made from the materials listed in table M which also gives details of the grinding wheels which were used.

Figure 3. Dynamometer passing under the grinding wheel as a cut is being made.

TABLE M
Materials

Grinding wheels		
7 in diameter aluminium oxide		
BA36 – L5 – VFBLU		
BA60 – L5 – VFBLU		
AA100 – L5 – VF8		

Steels		VPN
DAK(5)*	0.985%C steel hardened	890
DAK(5)	Hardened and tempered	710
DAK(5)	Hardened and tempered	598
DAK(5)	Hardened and tempered	441
DAK(5)	Hardened and tempered	315
DAK(5)	Hardened and tempered	250
Bright drawn mild steel		184
Annealed mild steel		112

Stellite	
Stellite 100	823

*DAK(5) provided by British Iron and Steel Research Association.

Experimental procedure

A period of half an hour was allowed for both the grinding machine and the measuring equipment to warm up. The dynamometer was calibrated at the start and finish of each series of tests. Wet grinding conditions were used for all the test and the transducers on the dynamometer were protected from the coolant. The grinding wheels were diamond dressed; a standard procedure was adopted to ensure consistent results. As a further check hardened DAK 5 specimens (p_m = 890 kg/mm²) were used as a control, force measurements being made with this material at the beginning and end of each series of experiments.

When a cut was put on the specimen both the dynamometer and the grinding machine deflected so that a number of cuts had to be taken before the down feed was equal to the material being removed. For the conditions used for most of the experiments (0.0003 in depth of cut, wheel speed 2800 rev/min and a table speed of ~ 50'/min) steady force readings were obtained after 40 passes. Force measurements were therefore taken after forty successive traverses; both the up and down cut forces were measured, a record being taken using the ultraviolet recorder. The

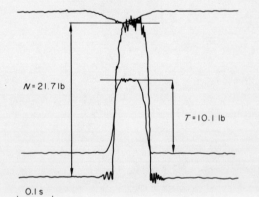

wheel speed 2780 rev/min

N = 21.7 lb

T = 10.1 lb

0.1 s

Figure 4. Typical ultraviolet recorder trace, material hardened DAK 5, wheel depth of cut 0.0003 in table speed 50 ft/min.

dynamometer is shown in action in figure 3; and a typical recorder trace in figure 4. A Talysurf trace was taken from each of the ground specimens using a ×500 horizontal magnification and an appropriate vertical magnification; the width/depth ratio of a typical scratch was then determined by sampling the results.

Another specimen of the same material which had just been ground was then fitted into the dynamometer; this specimen had a micropolished surface. The grinding wheel was clamped to prevent rotation and then brought into contact with the specimen; the wheel was lowered until the dynamometer indicated the same normal load as recorded in the grinding test. The table was moved past the wheel by hand, force measurements being taken with the ultraviolet recorder. Talysurf traces were taken and the scratches examined microscopically.

Effects of grit size and workpiece hardness

The first series of experiments was arranged to study the effects of abrasive grit size and workpiece hardness on grinding forces and grinding coefficient.

Tests were carried out using three grit sizes (36, 60 and 100 grit) and workpiece hardnesses ranging from 250 kg/mm² to 890 kg/mm² (DAK 5 steel hardened and tempered). Grinding conditions were kept constant throughout the tests (0.0003 in wheel depth of cut, ~ 50 ft/min table speed and a wheel speed of 2800 rev/min).

The complete set of results for hardened and tempered DAK 5 is shown in figure 5; normal force, tangential force and grinding coefficient are plotted

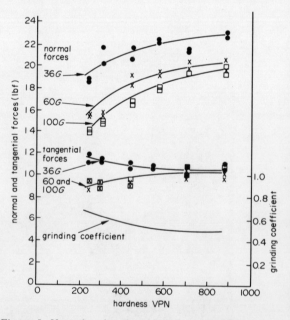

Figure 5. Normal and tangential forces grinding hardened and tempered DAK 5 steel. Conditions: wheel diameter 7 in; wheel speed 2800 rev/min; depth of cut 0.0003 in; table speed 50 ft/min.

as a function of hardness. The results show that grinding forces are only slightly dependent on hardness; reducing hardness by a factor of 3.5 reduces normal force by 15% with the 36 grit wheel, 20% with the 60 grit wheel and 30% with the 100 grit wheel. These reductions in normal force are partly associated with changes in grinding coefficient; reducing the hardness of the workpiece increases the grinding coefficient from ~ 0.5 to ~ 0.7, higher values of grinding coefficient reduce both wheel and table speed Although these parameters operate in opposite directions in equation 8 the net result is a drop in normal force.

Grit size can be seen to have some influence on normal force, the coarsest grit wheel giving the highest forces. These differences were found to be due mainly to variations in width/depth ratio for the three grinding wheels (determined from both the ground and scratched specimens as shown in table 2). Results from the two types of test show reasonable agreement and the mean width/depth ratio (although width/depth ratios are more difficult to determine from grinding than from scratching) can be seen to fall as grit size

TABLE 2
Width/depth ratios obtained from ground and scratched specimens

Wheel	Width/depth ratio		
	Grinding	Scratching	Average
36 grit	13.3:1	14.4:1	13.8:1
60 grit	13.7:1	12.8:1	13.3:1
100 grit	14.0:1	11.7:1	12.9:1

gets finer. This trend supports the experimental observations of normal grinding force; theory predicts lower forces with smaller width/depth ratios. Width/depth ratio measurements also help to explain the particularly large drop in normal force observed with the 100 grit wheel when grinding soft steel (250 kg/mm^2). The hardened DAK 5 specimen used as a control showed no change in the normal force with the 36 grit and 60 grit wheels. However, with the 100 grit wheel normal force obtained at the end of the test was 15% less than the original value. The width/depth ratio remained the same for all experiments with the 36 grit and 60 grit wheels but dropped by 12.5% with the 100 grit wheel. One would therefore expect lower normal forces from the wheel having the smallest width/depth ratio.

Force predictions based on dynamometer tests
The grinding conditions quoted in figure 5 are nominal and to calculate grinding forces the individual measurements appropriate to each experiment will be used.

Specimen calculations will be made for the grinding of fully hardened DAK 5 (p_m = 890 kg/mm^2) and hardened and tempered DAK 5 (p_m = 257 kg/mm^2), with the 60 grit wheel. In each calculation it is assumed that α and β are unity so that divergences between theoretical and experimental values should be capable of explanation by assuming other values of these parameters.

Experimental conditions
Wheel: BA60--L5--VFBLU
Wheel diameter: 6.765 in
Depth of cut: 3 x 10^{-4} in
Width of cut: 0.375 in

DAK 5 hardened (H = 890 VPN)
Wheel speed 2790 rev/min
Table speed 51.3 ft/min
Width/depth ratio (grinding) 14/1
Width/depth ratio (scratching) 14/1

Normal force measured 19.7 lb
Tangential force measured 9.7 lb
Grinding coefficient 0.49

DAK 5 soft (H = 250 VPN)
Wheel speed 2730 rev/min
Table speed 50 ft/min
Width/depth ratio (grinding) 13.8/1
Width/depth ratio (scratching) 12.2/1

Normal force measured 15.5 lb
Tangential force measured 9.0 lb
Grinding coefficient 0.58

The parameters listed above are substituted in equation 8 to give a predicted normal force in lb.

DAK 5 hard (H = 890):

$$N = \frac{10\,843 \times 51.3 \times 3 \times 10^{-4} \times 0.375 \times 890}{6.675 \times 2790 \times 1 \times 1 \times \frac{1}{7}}\,\text{lb}$$

= 20.7 lb   (experimental result 19.7 lb)

DAK soft (H = 257):

$$N = \frac{10\,843 \times 50 \times 3 \times 10^{-4} \times 0.375 \times 250}{6.765 \times 2730 \times 1 \times 1 \times \frac{1}{6.5}}\,\text{lb}$$

= 5.36 lb   (experimental result 15.5 lb)

A more complete comparison of predicted and observed forces is shown in table 3. The relationships between observed and predicted forces are very similar to those developed earlier when applying grinding theory to results avialable from the literature

TABLE 3
A comparison of observed and predicted normal forces for hard and soft DAK 5 steel using a range of abrasives

Wheel	Normal force (lb)			
	Hardened specimen 890 VPN		Soft specimen 250 VPN	
	Observed	Predicted	Observed	Predicted
36 grit	22.2	22.3	19.0	4.98
60 grit	19.7	20.7	15.5	5.36
100 grit	20.1	20.8	14.1	4.83

(namely predicted and observed forces in are good agreement for hard materials but with soft materials there is a marked divergence). As the hardness of the workpiece falls, the difference between observed and predicted force increases. With completely soft DAK 5 (p_m = 250 kg/mm^2), for example, the predicted force is only ~30% of the observed force.

Grinding of mild steel and stellite
Specimens of mild steel in the bright drawn (p_m = 184 kg/mm^2) and annealed (p_m = 112 kg/mm^2) conditions were ground to study the effects of cold working on grinding force. Using the standard conditions (0.0003

in depth of cut, 50 ft/min table speed, 60 grit wheel and plunge grinding) both the bright drawn and annealed specimens gave similar normal forces, ∿ 18 lb. These results are similar to those observed by Kruschov[4] who when abrading a range of materials on abrasive papers found that cold work did not increase wear resistance.

Experiments were carried out on Stellite as this is a material which is intrinsically hard (p_m = 823 kg/mm^2) unlike a steel which requires heat treatment. The normal grinding force was particularly high ∿ 90 lb when using the standard conditions. Grinding coefficient on the other hand was low, typically 0.19. It was noted that the grinding wheel exhibited slight glazing. The high forces were partly accounted for by the high value of the width/depth ratio 35:1. Normal force predictions were 42 lb, only about 50% of the observed value.

Effects of table speed on normal grinding force

The simplest way of achieving a large variation in metal removal rate on a grinding machine having a constant wheel speed is to vary table speed. The results of varying table speed on normal force are shown in figure 6. As would be expected from wear theory normal force shows an approximately linear variation with table speed for both hard and soft materials.

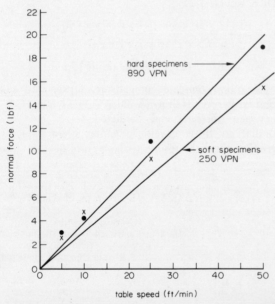

Figure 6. Variation of normal grinding force with table speed for hard and soft DAK 5 steel. Conditions: wheel diameter 7 in; wheel speed 2800 rev/min; depth of cut 0.0003 in.

DISCUSSION AND CONCLUSIONS

The dynamometer performed satisfactorily and the force measurements were in good agreement with those reported by other workers. When equation 8 was applied to the present results, force predictions were satisfactory for hard materials (assuming α and β = 1) but for soft materials were considerably in error.

This disparity between the observed and predicted forces for soft steels can be explained by reference to

the parameters making up the K factor. In the dynamometer experiments the value of the width/depth ratio for both ground and scratched specimens was substantially the same, so the validity of the Talysurf technique is not in doubt. The results also showed that in the same series of experiments the value of the width/depth ratio was the same with both hard and soft steel so this parameter cannot be responsible for the high forces observed when grinding soft steels. Adjusting either or both α and β in equation 8 to values less than unity would reconcile the practical results with theoretical predictions. Altering either parameter will considerably influence the basic cutting mechanism.

If it is assumed that only a proportion of the grits are cutting, the other merely ploughing a groove, it would imply that when only 10% were active, each grit will remove a chip ten times as large it as would have done had all the grits been cutting. The alternative explanation, that only a proportion of the groove volume is removed, appears to give a more acceptable result. For example, in surface grinding the assumption that only 10% of the groove volume is removed leads to the rather surprising conclusion that each grit will still remove a chip of the same size but this will of course be a proportion of a much larger groove. The mechanism by which a proportion of a groove can be removed must be fairly complex and probably very different from that of single-point cutting; the breakaway of the built-up edge of the groove may well be a significant factor.

It was noted that the Talysurf traces of the scratched specimens used in the dynamometer tests showed little pile-up at the edge of the groove with fully hardened steel (figure 7). A more detailed discussion of ploughing on the efficiency of individual grits

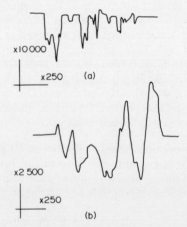

Figure 7. Talysurf traces of scratches made with a 100 grit grinding wheel during dynamometer tests. (a) Hardened DAK 5 (890 VPN) (b) B.D.M.S. (184 VPN).

has been reported elsewhere[9]. However, the precise effect of the variation of β upon the normal forces cannot be specified on the basis of present knowledge. Consider a comparison of the grinding of hardened steel and soft steel with the same grit depth of cut. For hardened steel β approaches unity and each element of material removed is involved in one individual abrasive process. Because for soft steels β is lower, each element of material is involved in a number of individual abrasive processes before removal. This can come about either by an increase in the number of active grits or by an increase in the grit depth of cut or by both of these changes in unknown proportions. Likewise these factors could affect tangential forces. However, in broad terms, it is clear that the fall of efficiency β with decreasing hardness is capable of explaining the relatively small effect of the hardness of heat treated steels upon grinding forces. Further experimental support for using a β term is provided by Stroud and Wilman[10] who, as a result of tests with abrasive papers, concluded that only a proportion of the groove volume was actually removed; a value of ~ 20% was quoted as a result of their work. Additional evidence is also provided by Grisbrook[11], who carefully determined the number of 'active grits' in the surface of a grinding wheel and then counted the number of chips produced when grinding with the same wheel. He deduced that the number of active grits was substantially the same as the number of chips formed so that each grit could be said to have removed a chip. These observations fully support the suggestion that in certain circumstances only a proportion of the groove volume is removed; if it is assumed that only a proportion of the grits are cutting there should be a corresponding reduction in the number of chips produced. Consequently one would have expected a significant difference between the number of chips and the number of grits.

Dressing techniques affect grinding forces and to obtain consistent results in grinding experiments a standard technique must be adopted. No very precise reasons have been advanced for these force variations except for the suggestion that in general terms they are a function of grit sharpness. The abrasive wear approach shows that the rate of metal removal depends critically on the width/depth ratio of the scratches formed by individual grits. If the method of dressing produces grits with a small included angle, the grinding forces will be low; grits with a very large included angle would give high cutting forces.

Grisbrook[6] has shown that grinding forces gradually change during a test; in the latter stages forces increase. Such changes are compatible with an abrasive wear treatment if it is assumed that θ gradually changes as grinding proceeds; it would be necessary to use a larger value of θ which implies the grits are getting blunter. In a process such as super-finishing the value of θ probably approaches $90°$.

When the theories presented in this paper are compared with conventional grinding theories, two facts emerge. Firstly, both rely on the accurate determination of the value of the width/depth ratio of a typical grinding scratch. Secondly, although the present theories have eliminated the controversial grit depth of cut, this has been replaced by two other factors α and β (proportion of active grits and proportion of groove volume removed respectively). In the author's opinion the factor α could be abandoned as it is to a large extent taken into account in the β term, since if a grit is not removing material (active) it will pile up a ridge. Such an inactive grit will be included in a mean value of B. As both theories are so critically dependent on the value of the width/depth ratio of a typical scratch and its geometrical relationship to the grit which produced it, it is this area which is considered to be most important for further study. In addition the nature of a surface produced by a ploughing—cutting mechanism may well have different qualities and characteristics to one formed by a pure cutting mechanism so an associated area of study would be to develop techniques capable of distinguishing such characteristics.

REFERENCES

1. A. G. WETTON, 1969. A review of theories of metal removal in grinding, *J. Mech. Eng. Sci.*, **11**, 412.
2. E. RABINOWICZ, 1965. Friction and Wear of Materials, Wiley, New York.
3. A. J. SEDRICKS and T. O. MULHEARN, 1963. Mechanics of cutting and rubbing in simulated abrasive processes, *Wear*, **6**, 457.
4. M. N. KRUSCHOV, 1957. Resistance of metals to wear by abrasion, as related to hardness, *Proc. Conf. Lubric. Wear*, 1952, (Institution of Mechanical Engineers, London), p. 655.
5. E. R. MARSHALL and M. C. SHAW, 1952. Forces in dry surface grinding, *Trans. ASME*, 51.
6. H. GRISBROOK, 1969. Precision grinding research, *Prod. Eng.*, **39**, 251.
7. P. LANDBERG, 1956. Experiments on grinding, *College International Pour l'Etude Scientifique de Techniques de Production Mechanique.*
8. R. S. HAHN, 1965. Some characteristics of controlled force grinding, *Proc. 6th M.T.D.R. Conf.*, p. 597.
9. T. C. BUTTERY and J. F. ARCHARD, 1971. Grinding and abrasive wear, *Proc. Inst. Mech. Engrs., London*, **185**, 537.
10. M. F. STROUD and H. WILMAN, 1962. The proportion of the groove volume removed as wear in the abrasion of metals, *Brit. J. Appl. Phys.*, **13**, 173.
11. H. GRISBROOK, 1962. Cutting points of the surface of a grinding wheel and the chips produced, *Proc. 3rd M.T.D.R. Conf.*, p. 155.

SIZE EFFECTS IN ABRASIVE PROCESSES

by

S. MALKIN*, K. L. WIGGINS*, M. OSMAN* and R. W. SMALLING*

SUMMARY

Size effects in abrasive processes are considered in terms of both the size of the abrasive grain and the size of the undeformed chip. On the basis of experimental results with grinding wheels and coated abrasives, it is shown that variations in specific cutting energy with abrasive grainsize and undeformed chip thickness can be attributed to the relative contributions of sliding, ploughing and chip formation energies. Experimental results are presented too for single-point cutting which show the effect of undeformed chip thickness on the extent of ploughing and on the fraction of cut material actually removed as a chip.

INTRODUCTION

Size effects which have been observed in abrasive processes fall into two categories. The first type, generally applicable to all metalcutting processes, is an increase in specific cutting energy, which is usually observed as the undeformed chip thickness is reduced. The second size effect is the decrease in metal removal rate which has been observed with coated and loose abrasives as the abrasive grain size is made smaller. It is the purpose of this paper to investigate the origin of these size effects and their inter-relationship.

A number of studies have been concerned with the effect of undeformed chip thickness on the specific cutting energy. Backer, Marshall and Shaw[1] found that the specific cutting energy became much larger as the undeformed chip thickness was decreased. The smallest undeformed chip thicknesses in their experiments were obtained by grinding. The increase in specific cutting energy for thinner chips was attributed to an increase in shear stress on the shear plane: this was thought to be due to the likelihood of finding fewer dislocations on a smaller shear plane. However, a recent analysis by Von Turkovich[2] predicts that dislocation sources at either end of the shear plane should provide a sufficient number of dislocations so that the shear stress would not increase. Experimental verification of this was achieved by Nakayama and Tamura[3] who found that, in orthogonal cutting, the shear stress did not become larger even when the undeformed chip thickness was only 2 μm (80 μin). More recently, it has been suggested by Kannappan and Malkin[4] that size effects in grinding can be attributed to the occurrence of relatively greater amounts of sliding and ploughing when the undeformed chip thickness is decreased. This latter work will be reviewed below in more detail.

Some surprising observations of the effect of undeformed chip thickness on specific cutting energy were made by Armarego and Brown[5], who found that the specific cutting energy sometimes decreased or remained virtually unchanged as the undeformed chip thickness was reduced. Their experiments were performed with single-point tools of various geometries. The specific cutting energy was calculated by dividing the force in the cutting direction by the intercepted area of the machined groove. This is correct only if all the material in the groove is removed as a chip, since the specific cutting energy is defined as the energy per unit volume of material removed. More recently, Shaw[6] has suggested that an abrasive grain can be modelled as a sphere. He presents an analysis which predicts that the force per unit area of the groove (considered to be the same as the specific cutting energy) will increase as the undeformed chip thickness is reduced. It will be seen below that, with a small undeformed chip thickness, a large portion of the groove volume is not removed. Consequently, the specific cutting energy is not equal to the cutting-force divided by the intercepted area of the groove.

There have been many investigations of size effects with coated and loose abrasives which indicate a decreasing abrasive wear rate with smaller abrasive grains. This work has been summarised by Larsen-Badse[7] and Finkin[8]. All these experiments were run as wear tests, which means that the material loss was determined for a given amount of sliding while the contact load was held constant. No measurements of the specific cutting energy were reported, so it cannot be determined whether, in addition to the decrease in abrasion rate in these experiments, there was also an increase in specific cutting energy.

The present paper is an attempt to elucidate the nature of size effects which occur in abrasive processes. As a first step, previous grinding experiments with wheels of different grain sizes are reviewed. These results are then compared with new experimental results obtained with coated abrasives. Some of the size effect phenomena are further clarified with simulated abrasive cutting experiments using a spherical single-point tool.

EXPERIMENTAL

Experiments with grinding wheels
All the grinding experiments described below were run on a straight surface grinder under plungecut conditions. For this condition, the total specific grinding energy u is given by

$$u = \frac{F_{\mathrm{H}} V}{bvd}$$

* Department of Mechanical Engineering, The University of Texas at Austin, Texas, U.S.A.

where F_H = the power force component
        V = wheel velocity
        v = workpiece velocity
        b = width of workpiece
        d = downfeed per pass.

It has recently been shown by Malkin and Cook[9] that, as the grinding wheel dulls by attritious wear, the increase in wear flat area causes a proportional increase in the grinding force components F_H and F_V, which act tangential and normal to the wheel surface, respectively. It was therefore postulated that each grinding force component can be considered as the sum of a sliding force, due to rubbing between the wear flats and the workpiece, and a cutting force

$$F_H = F_{HS} + F_{HC}$$

$$F_V = F_{VS} + F_{VC}$$

This is illustrated in figure 1. After subtracting the sliding component, the specific cutting energy u_c is given by

$$u_c = \frac{F_{HC} V}{bvd}$$

Values for u_c have been determined for a wide range of grain sizes and operating conditions[5, 9, 10].

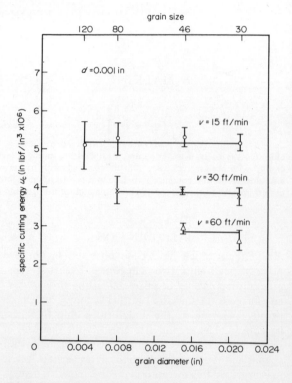

Figure 2. Specific cutting energy versus grainsize for different workpiece velocities. Grinding conditions[4]: plunge surface grinding, workpiece velocity V = 6000 ft/min, workpiece width b = 1/4 in, downfeed d = 0.001 in, 32A aluminium oxide abrasive, wheel grades G, I and K, AISI 1090 HR steel workpiece, no grinding fluid.

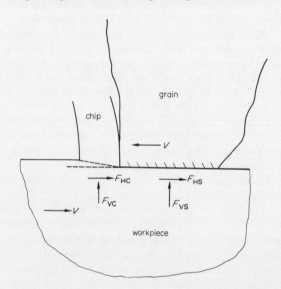

Figure 1. Illustration of cutting and sliding during grinding.

One set of results, presented in figure 2, shows that the specific cutting energy decreases with increased workpiece velocity but is independent of grainsize. From the results of these and other experiments with varying downfeed, it was concluded that a definite size effect exists whereby the specific cutting energy increases as the maximum undeformed chip thickness is reduced. In figure 2, the maximum undeformed chip thickness is made smaller by lowering the workpiece velocity. This size effect was attributed to an increase in the relative amount of ploughing which occurs as the undeformed chip thickness is decreased. It was also concluded that with finer grains, the undeformed chip thickness is unchanged. The result is that the specific cutting energy does not vary with

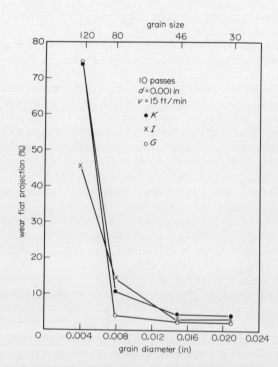

Figure 3. Wear flat projection versus grainsize for different wheel grades after 10 grinding passes. Grinding conditions[4]: plunge surface grinding, wheel velocity V = 6000 ft/min, workpiece velocity v = 15 ft/min, downfeed d = 0.001 in, 32A aluminium oxide abrasive, wheel grades G, I and K, AISI 1090 HR steel workpiece, no grinding fluid.

grainsize. More chips are produced with finer grains, but the chips are narrower.

Even though the specific cutting energy u_c was independent of grain size in the above experiments, the total specific grinding energy was u usually found to be larger with finger grains since smaller grains have a greater tendency to dull. This can be seen in figure 3 in which the wear flat projection is plotted versus grainsize for a given amount of grinding. The wear flat projection is determined by dividing the wear flat area per active grain by the projected area of a grain. Therefore, the larger specific energy usually observed with finer grainsizes in grinding can be attributed to a greater tendency for finer grains to dull and thereby contribute more sliding energy to the grinding process.

Experiments with coated abrasives

Size effects with coated abrasives were studied with a pin-on-disc wear machine. In these experiments a fixed normal load W was applied, and the tangential force component F measured with a strain-ring. For wear testing, it has been empirically found that

$$V_a = \frac{KWL}{H}$$

where V_a = wear volume
         K = wear constant
        W = normal load
         L = sliding distance
        H = hardness of wearing material.

From this equation, the wear constant is

$$K = \frac{V_a H}{WL}$$

The specific cutting energy for abrasion u_a is given by

$$u_a = \frac{FL}{V_a},$$

which is the total energy input divided by the volume of material abraded away.

Abrasive wear experiments were conducted to determine the effect of grainsize on K and u_a. The end of an AISI1018 cold rolled steel pin of 1/4 in diameter was run against a disk to which the coated abrasive paper was attached. The volume of abrasive was determined from the weight loss of the pin. Abrasive papers with aluminium oxide abrasive grains were used which included grainsizes of 36, 80, 100, 240 and 320. In these tests, the wear rate tended to decrease with increasing sliding length. This was attributed to the growth of wear flats on the grains as abrasion continued. Therefore, it was decided to determine values of K and u_a after a short sliding length before there was sufficient time for wear flats to develop. The value of u_a determined in this way neglects the sliding energy input due to rubbing between the wear flats and the workpiece, and is therefore analogous to the value of u_c determined for grinding.

In figure 4 experimental results are shown for K and u_a as a function of grainsize. As in previous experiments, a size effect was observed where the wear constant K decreased with finer grainsizes. A size effect was also observed for the specific cutting energy in abrasion u_a which increased with finer

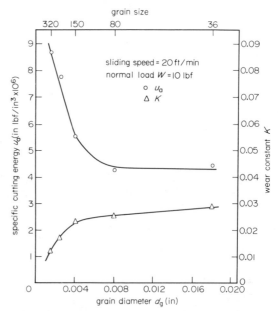

Figure 4. Specific cutting energy and wear constant versus grainsize.

grains. It is interesting to note that values of u_a with coated abrasives are comparable to those obtained with grinding wheels.

In view of the above results with grinding wheels, it was expected that size effects with coated abrasives may be due to either more dulling with finer grainsizes or to a decrease in the undeformed chip thickness associated with finger grains. In these experiments with coated abrasives, the effect of dulling was not a factor since the reported values of K and u_a were measured after a very short sliding distance, for which virtually no wear flat area could be detected on the abrasive grains. It was therefore believed that the observed size effect may have been due to a smaller undeformed chip thickness with finer grainsizes, which caused a relatively larger amount of ploughing. Another possibility was suggested by Larsen-Badse[7], who attributed the size effect with coated abrasives to increased amounts of elastic sliding during abrasion. With finer grainsizes, the relatively flexible backing of the coated abrasive allows some of the grains to come into elastic contact with the metal surface at a load per grain less than that required to cause ploughing or cutting.

In order to investigate the above ideas, a statistical analysis was made of the fluctuations in the tangential force F. It was felt that changes in the nature of contact between abrasive grains and the metal specimen should be reflected in the statistical variation of the measured force. For this analysis, the contact load was the same as in the previous experiments, but the sliding speed was greatly reduced in order to ensure a fast enough response for the force measuring system.

Force traces were recorded on a Sanborn model 321 stripchart recorder, and for each statistical analysis, 200 points were taken along the stripchart at a fixed interval corresponding to 0.008 in of sliding. The standard deviation σ was determined from

$$\sigma = [\frac{1}{N} \sum_{i=1}^{N} (F_i - \bar{F})^2]^{1/2}$$

where N = total number of readings
F_i = ith force reading
$\bar{F}$ = average of all force readings.

The relative amount of scatter in these experiments can be expressed by the coefficient of variation, defined as the ratio of σ to $\bar{F}$. Results for this ratio are presented in figure 5, where it can be seen that the

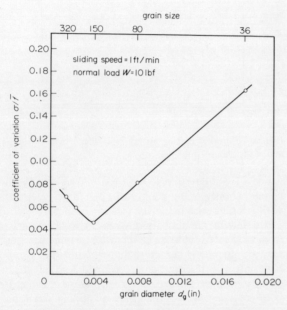

Figure 5. Coefficient of variation versus grainsize.

scatter decreased down to a grainsize of 150, and then started to increase again with finer grains. This result was somewhat surprising: with finer grains, a smoothing tendency would be expected in the data since many more grains are in contact with the metal specimen.

To eliminate the smoothing effect, the average scatter per grain can be obtained by multiplying the coefficient of variation by the square root of the number of abrasive grains in contact. Making the reasonable assumption that the number of abrasive grains in contact varies inversely with the square of the grain diameter d_g, a relative scatter factor per grain for different grain sizes can be obtained by dividing the coefficient of variation by d_g. Results for this scatter factor as a function of grain diameter are shown in figure 6, where a definite size effect can be seen analogous to those in figure 4. With grainsizes finer than 150, the scatter factor increases greatly, but it remains constant with coarser grains. With grains coarser than 150, the nature of the contact

between the abrasive grains and the metal specimen apparently does not change. With finer grains there is a transition to a different type of process which has inherently greater scatter. This is consistent with the idea that the decrease in wear constant and increase in specific cutting energy with smaller grains can be attributed to elastic contact between some of the grains and the metal surface. It can also be seen

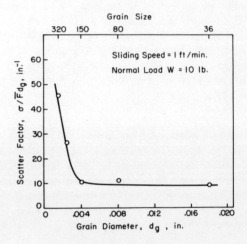

Figure 6. Scatter factor versus grainsize.

in figure 4 that, as the grain diameter approaches zero, K tends towards zero and u_a towards infinity. This would be the case of complete elastic sliding, where no abrasion occurs.

Experiments with spherical tools
In order to simulate abrasive action, experiments were conducted with single-point spherical tools (bearing balls). Grooves were cut along the surface of a block of commercially pure lead (Brinell hardness 9.2) which was held in a straingauge dynamometer. The dynamometer was mounted on the table of a milling machine, the spindle of the machine having been replaced by a stationary holder for the spherical tool. Force components F_c and F_q along and normal to the cutting direction were measured. The total length of the block along the cutting direction was 8 in. The depth-of-cut and the ball diameter were varied. The machine was not sufficiently rigid to maintain the nominal depth-of-cut, so the actual depth-of-cut was measured on completion of each cut.

During cutting, it was found that some of the groove material came off as a chip while the balance was ploughed aside along the groove. The chip appeared to separate from the ploughed material by tearing, which left a rough surface along the top of the ploughed ridges. The proportion of the groove volume actually removed was determined by weighing a chip to determine its volume V_c and dividing by the volume of the groove V_g, determined as the product of the intercepted area of the groove and the length-of-cut. The intercepted area of the groove A_g is given by[11]

$$A_g = \frac{\pi D_B^2}{4} - \left[\left(\frac{D_B}{2} - t\right)\left(D_B t - t^2\right)^{1/2}\right.$$
$$\left. + \frac{D_B^2}{4} \sin^{-1}\left(1 - \frac{2t}{D_B}\right)\right]$$

where D_B = ball diameter
        t = depth-of-cut.

Results for the fraction of the groove volume removed β as a function of cutting depth ratio t/D_B are presented in figure 7. The value of β varies

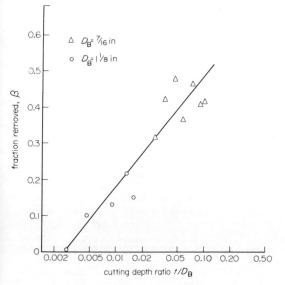

Figure 7. Fraction of material removed versus cutting depth ratio: lead workpiece, cutting velocity = 20 in/min.

from zero, when no chip is obtained, up to about 0.5 for the larger depth ratios.

For this process, the specific cutting energy can be calculated in two different ways. There is an apparent specific energy u^* for which it is assumed that all the material in the groove is removed

$$u^* = \frac{F_c L}{V_g} = \frac{F_c}{A_g}$$

where L is the length-of-cut. The actual specific cutting energy u is given by

$$u = \frac{F_c L}{\beta V_g} = \frac{F_c L}{Vc}$$

Results for u and u^* are shown in figure 8 as a function of cutting depth ratio. When the cutting energy per unit volume is based on actual chip volume, the size effect is very marked. Values of the specific cutting energy thus calculated approach infinity at very small depths because no chip is formed.

When the apparent specific cutting energy is calculated in terms of the intercepted area, the size effect is less evident. However, the specific energy is large for this material and a size effect is still observed.

Observations of subsurface deformations, made after cutting a previously sectioned workpiece along the top of the split, revealed that considerable plastic deformation occurs underneath the ball, even when ploughing only at the small depths-of-cut. Furthermore, the deformation energy for chip formation was

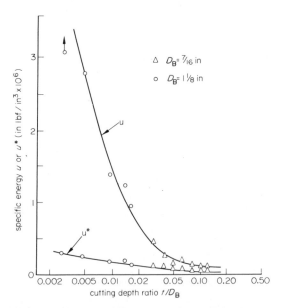

Figure 8. Specific cutting energy and apparent specific cutting energy versus cutting depth ratio: lead workpiece, cutting velocity = 20 in/min.

also very large, as evidenced by the very short chips that were generated. The cutting length was approximately 40 times the chip length for the smallest chip produced, and 10 times for the largest chip. These two factors, the proportionally larger plastic zone and greater chip deformation at smaller cutting depths, would seem to account for the large value u^* and its increase at smaller cutting depths.

DISCUSSION

The total energy in abrasive processes can be considered as the sum of sliding energy, ploughing energy and chip formation energy. Only chip formation results in material removal. Size effects in abrasion should be related to the relative amounts of sliding, ploughing and cutting that occur.

From the above results with grinding wheels and coated abrasives, it is apparent that sliding processes in abrasion are of two types. One of these is rubbing between wear flats on the abrasive grains and the workpiece surface, as illustrated in figure 1. As abrasive grains wear by attrition, the forces and energy associated with this rubbing increase in proportion to the wear flat area. This type of sliding is virtually always present in finish grinding, even at the start of grinding, since wear flats are dressed onto the abrasive grains[9]. This type of sliding also accounts for the larger grinding forces usually obtained with harder wheels, smaller abrasive grains and finer wheel dressing, because these factors tend to give larger wear flat areas[9].

Sliding of this type is not present with coated abrasives at the start of abrasion, since there is no initial wear flat area. As abrasion proceeds, however, wear flats develop and grow, and the sliding energy increases. Since wear flats grow relatively faster on finer grains than on coarser ones, the abrasion rates with finer grains decrease at a faster rate.

The second type of sliding that occurs during abrasion is due to elastic contact between the abrasive grains and the workpiece. This type of sliding, which was obtained above with the fine-grained coated abrasives, is responsible for the increase in specific cutting energy and the decrease in abrasive wear rate obtained with these fine abrasive grains. With coarser grains elastic contact does not occur, and the specific cutting energy remains constant. This agrees with the results obtained with grinding wheels which indicate, for given operating conditions, that the specific cutting energy is independent of grainsize. It is expected that elastic sliding would be a less significant factor with grinding wheels, because the grains in a vitreous bonded wheel are more stiffly held than in a coated abrasive and many grains near the wheel surface are broken out during wheel dressing.

Size effects which have been observed in abrasion can be partially attributed to the above sliding processes. However, when these sliding energy inputs are subtracted from the total energy of abrasion, the specific cutting energy which remains still increases substantially as the undeformed chip thickness decreases. This has been explained by considering the cutting energy to consists of chip formation energy and ploughing energy. This effect is clearly seen from the experiments with spherical tools. As the cutting depth is decreased, a larger fraction of the material to be removed is instead ploughed aside. It has been previously suggested that either ploughing or chip formation would occur during cutting, but not both[12]. The present experiments show that both ploughing and chip formation can occur simultaneously, with the chip separating from the ploughed ridge by a tearing action.

CONCLUSION

Size effects during abrasion can be attributed to the relative amounts of sliding, ploughing and chip formation that occur. Sliding can be classified as due either to elastic contact between the abrasive grains and the workpiece, or to rubbing between the wear flats and the workpiece. Ploughing causes metal to be displaced sideways rather than be removed as a chip. As the undeformed chip thickness is decreased, relatively more ploughing and less chip formation occur. From cutting experiments with a spherical tool, it was shown that both cutting and ploughing can occur simultaneously, and that the main contribution to the size effect is that much of the machined material is not removed as a chip.

ACKNOWLEDGMENT

The coated abrasives used in these experiments were provided by Norton Company, Coated Abrasives Division.

REFERENCES

1.   W. R. BACKER, E. R. MARSHALL and M. C. SHAW. The size effect in metalcutting. *Trans. ASME*, 74 (1952), p. 61.
2.   B. F. VON TURKOVICH. Shear stress in metalcutting. *Journal of Engineering for Industry*, *Trans. ASME*, 92 (1970), p. 151.
3.   K. NAKAYAMA and K. TAMURA. Size effect in metalcutting force. *Journal of Engineering for Industry*, *Trans. ASME*, 90 (1968), p. 119.
4.   S. KANNAPPAN and S. MALKIN. Effects of grain size and operating parameters on the mechanics of grinding. *Journal of Engineering for Industry, Trans. ASME*, 94 (1972), p. 833.
5.   E. J. A. ARMAREGO and K. H. BROWN. On the size effect in metalcutting. *International Journal of Production Research*, 1 (1962), p. 75.
6.   M. C. SHAW. Fundamentals of wear. *Annals of CIRP*, 19 (1971), p. 533.
7.   J. LARSEN-BADSE. Influence of grit size on groove formation during sliding abrasion. *Wear*, 11 (1968), p. 213.
8.   E. F. FINKIN. Abrasive wear. *Evaluation of Wear Testing*, ASTM STP 446, American Society for Testing and Materials, 1969, p. 55.
9.   S. MALKIN and N. H. COOK. The wear of grinding wheels, Part 1 – attritious wear. *Journal of Engineering for Industry, Trans. ASME*, 93 (1971), p. 1120.
10. S. OSMAN and S. MALKIN. Lubrication by grinding fluids at normal and high speed wheels. *Trans. American Society of Lubrication Engineers* (to be published).
11. *Standard mathematical tables*, Chemical Rubber Publishing Company, 12th Ed., 1959, p. 399.
12. A. J. SEDRIKS and T. O. MULHEARN. Mechanics of cutting and rubbing in simulated abrasive processes. *Wear*, 6 (1963), p. 457.

'APPARENT' RUN-OUT OF THE GRINDING WHEEL PERIPHERY AND ITS EFFECT ON SURFACE TOPOGRAPHY

by

H. KALISZER* and G. TRMAL*

INTRODUCTION

The number of variables in production processes where abrasives are applied can be almost infinitely large. An abrasive machining can be achieved by means of grains bonded together to form abrasive tools (grinding wheels, etc.) or by means of abrasive grains which are free. Although there are many types of abrasive machining, the conventional grinding with bonded abrasives (grinding wheels) plays the most important part. The diagram in figure 1 shows the

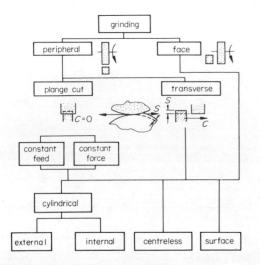

Figure 1. Application of grinding wheels in various grinding operations.

possible application of grinding wheels in various grinding operations. From those shown in the diagram *traverse* and *plunge cylindrical* grinding are most often used.

The purpose of a grinding operation is to achieve the required shape, size and surface topography of a finished product in the most efficient way and in the shortest time possible. From all the parameters resulting from a grinding process, the surface topography parameters are by far the most important, because they largely influence the wear and, therefore, the future functioning of various machining parts. The surface topography parameters and in particular the the surface roughness and waviness height is usually associated with the motion of the wheel relative to the workpiece due to the presence of vibration which may appear during grinding. These vibrations can be conveniently divided into the following three groups.

(a) Forced vibration of the machine tool strucutre or its various parts caused by the action of known excitation forces.

(b) 'Passive' vibration transmitted through foundations from other machines or resulting from random changes in the workpiece material, non-uniform wear, etc.

(c) Self-excited vibration (or chatter) generated by internal forces formed by the cutting action itself without presence of any external forces.

The object of this paper, which forms a part of a very comprehensive inter-university coordinated program in grinding (sponsored by the Science Research Council), is to analyse the wheel motion relative to the workpiece due to the wheel unbalance and its effect on surface topography. In addition an attempt is also made to establish permissible limits of wheel unbalance in conjunction with other rotating parts of the wheel—spindle assembly. This part is investigated by considering (a) the surface topography parameters, (b) the rate of smoothness in relation to the threshold of human perception and (c) the wear of bearing spindle assembly due to the presence of centrifugal forces. The graphical representation of all variables affecting the above is shown in figure 2.

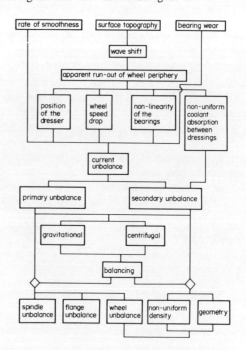

Figure 2. Variables due to wheel unbalance affecting the surface topography.

*Department of Mechanical Engineering, University of Birmingham

APPARENT RUN-OUT OF THE WHEEL PERIPHERY

In order to analyse the relative motion an assumption has been made that the workpiece holding system has no motion in relation to the machine bed and that the wheel spindle is subject to a motion in the radial plane (plane passing through the axis of the workpiece or the dressing tool and the wheel spindle) so that

$$a_S = f(\phi)$$

where a_S = displacement of the spindle in the radial plane during grinding
ϕ = angular position of the wheel in relation to its vertical axis.

In general as a result of the wheel motion the dressing tool will generate a non-circular shape on the wheel periphery. The form of this periphery in any given cross-section of the wheel can be described by a

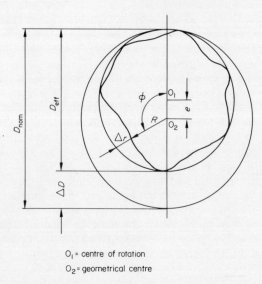

O_1 = centre of rotation

O_2 = geometrical centre

Figure 3. Deviation from a true cylindrical cross-section of a ground workpiece.

variable radius R from a fixed geometrical centre O_2 (figure 3). If the work-holding system has no motion then

$$a_d = \Delta r = R - r = f(\phi) \qquad (1)$$

where a_d = motion of the wheel in the radial plane during the dressing action
Δr = error of form in the wheel periphery
r = nominal radius of the wheel periphery.

The expression $f(\phi)$ represents a periodic function with a period $T = 2\pi$.

Such function can be described by a Fourier series

$$r = f(\phi) = \frac{a_0}{2} + \sum_{n=1}^{n=\infty} a_n \cos(n\phi + \phi_n) \qquad (2)$$

where a_0 and a_n = amplitudes
ϕ and ϕ_n = phase angles.

As shown previously[1] the term $a_0/2$ represents the deviation of size (ΔD) due, for example, to an error in axial location of the dressing tool.

Higher harmonics will describe waviness and surface roughness (figure 3), i.e.

$$\Delta r = a_n \cdot \cos(n\phi + \phi_n) \qquad (3)$$

The apparent diametral run-out of the wheel periphery in relation to the axis of the workpiece (assumed to have no motion in the radial plane) depends upon the difference in the wheel motion during dressing and grinding and can be described as follows:

$$X = a_g - \Delta r$$

a_g = motion of the wheel in the radial plane during the actual grinding.

If a perfectly balanced wheel is dressed and there is no motion of the wheel spindle, a true cylindrical shape of the wheel periphery will be formed. When grinding with such a wheel, there will be no relative motion between the wheel and the workpiece. Hence practically no waviness will be generated on the workpiece periphery (figure 4(a)).

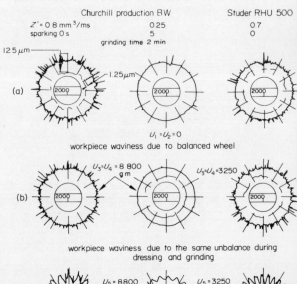

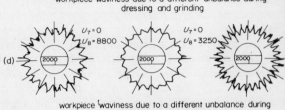

Figure 4. Workpiece waviness due to unbalance during dressing and grinding.

As shown previously[1] the difference in the wheel motion is most pronounced when the state of un-balance during dressing is different from the state of unbalance during grinding (figure 4(c), 4(d)). In the figure U_1, U_3, U_5, U_7 denote a primary unbalance (U_p) at the end of the dressing operation and

U_2, U_4, U_6, U_8 a current unbalance (U_c) present during grinding. (The current unbalance represents the vectorial sum of primary unbalance and secondary unbalance (U_s) developed during grinding.)

If the unbalance remains the same during dressing and grinding and the dresser is mounted on the table or tailstock, the sinusoidal shape of the wheel periphery generated by the dresser will be compensated by the periodic motion of the wheel, and the apparent run-out will approach zero (figure 4(b)).

If the current unbalance develops during grinding, the apparent run-out can be determined as follows:

$$X_c = 2\frac{U_c}{g}\omega^2 \, (J_S + J_F) \qquad (4)$$

where ω = rotational speed

J_S, J_F = compliances of the spindle assembly and—the infeed system correspondingly (figure 5).

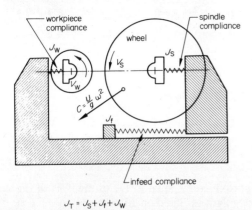

Figure 5. Compliances of a grinding machine system.

In the case when the dresser is fixed on the wheelhead and the state of unbalance remains unchanged during dressing and grinding the run-out of the wheel periphery depends only upon the compliance of the spindle assembly and can be calculated as follows:

$$X_{p\alpha} = 2\frac{U_c}{g}\omega^2(J_S + J_F)\frac{J_F}{J_S + J_F} \qquad (5)$$

where

$$\alpha = \frac{J_F}{J_S + J_F}$$

then

$$X_{p\alpha} = \alpha X_c \qquad (5a)$$

An apparent run-out may also develop owing to the difference in the magnitude of forces developed during dressing and grinding. Because of this difference two conditions may arise, i.e. a speed drop caused by the motor load and an increase in the spindle rigidity due to some nonlinearity in the bearing characteristics. Both these conditions can be described by a coefficient β, which is normally very small. For the new universal grinder (wheel diameter D_S = 300 mm) which we investigated for example[2] β = 0.02.

In general

$$X_{p\beta} = \beta X_C \qquad (5b)$$

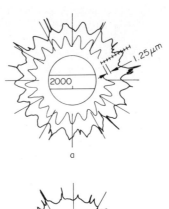

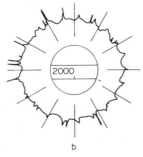

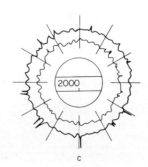

workpiece waviness as a function of wave shift (0.13π; 0.33π; 1.1π)

(a)

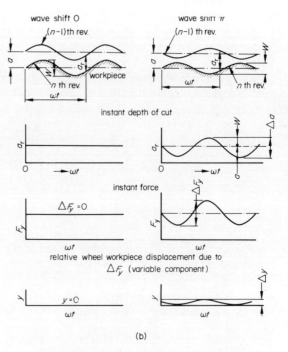

(b)

Figure 6. (a) Workpiece waviness as a function of a wave shift (b) Variation of grinding parameters due to a wave shift.

For (larger) older machines where the bearing play may be relatively large, the coefficient β may be much larger.

As explained in detail earlier[1] the amplitude of workpiece waviness generated by the apparent run-out condition depends to a large extent upon the superposition of the generated waves during subsequent revolutions of the workpiece. The effect of superposition of waviness can be best explained by considering two extreme conditions. One condition is when the number of waves forms an integral number, and the generated waves will reinforce each other at subsequent revolutions (wave shift $\phi = 0$). In this situation the maximum waviness generated on the workpiece periphery cannot exceed the amount of apparent run-out. The second condition will take place when the number of waves is not an integer. This will result in a wave shift and a variation in the depth of cut during one revolution of the wheel. The above will become clear by comparing the polarographs shown in figure 6(a). All the various waveforms shown in the figure were generated by the same wheel unbalance, all other conditions remaining constant. As can be seen the above effect is most pronounced when the wave shift $\phi = \pi$. For such a wave shift the instantaneous depth of cut (a_r) will vary as shown in figure 6(b). This will produce a variable force increment ΔF_y and consequently a variation in the deflection of the system (Δ_y). All this will lead to a reduction of the waviness height (W) generated on the workpiece periphery.

As a result

$$W_\pi = X - \Delta_y$$

where $\Delta y = \Delta F_y \times J_T$ (J_T = total grinding machine compliance.

It can be shown that for a wheel width (B) and workspeed (V_w)

$$W_\pi = \frac{X}{1 + (2\pi V_w B/\Lambda)J_T} = \frac{X}{\epsilon} \qquad (6)$$

As can be seen the coefficient ϵ depends upon the total compliance (J_T) of the grinding machine system,

and the ratio between the grinding force and the depth of cut expressed in terms of the metal removal parameter[3] (Λ).

SELECTION OF LIMITS FOR WHEEL UNBALANCE

To establish the limits of unbalance from the technological point of view a ground surface profile was recorded axially and circumferentially (figure 7). An analysis of the graphs shows that waviness cannot be effectively measured if its double amplitude is

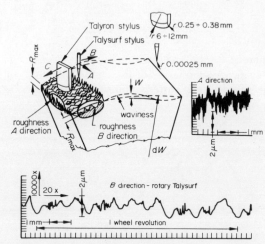

Figure 7. Surface topography of a ground workpiece.

approximately equal to 1/7 of the PVH of the surface irregularities. Based on this assumption and considering that R_a from 0.3 μm to 0.4 μm should form a permissible limit for conventional grinding operations an apparent run-out of approximately 0.3 μm has been established as an acceptable technological basis for determining the grinding wheel unbalance. On the basis of the assumed run-out, the permissible unbalance according to equations (4), (5a) and (5b) is given in table 1.

Besides the effect of the wheel unbalance upon the surface topography it is also necessary to consider the threshold of human perception. According to VDI such threshold corresponds to a wheel unbalance of

TABLE 1
Permissible unbalance (g mm)

Type of the grinding machine and size of the grinding wheel	Wheel speed (m/s)	Secondary	Primary unbalance	
		Dresser on the table		Dresser on the wheel head
		$U_s = \dfrac{X_g}{2\,W^2\,(J_S + J_F)}$	$U_{p\beta} = \dfrac{1}{\beta}U_s$	$U_{p\alpha} = \dfrac{1}{\alpha}U_s$
Production cylindrical grinder wheel diameter 500 mm	30	610		2350
	60	152		590
Small surface grinder wheel diameter 175 mm	30	17.4		33
	60	4.3		8.2
Universal cylindrical grinder wheel diameter 300 mm	30	92	4600	315
	60	23	1150	79

250 g on a universal grinder with a wheel diameter $D_S = 300$ mm. Such unbalance has practically no effect upon bearing wear.

With the aim of determining the permissible limits for wheel unbalance it is necessary to analyse four sources of unbalance in the wheel–spindle assembly, i.e. (a) the non-uniform distribution of grinding wheel mass; (b) the eccentric mounting of the wheel in flanges; (c) mounting accuracy due to the run-out of the spindle nose; (d) the unbalance of the flange–spindle assembly excluding the wheel.

Wheel unbalance

The permissible amount of wheel unbalance depends upon the commercial limits of unbalance set by the grinding wheel manufacturers or national standards. The change of unbalance due to wheel wear can be determined by considering figure 8 where it was

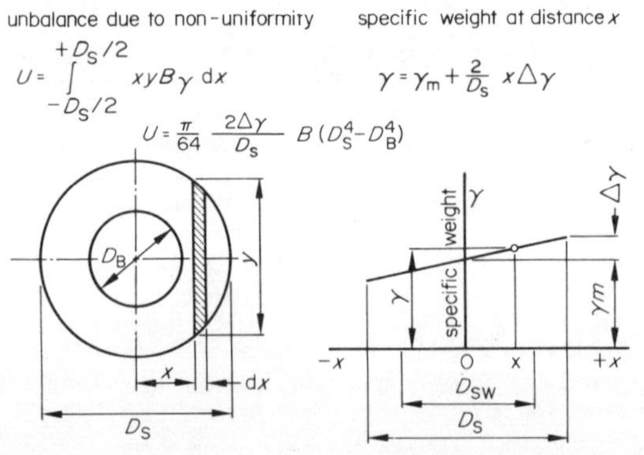

unbalance due to non-uniformity

$$U = \int_{-D_S/2}^{+D_S/2} xyB\gamma\, dx$$

specific weight at distance x

$$\gamma = \gamma_m + \frac{2}{D_S} x \Delta\gamma$$

$$U = \frac{\pi}{64} \frac{2\Delta\gamma}{D_S} B (D_S^4 - D_B^4)$$

Figure 8. Non-uniform distribution of mass in grinding wheels.

assumed that a non-uniform distribution of mass is present in the body of the grinding wheel.

As can be seen from the figure by considering an elementary unbalance

$$dU = Xy dx B$$

the unbalance for a new wheel can be determined as follows:

$$U = B\gamma_m \int_{-D/2}^{+D/2} xy\, dx + \frac{2\Delta\gamma}{D} B \int_{-D/2}^{+D/2} x^2 y\, dx \tag{7}$$

The solution of the last equation gives the following expression:

for a new wheel $U_N = \dfrac{2\Delta\gamma}{D_S} B \dfrac{\pi}{64} (D_S^4 - D_B^4)$ (7a)

for a worn-out wheel $U_W = \dfrac{2\Delta\gamma}{D_S} B \dfrac{\pi}{64} (D_{SW}^4 - D_B^4)$ (7b)

where D_S = diameter of a new wheel
D_{SW} = diameter of a worn wheel
D_B = diameter of the wheel hole

By considering the last two equations the change of unbalance for a worn-out wheel

$$\Delta U = U_N \frac{D_S^4 - D_{SW}^4}{D_S^4 - D_B^4} \tag{7c}$$

From equation (7a), the mass non-uniformity is given by

$$\Delta\gamma = U_N \frac{32 D_S}{\pi B (D_S^4 - D_B^4)} \tag{8}$$

and finally the relative non-uniformity is

$$S = \frac{\Delta\gamma}{\gamma_m} 100 = \frac{32 D_S}{\pi B \gamma_m} \frac{U}{D_S^4 - D_B^4} 100 \tag{8a}$$

If at the end of a wheel life its diameter is reduced from $D_S = 300$ mm to $D_{SW} = 190$ mm, the permissible wheel unbalance will be reduced from 1530 g mm (DIN standard) to 205 g. For the same wheel $S = 0.71\%$, which corresponds roughly to $\pm\frac{1}{2}$ hardness grade.

Eccentric mounting of the wheel

According to the existing safety regulations there should be a clearance between the grinding wheel hole and the mounting spigot of the flanges. This clearance must always be present to prevent the spigot expanding owing to grinding and bearing heat.

If a recommended fit f8/H12 is considered then for a hole diameter $D_B = 127$ mm the maximum diametral clearance will amount to 0.506 mm.

Owing to this clearance the resulting unbalance can be calculated by considering the eccentricity (e) and the wheel weight (W_S)

$$Ue = eW_S = \frac{0.506}{2} 5100 = 1300 \text{ g mm} \tag{9}$$

After the first dressing operation the outer diameter becomes concentric and the unbalance will result from the eccentricity of the hole, i.e.

$$U_{eb} = eW_S \frac{D_B^2}{D_S^2 - D_B^2} = 286 \text{ gm} \tag{9a}$$

Mounting accuracy

By considering the recommended maximum permissible run-out of a grinding wheel spindle which amount to 0.01 mm the resulting unbalance can be determined by considering the spindle eccentricity (e_s) and flange weight (W_F)

$$U = e_S (W_S + W_F) = 51 \text{ gm} \tag{10}$$

The unbalance due to the flange–spindle assembly does not need to be considered, since a permissible state of balance can be obtained by the grinding machine manufacturer during the production stages.

By considering equations (7a), (7b), (9), (9a) and (10) and by assuming that an average accuracy of conventional gravitational balancing is approximately 200 gm it is possible to predict the effect of wheel unbalance at various stages of wheel life. Such a comparison is given in table 2.

CONCLUSIONS

On the basis of the results given in tables 1 and 2 the following conclusions can be made.

(1) From the technological (surface topography) point of view, the balancing of grinding wheel is not absolutely essential.

TABLE 2

Wheel 300 x 127 x 40 (mm)	Max. unbalance before first dressing (g mm)	After first dressing (g mm)	End of wheel life (min. wheel diameter) (g mm)
Wheel balanced after fitting on the flange	250	1850	3150
Wheel balanced after fitting on the flange and again after first dressing	250	250	1550
Norton method	2880	1550	330
No balancing	2880	1850	550

(2) To achieve a high rate of smoothness several balancing procedures will be required during the wheel life.

(3) In the case when the dresser is mounted on the wheelhead, to achieve satisfactory technological results the unbalance must be kept within the same limit as for high rate of smoothness.

In general it may also be stated that with reasonable care in maintaining the balancing equipment the accuracy of gravitational balancing[4] can be considered satisfactory for wheel speed of 30 m/s. In the case of high-speed grinding the gravitational method would not provide the required accuracy. The disadvantage of gravitational method includes the balancing time and the necessity to remove the wheel assembly from the machine.

ACKNOWLEDGEMENTS

The authors wish to thank Professor S. A. Tobias for providing facilities and the Science Research Council for the financial support.

REFERENCES

1. H. KALISZER and G. TRMAL, 1972. 'Force vibration during plunge grinding and its effect on surface topography', *Intern. Grinding Conf., Pittsburgh.*
2. H. KALISZER and G. TRMAL, 1972. 'Some aspects of unbalance in the wheel–spindle assembly of cylindrical grinding machines', *C.I.R.P. 22 Gen. Assemb., Sweden.*
3. R. P. LINDSAY and R. S. HAHN, 1970. 'On the basic relationship between grinding parameters', *Ann. C.I.R.P.,* 18.
4. H. KALISZER, 1963. 'Accuracy of balancing grinding wheels by using gravitational and centrifugal methods', *Proc. 4th M.T.D.R. Conf., Manchester.*

CORRELATION ANALYSIS OF THE STRUCTURE OF A GROUND SURFACE

by

T. R. THOMAS*

SUMMARY

Autocorrelation functions of profiles of a ground surface are measured in a number of directions at different angles to the lay, using a correlating computer on-line to a stylus instrument. The correlation length is found to vary in a systematic way with orientation and the average length and breadth of a grinding scratch can be deduced. Measurements on three ground surfaces yield ratios of scratch length to breadth between 5 and 7, compared with values of between 20 and 30 calculated from existing theories.

NOTATION

b = mean breadth of grinding scratch
C = number of cutting edges per unit area of grinding wheel
d = mean diameter of abrasive particle
D = grinding wheel diameter
ℓ = mean length of grinding scratch
L = profile length
m = separation of pairs of correlated points
r_m = autocorrelation coefficient at separation m
v = work speed
V = peripheral wheel speed
x = distance along profile

INTRODUCTION

The continuing effort to improve the efficiency of grinding processes has led to recent studies of the topography of ground surfaces[1,2] being carried out in the hope of relating surface finish to properties and movements of the grinding wheel. A ground surface is the resultant of a very large number of individual scratches, each caused by the impact of a single grain of abrasive. It would be helpful to be able to obtain quantitative information on the average shape and size of grinding scratches, but this is not easy because of their repeated random superposition. Attempts have been made[2] to measure individual scratches on a surface sufficiently lightly ground for superposition not to occur, but it is not clear whether the results can be generalised to apply to more heavily ground surfaces.

Attention has recently been given to the description of surfaces as random processes. Several geometrical properties of a surface profile have been shown to depend on the autocorrelation function of the profile[3]. This suggests the possibility of using correlation techniques to investigate the variation of finish with orientation on anisotropic surfaces. Kubo and Peklenik[4] have applied this method to spark-eroded surfaces with some success. By making a digital recording on paper tape of the signal from a stylus drawn over the surface and processing the recording off-line on a digital computer they were able to show clearly the existence of anisotropy in the surface texture.

* Mechanical Engineering Department, Teesside Polytechnic.

This method is suitable for a weakly anisotropic surface where a small number of measurements will suffice, but the labour and computational time involved is rather excessive for the detailed scrutiny of a surface with strongly directional properties. Recently, however, compact hybrid computers have become available which are capable of the quick calculation of many statistical properties of a varying signal while connected on-line. The present paper describes the application of an on-line correlating computer to the investigation of a ground surface, with special reference to the dimensions of an average grinding scratch.

THEORY

The finish of a surface can be regarded as the result of a large number of small discrete similar random events[5]. The *central limit theorem* predicts that a process of this kind should yield a Gaussian distribution of surface heights, and experiments[3,5] have confirmed that many surfaces do in fact conform to this distribution. A profile of such a surface thus reveals a random structure with a Gaussian amplitude distribution. This satisfies the definition of a Gaussian random process, and attempts[3] to describe surface profiles in terms of random process theory have met with considerable success.

Consider a surface profile whose height at any distance x along the profile is $f(x)$. Pairs of points on the surface whose separation is less than the average size of a unit event will tend to be more or less highly correlated; this simply means to say that there is a good chance that they will occur on the same peak or valley. Pairs of points whose separation is large compared with a unit event, however, are unlikely to belong to the same peak or valley and so will tend to be poorly correlated. The degree of correlation between pairs of points at any separation m is expressed by the autocorrelation coefficient

$$r_m = \frac{1}{L} \int^L f(x)f(x+m)\,\mathrm{d}x$$

where L = length of the profile.

The variation of r_m with separation m is called the

autocorrelation function of the profile. It is found that the autocorrelation functions of the profiles of many surfaces decrease progressively with m from an initial high value to a value showing no significant correlation. The length taken for the function to fall to a certain fraction of its initial value is called the correlation length.

In the case of a ground surface the unit event responsible for its production is the scratch made by a single particle of abrasive. There is general agreement[2,6] that a typical grinding scratch is needle-shaped, long and narrow. As a consequence of the above argument one would expect to find a high correlation length in directions parallel to or nearly parallel to the lay, falling off to a much smaller value at right angles to the lay. The magnitudes of these two extreme values should be related to the average length and breadth of a grinding scratch.

EXPERIMENT

The ground finishes investigated were produced by reciprocating horizontal surface grinding. Three specimens were prepared in the form of mild steel discs of 2 cm nominal diameter. Three vitrified,

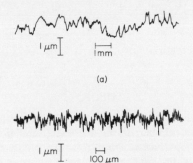

(a)

(b)

Figure 1. Profiles of ground specimen II (a) parallel to lay; (b) across lay. Profiles of the other ground surfaces were similar.

TABLE 1
Surface parameters

Surface	I	II	III
Grinding wheel	46K8	60K8	100M7
d (μm)	550	420	250
Roughness along lay (μm)	0.52	0.26	0.62
Roughness across lay (μm)	0.65	0.30	0.66
ℓ, measured (μm)	240	252	268
ℓ, equation (2) (μm)	870	940	1060
b, measured (μm)	48	34	42
b, equation (3) (μm)	46	43	38

aluminium oxide wheels of 20 cm (8 in) diameter (D) and different grain sizes were dressed and each wheel was used to grind a different specimen (table 1, figure 1). Wheel speed (V) was 52 m/s (10 300 ft/min) and work speed (v) was 0.3 m/s (60 ft/min). A fourth surface was introduced as a control. This had been grit-blasted to a roughness of 6.5 μm and was believed to possess no directional properties. Each disc in turn was mounted on a ruling table, capable of positioning to one minute of arc, on

the bed of a stylus surface measuring instrument (Talysurf 3, Rank Precision Industries). The output from the stylus transducer was conditioned to remove the 3 kHz carrier and connected on-line to a correlating computer (Hewlett-Packard 3721A).

The correlating computer delayed the signal in a large number of equal increments, averaged the product of each set of delayed impulses over a selected period and stored the averages in registers. Each averaged product was an autocorrelation coefficient; when the contents of all the registers were displayed simultaneously on an oscilloscope screen the resulting picture was the autocorrelation function (figure 2). As

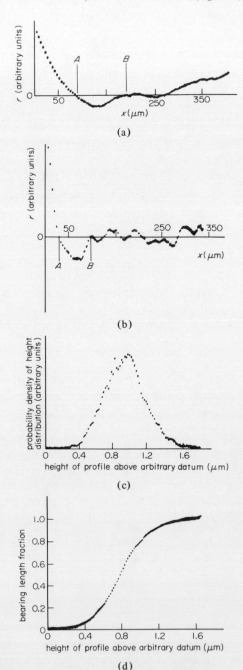

Figure 2. Oscillograms computed from the profiles of Figure 1 for ground specimen II. (a), (b) Autocorrelation functions parallel to lay and across lay respectively, to same horizontal scale. A = correlation length as plotted in figure 3; B = second intercept. (c) Height distribution across lay. (d) Bearing length across lay.

the Talysurf gearbox traversed the stylus across the surface at a constant speed of about 1 mm/s, a time delay of say 10 ms on the oscilloscope screen was equivalent to a 10 μm separation of points on the surface. The correlator would also compute and display height density distributions and bearing area curves (figure 2).

It was assumed that the shape of a grinding scratch possessed symmetry about two axes at right angles[2], in which case the angular rotation necessary to measure the scratch completely could be reduced to a 90° sector starting parallel to the lay and finishing at right angles to the lay, or vice versa. In practice this was extended to up to 120° where necessary to emphasise peaks in the results. The ruling table was rotated in increments of 5°, or less in a region where measurements were varying rapidly. At each angular setting the stylus was traversed over its full range of about 8 mm at its fastest speed, taking about 8 s. The computer was started after the beginning of the traverse and stopped before the end to avoid spurious signals from the limit switches of the gearbox. One or two trial runs were needed initially to select the most appropriate timebase. The autocorrelation function took shape on the oscilloscope screen almost immediately and settled down to its final form after three or four seconds. The display could then be frozen indefinitely for subsequent inspection or recording. For convenience the correlation length was taken as the distance (time) required for the autocorrelation function to decay to zero.

RESULTS AND DISCUSSION

The above definition of the correlation length was not entirely satisfactory, for two reasons. Firstly, the autocorrelation function sometimes approached zero asymptotically. This effect was responsible for some of the unexpectedly large values found in figure 3. The second reason is more interesting. The autocorrelation function of an ideally random surface should decay progressively to zero and remain there. In general, however, the functions obtained on the ground surfaces went steeply through zero and reached a pronounced negative trough before climbing back to zero at roughly twice the correlation length. It is suggested that the initial positive region of the function represents points on the same downward (or upward) slope; the subsequent negative region represents points on the adjacent upward (downward) slope. On an ideally random surface adjacent slopes would be of unrelated lengths and points on them would not be correlated; on a ground surface, however, the unit event is a scratch whose sides are generally of equal length, leading to twin correlation lengths of similar magnitude but opposite sign. On this hypothesis a more appropriate measure of the average dimension of a scratch in a given direction would be twice the value of the first intercept of the autocorrelation function on the x axis. In practice the first and second intercepts generally occurred at roughly equal intervals, but the second intercepts were not always unambiguous. For the sake of consistency, therefore, values of the first intercept

have been plotted in figure 3 as correlation lengths The mean scratch length ℓ for each surface has been taken as twice the highest correlation length for that surface, and similarly the mean scratch breadth b has been taken as twice the lowest correlation length (table 1).

The behaviour of the three ground surfaces was basically similar (figure 3). A low value of the correlation length across the lay increased more or

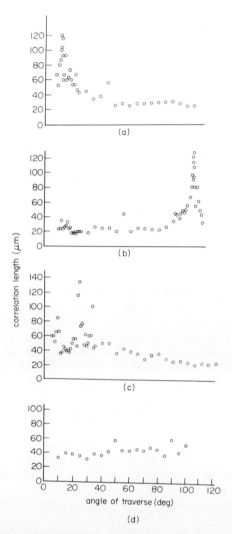

Figure 3. Variation of correlation length with angle of traverse measured from an arbitrary datum. (a) Ground surface I. (b) Ground surface II. (c) Ground surface III. (d) Grit-blasted surface.

less slowly until a peak was reached parallel to the lay. This peak was rather ill-defined for Specimen III but quite pronounced on specimen II. On this latter specimen an angular shift in the region of the peak of as little as 10 minutes of arc produced a measurable change in the correlation length. By contrast the grit-blasted surface showed no large variations in correlation length, though there seemed to be a small but progressive increase with rotation.

The ratio of scratch length to breadth is given by[2]

$$\ell/b = (D/d)^{\frac{1}{2}} \qquad (1)$$

This leads to ratios of between 20 and 30 for the present surfaces, compared with measured ratios of between 5 and 7. In view of this large discrepancy it seemed worth while to attempt calculations of scratch length and breadth individually.

Again according to Nakayama and Shaw[2],

$$\ell^2 = \frac{2v}{VC} \left(\frac{D}{d}\right)^{\frac{1}{2}} \qquad (2)$$

and

$$b^2 = \frac{2v}{VC} \left(\frac{d}{D}\right)^{\frac{1}{2}} \qquad (3)$$

C is the number of cutting edges per unit area of the grinding wheel. Rather large uncertainties are involved in its determination, but its value is not very critical as ℓ and b do not vary very rapidly with it. Taking an approximate figure of 30 cm^{-2} (200 in^{-2}) for a 60H wheel[3] and using it for all three wheels, values of ℓ and b were calculated from equations (2) and (3). The calculated values for the average breadth were in good agreement with measured values (table 1), but the calculated and measured lengths differed by a factor of about 4. This disagreement is rather difficult to explain, as it seems unlikely that the present technique could have missed peaks in the correlation lengths four times as high as those actually found.

CONCLUSIONS

Correlation analysis is a powerful technique for investigating surface finish, and the correlating computer connected on-line to a stylus instrument has shown itself to be a convenient and versatile tool. Its application to the measurement of a ground surface has yielded values for the mean breadth of a grinding scratch which are in good agreement with an existing theory. In view of the number of assumptions made in the application of the theory the extent of the agreement is probably fortuitous. The large discrepancy between calculated and measured scratch lengths, however, is rather puzzling. In the absence of further evidence it seems possible that grinding scratches may be much shorter than had hitherto been supposed, though the reason for this is by no means clear.

ACKNOWLEDGEMENTS

The advice and help with instrumentation of Mr. J. Coulthard of the Department of Electrical Engineering is gratefully acknowledged.

REFERENCES

1. D. A. FARMER, J. N. BRECKER and M. C. SHAW, 1967–8. 'Study of the finish produced in surface grinding, Part 1', *Proc. Inst. Mech. Engrs. (London)*, **182,** Part 3K, 171.
2. K. NAKAYAMA and M. C. SHAW, 1967–8. 'Study of the finish produced in surface grinding, Part 2', *Proc. Inst. Mech. Engrs. (London)*, **182,** Part 3K, 179.
3. D. J. WHITEHOUSE and J. F. ARCHARD, 1970. 'The properties of random surfaces of significance in their contact', *Proc. Roy. Soc., Ser. A*, **316,** 97.
4. M. KUBO and J. PEKLENIK, 1968. 'An analysis of micro-geometrical isotropy for random surface structures', *Ann. C.I.R.P.*, **16,** p. 235.
5. J. B. P. WILLIAMSON, J. PULLEN and R. T. HUNT, 1970. 'The shape of solid surfaces', in *Surface Mechanics*, A.S.M.E., New York, p. 24.
6. J. DYSON and W. HIRST, 1954. 'The true area of contact between solids', *Proc. Phys. Soc.*, **67,** 309.

INFLUENCE OF THE ABRASIVE GRAIN ON THE SURFACE INTEGRITY OF HIGH SPEED STEEL

by

NATHAN P. NAVARRO*

SUMMARY

The surface integrity of high-speed steels ground with aluminium oxide and cubic boron nitride abrasive is examined. Experimental tests demonstrate that the particular type of abrasive used in grinding is one of the most important factors influencing the surface integrity of high-speed steels. Even under moderate feedrate conditions, aluminium oxide wheels create disturbed material zones which differ in microstructure and hardness from the parent metal. There is significantly less tendency to disrupt the workpiece when using cubic boron nitride abrasive under similar grinding conditions. The results of these metallurgical studies are substantiated by drill-life studies where drills pointed with CBN abrasive exhibited better tool-life than drills similarly ground with aluminium oxide.

INTRODUCTION

It is generally accepted that grinding is a metal removal operation which may seriously disturb the workpiece surface. In the grinding of high-speed steel tool materials, where maintaining the hot hardness properties is particularly important, any change in tool surface properties as a result of such a disturbance may strongly influence the performance of the tool itself. The extent of this damage can depend on the workpiece being ground, the feedrate conditions and the length of time between wheel dressing. Most often, the damage can be predicted by a discoloration of the ground surface. However, in some instances, considerable workpiece damage occurs even though the surface may appear to be virtually burn free.

A new General Electric Company abrasive, Borazon (cubic boron nitride), has been found particularly effective on high-speed steels, in that relatively little wheelwear occurs when grinding these hardened steels.

The study was conducted to evaluate the influence of these two abrasives on the properties of the workpiece when used under identical grinding conditions.

EXPERIMENTAL PROCEDURE

Three typical high-speed steels were chosen as test materials, two molybdenum types (M-2 and M-42) and one tungsten type (T15). The hardness and principal alloy elements of these steels and two other steels used in this study are shown below.

The specimens measured 1/4 x 3/4 x 1 in long (6.35 x 19.65 x 25.4 mm). Each specimen was mounted into a fixture which was similar to an inserted tooth milling head capable of holding 16 inserts. This was done to provide a reasonable heatsink for the specimen.

The 1/4 in (h) x 3/4 in (b) dimension was ground under the following conditions:

wheel speed:	20.3 m/sec (4000 s ft/min)
table speed:	1.83 m/min (6 ft/min)
infeed per pass:	0.076 mm (0.003 in) on M-2 and M-42
	0.025 mm (0.001 in) at 15 passes on T-15
	0.051 mm (0.002 in) at 5 passes on T-15
time between grinds:	5 s

The wheels used were Type II flaring cup wheels, and their specifications were:

abrasive	wheel description
aluminium oxide wheel	4/3 − −A60J5− white aluminium oxide, vitrified bond, most commonly used for toolroom grinding.
cubic boron nitride	80/100 CBN Type II*, 25 volume per cent crystal content, 3−3/4 in OD. 1/8 in rim THK, resin bond. (CBN Type II is a metalcoated cubic boron nitride crystal. The metalcoated crystals are used in resin bonds.)

The aluminium oxide wheel was freshly dressed with a boron carbide stick before each change in test conditions. The CBN Type II wheel was not dressed during the entire test period. The test procedure involved varying the grinding time (number of passes) between dresses with the aluminium oxide wheel on M-2 and M-42.

On T-15 steel, which is more difficult to grind, the infeeds were reduced to 0.001 and 0.002 in per pass, and the number of grinding passes changed to 15 and 5 passes, respectively.

material	hardness	C	Cr	V	Mo	W	Co	
M-2	R_c 66	0.85	4.00	2.00	5.00	6.00	−	
M-42	R_c 66	1.10	3.75	1.15	9.50	1.50	8.00	
T-15	R_c 67	1.50	4.00	5.00	−	12.00	5.00	
M-10	R_c 64	0.85	4.00	2.00	8.00	−	−	
4340	BHN 341	0.40	0.80	−	0.25	−	−	(1.8 Ni)

*General Electric Company, PO Box 568, Worthington, Ohio 43085, USA.

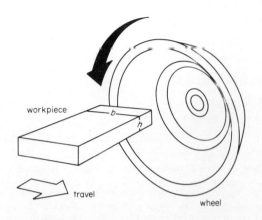

Figure 1. Schematic of test arrangement.

After the specimens were ground, the surface appearance was noted, and a side perpendicular to the ground surface was polished and nital etched for examination.

Following the examination of the area perpendicular to the ground surface, the ground surface was also etched for 20 min in a 50% hydrochloric acid and water solution heated to 140°F. This surface was then examined for cracks that may have been caused by grinding.

RESULTS AND DISCUSSION

Surface appearance and depth of damage
The appearance of each ground surface was recorded in order to establish whether a correlation existed between surface appearance and sub-surface workpiece damage. Based on the discolorations of the ground surfaces, one would have predicted that the aluminium oxide wheels had caused sub-surface damage to the M-42 and T-15 workpieces, while the CBN abrasive wheel would have caused similar damage to the M-2 specimens.

After polishing and nital etching the test pieces, changes in microstructure could be observed on certain specimens as a white layer directly below the

ground area. This white layer or rehardened martensite was readily measured from magnified photographs. The surface discoloration, along with any measured change in microstructure, is shown in figure 2.

The examination of the microstructure was rather surprising in that our predictions, based on surface discoloration, would not have been totally accurate. Indeed, those surfaces that were significantly discoloured when ground with aluminium oxide did show varying degrees of sub-surface damage. However, those workpieces that were ground with aluminium oxide, where only a slight straw-like surface discoloration occurred, also contained a significant layer of disturbed material.

More surprising were the results on the workpieces ground with cubic boron nitride abrasive, where no change in microstructure was observed on those workpieces that appeared severely burnt. It would seem that the discoloration caused by the CBN wheel is quite superficial.

Why this discoloration of workpiece when using CBN abrasive wheels does not correlate with depth-of-damage to the workpiece is presently unexplainable. However, the phenomenon had been observed in some of our previous work, and these studies seem to confirm earlier suspicions that when using CBN abrasive in resin bonds, discoloration of the workpiece is rarely (if ever) indicative of sub-surface damage.

Microstructure of sub-surface layers
In figures 3, 4 and 5, photographs of the etched surfaces perpendicular to the ground areas show the extent of damage caused by grinding. The white layers observed show that sufficient heat was generated in grinding to cause rehardening of the material. In order to reharden these steels, the temperatures at the surface must have exceeded 1700°F[1].

In addition to the rehardened layers, a transition zone of overtempered material is sometimes visible in the photographs since it is darkened by the etchant.

Abrasive	Material	Infeed (in)	No. of grinding passes	Surface appearance (degree of burn)	Depth of damage (in)
ALOX	M-2	0.003	5	light	0.0003
CBN	M-2	0.003	5	light	none
ALOX	M-2	0.003	15	light	0.0017
CBN	M-2	0.003	15	medium	none
ALOX	M-2	0.003	35	light	0.0020
CBN	M-2	0.003	35	dark	none
ALOX	M-42	0.003	5	medium	0.0020
CBN	M-42	0.003	5	light	none
ALOX	M-42	0.003	15	medium	0.0040
CBN	M-42	0.003	15	light	none
ALOX	T-15	0.001	15	dark	0.0010
CBN	T-15	0.001	15	light	none
ALOX	T-15	0.002	5	dark	0.0020
CBN	T-15	0.002	5	light	0.0001

Figure 2. Tabulation of grinding conditions, surface appearance and depth of damage observed on the ground workpieces.

This overtempered material is caused by heating the material above the original tempering temperatures. Both the rehardened and overtempered material are normally softer than the parent metal. Also, the rehardened layers are extremely brittle and catastrophic failure of the cutting-edge easily occurs when attempting to use such a tool.

In figure 3, it is seen that by stopping the grinding after five passes, no damage was done to the M-2 specimen ground with CBN abrasive, while a thin layer of rehardened martensite approximately 0.003 in

material: M-2, R$_c$66

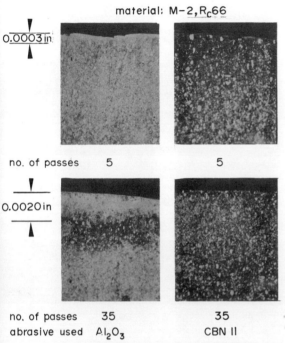

no. of passes   5   5

no. of passes   35   35
abrasive used   Al$_2$O$_3$ CBN II

Figure 3. Effect of the abrasive and increased grinding time on the sub-surface microstructure of M-2 steel.

material : M 42, R$_c$66

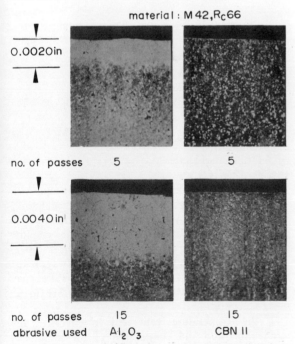

no. of passes   5   5

no. of passes   15   15
abrasive used   Al$_2$O$_3$ CBN II

Figure 4. Effect of the abrasive and increased grinding time on the sub-surface microstructure of M-42 steel.

material: T-15, R$_c$67

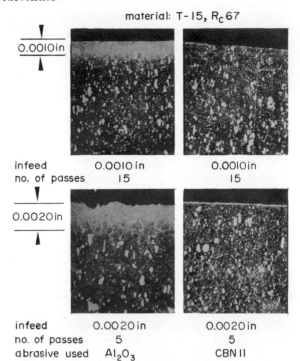

infeed   0.0010 in   0.0010 in
no. of passes   15   15

infeed   0.0020 in   0.0020 in
no. of passes   5   5
abrasive used   Al$_2$O$_3$ CBN II

Figure 5. Effect of the abrasive and depth-of-cut on the sub-surface microstructure of T-15 steel.

is visible on the specimen ground with aluminium oxide. After the aluminium oxide wheel was dressed and 35 passes taken before the grinding was stopped, the damage to the sub-surface material increased to 0.002 in. The damage consisted of 0.001 in depth of rehardened martensite and an additional 0.001 in layer of overtempered material. It is apparent that, by not stopping frequently to dress the aluminium oxide wheel, the depth of damage will increae substantially.

Although the surface of the M-2 specimen ground for 35 passes with CBN abrasive was appreciably discoloured, the microstructure was virtually unchanged, indicating that the increased grinding-time has little effect when using this abrasive.

In figure 4, the effect of the abrasive and grinding time on M-42 steel is similar to that observed on M-2 steel. However, on M-42, which is more difficult to grind than M-2, the depth of damage when using the aluminium oxide wheel is much greater. Therefore, when grinding M-42 with aluminium oxide, much more care must be taken to avoid damaging the workpiece. Reasonable precautions would be able to reduce the depth-of-cut and more frequent wheel dressing.

The tungsten type high-speed steel T-15 is the most difficult to grind high-speed steel. As seen in figure 5, even at 0.001 in depth of cut, damage to the workpiece occurs when using aluminium oxide wheels. When increasing the infeed to 0.002 in with the aluminium oxide wheel, the depth of damage increased, even though the grinding time was reduced. When closely examining the photographs of the T-15 steel ground with aluminium oxide at 0.001 in and 0.002 in depths-of-cut, it can be seen that not only have sub-surface layers been rehardened but that considerable grain growth has occurred. Such significant grain growth and distortion indicates that the surface of the steel was plastic during grinding[2].

The difficulty in grinding T-15 steel is probably best evidenced by the fact that at 0.002 in depth-of-cut, a very thin (about 0.001 in thick) layer of rehardened martensite is visible even when using the CBN abrasive wheel. This was the only material where, when using CBN abrasive, any degree of microstructure change could be observed.

Hardness measurements

Knoop hardness tests were conducted on each of the specimens and the results converted to the Rockwell C scale.

As is witnessed by the tabulated results in figure 6, a significant reduction in hardness is noted on the materials ground with aluminium oxide. This reduction in hardness extends through the rehardened and overtempered layers, and the degree of softening is a function of grinding time and depth-of-cut.

On those specimens ground with CBN abrasive, some very slight change in hardness is noted very close to the surface. This is indicative of retempering, which occurs even when steels are heated to temperatures as low as 200°F. This slight reduction in hardness caused by grinding with CBN should not significantly affect tool life.

resultant photograph shown in figure 7 is typical of the cracks found on the specimens ground with

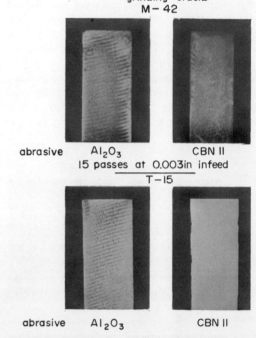

grinding cracks
M − 42

abrasive   Al₂O₃   CBN II
15 passes at 0.003in infeed
T−15

abrasive   Al₂O₃   CBN II
15 passes at 0.0010in infeed

Figure 7. Cracks resulting from grinding with aluminium oxide abrasive.

Abrasive	Infeed (in)	No. of passes	Hardness (R_C) Depth below surface (in)					
			0.0005	0.0010	0.0015	0.0020	0.0030	Parent metal
A. MATERIAL M-2								
ALOX	0.003	5	55	58	61	63	65	66
CBN	0.003	5	63	64	66	66	66	66
ALOX	0.003	15	55	62	64	63	65	66
CBN	0.003	15	63	64	65	66	66	66
ALOX	0.003	35	55	60	64	65	65	66
CBN	0.003	35	64	64	66	66	66	66
B. MATERIAL M-42								
ALOX	0.003	5	62	65	66	64	65	66
CBN	0.003	5	66	66	66	66	66	66
ALOX	0.003	15	56	60	65	66	66	66
CBN	0.003	15	66	67	66	66	66	66
C. MATERIAL T-15								
ALOX	0.001	15	54	64	67	67	67	67
CBN	0.001	15	64	66	67	67	67	67
ALOX	0.002	5	54	58	62	65	66	67
CBN	0.002	5	64	66	67	67	67	67

Figure 6. Hardness measurements on sub-surface layers (converted from Knoop hardness tests, 500 g load).

All the materials having changes in microstructure as a result of grinding had sub-surface layers that were softer than the parent metal.

However, for a cutting tool, significant reductions in hardness, as those seen on the specimens ground with aluminium oxide, would render it useless. Unfortunately, the discoloration that occurs in grinding is easily removed with a light cutting pass and too often grinding damage is hidden in this manner.

Surface cracks

The ground specimens were hot-etched and the ground surfaces examined for grinding cracks. The

aluminium oxide. None of the specimens ground with CBN abrasive showed evidence of surface cracking.

When excessive temperatures are generated in grinding, the surface near the grind area rapidly expands, and then contracts as the wheel moves away and heat is conducted or converted away. If the temperatures are not too high, residual stresses may be set up which, at best, leave the surface in an unstable condition.

If the temperatures generated are excessive, the rapid expansion and contraction of the metal can cause surface cracking. Except in extreme cases, these very fine cracks are normally invisible even with the aid of a microscope. They can act as stress points and under load cause a cutting tool to chip or break.

Drill life

Since it is apparent that the abrasive used can effect the metallurgical properties of high-speed steels, an investigation was conducted as to the influence of the abrasive on tool life.

In this study, a prominent US drill manufacturer produced sets of both M-10 and T-15, 1/4 in diameter drills. The drills were manufactured in a conventional manner except that some were pointed with an aluminium oxide wheel and others with a CBN Type II abrasive wheel.

In order to simulate actual regrinding practice, these sets of drills were further subdivided: some were pointed with the use of a grinding fluid while others were pointed dry. The grinding conditions were established by the manufacturer so that grinding times with both types of abrasive wheel were

considerable care must be taken to prevent tool damage.

Another interesting aspect of this investigation is to compare the performance of the drill materials. Since T-15 steel is much more expensive than M-10, one would expect more holes to be produced with drills made of it. The results with the M-10 drills ground with aluminium oxide appear equal (if not superior) to those obtained with T-15. Based on this data, it would be concluded that, for drilling 4340 steel, the cheaper M-10 steel drills are more economical.

When comparing the number of holes obtained with M-10 and T-15 drills that had been pointed with CBN abrasive, the conclusion is that the use of T-15 material would be economically justifiable. Based on these results, it is apparent that in order to justify the purchase of the more expensive T-15 drills, it is necessary to ensure that they are ground with CBN or at least very carefully ground with aluminium oxide abrasive.

Drills: 1/4 in dia., Jobbers length, M-10 and T-15 HSS

point angle 115°
helix angle: 29°
clearance: 12°
point: plain

Feed: 0.005 in/rev
Cutting speed: 60 s ft/min (920 rev/min)

Depth of hole: 1/2 in through

Cutting fluid: chlorinated oil
Tool life end point: 0.015 in wear

| | | Number of holes | |
| | | drill material | |
Abrasive	Grinding fluid	M-10	T-15
Aluminium oxide	NO	150[a]	120
CBN type II	NO	219	247
Aluminium oxide	YES	220	208
CBN type II	YES	203	350[b]

(a) considerable test scatter: number of holes per drill ranged from 5 to 300
(b) interpolated from data at higher speeds.

Figure 8. Drill-life test results showing influence of abrasive on drill-life.

equivalent. The drills were then sent to an independent machining research centre where they were evaluated by drilling AISI-4340 steel. The criterion for determining the drill-life end point was a 0.015 in wear land on the drill corners.

The drills were run at various speeds, and figure 8 shows the results at 60 s ft/min (920 rpm).

The results of the drill-life test confirm earlier suspicions that the abrasive has a considerable influence on tool-life. When analysing the results with the tough-to-grind T-15 drills, the results clearly indicate that many more holes can be produced with these drills pointed with CBN abrasive. On the M-10 drills, the first inclination may be to state that the abrasive did not influence drill-life, but the considerable data scatter observed when using the M-10 drills that were pointed dry with aluminium oxide indicates that, even on this relatively easy-to-grind material,

CONCLUSION

The results of the surface integrity studies, together with the drill-life studies, substantiate that the abrasive plays a most important role in grinding high-speed steels. What is readily apparent is that unless considerable care is used when grinding, aluminium oxide wheels can seriously damage high-speed steels, with subsequent detrimental effects on tool life.

When using cubic boron nitride abrasive wheels, there is less tendency to damage the workpiece even under conditions that would be considered 'abusive' with aluminium oxide.

REFERENCES

1. Latrobe Steel Company. Metallurgical factors affecting the service life of tool steels, *Tech Topics, Bulletin 108.*
2. P. TECHIE-EWING, 1965. Improper grinding can change the surface properties of HSS tools, *Cutting Tool Engineering,* June.

ELECTROCHEMICAL GRINDING
AND
ELECTRO-DISCHARGE MACHINING

PERIPHERAL ELECTROCHEMICAL GRINDING WITH A FORMED WHEEL

by

A. GEDDAM* and C. F. NOBLE*

SUMMARY

An important application of electrochemical grinding is component production using a formed wheel. The paper presents an introduction to features peculiar to a peripheral configuration and discusses some experimental results obtained when machining a trapezoidal slot in Nimonic 105 using $NaNO_2$ aqueous electrolyte solution and a Norelek formable wheel. The results indicate the most desirable operating conditions for rough forming by single-pass plunge grinding.

A method is illustrated for presenting results in a form suitable for industrial use and is shown to have fundamental significance.

NOTATION

a	=	feedrate
A	=	atomic rate
d_s	=	set depth of cut
F	=	the Faraday
I	=	current
J	=	current density
J_0	=	current density corresponding to the leading edge
k	=	electrolyte conductivity
m	=	mass of material removed
r	=	wheel radius
t	=	time
u	=	a constant, $\epsilon/\rho_m F$
V	=	applied voltage
ΔV	=	overvoltage
y	=	working gap
y_0	=	gap corresponding to the leading edge
z	=	valency
ϵ	=	chemical equivalent
θ	=	angle between the feed direction and the normal to workpiece surface
θ_0	=	angle between the feed direction and the normal to workpiece surface corresponding to the leading edge
ρ_m	=	workpiece material density

INTRODUCTION

Electrochemical grinding (ECG) is basically an electrochemical process as the bulk of the material is removed by electrolytic dissolution. The grinding machine is more-or-less of conventional form but employs an abrasive bearing grinding wheel with an electrically conducting bond and is provided with electrolyte solution instead of coolant. The purpose of the non-conducting abrasive grit is to maintain a gap between the wheel and workpiece, and to remove any oxide film which might impede electrolysis. The rotation of the wheel combined with the continuous flow of electrolyte is responsible for the removal of reaction products from and renewal of electrolyte into the machining zone.

Peripheral electrochemical grinding is a particular application of ECG, and has the same configuration as in conventional grinding. The working gap formed will be smaller at the leading edge than at the trailing edge and the dimensional variation is a time-dependent factor[1,2]. When using a formed wheel the shape imparted in the workpiece resembles that of the wheel but separated by a working gap. The gap formed may be assumed to be constant at any one section but as in plain peripheral grinding it increases from the leading edge to the trailing edge and, therefore, the same process considerations may be presumed applicable.

Peripheral electrochemical grinding with a formed wheel (electrochemical form grinding) has been viewed as a potential manufacturing technique for component production. Low wheel breakdown during ECG and the ability to hold wheel form for long periods of time make it an ideal process for form grinding, whereas wheel loading in conventional form grinding causes rapid wheel breakdown necessitating frequent dressing in order to maintain accurate profiles. The early publications only refer to forming cutting tools[3,4] although more recent reports mention a wide range of profiles in 'difficult-to-machine' materials[5,6].

The general process information available so far in peripheral electrochemical grinding[1,2,7] has largely been under electrochemical removal conditions and any mechanical abrasive action has been limited to the removal of any oxide layer. For reproducing a shaped configuration it is important to combine removal under both electrochemical and mechanical conditions, the mechanical proportion being varied to accomplish the desired tolerances. There is, as yet, very little information on the operating characteristics of combined removal.

It is difficult to discern the exact boundary between purely electrochemical removal and the commencement of mechanical contribution. However, the machining variables can be adjusted so as to accomplish the desired removal condition[8], since the extent of grit participation, which will determine the mechanical proportion, can be increased or decreased by varying applied voltage, set depth of cut and feed rate. The experimental work carried out has been aimed at assessing the influence of the set operating variables and to arrive at desirable machining conditions for single-pass plunge grinding with particular reference to a shaped configuration. The 6 in diameter, $\frac{1}{2}$ in wide wheel used is a formable

* The University of Manchester Institute of Science and Technology

type (Norelek) and was form dressed to bear a trapezoidal shape having 60° included angle and 0.2 in peripheral width. Nimonic 105 was machined with 10% (by weight) aqueous electrolyte solution of sodium nitrite. Other operating variables, such as wheel size and speed, relative direction of table feed, workpiece material, type of electrolyte solution, have been kept constant throughout the investigation.

The experimental equipment[9] consisting of a modified grinding machine[1,2] a d.c. power unit, an electrolyte supply system with a filtering unit, an exhaust gas extraction unit, a form dressing unit and instrumentation, is shown in figure 1.

The machining performance has been evaluated by measuring (i) current developed; (ii) metal removal

Figure 1. Experimental equipment.

rate; (iii) shape reproduction under the following operation conditions: (a) applied voltage: 5, 10 and 15 V; (b) feed rate: 0.5, 1 to 5 in/min at intervals of 1 in/min; (c) set depth of cut: increased from 0.010 in by steps of 0.010 in until failure occurred.

METAL REMOVAL RATE

Faraday's laws may be used to calculate a theoretical removal rate of both pure metals and alloys if the valency of the ions produced is known and if it is assumed that the whole of the current is used for metal removal. However, in practice, an exact estimation is difficult because it has not so far been clear how the composition of individual work materials affects the stock removal. It is also not certain how much of the current passed is actually used to remove metal and how much is used by side reactions and gas evolution. As most of the engineering materials are alloys consisting of many elements, some

of which have the possibility of multiple valency state with which they can react electrochemically, it is, therefore, difficult to calculate the theoretical current required to effect stock removal. In electrochemical grinding, mechanical abrasive action of the wheel could significantly alter the machining characteristics: (i) by effectively removing any passivating oxide film that may have formed; (ii) by effecting some stock removal by abrasive grit action.

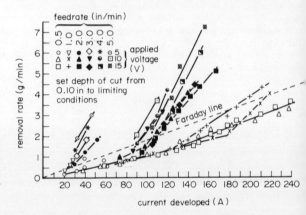

Figure 2. Removal rate versus current.

The experimental data shown in figure 2 signify the influence of the operating parameters in determining the removal conditions. It is apparent that substantial stock removal far in excess of the predicted Faraday removal is possible if the operating parameters are stretched to their limits. The Faraday line shown represents predicted values based on 'percentage' chemical composition of the workpiece material assuming 100% current efficiency. The nature and extent of deviation of the actual removal from that predicted by Faraday's laws could have been influenced by factors both electrochemical and mechanical in nature. The significant features appear to be (i) the removal for varying set depth of cut at particular feed rates and applied voltages lies on the same straight line suggesting that the electrochemical nature of the process is unaffected by the set depth of cut; (ii) the mechanical contribution (the term used to effect removal in excess of the predicted Faraday values) is dependent on applied voltage and feedrate and can be increased significantly by increasing set depth of cut.

The following table represents total removal as a percentage of Faradaic removal, i.e. 100 represents removal according to Faraday's laws, below and above 100 represent electrochemical removal at low efficiency and removal due to mechanical contribution respectively. The 'minimum' values correspond to 0.010 in set depth of cut (lowest used in the experiments) and the 'maximum' to those at limiting operating conditions.

With reference to the 'maximum' values, i.e. limiting operating conditions, the mechanical contribution as a percentage of predicted Faradaic removal is of the order of 30 to 250% for 5 V, 0 to 90% for 10 V, and 0 to 150% for 15 V (the values in brackets indicate mechanical removal as a percentage of total removal). These values appear to be far in excess of those

Applied voltage	Feedrate (in/min)					
	0.5	1.0	2.0	3.0	4.0	5.0
	Material removal (% Faraday, in parentheses % total)					
5 min. max.	56 131 (24)	80 170 (41)	122 194 (49)	181 243 (58)	230 316 (68)	320 356 (72)
10 min. max.	53 77	53 104 (4)	73 161 (38)	80.5 153 (35)	101 194 (48)	126 188 (47)
15 min. max.	53 85	53 114 (12)	74 170 (41)	85 167 (40)	98 218 (54)	126 254 (60)

reported in face grinding, e.g. Hughes and Notter[10] quote 75 to 95% electrochemical removal for carbides, and Cole[11] quotes 99.5% electrochemical removal for tool steel, but they make no reference to limiting conditions.

FARADAIC REMOVAL CONDITIONS

The Faradaic operating conditions are defined here as those set variable operating conditions, viz. set

depth of cut and feedrate at a particular applied voltage, that can yield metal removal according to predicted Faraday values. These conditions can be found by relating current output and removal rate according to Faraday's laws as shown in figure 3. The intersecting points with actual removal curves represent the Faradaic removal conditions of feedrate and set depth of cut at constant voltage and are shown separately in the adjacent figures which also

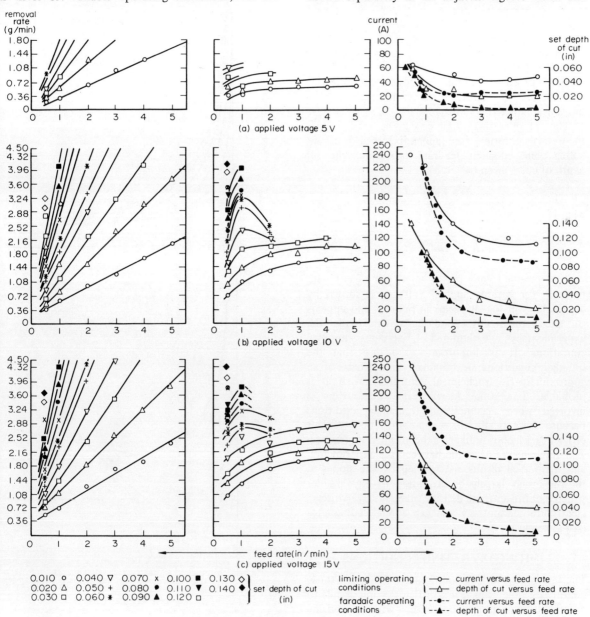

Figure 3. Determination of Faradaic removal conditions.

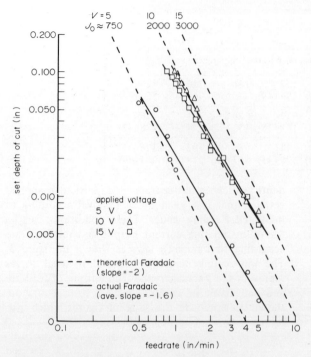

Figure 4. Assessment of Faradaic operating parameters.

or mechanical in nature — failure by short-circuiting (predominantly at high applied voltages) and failure by wheel seizure (mainly at low applied voltages). Each form of failure must be avoided for safe and trouble-free operation as the consequent wheel/workpiece damage is detrimental to both wheel life and component accuracy.

If the machining is continued beyond the Faradaic level, the proportion of mechanical contribution will increase. A reduction in applied voltage, increase in feedrate and increase in set depth of cut individually or together will demand more and more mechanical abrasive action resulting in severe pressure between the wheel and workpiece. This could cause premature breakdown of the grit, sparking and arcing finally culminating in short-circuiting or wheel seizure or both. For constant voltage operation, the limiting conditions are maximum operating feedrate at a specified depth of cut or maximum set depth of cut that allows a particular operating feedrate. Under

show the practical limiting conditions. It is interesting to note that the nature of curves for Faradaic and limiting conditions is similar.

The set operating variables can be related by a theoretical criterion (see appendix). For constant voltage and constant feedrate operation, the set depth of cut is given by

$$ds = Ka^{-2}$$

where

$$K = \frac{r}{2}(uJ_0)^2$$

The expected variation of set depth of cut with feedrate for various applied voltages is shown by dotted lines in log–log scale in figure 4 and appears to show reasonably good agreement with the experimentally determined Faradaic removal conditions. It is important to realise that the Faradaic conditions are obtained on the basis of an approximate electrochemical equivalent. Also, it is probable that some mechanical contribution is involved when accomplishing Faradaic conditions because theoretically these represent 100% current efficiency which is unlikely to be attained in practice. The apparent net deviation between the experimental and theoretical results, as signified by the slopes of the lines, may largely be attributed to the variables which control current density, namely, overvoltage, electrolyte conductivity and gap at the leading edge (see appendix).

LIMITING OPERATING CONDITIONS

The limiting operating conditions are defined as those conditions under which it is difficult or impossible to extend the machining operations without causing imminent failure. This failure could be either electrical

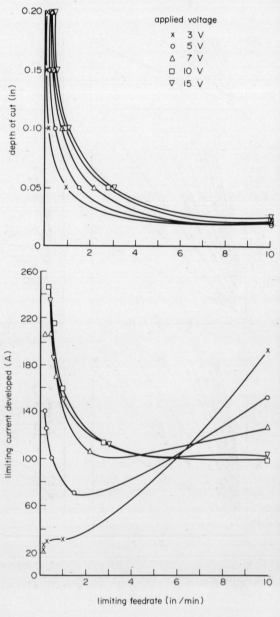

Figure 5. Determination of limiting operating conditions.

these conditions, the current developed will be the limiting current and the stock removal will be maximum. Figure 5 represents the variables concerned covering applied voltages from 3 to 15 V, depths of cut up to 0.200 in and feedrates up to 10 in/min.

The limiting operating conditions may be represented on a log–log scale together with the theoretical Faradaic conditions as shown in figure 6 where the dotted lines indicate the theoretical Faradaic values and the solid lines denote the

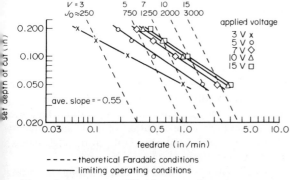

Figure 6. Assessment of limiting operating parameters.

limiting conditions. The effective deviation as represented by the slope of −0.63 compared with −2 of the theoretical Faradaic lines indicate the extent of mechanical contribution. It is worth noting that the slopes for different operating voltages are similar in spite of the fact that varying degrees of mechanical to electrochemical proportions of the removal are

presumed to have been involved. This method of illustrating the results offers a particularly useful approach for industrial needs.

SHAPE REPRODUCTION

The accuracy of shape reproduction in the present experimental work has been assessed by measuring form included angle, form corner radius and form oversize and comparing with those of the wheel. The features of particular concern are those relating to the Faradaic and limiting conditions which are shown in figure 7. The wheel profile deviations in terms of enlargement of included angle and rounding-off of profile corners are also indicated at specific intervals. (The sequence of experimentation is as indicated and the wheel profile was recorded after completing the tests at a particular feedrate.) This method, incidentally, provides a rough guide as to the nature of cumulative wheel profile deviation and the form-holding capability of the wheel. Re-form dressing after each test run would have provided only an absolute measure instead of a relative assessment.

The results of figure 7 support the following deductions.

(i) Form angle reproduction shows better accuracy at limiting conditions than at the Faradaic conditions because larger depths of cut are associated with the limiting conditions. The fact that the angle reproduced approaches the actual (or nominal) value with increasing set depth of cut suggests that the deeper the machining area the less it is affected by stray

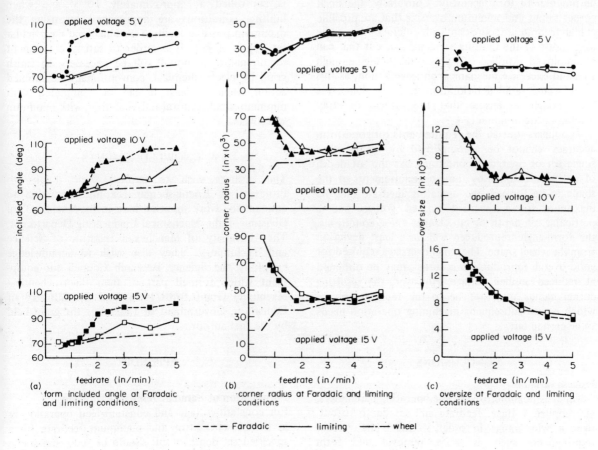

(a) form included angle at Faradaic and limiting conditions.

(b) corner radius at Faradaic and limiting conditions

(c) oversize at Faradaic and limiting conditions

- - - - Faradaic   ——— limiting   —·—· wheel

Figure 7. Assessment of shape reproduction.

machining near the component surface. This also suggests that a small depth of cut and a high voltage results in enlarged included angle.

(ii) Corner radius appears to show similar variation under both Faradaic and limiting conditions. Low feedrates seem to result in large corner radius and this is particularly noticeable with increasing voltage. This suggests that the main source of error is stray machining which will have its greatest influence at highest applied voltage and lowest feedrate. In terms of reproduction accuracy, the higher the feedrate and lower the voltage the better will be the radius reproduced.

(iii) Form oversize − the final gap separating the shape produced in the workpiece and the wheel shape − represents the overall reproduction characteristics and general tolerances. As expected from fundamental principles, the results show an increasing oversize with applied voltage and hyperbolic variation with feedrate. The results corresponding to the Faradaic and limiting operating conditions are remarkably similar in spite of the fact that substantial mechanical removal is involved at limiting conditions. As the best reproduction accuracy is associated with the smallest oversize, this can be achieved by reducing applied voltage and increasing feedrate.

In practice the aspects of form accuracy must be accounted together. The results reported above appear to show diverse trends for each item. For instance, while large set depths of cut are desirable for good angle reproduction the slow feedrates associated result in large corner radius and oversize which are detrimental to form accuracy. Conversely, the small corner radius and minimum oversize that are possible at high feedrate are coincident with enlarged form-angle as a result of the limiting set depths of cut that can be employed. Applied voltage should be high enough to avoid excessive mechanical abrasive action, as this would cause rapid deterioration in wheel shape, yet low enough to ensure that the effects of stray machining are minimum.

It appears that all the requirements for good form accuracy cannot be accomplished by means of a single set of operating conditions. As the set depth of cut is determined by the size requirements of the shape to be formed, the operating feedrate may be maximum for a specified applied voltage and a particular set depth of cut. Under these conditions, the form-angle reproduced will have least deviation from the wheel shape. The small oversizes required for good overall reproduction accuracy may be obtained at reduced applied voltage. However, the resulting corner radius may not be to the requirements in which case a subsequent machining operation needs to be carried out.

CONCLUSIONS

Process variables

(a) It is possible to vary the operating parameters, viz. applied voltage, feedrate and set depth to cut, over a wide range in order to meet the various requirements such as stock removal and form accuracy.

(b) A Faradaic removal condition exists for each set of operating parameters. These conditions show reasonably close correlation with theoretically maximum Faradaic removal under ideal conditions.

(c) Machining carried out beyond the Faradaic level results in imminent failure at some stage by either sparking, arcing, short-circuiting or excessive loading on the wheel causing wheel seizure.

(d) Diagrams similar to figures 4 and 6 could be used to forecast the boundary conditions for practical ECG. The user would simply be required to accommodate his own particular wheel diameter and form configuration.

Form accuracy

(a) Form accuracy as measured by angle reproduction, corner radius and oversize shows improvement with decreasing voltage and increasing feedrate. However, maximum operating feedrates are limited by the depths of cut and thereby affect the form accuracy. Only form-angle shows good reproduction accuracy at deeper cut.

(b) Results at Faradaic and limiting operating conditions show similar trends. The substantial mechanical proportion involved under limiting conditions does not appear to alter the trend significantly. This fact suggests the desirability of working near limiting operating conditions for maximum stock removal as well as for good shape reproduction.

The most desirable conditions for rough forming in a single pass appear to be when applied potential is controlled at approximately 10 V and when limiting conditions are approached by using the maximum possible feedrate obtainable at a particular set depth of cut. If in order to satisfy component requirements it should still be necessary to finish grind by electrochemical, conventional or combined action, then at least it can be completed with minimum stock removal and, therefore, with minimum wheelwear.

ACKNOWLEDGEMENTS

The authors wish to express their thanks to Professor F. Koenigsberger for his permission to conduct the work reported here in the Machine Tool Division of the Mechanical Engineering Department, The University of Manchester Institute of Science and Technology. They also wish to acknowledge gratefully the Science Research Council for equipment which formed part of their Electrochemical Machining Grant B/SR/1096, and the Ministry of Education, Government of India, for the grant held by the first author.

APPENDIX

Formulation of Faradaic removal conditions

For constant-voltage and constant-feed operation as in the present setup, the maximum feedrate for a specified set depth of cut should be such that stable machining conditions exist. In peripheral configura-

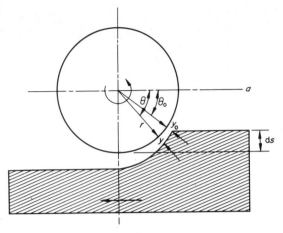

tion, this establishes a variable equilibrium gap[1,2].

Considering an elemental area 'dA' at an angle θ to the direction of tool feed, metal removal in time 'dt' according to Faraday's laws may be given by

$$\frac{(V - \Delta V)dA\,k}{y} \frac{\epsilon}{\rho_m F} dt \tag{1}$$

The rate at which the workpiece being eroded, at an angle θ, is

$$\frac{(V - \Delta V)k\epsilon}{y\rho_m F} \tag{2}$$

Since the wheel is fed towards the workpiece surface at a rate '$a\cos\theta$' in the opposing direction, the rate of change of position of the gap is given by

$$\frac{dy}{dt} = \frac{(V - \Delta V)k\epsilon}{y\rho_m F} - a\cos\theta \tag{3}$$

In a peripheral configuration, the gap at the leading edge (at $\theta = \theta_0$) will be the least for any set of operating conditions. and, therefore, the stable conditions will be determined by the minimum gap requirements corresponding to this position. Hence, for operating criterion,

$$\frac{dy}{dt} = \frac{(V - \Delta V)k\epsilon}{y_0\rho_m F} - a\cos\theta_0 \tag{4}$$

Under equilibrium conditions, $dy/dt = 0$, i.e.,

$$\frac{(V - \Delta V)k\epsilon}{y_0\rho_m F} = a\cos\theta_0 \tag{5}$$

If θ_0 is such that the gap formed at leading edge is at height ds, the set depth of cut is

$$\cos\theta_0 \approx \left(\frac{2ds}{r}\right)^{\frac{1}{2}} \tag{6}$$

from which

$$ds = \frac{r}{2}\left\{\frac{(V - \Delta V)k}{y_0} \frac{\epsilon}{\rho_m F}\right\}^2 a^{-2} \tag{7}$$

$$= Ka^{-2}$$

where

$$K = \frac{r}{2}(uJ_0)^2$$

$$u = \frac{\epsilon}{\rho_m F}$$

and

$$J_0 = \frac{(V - \Delta V)k}{y_0}$$

The graphs represented in figures 4 and 6 have been calculated for the following data:

the Faraday constant $F = 96\ 500$ C
for 6 in diameter wheel $r = 3$ in
for Nimonic 105 $\epsilon = 29$ g
                     $\rho_m = 7.99$ g/cm^3

for 10% (by weight) sodium nitrite electrolyte solution

         $k = 0.10\Omega^{-1}$ cm^{-1}

from which $u = 1.375 \times 10^{-4}$ in^3/A min
and $J_0 = 254\,(V - 2)$

assuming $\Delta V = 2$ V and $y_0 = 0.001$ in (for wheels of 100 grit size[12]).

REFERENCES

1. A. PAINTER, 1965. 'Electrolytic grinding', *Thesis*, College of Aeronautics, Cranfield.
2. D. C. PERRY, 1967. 'A study of peripheral electrolytic grinding', *Thesis*, College of Aeronautics, Cranfield.
3. C. R. STROUP, 1958. 'Applications of electrolytic grinding', *Metalworking Prod.*, Feb.
4. F. PEARLSTEIN, 1958. 'Test results in electrolytic grinding', *Metalworking Prod.*, Mar.
5. R. B. MASSAD, 1966. 'Electro-abrasive machining', *5th Natl. Tech. Conf. Am. Soc. for Abrasive Methods.*
6. H. HAUSFATER and I. WEBER, 1969. 'Electrochemical grinding of screw threads', *Tool Mfg. Engr.*, April.
7. A. GEDDAM, 1969. 'A study of peripheral electrochemical grinding with some wheel variables', *M.Sc. Thesis*, University of Manchester Institute of Science and Technology.
8. A. GEDDAM, 1971. 'Some studies in electrochemical form grinding of Nimonic 105', *Ph.D Thesis*, University of Manchester Institute of Science and Technology.
9. A. GEDDAM and C. F. NOBLE, 1970. 'Equipment for peripheral electrochemical grinding research', *Rept.*, No. 34, University of Manchester Institute of Science and Technology, Oct.
10. F. HUGHES and A. NOTTER. 'Evaluation of the electrolytic grinding process', *Diamond Information L1*, De Beers Industrial Diamond Division.
11. R. R. COLE, 1961. 'An experimental investigation of the electrolytic grinding process', *Trans. A.S.M.E.*, *Ser. B, J. Eng. Ind.*, May.
12. M. M. SFANTSIKOPOULOS, 1971. 'Electrochemical vertical spindle surface grinding', *M.Sc. Thesis*, University of Manchester Institute of Science and Technology.

DYNAMIC AND GEOMETRIC ASPECTS OF VERTICAL SPINDLE ECG

by

M. M. SFANTSIKOPOULOS* and C. F. NOBLE*

SUMMARY

The critical parameter for ECG, as for ECM, is that of *electrochemical gap.* Four main factors influence this gap when surface grinding with a vertical spindle configuration: the principle of operation, the grinding wheel conditioning, the geometrical accuracy of the machine tool and the forced vibration level. The paper combines a short fundamental analysis and related experimental work to establish the contributing role of each of the above factors for process control.

NOTATION

A, A_i	equivalent vibration peak-to-peak (PTP) amplitude
b	radial width of wheel rim
b_i	electrochemical vibration
B, B_i	electrochemical vibration PTP amplitude
ΔB_{max}	maximum wheel bond surface runout
C	ECM constant
C'	grinding conditions constant $= f_2/C$
D	wheel diameter
e	grit protrusion height
f	wheel frequency
f_1	set depth-of-cut or upfeed
f_{1a}	actual depth-of-cut
f_2	worktable feedrate
ΔG_{max}	maximum grit wear between two successive dressings
$K_1, K_2,$	bond ratio, grit ratio, depth-of-cut ratio,
K_3, K_4	forced vibration ratio, respectively
t	time
V_i, V_w	PTP amplitudes of the components of the equivalent vibration
V_H	horizontal spindle vibration PTP amplitude
y	electrochemical gap
y_0	initial electrochemical gap
Y_0	average electrochemical gap
α	angular error of worktable motion
ω	wheel angular frequency

INTRODUCTION

In a previous paper[1] the authors discussed some fundamental characteristics of vertical spindle electrochemical surface grinding and reported that phenomena directly connected with in-process variation of the electrochemical gap were

(1) excessive arcing and sparking between the wheel and workpiece, especially at high table feedrates, coupled with a considerable drop in electrochemical contribution to removal at high voltage when total removal was observed to increase,

(2) considerable deviation between actual and theoretically expected current waveforms,

(3) peculiarities in the material removal-table feed relationship.

The special and predominant feature of ECG is the continuous electrochemical action which affects an electrochemical gap. Compared with conventional machining this gap is an important consideration when examining the influence of the process on accuracy. ECG practice has established for most grinding wheels a gap of about 0.001 in but it should vary only within a close tolerance band. Four main factors of primary importance in this connection can be distinguished

- process configuration
- grinding wheel conditioning
- the geometric accuracy of the machine tool
- the forced vibration level.

The change in gap due to the process principle in vertical spindle ECG is covered in the aforementioned work[1]. This paper deals exclusively with the three remaining factors to assess their role in controlling the shallow depth-of-cut grinding technique.

GRINDING WHEEL CONDITIONING

Figure 1(a) shows the ideal grinding condition where the wheel-bond surface is perfectly flat and the abrasive grits protrude from it in a uniform manner with height e. This model has been used for a theoretical study of the method[1]. In fact, however, the wheel-bond surface is not flat but has an irregular shape with blunt and worn abrasive grits and freshly fractured or freshly revealed grits, some of which will contact the workpiece surface, as in figure 1(b). In accordance with the bond surface topography, the actual electrochemical gap within the active area of the wheel, figure 1(c), will vary in both the radial and circumferential directions. The effective grinding condition is given diagrammatically in figure 2(a), where the grit envelope is determined by the average grit exposure e. For a typical ECG wheel the electrochemically active proportion of the total working surface is about 85 per cent.

Bond surface irregularity represents a quite phenomenal gap vibration, which may alternatively be termed an electrochemical vibration, and will have the same fundamental frequency as that of the wheel. In figure 2(b), the wheel-bond waves correspond to a sequence of adjacent wheel reference diameters $D - 2b \leqslant D_i \leqslant D$, $i = 1.2 \ldots$ which the

* Machine Tool Division, Mechanical Engineering Department, UMIST.

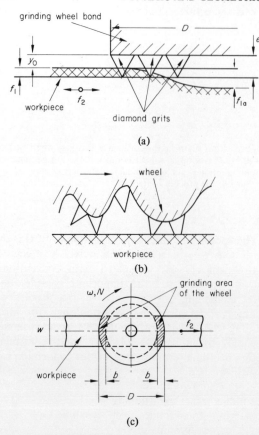

(a)

(b)

(c)

Figure 1. Vertical spindle surface grinding configurations.

workpiece, being fed by the table at rate f_2, will meet in varying phases. The nature of the bond surface demands that $(f/2f_2)(D_{i-1} - D_i) \gg 1$ and for each reference diameter D_i the electrochemical vibration can be represented by a Fourier series:

$$b_i = B_{i,o} + \frac{B_{i,1}}{2} \sin(\omega t + \phi_{i,1}) + \frac{B_{i,2}}{2} \sin(2\omega t + \phi_{i,2})$$

$$+ \frac{B_{i,3}}{2} \sin(3\omega t + \phi_{i,3}) + \ldots \qquad (1)$$

where $B_{i,n}$, $n = 0, 1, 2 \ldots$ is the PTP (peak-to-peak) amplitude of the nth harmonic and $\phi_{i,n}$ its phase angle.

The eventual grit wear between two successive wheel dressings can be included in the new grinding model by reference to figure 2(c), from which the following ratios are revealed

$$\text{bond ratio } K_1 = \frac{\Delta B_{max}}{e} \qquad (2A)$$

$$\text{grit ratio } K_2 = \frac{\Delta G_{max}}{e} \qquad (2B)$$

$$\text{depth-of-cut ratio } K_3 = \frac{f_{1,max}}{e} \qquad (2C)$$

If $K_1 \geqslant 1$, it becomes apparent that periodic arcing is assured, its duration and severity being dependent on the specific bond wave. If the reference diameter interval for which $K_1 \geqslant 1$ is comparable with the radial width of wheel rim b, then ECG is definitely impossible.

Grit ratio $K_2 \leqslant 1$ and is increased by grit wear. This increase is limited because it will restrict electrolyte supply to the gap and will tend to initiate arcing and sparking.

$K_3 < 1$ due to the grinding technique used.

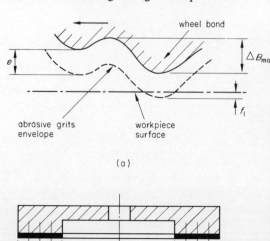

(a)

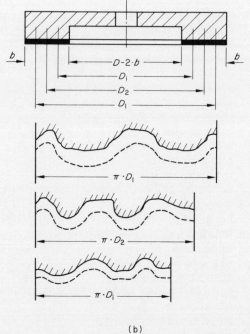

(b)

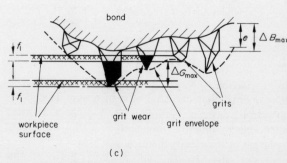

(c)

Figure 2. The actual ECG-model.

Figure 3(a) shows a series of continuous bond-wave records from an 8 in OD x 1 in rim ECG metal bond, diamond grit cup wheel as fitted to an Abwood vertical spindle surface grinding machine, model SG4 HE. The records were obtained using capacitive transducers having a 0.1 in diameter active area. All the measurements were taken following trueing and dressing to industrial standards and following a

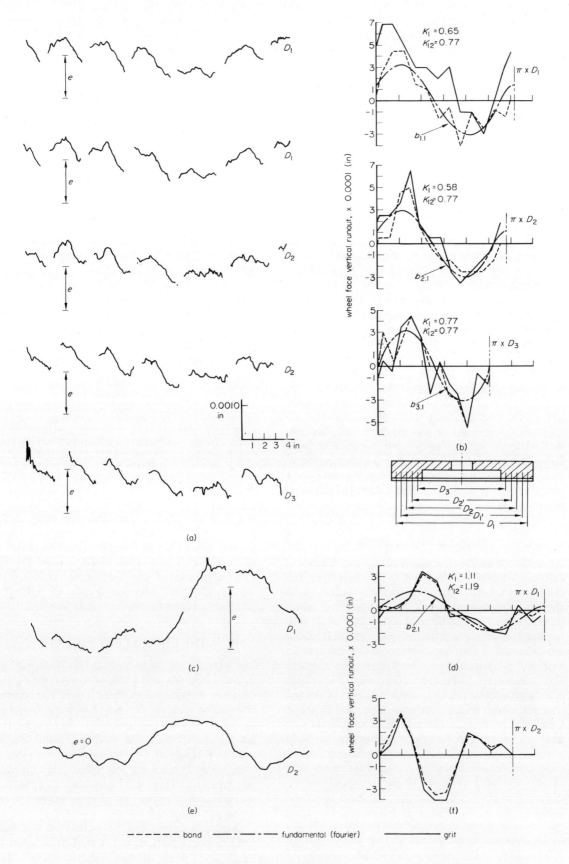

Figure 3. Wheel bond waves.

grinding operation on a few tungsten carbide tool tips to stabilise the condition of the wheel. For all five reference diameters $K_1 \geqslant 1$ and periodic arcing must therefore take place.

In figure 3(b) the average of several dial test indicator readings taken at discrete positions is compared with discrete bond measurements. Here K_1 is indicated as having values always less than unity and in fact very much lower than in figure 3(a). The apparent lowest-bond-to-highest-grit dimension is also lower than unity and gives further emphasis to the inappropriate use not only of discrete measurements but also of measuring the grit envelope runout. The latter is due to the uncertainty created by the obvious difficulty of locating the stylus of the indicator on the outer grit envelope.

Figures 3(c) and (d) show results from a 10 in OD x $\frac{3}{4}$ in rim ECG metal bond, diamond grit wheel, but here the wheel was not trued or dressed after mounting on the grinding spindle adaptor. The K_1 values are high and although the previous conclusions are equally relevant with regard to the relative measurements the results reveal the following observations: the local smoothness of the bond waviness as compared with figure 2(a), and the grit envelope more closely follows the bond wave than in figure 2(b). Both observations seem to justify special care in selecting the dressing technique. This is further confirmed from the results on an 8 in OD x $\frac{3}{4}$ in rim ECG metal bond, aluminium oxide grit wheel, figures 3(e) and (f), where no grit exposure is apparent or was measured. In this case the relatively soft abrasive grit (compared with diamonds) presents a trueing and dressing problem; small wonder that such wheels were found to have a serious tendency to short circuit the electrical power-supply[2].

FORCED VIBRATIONS

As with conventional grinding the main sources of forced vibrations are the grinding spindle system and the masses rotating with it, particularly unbalance in the grinding wheel from inhomogeneity, wheel mounting and general thermal distortion. As a result of this unbalance excitation, two parallel effects are observed: a whirling displacement of the spindle axis and a synchronous vibration of the machine tool structure[3,4,5]. In connection with ECG both effects produce a time-dependent variation in electrochemical gap which has the same fundamental frequency as that of the spindle. The existence of nonlinear systems in a machine tool usually result in a vibration response which includes several higher-order harmonics of their fundamental frequency and, in the case of a vertical spindle configuration, the variation of electrochemical gap will have two main components: a component of amplitude V_w from the relative displacement of wheel and workpiece irrespective of table travel; and a component of amplitude V_i which will vary with table travel.

Vibration analysis of the Abwood machine showed a predominance for the wheel frequency vibration component (3180 rpm) over the other disturbance sources (hydraulic, electric and exhaust system). Table 1 gives the horizontal and vertical PTP amplitudes of the fundamental and harmonic frequencies from the 8 and 10 in diamond grit wheels. As expected, vertical amplitudes are considerably lower than the horizontal ones but, although

TABLE 1
Horizontal and vertical vibration PTP amplitudes (μin)

diamond wheel OD	harmonics / vibration	1st	2nd	3rd	resultant
8 in	horizontal	520	60	84.5	600
	vertical	80	27	34	109
10 in	horizontal	480	80	120	617
	vertical	120	19	–	145

both wheels have vibration levels of a similar order of magnitude, the 8 in wheel surprisingly has the somewhat higher unbalance. Since metalbonded ECG wheels possess no porosity for absorbing electrolyte, and the wear rate is very low, this feature must be attributed to the trueing, dressing or cleaning method used. (Cleaning refers to the removal of any wheel layer which is a particular characteristic when grinding tungsten-carbide workpieces.) In table 2

TABLE 2
Vibration ratios

ratio	8 in diamond wheel	10 in diamond wheel
$V_w/f_1,\text{min}$	109/100 = 1.090	145/100 = 1.450
$V_w/f_1,\text{max}$	109/500 = 0.218	145/500 = 0.290
V_w/e	109/1300 = 0.084	145/1850 = 0.078
K_4	600/1300 = 0.461	617/1850 = 0.334

the vertical vibration amplitudes are compared with grit protrusion height and the available range of upfeed values. A horizontal forced-vibration ratio $k_4 = V_H/e$ is also introduced, where V_H is the PTP amplitude of the horizontal (radial) grinding spindle vibration. The significance of upfeed value on vibration level becomes obvious from this table.

THE EQUIVALENT VIBRATION

The coexistence of wheel surface topography and forced vibrations will produce a resultant equivalent vibration of the electrochemical gap. The representative vector diagram for any harmonic frequency is shown in figure 4 (equation 1). Amplitude A_i is to be compared with the ECG gap value for every reference diameter D_i, the most unfavourable value being $D_i = D$ since for any other value the gap will be greater[1]. This will produce a criterion for acceptable ECG operation, that is, without excessive arcing and sparking.

Arcs and sparks can be caused by high voltage, sudden gap reduction, direct contact between wheel bond and work surface, shortcircuiting due to mechanical grinding debris, defective electrolyte supply. As well as damaging both the workpiece surface and the wheel-bond surface, arcing and sparking disrupt the continuity of electrolyte film in the inter-electrode gap. The localised high current density with which they are related can cause local boiling, evaporation or even ejection of the electrolyte from the gap, so creating a breakdown in the electrochemical process. Work material may also be

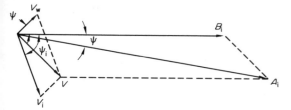

Figure 4. Equivalent vibration components.

removed in a similar manner to the spark erosion process. The K_1 ratio values quoted in figure 2 for the 8 in diamond grit wheel which was used for the experimental work[1] and the phenomena described in the Introduction to this paper, support these remarks. In addition, the local damage spots on the wheel will act to initiate further arcing and sparking until the next wheel conditioning operation. This situation tends to nullify the value of expensive spark-detection equipment[6].

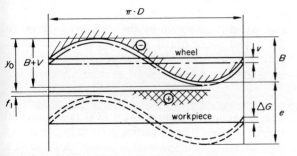

Figure 5. The effect of machine vibration and grit wear on a single bond wave.

For PTP amplitudes of the resultant vibrations, it can be seen from figure 5 that

$$\frac{y_0}{A} \geqslant \frac{y_0}{B+V} \geqslant \frac{y_0}{B+V_w + V_D} > 1 \qquad (3)$$

If $y_0 = e - f_{1\,max} - \Delta G_{max}$
     (after dressing $\Delta G_{max} = 0$)
and $\Delta B_{max}(B_1)$

then, by replacing $V_w + V_D$ by V_H, we may establish from equation (3) a stricter criterion for acceptable grinding operations, namely

$$K_1 + K_2 + K_3 + K_4 < 1 \qquad (3A)$$

The theoretical expression for the average gap corresponding to the critical table feed[1] is given by

$$Y_0 = \frac{2}{3}\left[1 - K_2 - K_3 + \frac{(1-K_2)^2}{2(1-K_2) - K_3}\right]e \qquad (4)$$

$$\cong (1 - K_2 - K_3/2)e \qquad (4A)$$

Now, equation (4A) only *expresses a limitation for the abrasive grit size*, but combining it with the criterion of equation (3A) yields

$$\Delta B_{max} < Y_0 - \frac{f_{1\,max}}{2} - V_H \qquad (5)$$

and this represents a more informative *wheel trueing criterion* 'directly related to wheel *bond-surface topography*. This simple principle has been neglected by industrial ECG users who instead base their trueing *on measurements of grit envelope runout* and take account only *of the average Y_0 in the criterion of equation* (5).

Table 3 gives the measured amplitude A_i of the equivalent vibrations for the same reference diameters as figure 3. The corresponding frequency analysis record for a reference diameter of the 8 in wheel is reproduced in figure 6. Amplitudes A_i for the fundamental wheel frequency (53 Hz) are compared

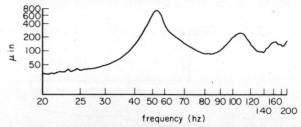

Figure 6. Equivalent vibration-frequency-analysis record.

TABLE 3
Equivalent vibration PTP amplitudes, μin

harmonics ref. diameter	8 in diameter wheel				10 in diameter wheel			
	1	2	3	4	1	2	3	4
D_1	750	240	150	—	1900	600	240	375
D_2	600	106	120	—	1345	600	300	375
D_3	600	75.5	—	—	860	600	150	375

TABLE 4
Fundamental PTP amplitudes (μin) of equivalent and bond vibrations

ref. diameter	8 in diamond wheel		10 in diamond wheel	
	A_i	B_i	A_i	B_i
D_1	750	630	1900	1950
D_2	600	590	1345	1408
D_3	600	648	860	1088

in table 4 with amplitudes of bond wave-form B_i obtained from figure 3 by a graphical Fourier integration technique. In this table it can readily be seen that bond waviness predominates over the time-dependent electrochemical gap. Therefore, forced vibrations are of secondary importance provided a reasonably balanced wheel assembly is used and this conclusion applies even if the rather high level of radial vibrations (K_4) is incorporated in equation (3A).

Within a satisfactory degree of approximation it can be stated that the frequency analysis spectrum of table 3 represents the harmonic components of the wheel-bond waveforms. This is especially important for ECG because it indicates the desirability of higher-order harmonic bond waviness which must have an amplitude to satisfy the limitation of equation (5) in order to assure supply of electrolyte into the inter-electrode gap. This applies equally well to non-grit wheels and could apply to other ECG configurations.

GEOMETRY OF THE WORKTABLE MOTION

The pitch of the machine tool worktable, represented by the angle α of figure 7, will produce an electrochemical gap change whose effect is described by the differential equation

$$\frac{dy}{dt} = \frac{C}{y} - f_2 \sin \alpha \qquad (6)$$

The existence of the angle α, which may (or may not) be constant throughout the whole table travel, is due either to inherent errors of the machine tool, or to several functional effects such as weight

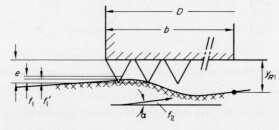

Figure 7. Modification of ECG-model by angular error of the worktable feed.

deformation, deformations due to clamping, thermal distortions or elastic deflections caused by the cutting forces[3].

For a very small and constant angle α the solution of equation (6) yields a value for electrochemical gap y_{R1} at the end of the first half of the grinding wheel as given by

$$e - f_1 + \alpha\left(\frac{D}{2} - b\right) - y_{R1} +$$

$$+ \frac{1}{\alpha C'} \ln \frac{e - f_1 + \frac{D\alpha}{2} - \frac{1}{\alpha C'}}{y_{R1} - \frac{1}{\alpha C'}} = 0 \qquad (7)$$

For a typical set of arithmetical parameters (taken from the experimental work[1]) a graphical representation of equation (7) is given in figure 8. It can be seen that generally the existence of table pitch results in reduced material removal, except perhaps for a very narrow band of very small angles.

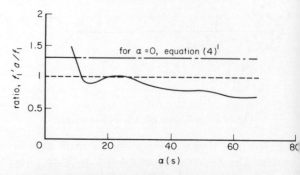

Figure 8. Effect of worktable angle α on depth-of-cut.

If we now consider that angle α is likely to change with worktable position, feedrate and the process forces involved and further, that for higher feeds the mechanical contribution (and hence the forces) are greater, it is reasonable to expect larger angular total deviations. In consequence there will be a decrease in the total material removed. This conclusion is supported by the similarity in shape of the actual depth-of-cut ratio versus feedrate curve previously published[1] with that of figure 8.

CONCLUSION

The foregoing has shown that if ECG (and a vertical spindle configuration in particular) is to be conducted for attaining component accuracy and surface integrity, rather than simply rapid removal with low wheel wear, wheel conditioning is an extremely important aspect of the process, and that machine tool deflection is no less important than when grinding by conventional means. The work has highlighted cause and effect of dynamic and geometric problems and it is hoped that further study will reveal how the indicated solutions can be practised in an industrial environment. Meanwhile, the user should apply the information previously published[1] and at least pay heed to wheel balance, table feed and table inclination limitations, as discussed in this paper.

ACKNOWLEDGEMENTS

The authors are pleased to thank Messrs. Abwood Machine Tools Ltd. for the loan of an electrochemical vertical spindle surface grinding machine for research covering aspects of this type of grinding. UMIST is acknowledged for permitting the experimental work to be carried out in the Royce Laboratory and Professor F. Koenigsberger is thanked for his encouraging interest and support. The National Technical University of Athens is also acknowledged for the S. Niarchos Scholarship awarded to the first-named author.

REFERENCES

1. M. M. SFANTSIKOPOULOS and C. F. NOBLE. Vertical spindle electrochemical surface grinding. *12th International MTDR Conference,* Macmillan, 1972, p. 265.
2. A. GEDDAM and C. F. NOBLE. An assessment of the influence of some wheel variables in peripheral electrochemical grinding. *International Journal for Machine Tool Design & Research,* Vol. 11, pp. 1–12, 1971.
3. Specifications & tests of metal cutting machine tools. *Proceedings of the Conf.,* 19th–20th February, 1970, UMIST.
4. J. E. STEVENS. An investigation into a procedure for dynamically testing grinding machines. M.Sc. Diss., UMIST, 1970.
5. R. S. BENETT and C. F. MAY. Performance studies on a typical centreless grinding machine with reference to truing and balancing of the grinding wheel. *Int. J. Mach. Tool Des. & Res.,* Vol. 6, pp. 47–101, 1966.
6. L. A. WILLIAMS. Electrochemical grinding. *ASTME Technical Paper* MR67–649.

RESIDUAL STRESSES AND SURFACE EFFECTS IN ELECTRO-DISCHARGE MACHINING

by

J. R. CROOKALL* and B. C. KHOR*

SUMMARY

Electro-discharge machined specimens were obtained from two different relaxation-circuit machines and one pulse-generator machine. Determination of residual stresses was by a bending-deflection method, involving a very sensitive technique for continuously recording the deflections due to stresses released by an electro-chemical cell. Such a method is necessary for measuring stresses in the often very thin sub-surface layers due to electro-discharge machining.

Very high residual tensile stresses are induced by electro-discharge machining, which near the surface approach the ultimate tensile strength of the material, but fall rapidly below the surface to a relatively low value or to zero. Residual stresses in different materials are related to their thermal and other properties, and are largely unaffected by the tool electrode material and dielectric. In general, the maximum stress for a given material is less affected by machining conditions (discharge energy and duration) than is the distribution and depth of penetration of stress.

INTRODUCTION

The applications of electro-discharge-machining (EDM) stem partly from the fact that it is basically a 'forming' process, i.e. the tool electrode can be made to impose its complementary shape upon a workpiece with a constant small overcut, typically of 0.08 mm. As metal removal is primarily on a thermal basis the physical strength of the workpiece material is of minor consequence, and it has been widely found that the thermal properties in general correlate with metal-removal ability. Also the erosive action is essentially force-free, hence electrode sections may be thin and relatively complex, and yet may be used to machine very hard materials such as tungsten carbide, for example. Whilst die sinking is a common general engineering application, the drilling of fine holes, and even curved holes, in 'high-strength temperature-resistant' materials is typical of applications in which it has few, if any, competitors.

The 'force-free' nature of the process should not, however, be taken to imply that the surface produced is stress-free. Furthermore, in common with processes producing intense surface heating, the stresses induced are likely to be of tensile nature, and hence detrimental to fatigue life. They may also cause distortion in thin components particularly, and increase susceptibility to stress corrosion. This, together with the peculiar advantages for form generation which lead to its employment in conventionally difficult situations, implies that a proportion of products on which it is used are likely to be highly stressed. In this combination of circumstances, the study of residual stresses due to EDM assumes some importance.

What has come to be known as 'surface integrity' involves all aspects of the surface and near-surface regions which may ultimately affect the functional behaviour. Such aspects as microgeometry, hardness and microstructure are involved, in addition to residual stresses. However, quite different techniques of investigation are involved for each, and furthermore their relative importance, and possibly their interaction when present simultaneously, is somewhat uncertain at present. The present paper is therefore concerned with the study of residual stresses, with some observations on the mechanism of surface formation and its effects.

ELECTRICAL DISCHARGES AND THEIR EFFECTS

In EDM a rapid succession of discrete discharges are introduced between the electrodes, each discharge taking the path at which breakdown can be most readily achieved. For each discharge, the application of pulse voltage causes the field intensity at some local area within the gap to produce sufficient current flow to vaporise and then ionise a very narrow channel within the dielectric[1]. The discharge then takes place as an electron avalanche towards the anode. The cathode is later bombarded by ions from the decomposed dielectric and later still by ions, and possibly particles, from the vapourised anode. The anode is heated sooner and more rapidly than the cathode[2]. Liquid dielectrics assist in confining the discharge to a very narrow channel, thus maintaining a high current density, as also does the magnetic 'pinch' effect. However, if pulse current duration is prolonged the discharge will assume the characteristics of an arc, and current density falls.

At the anode, comparatively little erosion occurs during the initial breakdown. The electron avalanche, of intense current density ($\approx 10^6$ A/mm^2) and accelerated by the electro-dynamic forces[3], heats the impact area probably to boiling point and melts a volume of metal below it; the extent of such boiling and melting depends upon the properties of the anode and the characteristics of the discharge. As the discharge continues, and the current density falls, the amount of boiling is reduced, but the melted volume continues to increase owing to the penetration of heat with time, and the heated zone below this experiences considerable compressive forces due to the high temperature gradients. During the discharge the pressure in the dielectric vapour bubble remains high[4], but falls suddenly when the discharge ceases, causing melted metal in the crater to be ejected, accompanied by an

*Imperial College of Science and Technology

evolution of gas within the metal itself. The ejected metal forms microscopic solid and hollow spheres of 3 to 5 μm diameter[5-8]. The small quantity of liquid metal remaining in contact with the crater base then undergoes the physico-chemical changes associated with an extremely high cooling rate. Cermets and a few other materials are exceptions to this general rule since they readily fracture under the action of the current impulse[9].

Very similar events occur at the cathode, but as bombardment is by ions rather than by electrons, the rate of transfer is slower, and erosion is consequently generally less. The cathode can be bombarded also by particles expelled from the anode. Oscillatory discharges may occur with certain circuits, in which case the cathode will be subjected to periodic anodic bombardment on current reversals.

Surface and metallurgical effects in EDM have been the subject of contributions by Wilms and Wade[10], Opitz[11], Barash[12-15], Lloyd and Warren[16], and others. In a comprehensive review, Bucklow and Cole[17] emphasise the metallurgical aspects, and categorise existing investigations on the nature of electro discharge-machined surfaces by investigational techniques of optical metallography, electron microscopy, X-ray diffraction, dislocation etching and mechanical and temperature measurements.

RESIDUAL STRESSES

Residual stresses due to EDM have been investigated by Aleksandrov[18] and Aleksandrov and Zolotykh[19], Goldsmidt[20], Barash[12-15], and Lloyd and Warren[16]. Barash[12] used an RC generator to prepare specimens which were tested for residual stresses by the discrete layer removal bending-deflection method, and found high residual tensile stresses. His later work[13-15] referred to residual stresses only qualitatively, and was mainly concerned with the effects on mechanical performance, notably fatigue life. Spark-turned specimens of tool steels showed a reduction in endurance limit[13] when compared with specimens which had been turned conventionally to give the same surface finish. However, the detrimental effects of the tensile residual stresses were found to be minor when compared with the effects of permanent damage, such as cracks and microcracks.

Lloyd and Warren[16] also employed an RC generator and the bending-deflection method. Tensile residual stresses were found to approach the upper tensile strength of the material at the immediate surface, and then fall rapidly to a relatively low value before giving way to small residual compressive stresses in the core of the material. This general pattern was found by Barash also. This smooth transition in the stress pattern was not maintained for copper, but no explanation was forwarded for the difference.

The effect of discharge energy on the stress distribution in EN2 steel machined with a copper electrode was found to be small near the surface, but the depth of the stress distribution increased with increasing energy. It was considered that is should be possible to reduce the near-surface tensile stress level by introducing metals with a high atomic volume through the alloying effect with the tool electrode. Using lead, the stresses near the surface were slightly reduced, but those further within the material were unchanged.

It was expected that heat treatment at around 300 °C would reduce the surface stresses for carbon and low alloy steels, which would have an austenitic and martensitic surface after spark machining. The martensite would be tempered and the austenite transformed. Also, the surface hardness would increase owing to the austenite transforming with an expansion to high carbon bainite. Heat-treatment was found to reduce markedly the surface stress, although the stresses due to plastic deformation below the fused layer were unaffected and only became lower when the heating temperature was increased to 500 °C.

From fatigue tests on annealed specimens of a 0.2% C steel they inferred that the tensile fatigue properties depended more upon the sensitivity of the material to spark damage (cracks and microcracks) than to the tensile residual stresses. This is also in agreement with Barash's result[13].

In the Soviet Union, Aleksandrov[18] studied the effects of pulse duration and pulse energy on the formation of residual stresses. Results by Aleksandrov and Zolotykh[19] indicated that surface finish and depth of surface layer affected depended to a considerable extent upon the pulse duration; hence it was expected that the residual stresses would be so influenced also. The observed residual stresses, determined by Davidenkov's method, clearly supported the expected trend, being increased by both the pulse duration and pulse energy used.

Similar tentative explanations for the formation of residual stresses were forwarded by Aleksandrov, and by Lloyd and Warren. The stresses were deemed to arise mainly as a result of the thermal contraction of the resolidified metal which was not expelled from the craters, onto the relatively unaffected parent metal, inducing plastic deformation and biaxial tensile stresses. It was also suggested that this effect, together with the occurrence of slip, twinning, and work hardening below the fused layer, might indicate the passage of shock waves into the bulk of the metal.

However, a rapid thermal cycle at the surface producing temperature gradients steep enough to cause yielding of the parent material would also result in subcutaneous tensile stresses on restoration of equilibrium. Hence stresses of this nature must be expected in such circumstances irrespective of the presence and extent of re-solidified surface metal.

METHODS OF EXPERIMENTAL DETERMINATION OF RESIDUAL STRESSES

The subject of experimental determination of residual stresses has received considerable attention, and a number of qualitative and quantitative methods can be found in the literature. Contributions to this field appear in the symposia on internal stresses in metals and alloys[21, 22]. Books by Heindlhofer[23], Almen and Black[24], American Society for Metals[25] and National

Research Council (U.S.A.)[26] contain detailed discussions of the subject. Also Denton[27] and Rao[28] have reviewed the existing experimental techniques.

The two main methods of quantitative determination of residual stresses are the X-ray method, and the mechanical 'bending-deflection' method based on successive removal of stressed layers.

However other methods include qualitative assessments using brittle lacquers[29, 30] and strain etching techniques[31]. Also Knoop[32] and Hertz[33, 34] hardness tests have been employed for quantitative determination of residual stresses. Hole-drilling methods proposed by Mathar[35] and further studied by Lake et al.[36] can be employed if a quick assessment of the surface stresses is required to an accuracy of only 20%. The use of ultrasonic sound waves has been proposed[37], but it appears that further development is necessary before this can be employed for precise quantitative determination of residual stresses.

Several methods based on the modification of the electro-chemical potential[38], and the magnetic properties of ferromagnetic materials[39], were recently explored as possible means of stress detection. Oliver[40] suggested new methods involving electron emission, proton annihilation and nuclear resonance. Again, the value of these methods in the precise determination of residual stresses is yet to be established.

X-ray methods have been treated in detail by several investigators[26, 27, 41-46] but a careful comparative experimental assessment by Rao[28] has indicated no advantage over the bending-deflection method developed by him.

The bending-deflection method and analysis

Residual stresses are those existing within the body of a material when no external forces act upon it. If a layer of material is removed, the stress previously in that layer cannot of course be sustained, and, to restore the equilibrium of the remaining material, a deflection takes place. This deflection can be measured and related to the stress originally in the removed layer, and by successive or progressive removal of layers the distribution of residual stress in the material can be determined.

For the derivation of residual stresses from experimental data in the bending-deflection method, the first accurate expressions for the circumferential stresses in thin-walled tubes were provided by Davidenkov[47], who improved an earlier approximate theory. Sachs and Espey[48] modified the expressions later to effect an improvement in the computation time. More recently Denton and Alexander[49] have removed approximations in both Davidenkov's and Sachs and Espey's work to provide an exact method for use with a digital computer.

The relationship between stress in a removed layer and the change in curvature of plates can be found in standard works on elasticity and has been quoted by Treuting and Read[50], Lenter[51], and others. The change in the stress in the remaining layers, however, becomes an implicit integral function when expressed analytically, requiring an approximation for its completion. Treuting and Read[50] attempted to overcome

this difficulty by returning the plate to the flat position before each layer was removed, thereby reducing the change which affects the stress in the underlying layers. The plate would also require stretching to its original size for complete avoidance of the change in other layers, but this was not done. The error incurred was shown by Alexander and Denton[49] to be approximately 12% after one-third of the plate thickness had been removed layer by layer. They advocated the use of the simple bending theory instead.

The bending-deflection method is applicable to various configurations of workpiece. However, with discrete layer removal, if the layers are too thin accuracy of determination is poor, and if they are too thick definition is poor particularly in the important region just below the surface itself (where maximum stresses occur in EDM). This problem becomes particularly acute for EDM because the total depth in which the major residual stresses occur can be very small, and stress gradients (which are averaged across a layer) can be very high.

By careful development of the method for measurement of residual stresses due to conventional machining, Rao[28] raised the sensitivity and definition of the technique, and made both layer removal and the recording of the resulting deflections a continuous process. Thus information can be obtained comparatively near the surface, and the technique gives reproducible results with an accuracy of determination of ± 4.6 MN/m^2 in a layer 5 μm thick.

The specimen is normally in the form of a beam (of section I and Young's modulus E), the deflection of which occurs as a change of radius δR, which is related to the moment M producing it by the simple bending relationship

$$M = EI\left(\frac{1}{R_2} - \frac{1}{R_1}\right) = \frac{EI\delta R}{R_m^2} \qquad (1)$$

This in turn is related to the stress σ_m in the layer removed (assuming uniform distribution of stress across the layer) by

$$M = \sigma_m \frac{a(a+b)w}{2} \qquad (2)$$

Thus,

$$\sigma_m = \frac{Eb^3}{6a(a+b)} \cdot \frac{\delta R}{R_m^2}$$

where a = thickness of layer removed
b = current thickness of the specimen
w = width of specimen
R = mean radius of curvature = $\frac{1}{2}(R_1 + R_2)$.

However, with the removal of each layer, a re-distribution of stress takes place during the restoration of equilibrium within the remaining material. Thus stresses observed in successive layers after the first are not the original residual stresses, which must be found by correcting for the redistribution of stress on each layer removal. In terms of forces and moments on the remaining beam, these are

$$\sigma_1 = \frac{a}{b}\sigma_m \tag{3}$$

$$\sigma_2 = \frac{6a(a+b)y\sigma_m}{b^3} \tag{4}$$

where y = distance of each layer from the neutral axis.

Thus in all layers 1, 2, 3, . . . i, the original residual stresses σ_i, may then be found from the observed stresses σ_{mi} in each layer from

$$\sigma_i = \sigma_{mi} - \sum_2^i (\sigma_1 + \sigma_2) \tag{5}$$

The stress within a layer is not normally uniform, and a better estimate of the stress distribution is obtained by making a correction assuming stress to vary linearly across the layer. If the total variation within the layer is $\pm\delta\sigma$ and the mean stress is σ_m, then the component of bending moment per unit length to be accounted for is $\delta\sigma a^2/6$. Hence the corrected value of the residual stress is

$$\sigma = \sigma_m \pm \frac{\delta\sigma a^2}{3a(a+b)} \tag{6}$$

The positive sign is used in the equation when σ_m is increasing towards the interior of the specimen.

Apparatus and specimens
Figure 1(a) shows the experimental arrangement with the specimen supported in a glass holder positioned in the tank containing an electrolyte. A cathode of stainless steel is also fixed to the glass holder, with a uniform gap of about 25 mm between the electrodes. The surfaces of the specimen other than the spark-eroded portion over which layer removal is to be effected is protected with a cellulose 'dope'.

The electro-polishing behaviour of the specimen material is obtained experimentally as illustrated in figure 1(b). In the linear region up to voltage V_1 a matt surface is produced, but in the region V_2-V_3 current density changes little with voltage and a good electro-polished surface is produced. Pitting tends to occur beyond V_3.

During continuous layer removal by electro-polishing, the resulting deflection is continuously monitored by a sensitive capacitive transducer and an ultraviolet recorder (figure 1(a)).

Theoretically the volume of metal removed by electropolishing can be related to the electrical charge passed, and the layer thickness a, removed from area A of the specimen in time dt is

$$a = \frac{K}{A} I \, dt \tag{7}$$

where I is the current, and K is a constant involving the equivalent weight of the metal. However, uncertainty of the effective valency of Fe ions (which can be divalent or trivalent) during the process, and also

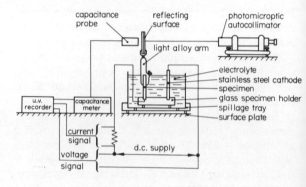

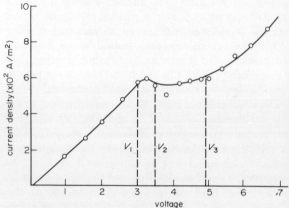

Figure 1. Experimental details of residual stress measurement. (a) Schematic diagram of apparatus for measuring residual stresses. (b) Conditions for the electro-polishing of steel using perchloric acid based electrolyte

the existence of anode reactions involving gas evolution, renders the direct application of Faraday's laws untenable in practice on grounds of accuracy. Recourse was therefore made to direct calibration of material removal against magnitude of charge, and during electro-polishing current was also continuously recorded on the ultraviolet trace for subsequent integration with respect to time. This, and the derivation of residual stresses from the deflections, were performed by digital computer. A linear relationship between the translatory displacements recorded by the capacitive transducer and an angular change $\delta\theta$ of the reflector due to a change in specimen radius δR was established experimentally, and it can be shown[52] that the latter two parameters are related by

$$\delta R = \frac{R_1(b/6 - R_1)\delta\theta}{(S_1 + R_1\delta\theta)}$$

where R_1 and S_1 are respectively the initial radius and length of specimen subject to electro-polishing.

Three different workpiece materials were used for this investigation, namely commercially pure copper, EN1 steel and Darwin HW5 tool steel. Copper was selected as an example of a non-transforming material, in contrast to EN1 steel, and the tool steel as representative of materials which are frequently machined by EDM and where residual stresses may be of significance.

The compositions (per cent by weight) of the steels were as follows:

	C	Si	Mn	S	P
EN1 steel	0.07–0.15	0.10 max.	0.80–1.20	0.20–1.30	0.07 max

	C	W	Cr	Mo	V
Darwin HW5 tool steel	0.3	1.0	5.0	2.0	0.25

The specimens, measuring 50 mm x 10.2 mm by 2.04 mm, were first given an annealing/stress relieving treatment in a vacuum furnace to eliminate existing stresses prior to electro-discharge machining.

Two different relaxation-circuit generator machines, and one pulse-generator machine were used in the tests. Normally, brass tool electrodes and paraffin dielectric were used, although graphite and lead electrodes were employed for certain tests, and also water as a dielectric fluid.

RESULTS AND DISCUSSION

Pulse-generator machine
Residual stresses were determined for the tool steel HW5 after electro-discharge machining at three different energy levels, and with pulse duration varying

between 75 μs and 500 μs. The results are presented in figure 2.

At the highest energy level used (1.1 J, figure 2(a)) a maximum tensile residual stress of 420 MN/m² occurs at about 0.04 mm below the surface, but the stresses fall away relatively gradually, and would apparently reach zero at a depth of around 0.2 mm. At a lower energy of 0.5 J but the same pulse duration of 250 μs (figure 2(b)), the general extent and penetration of residual stress has decreased, although a similar value of maximum stress occurs at a decreased depth of 0.015 mm.

At the same energy level (0.5 J) a twofold increase in pulse duration to 500 μs (figure 2(b)) produces a steeper stress gradient, without evidence of a maximum in the sense of a turning point, the highest observed stress level being 640 MN/m².

The two values of pulse duration used at the lowest energy level (0.1 J) namely 75 μs and 188 μs (figure 2(c)) show a somewhat similar effect of pulse duration. However, the general magnitude of stress at this, the lowest energy level, lies between those of the levels discussed previously; the effect is evidently genuine in view of the two independent tests involved. The existence of fairly narrow 'optimal' conditions from the viewpoint of metal removal rate is comparatively well known, and it appears likely that such a situation occurs also with respect to thermal penetration, and hence residual stress levels, so that monotonic trends should not necessarily be expected.

Relaxation-circuit machines
The experimentally determined residual stress distributions obtained with the two relaxation-circuit machines are shown in figure 3 for EN1 steel. The

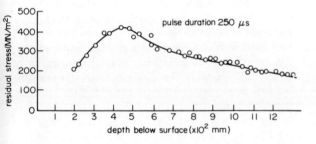

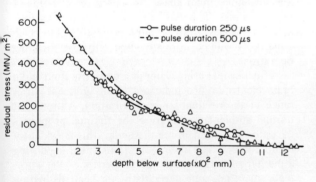

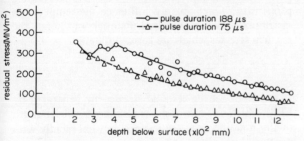

Figure 2. Residual stresses in HW5 tool steel using a pulse-generator machine (Agiepuls). Tool electrode, brass; dielectric paraffin. (a) Pulse energy 1.1 J. (b) Pulse energy 0.5 J. (c) Pulse energy 0.1 J.

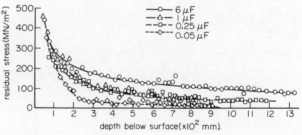

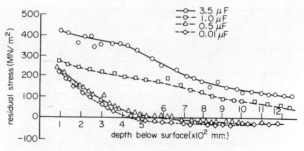

Figure 3. Residual stresses in EN1 steel using relaxation-circuit machines. (a) Servomet SMD. (b) Agemaspark F500

general feature is the high near-surface stresses, which decrease with distance from the surface, becoming low tensile or even compressive at a depth of about 0.1 mm. Also, with increasing capacitance in the *RC* circuit (and hence spark energy) the general level and penetration of the stresses increase. However, acknowledging the difficulty of obtaining stresses very near the surface, and the necessary caution consequent upon extrapolation, it appears that as a rough approximation the highest stresses which occur are fairly similar, irrespective of the spark energy used.

Thereafter, similarity in the residual stresses from the two machines ceases, those from the Servomet (figure 3a) being characterised by a very rapid initial decrease with depth, this being much less evident in the case of the Agemaspark, figure 3(b). With the latter a 'plateau' region occurs for the higher values of capacitance investigated, 3.5 μF and 1.0 μF, whereas at values of 0.5 μF and 0.01 μF, stress distributions show greater similarity to those of figure 3(a). The possible reasons may only be surmised in the absence of detailed evidence of the discharge performance of the two machines. However, it is considered that a variation in the effective duration of the discharges supplied by the two machines is the probable cause of the observed difference in residual stresses. Evidence of the significant effect of pulse duration upon the distribution of residual stress has been noted already for the pulse-generator machine (figure 2).

The remaining tests were conducted with the Servomet machine.

Effect of workpiece material
A comparison of the residual stresses obtained in HW5 tool steel and copper is given in figure 4, and the

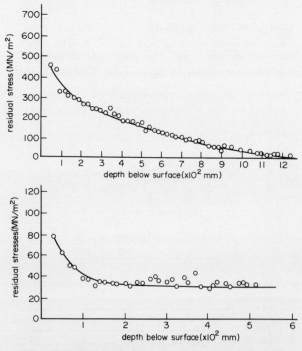

Figure 4. Residual stresses in different workpiece materials using a relaxation-circuit machine (Servomet). Tool electrode, brass; dielectric, paraffin; *RC* generator capacitance: 6 μF (a) workpiece HW5 tool steel, (b) workpiece copper.

result for EN1 steel at the same capacitance of 6 μF has been given already in figure 3(a).

The stresses in the tool steel are generally somewhat higher than in EN1 steel, with a lower rate of decrease with distance from the surface. However, the difference is not marked, and it is noted that their thermal properties (fusion temperature and thermal diffusivity) are similar. In the case of copper, stresses are considerably lower, and decrease rapidly to a low level within about 0.02 mm. The lower yield strength will produce lower values of maximum stress but the principal difference in distribution of stress is attributed to the effect of the thermal properties. The thermal conductivity of copper is approximately eight times that of steel, the effect of which is to reduce the penetration of isotherms and also to steepen the temperature gradient within the metal resulting from a thermal pulse of finite duration and energy[53].

Effect of tool material and dielectric
Some inter-electrode alloying is known to occur in EDM, although tool electrode metal which is transferred across the gap appears chiefly in the re-solidified layer. Also with a hydrocarbon dielectric a steep carbon concentration gradient can occur by diffusion in the unmelted material as well as in the resolidified metal. To examine the possible effects of such alloying upon residual stresses, further specimens of HW5 tool steel were machined with graphite and lead electrodes in paraffin dielectric, using a capacitance of 1 μF (RC circuit). The differences in the resulting residual stress distributions were negligible (and are therefore not presented here). These results therefore largely confirm the findings of Lloyd and Warren[16] for EN2 steel, and indicate that such diffusion effects are negligible under these circumstances in terms of residual stress. This is likely to be less true of long duration pulses from a pulse generator, but such conditions will be normally used only for roughing operations.

A further test using water as a dielectric fluid and a brass electrode also indicated an insignificant difference in the residual stress distribution. Although at normal ambient temperature the thermal properties of water and paraffin are dissimilar, this is evidently of no consequence under conditions obtaining during a discharge. It is the thermal properties of the workpiece material, and particularly its heat dissipating propensity, which mainly influence the temperature distribution in the unmelted material, and hence the residual stresses which are formed thereby.

CONCLUSIONS

(1) High tensile residual stresses are generated by EDM, which can approach the ultimate tensile strength of the material near the surface, but fall rapidly with depth to low values, or to zero.

(2) Some consistent differences have been found between residual stresses produced by a pulse-generator machine and two different *RC* circuit machines. However, the highest occurring stress is less dependent

upon either the type of EDM generator or the machining conditions used than is the distribution and penetration of stresses.

(3) The workpiece material, and particularly its thermal properties, influence the distribution of residual stress. However, neither the tool electrode material nor the dielectric fluid appear to affect significantly the residual stress distribution in the workpiece.

(4) Residual stresses cannot be avoided altogether, and machining conditions for minimising their extent almost inevitably conflict (as in the case of surface roughness) with rapid metal removal requirements. The comparatively thin stressed surface layers could be removed (e.g. by chemical milling); alternatively the stresses could be eliminated by annealing, or by mechanical treatment (e.g. shot peening) residual compressive stresses could be induced in the surface following electro-discharge machining.

(5) The work of others has shown that reductions in fatigue life of electro-discharge machined metals arise more from the surface irregularities, and the possible occurrence of cracking, than from residual stresses.

REFERENCES

1. J. A. KOK, 1963. *Conf. Electronic Processes in Dielectric Liquids, Durham,* Institute of Physics, London, Paper 8.5.
2. F. LLEWELLYN JONES, 1950. 'Electrode erosion by spark discharges', *Brit. J. Appl. Phys.,* 1, 60.
3. V. M. ZIMIN, 1961. *Izv. VUZ, Fiz.,* 6, 164–71.
4. B. N. ZOLOTYKH, 1957. *Izv. Akad. Nauk SSSR,* 1, 38.
5. B. N. ZOLOTYKH, 1959. *Radio Engng. Electron.,* 4, 8.
6. B. N. ZOLOTYKH, et al., *Probl. Elektr. Obra. Mat. Tr. TsNIL-Electrom. Akad. Nauk SSSR,* 58–64.
7. Y. KIMOTO, 1962. *Inst. Elect. Eng. Japan J.,* 82, 530–6.
8. I. Z. MOGILVESKII and A. R. KUTSAR, 1962. *Probl. Elektr. Obra. Mat. Tr. TsNIL-Electrom, Akad. Nauk SSSR,* 59–65.
9. A. S. ZINGERMAN, 1955. *Zh. Eksperim. i Tear. Fiz.,* 25, 1931–43.
10. G. R. WILMS and J. B. WADE, 1956. 'Some observations with the electric spark machining processes', *Metallurgica,* 54, 263.
11. H. OPITZ, 1960. 'Metallurgical aspects and surface finish', *Spark Machining Symp., Metal Treatment and Drop Forging,* pp. 237–50.
12. M. M. BARASH, 1958. 'Investigation into some aspects of the spark erosion process', *Ph.D. Thesis,* University of Manchester.
13. M. M. BARASH, 1961. 'Electric spark machining', *Proc. 2nd Intern. M.T.D.R. Conf., Manchester.*
14. M. M. BARASH and M. G. SRI-RAM, 1962. 'Some properties of the spark-machined heat-treated steels', *Proc. 3rd Intern. M.T.D.R. Conf., University of Birmingham.*
15. M. M. BARASH, 1965. 'Effect of EDM on the surface properties of tool and die steels', *Metals Eng. Quart., ASM,* Nov. 48–51.
16. H. K. LLOYD and R. H. WARREN, 1965. 'Metallurgy of spark machined surfaces', *J. Iron Steel Inst.,* March, 238–47.
17. I. A. BUCKLOW and M. COLE, 1969. 'Spark machining', *Met. Rev.,* Review No. 135, 3, No. 6, June, 103–13.
18. V. P. ALEKSANDROV. 'Residual stresses and the long term and fatigue strengths of heat-resistant materials after electro-spark machining', *Electrospark Machining of Metals,* Vol. 3 (Ed. B. R. Lazarenko), Consultants Bureau, New York.
19. V. P. ALEKSANDROV and B. N. ZOLOTYKH, 1958. 'On the selection of the optimum conditions in machining nickel-based heat-resistant alloys by the electrospark method', *Izv. Akad. Nauk SSSR, Otd. Tekhn. Nauk.,* No. 6.
20. GOLDSCHMIDT, 1959. *Iron Steel,* 469–71.
21. *Symp. Internal Stresses in Metals and Alloys,* 1948. Institute of Metals, London.
22. *Symp. Internal Stresses and Fatigue of Metals,* 1958. Detroit and Warren, Michigan.
23. K. HEINDLHOFER, 1951. *Evaluation of Residual Stresses,* McGraw-Hill, New York.
24. J. O. ALMEN and P. H. BLACK, 1963. *Residual Stresses and Fatigue in Metals,* McGraw-Hill, New York.
25. R. J. TREUTING et al., 1951. *Residual Stress Measurement,* ASM, Pittsburg, Pa.
26. W. R. OSGOOD, 1954. *Residual Stresses in Metals and Metal Construction,* Reinhold, New York.
27. A. A. DENTON, 1966. 'Determination of residual stresses', *Met. Rev. Inst. Metals,* 11, 1–23.
28. U. R. K. RAO, 1971. 'An analysis of residual stresses in the basic cutting process', *Ph.D. Thesis,* University of London.
29. C. W. GADD, 1946. 'Residual stress indications in brittle lacquer', *Proc. Soc. Exptl. Stress Analysis,* 4, No. 1, 74.
30. A. J. TOKARCIK and POLZIN, 1952. *Proc. Soc. Exptl. Stress Analysis,* 10, No. 1, 237.
31. D. M. TURLEY, 1968. 'Deformed layers produced by machining 70/30 brass', *J. Inst. Metals,* 96, March, 82–5.
32. G. U. OPPEL, 1946. *Proc. Soc. Exptl. Stress Analysis,* 21, No. 1, 135.
33. M. R. LECLOUX, 1965. 'Evaluation of principal surface stresses by determination of Hertz's hardness', *Rev. Franc. Mecan.,* No. 13, 75–81.
34. B. STENGEL and Th. GAYOMANN, 1970. 'Determination of residual stresses by indentation hardness testing', *AGARD Conf. Proc.,* No. 53, p. 16.
35. J. MATHAR, 1934. *Trans. A.S.M.E.,* 56, 259.
36. B. R. LAKE et al., 1970. 'An investigation of the hole drilling technique for measuring planar residual stresses in rectangular orthotropic materials', *Exptl. Mech.* 10, No. 6, 233–9.
37. D. I. CRECRAFT, 1967. 'The measurement of applied and residual stresses in metal using ultrasonic waves', *Sound and Vibration,* 5, No. 1, Jan., 173–92.
38. 'Electrochemical technique for residual stresses', 1969, abstract reported in *Machine Design (U.S.A.),* 16th Oct., 153.
39. J. IWAYANAGI and S. ABUKU, 1968. 'A contribution to the magnetic measurement of stress in plastically deformed carbon steel', *Proc. 11th Japan Congr. on Mat. Res.*
40. R. B. OLIVER, 1970. 'A review of non-destructive methods for evaluation of residual stresses and stress corrosion', *AGARD CP*-53, paper 17.
41. C. S. BARRETT and T. B. MASSALSKI, 1966. *Structure of Metals,* 3rd edn., McGraw-Hill, New York.
42. E. MACHERAUCH, 1966. 'X-ray stress analysis', *Exptl. Mech.,* 6, Mar., 140–53.
43. D. N. FRENCH and B. A. MACDONALD, 1969. 'Experimental methods of X-ray stress analysis', *Exptl. Mech.,* 9, Oct., 456–62.
44. A. L. CHRISTENDON (Ed.) et al., 1960. 'Measurement of stress by X-ray', *SAE Rep.,* No. TR-182, 36.
45. S. TAIRA and Y. YOSHIOKA, 1965. 'X-ray investigation on the residual stress of metallic materials', *Bull. J.S.M.E.,* 5, No. 31, 307–14.
46. B. D. CULLITY, 1956. *Elements of X-ray Diffraction,* Addison-Wesley, Reading, Mass.
47. N. DAVIDENKOV, 1932. 'Bending deflection method', *Z. metallku.,* 24, 25.
48. G. SACHS and G. ESPEY, 1941. 'The measurement of residual stresses in metals', *Iron Age,* 148, 18th Sept. 63–71; 24th Sept. 36–42.

49. A. A. DENTON and J. M. ALEXANDER, 1963. 'On the determination of residual stresses in tubes', *J. Mech. Engng. Sci.*, **5**, No. 4, 75.

50. R. G. TREUTING and W. T. READ, 1951. 'A mechanical determination of biaxial residual stress in sheet materials', *J. Appl. Phys.*, **22**, 130–4.

51. H. R. LETNER, 1953. *Proc. Soc. Exptl. Stress Analysis*, **10**, No. 2, 23.

52. B. C. KHOR, 1971. 'A study of residual stresses and surface effects in electro-discharge machining', *M.Sc. Thesis*, University of London.

53. F. VAN DYCK, and R. SNOEYS, 1971. 'Thermo-mathematical analysis for electro-discharge machining operations', *Proc. Natl. Conf. Tech. of Non-Traditional Machining of Metals, Timisoara, Roumania*, Paper B1, pp. 21–38.

HIGH VELOCITY FORMING

THE EFFECT OF IMPACT SPEED AND LUBRICANT IN HOT FORGING: PART 1. INTERFACE FRICTION AND DIE CAVITY PRESSURE

by

A. D. SHEIKH,* T. A. DEAN,* M. K. DAS* and S. A. TOBIAS*

SUMMARY

The effectiveness of lubricant conditions on the hot forging of steel has been investigated using four machines with ram speeds ranging from 0.4 in/s to 62 ft/s. Four conditions of lubrication have been used, namely Copaslip, colloidal graphite in water, clean dry dies and lightly oxidised billets and clean dry dies with heavily oxidised billets.

Ring tests were employed to determine the frictional restraints at the billet–die interface, during metal flow. Subsequently a cavity die was used to measure gas pressures in the unfilled portion of the die cavity, resulting from each deformation speed and condition of lubrication.

INTRODUCTION

The main reason for using a lubricant in hot forging processes is to eliminate contact between the stock and tools; and any medium which promotes this is, in drop forging vernacular, a lubricant. Metal-to-metal contact can result in 'pick-up', i.e. welding, between the product and the dies. This leads to the following.

(a) Damaged and perhaps unusable tools.
(b) Blemished components.
(c) Components which cannot be removed from the tools.

To obviate 'pick-up' a lubricant must form an unbroken barrier between tools and the stock, the surface area of which may increase by an order of magnitude during a forming operation.

Two other useful features of a lubricant are as follows.

(a) *Lubricity.* This term is used to describe the property of lowering the frictional drag at the billet–die interface. It is of importance when components having large surface areas and thin sections are made. In such cases high friction coefficients can raise required forging loads prohibitively.

(b) *Repellent action.* The ejection of hot forgings from deep die cavities is facilitated by the use of lubricants which decompose. Oil or saw dust is often used for this purpose. The pressure of the resulting gases, trapped by deforming stock in unfilled portions of the die cavity, can reach a value sufficient to force the component from the tools without the aid of mechanical ejectors. Repellents are widely used for normal drop forgings, where the dimensional tolerances on components are fairly generous. However, their use often results in underfilling and they are not recommended for precision forgings.

The performance of lubricants under conditions of bulk plastic deformation has been the subject of various types of investigation. Most of these have been directed to the determination of friction coefficients.

Direct measurements of coefficients of friction have been made using pin load cells. A notable early use of these to measure friction in strip forming rolls was by Siebel and Lueg[1]. Similar uses have been found for these devices by Smith[2] *et al.* and Van Rooyen and Backofen[3]. Subsequent developments in the use of pin load cells have been well documented recently by Cole and Sansome[4].

The direct measurement of friction in the deformation of hot metals has been made by Peterson and Ling[5] using apparatus in which metal foil was deformed between a slender anvil and a flat tool. Coefficients of friction were determined by the ratio of the compressive load to the load required to move the foil and anvil across the tool. For a number of pure metals friction was significantly affected by the tool material, its surface roughness, interface temperature and the amount of oxidation existing initially on the surface of the test piece. The disadvantage of these tests is that the contact areas between tools and test pieces were necessarily small; a condition which does not arise in most metal working operations. Pawelski[6] and co-workers have measured frictional constraints in hot working by extruding square sectioned specimens through flat open-sided tools. Coefficients of friction were determined from measurements of punch pressures and lateral loads on the tool faces. It was found that for unlubricated surfaces friction was markedly dependent on the degree of surface oxidation. Colloidal graphite was shown to be a stable lubricant up to temperatures of about 840 °C.

A most useful method of indirectly measuring the coefficients of friction during metal deformation is the ring test first suggested by Kunogi[7] and Kudo[8] and developed by Male and Cockroft[9]. The advantages of this test are that specimens are deformed in a manner similar to that of practical working conditions, tool requirements are simple and the only measurements needed are those of geometrical changes of the specimens. Male[10] has used this test to determine the effect of temperature on frictional behaviour and concluded that in the absence of a lubricant the coefficients of friction at high temperatures depends largely on the mechanical properties of the oxide films formed. Jain and Bramley[11] have also used the same test to show that the lubricity of lubricants increases with forging speed during the hot upset forging process.

* Department of Mechanical Engineering, University of Birmingham

Tolkien[12] has performed a practical investigation of several properties of various lubricants and propellants in the drop forging of steel. Of the commonly used lubricants, graphite was shown to be particularly effective in reducing required forging loads and adhesive forces. Sawdust was demonstrated to produce very high pressures in die cavities, hence its good propellant action.

The object of the experiments reported in this paper is (by using four machines having a wide range of impact velocity) to study the effect of forging speed on two facets of lubrication which are important to the filling of die cavities. They are the following.

(a) The coefficient of friction between stock and tools.

(b) The pressure of gases trapped in initially unfilled portions of the cavity.

EXPERIMENTAL MACHINES

Four forging machines were used in the experiments.

They were as follows.

(a) A Petro-Forge Mk II D–10K[13] with a range of ram speeds of 22–62 ft/s.

(b) A slow-speed Petro-Forge[14] with impact speeds ranging between 5 and 22 ft/s.

(c) A 200 ton eccentric press using impact speeds of 0.4–1.5 ft/s.

(d) A 150 ton hydraulic press with ram speeds of 0.4–2.5 in/s.

EXPERIMENTAL TOOLING

A common set of bolsters was used for all the experiments. Two types of tooling were employed.

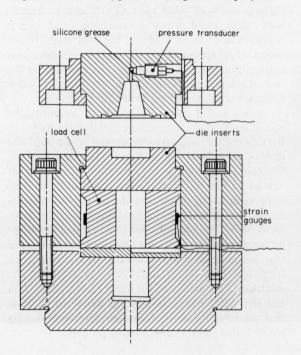

Figure 1. Cavity die tool set.

(a) For the friction ring tests, flat dies made from a 5% chromium hot working steel were used.

(b) A die of the form shown in figure 1 was used for the cavity forging tests. The tool material and heat treatment were as for the flat dies. A strain-gauged load cell was positioned under the lower die to obtain measurements of forging loads reported in part 2. The height of the tapered boss in the upper die was 1.55 in. A small bore hole was drilled to lead from this cavity to the diaphragm of a pressure transducer, which was mounted horizontally in the top die, to ensure that the shock loads due to forging did not interfere with its output. This fine hole was filled with silicone grease.

IMPACT VELOCITY AND RAM DISPLACEMENT

A different method of measuring impact velocity was used on each of the machines.

On the Mk II Petro-Forge the velocity was determined by differentiating the record of ram displacement against time.

An Optron tracker model 680, an optical electronic instrument, which requires no physical contact with the ram, was used to measure displacement.

The impact velocity of the slow-speed Petro-Forge was measured directly using an inductive velocity transducer.

The slide velocity of the eccentric press was determined from a knowledge of the rotational speed of the flywheel and the geometry of the links.

The speed of the hydraulic press was determined from records of displacement obtained from a capacitance displacement transducer.

DIE CAVITY PRESSURES

Pressure measurements were obtained by means of a Kistler piezo-electric transducer type 6201.

EXPERIMENTAL DATA

Readings of velocity, displacement, and gas pressure were recorded on a Tektronix type 564 storage oscilloscope.

TEST PIECES

All test pieces for the hot forging experiments were machined from bright drawn En 8 bar stock. Specimens for the ring tests were made to the geometry originally suggested by Male and Cockroft[9], as shown in figure 2. Also shown in this figure are the billets used for the cavity die forging tests.

Preformed billets were employed to obtain larger amounts of die filling and commensurate high pressures in the cavity.

Cylindrical and preformed billets were also made in commercially pure aluminium for cold cavity die forging tests.

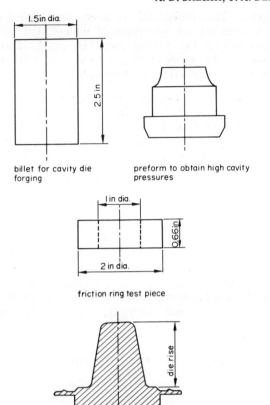

billet for cavity die
forging

preform to obtain high cavity
pressures

friction ring test piece

typical section of forging

Figure 2. Test pieces.

LUBRICANTS

Two lubricants were used.

(a) Copaslip, a mixture of copper and lead particles in a base of bentone grease. This substance is normally used as an anti-seize compound.

(b) Colloidal graphite in water, a normal hot working lubricant.

BILLET HEATING

All steel test pieces were heated to 1100 °C in an electric muffle furnace. Two conditions of heating were employed for the friction rings.

(a) Specimens were protected with a steel cover and heated for 8 min. This produced the minimum amount of surface oxidation obtainable. Measurements based upon those described elsewhere[15] indicate that an oxide layer 0.001–0.0015 in thick was formed.

(b) Specimens exposed to the furnace atmosphere were heated for 20 min. An oxide layer between 0.006 in and 0.007 in was obtained in this manner. Billets, heated thus, are later referred to as having the maximum oxide layer.

All ring specimens were placed on edge in the furnace to ensure similar oxidation of top and bottom surfaces.

The billets for the cavity die forgings were unprotected in the furnace and heated for 20 min.

TEST CONDITIONS

For all the hot deformation tests the tools were preheated to approximately 100 °C. They were then cleaned with acetone and the appropriate lubricant brushed over all the working surfaces.

The time intervals between the removal of the test pieces from the furnace and deformation were between 3 and 6 s.

Friction ring tests

For these tests four conditions of test piece and die surfaces were used.

(a)   Copaslip lubricant maximum oxide thickness.
(b)   Graphite lubricant maximum oxide thickness.
(c)   No lubricant maximum oxide thickness.
(d)   No lubricant minimum oxide thickness.

Two methods were used to obtain various amounts of deformation on the Petro-Forge machines.

(a) The impact velocity of each machine was varied and the blow energy wholly absorbed by the deforming test piece. Deformation was thus dependent on the impact velocity.

(b) Clash rings of various heights were placed on the lower die which arrested the ram at a predetermined position irrespective of its impact velocity.

Clash rings were used to obtain different test piece reductions on the hydraulic press. On the eccentric press test piece deformation was varied by adjusting the shut height.

Cavity forging tests

Three machines were used for the hot forging tests. They were the Mk IID Petro-Forge, the slow-speed Petro-Forge and the eccentric press. All billets were coated with the maximum thickness of oxide. Different amounts of deformation were obtained by varying the blow velocity on the impact machines and by varying the shut height of the eccentric press.

Cold aluminium forging

These tests were conducted on the slow-speed Petro-Forge only. No lubricant was used and the dies were not pre-heated.

EFFECT OF IMPACT SPEED ON FRICTION COEFFICIENT

The results obtained from the friction ring tests are summarised in figures 3, 4 and 5. No sensible differences in coefficients of friction could be detected, from tests conducted at the same impact speeds, whether or not crash rings were used. All these figures show the coefficient of friction to decrease with impact speed and to increase with deformation. At all speeds except the lowest (that of the hydraulic press) the lubricity of Copaslip proves superior to that of colloidal graphite and the unlubricated condition. At the speed of the hydraulic press the measured coefficients of friction are lowest for the graphite lubricant (figure 4). A probable reason for the poor performance of Copaslip in the hydraulic press is that owing to the longer process times occasioned on this machine the grease base of the Copaslip decomposes. Indeed it was noted that burning rapidly occurred when hot billets were placed in the dies.

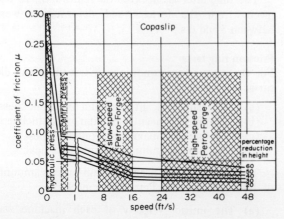

Figure 3. The relation of friction and speed using Copaslip.

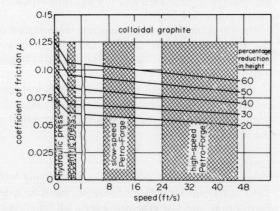

Figure 4. The relation of friction and speed using graphite.

It can be seen in figure 5 that the friction coefficients for billets with thick oxide layers are lower at all speeds and all deformations than for those with thin oxide layers. Complete sticking friction ($\mu = 0.577$) was recorded for the thinly oxided billets deformed to 60% on the hydraulic

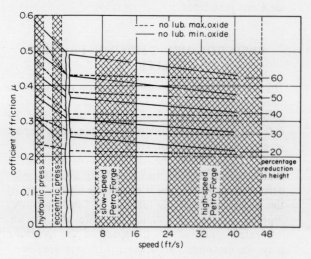

Figure 5. The relation of friction and speed – no lubricant.

press. The differences between the coefficients recorded for the two types of unlubricated billets decrease as the forging speed increased. It thus appears that oxide coatings on hot steel billets can have two major effects on forgeability.

(a) They can physically separate the billet and die surfaces and prevent 'pick-up'.

(b) If the shear strength of the oxides is lower than that of the billet, slip will occur within them rather than in the billet and they will serve as a 'lubricant'.

It is notable that, over a wide range of deformation, the decrease of friction coefficient, for all forging conditions, is virtually linear. However, the most rapid decrease occurs between the speed ranges of the hydraulic and eccentric press. It should be noted that the calibration for the friction ring test is strictly valid when measuring coefficients which do not vary with deformation. The values plotted in the curves under discussion will represent cumulative results of varying frictional restraints up to the final deformation.

GAS PRESSURES IN THE DIE CAVITY

Figures 6 and 7 show the recorded gas pressures as a function of the volume compression ratio within the unfilled portion of the die cavity.

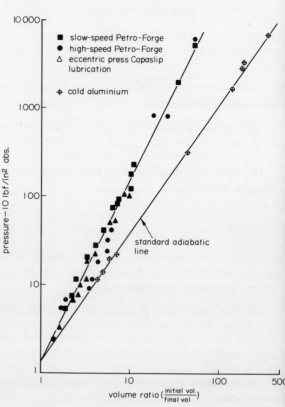

Figure 6. Pressure rise in die cavity using Copaslip.

The results obtained from cold forging aluminium billets on the slow-speed Petro-Forge are shown in both diagrams. It can be readily seen that these points fall substantially on the line drawn for adiabatic compression. It can thus be inferred that, not only did the entrapped air behave adiabatically, but also no leakage occurred either around the transducer or past the deforming billet.

In figure 6 it is seen that no substantial differences arose in the pressure rise, for a given compression on each of the machines, when Copaslip was used for a lubricant. A log–log plot of the data is sensibly linear and corresponds to a compression index of 2.

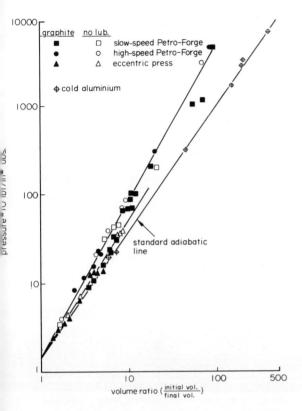

Figure 7. Pressure rise in the die cavity using (a) graphite; (b) no lubricant.

The results plotted for the graphite and no lubrication conditions fall into two groups. Those obtained at low speeds (on the eccentric press and for low compression ratios of the slow-speed Petro-Forge, when the blow energy was low) lay substantially on a straight line, close to but a little above that for adiabatic compression, having a compression index of about 1.5. The pressures obtained at higher speeds (high-speed Petro-Forge and large compressions on the slow-speed Petro-Forge) are greater for a given compression ratio. Again the plotted points are substantially linear with a compression index of approximately 1.7.

As the pressure rises were greater than adiabatic in the case of the unlubricated billets it appears that a significant amount of heat was transferred from the hot billets to the entrapped air.

The fact that the results, for a particular machine, obtained for both the graphite and unlubricated condition, are similar, indicates that the graphite remained inert throughout the forging process.

The high pressures caused by the Copaslip lubricant can be attributed to burning of the grease base by the hot billets. The resulting high pressures were noted to eject forgings with an audible report.

LUBRICANTS AS PROTECTIVE MEDIA

No gross pick-up was observed in any of the cavity die forging tests, although signs of abrasion were noted on the flash lands and the transition radius around the tapered boss. It appears then that for the degree of flow encountered in the cavity used even a thin layer of surface oxide is sufficient to separate tool and billet surfaces.

CONCLUSIONS

From the results of the experiments reported in this paper the following conclusions have been made.

(a) For a given condition of lubrication the coefficient of friction decreases with increased impact velocity.

(b) The coefficient of friction between the billet–die interface is dependent more on impact velocity than the change in velocity during deformation.

(c) Copaslip provided the lowest frictional forces on all machines except the hydraulic press. Colloidal graphite was superior on this machine.

(d) In the absence of a lubricant a lower friction coefficient is obtained with a thicker oxide layer.

(e) For all lubricant conditions the pressure rise in the die cavity is greater than adiabatic, when forging hot billets.

(f) The highest pressure rises are obtained using Copaslip.

(g) No significant differences in pressure rise for a given die filling, are obtained using Copaslip on each of the three machines.

(h) The same pressure rise, for a given die filling, is obtained using graphite as for no lubricant.

(i) Gas pressures considerably higher than the flow stress of the forging stock, can occur.

(j) All the surface conditions employed are effective in preventing 'pick-up' between stock tools.

(k) The high gas pressures formed within the die cavity indicate the need for well-vented dies for both low- and high-speed forging processes.

REFERENCES

1. E. SIEBEL and W. LUEG, 1933. 'Untersuchungen über die Spannungsverteilung im Waltzspalt', *Mitt. K. W. Inst. Eisenforschung*, Dusseldorf, **15**, 1–14.
2. C. L. SMITH, F. H. SCOTT and W. SYLWESTROWIZ, 1952. 'Pressure distribution between stock and rolls in hot and cold flat rolling', *J. Iron Steel Inst.*, **170**, 347–59.
3. G. T. VAN ROOYEN and W. A. BACKOFEN, 1957. 'Friction in cold forging', *J. Iron Steel Inst.*, **186**, 235–43.
4. I. M. COLE and D. H. SANSOME, 1968. 'A review of the application of pin load cell pressure measurement techniques to metal deformation processes', *Proc. 9th Intern. M.T.D.R. Conf.*, pp. 271–86.
5. M. B. PETERSON and F. F. LING. 'Friction and Lubrication in Hot Metal Deformation'. Mechanical Technology Incorporated, Luttam, New York, AD 630–204 March 20: 1966.
6. O. PAWELSKI, G. GRAVE and D. LOHR, 1969. 'Coefficient of friction and temperature distribution during hot forming of steel with different lubricants', *Proc. Iron Steel Inst.* and *Inst. Mech. Engrs. Tribol. Conf.*

7. M. KUNOGI, 1954. 'On the plastic deformation of the hollow cylinder under axial load', *J. Sci. Res. Inst. Japan*, **30**, No. 2, 63–92.

8. H. KUDO, 1955. 'An analysis of plastic compression deformation of a lamella between rough plates by the energy method', *Proc. 5th Japan Natl. Congr. Appl. Mech.*, Vol. 5, pp. 75–8.

9. A. T. MALE and M. G. COCKROFT, 1964. 'A method for the determination of the coefficient of friction of metals under conditions of bulk plastic deformation', *J. Inst. Metals*, **93**, 38–46.

10. A. T. MALE, 1964. 'The effect of temperature on the frictional behaviour of various metals during metal working', *J. Inst. Metals*, **93**, 489–94.

11. S. C. JAIN and A. N. BRAMLEY, 1967–8. 'Speed and frictional effects in hot forging', *Proc. Inst. Mech. Engrs.*, **182**, pt. 1, 783–95.

12. H. TOLKIEN, 1961. 'Lubricant effects in drop forging dies', *Werkstattstechnik*, **51**, No. 2, 102–5.

13. S. A. TOBIAS, 1970. 'The Petro-Forge forming system', *A.S.M.E. Tech. Paper*, No. MF–70–184.

14. A. D. SHEIKH, M. K. DAS and S. A. TOBIAS, 1971. 'The development of the slow-speed Petro-Forge machine', *Intern. J. Machine Tool Design Res.*, **11**, 13–29.

15. T. A. DEAN, 1970. 'A preliminary warm forging essay', *Proc. 11th Intern. M.T.D.R. Conf.*, pp. 779–801.

THE EFFECT OF IMPACT SPEED AND LUBRICANT IN HOT FORGING: PART 2. METAL FLOW AND FORGING LOADS

by

A. D. SHEIKH,* T. A. DEAN,* M. K. DAS* and S. A. TOBIAS*

SUMMARY

The effect of lubrication on the flow of metal in a cavity die and the concomitant die loads has been studied. The die set was provided with a flash and gutter and allowed both lateral and vertical flow of stock.

Three forging machines having impact speeds ranging from 0.4 ft/s to 62 ft/s were used.

Two different treatments of the die surface were employed. They were a coating of 'Copaslip' and a covering of colloidal graphite.

INTRODUCTION

The coefficient of friction at the billet—die interface was shown, in part 1 of this paper, to be markedly affected by the deformation speed.

The influence of friction in an upsetting operation is manifested in two major ways.

(a) The deformation of the billet is rendered inhomogeneous and barreling of the sides of billets occurs.

(b) The forging loads required for a given deformation are increased as the friction coefficient increases.

The load dependency on friction in an upset forging operation was first illustrated by the approximate analysis of Siebel[1]. Later work by Schroeder and Webster[2] has shown how the effect of friction on forging load increases significantly with the amount of deformation.

Frictional restraints similarly affect the forging loads for an extrusion operation and cause uneven deformation within the product.

The flow of metal in most drop forging processes is complicated and defies theoretical analysis. A rather gross but useful simplification of deformation can be made by considering a flow model in which material in a forging die flows either axially (extrusion into a loss of flange) or laterally as in an upsetting process. In most operations these forms of flow will occur simultaneously and in degrees which vary throughout the process.

Previous studies of metal flow in a forging operation have been made in which concurrent extrusion and spreading forms of flow were obtained in test pieces. Dean[3], using cold unlubricated aluminium billets, concluded that forging speed had no significant effect on the mode of deformation. Later work[4] has demonstrated that important factors influencing the relation between lateral spread and vertical extrusion are the size of billet compared with the dimensions of the die cavity and the lubricity between the stock and die surfaces. It was also found that the forging loads depended solely on the upset area of the forged component and were independent of the amount of extruded material.

The intention of the work described in this paper is to determine the effect of the two lubricants and three of the machines used in part 1 on forging loads and metal flow. Forging loads required to upset rings are first examined. The loads and consequent deformations within the cavity die, which allows both axial and lateral movement of the billet are then determined.

EXPERIMENTAL MACHINES

Three of the machines used in part 1 were used.

(a) A Petro-Forge Mk II D—10K[5] with a range of ram speeds of 22—62 ft/s.

(b) A slow-speed Petro-Forge[6] with ram speeds between 5 and 22 ft/s.

(c) A 200 ton eccentric press with a slide speed varying between 0.4 and 1.5 ft/s.

EXPERIMENTAL TOOLING

The tool sets were the same as used in part 1 of this paper. The cavity die is shown in figure 1.

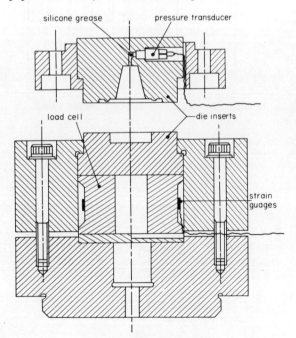

Figure 1. Cavity die tool set.

* Department of Mechanical Engineering, University of Birmingham

The height of the boss in the upper die was 1.55 in and the walls were provided with a draft of 10°. A flash and gutter was incorporated to constrain lateral material flow. The ratio of the flash land width to flash thickness, with dies fully closed, was 5 and the minimum flash thickness was 0.04 in.

The cavities were replaced by flat dies for upsetting rings. A load cell was placed beneath the lower die in each of the two configurations used.

FORGING LOADS

Eight Hawker-Siddeley paper-backed strain gauges, of nominal resistance 250 ohms, were attached to the load cell. It was calibrated using a 300 ton hydraulic testing machine. The calibration was subsequently checked dynamically by comparing its output with that of a piezo-electric accelerometer attached to the ram of the slow-speed Petro-Forge.

The output of the load cell was transmitted through a Hottinger strain gauge bridge of 50 kHz carrier frequency and recorded on a Tektronic type 564, storage oscilloscope.

TEST PIECES

Test pieces were machined from bright drawn En8 bar stock. The ring test pieces are described in part 1. Their dimensions are shown in figure 2. Billets for the cavity die forging tests were machined to the size also shown in figure 2.

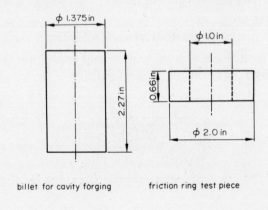

billet for cavity forging   friction ring test piece

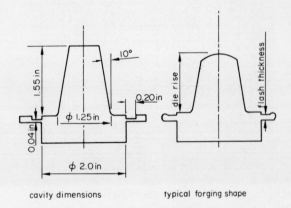

cavity dimensions   typical forging shape

Figure 2. Forging test pieces.

LUBRICANTS

The two lubricants employed in part 1 of this paper were used.

(a) Copaslip, a mixture of copper and lead particles in a base of bentone grease.

(b) Colloidal graphite in water.

For a particular test one of these lubricants was brushed over the whole surface of the pre-heated (to approximately 100 °C) tooling.

BILLET HEATING

All test pieces were heated to 1100 °C in an electric muffle furnace. The heating time was maintained at 20 min. An oxide layer of thickness between 0.006 in and 0.007 in was obtained on the billet surfaces. The friction rings were placed in the furnace on edge so as to obtain even oxidation on top and bottom faces.

TEST PROCEDURE

The transport time between the removal of the test pieces from the furnace and the start of deformation varied between 3 and 6 s.

Different degrees of deformation were obtained, on the Petro-Forge machines, by varying impact velocity and on the press by adjusting the shut height.

FORGING LOAD REQUIREMENTS FOR UPSETTING RINGS

The curves plotted in figure 3 show the relationships between maximum forging load and deformation for two conditions of lubrication on the three machines. For a particular machine the relative disposition of each curve is as expected from the measured

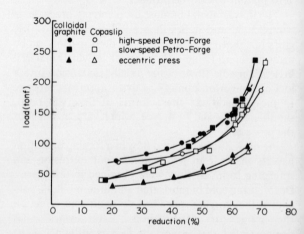

Figure 3. Deformation loads versus reduction – friction rings.

coefficients of friction reported in part 1. The required load at a given deformation is less for Copaslip than for the graphite lubricant. The differences occasioned by the two lubricants, when used on the eccentric press, are small and insignificant at low deformations. This reflects the fact that the friction coefficients for the two lubricants are similar when used on this machine.

The greater load requirements on the Petro-Forge machines are due to the increased flow stresses caused by the higher deformation velocities. The relation of the curves for the Petro-Forge machines is the result of three contributory factors.

(a) Interface friction, which judging from the curves in part 1 will tend to reduce the loads on the Mk IID machine. This reduction in friction between the slower- and higher-speed machine is greater for greater deformations. Thus the contribution of the lubricant in reducing loads will increase with reduction.

(b) Strain rate, which will cause the flow stress of the billet material to be greater on the higher-speed machine than on the slow-speed Petro-Forge.

(c) Billet temperature; this rises during the process due to the deformation work. A greater rise is to be expected on the Mk IID Petro-Forge because of the initially higher flow stress. This increase in temperature can cause thermal softening and, particularly in the latter stages of deformation, can tend to reduce the flow stress.

RELATION OF FLASH THICKNESS TO DIE RISE

Figure 4 shows the relation between die rise and the thickness of flash formed for a particular billet deformation. No significant differences are discernable between the results obtained using Copaslip and those for graphite lubrication, on each of the Petro-Forge

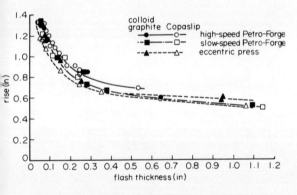

Figure 4. The relation of die rise to flash thickness

machines. It thus appears that the mode of metal deformation was unaffected by lubricant on these machines.

The results obtained from the eccentric press are represented by two distinct curves, showing that a particular rise was obtained with a thicker flash using graphite than with Copaslip, indicating that graphite inhibited lateral flow to a greater degree than the other lubricant.

Differences in the mode of flow obtained on each machine are also readily apparent in these curves. At higher values of rise a given die filling is obtained with a thicker flash on the machine with the highest impact speed than on the other two machines. A thinner flash (that is, more deformation) is required on the eccentric press, to obtain a particular rise, than on the two Petro-Forge machines. This result

provides contrary evidence to a common assumption that flash and gutters are ineffective at high forging speeds.

RELATION OF FLASH THICKNESS TO MAXIMUM FORGING LOAD

The curves of maximum forging load plotted on an axis of flash thickness, figure 5, are related in a manner similar to those of the previous figure.

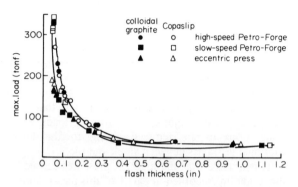

Figure 5. The relation of forging load to flash thickness.

Differences ascribable to the two lubricants when used on the eccentric press, noted in figure 4, are not apparent in the plots of figure 5. It is evident from these curves that the loads required to close fully the dies (the flash thickness at this stage is 0.04 in) increase with increase in speed of the forging machine. Thus, although the flash is more effective in increasing die rise on the highest-speed machine as evidenced in figure 4, the forging loads are commensurately increased.

EFFECT OF MAXIMUM FORGING LOAD ON DIE RISE

Figure 6 compares the die rise obtained for the different forging conditions with required maximum forging loads. Two discrete curves are identifiable for the two lubricants used on each machine. As expected

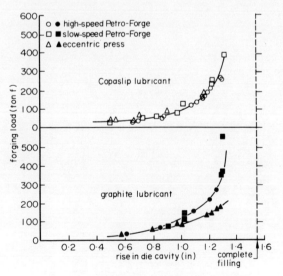

Figure 6. The relation of die rise to forging load.

from consideration of figure 4, rise was more readily obtained on the eccentric press using graphite lubricant than on the two higher-speed machines.

The relatively poor performance of Copaslip on the press is due to the fact that this lubricant promotes a greater spread of material (and in consequence a thinner flash) for a given die rise. The increased load requirements are due to this excess flash formation. A reason for the relative ease of obtaining rise on the two Petro-Forge machines using Copaslip can be proposed by consideration of figure 4 and the results of the friction ring tests reported in part 1. Figure 4 shows that the mode of metal flow is similar for the two conditions of lubrication. Therefore disregarding other process variables the deformation loads should be the same. However, it is shown in part 1 that the coefficient of friction occasioned by Copaslip is appreciably less than that associated with graphite. Thus forging loads when using Copaslip are reduced owing to the lower frictional restraints between stock and die.

It is notable that the high gas pressures caused by the 'Copaslip', recorded in part 1 of this paper, appear to have an insignificant effect on forging loads.

CONCLUSIONS

The results obtained from the tests described in this paper justify the following conclusions.

(a) Deformation loads required to upset ring specimens increase with interfacial friction.

(b) Lower upsetting loads are required using 'Copaslip' than with graphite.

(c) Higher upsetting loads are required for a given deformation on the Petro-Forge machines than on the eccentric press.

(d) The mode of flow is similar when using the two different lubricants on each of the Petro-Forge machines. On the eccentric press less lateral flow is obtained using graphite.

(e) Greater rise is obtained for a particular deformation on the Petro-Forge machines than on the slower-speed eccentric press.

(f) Higher loads are required on the Petro-Forge machines for a given die rise than on the eccentric press.

REFERENCES

1. E. SIEBEL, 1923. *Stahl und Eisen,* **43,** 1295.
2. W. SCHROEDER and D. A. WEBSTER, 1949. 'Press forging thin sections: effect of friction, area and thickness on pressures required', *J. Appl. Mech.,* 289–94.
3. T. A. DEAN, 1966. 'A comparison of high rate and conventional forging machines', *Proc. 7th Intern. M.T.D.R. Conf.,* pp. 25–40.
4. S. C. JAIN, A. N. BRAMLEY, C. H. LEE and S. KOBAYASHI, 1970. 'Theory and experiment in extrusion forging', *Proc. 11th Intern. M.T.D.R. Conf.,* pp. 1097–115.
5. S. A. TOBIAS, 1970. 'The Petro-Forge forming system', *A.S.M.E. Tech. Paper,* No. MF–70–184.
6. A. D. SHEIKH, M. R. DAS and S. A. TOBIAS, 1971. 'The development of the slow-speed Petro-Forge machine', *Intern. J. Machine Tool Design Res.,* **11,** 13–29.

THE EFFECTS OF TEMPERATURE AND SPEED ON THE WARM EXTRUSION OF STEEL

by

A. SINGH,* T. A. DEAN* and R. DAVIES*

SUMMARY

An investigation has been made of the effects of temperature and speed at various extrusion ratios in warm forward extrusion of steels. The steels selected for warm extrusion were En2E, En8, En19 and En30B. The extrusions were performed in the temperature range of 20 °C to 800 °C. The properties of the extruded products have also been examined.

INTRODUCTION

In recent years there has been an increasing interest in the warm forming of steels, in which the workpiece is initially heated to temperatures below those of recrystallisation. Warm forming allows the deformation of high-strength metals which cannot be formed cold, because of the excessive tool stresses encountered. Warm extrusion thus makes it possible to produce a new range of components, which have some of the shape and complexity of hot forged components, and at the same time retain a large proportion of the mechanical and physical properties of cold forged components.

The warm forming process is based on the fact that resistance of metal to deformation usually decreases as the temperature at which it is worked is raised. Workers at N.E.L.[1] and P.E.R.A.[2] have made intensive investigations into the warm working of steels. Dean[3] has reviewed some of the literature relating to the deformation behaviour of steels at elevated temperatures. Useful information on the flow stress at various strain rates and temperatures have been obtained by Cook[4] and Suzuki[5], where it was observed that yield stress of many highly alloyed steels was reduced at high strains. Generally the yield stress at a particular strain increases with strain rate up to a point when thermal softening due to plastic working predominates over work hardening, beyond which there is sharp drop of the yield stress. Depending on the strain rate, many common materials show a maximum flow stress at a temperature of about 400 °C, which is regarded as being due to strain age hardening[6].

One of the main effects of the high-speed deformation processes is to reduce friction. The effect of temperature and speed on friction during forward extrusion of steels has been investigated by Singh[7] by directly measuring billet and tool interface temperatures using surface thermocouples. The temperatures were related to the frictional work expended at the billet—tool interface. The coefficient of friction in the container was found to increase from 0.04 to 0.061 with an increase of billet temperature from ambient to 400 °C, and to decrease to a value of 0.054 at a temperature of 600 °C. It then remained fairly constant up to 800 °C. At room temperature the coefficient of friction was found to decrease with increase of ram velocity. For a billet surface roughness value of 90 CLA, and using graphite lubricant, the decrease of the coefficient of friction for a velocity range of 1 to 30 ft/s was found to be from 0.056 to 0.038 for a reduction ratio of 2.56. This decrease in frictional resistance at high speed was found to improve some of the mechanical, physical and metallurgical properties of the extruded product.

The object of the work reported in this paper was to investigate the effects of temperature in the range, ambient to 800 °C, and forming speeds of about 0.7 and 25 ft/s, on the extrusion pressures of a range of steels. The steels chosen for this work were En2E, En8, En19 and En30B. The effect of the forming condition on material flow, grain structure and hardness has also been assessed.

EQUIPMENT

Forging machines
High-speed work was conducted on a Petro-Forge Mk IIB machine[8]. The velocity at impact was in the range of 21 to 27 ft/s. The slow-speed work was carried out on a 200 tonf eccentric press with a platen velocity of about 0.7 ft/s at the beginning of extrusion.

Tooling
The general layout of tooling used for warm forward extrusion is shown in figure 1. The container A and die B were located in the main body C and clamped by ring D. An ejector guided by the pressure pad E was operated by a hydraulic cylinder mounted under the machine bed F. The extrusion punch G backed by pressure pad H was clamped by the holder I fixed to the moving platen J.

A duplex-type container made of En30B steel with a nominal internal diameter of 1.010 in was used for the warm forming tests. The dies were provided with a 120° die entry angle and reduction ratios of 1.52, 1.78, 2.56, 3.3 and 4.0. Dies and punches were made from 5% chromium steel.

Test materials
Test pieces 1 in diameter and $1\frac{1}{4}$ in long were machined from bars of En2E, En8, En19 and En30B steels. All billets were furnace annealed before use.

* Department of Mechanical Engineering, University of Birmingham

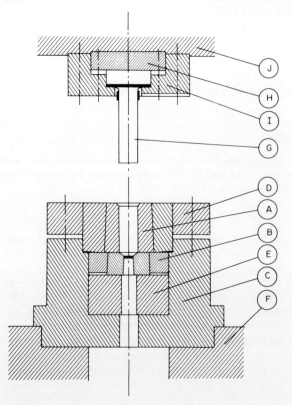

Figure 1. General layout of tooling for forward extrusion.

Lubrication

Zinc phosphate and Bonderlube lubricant was used for tests at room temperature. For other tests graphite was sprayed onto the billet prior to heating. At a temperature of 800 °C, the dies were also sprayed with lubricant.

Billet heating

The heating times necessary to bring the billet to test temperatures were established by Dean[3] for black and bright billets. The billets for extrusion tests were heated for 30 min at 800 °C and 1 h at 400 °C.

RESULTS AND DISCUSSION

Effect of temperature on extrusion pressure

The extrusion pressure and punch stroke characteristics for En2E at a reduction ratio of 3.3 are shown in figure 2. The effect of temperature on the extrusion pressure is comparatively more noticeable at high temperatures than at lower temperatures, particularly for En2E and En8 steels. The drop of pressure with displacement increases as the temperature is increased; this is discussed later. Figure 3 shows the effect of temperature on maximum extrusion pressure at extrusion ratios of 1.52, 1.78, 2.56, 3.3 and 4.0 for En2E, En8, En19 and En30B steels. Increases of billet temperature up to 600 °C had less effect on the extrusion pressure for En2E and En8 than other materials. The decrease of extrusion pressure is more noticeable in the temperature range of 600 to 800 °C for all steels. The curves for billet temperatures of 20 °C and 400 °C follow

sensibly a straight line. These curves conform to the results of previous workers which show that extrusion pressure is proportional to the area reduction. This implies that strain rate effects were negligible. The curves for 600 °C and 800 °C tend to

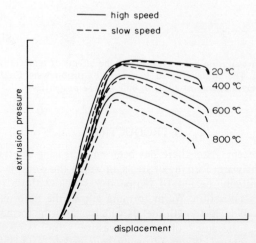

Figure 2. Typical extrusion pressure and displacement characteristics for En2E steel at a reduction ratio of 3.3.

depart from linearity as the strain is increased, which could be due to reducing friction or reducing yield stress.

The maximum extrusion pressures, which are reached as soon as the die is filled with the workpiece material are plotted in figure 3. Friction during this part of the process would play a less important role than the yield stress. For a given velocity an increase of extrusion ratio (or strain) increases the strain rate. Increasing strain rate normally increases the flow stress only slightly at 600 °C and more so at 800 °C, from which it would be expected that the lines of figure 3 should curve upwards. However, increasing yield stress would require more work input. This increased work input raises the temperature of the plastically deforming zone, which causes thermal softening, thus causing the yield stress to drop. Cook[4] found similar results in compression, and concluded that at high rates of strain thermal softening predominates over strain hardening as compression proceeds.

If the curves in figure 3 are plotted for a reduction ratio of 2.56, it can be shown that the increase of billet temperature had comparatively less effect on En2E than on the other steels. The effect of the increase of carbon content is to increase the flow stress. With the increase of billet temperature above 400 °C this effect of carbon content drops rapidly at both slow and high speeds.

An attempt was made to extrude En30B at room temperature and high speed. Cracking was observed in the extruded billet for a reduction ratio of 1.52 and for a reduction ratio of 1.78. Although the extrusion pressure for the latter reduction was greatly in excess of the allowable stress for the tools, no failure of the punch or die was observed at this reduction. This may be a characteristic of high-speed forming where tools have to sustain loads for a short interval. At 600 °C

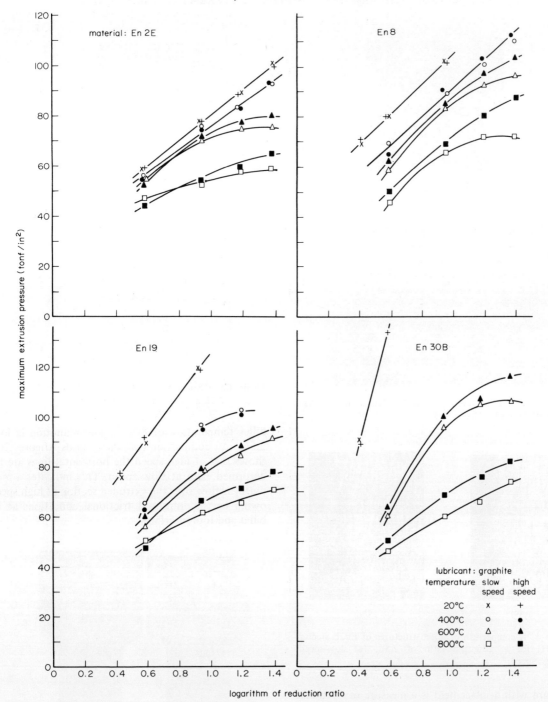

Figure 3. Effect of temperature on extrusion pressure for various reductions for En2E, En8, En19 and En30B steels.

it was still difficult to extrude En30B, and the high-speed extrusion at this temperature failed in a typical way, as shown in figure 4. This type of failure is known as internal cracking or central burst which is thought to be due to the extruded part moving faster than the plastically deformed zone, in materials of low ductility.

A metallurgical examination of the En2E billets at slow and high speed for a reduction ratio of 2.56, figure 5(a) and (b), shows that the material near the surface of the billet extruded at 800 °C had started to recrystallize. This is probably due to the increase in temperature caused by deformation work and friction at the billet die interface.

Figure 6 shows the effect of billet temperature on the hardness of extruded and undeformed parts for various steels extruded at a reduction ratio of 2.56. These curves show that a maximum value occurs around 400 °C which agrees with the results of Burgdorf[6].

Effect of speed on extrusion pressure
One of the most important effects of high ram speeds when compared with slow speeds is to reduce the coefficient of friction. This is shown to some extent in figure 2, where it is seen that the slope of the high-speed curves or the drop of pressure with displacement is less than in the slow-speed case. This reduction of

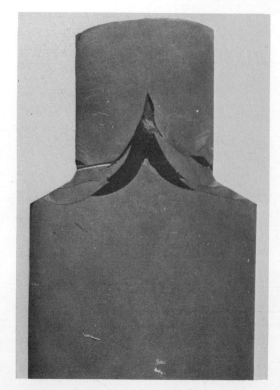

Figure 4. A typical failure of En30B steel at 600 °C and reduction ratio of 2.56.

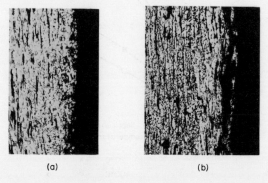

(a)   (b)

Figure 5. A comparison of microstructures of En2E steel at 800 °C for a reduction ratio of 2.56: (a) slow-speed; (b) high-speed.

pressure with displacement is a measure of frictional conditions between billet and container during the process. Thus at high speeds the coefficient of friction appears to be considerably less than that at low deformation speeds.

As shown in figure 3, the extrusion pressure for temperatures up to 400 °C are less affected by speed than for temperatures of 600 °C and 800 °C. The high-speed extrusion pressures at 600 °C and 800 °C were considerably higher than those for the low-speed deformation, particularly at high reductions. This increase of extrusion pressures at high speed is due to the increase of strain rate for the same reduction ratio, the flow stress being increasingly strain rate sensitive at higher temperatures.

Flow patterns were studied by using grids on split billets. Figure 7(a) and (b) show typical results obtained at slow and high speeds for En2E steel at a

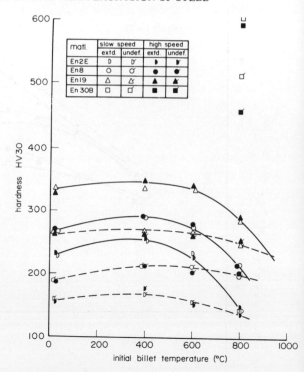

matl.	slow speed		high speed	
	extd.	undef.	extd.	undef.
En2E	▷	▷′	▶	▶
En8	○	○′	●	●
En19	△	△	▲	▲
En 30B	◻	◻′	◼	◼

Figure 6. Effect of temperature on hardness for a reduction ratio of 2.56 at slow and high speeds.

billet temperature of 800 °C and reduction ratio of 2.56. A comparison of these grids, figure 7(c), shows that at high speed the horizontal lines are less deformed than at slow speed. This indicates a more uniform flow over the extruded section at high speed, owing to the improved frictional conditions at the billet and tool interface.

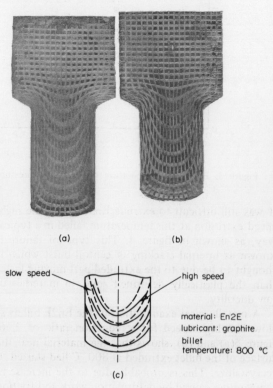

(a)   (b)

slow speed   high speed

material: En2E
lubricant: graphite
billet temperature: 800 °C

(c)

Figure 7. A comparison of material flow: (a) slow-speed; (b) high-speed.

Material starts recrystallising near the extruded surface at a billet temperature of 800 °C as shown in figure 5(a) and (b). The process time at slow speeds is about 120 ms compared with 3 ms at high speeds. Owing to this prolonged contact of the workpiece and tools at slow speed the billet surface cools relatively more than at high speed. This means that at high speeds the extrusion surfaces are deformed at higher temperatures than at slow speeds. Because of the higher temperatures at high speed the recrystallisation process has gone a stage further than at slow speed.

At high speeds a dead metal zone was formed during the extrusion of En2E at a temperature of 400 °C and reduction ratio of 2.56, as shown in figure 8. The material in the region of the surface of

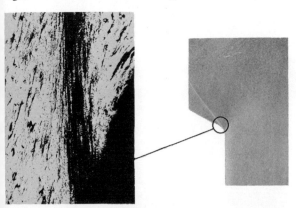

material: En2E

Figure 8. A typical dead metal zone at high speed for a reduction ratio of 2.56 and a billet temperature of 400 °C.

the dead metal zone showed a tendency to recrystallise which indicates that the rates of strain were high enough to increase the temperatures significantly. The dead metal zone is usually associated with high billet-tool friction. Friction varies with temperature at the billet–tool interface and is maximum at about[7] 400 °C. No dead metal zone was observed at 200 °C and 600 °C or at other reductions. The flow stress maximum occurs at about 400 °C owing to strain hardening as described earlier. The dead metal zone is, therefore, probably the combined result of friction and increase of flow stress at this temperature.

The effect of high speed on the deformation of grains in the plastically deformed region is shown in figure 5(a) and (b). At high speed the grains near the extruded surface are less deformed than at slow speed. This is again due to the reduction of frictional forces at high speed.

CONCLUSIONS

(1) The extrusion pressure against reduction curves follow a straight line at billet temperatures of 20 °C and 400 °C. At 600 °C and 800 °C the results for higher reductions were thought to be affected by thermal softening.

(2) An increase of temperature had less effect on the extrusion pressure required to extrude En2E than on that required for other steels. The effect of carbon content is to increase the extrusion pressure, but its effect is reduced considerably as the temperature is increased to 800 °C.

(3) En30B steel was impossible to extrude successfully at temperatures up to 600 °C. At 800 °C a reduction ratio of 4, the highest attempted, was well within the allowable stress of tools.

(4) Material near the extruded surface recrystallises slightly more at high speeds than at low.

(5) The hardness of the extruded billet was found to increase slightly with an increase of temperature up to 400 °C. Above that temperature it decreases markedly.

(6) Extrusion pressures are little affected by speed up to 400 °C, but at 600 °C and 800 °C a considerable difference was observed at higher reductions. In the latter case the extrusion pressures were higher at high speed than at slow speed.

(7) Material at high speed deforms more uniformly at all temperatures than at slow speed. For a billet temperature of 400 °C a dead metal zone was observed at high speed and a reduction ratio of 2.56.

ACKNOWLEDGEMENTS

Part of this work was financed by a Science Research Council grant awarded for investigations of warm forming processes.

REFERENCES

1. *N.E.L. Rept.*, No. 456. 'Extrusion pressures, press loads and tool stresses for warm extrusion of steel'.
2. *P.E.R.A. Rept.*, No. 147. 'Elevated temperature extrusion of steels, Part 2 – Forward Extrusion'.
3. T. A. DEAN, 1970. 'A preliminary warm forging essay', *Proc. 11th Intern. M.T.D.R. Conf.*
4. P. M. COOK, 1957. 'True stress strain curves for steel in compression at high temperatures and strain rates', *Proc. Conf. on the Properties of Materials at High Rates of Strain.*
5. H. SUZUKI *et al.*, 1968. 'Studies of the flow stress of metals and alloys', *Rept. Inst. Ind. Sci. Univ. Tokyo*, **18**, No. 3.
6. M. BURGDORF, 1971. 'Extrusion of steel in the temperature range of 20–700 °C', *Metal Forming*, **38**, No. 3.
7. A. SINGH, 1971. 'High-speed forming or metals, Part II – Lubrication and friction in forward extrusion', *Ph.D. Thesis*, University of Birmingham.
8. L. T. CHAN and S. A. TOBIAS, 1968. 'Performance characteristics of Petro-Forge Mk I and Mk II machines', *Proc. 9th Intern. M.T.D.R. Conf.*

MECHANICAL PROPERTIES OF MILD STEEL AFTER COLD AND WARM HIGH-SPEED FORGING

by

K. OSAKADA,* M. OYANE† and H. TANAKA†

SUMMARY

A 0.03% carbon steel was forged in high- and low-speed forging machines at temperatures between –78 and 600°C. Results of tensile, hardness and Charpy impact tests are presented. The effect of deformation speed and temperature on workhardening and the optimum temperature of high-speed warm forging are discussed.

INTRODUCTION

Considerable attention has been focused on the various types of high-speed forging machine, and many uses of these machines have been developed. Since a high-speed forging machine produces a large impact energy proportional to the square of the hammer speed per blow, it enables the forging of products with thin flanges or fins. It has not been possible to forge these articles with low-speed forging presses or with multi-blow hammers because of the cooling of the material that occurs during forging.

It is generally accepted that it is advantageous to use the cold forging process for the production of workhardened articles with a good surface finish and accurate dimensions. The application of the high-speed forging process to cold forging is beneficial for forging larger scale products. Another advantage of high-speed cold forging is that the coefficient of friction decreases with increasing deformation speed for many lubricants.[1] However, it may not be simple to realise the wide use of high-speed cold forging because the flow stresses of steels increase considerably as deformation speed increases.

The warm forging process, in which a material is deformed at an elevated temperature below its recrystallisation temperature, may be a solution for lowering the flow stress in the high-speed forging whilst maintaining a workhardening capacity in the material. The authors[2] have reported that the flow stress of carbon steel under conditions of high-speed deformation decreases with temperature from room temperature to 400°C, at which temperature it is a minimum, and then increases up to 600°C: this is the blue brittleness temperature at high strain rates ($\dot{\varepsilon} \doteqdot 500$/s). Because of this property, it is thought that high-speed warm forging at about 400°C is desirable. Thus another important advantage of high-speed warm forging is that the mechanical properties are utilised in the process. The mechanical properties of the material will be affected by forging speed and temperature.

In this report, the mechanical properties of a mild steel after cold and warm high-speed forging are presented. The results are compared with those after low-speed forging, and the effect of deformation speed and temperature are discussed. The optimum temperature for high-speed warm forging is also discussed.

*Department of Production Engineering, Kobe University, (Japan)
†Department of Mechanical Engineering, Kyoto University

EXPERIMENTAL PROCEDURE

Material

The material was a mild steel with a chemical composition of: 0.03% C; <0.01% Si; 0.23% Mn; 0.012% P and 0.02% S. Although the carbon content of this material was lower than that of ordinary forging steels, it was considered to show more clearly the effect of deformation speed and temperature than higher carbon steels.

The forging billets were annealed at 950°C for 30 min with subsequent furnace cooling. The grain size was approximately 70 μm.

TOOLING

Billets with a rectangular cross-section were compressed in a die shown in figure 1. Longitudinal deformation of the billets was restricted and deformation occurs in plane strain. The dimensions of the billets were: height 20 mm; width 12 mm; length 100 mm (for tensile test specimen) or 55 mm (for Charpy impact test specimen).

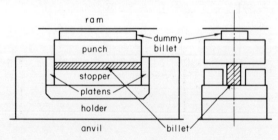

Figure 1. Forging die for plane strain compression.

In order to obtain the same strain for different forging temperatures, a stopper and a dummy billet, which absorbed the remainder of the hammer energy, were used.

The billets were heated in an electric furnace or cooled in a mixture of ethyl alcohol and dry ice together with the die before forging.

The high-speed forging machine was a Hi-fomac with a maximum impact energy of 1.1 ton m driven by nitrogen gas of up to 150 kg/cm² pressure. The forging speed in these experiments was about 15 m/s. The average strain-rate during high-speed forging was about 500/s.

The low-speed forging was performed in a 100 ton hydraulic press with a ram speed of 0.3 mm/s. The strain rate was about 0.015/s.

The lubricant was a mixture of mineral oil and molybdenum disulphide for forging temperatures up to 200°C; powdered graphite was added to the mixture for use at higher temperatures.

Testing

The forged materials were machined into tensile test specimens and Charpy V-notch impact tests specimens. The tensile test specimens had diameters of 6 mm and gauge lengths of 20 mm. The tensile tests were conducted at room temperature in a 5 ton universal testing machine with a crosshead speed of 1 mm/min.

Charpy impact tests were carried out in a Charpy testing machine with a maximum energy of 30 kg m. The notch radius of the specimens was 0.25 mm.

The Vickers hardness was measured with a load of 30 kg, and the hardness values given are the mean of five impressions.

RESULTS

Forging at room temperature

At room temperature, the billets were forged to strains of up to about 0.4 in the high- and low-speed forging machines. The logarithmic strain of height reduction $\epsilon_h = \ln(h_0/h)$. The forged materials were subjected to tensile tests. Charpy impact tests and Vickers hardness tests after machining within two weeks of forging.

Figures 2(a), (b) and (c) show the tensile and hardness tests results of mild steel after high- and low-speed forging. The values of yield-stress, ultimate tensile stress and hardness of high-speed forgings are

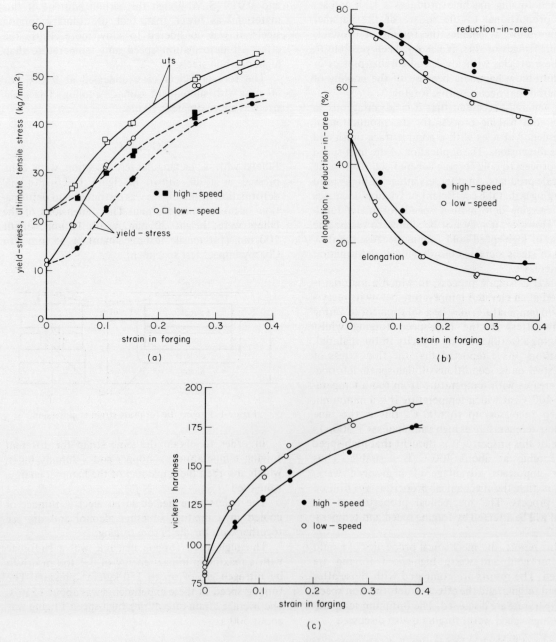

Figure 2. Effect of forging speed on the mechanical properties of mild steel deformed at room temperature.

lower than those of low-speed forgings. The difference between these values increases with strain up to a value of about 0.1, and then tends to be constant. It is obvious that the difference in the rate of work-hardening between high- and low-speed forgings was produced during the initial stage of deformation. The true fracture stress (load at fracture/final area in neck) was approximately 85 kg/mm^2 irrespective of the deformation speed and strain of forging operation.

The elongation and reduction in area produced by high-speed forging are higher than those produced by low-speed forging. Whilst the decrease in elongation is noticeable even after a small amount of deformation, the reduction in area is almost constant for small strains up to 0.05.

Two series of Charpy impact tests were conducted for the materials forged at room temperature. Firstly, the brittle—ductile transition temperature of annealed steel and high-and low-speed forgings were investigated. The strain produced by forging was 0.12, and the test temperatures were between −78 and 130°C.

Figure 3 shows the dependence of the Charpy impact value on the test temperature. The annealed

elongation and reduction in area were caused by differences in the degree of workhardening. However, as shown in figure 5, the Charpy impact results for high-speed forging are numerically greater than those for low-speed forging when they are plotted against Vickers hardness. It was thought that some micro-structural differences, as well as the differences in the degree of workhardening, were caused by differences in strain rate, since the impact values are sensitive to changes in microstructure.

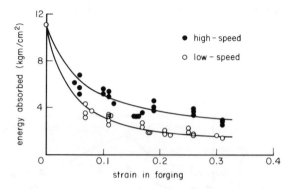

Figure 4. Effect of forging speed on the Charpy impact strength (at 50°C) of mild steel deformed at room temperature.

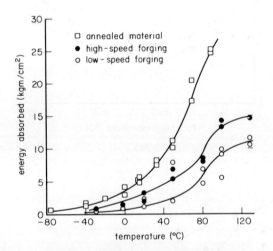

Figure 3. Effect of forging speed on the Charpy impact property of mild steel deformed at room temperature.
Strain in high- and low-speed forgings is 0.12.

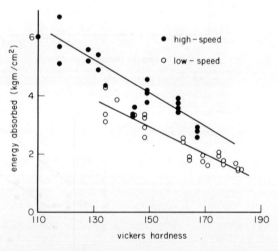

Figure 5. Effect of forging speed on the relationship between hardness and Charpy impact strength (at 50°C) of mild steel deformed at room temperature.

material shows the lowest transition temperature, and the transition temperature of high-speed forgings is lower than that of low-speed forgings. For the forged material it is found that the difference in impact values is great at temperatures above 50°C.

For the second series of impact tests, the effect of strain on the values of impact strength was examined at 50°C, and the results shown in figure 4. High-speed forging produces a higher ductility than low-speed forging. In tensile tests, carried out on the forged materials, the same tendency to decrease with forging strain was obtained for the values of reduction in area and elongation.

It is obvious that the lower rate of workhardening in high-speed forging leads to a higher elongation, reduction in area and impact value. Both the relation-ships between elongation and hardness, and the reduction in area and hardness did not change with forging speed. This may mean that the differences in

The observation of slip bands revealed that those produced by high-speed deformation were very fine compared to the slip bands obtained from low-speed deformation. The high impact values in high-speed forging may have been caused by microscopically uniform deformation, as was indicated by slipbands.

The other metallurgical variables studied was strain ageing. High- and low-speed forgings ($\epsilon_h = 0.16$) were aged at 80°C and the Vickers hardness variations were obtained for various ageing times up to 120 min. It was found that the increases in hardness due to ageing were the same in both materials, and the hardness differences remained almost constant with varying ageing time.

WARM FORGING

High- and low-speed forgings were performed at elevated temperatures up to 600°C. For comparative purposes, forgings at −30 and −78°C were carried out too. The strain in the forgings was 0.14 for all forging temperatures. The materials were machined into tensile test specimens, and tested at room temperature.

Figure 6 shows the relationship between forging temperature and tensile test properties and Vickers hardness. High-speed forgings exhibit a lower yield-stress than low-speed forgings at forging temperatures

that the dislocation density of a mild steel deformed at the blue brittleness temperature was significantly higher than steel deformed at room temperature to the same strain.

For high-speed forging, however, the peak occurs at about 400°C whilst the peak of flowstress, which occurs at the blue brittleness temperature, is at about 600°C. Referring to the recovery tests carried out on iron and carbon steel,[4] it is considered that the recovery of the forgings has taken place during air cooling after forging at between 400 and 500°C.

The elongation and reduction in area of high-speed

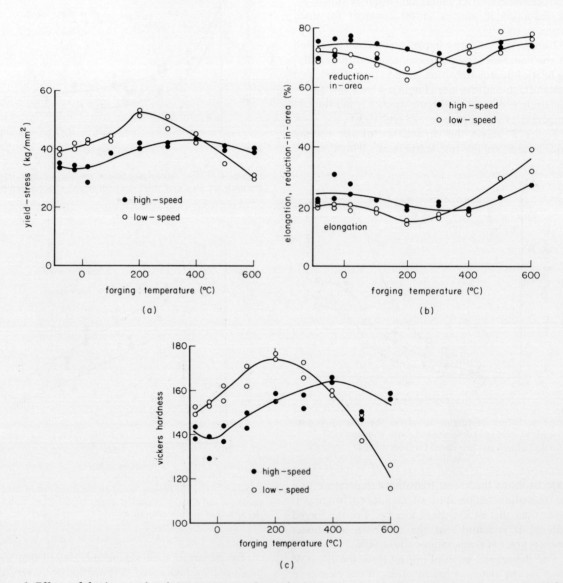

Figure 6. Effects of forging speed and temperature on the mechanical properties of mild steel deformed to a strain of 0.14.

of up to 400°C, and the relationship is reversed at temperatures above 400°C. The hardness follows the same tendency as the yield-stress.

Low-speed forgings show a peak in the yield-stress at a forging temperature of about 200°C, which is nearly the blue brittleness temperature for low-speed deformation. The result of peak strength at the blue brittleness temperature appears to correlate with the electron micrographic observations,[3] which showed

forgings are greater than those of low-speed forgings for temperatures up to 400°C and the relationship was reversed at higher temperatures.

These results, it is clear, show that the elongation and reduction in area vary with temperature in the opposite sense to the yield-stress and hardness.

Figure 7 shows the relationship between elongation and hardness: yield-stress and hardness are not affected by forging speed.

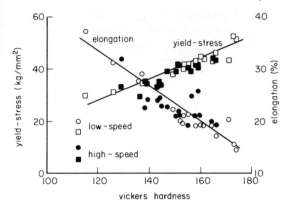

Figure 7. Relationships between hardness and yield-stress, and hardness and elongation of mild steel deformed to a strain of 0.14, in high and low-speed machines between −78 and 600°C.

No difference was found between warm and cold forgings when they were subjected to optical microscopic investigation, but the high-speed forgings formed at temperatures between −30 and −78°C exhibited a large number of mechanical twins.

DISCUSSION

Effect of forging temperature and forging speed on the workhardening of mild steel

As shown in the experimental results, the variation in the rate of workhardening is the most important effect by appropriate variation of temperature and speed of forging. Further, the other results, except for the Charpy impact test results, appear to be related to this difference in workhardening characteristics.

At room temperature, a number of experimental investigations have been carried out concerned with the effect of deformation speed on hardness and strength after deformation. It has been reported that Cu^5, Al^6 and 18-8 stainless steel[6] showed higher hardness and strength for high-speed than for low-speed deformation. However, pure iron and mild steel[6,7] showed lower hardness and strength for high-speed deformation.

One reason considered for this variation in workhardening characteristics was the differences in the crystallographic structures of the metals, that is, the difference between f.c.c. metals (Cu, Al and 18-8 stainless steel) and b.c.c. metals such as mild steel. To check this difference, a 13% Cr stainless steel (b.c.c.) was deformed in both high- and low-speed forging machines to the same strain. The hardness of the high-speed forgings was greater than the low-speed forgings at forging temperatures between zero and 600°C, which differed from the results obtained for mild steel. Though the number of examples of b.c.c. metals was limited, it was considered that the difference in the strain rate dependence of the workhardening characteristics between mild steel and f.c.c. metals might not be due to a crystallographic difference.

As shown in figure 6, the hardness of mild steel tends to increase when approaching the blue brittleness temperature, if recovery does not take place after

deformation. Since the blue brittleness phenomenon is a result of dynamic strain ageing, dynamic strain ageing is considered to be the controlling factor of speed and temperature dependence of the workhardening characteristics of mild steel. The shift of the blue brittleness temperature to a higher temperature for high-speed deformation may reduce the effect of dynamic strain ageing on the workhardening rate at room temperature, and may bring about a lower hardness in high-speed forging than in low-speed forging.

Optimum temperature of high-speed warm forging

The effect of high- and low-speed deformation on the flow stress of mild steel is shown in figure 8. A minimum value of flow stress for high-speed deformation is observed at about 400°C. The flow stress at

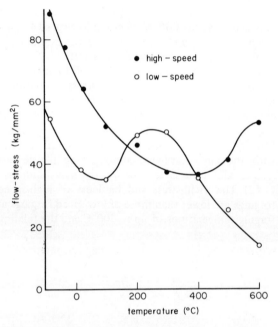

Figure 8. Effects of deformation speed and temperature on the flow stress of mild steel at a strain of 0.16. Strain rate is 380/s for high-speed; 0.003/s for low-speed.

400°C is about half that at room temperature and is even lower than the flow stress for low-speed deformation at room temperature. Since the main purpose of warm forging is to decrease the flow stress, this temperature appears to be the optimum: the forged material is then heavily workhardened and has a high strength, as shown in the experimental results.

The main advantage of both high-speed hot and warm forging is the prevention of an increase in flow stress caused by cooling during contact with the die.

The other factors to be considered are: cracking during forging and lubrication. It is obvious that the blue brittleness temperature should be avoided because of the increased tendency for cracking. This optimum forging temperature of 400°C in high-speed forging is sufficiently lower than the blue brittleness temperature of about 600°C.

Although the lubricants for warm forging are not well developed yet, it is possible to say that many lubricants deteriorate at high temperatures. If direct contact between the lubricant and the heated material is avoided before die impact, high-speed forging will provide better lubrication because the contact time is so short that the lubricant will not have completely deteriorated.

To ascertain the validity of the optimum temperature of high-speed warm forging, a spur gear was forged at 400°C in the high-speed forging machine. It was found that the gear (module = 3, number of teeth = 10) was satisfactorily forged without any defects. Because of the limitation set by the die strength, it was not possible to forge the gear at room temperature. The forged gear exhibited a reasonable hardness distribution, and the fatigue strength of the gear was higher than that of a machined gear.

CONCLUSIONS

A 0.03% carbon steel was forged at temperatures between −78 and 600°C in high- and low-speed forging machines, and the mechanical properties examined at room temperature. The conclusions are:

(1) At room temperature, high-speed forgings show lower yield-stress, ultimate tensile stress and hardness than low-speed forgings. Consequently higher values of elongation, reduction in area and Charpy impact values are obtained. The Charpy impact values of high-speed forgings are greater than those of low-speed forgings with the same hardness, although the tensile test results exhibit the same values for the same hardness.

(2) The yield-stress and hardness of high-speed forgings are lower than those of low-speed forgings at forging temperatures of up to 400°C and the relation is reversed at temperatures above 400°C. A maximum in yield-stress and hardness for high-speed forgings is observed at 400°C for a forging strain of 0.14.

(3) From the viewpoint of flow stress and the subsequent properties of the material after forging, 400°C is concluded to be the optimum temperature of high-speed warm forging.

ACKNOWLEDGEMENT

The authors would like to thank Mr. B. Dodd of the Department of Industrial Metallurgy, The University of Birmingham for his advice during the preparation of the manuscript.

REFERENCES

1. K. OSAKADA and M. OYANE, 1970. The effect of deformation speed on friction and lubrication in cold forging, *Bull. JSME,* **13**, p. 1504.
2. M. OYANE, F. TAKASHIMA, K. OSAKADA and H. TANAKA, 1967. The behaviour of some steels under dynamic compression, *The 10th Japan Congress on Testing Materials,* p. 72.
3. B. J. BRINDLEY and J. T. BARNBY, 1966. Dynamic strain ageing in mild steel, *Acta Metallurgica,* **14**, p. 1764.
4. A. S. KEH, 1962. Direct observation of imperfections in crystals, Interscience, p. 213.
5. H. J. LIPPMAN, 1964. Zur physik der metallumformung mit hochen magnetfeldimpulsen, *Zeitschrift für Metall · kunde,* **56**, p. 737.
6. T. WILLIAMS, 1962. Some metallurgical aspects of explosive forming, *Sheet Metal Industries,* **39**, p. 487.
7. J. D. CAMPBELL and J. DUBY, 1956. The yield behaviour of mild steel, *Proc. Roy. Soc. (London),* **A236**, p. 24.

DYNAMIC EFFECTS IN HIGH-VELOCITY COMPRESSION TESTING

by

C. E. N. STURGESS* and T. A. DEAN*

SUMMARY

Plane strain and axisymmetric compression tests have been conducted on electrolytically pure copper at strain rates of approximately $4000 \, s^{-1}$, to final strains of 1.6, at ambient temperature.

The natural strain rate variations and inertial effects arising in high-velocity plane strain compression have been analysed theoretically.

The experimental strain rate variations are in reasonable accord with the analysis. However, the results of the high-velocity plane strain and axisymmetric compression tests on copper are only in qualitative agreement with the inertia theory developed.

NOTATION

A	= constant
$2b$	= overall breadth of plane strain billet
CG	= centre of gravity
D	= constant
h	= current billet height
H	= initial billet height
k	= constants
L	= overall length of plane strain billet
m	= strain rate index
M	= mass of platen
m_b	= mass of billet
n	= strain index
t	= time
T	= total process time
V_x, V_y, V_z	= velocity components; subscripts denote direction
$\dot{V}_x, \dot{V}_y, \dot{V}_z$	= acceleration components; subscripts denote direction
V	= current platen velocity
$\dot{V}$	= current platen deceleration
V_0	= platen impact velocity
w	= width of plane strain billet
x, y, z	= cartesian coordinates
γ	$= \dfrac{\rho \pi R^4 H^2}{8M(1 + m_b/4M)}$
ϵ	= natural strain
ϵ_E	= equivalent strain
ϵ_A	= axisymmetric strain
ϵ_{PS}	= plane strain
ϵ_f	= final natural strain
$\dot{\epsilon}$	= current natural strain rate
$\dot{\epsilon}_0$	= initial natural strain rate
$\dot{\epsilon}_E$	= equivalent natural strain rate
$\dot{\epsilon}_A$	= axisymmetric natural strain rate
$\dot{\epsilon}_{PS}$	= plane natural strain rate
$\eta = \dfrac{1}{1 + \rho b H/M}$	= impact efficiency factor
ι	$= \dfrac{4wb}{\sqrt{(3)}M}$
ξ	$= \dfrac{2m_b b^2}{M}$
ρ	= density
σ	= stress
σ_E	= equivalent stress
σ_A	= axisymmetric stress
σ_{PA}	= plane strain stress
σ_D	= dynamic flow stress
σ_S	= quasi static stress
σ_{XI}	= inertial stress in x direction

INTRODUCTION

Material properties are required to estimate the loads and energies required for a given forming operation. Tensile[1] and compression[2] tests have been used to obtain material data.

Compression tests are favoured for determining material properties pertinent to metal forming, as tensile tests terminate at low strains.

The axisymmetric compression test has been used at high speeds[3,4] to obtain stress strain data applicable to high-velocity metal working. The attention of many workers was drawn to the possibility that workpiece inertia would significantly affect the stresses measured from a high strain rate upsetting test[5,6]. Theoretical treatments[7,8] have shown that in axisymmetric compression inertial effects are generally of little consequence.

The plane strain compression test[9,10] has been utilised, of late, to provide high strain rate material properties[11,12]. In the plane strain test a constant area of the billet is in contact with the tools and relatively low deformation forces are required. Because of this feature plane strain compression has been employed to expand the working range of load-restricted cam plastometers[12]. The low loads encountered in the plane strain test also appear to make this test a suitable choice for work-restricted free flight impact testing machines.

As yet no investigations have been conducted into the dynamic aspects, such as strain rate variations and inertial effects, of high-velocity plane strain compression testing. The large specimen sizes commonly employed in the plane strain compression test would appear conducive to the appearance of appreciable inertial forces if used at high velocities.

*Department of Mechanical Engineering, University of Birmingham.

The object of these investigations was to compare the stress–strain curves obtained from high strain rate plane strain and axisymmetric compression tests and, if possible, to isolate inertial effects.

THEORETICAL ANALYSIS OF THE HIGH-SPEED PLANE STRAIN COMPRESSION TEST

To identify and compare inertial forces in axisymmetric and plane strain compression it is necessary to eliminate the effects of strain rate and friction; this requires (1) low friction, (2) equality of equivalent strain, (3) identical initial equivalent strain rates and strain rate histories in both the plane strain and upsetting tests.

Of these conditions (1) and (2) are matters of experimental technique. The strain rate variation in upsetting has been analysed elsewhere[3, 13].

Appendix 1 contains an analysis of the strain rate variations in the impact plane strain compression of a strain and strain rate sensitive material, whose properties can be described by

$$\sigma = k\epsilon^n \dot{\epsilon}^m \qquad (1)$$

To assess inertial effects in plane strain compression the process was visualised as one where two fixed masses are moved horizontally outwards from the anvils by a 'light' extending rod, figure 2, the motion of the rigid masses inducing inertial forces within the 'rod', which is the deforming region of the test specimen.

The assumptions embodied in the analysis are as follows.

(1) The moving platen is in free-flight prior to impact.

(2) The billet material is homogeneous, incompressible, and obeys the Von Mises yield criterion.

(3) The undeformed regions of the billets are of constant mass.

(4) Strain rate effects are ignored.

(5) Plastic wave propagation effects are neglected.

The material between the anvils is progressively transferred to the 'fixed' masses during the test. This effect is small and, as it unduly complicates the analysis, is not included.

Inertial effects arise as a result of the velocity variations occurring in an impact operation; these are affected by material strain hardening and strain rate sensitivity. It was possible to include material work hardening in the analysis by assuming the simple relationship

$$\sigma = A - Dh \qquad (2)$$

This expression provides a good approximation to the behaviour of a real material.

The inclusion of material strain rate sensitivity in the analysis would have rendered an analytical solution impractical.

Plastic wave propagation effects are not considered as the process times are considerably longer than the transit time of stress waves through the billets.

Utilising these assumptions expressions have been derived in Appendix 2 for the effects of inertial forces in high velocity plane strain compression.

EXPERIMENTAL EQUIPMENT

Test materials

Fully annealed electrolytically pure copper was used for the test pieces. Billets of rectangular section 0.250 in × 1.0 in and lengths 1.0 in, 2.0 in, 3.0 in and 4.0 in were prepared for the plane strain compression tests, and cylindrical billets 0.375 in diameter 0.250 in long for upsetting.

To eliminate the effects of anisotropy, the billets were machined with the axes of deformation of both the upsetting and plane strain billets in the same direction relative to the structure of the stock material.

The lubricant employed was Copaslip, a suspension of copper and lead globules in a bentone grease base.

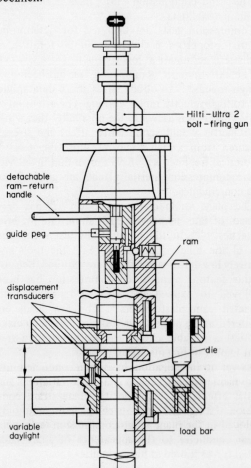

detachable
ram–return
handle

guide peg

displacement
transducers

variable
daylight

Hilti – Ultra 2
bolt – firing gun

ram

die

load bar

Figure 1. Explosively driven compression testing machine.

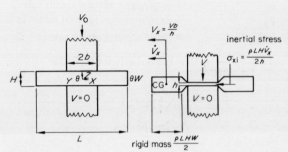

Figure 2. Inertial effects in high-velocity plane strain compression.

Compression testing machine

An explosively drive compression testing machine, embodying a Hulti-Ultra Mk II bolt firing gun as a power source, capable of 300 ft lb at velocities up to 100 ft/s (fig. 1) was used for the trials.

The gun incorporated a co-axial capacitance displacement transducer, used in conjunction with a Southern Instruments M 700L gauge oscillator and MR 220F F.M. amplifier.

Deformation loads were measured using a 2 in diameter, 8 ft long strain-gauged En8 Hopkinson bar. With this device records of deformation loads from tests up to 0.8 ms in duration can be obtained without interference from reflected stress waves.

The bar was strain gauged with four 1300 Ω Hawker Siddeley KW8/1300 gauges. The bridge was supplied with 36 V d.c. and the output measured using an Ellis B.A.M.1 d.c. amplifier with a 20 kHz frequency response. The basic signal frequency expected was approximately 1 kHz and the 20x margin existing between this and the bridge cut-off frequency was thought adequate to avoid attentuation of the output.

The interchangeable plane strain anvils and upsetting dies were assembled with the bar with a thin film of light machine oil. This tooling arrangement has been shown[15] not to interfere with the passage of stress waves, and hence measurement of the deformation loads. The interchangeable dies were produced from Orvar, a 5% chromium die steel, heat treated to 45 Rc.

The Hopkinson bar was calibrated by comparing the area under the load—time traces with the momentum change of the ram, determined from measurements of impact and rebound velocities. For these tests billets of various diameters were employed, the output of the bar was independent of the loaded area.

The displacement—time and load—time signals were recorded using a Tektronix 564 storage oscilloscope.

The punch velocity was obtained from the punch displacement—time trace by utilising a five-point numerical differentiation technique based on a Taylor series expansion.

RESULTS AND DISCUSSION

Experimental strain-equivalence

In the absence of inertial effects the stress—strain curves obtained from axisymmetric and plane strain compression should only agree if compared on the basis of equivalent stress, strain and strain rate[10], which for ideal plane strain conditions give

$$\sigma_E = \sigma_A = \sigma_{PS} \times \frac{\sqrt{3}}{2} \qquad (3)$$

$$\sigma_E = \epsilon_A = \epsilon_{PS} \times \frac{2}{\sqrt{3}} \qquad (4)$$

$$\dot{\epsilon}_E = \dot{\epsilon}_A = \dot{\epsilon}_{PS} \times \frac{2}{\sqrt{3}} \qquad (5)$$

In the plane strain compression tests reported in this paper some lateral spreading was observed, figure 3, this provided an experimental correction factor for equivalent strain.

Equivalent strain is defined as

$$\epsilon_E = \tfrac{2}{3}[\tfrac{1}{2}\{(\epsilon_1 - \epsilon_2)^2 + (\epsilon_2 - \epsilon_3)^2 + (\epsilon_3 - \epsilon_1)^2\}]^{1/2} \qquad (6)$$

but $\epsilon_1 = \epsilon_1$ and $\epsilon_3 = -0.06\ \epsilon_1$ (figure 5). Therefore, from constancy of volume

$$\epsilon_1 + \epsilon_2 + \epsilon_3 = 0 \qquad (7)$$

$$\epsilon_E = 1.1\epsilon_1 \qquad (8)$$

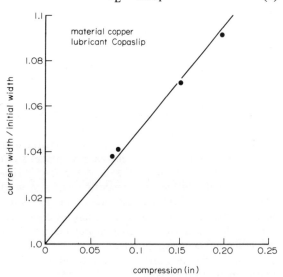

Figure 3. Lateral spreading in high-velocity plane strain compression test.

In analysing the test results the experimental correction factor 1.1 was used instead of $2/\sqrt{3}$.

Strain rate variations in upsetting and plane strain compression

The strain rate variations encountered during the experimental high-velocity upsetting and plane strain compression tests are shown in figure 4.

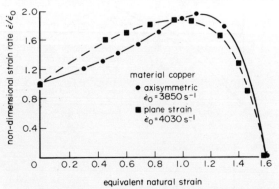

Figure 4. Comparison of the experimental natural strain rates in high-velocity axisymmetric and plane strain compression.

The initial strain rates are different by approximately 8% and maximum differences of 15% occur in the strain rate histories.

In figure 5 the experimental strain rate histories in high-velocity plane strain compression is compared with the theoretical results for work hardening and rigid plastic materials of the same mean flow stress.

The strain hardening index $n = 0.4$ is a good approximation to the work hardening exhibited by copper when deformed at ambient temperature.

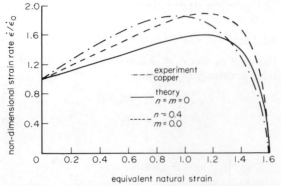

Figure 5. Experimental and theoretical natural strain rate variations in high-velocity plane strain compression testing.

As can be seen from figure 5 good agreement was obtained between the theoretical expressions, incorporating work hardening, and experiment.

Work hardening acts to increase the instantaneous strain rates observed in high-velocity plane strain compression in comparison with deforming a rigid plastic material of the same mean flow stress. A work hardening material extracts less energy from the punch in the early stages of deformation than the equivalent rigid plastic material; this gives higher instantaneous velocities and hence strain rates.

These effects are similar to those found to apply to axisymmetric compression[13].

Inertial effects in plane strain and axisymmetric compression

To illustrate the results of the theoretical analysis of inertial effects in plane strain compression equation (15) appendix 2 has been evaluated in figure 6(a) for

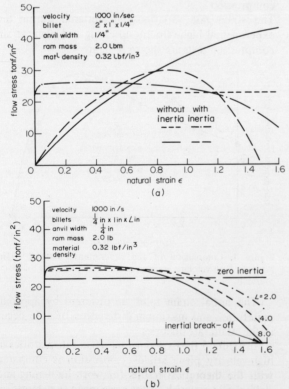

Figure 6. (a) Theoretical inertial effects in high-velocity plane strain compression testing. (b) Theoretical variations of inertial effects in high-velocity plane strain compression with different billet sizes.

a rigid perfectly plastic material, and a strain hardening material of the same mean flow stress, and in figure 6(b) for a rigid plastic material of different billet sizes.

The major effects of inertia are shown to be the following:

(a) The measured flow stress in the early stages of compression is increased, and the final stress is lower than in the absence of inertia.

(b) The magnitude of inertial forces increases with increase in billet size.

(c) Work hardening reduces the positive inertial effects in the initial stages of the process, and increases the final inertial stress reductions.

In figure 6(b) the measured stress for the case of an 8 in long billet demonstrates inertial break-off. As a rigid palstic material was assumed, inertial break-off was taken to occur when the inertial stress exceeded the yield stress of the material (zero punch pressure). This is the same criterion as applied by Cole and Bakhtar[14] for inertial break-off in high-velocity extrusion.

To compare the theoretical inertial effects in high-velocity plane strain compression with the inertial effects in upsetting, the relationship

$$\sigma_D = \frac{Y}{(1 + \gamma/h^2)^2} \left\{ 1 + \frac{3\gamma}{(h^3 + \gamma)} \ln \frac{h}{hf} \right\} \quad (9)$$

was taken, depicting the inertial effects in the upsetting of a rigid plastic material with zero friction[8]. This expression, evaluated for the conditions outlined in figure 6 and a billet size 0.375 in x 0.250 in did not depart from the flow stress in the absence of inertia by more than 2%.

The stress—strain curves obtained from the experimental high velocity upsetting and plane strain compression tests, displayed as equivalent stress and strain are shown in figure 7(a).

The levels of stress exhibited by copper, when deformed under axisymmetric and plane strain conditions, at ambient temperature are seen to be similar; however, the curves are markedly different in shape.

The results of the plane strain compression tests, when compared with upsetting demonstrate (1) increased stresses at low strains; (2) lower stresses at intermediate strains; (3) a rapid fall-off of stress at the end of the test.

These effects become slightly more pronounced as the billet size increases from 0.5 in to 4 in.

The results for the 1 in and 2 in billets, although generally intermediate to the 0.5 in and 4 in billets, are confused, the resolution not being sufficient to separate these results.

These effects are broadly in agreement with the inertial theory for high-velocity plane strain compression developed in appendix 2.

That the experimental results shown in figure 7(a) are not unduly influenced by strain rate can be demonstrated by reference to figure 7(b).

In this figure the high strain rate stress—strain curve obtained from the upsetting tests is compared with a low strain rate result obtained by Suzuki et al.[16]. There is little difference between the two curves for widely different values of strain rate, hence it is obvious that the 8% strain variations in strain rates obtained in these present tests (figure 6(a)) are insignificant.

To compare the inertial theory with the experimental results, the stress—strain curve obtained for copper from the axisymmetric compression test was described as

$$\sigma = 37.5 - 100h \tag{11}$$

This expression, within 5% of the experimental axisymmetric stress—strain curve, was inserted in the analysis of inertial effects in plane strain compression as depicting the stress—strain curve in the

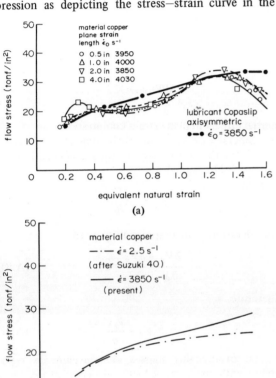

(a)

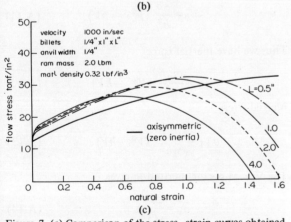

(b)

Figure 7. (a) Comparison of the stress—strain curves obtained from high-velocity plane strain and axisymmetric compression. (b) Comparison of high and low strain rate stress—strain curves for copper. (c) Theoretical inertial effects in the high-velocity plane strain compression of copper.

absence of inertial effects. The results are shown in figure 7(c).

By comparing figures 7(a) and 7(c) it can be seen that theoretically the positive inertial effects at low strains are similar for all billet sizes. Experimentally the 4 in long billet produced a considerably higher stress than the other billets.

The stress reductions evident in the plane strain results at the end of the process are overestimated by the inertial theory for all the billets except the 0.5 in long specimen. The 0.5 in billet displays a considerably higher stress reduction than anticipated.

That the plane strain compression test should yield lower stresses than upsetting at medium strains is not in agreement with the inertial theory. The inertial effects at moderate strains should be small and positive, figure 7(c).

The plane strain compression tests conducted did depart slightly from ideal plane strain conditions, resulting in some lateral spreading. Inertial effects are theoretically considerably higher in plane strain than in axisymmetric compression. Therefore, the departures from ideal plane strain conditions obtained in these tests would be expected to reduce the inertial effects observed. However, such discrepancies could not account for the reduced stresses at intermediate strains.

Effects which have not been considered are the following.

(a) Anisotropy — differences do occur between plane strain and upsetting[10] due to anisotropy, but the billet preparation should have avoided such difficulties.

(b) Friction — the same lubricants and velocities were employed for both plane strain and upsetting; therefore the frictional conditions were probably similar in all tests.

(c) Heat generation — the deformation in plane strain is delineated by bands of intense shearing. At high velocities, once the deformation occurs in narrow zones, heat generation may locally reduce the flow stress. However, it is difficult to visualise such effects producing the particular discrepancies observed.

CONCLUSIONS

Plane strain and axisymmetric compression tests have been carried out on electrolytically pure copper at ambient temperature. The tests were conducted at strain rates of approximately 4000 s^{-1} to final strains of 1.6. From the results of these investigations the following conclusions may be drawn.

The experimental strain rate variations observed were in reasonable agreement with theory. Copper deformed at ambient temperature work hardens, this produces higher strain rates than a rigid perfectly plastic material.

The stress—strain curves, for copper, obtained from high-velocity plane strain and axisymmetric compression, although similar in level, were markedly different in shape. Compared with upsetting high-velocity plane strain compression produces (a) increased stresses at low strains; (b) lower stresses at intermediate strains; (c) a reduced final stress.

These effects tend to increase with increase in billet size.

The inertial theory, although predicting increased initial stresses, and reduced final stresses in plane strain, as opposed to axisymmetric, compression over-estimated these effects, and was only in qualitative agreement with theory.

It has been demonstrated that the effects of billet inertia on flow stress are considerably more pronounced in plane strain than in axisymmetric compression. Therefore, when conducting high strain rate plane strain compression tests, caution should be exercised lest inertial effects occur.

ACKNOWLEDGEMENTS

The authors would like to thank Professor S. A. Tobias for use of departmental facilities and financial support during these investigations. One of the authors (C.E.N.S.) is CHUBB Research Fellow.

APPENDIX 1

Strain rate variation in high velocity plane strain compression

Assuming a double-power constitutive equation

$$\sigma = k\epsilon^n \dot{\epsilon}^m \tag{A.1}$$

ignoring inertial effects the equation of motion is

$$-M \frac{d^2 h}{dt^2} = kA\epsilon^n \dot{\epsilon}^m \tag{A.2}$$

as $V = dh/dt$, therefore $V \dfrac{dV}{dh} = \dfrac{d^2 h}{dt^2}$

$$\epsilon = \ln \frac{H}{h}, \text{ therefore } dh = -hd\epsilon$$

and $$h^m = H^m \exp(-m\epsilon)$$

Substituting these expressions in equation (A.2) gives

$$\int_{V_0}^{V} V^{(1-m)} dV$$

$$= \int_{0}^{-\epsilon} \frac{KA\epsilon^n H^{(1-m)}}{M} \exp\{\epsilon(m-1)\} d\epsilon \tag{A.3}$$

Now expanding the exponential and integrating term by term gives

$$V^{(2-m)} = V_0^{(2-m)} - \frac{(2-m)K AH^{(1-m)}}{M}$$

$$\times \left\{ \frac{\epsilon^{(n+1)}}{(n+1)} + \frac{(m-1)\epsilon^{(n+2)}}{(n+2)} + \frac{(m-1)^2 \epsilon^{(n+3)}}{(n+3)} \right\} \tag{A.4}$$

As $\dot{\epsilon} = 2V/\sqrt{3}h$ and $\dot{\epsilon}_0 = 2V_0/\sqrt{3}H$ in plane strain, then the non-dimensional strain rate variation is

$$\frac{\dot{\epsilon}}{\dot{\epsilon}_\sigma} = \left[1 - \frac{\epsilon^{(n+1)}}{\epsilon_f(n+1)} \right.$$

$$\left. \times \left\{ \frac{\dfrac{1}{(n+1)} + \dfrac{(m-1)\epsilon}{(n+2)} + \dfrac{(m-1)^2\epsilon^2}{2(n+3)} + \dots}{\dfrac{1}{(n+1)} + \dfrac{(m-1)\epsilon_f}{(n+2)} + \dfrac{(m-1)^2 \epsilon_f^2}{2(n+3)} + \dots} \right\}^{1/(2-m)} \right]$$

$$\times \exp \epsilon \tag{A.5}$$

for $m = n = 0$, $k = \sigma$, equation (A.5) reduces to

$$\frac{\dot{\epsilon}}{\dot{\epsilon}_0} = \exp \epsilon \left[1 - \left\{ \frac{1 - \exp(-\epsilon)}{1 - \exp(-\epsilon_f)} \right\} \right]^{1/2} \tag{A.6}$$

APPENDIX 2

Inertial effects in plane strain compression

The general case of a free flight impact test is considered for the configuration shown in figure 1. The velocity field assumed was

$$V_z = -Z\frac{V}{h} \tag{A.7}$$

Consideration of incompressibility gives

$$\frac{\partial V_x}{\partial X} + \frac{\partial V_z}{\partial Z} = 0 \tag{A.8}$$

Therefore

$$V_x = \frac{V.x}{h} \tag{A.9}$$

The overhanging masses are travelling as rigid bodies with the same velocity as occurs at the edge of the anvil, i.e.

$$V_b = \frac{Vb}{h} \tag{A.10}$$

hence

$$\dot{V}_b = \frac{-b}{h^2} (V^2 - h\dot{V}) \tag{A.11}$$

Thus we have inertial force

$$= \frac{m_b b}{h^2} (V^2 - h\dot{V}) \tag{A.12}$$

inertial stress

$$\sigma_{XI} = \frac{m_b b}{wh^3} (V^2 - h\dot{V}) \tag{A.13}$$

The Von Mises yield criterion gives

$$\sigma_x = \frac{2Y}{\sqrt{3}} + \sigma_z \tag{A.14}$$

$$\sigma_x = \frac{2Y}{\sqrt{3}} + \sigma_{XI} \tag{A.15}$$

But deformation force $= M\dot{V}$ $\qquad\qquad$ (A.16)

and

$$Y = A - Dh \qquad\qquad (A.17)$$

Substituting from (A.13), (A.15), (A.16) and (A.17) gives

$$\left(1 + \frac{\xi}{h^2}\right)\frac{d^2 h}{dt^2} - \frac{\xi}{h^3}\left(\frac{dh}{dt}\right)^2 - \iota(A - Dh) = 0 \tag{A.18}$$

assuming

$$\lambda = \left(\frac{-dh}{dt}\right)^2 \tag{A.19}$$

$$\left(1 + \frac{\xi}{h^2}\right)\frac{d\lambda}{dh} - \frac{2\xi\lambda}{h^3} - 2\iota(A - Dh) = 0 \tag{A.20}$$

Evaluating this expression for λ and differentiating gives

$$\sigma_D = \frac{M}{2bw(1 + \xi/h^2)}\left\{\frac{(1 + \xi/h^2)}{h}\left(1 - \frac{1}{(1 + \xi/h^2)}\right)V_0^2\right.$$

$$+ \iota\left[A - Dh + 2(A - Dh/2)\left(1 - \frac{1}{(1 + \xi/h^2)}\right)\right.$$

$$\left.\left. - \frac{2H}{h}\left(\left(A - \frac{Dh}{2}\right)\left(1 - \frac{1}{1 + \xi/h^2}\right)\right)\right]\right\} \tag{A.21}$$

REFERENCES

1. M. MANJOINE and A. NADAI, 1950. High-speed tension tests at elevated temperatures, Part 1, *Proc. A.S.T.M.*, **40**, 822.

2. P. M. COOK, 1957. 'True stress–strain curves in compression at high temperatures and strain rates for application to the calculations of load and torque in hot rolling', *Proc. Conf. the Properties of Materials at High Rates of Strain*, Institution Mechanical Engineers, London.

3. R. A. C. SLATER, W. JOHNSON and S. Y. AKU, 1968. 'Experiments in the fast upsetting of short pure lead cylinders and a tentative analysis', *Intern. J. Mech. Sci.*, **10**, 169.

4. S. K. SAMANTA, 1968. 'Resistance to dynamic compression of low carbon steel and alloy steels at elevated temperatures and high strain rates', *Intern. J. Mech. Sci.*, **10**, 613.

5. H. KOLSKY, 1949. 'An investigation of the mechanical properties of materials at very high rates of loading', *Proc. Phys. Soc., Ser. B*, **62**, 676.

6. E. D. H. DAVIES and S. C. HUNTER, 1963. 'The dynamic compression testing of solids by the method of the split Hopkinson bar', *J. Mech. Phys. Solids*, **11**, 155.

7. T. A. DEAN, 1970. 'The influence or billet inertia and die friction in forging processes – a simple energy approach', *Proc. 11th M.T.D.R. Conf., Birmingham, B*, p. 761.

8. C. E. N. STURGESS and M. G. JONES, 1971. 'Estimation of dynamic forces in high-speed compression using a free-flight impact forging device', *Intern. J. Mech. Sci.*, **13**, 309.

9. A. B. WATTS and H. FORD, 1952. 'An experimental investigation of the yielding of strip between smooth flat dies', *Proc. Inst. Mech. Engrs. (London), Ser. B*, **166**, 448.

10. A. B. WATTS and H. FORD, 1955. 'On the basic yield stress curve for a metal', *Proc. Inst. Mech. Engrs.(London)*, **159**, 1141.

11. J. A. BAILEY and A. R. E. SINGER, 1963–4. 'Effects of strain rate and temperature on the resistance to deformation of aluminium, two aluminium alloys and lead', *J. Inst. Metals*, **92**, 404.

12. P. F. THOMASSON, B. FOGG and A. W. J. CHISHOLM, 1968. 'A cam-plastometer for investigation into the cold and warm working characteristics of alloy steels', *Proc. 9th M.T.D.R. Conf., Birmingham*.

13. C. E. N. STURGESS and A. N. BRAMLEY, 1970. 'The use of impact forming devices to obtain dynamic stress–strain data', *Proc. 11th M.T.D.R. Conf., Birmingham*.

14. B. N. COLE and F. BAKHTAR, 1963. 'Dynamic effects in very high speed impact extrusion', *Intern. J. Machine Tool Design Res.*, **3**, 77.

15. R. M. DAVIES. 'A critical study of the Hopkinson pressure bar', *Phil. Trans. Roy. Soc. London, Ser. A*, **240**, 375.

16. H. SUZUKI, S. HASIZUME, Y. VABUKI, Y. ICHIBARA, S. NAKAJIMA and K. KENMOCHI, 1968. 'Studies on the flow stress of metals and alloys', *Inst. Ind. Sci., Univ. Tokyo*, **18**, No. 117.

APPLICATION OF A COMPUTER SIMULATION TECHNIQUE TO ESTIMATE LOAD AND ENERGY IN AXISYMMETRIC CLOSED DIE FORGING

by

S. K. BISWAS* and B. W. ROOKS*

SUMMARY

A high-speed forming hammer (Petro-Forge) is used to forge two medium carbon steel (EN8) axisymmetric parts with longitudinally symmetric and longitudinally asymmetric flash openings under four different conditions of temperatures. The experimental results are used to verify a step-by-step simulation technique used to predict the forging characteristics of a work restricted forging machine. The technique analyses forging in small steps of deformation such that the stress distribution, the load, the energy and die filling are estimated at each deformation step.

Close agreement is reached between the experimental and the theoretical results and it is concluded that the simulation technique is appropriate for predicting the load and energy characteristics of a work-restricted machine forging axisymmetric parts.

NOTATION

A_1 = thermal diffusivity
A_2 = structural constant of the deforming material
b = actual flash width
C_p = specific heat
ΔE = incremental energy input for a deformation stroke of Δh
f_f = friction factor at the flange–die interface
f_{fl} = friction factor at the flash–die interface
f_s = friction factor at the shaft–die interface
h_1 = instantaneous flange height
H_0 = original billet height
J = mechanical equivalent of heat
m = mass of the platen
M = total mass of the forging machine
n_2 = structural constant of the deforming material
Q = structural constant of the deforming material
t = flash thickness
T_i = absolute temperature of any deformation zone i
V = volume of the specimen
V_0 = impact velocity
V_p = instantaneous platen velocity
w = flash width on the die
α = structural constant of the deforming material
$\dot{\epsilon}_i$ = mean strain rate in any deformation zone i
ρ = density of the deforming material
$\bar{\sigma}_i$ = flow stress in any deformation zone i
τ = frictional shear stress at die–material interface

1. INTRODUCTION

The primary concern in any kind of hot forging operation is to produce the desired component by the expenditure of minimum time, labour and wastage of material whilst maximising die life. Principally these requirements are met when the energy applied to the operation is just sufficient to form the component, assuming that the billet size and temperature are the correct ones. In a stroke-restricted machine such as a crankpress, the platen stroke is set and for a given operation energy sufficient to form the component is extracted from the flywheel. In contrast, a work-restricted machine such as a hammer, expends all the energy available irrespective of the component requirements. Therefore, if the energy is in excess of that required to form the component, excessive loading of the machine and dies results, whereas if there is insufficient energy the production efficiency is impaired. Hence there is a need for a method that will enable the energy to be set to give the desired effect by consideration of the initial billet size, the component shape, the billet material properties, the billet temperature, interface friction and the speed of forging. It is with such a method that this paper is concerned.

The principal problem in such estimation procedures is the determination of the stress developed at each stage of the forging operation. Kobayashi et al.[1] adopted a 'slab method' of analysis to predict mean forging pressures by estimating the stress generated due to shearing of a fictitious disc having the same thickness as that of the component flash. This assumption was shown to be in error by visioplastic experiments conducted by Unksov[2]. He found that the above shear zone was lens-shaped. Equations relating the optimised geometry of this lens-shaped zone have been derived by Altan[3].

Once the stresses at each stage have been established, a step-by-step simulation technique can be employed as proposed by Altan[4]. His technique attempts to establish the load–displacement characteristics of a forging component by estimating the stress distribution at various positions of the deforming stroke. The metal flow in the workpiece is simulated by considering the latter to be made up of zones for each of which the flow pattern and the flow stress can be established. Each step of deformation alters the preceding zonal geometry and flow stresses, thus altering the overall stress distribution in the flow model. The changed

* Department of Mechanical Engineering, University of Birmingham.

geometry and stresses give rise to a new workpiece resistance which is a measure of the corresponding load. The energy requirement follows from the sum of this load and the amount of deformation.

A technique similar to the above has also been developed by Biswas[5]. This paper is concerned with an experimental verification of this theoretical method by comparing the predicted load and energy characteristics with those obtained in forging the two components shown in figure 1. These tests

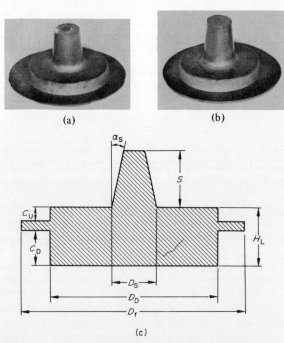

(a)   (b)

(c)

Figure 1. Forged components: (a) specimen 'A' (symmetric flash, $C_U = C_D$); (b) specimen 'B' (asymmetric flash, $C_D = 0$); (c) generalised component used for analysis.

were carried out on a Petro-Forge high-speed hammer (a work-restricted machine) under varying conditions of strain rate and forging temperature.

2. FLOW STRESS ESTIMATION

The main feature of hot working is that large strains are applied to materials at high rates of strain and at temperatures above half the material's melting temperature. Hot working is similar to steady-state creep in that it is temperature and strain rate sensitive and an equation derived from creep theory has been found to fit hot working data reasonably well[5,6], given by equation (A.1) in appendix 1. A statistical method to determine the temperature and strain rate independent material constants has been developed by Biswas[5].

However, the mean strain rate and temperature are not constant during forging and vary from one zone of deformation to the other, and also change continuously as the deformation proceeds. In the simulation technique developed the instantaneous zonal flow stress is estimated from equation (A.1) (appendix 1(a)), on substitution of the current values of zonal strain rate and zonal temperature.

Because of the high volume to surface area ratio in the cavity and the time of forging, the heat loss

due to conduction is assumed to be small in high-speed forging and is neglected in the present analysis. Therefore, the only change in cavity temperature during deformation is due to adiabatic heating as given by equation (A.2) in appendix 1(b).

Owing to the low volume to surface area ratio at the flash the heat loss due to conduction is considerable and cannot be neglected as above. Consequently the temperature drop during one deformation step is estimated by considering the flash to be a thin plate with an average uniform temperature, cooled symmetrically from both sides, as described in the appendix 1(c), and given by equation (A.3).

Although the platen velocity varies throughout the deformation stroke, the mean strain rate ($\dot{\epsilon}_i$) in the zones of lateral flow at the flange, as given by equation (A.4) in appendix 1(d), is assumed constant during the whole process of deformation. The mean strain rate in the flash varies with deformation according to equation (A.5) in appendix 1(d). At the centre of the flange, the material changes its direction of flow from a lateral one to extrude longitudinally into the shaft. This implies a change in velocity so that the longitudinal velocity in this zone varies from one radial plane to another. The assumptions regarding the estimation of an average velocity (equation (A.7)) and mean strain rate (equation (A.8)) are explained in appendix 1(d).

3. THEORETICAL FLOW MODELS

Three main stages of deformation, as shown in figure 2(a), 2(b), and 2(c), can be distinguished during deformation of the model workpiece shown in figure 1(c).

The first stage signifies the metal flow in the cavity corresponding to a decrease in the total forging height (H) (figure 2(a)), while stage II represents the flow in the cavity and flash associated with any increase in the said height. Stage III describes the flow in the cavity and flash after complete filling of the shaft.

(a) Stage I, as shown in figure 2(a). At the commencement of forging, the axisymmetric billet is compressed between the upper and lower dies, the material flows out laterally and there is a decrease in the total forging height (H). There is an outward flow in zone 1 and backward extrusion in zone 4. At the centre of the flange the inwardly flowing metal changes velocity and direction to flow into the shaft. This transition zone, the existence of which depends upon sufficient resistance to outward lateral flow, is defined by radius R_c and angle γ. The magnitudes of the above parameters in relation to current flange height h_1 and radius R_S are given in appendix 2(a) (equations (A.9), (A.10) and (A.11)). The axial stress in this zone of transition varies from one radial plane to the other

During upsetting a neutral surface of radius R_N is established. The latter is defined as the surface on both sides of which the metal flows in opposite directions. This radius is given in the appendix 2(d) by equation (A.15). At the beginning of deformation, the resistance to flow into the transition zone 3 is large compared with the resistance to the outward flow (zone 1), and hence the radius R_N is contained within the cylinder of radius R_S. This signifies no lateral inward flow from the flange and limits the outside radius of zone 3 to R_N. Thus in stage I, with reference to figure 2(a), for a platen descent of Δh extrusion takes place by a

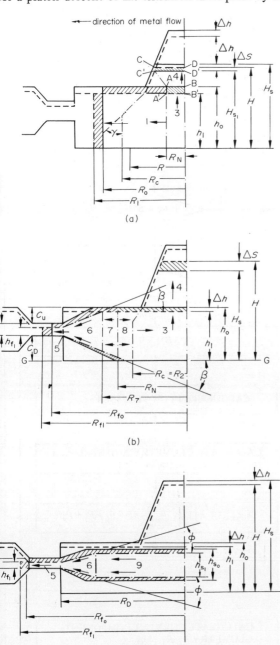

(a)

(b)

(c)

Figure 2. Flow models showing deformation zones: (a) stage I; (b) stage II; (c) stage III.

layer of material AA'B'B being pushed by a longitudinal thrust generated in the cylindrical zone 3, into the shaft as CC'D'D with no inward lateral pressure. This brings down the total forging height H_S to H as DD' < BB'. As a result of deformation Δh, the flange radius increases from R_0 to R_1.

The axial stresses in zones 1, 3 and 4 are estimated by Altan[4] and Biswas[5]. The total load to deform the

workpiece corresponding to each deformation step is $P = P_1 + P_3 + P_4$, where the subscripts 1, 3 and 4 refer to the zones of deformation.

(b) Stage II, as shown in figure 2(b). With the progress of the upsetting stage the resistance to outward flow increases and the radius R_N exceeds R_S signifying a positive inward flow. The flow model at the centre of the flange is a real transition zone 3. At this stage of deformation the material reaches the cavity wall and starts to extrude into the flash by shearing along a tapered surface as shown in figure 2(b). The geometry of the shear zone 6, in a cavity with symmetrically placed flash, was optimised to yield minimum energy consumption. The optimised value of angle β is given in appendix 2 by equation (A.12). There are several possible configurations of this shear envelope for an asymmetrically placed flash, the one adopted here being considered to be made up of two zones, 6 and 7. Zone 6 is bounded by two shear surfaces both inclined at angle β to the radial plane of the flange. Zone 7 is bounded on one side by a shear surface inclined at angle β and the die material interface on the other. For a component with C_U or $C_D = 0$, zone 6 disappears leaving only a deformation zone 7. The equations relating the stresses in zones 8, 3, 4, 5, 6 and 7 to the other forging parameters are in references 4 and 5. These stresses are integrated over their respective areas to give the transient load.

$$P = P_8 + P_3 + P_4 + P_5 + P_6 + P_7$$

(c) Stage III, as shown in figure 2(c). The commencement of this stage is characterised by the complete filling of the shaft. The metal from the cavity extrudes into the flash by shearing along the surfaces of a lens-shaped zone 9 and 6. The optimised geometry of this zone, to yield minimum energy consumption, is given in appendix 2(c) by equations (A.13) and (A.14). The friction coefficient at the shear surface of zone 9, in this case, is 0.577 from Von Mises' flow rule. The total load at this stage of deformation is thus given by

$$P = P_9 + P_5 + P_6$$

4. COMPUTER SIMULATION

The flow models described in section 3 were used for formulating the equations relating the stress distribution, forging load and forging geometry at each position of the platen stroke. These equations were programmed for a digital computer (KDF9) in simulating the forging process.

In simulating the complex mechanics of deformation in closed die forging, the following assumptions were made.

(1) Plane surfaces are assumed to remain plane on deformation, implying no barrelling.

(2) The frictional shear stress τ, at the die–material interface is expressed[4,5] by $\tau = f\bar{\sigma}$, where f is the friction factor at the interface.

(3) The calculation of the forging geometry after each deformation step is based on the radius of the neutral surface calculated prior to the application

TABLE 1

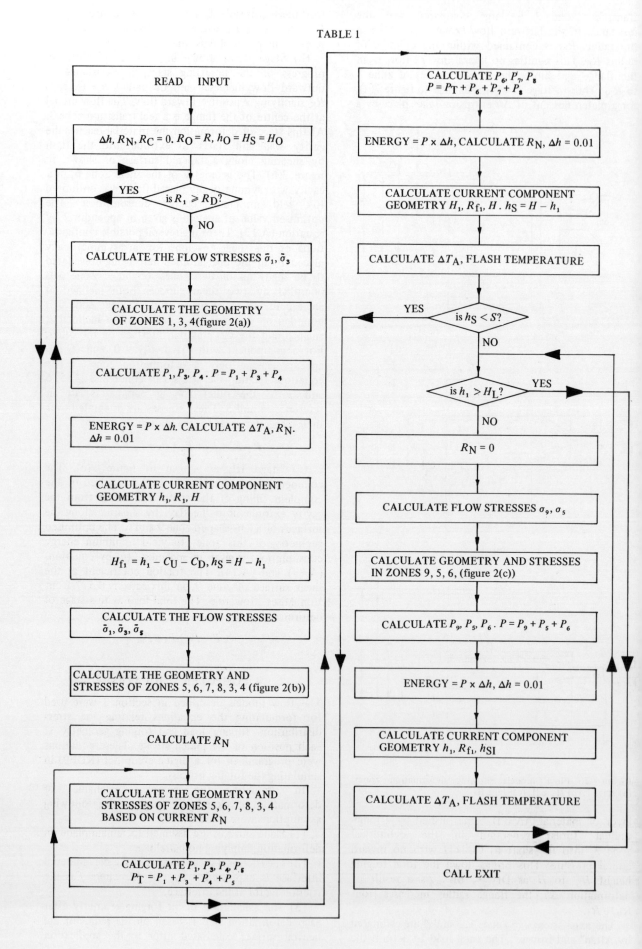

of that step, implying a negligible variation in the neutral radius over each step.

(4) The deformation, metal flow and instantaneous flow stress are assumed to remain uniform inside any zone at one position of the deformation stroke.

The input data to the computer programme broadly fall into the following categories.

(i) Initial billet size.

(ii) Dimensions of the component to be forged.

(iii) Material properties of the billet in terms of flow stress and heat transfer constants.

(iv) Nominal forging temperature and impact velocity.

A simulation cycle starts by the introduction of a platen descent, Δh. This changes the billet geometry depending on the neutral radius calculated at the end of the last cycle. A test subroutine is introduced at this stage to determine the particular stage of deformation (i.e. I, II or III) to which the current workpiece geometry belongs. Subroutines are brought into the main programme thereby to construct appropriate flow models. The zonal mean strain rates and zonal temperatures are estimated to determine the zonal flow stresses. The stress distribution at various zones are calculated to predict load and energy requirements corresponding to the current workpiece geometry. The radius of the neutral surface is computed at this stage from the current flow model configurations and stresses in order to start the next cycle.

A flow diagram of the computer programme is shown in table 1.

5. EXPERIMENTAL DETAILS

All the tests were carried out on a Mk IIA Petro-Forge high-speed hammer[7] which has a nominal output rating of 10 000 ft lbf given by a platen impact speed of 50 ft/s. Correspondingly lower energies are obtained at lower impact speeds. In this investigation the impact speeds covered the range 15–45 ft/s (1050–9450 ft lbf).

Instrumentation fitted to the forging machine included a strain gauge load cell[8] placed under the bottom die to measure transient die load, a capacitance type displacement transducer to measure platen position and a differential transformer transducer[9] to indicate the platen velocity.

The two components (A and B) forged in the tests are shown in figure 1(a) and the corresponding billet sizes are given in table 2. Both sets of billets

TABLE 2

Component	Height (in) (H_0)	Diameter (in) (D_0)
A	1.5	1.0
B	1.125	1.0

were from 'bright' EN 8 bar, those of component 'B' being cropped to length whilst those of component 'A' were sawn.

The tests were carried out at four nominal forging temperatures, 900, 1000, 1100 and 1200 °C, the billets being heated in an electric muffle furnace. As billets were transferred manually to the hammer some drop in billet temperature occurred and this was estimated to be 50 °C on average.

No lubrication was applied either to the billets or to the dies and hence for estimation of the die–billet interface friction metal-to-metal contact was assumed at the interface.

For each test the transient load, displacement and velocities were measured and recorded on a storage tube oscilloscope. The displacement signal was fed through a differential amplifier which was set to be triggered at impact, recording the outputs of the load cell and the velocity transducer as a function of platen displacement. The voltage output from the

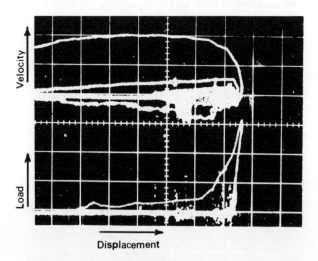

Figure 3. Typical oscilloscope record showing the variation of load and velocity with deformation.

velocity transducer at impact was computed to yield the kinetic energy available while the area of the load–displacement curve provided the energy absorbed in the billet. A typical recorded trace is shown in figure 3.

6. ESTIMATION OF THE LOAD AND ENERGY CHARACTERISTICS FOR WORK RESTRICTED FORGING

The characteristics of 'work-restricted' machines such as Petro-Forge is that the impact energy is related to the impact velocity and platen mass by a fixed relationship expressed by equation (A.16) in appendix 3. On the other hand the impact velocity influences the flow stress of the billet material and hence the required deformation energy through the strain rate effects. When a workpiece is subjected to a given energy, the simulation technique outlined above is employed to predict the die load and the total amount of deformation suffered by that work-piece at an assumed deformation velocity. This velocity in turn is related to the required energy by equation (A.16). To overcome this problem the following procedure has been adopted to predict the energy deformation and the load–deformation characteristics of the workpiece compatible to the deformation mechanics of the process and the forging machine characteristics.

(i) The computer simulation technique is employed to predict the energy–deformation and load–deformation characteristics of a component for a suitable range of impact velocities, assuming the velocities to be independent of the predicted energies.

(ii) From the above, characteristic curves of constant deformation energy–impact velocity can be constructed for a number of final deformation values. The related energy–velocity characteristic of the machine given by equation (A.16) is then superimposed on the above constant deformation energy–impact velocity characteristic.

The points of intersection give the compatible values of velocity and energy for the chosen values of deformation. These values of energy constitute the energy–deformation characteristics of the 'work-restricted' process.

(iii) In a similar way to (ii) curves of constant deformation load–impact velocity are constructed. The value of the load compatible to the 'work-restricted' process is the load on this curve which corresponds to the impact velocity and resulting final deformation given by the energy–velocity characteristics obtained above.

7. EXPERIMENTAL RESULTS

Figure 4 shows the theoretical and experimental relationships between the compatible peak load and percentage flange reduction $\{(H_0 - H_L)/H_0\} \times 100$ for the two components at the four forging temperatures.

Both components show a similar characteristic

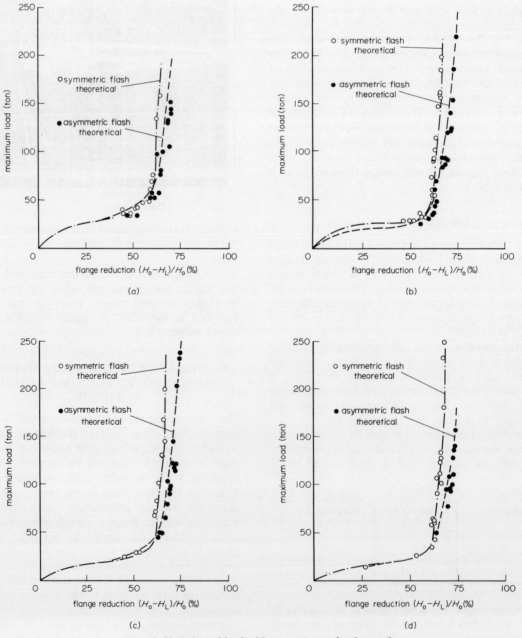

Figure 4. Variation of load with percentage reduction at the four forging temperatures: (a) 900 °C; (b) 1000 °C; (c) 1100 °C; (d) 1200 °C.

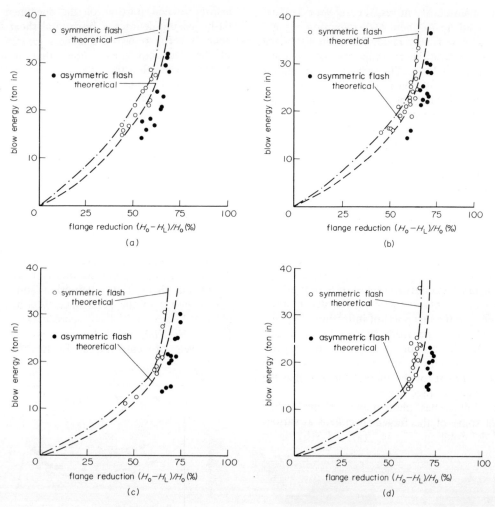

Figure 5. Variation of energy with percentage reduction at the four forging temperatures: (a) 900 °C; (b) 1000 °C; (c) 1100 °C; (d) 1200 °C.

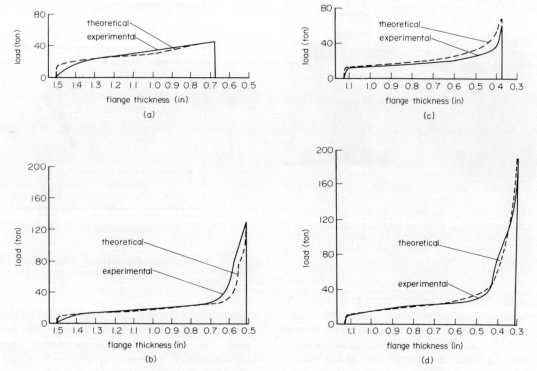

Figure 6. Variation of transient load with flange height: (a) symmetric flash at 900 °C; (b) symmetric flash at 1200 °C; (c) asymmetric flash at 1000 °C; (d) asymmetric flash at 1100 °C.

as do the theoretical curves, i.e. a very gradual increase in load up to about 60% reduction beyond which the increase is very rapid. The main difference is that the symmetric flange component (A) exhibits a steeper rise in load. The effect of increasing forging temperature is seen to cause a proportionate decrease in peak load for both components. It is also interesting to note the better agreement with theory for component 'A' than for component 'B'.

Figure 5 shows the estimated energy–percentage reduction characteristics for the two components at the four forging temperatures and the corresponding experimental points. Whilst the general characteristics for both components are similar to the load, the quantitative agreement between theory and experiment is less, particularly in the case of component 'B'.

Experimental transient load characteristics obtained from oscilloscope traces and the theoretical transient load characteristics are shown plotted for both the components in figure 6. The deformation corresponding to the first point of inflection, although occurring at the same load, is seen to be underestimated by theory for component 'B'. A pronounced second point of inflection is noticeable in figure 6(b) and 6(d) for the theoretical as well as the experimental curves. (See figure 3.)

Figure 7 illustrates photographs of the etched sections of some of the forged specimens at various stages of deformation. Figure 7(a) and 7(b) shows

Figure 7. Macrophotographs showing grain flow in sectioned and etched specimen at different forging temperatures (and reductions): (a) 1000 °C (45.4%); (b) 1000 °C (62.1%); (c) 1200 °C (68.1%); (d) 900 °C (47.3%); (e) 1100 °C (67.1%); (f) 1100 °C (75%).

component 'A' just before and just after the commencement of the flash extrusion stage, respectively. The formation of a dead metal zone at the die flange corner with the commencement of flash and a triangular dead metal zone at the bottom face of the component flange are noteworthy features of these specimens. Referring to figure 7(c) the

heavily etched portion of the specimen near the flash opening denotes a horizontal shear zone. This zone is seen to be of thickness greater than that of the flash but smaller than that of the flange. Moving towards the flange centre this shear zone is seen to be deflected at an angle to the radial plane of the flange. Component 'B' shows a shear corridor created between the dead metal zone at the corner and the dead metal zone adjacent to the component flange base, as seen in figure 7(f) and 7(e). This corridor is seen to be consistently inclined to the radial plane of the component starting at the flash corner. These specimens indicate that flash commences to extrude at about 60% reduction for both the components. The maximum shaft height achieved in this investigation is that given by the specimens shown in figure 7(c) and 7(f).

The change in component geometry with deformation is illustrated by means of figure 8 which shows the theoretical relationship and the experimental points of the percentage total height (H/H_0) plotted against the percentage reduction $\{(H_0 - H_L)/H_0\} \times 100$ for the two components at two levels of temperature. Figure 8(a) and 8(b)

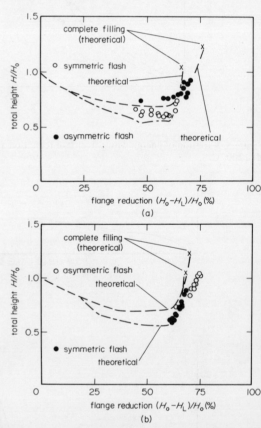

Figure 8. Variation of percentage total height with deformation at two forging temperatures: (a) 900 °C (b) 1200 °C.

shows that complete die filling is not achieved in practice under the conditions of the present investigation. The existence of three distinct stages of deformation is observed in Figure 8(a) and 8(b) which also show a theoretical underestimation in stages I and II.

8. DISCUSSION

The maximum change in slope of the experimental load characteristics observed to occur between 55–60% reduction in figure 4 is seen to be closely estimated by the theoretical model. This change in slope starts at about 45–50% which corresponds to the commencement of the predicted stage II deformation, where the introduction of zone 3 of transition and zone 2 of inward flow predict an increase in the slope of the load curves. Figure 8 shows the experimental forging height to be reduced to about 60% of the initial billet height at 47% reduction of the flange, beyond which it remains steady until about 60% reduction. The theoretical model of stage I deformation predicts a reduction in total height until the introduction of stage II, which is predicted to commence at about 45%. The above agreements suggest the assumed models at this stage of deformation to be fairly realistic, although the 5% underestimation of the total height indicates an overestimation of the resistance to inward flow.

A steep increase in the experimental load is observed in figure 4 for both components beyond 60% flange reduction. This closely corresponds to the first inflection of the estimated curve, which is predicted to occur at the commencement of flash extrusion in the theoretical model of stage II. The increased strain rate and reduced temperature gives rise to a high flash flow stress. This in addition to the large stresses at zone 6 (figure 2(b)) accounts for the steep rise in load at this stage of deformation. The validity of the theoretical model is further confirmed by the close agreement observed in figure 8, between the experimental and theoretical total heights. The steep rise in total height, in that case, could be due to the 'jump' of the neutral surface initiated by the greatly increased resistance to outward flow caused at this stage by large stresses in zone 5 and 6 (figure 2(b)). The actual existence of zone 6 is confirmed by the dead metal zone seen at the flange die corner of the etched specimen shown in figure 7(b).

The large overestimation of the energy requirement for component 'B' and in the 60–67% range of flange reduction is reflected to an extent in figure 4, but much more so in the transient load characteristics of figure 6. The large dead metal zone adjacent to the component flange base, unaccounted for in the stage II theoretical flow model and seen in figure 7(e), is considered responsible for this discrepancy. The ratio of the volume of this dead metal zone to that of the component cavity is larger for component 'B' than for 'A' which can be estimated by comparing figure 7(b) and 7(e). This indicates a larger influence of this zone on the actual stress distribution in component 'B'. The load to deform the shear corridor seen in figure 7(e) is less than that to deform the converging shear zone, as suggested by the zone 6 (figure 2(b)) of the theoretical model, suggests a possible explanation of the energy overestimation.

The second inflection of the transient load observed in practice is seen to be predicted by theory; by reference to figure 6, angle β (figure 2(b)) defined by equation (A.12) in appendix 2(b) decreases with deformation increasing the area of zone 6 and thus the load P_6. At one point of the deformation the advancing neutral surface enters zone 6, and any further deformation reduces the area of zone 6 thus decreasing P_6. This point, the point of second inflection marks a retardation of load with respect to deformation. The agreement observed in figure 6 suggests that the assumed model to be a fair approximation of the actual deformation mechanism. This suggestion is also confirmed by the oscilloscope trace shown in figure 3.

The very close agreement between the theory and practice at the last stages of deformation, as observed in figure 4 and figure 5, suggests the theoretical model, of stage III, to be a valid one. This model predicts a steep rise in load and energy because of the high flow stresses and greatly reduced thicknesses of zones 5, 6 and 9 (figure 2(e)). Further confirmation of the predicted mode of deformation is seen in figure 7(e) and 7(f), which shows the thin shear zone extending from the centre of the flange to the flash. The observed deflection of the shear zone although not influencing the forging characteristics is due to barrelling induced by the dead metal zone at the flange base.

The inability to achieve full shaft height at large reductions as seen in figure 8(a) and 8(b) could be due to the following reasons.

(a) The pressure of the gas trapped in the shaft cavity restricts backward extrusion[11].

(b) The top part of the extruding shaft, which undergoes little deformation since entering the extrusion cavity, becomes sufficiently cool by this stage of deformation to raise the flow stress and reduce flowability.

9. CONCLUSIONS

(1) A step-by-step simulation technique has been shown to predict with good accuracy the load–reduction characteristics of a work-restricted forging machine; a component with a longitudinally symmetric flash showing better agreement than one with a longitudinally asymmetric flash.

(2) The energy–reduction characteristics given by the step-by-step technique generally overestimated the actual characteristics by approximately 10% and 20% in the cases of the symmetric flash component and asymmetric components, respectively.

(3) Whilst some limitations are apparent in the flow model chosen, particularly with respect to the height of the extruded portion of the component, the agreement between the theoretical and actual component geometries is good.

REFERENCES

1.   S. KOBAYASHI, V. McDONALD and E. G. THOMSEN, 1960. 'Some aspects of press forging', *Intern. J. Mech. Sci.*, **1**, 282–300.
2.   E. P. UNKSOV, 1961. *An Engineering Theory of Plasticity*, Butterworths, London.

3. T. ALTAN, H. J. HENNING and R. J. FIORENTINO, 1969. 'The use of analytical methods in predicting loads and stresses in closed die forging. A study of mechanics of closed die forging', *Final Rept. Battelle Mem. Inst.*, Chapter 3, April 30.

4. T. ALTAN and R. J. FIORENTINO, 1969. 'Computer simulation to predict load, stresses and metal flow in an axisymmetric closed die forging', *Final Rept. Battelle Mem. Inst.*, Chapter 4, April 30.

5. S. K. BISWAS, 1971. 'Development towards automation of hot forging process', *M.Sc. Thesis*, University of Birmingham.

6. S. K. SAMANTA, 1970. 'The dynamic compression of steels at elevated temperatures', *Proc. 11th M.T.D.R. Conf.*, Pergamon Press, Oxford.

7. L. T. CHAN, 1967. 'The development of combustion driven high energy rate forming machines', *Ph.D. Thesis*, University of Birmingham.

8. S. C. JAIN and E. AMINI, 1968. 'Development of a short load cell for metal forming application', *Proc. 9th M.T.D.R. Conf.*, Pergamon Press, Oxford.

9. A. J. ORGAN. 'The calibration of high energy rate forming machines', *Intern. J. Mech. Tool Design Res.*, 7, 325–49.

10. J. M. ALEXANDER and R. C. BREWER, 1963. *Manufacturing Properties of Material*, London.

11. A. SHIEKH, 1971. *Ph.D. Thesis*, University of Birmingham.

APPENDIX 1

(a) *Flow stress.* The zonal flow stress $\bar{\sigma}_i$ is given by

$$\dot{\epsilon}_i = A_2 (\sinh\alpha\bar{\sigma}_i)^{n_2} \exp\left(-\frac{Q}{RT_i}\right) \qquad (A.1)$$

where i denotes the zone of deformation

(b) *Adiabatic temperature rise.* Most of the energy expended during deformation is converted into heat. At high strain rates such as used in the present investigation there is insufficient time for the heat to transfer to the surrounding medium. The adiabatic temperature rise at the workpiece cavity ΔT_A is given by

$$\Delta T_A = \frac{\Delta E/V}{\rho C_p} \times \frac{1}{J} \qquad (A.2)$$

(c) *Cooling of the flash.* In the time 't' for a platen descent of h, the original flash temperature T_{fo} decreases owing to conduction of heat to the adjacent die surface. If the die temperature is T_d, the new flash temperature T_{fi} is given by

$$T_{fi} = T_d + (T_{fo} - T_d) \exp\frac{A_1 t 2w}{bt} \qquad (A.3)$$

(d) *Mean strain rates.* It is assumed that the mean strain rate in the flange $\dot{\epsilon}_i$ remains constant in zones of parallel lateral flow and is given by

$$\dot{\epsilon}_i = \frac{V_0}{H_0} \qquad (A.4)$$

The mean strain rate in the flash is given by

$$\dot{\epsilon}_5 = \frac{V_5}{h_f} \qquad (A.5)$$

where

$$V_5 = V_p = \dot{\epsilon}_1 h_1$$

and

$$h_f = h_1 - C_u - C_d \qquad \text{(figure 2(b))}$$

Assuming the metal to be deforming at velocity V_p in the cylindrical part of zone 3 (figure 2(b)), the velocity V_4 at the entrance to zone 4 is given by

$$V_4 = \frac{R_c^2}{R_s^2} V_p \qquad (A.6)$$

The average velocity V_3 in zone 3 is assumed to be given by

$$V_3 = \frac{V_4 l_1 + V_p l_2}{h_1} \qquad (A.7)$$

The average mean strain rate in zone 3, $\dot{\epsilon}_3$ is given by[10]

$$\dot{\epsilon}_3 = \frac{6 V_3 D_c^2 \tan\gamma \ln(D_c/D_s)}{(D_c^3 - D_s^3)} \qquad (A.8)$$

APPENDIX 2

(a) *Geometry of the flow model in zone 3* (figure 2(a), 2(b)). To determine the geometry of transition zone 3 the stress in this zone at plane GG was optimised[4,5] to yield the following equations.

$$\tan\gamma = \pm \left(1 - \frac{\rho - 1}{\rho \ln \rho}\right)^{1/2} \qquad (A.9)$$

where

$$\rho = D_c/D_s \quad \text{when } \tan\gamma \geqslant 0$$

and

$$\rho = D_s/D_c \quad \text{when } \tan\gamma \leqslant 0$$

The diameter D_c is given by

$$D_c/D_s \approx 0.8\left(\frac{h_1}{D_s}\right)^{0.92}, \quad \text{for } \frac{h_1}{D_s} \geqslant 2 \quad (A.10)$$

and

$$D_c/D_s \approx 0.333 + 0.584\frac{h_1}{D_s}, \text{ for } 0.5 \leqslant \frac{h_1}{D_s} < 2 \quad (A.11)$$

(b) *Geometry of the model in zone 6* (figure 2(b)). To determine angle β the maximum stress in the flange was optimised[4,5] to yield the following equation:

$$\tan\beta = \left\{1 - \frac{(h_1/h_f - 1)\sqrt{3} f_f}{(h_1/h_f) \ln(h_1/h_f)}\right\}^{1/2} \qquad (A.12)$$

(c) *Geometry of the shear model in figure 2(c).* Angle ϕ is obtained by optimising[4,5] the maximum stress on the lens-shaped shear zone.

$$\tan\phi = \pm \left\{1 - \frac{h_{s1}/h_{f1} - 1}{(h_{s1}/h_{f1}) \ln(h_{s1}/h_{f1})}\right\}^{1/2} \qquad (A.13)$$

h_{s1} is given by

$$\frac{h_{s1}}{h_f} = 0.8\frac{R_D}{h_f} 0.92 \qquad (A.14)$$

(d) *Neutral surface.* The radius of the neutral surface, R_N is determined by equating the boundary stresses in the two zones of deformation adjacent to the neutral surface.

$$R_N = \frac{1}{\bar{\sigma}_x f_x + \bar{\sigma}_y f_y} \times$$

$$\left[\bar{\sigma}_x f_x R_x + \bar{\sigma}_y f_y R_y + \frac{h_0}{2} (\sigma'_{zx} - \sigma'_{zy}) \right] \qquad (A.15)$$

where subscripts x and y refer to the zones of deformation with neutral surface as one common boundary, the other boundary radii being R_x and R_y, and where σ'_{zx} is σ_{zx} at $r = R_x$ and σ'_{zy} is σ_{zy} at $r = R_y$.

APPENDIX 3

The energy delivered to the workpiece E is given by

$$E = \frac{1}{2} m \left(\frac{m}{m + M} \right) V_0^2 \qquad (A.16)$$

ACKNOWLEDGEMENTS

The present investigation forms a part of the programme to develop an 'automatic multistation forging unit' sponsored by the Wolfson Foundation and carried out under the overall supervision of Professor S. A. Tobias.

The authors wish to thank Professor S. A. Tobias for his friendly help and guidance and the trustees of the Wolfson Foundation for their financial support.

HOT KINETIC FORMING OF METALS

by

R. BALENDRA* and F. W. TRAVIS*

SUMMARY

Tests are described in the use of a large tool for the hot kinetic forming of metals. The tool basically comprises (i) a barrel and breech assembly, used to accelerate a hot metal billet to a velocity of the order of 100 m/s, and (ii) a closed die-set into which the billet is projected so that it takes up the form of the die-set under the heavy inertia stresses generated upon its arrest. Hot formed aluminium and brass specimens are presented and discussed, and general features encountered in forming metals at this order of speed are considered.

INTRODUCTION

In work reported earlier[1,2] on the cold forming of metals entirely under their own inertia forces, using a commercial stud-driving tool to accelerate the billet, a major parameter was found to be the ratio of the impact kinetic energy of the billet to its yield stress, represented by $\rho V^2/Y$, where ρ and Y are the billet density and yield stress and V is the impact velocity; the more deformation required of the billet to fill the die-set, then the higher the relevant value of $\rho V^2/Y$. Thus, to form a component from hard drawn copper was found to require only the same impact velocity as that to form the same component from soft aluminium, as the ratio of the densities of the two materials was approximately equal to the ratio of their yield stresses.

Having established this basis, it became possible thereby to predict to what extent different materials could be formed by the process; in some cases, particularly where considerable deformation of a low-density billet was required, a 'driver' of some dense material was set behind it to increase its effective density and thus to lower the impact velocity to within the range of the tool employed.

Whilst in early work only cold billets were used, it was apparent that the forming capabilities of the process for any material would be increased by a reduction in the yield stress, secured by heating the billet. However, although it is well appreciated that a very large reduction in forming forces is secured by hot working at conventional forming speeds, at the level of forming speed presently employed it is relevant to consider the work of Johnson and Slater[3,4] on the combined effects of temperature and strain rate. These authors found, for example, that in the cold† blanking of aluminium at both high and low strain rates, the ratio of the respective blanking energies was approximately 2, whereas in the hot blanking operation the ratio rose to approximately 18.

A similar trend was observed for copper and mild steel, and the indications are that at very high strain rates, where the forming time is less than that required for the recrystallisation reaction, material is unable to effect complete thermal recovery, and thus behaves, to some degree, effectively as a 'cold' material. For discussion of these factors, the reader is referred to references 5 and 6.

Whilst the advantages of hot forming are diminished therefore at high strain rate, this aspect forms the basis of the present investigation. The billet is heated in an adjacent furnace and transferred to the tool immediately prior to forming.

FORMING TOOL

Sectioned views of the breech assembly and of the die-set assembly of the tool are presented as figures 1 and 2 and a photograph looking from the die-set end is presented as figure 3. The 1 in diameter billet is slipped into a mild steel sleeve prior to heating in the

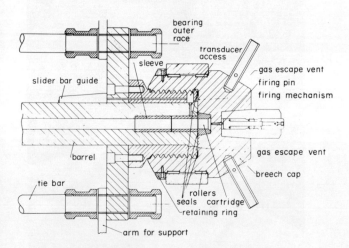

Figure 1. Cross-sectional view of breech assembly.

furnace, and, upon reaching the required temperature, the assembly is inserted intact into the barrel of the tool, using tongs. A second sleeve containing a lead 'driver' is then inserted by hand and the breech closed. To protect the cartridge from the heat of the billet, it is mounted in the breech cap, and only approaches the vicinity of the billet and driver as the breech cap is finally screwed home. Rubber ropes support the weight of the breech cap via the outer race of a roller bearing, thus enabling it to be swung

* Production Engineering Department, University of Strathclyde, Glasgow

† In the work of references 3 and 4, 'hot' or 'cold' are taken as meaning above or below the recrystallisation temperature of the primary constituent of the material.

into place on a hinged mounting, after insertion of the billet and driver, and the breech to be quickly closed.

A frame is used to secure alignment between the breech assembly and the die-set assembly, consisting of two $2\frac{1}{2}$ in thick end plates held together by four $1\frac{3}{4}$ in diameter steel tie-bars.

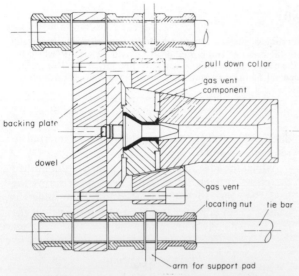

Figure 2. Cross-sectional view of die-set assembly.

Figure 3. Forming tool, viewed from die-set end.

A tubular protective shroud, shown in figure 3, runs between the end of the barrel and the mouth of the die-set.

For full details of the manufacture, operation and development of the forming tool, the reader is referred to reference 7.

RESULTS AND DISCUSSION

The cross-sectional form of the component to be produced is shown, filled-in, in figure 2. The billets to be used are of a slightly greater volume than that of the die cavity, so that in producing a fully formed component the excess material leaves a slight stub, beyond the peripheral lip at the top of the cylindrical section, which is required to be removed subsequently. As the billet material passes in entirety beyond the nose of the male die, then, should a 'driver' be used, this has to be of a non-rigid material to avoid damage to the male die; having a high density and low strength, lead is the ideal material for this purpose.

Preliminary tests

As a preliminary investigation, to provide an indication of the level of operation necessary with the higher-strength materials, model tests were carried out with lead billets at room temperature. Figure 4 shows an 'as-formed' lead specimen, where the

Figure 4. Cold-formed lead specimen as removed from die-set.

estimated impact velocity was 100 m/s, and the billet length was $5\frac{1}{2}$ in, and figure 5 shows the same specimen after axial sectioning. The material has completely filled the die-set, and there is extensive flash, of the order of 0.010 in thickness, indicating the very high level of pressure developed.

An evident minor defect of the component is the presence of small surface blisters, as indicated by the arrows of figure 4, resulting from entrapment of small pockets of die lubricant — Shell Barbatia Grease 4. This particular defect arises owing to unsuitable die design; at the sharply angled transition from the cylindrical to the conical walls, axial inertia pulls material away from the walls of the female die, despite the compressive forces imposed by the rearward material, and die lubricant collects in this region. As the operation progresses, and the level of speed reduces, material commences to flow into this region, and displaces the lubricant, which then becomes entrained with the forming material moving down the conical walls, as indicated by the lower arrow of figure 5. The pockets of lubricant are thus subjected

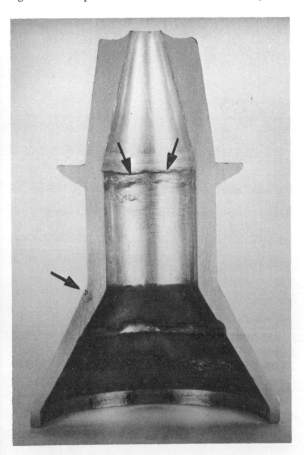

Figure 5. Sectional view of specimen of figure 4.

to the high forming pressure experienced by the material and, at the conclusion of the operation, the relief of this pressure by expansion produces the evident blisters.

A further defect arises at the peripheral lip. After piercing, material flows down the parallel walls of the cylindrical region, and proceeds to fill up the conical region, without any flow into the lip. Having filled both the conical and the cylindrical regions, material then commences to flow into the lip by a shearing action across the top of the tubular walls. Subsequent metallurgical inspection of sectioned and etched specimens reveals a region of intense deformation, running from the arrowed defect downwards and outwards at 45°, across to the junction of the bottom of the lip and the commencement of the cylindrical walls. This feature, which is contrary to that normally experienced in slow speed forging of such sections where some degree of continuity of the flow pattern is maintained throughout the component, produces

an inherent weakness, which in a number of cases, in subsequent tests with harder materials, resulted in rupture there upon cooling of the hot formed component within the cold die-set.

Aluminium billets

Figure 6 shows an aluminium component, hot formed at an estimated temperature of 550 °C, and an impact velocity of 100 m/s, using a $3\frac{1}{2}$ in billet supplemented by a $2\frac{1}{2}$ in long driver. The component is 'as-formed' excepting that the stub of excess material has been turned off. A modified male die was used to restrict the length of the conical walls, and a flash gap of approximately 0.010 in was allowed between the male and female die at the bottom of the die cavity; the flash produced is shown in figure 6.

Figure 6. Hot-formed aluminium component.

A defect not previously encountered in cold tests is the general surface roughness, arrowed, over the lower regions of the conical walls. Upon sectioning and polishing a specimen from this vicinity, there was found to be a general distribution of small cavities extending throughout the whole thickness of the wall, as shown in figure 7. A similar phenomenon has been observed and reported by Dower[8,9] with components

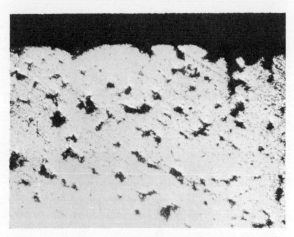

Figure 7. Sectioned and polished specimen from lower wall of conical region of component of figure 6 (magnification 10x)

formed by 'extrusion moulding'; this operation is similar to kinetic forming, excepting that the high velocity required of the billet material is secured by its high-speed impact extrusion at an area ratio of 40:1, to provide an exit velocity of the order of 300 m/s. Dower[8] states that 'if the billet temperature or the extrusion ratio is too high, excessive turbulence can occur in the die, leading to porosity in the product'. In extrusion moulding, the cross-sectional area of the extruded rod is small in relation to that of the die-set, so that 'the die filling mechanism appears to be a combination of splashing followed by buckling and upsetting of the extruding rod against the closed die. Welding occurs where the metal impinges on itself.'[8] Thus the forming material is not subjected to heavy lateral constraint, and an (oversimplified) analogy might be that of filling a cup from a water tap. With the parallel walls of the present die-set, the cross-sectional area of the die cavity encountered by the advancing material is continually increasing, so that, as in the extrusion moulding operation, the forming material is without lateral constraint, and conditions leading to turbulence are promoted. A further result of the absence of lateral constraint was encountered in earlier reported work, where in the forming of parallel-walled cones, excessive hoop rupture of the lower walls occurred. This was elimin-ated by providing a 5° convergence of the die walls, so that slight reduction of the cross-sectional area of the die cavity was maintained as the material advanced, thus producing compressive stresses in the 'thickness' direction to offset the tensile hoop stresses causing hoop rupture, and perfectly sound components were then produced. A similar modification of the present die-set would be advisable to constrain the material laterally over the entire forming operation. The com-ponent of figure 6 also suffered partial cracking at the peripheral lip, arrowed, along the line of inherent weakness discussed earlier, but this could have been avoided by heating the die-set prior to forming. Similarly, the hot formed billet cooled to a tight shrink fit over the cylindrical part of the cold male die, which made removal difficult, and emphasised the necessity for provision of adequate draught.

Brass billets

A trimmed and polished brass component is shown in figure 8, hot formed at an estimated temperature of 800 °C and an impact velocity of 100 m/s, using a $3\frac{1}{2}$ in long billet and a $2\frac{1}{2}$ in long driver.

The main defect of this specimen is the development of hoop rupture of the material as it proceeds down the conical part of the die cavity, but it will be noted, however, that over the forward end of the component the material has welded together, upon impact with the bottom of the die cavity, to produce a coherent mass. By designing slight convergence into the die walls, as discussed earlier, this defect could have been avoided in the first instance or, alternatively, by use of a slightly higher forming temperature and/or impact velocity, the welding-up of the ruptures could have been carried to completion. The absence of any flash, and the incomplete filling of the

peripheral lip, indicate that the initial forming conditions were inadequate, and that there was not the heavy compressive stress developed throughout the component, such as indicated in figure 6, which would be required to eliminate these flaws.

Figure 8. Trimmed and polished hot-formed brass component.

Venting of the die-set

In kinetic forming, venting of the die-set is found to be essential, particularly to avoid entrapment of pockets of air. When forming a hemispherical dome in the work of reference 1, using a solid die, the en-trapped pocket of air caused further deformation of the component as it expanded at the end of the forming operation, and the dome was not to form. In addition, the adiabatic compression of the pocket of air caused local overheating of the surface of the component. When the same die was used in the cracked condition, perfect components were produced. In view of the high level of die pressure developed in kinetic forming, the flash gap does not need to be large; an analysis of the kinetic forming of a typical component[10] revealed that, when forming a thin flange or section, the kinetic energy acquired by the advancing material becomes sufficiently high for the material to form into the flange or section virtually without need of rearward applied forces, so that pressure levels in this region are extremely high. Thus, as shown in figures 4 and 6, the considerable volume of air entrapped within the die-set by the advancing billet is discharged through the flash land, and a considerable flash formed, although the flash gap is only of the order of 0.010 in. In general, a flash gap of the order of 0.005 in was found adequate.

For the extrusion moulding operation, Dower[8] advises evacuation of the air from within the die, to avoid its entrapment within the material, which is found to produce internal flaws and to give the product a blistered appearance. For kinetic forming, evacuation does not appear to be necessary. In this respect, however, it is worth pointing out that a further advantage of ensuring that the advancing material completely fills the cross-section of the die-cavity is that the air therein is progressively driven ahead of the material, towards the flash gap, and the possibility of its entrapment is thus excluded.

Performance of the tool

In operation, the tool is found to be fairly quiet, this being attributed to the protective shroud, which allows the propellant gases to expand and reduce in velocity after emerging from the barrel, prior to escaping to atmosphere. The propellant employed, 'Rifle Neonite', produces little smoke and, by having sited the tool in the vicinity of an extractor, fumes are not particularly noticeable.

As the only physical result of firing the tool is to displace the billet and driver from the breech end of the barrel to the die-set, there is no 'recoil' as such, and the short displacement of the frame to maintain the centre of gravity of the tool is readily accommodated by the 4 in thick resilient mountings.

A modification to be incorporated is the replacement of the mild steel sleeve in which the billet is heated, by one of some ceramic or asbestos materials, so that heat losses prior to firing can be minimised. It is envisaged that the tool will then be able to form steel billets successfully; with the present arrangement, the heat losses through the metal sleeve to the barrel of the tool are excessive and deformation of steel billets, at the present level of impact velocity, is limited to little more than the initial piercing operation. Whilst it would be quite simple, by use of a larger propellant charge, to increase the energy available, possibly by an order of magnitude — in the work of references 1 and 2, impact velocities of

up to 600 m/s were used — it was found that higher impact velocities resulted in damage to the male die when using steel billets.

ACKNOWLEDGEMENTS

The authors wish to acknowledge the financial support of the Department of Trade and Industry, under agreement reference AT/1065/09/NEL.

REFERENCES

1. F. W. TRAVIS, 1969. 'An investigation into kinetic forming', *Intern. J. Mach. Tool Design Res.*, **9**, 51.
2. R. BALENDRA and F. W. TRAVIS, 1970. 'An investigation into the kinetic forming of tubular and other thin-walled components from solid cylindrical aluminium and copper billets', *Intern. J. Prod. Res.*, **8**, 345.
3. R. A. C. SLATER and W. JOHNSON, 1967. 'The effects of temperature, speed and strain-rate on the force and energy required in blanking', *Intern. J. Mech. Sci.*, **9**, 271.
4. W. JOHNSON and R. A. C. SLATER, 1967. 'A survey of the slow and fast blanking of metals at ambient and high temperatures', *Proc. Intern. Conf. Manuf. Tech.*, C.I.R.P./A.S.T.M.E., Ann Arbor, p. 825.
5. W. JOHNSON and R. A. C. SLATER, 1968. 'The dynamic blanking, forging, indenting and upsetting of hot metals', *Proc. I.U.T.A.M. Symp, East Kilbride*, Springer-Verlag, Wien, p. 120.
6. R. A. C. SLATER, S. Y. AKU and W. JOHNSON, 1971. 'Strain rate and temperature effects during the fast upsetting of short circular cylinders of 0.55% plain carbon steel at elevated temperatures', *Ann. C.I.R.P.*, **19**, 513.
7. R. BALENDRA, 1971. 'Studies in bulk metal forming under inertial forces', *Ph.D. Thesis*, University of Strathclyde.
8. R. J. DOWER, 1967. 'A preliminary investigation of the extrusion moulding process', *Proc. 7th Intern. Mach. Tool Design Res. Conf.*, Birmingham, p. 41.
9. R. J. DOWER, 1966. 'Preliminary assessment of the extrusion moulding process', *NEL Rep. No. 214*.
10. R. BALENDRA and F. W. TRAVIS, 1972. 'An analysis of the kinetic forming of cones', *Intern. J. Mech. Sci.*, **14**, 247.

HIGH-VELOCITY HYDROSTATIC EXTRUSION – A FEASIBILITY STUDY

by

C. E. N. STURGESS* and T. A. DEAN*

SUMMARY

High velocities have been applied to small-scale hydrostatic extrusion to dispense with ram seals.

An unsealed punch follower has been employed, at high impact velocities (20 ft/s), to extrude hydrostatically 1 in diameter billets of materials ranging from aluminium to high-speed steel, at an extrusion ratio of 2 through 90° dies, using various viscosity blends of Shell low viscosity index oil as pressure-transmitting media.

An elementary analysis of the process was in qualitative agreement with the results, and highlighted the essential mechanics of the process.

NOTATION

C	=	velocity of sound in the fluid
D	=	diameter of container
D_b	=	diameter of billet
h	=	gap width
L	=	length of punch follower
M	=	mass of moving platen
P	=	pressure
P_w	=	amplitude of pressure wave
$\dot{q}$	=	flow rate/unit width
$\dot{Q}$	=	flow rate
Q	=	leaked volume
Q_p	=	pressurisation leakage
Q_T	=	total leaked volume
S	=	punch stroke
t_p	=	time of pressurisation
T	=	total process time
V	=	velocity
V_0	=	impact velocity
V_p	=	punch velocity
V_b	=	billet velocity
$V_c = \dfrac{Ph^2}{6\eta L}$	=	critical velocity for no leakage
$\overline{V}$	=	mean velocity
$\overline{V}_b$	=	mean billet velocity
δV	=	relative volume
x, y, z	=	Cartesian coordinates
η	=	dynamic viscosity
μ	=	particle velocity
ρ	=	density
τ	=	shear resistance
ψ_S	=	volume swept by punch
ψ_E	=	volume extruded

INTRODUCTION

Hydrostatic extrusion, in which the billet is expelled from the die by a high-pressure fluid, was originally explored by Bridgman[1]. Subsequently this process has been extensively investigated. Summaries of the current information have been published by Pugh[2,3] and Beresnev et al.[4].

A number of obvious differences exist between conventional and simple hydrostatic extrusion, these may be summarised in the following way.

Advantages

(1) Friction between billet and container is absent, permitting the use of long billets.

(2) Die friction is low enabling high-strength low-ductility materials to be extruded without cracking.

(3) Thin-walled dies can be used, supported by the hydrostatic pressure – this reduces die costs.

Disadvantages

(1) Elastic energy, stored in the fluid during compression, can constitute a hazard.

(2) The billet has to be profiled to suit the die to afford a primary seal during compression of the fluid.

(3) As the billet is not in contact with the ram, variations in die friction can cause pressure fluctuations in the container, resulting in intermittent extrusion of the product (stick-slip).

(4) Sealing of the high-pressure fluid presents production difficulties particularly with respect to the moving ram seals.

The advantages of the process have resulted in two major applications.

(1) Materials can be extruded at higher reductions than by conventional means

(2) Materials having high strength and low ductility may be extruded.

Developments in hydrostatic extrusion have been directed towards the large-scale application of the process. However, if hydrostatic extrusion is to be used to produce small components from difficult materials, then higher production rates are required than are possible with existing equipment.

The investigations reported here centre on the application of high velocities to hydrostatic extrusion in an attempt to dispense with ram-seals, thereby simplifying tooling and so achieving higher cycling rates.

SPEED EFFECTS IN HYDROSTATIC EXTRUSION

The two major speed effects in hydrostatic extrusion are (a) suppression of stick-slip instabilities; (b) promotion of hydrodynamic lubrication.

Green[5] designed and developed an experimental cross-bore high-speed machine (20 ft/min) for the

* Department of Mechanical Engineering, University of Birmingham.

practical exploitation of hydrostatic extrusion. The amplitude of the pressure fluctuations during stick-slip were reduced by 50–80% in comparison with those obtained at slow speeds.

Low and Donaldson[6] investigated various methods for reducing stick-slip instabilities. They found that increasing the working speed reduces the amplitude of the stick-slip, but results in high breakthrough pressures. An analysis indicated that the process would be stable above ram velocities of 5–10 in/min.

Low[7] continued the investigations of speed effects in hydrostatic extrusion using a press capable of ram speeds of 90 in/min. He found that in general the amplitude of the pressure variations encountered decreased with increase in speed, and at product speeds above 100 in/min no stick-slip occurred.

Stick-slip resulted in corresponding variations in extrudate diameter. The extrusion products decreased in diameter as the ram velocity increased owing to hydrodynamic lubrication.

Hillier[8] investigated the conditions necessary to promote hydrodynamic lubrication in hydrostatic extrusion, and has shown that the onset of hydrodynamic lubrication is a function of

$$N = \frac{\eta V}{\sigma r} \qquad (1)$$

Hillier utilised Pugh's[9] experimental results and found that N is of the order of 10^{-4} for full-fluid lubrication.

SIMPLIFIED ANALYSIS OF HIGH-VELOCITY HYDROSTATIC EXTRUSION

The process considered here is one where an unsealed punch enters a hydrostatic extrusion chamber at sufficiently high velocities to accomplish the extrusion before the fluid leaks past the punch (figure 1).

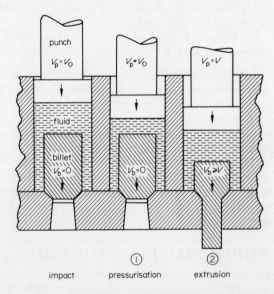

Figure 1. Stages of impact hydrostatic extrusion.

To analyse high-velocity hydrostatic extrusion the following assumptions were made.

(1) Flat plate analogue. The gap between the punch and container is small in comparison with the diameter of the punch. Therefore the conditions occurring between two flat plates in relative motion were applied to the gap between the punch and container.

(2) Incompressible flow. Fluids are appreciably compressible at the pressures encountered in hydrostatic extrusion[10]; however this assumption considerably simplifies the analysis.

(3) Steady-state laminar flow. The flow is of small volumes of highly viscous fluids.

(4) Newtonian fluids. This assumption will be shown to be reasonable for the fluids and conditions encountered.

(5) Isoviscous fluid. Fluid viscosity will change owing to the opposing influences of temperatures generated and pressured developed in the process[11].

(6) Constant gap width. The gap between the punch follower and container is assumed to be parallel[12]. The gap would probably be subject to considerable variations along the length of the punch[13]; it was considered unprofitable to incorporate these effects.

(7) A constant length punch follower.

Utilising these assumptions, expressions have been derived in the appendix which enable assessment of the relative importance of the process parameters; these will now be discussed.

Two factors influence the flow past the unsealed punch follower.

(1) Pressure produces outward flow.

(2) Viscous drag retards the outward flow (see equation (7) of the appendix).

Therefore, there exists a certain critical velocity at which the pressure flow is counterbalanced by the viscous drag; above this critical velocity no fluid leaks past the punch.

The critical velocity (equation (A.9)) is given by

$$V_c = \frac{Ph^2}{6\eta L} \qquad (2)$$

In the following discussion the parameters of interest are expressed in terms of this characteristic quantity.

High-velocity hydrostatic extrusion, as depicted in figure 1 was visualised as proceeding in two distinct stages.

(1) Pressurisation. To achieve 'pure' hydrostatic extrusion (unaugmented), the volume of fluid in the container must be sufficient to allow for compressibility and leakage without the punch contacting the billet.

The height of fluid above the billet necessary to achieve this condition may be found from.

$$h_f = \frac{4\delta V}{\pi D^2} \left\{ 1 + \frac{2h}{D}\left(\frac{V_c}{V_0} - 1\right) \right\} \qquad (3)$$

(see equation (A.12)).

(2) Extrusion. Once the fluid has been pressurised, and extrusion starts, the velocity of the billet must be equal to, or greater than, the punch velocity to maintain pure hydrostatic extrusion.

This condition is given by

$$V_0 \geqslant \frac{2V_c}{1 + \dfrac{D}{2h}\left(1 - \dfrac{D_b{}^2}{D^2}\right)} \qquad (4)$$

Equation (21) of the appendix).

However, in a free-flight impact forming device the deformation energy, and hence the volume extruded, is directly related to the punch velocity. Therefore it is possible to relate machine and process variables as

$$\sqrt{\frac{P\psi_E}{M}} \geqslant \frac{2V_c}{1 + \dfrac{D}{2h}\left(1 - \dfrac{D_b{}^2}{D^2}\right)} \qquad (5)$$

EXPERIMENTAL EQUIPMENT

Forging machine

To apply high velocities to hydrostatic extrusion a Petro-Forge machine was used as a power source.

Petro-Forge is a commercially available combustion actuated high-velocity forging machine, which has been fully described elsewhere[14,15]. The particular Petro-Forge utilised in these investigations was characterised by an anvil weight of 300 lbf, and operating velocities between 10 and 50 ft/s.

In use the machine was instrumented with displacement- and load-measuring systems.

Hydrostatic extrusion tooling

The hydrostatic extrusion tooling used is illustrated in figure 2.

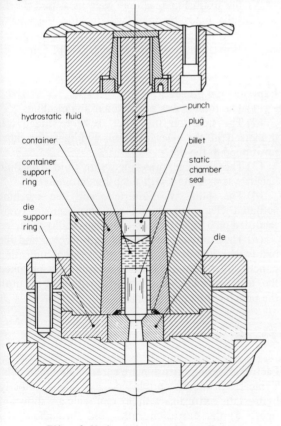

Figure 2. Hydrostatic extrusion tooling.

The tool material used was E.S.C. Super C.12, a high carbon, high chromium cold working die steel, heat treated to 60 Rc to provide high strength, wear resistance and good impact properties.

The die and container were optimised duplexed cylinders[12] produced to the following sizes.

Component	I/D (in)	O/D (in)	Optimum interface diameter (in)	Interference in/in of diameter	Pressure capability (tonf/in²)
die	0.707	6.5	2.14	0.005	100
container	1.250	6.5	2.86	0.005	100

The interface fits were achieved by press-fitting with 2° inc. tapers at the interfaces.

The presence of high-pressure fluids has a deletereous effect on container fatigue life[16]. However, owing to the restricted nature of the trials, fatigue was not considered.

A punch follower, or plug, was incorporated in the tool set for experimental convenience. The plug was ground to suit the container bore; subsequently the ambient clearance was assessed as 0.0002 in.

The gap between the plug and container at pressure was calculated according to the previously mentioned assumptions and is shown in figure 3.

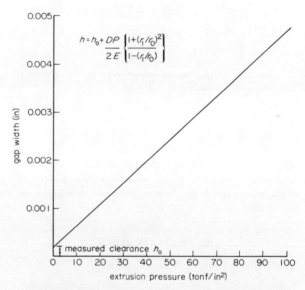

Figure 3. Plug–container gap width versus extrusion pressure for hydrostatic extrusion.

To investigate the volume of fluid leaked during pressurisation the lower face of the plug was provided with a cone of 130° inc. angle. Therefore punch/billet contact was indicated by a conical impression in the rear of the billet.

The extrusion punch having a high length-to-diameter ratio was produced from Jessop Saville G.110 Vacumelt Maraging Steel, to provide high strength and good ductility.

The tool set was equipped with a static seal between the die and container to avoid fluid leakage. The seal employed was a 'standard'[3] 'O'-ring and beryllium–copper bevel mitre anti-extrusion ring.

Fluid selection

High viscosity is the major requirement of a fluid for the application of high working speeds to hydrostatic extrusion in order to dispense with ram seals. Further desirable fluid attributes are good lubricating properties, low compressibility and low cost.

Fluid viscosity increases with increase in the applied pressure and decreases as temperature is raised[11]. The considerable viscosity increases at high pressures often result in fluids freezing[2]; this has led to many diverse fluid mixtures being used for hydro-static extrusion[2–4]. However, fluid freezing is a time-dependent phenomenon[17] and should not be a problem at high velocities where process times are approximately 5 ms. Nevertheless a fluid possessing a low pressure coefficient of viscosity linked with a high controlling viscosity is desirable.

The oils used in these tests were those recommended and supplied by Shell Thornton Research Ltd. They were low viscosity index oils (L.V.I. oils) of high and low controlling viscosities. These fluids are typified by pressure coefficients of viscosity around 30×10^{-5} in^2/lbf. From the two basic oils the following blends were produced.

Designation	Blend	Viscosity (lbf s/in^2)	Temperature ($^\circ$F)
A	100/0	4.45×10^{-3}	60
B	80/20	0.97×10^{-3}	60
C	60/40	0.23×10^{-3}	60
D	40/60	0.068×10^{-3}	60
E	20/80	0.024×10^{-3}	60
F	0/100	0.0097×10^{-3}	60

The viscosities were measured by a Ferranti rotating-cylinder portable viscometer.

The temperature of 60° F was the ambient laboratory temperature when the trials were conducted.

L.V.I. oils are Newtonian up to shear rates of approximately[17] 10^6 s^{-1}. Typical shear rates for this application were 10^5 s^{-1} (equation (A.16)). Therefore the assumption of Newtonian fluids should be reasonable.

BILLET MATERIALS

The following materials were extruded.

Material	Hardness (VPN)
commercially pure aluminium	29
HE30WP	64
electrolytically pure copper	66
low residual En2E	200
6:5:2 high-speed steel (M2)	300

These materials were machined into billets of the sizes shown in figure 4. The provision of a parallel nose was necessary to ensure a straight extrusion, no container supports being utilised.

Instrumentation

A co-axial capacitance displacement transducer[18], used in conjunction with a Southern Instruments

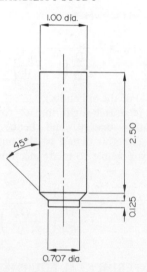

Figure 4. Billet dimensions.

M.700.L gauge oscillator and M.R.220F F.M. gauge amplifier, was employed to monitor punch movement.

It was impractical to measure pressures inside the chamber by a Manganin gauge; therefore the extrusion pressures were measured by a strain gauged punch. The punch was strain gauged with four (two axial, two circumferential) Hawker Siddeley K250 Ω gauges. The bridge was arranged to eliminate bending effects in the punch.

The bridge output was amplified by a Hottinger 50 Hz carrier frequency impedance bridge. The Hottinger possessed a flat frequency response to 5 kHz which was considered adequate to avoid attenuation of the bridge output.

As the extrusion pressures were measured on the punch, the loads recorded will include the viscous drag along the plug. The viscous drag on the punch was estimated as approximately 1000 lbf (equation (A.18)).

Experimental procedure

(1) The tooling was assembled in the machine.

(2) The nose on the billet to be extruded was immersed in the relevant fluid, and then the billet positioned in the die.

(3) The container was filled with fluid. The fluid height was measured with a depth micrometer.

(4) The plug was inserted, and the machine actuated. The punch load and displacement were recorded on a Tektronix 564 Storage oscilloscope.

(5) The formed components were ejected and the billets checked for a conical impression indicating punch–billet contact. If no impression was obtained the test procedure was repeated with less oil until the punch contacted the billet.

RESULTS AND DISCUSSION

Factors affecting extrusion pressures – billet materials

Typical products produced by high velocity hydrostatic extrusion without ram seals are shown in figure 5; the extrusion ratio used throughout the trials was 2:1.

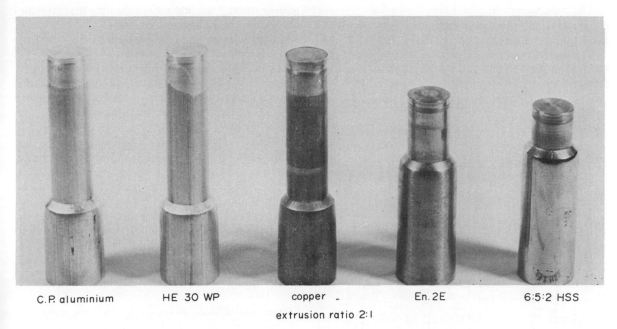

C.P. aluminium   HE 30 WP   copper   En.2E   6:5:2 HSS

extrusion ratio 2:1

Figure 5. Typical products of high-velocity hydrostatic extrusion.

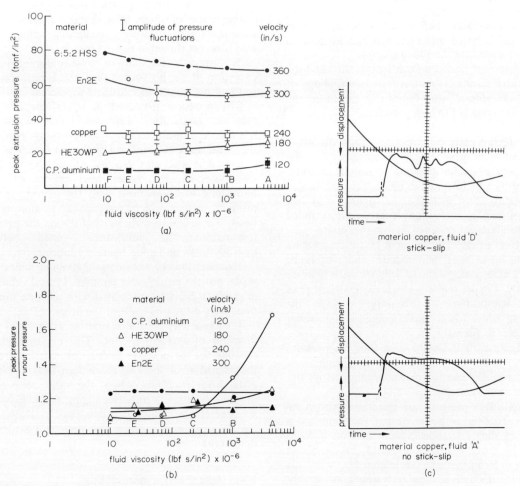

Figure 6(a). Effect of fluid viscosity and material on the peak extrusion pressure in high-velocity hydrostatic extrusion; (b) Effects of fluid viscosity on the ratio of peak to runout pressure for the high-velocity hydrostatic extrusion of various materials; (c) Typical pressure and displacement−time traces for the high-velocity hydrostatic extrusion of copper.

The pressures required to extrude these components are displayed in figure 6(a), plotted against fluid viscosity.

To compare the general level of extrusion pressures required in high-velocity hydrostatic extrusion with pressures required for conventional low-speed hydrostatic extrusion the expression

$$P = (0.374H + 4) \times \ln R \text{ tonf/in}^2 \qquad (6)$$

obtained by Pugh[3] has been used. This expression has been found to be accurate to 25%, for a wide range of materials.

The comparison is shown in the following table.

Materials	Hardness (VPN)		Extrusion pressure (tonf/in²)	
	Initial	Final	Mean experimental	Predicted
C.P.A	29	47	11	10.5
HE30WP	64	84	24	19.5
copper	66	130	32	20.0
En2E	200	265	57	54.8
6:5:2 HSS	300	350	73	80.8

The agreement between the Pugh's empirical expression (equation (6)) and experiment is good except for the extrusion of copper, where the actual pressure is considerably underestimated.

As can be seen from the table, copper work hardens considerably during the process, and as equation (6) is based on initial material hardness it cannot accommodate this phenomenon.

The agreement between Pugh's expression and the experimental results is contrary to the findings of Low and Donaldson[6], who found that increasing the pressurisation rate led to high pressures.

Factors affecting extrusion pressure – fluid viscosity
The variation of peak extrusion pressure, and the ratio of peak extrusion pressure to runout pressure, with viscosity are illustrated in figures 6(a) and 6(b).

No results are included in figure 6(b) for high-speed steel, as insufficient lengths were extruded to obtain reliable estimates of runout pressures.

The effect of viscosity on peak extrusion pressure shown in figure 6(a) demonstrates that for En2E and 6:5:2 HSS peak pressures decrease with increase in viscosity. Copper requires a virtually constant peak pressure irrespective of viscosity, whereas the aluminium alloys show peak pressures increasing with increase in viscosity.

For figure 6(b) it can be seen that the pressure ratios for En2E and copper are independent of viscosity, whilst those for the aluminium alloys rise with increase in viscosity.

Peak extrusion pressures are associated with static billet–die friction, whereas runout pressures are a function of sliding friction. Sliding friction in high-velocity hydrostatic extrusion could involve either boundary or hydrodynamic lubrication.

The velocities used in these tests were considerably less than those required to achieve hydrodynamic lubrication according to equation (1). However, the billet surface finish was substantially unmodified on

passing through the die zone, indicating that intimate die–billet contact was not occurring.

Lubrication should improve with increase in fluid viscosity and decrease in material strength. For the high-strength materials increased fluid viscosity appears to reduce both static and sliding friction; these effects could be due to improving boundary lubrication with increase in fluid viscosity.

For the aluminium alloys the high peak pressures may have been due to the shear resistance of the high-viscosity fluids surrounding the punch nose being appreciable compared with the strength of the billet material. That the runout pressures for the extrusion of aluminium were independent of fluid viscosity would indicate full-fluid lubrication.

On the basis of the above arguments the extrusion of copper would appear to have been conducted under conditions representing a transition between boundary and hydrodynamic lubrication.

Pressure instabilities
It has been found by Low and Donaldson[6] that increasing the pressurisation rate in hydrostatic extrusion reduces or eliminates stick-slip instabilities. It was hoped that pressure instabilities would not be encountered during the high-velocity hydrostatic extrusion process. However, as can be seen from figure 6(a) considerable pressure fluctuations occurred during some of these tests. Typical pressure traces demonstrating pressure oscillations and stable extrusion are shown in figure 6(c).

No pressure fluctuations were observed in the extrusion of the high-speed steel; this is probably due to the short lengths of extrudate produced.

The products produced in operations involving pressure oscillations displayed bands where the lubrication conditions were intermittent. However, the conditions encountered could not be adequately described as stick-slip. No 'galling' on the die lands was observed; the zones were regions of slight burnishing in comparison with the majority of the product surface which was similar to that of the billet. These results indicate that during the pressure fluctuations the lubrication varied between hydrodynamic and good boundary lubrication.

Examinations of extrudates containing zones were made, but the results were tenuous. Typical variations in product diameter associated with pressure fluctuations were estimated to be of the order of 0.001 in.

It is generally held[2] that pressure instabilities are a function of starting friction. This explanation can account for the pressure fluctuations encountered in the extrusion of the two aluminium alloys, pressure instabilities occurring where the ratio of peak to runout pressures is high (see figure 6(a)). However, no general correlation between peak extrusion and the occurrence of pressure fluctuations is evident in figure 6(a).

Fluid leakage and compressibility
To investigate theoretically the amount of fluid leakage during high-velocity hydrostatic extrusion, it is necessary to know both the compressibility of the

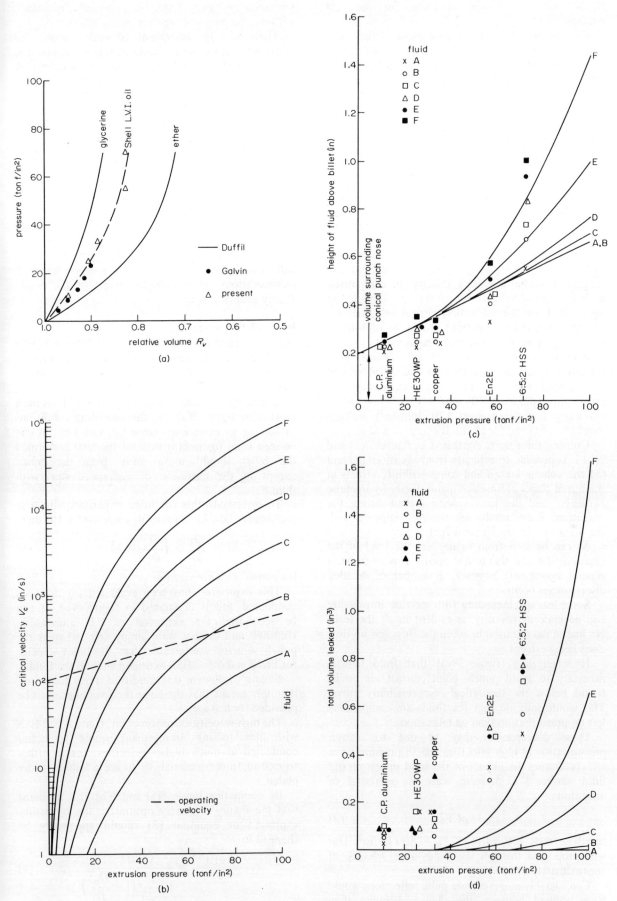

Figure 7(a). Compressibility of Shell L.V.I. oil. (b) Effects of extrusion pressure and fluid viscosity on the critical velocity for no leakage in high-velocity hydrostatic extrusion. (c) Fluid height necessary to avoid punch–billet contact in high-velocity hydrostatic extrusion. (d) Total volume leaked versus extrusion pressure in high-velocity hydrostatic extrusion.

fluids and the critical velocities for the test configurations.

For the tests reported here compressibility data up to pressures of 70 tonf/in^2 were required. The magnitude of the compressibility of Shell L.V.I. was assessed by considering the oil relaxation at the end of the extrusion operation; it was presumed that no leakage occurs during this phase of the operation.

As the compressibility of L.V.I. oil is but little affected by viscosity[19], the compressibility results shown in figure 7(a) are the means of the compressibilities of all the fluids used at a particular pressure.

The adiabatic compressibility curve for Shell L.V.I. oils shown in figure 7(a) is compared with Galvin's[19] isothermal results; it can be seen that good agreement is obtained.

The amount of fluid leaked in high-velocity extrusion is theoretically directly related to the magnitude of the operating velocity, in comparison with the critical velocity for no leakage. Depicted in figure 7(b) are the theoretical critical velocities for the tool geometries and fluid properties of interest here; see section on experimental equipment.

Also included in figure 7(b) are the operating velocities. It can be seen that no leakage should be encountered for any fluid when extruding C.P. aluminium (pressure 10 tonf/in^2), whereas when extruding high-speed steel (70 tonf/in^2) leakage should be obtained for all fluids except A and B.

Utilising the results contained in figures 7(a) and 7(b) it is possible to estimate from equations (4) and (5) the volume leaked and compressibility effects in the initial stages of unsealed high-velocity hydrostatic extrusion, and the total volume leaked during the operation. These results are shown compared with the theory in figures 7(c) and 7(d).

As can be seen from figure 7(c) the level of the experimental and theoretical results appear to be in general agreement; however, a number of detailed discrepancies occur.

Some leakage, increasing with increase in pressure and decrease in viscosity, is evident in all the tests. No line of demarcation between the flow and no-flow cases being evident.

It is puzzling (figure 7(c)) that fluid heights necessary to avoid punch–billet contact are to be found below the theoretical compressibility curve. This would indicate that the fluids are compressing less on pressurisation than on relaxation.

These discrepancies may be due to uneven pressurisation at high velocities due to pressure wave effects. During the impact of the solid punch on the fluid surface[20] a pressure wave is generated of magnitude

$$P_W \approx \rho C V_0 \tag{7}$$

for these tests 250 lbf/in$^2 < P_W < 800$ lbf/in^2. This will propogate through the fluids at a velocity of approximately 2000 ft/s.

Considerably more pressure pulse reflections would have occured between the plug and billet, than between the plug and the bottom of the container (figure 1). Therefore the initial volume subjected to pressurisation may have been sensibly contained between the plug and top face of the billet.

Therefore the agreement between theory and experiment apparent in figure 7(c) is a combination of (a) leakage volume being underestimated; (b) compressibilities being overestimated.

When total volumes are compared with theory (figure 7(d)), it can be clearly seen that the theory considerably underestimates the volumes of fluid leakage occurring for the high viscosity fluids. However, the total leaked volumes observed experimentally are again in general agreement with theory in that the leaked volume increases with increase in pressure and decrease in viscosity.

The major assumption contained in the analysis is that of isoviscous fluids. If pressure–viscosity effects are included, then the theoretical leakages are many orders of magnitude too low. Temperature effects will offset viscosity rises owing to pressure but such considerations are outside the scope of the simple theory proposed here.

Punch–billet separation

Having investigated fluid leakage, the other criterion for successful high-velocity hydrostatic extrusion is that the punch shall not unduly encroach upon the billet.

The inequality contained in equation (5) has been plotted in figure 8(a) for the operating conditions presuming an extruded volume (ψ_E) of 1 in^3. It can be seen from figure 8(a) that for the tests performed the billet should move away from the punch except in the extrusion of high-speed steel with fluid F.

To investigate this condition experimentally it is possible to show by rearranging equation (A.19) that

$$\frac{Q_T}{\psi_E} \geqslant \left(\frac{D^2}{D_b{}^2} - 1 \right) \tag{8}$$

for punch–billet separation.

This expression has been evaluated for the tests conducted and is illustrated in figure 8(b). It can be seen that the extrusion of C.P. aluminium, HE30WP and copper were amply safe, whereas the punch velocity was greater than the billet velocity for En2E and 6:5:2 HSS extruded with all the fluids.

Failure to observe this condition is not disastrous, it simply means that the large fluid volumes must be provided for leakage.

The high-velocity extrusion of En2E and 6:5:2 HSS with this tooling arrangement would be better conducted at much higher velocities, that is either to extrude more material, or to use a lighter, faster platen.

By comparing figure 8(a) and 8(b), it is apparent that the theory is unduly optimistic; the inequality expressed in equation (8) should apparently be changed to

$$\frac{(P\psi_E)^{1/2}}{E} \geqslant \frac{(\sqrt{2})V_c}{1 + \dfrac{D}{2h}\left(1 - \dfrac{D_b{}^2}{D^2}\right)} \tag{9}$$

where the difference is of the order of 100 times the velocity term.

Evaluation of tooling

Within the restricted nature of the trials it is possible to make the following comments on the tooling.

In general the tooling performed well; no die wear was observed.

The punch follower was cumbersome and displayed a tendency to 'pick up'. The plug need not

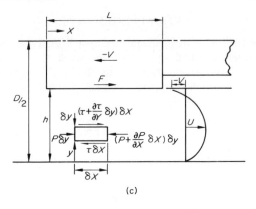

(c)

Figure 8(a). Theoretical operating criteria for punch–billet closure in high-velocity hydrostatic extrusion. (b) Experimental determination of punch–billet closure in high-velocity hydrostatic extrusion. (c) Conditions assumed to exist in the annular gap between an unsealed punch and container in hydrostatic extrusion.

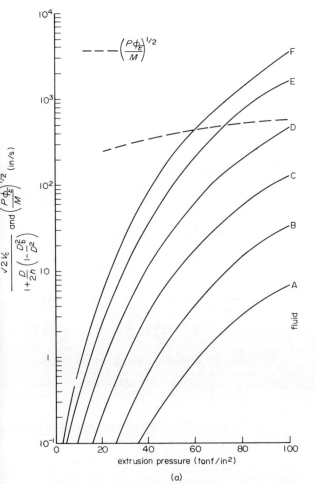

(a)

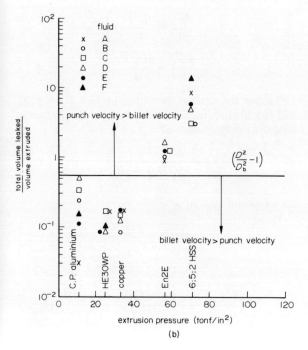

(b)

be used in future if small (0.001–0.002 in) clearances are maintained between punch and container.

If air is trapped between the punch and fluid, combustion takes place, resulting in an apparent stick-slip phenomenon; however, this condition is not serious.

Hydrostatic pressure acting on the annular area containing the 'O'-ring seal exerts considerable separation forces between the die and container. At high extrusion pressures these separation forces resulted in extensive extrusion of the bevel mitre ring.

CONCLUSIONS

From the tests reported here the following conclusions may be drawn.

(1) Small-scale hydrostatic extrusion can be accomplished without ram seals by utilising a close fitting punch follower and high punch velocities.

(2) Unsealed high-velocity hydrostatic extrusion can be used to produce satisfactory extrusions from aluminium, copper, En2E and 6:5:2 high-speed steel.

(3) Impact hydrostatic extrusion can be successfully conducted using a diverse range of fluid viscosities. High fluid viscosities minimise the leakages occurring during the process.

(4) The pressures required for high-velocity hydrostatic extrusion are comparable with those required for hydrostatic extrusion at conventional speeds.

(5) Increasing fluid viscosity reduces the extrusion pressures required for En2E and high-speed steel, but increases the pressure required to extrude aluminium and HE30WP. Variation of fluid viscosity has no marked effect on the pressures required for the extrusion of copper.

(6) Pressure fluctuations were observed in high-velocity hydrostatic extrusion, which resulted in variations in the product surface finish.

(7) The elementary theory developed considerably underestimated the volumes of fluid leaked during extrusion, the discrepancies being marked for the high-viscosity fluids.

ACKNOWLEDGEMENTS

The authors wish to thank Professor S. A. Tobias for his interest in, and support for, this work.

The authors are indebted to Mr. G. D. Galvin of Shell Research Ltd., for helpful discussions, and supplying the fluids.

Thanks are extended to Mr. C. Anderton for his help in conducting the trials and preparing the figures for the manuscript.

One of the authors (C.E.N.S.) is CHUBB Research Fellow.

APPENDIX

Elementary analysis of flow past an unsealed hydrostatic punch

Considering the flow through a parallel gap with one moving plate gives (figure 8(c))

$$P\delta y - \left(P + \frac{\partial P}{\partial x}\delta x\right)\delta y - \tau\delta x + \left(\tau + \frac{\partial \tau}{\partial y}\delta y\right)\delta x = 0 \quad \text{(A.1)}$$

Therefore

$$\frac{dP}{dx} = \frac{d\tau}{dy} \quad \text{(A.2)}$$

as $\tau = \eta(du/dy)$ for a Newtonian fluid; then integrating (A.2) twice with the bounds

$$\left.\begin{array}{l} u = 0 \text{ at } y = 0 \\ u = V \text{ at } y = h \end{array}\right\} \quad \text{(A.3)}$$

then

$$u = -\frac{Vy}{h} + \frac{1}{2\eta}\frac{dP}{dx}(y^2 - hy) \quad \text{(A.4)}$$

Therefore the flow rate per unit width is

$$\dot{q} = \int_0^h u\,dy = -\frac{Vh}{2} - \frac{h^3}{12\eta}\frac{dP}{dx} \quad \text{(A.5)}$$

Integrating (A.5) with

$$\left.\begin{array}{l} P = P \text{ at } x = 0 \\ P = 0 \text{ at } x = L \end{array}\right\} \quad \text{(A.6)}$$

provides

$$\dot{q} = \frac{Ph^3}{12\eta L} - \frac{Vh}{2} \quad \text{(A.7)}$$

Therefore the flow is

$$Q = \pi Dt \left|\frac{Ph^3}{12\eta L} - \frac{Vh}{2}\right| \quad \text{(A.8)}$$

For no flow $\dot{Q} = 0$ and $V = V_c$, therefore

$$V_c = \frac{Ph^2}{6\eta L} \quad \text{(A.9)}$$

To estimate the volume of fluid leaked during pressurisation, the punch stroke is

$$S = \frac{4\delta V}{\pi D^2} \quad \text{(A.10)}$$

therefore

$$t_p = \frac{S}{V_0} = \frac{4\delta V}{\pi D^2 V_0} \quad \text{(A.11)}$$

Substituting these values in (A.8) gives

$$Q_p = \frac{2\delta Vh}{D}\left(\frac{V_c}{V_0} - 1\right) \quad \text{(A.12)}$$

The total process time is

$$T \approx \frac{4MV_0}{\pi D^2 P} \quad \text{(A.13)}$$

Therefore the total volume of fluid leaked is

$$Q_T = \frac{MV_0^2 h}{DP}\left(\frac{2V_c}{V_0} - 1\right) \quad \text{(A.14)}$$

To estimate the shear rate at the punch, equation (A.4) becomes

$$\left.\frac{du}{dy}\right|_{y=h} = -\frac{V}{h} + \frac{h}{2\eta}\frac{dP}{dx} \quad \text{(A.15)}$$

substituting from (A.5) and (A.7) gives

$$\left.\frac{du}{dy}\right|_{y=L} = \frac{Ph}{2\eta L} - \frac{V}{h} \quad \text{(A.16)}$$

The fluid shear on the punch at any x is given by

$$\tau_x = \eta\frac{du}{dy} = \frac{Ph}{2L} - \frac{V\eta}{h} \quad \text{(A.17)}$$

Therefore the total viscous drag on the punch is

$$F = \int_0^L \pi D\tau_x\,dx = \pi D\left(\frac{Ph}{2} - \frac{V\eta L}{h}\right) \quad \text{(A.18)}$$

To assess the relative velocities of punch and billet, then

$$\psi_S = Q_T + \psi_E \quad \text{(A.19)}$$

but

$$\left.\begin{array}{l} \dfrac{d\psi_S}{dt} \approx \dfrac{\pi D^2\bar{V}}{4} \\[2mm] \dfrac{d\psi_E}{dt} \approx \dfrac{\pi D_b^2\bar{V}_b}{4} \end{array}\right\} \quad \text{(A.20)}$$

Taking the critical condition as $\bar{V} = \bar{V}_b = V_0/2$ and substituting from (A.8) and (A.9), the following inequality may be derived:

$$V_0 \geqslant \frac{2V_c}{1 + \dfrac{D}{2h}\left(1 - \dfrac{D_b^2}{D^2}\right)} \quad \text{(A.21)}$$

for successful operation.

However,

$$\frac{MV_0^2}{2} = P\psi_E \quad \text{(A.22)}$$

therefore

$$\frac{(P\psi_E)^{1/2}}{M} \geqslant \frac{(\sqrt{2})V_c}{1 + \dfrac{D}{2h}\left(1 - \dfrac{D_b^2}{D^2}\right)} \quad \text{(A.23)}$$

REFERENCES

1. P. W. BRIDGMAN, 1931. *The Physics of High Pressure*, G. Bell, London.

2. H. LL. D. PUGH, 1969. 'Hydrostatic extrusion', *N.E.L. Rept.*, No. 416.

3. H. LL. D. PUGH, 1969. 'Hydrostatic extrusion', *9th Commonwealth Mining and Met. Congr.*, London.

4. B. I. BERESNEV, L. F. VERE SHCHAGIN, Yu. N. RIABIN and L. D. LIVSHITS, 1963. *Some Problems of Large Plastic Deformation of Metals at High Pressures*, Pergamon Press, Oxford.

5. D. GREEN, 1964–5. 'An experimental high-speed machine for the practical exploitation of hydrostatic extrusion', *J. Inst. Metals*, **93**, 65.

6. A. H. LOW and C. J. H. DONALDSON, 1967. 'An investigation of speed instability in hydrostatic extrusion', *N.E.L. Rept.*, No. 289.

7. A. H. LOW, 1968. 'Some effects of varying the pressurising speed in hydrostatic extrusion', *N.E.L. Rept.*, No. 358.

8. M. J. HILLIER, 1966. 'A hydrodynamic model of hydrostatic extrusion', *Intern. J. Prod. Res.*, **5**, 171.

9. H. LL. D. PUGH, 1964. 'Redundant work and friction in the hydrostatic extrusion of pure aluminium and an aluminium alloy', *J. Mech. Engng. Sci.*, **6**, 362.

10. A. W. DUFFILL, 1967. 'Hydrostatic extrusion', *Ph.D. Thesis, University of Birmingham.*

11. D. BRADBURY, M. MARK and R. V. KLEIN-SCHMIDT, 1951. 'Viscosity and density of lubricating oils from 0–150 000 p.s.i. and 32 to 425 °F', *Trans. A.S.M.E.*, **73**, 667.

12. S. J. BECKER and L. MOLLICK, 1960. 'The theory of the ideal design of a compound vessel', *Trans. A.S.M.E.*, **82**, 136.

13. A. S. NIKOLICH. 'Stresses in cylinders subjected to internal pressures on portions of their length', *Russian Engng. J.*, **47**, No. 1.

14. L. T. CHAN, F. BAKHTAR and S. A. TOBIAS, 1965–6. 'Design and development of Petro-Forge high energy rate forming machines', *Proc. Inst. Mech. Engrs.*, **180**, pt. 1.

15. L. T. CHAN and S. A. TOBIAS, 1968. 'Performance characteristics of Petro-Forge Mk I and Mk II machines', *Proc. 9th M.T.D.R. Conf.*, Birmingham.

16. J. L. M. MORRISON, B. CROSSLAND and J. S. C. PARRY, 1960. 'Strength of thick cylinders subjected to repeated internal pressure', *Proc. Inst. Mech. Engrs.*, **174**, No. 2, 95.

17. G. D. GALVIN, private communication, Shell Research Ltd., Thornton Research Centre.

18. A. J. ORGAN, 1967. 'The calibration of high energy rate forming machines', *Intern. J. Machine Design Res.*, **7**, 325.

19. G. D. GALVIN, H. NAYLOR and A. R. WILSON, 1963–4. 'The effect of pressure and temperature on some properties of fluids of importance in elastohydrodynamic lubrication', *Proc. Inst. Mech. Engrs.*, **178**, Pt.3N, 283.

20. F. J. HEYMANN, 1968. 'On the shock wave velocity and impact pressure in high-speed liquid–solid impact', *Trans. A.S.M.E., Ser. B*, **90**, 400.

EXPLOSIVE AND
ELECTROHYDRAULIC FORMING

THE DESIGN AND ANALYSIS OF EXPLOSIVE FORMING THIN SHELL DIES

by

S. B. KULKARNI* and A. A. EZRA*

SUMMARY

Thin shell dies backed up by water have been found to be a very economic solution for the explosive forming of metal domes. A method of analysis is presented here that will permit the rational design of such dies. Experimental observations of strain in the die are shown to agree very well with the theoretical predictions.

Based on these results, it is possible to analyse and design thin shell dies with complete confidence for the explosive forming of domes.

NOTATION

C_R	=	wave velocity in back-up media (in/s)
C_w	=	sonic velocity of water (in/s)
e	=	specific energy of explosive (in lb/lb)
E	=	Young's modulus (lb/in^2)
F_1	=	pressure force on the internal surface of shell (lb s/in^3)
F_2	=	reactive pressure force on the external surface of shell (lb s/in^3)
h	=	shell thickness (in)
m	=	mass per unit area of shell wall = ρh (lb s^2/in^3)
R	=	shell radius (in)
t	=	time (s)
V	=	velocity of shell dw/dt (in/s)
W	=	weight of explosive (lb)
w	=	radial displacement (in)
ϵ	=	hoop strain (in/in)
ν	=	Poisson's ratio
ρ	=	mass density of shell material (lb s^2/in^4)
ρ_R	=	mass density of die back-up media (lb s^2/in^4)
ρ_w	=	mass density of water (lb s^2/in^4)
$\sigma_\theta, \sigma_\phi, \sigma$	=	normal stress (lb/in^2

1. INTRODUCTION

The present study is part of a continuing investigation of the explosive forming process. In this paper we shall investigate the design of one of the largest single items of cost in the process, namely the die.

Large dies result in significant expenditures and, therefore, various die materials have been used in minimising this expense. Because of the large pressures involved in explosive forming, a reliable way must be found to reduce the resulting stresses in the die and the die material required. The use of a thin shell die backed by water has been found to be very effective. The purpose of this paper is to provide the analytical capability for designing such dies. The concept of supporting the die with water was first introduced by Beyer[1]. The double die configuration wherein two lightweight shell dies are placed face-to-face would also alleviate this problem because of opposite reacting forces during forming.

The critical shots in explosive forming are the initial and final sizing shots. A large part of the energy of the initial shot is absorbed by the plastic deformation of the blank into the die. Therefore, it is in the final sizing shot that the die has to withstand almost directly the incident explosive pressures. While it is better to avoid the necessity of a final sizing shot, the die should be designed with this eventuality in mind.

Die stresses have been investigated theoretically by Reismann and Jurney[2]. They have shown that the stresses may be substantially reduced by water immersion. In preparation for an experimental investigation of this case, theoretical calculations similar to Reismann's have been carried out and stresses in thin shell dies subjected to internal blast loads have been measured. The comparison of theoretical and experimental results is summarised herein.

2. THEORY

The physical limitation on the die under repeated final sizing shots is that it should remain elastic. In this section we shall develop the theory to predict the response of a thin spherical shell subjected to pressures from underwater explosions. This pressure will be assumed to act uniformly over the internal surface of the shell.

2.1. Underwater explosion assumptions

Some well-known underwater explosion formulae will be stated. The pressure history of the major portion of the shock can be approximated by

$$P(t) = P_m e^{-t/\theta} \tag{1}$$

where P_m is the peak pressure from an underwater explosion and θ is known as the exponential decay constant. The peak pressure and the decay constant can be represented by the empirical equations

$$\left. \begin{array}{c} P_m = B\left(\dfrac{W^{1/3}}{R'}\right)^\alpha \\[2ex] \theta = CW^{1/3}(W^{1/3}/R')^\beta \end{array} \right\} \tag{2}$$

where W = weight of the explosive (lb)
R' = distance from the explosive (ft)

* Department of Mechanical Sciences and Environmental Engineering, University of Denver, Denver, Colorado, U.S.A.

The values of the empirical constants B, C, α and β have been experimentally determined for each explosive and are given in the table below.

Explosive	Peak pressure (lb/in²)		Decay constant (ms)	
	B	α	C	β
PETN	23 100	1.13	0.06	−0.18
Pentolite	22 500	1.13	0.06	−0.18
Comp C4	22 500	1.13	0.06	−0.18
Comp A3	23 500	1.13	0.06	−0.18
HBX-1	28 400	1.15	0.049	−0.29
HBX-3	20 900	1.14	0.06	−0.218

2.2. Shell response

For a thin shell die the die radius is normally very large in comparison with the die thickness. Therefore, membrane theory can be used to analyse the die stresses. Further, because of spherical symmetry, the die is in a biaxial state of stress with $\sigma_\theta = \sigma_\phi = \sigma$ (figure 1). The hoop strain ϵ is given by

$$\epsilon = \frac{w}{R} = \frac{\sigma}{E}(1 - \nu) \tag{3}$$

where E is Young's modulus and ν is Poisson's ratio.

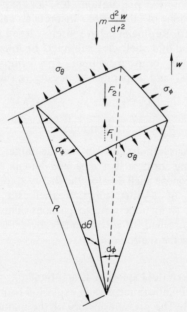

Figure 1. Spherical shell element.

The radial equation of motion can be written as

$$m\frac{\mathrm{d}^2 w}{\mathrm{d}t^2} = -\frac{2\sigma h}{R} + F_1 - F_2 \tag{4}$$

where F_1 is the pressure force on the internal surface of the shell and F_2 is the reactive pressure force exerted by the water. If a dome is being formed in the final sizing shot, then the mass m per unit area should include the blank.

In equation (1) the pressure P is doubled owing to the reflection of the pressure wave from the die wall. Therefore, we can write

$$F_1 = 2P_\mathrm{m}e^{-t/\theta} \tag{5}$$

The rarefaction wave due to the outward motion of the die wall causes a suction pressure acting radially inward given by

$$F_2 = \rho_\mathrm{w}C_\mathrm{w}V \tag{6}$$

where ρ_w is the density of water, C_w is the sonic velocity of water and V is the velocity of the die wall ($\mathrm{d}w/\mathrm{d}t$). The pressure force resisting the die wall movement has the same form as equation (6). This resisting pressure force and the effect of the rarefaction wave (6) can be combined to give the total reactive pressure force F_2 as

$$F_2 = (\rho_\mathrm{R}C_\mathrm{R} + \rho_\mathrm{w}C_\mathrm{w})V \tag{7}$$

If the die were backed by water, then $\rho_\mathrm{R} = \rho_\mathrm{w}$ and $C_\mathrm{R} = C_\mathrm{w}$, respectively, so that

$$F_2 = 2\rho_\mathrm{w}C_\mathrm{w}\frac{\mathrm{d}w}{\mathrm{d}t} \tag{8}$$

Substituting equations (8), (5) and (3) in (4), we have

$$m\frac{\mathrm{d}^2 w}{\mathrm{d}t^2} + 2(\rho_\mathrm{w}C_\mathrm{w})\frac{\mathrm{d}w}{\mathrm{d}t} + \frac{2hE}{R^2(1 - \nu)}w = 2P_\mathrm{m}e^{-t/\theta} \tag{9}$$

Equation (9) is easily recognised to be the differential equation of a mass-spring model with damping and a disturbing force present. Written in the standard form, it is

$$\frac{\mathrm{d}^2 w}{\mathrm{d}t^2} + 2\xi\omega\frac{\mathrm{d}w}{\mathrm{d}t} + \omega^2 w = \frac{2P_\mathrm{m}e^{-t/\theta}}{m} \tag{10}$$

where

$$\omega^2 = \frac{2hE}{mR^2(1 - \nu)}$$

and

$$\xi = \frac{\rho_\mathrm{w}C_\mathrm{w}}{m\omega} = \rho_\mathrm{w}C_\mathrm{w}R\left\{\frac{(1 - \nu)}{2mEh}\right\}^{1/2}$$

For $\xi > 1$, the solution of (10) is

$$\frac{w}{\psi} = e^{-t/\theta} - e^{-\xi\omega t}\left[\cosh\left\{(\xi^2 - 1)^{1/2}\omega t\right\}\right.$$
$$\left. + \frac{1}{\omega\theta}\frac{(\xi\omega\theta - 1)}{(\xi^2 - 1)^{1/2}}\sinh\left\{(\xi^2 - 1)^{1/2}\omega t\right\}\right] \tag{11}$$

If $\xi < 1$, the solution is

$$\frac{w}{\psi} = e^{-t/\theta} - e^{-\xi\omega t}\left[\cos\left\{(1 - \xi^2)^{1/2}\omega t\right\}\right.$$
$$\left. - \frac{1}{\omega\theta}\frac{(1 - \xi\omega\theta)}{(1 - \xi^2)^{1/2}}\sin\left\{(1 - \xi^2)^{1/2}\omega t\right\}\right] \tag{12}$$

If $\xi = 1$

$$\frac{w}{\psi} = e^{-t/\theta} - e^{-\omega t}\left(1 - \frac{t}{\theta} + \omega t\right) \tag{13}$$

In equations (11) to (13)

$$\psi = \frac{2P_\mathrm{m}\theta^2}{m(1 - 2\xi\omega\theta + \omega^2\theta^2)}$$

The time τ to reach the maximum deflection is given by the lowest non-zero roots of the following three equations:

For $\xi > 1$

$$-e^{-\tau/\theta} + e^{-\xi\omega\bar{\tau}}\left[\cosh\left\{(\xi^2 - 1)^{1/2}\omega\tau\right\}\right.$$
$$\left. + \frac{(\omega\theta - \xi)}{(\xi^2 - 1)^{1/2}} \sinh\left\{(\xi^2 - 1)^{1/2}\omega\tau\right\}\right] = 0 \quad (14)$$

For $\xi < 1$

$$-e^{-\tau/\theta} + e^{-\xi\omega\tau}\left[\cos\left\{(1 - \xi^2)^{1/2}\omega\tau\right\}\right.$$
$$\left. + \frac{(\omega\theta - \xi)}{(1 - \xi^2)^{1/2}} \sin\left\{(1 - \xi^2)^{1/2}\omega\tau\right\}\right] = 0 \quad (15)$$

For $\xi = 1$

$$-e^{-\tau/\theta} + e^{-\omega\tau}\left\{1 - \omega\theta(1 - \omega\theta)\frac{\tau}{\theta}\right\} = 0 \quad (16)$$

Computations have been made for various values of ξ for τ/θ versus $\xi\omega\theta$. These are shown in figure 2. The curves represent the solutions of the transcendental equations (14) to (16).

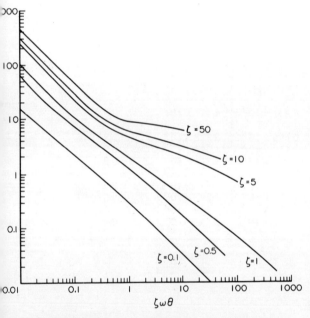

Figure 2. Lowest non-zero roots of transcendental equation for calculating maximum radial wall deflection.

3. COMPARISON WITH EXPERIMENTS

A series of tests were conducted to compare the theoretical response with strains actually measured on shell dies subjected to explosive loading on the inside. The die shells were segments of spheres fabricated from sheet metal by explosive forming. Two Bean-type BAE-06-187BB-350TE strain gauges were mounted in the conventional manner using an epoxy cement. The gauges were waterproofed with No. 5 Gagekote rubber and excited by constant-current d.c. sources. A Tektronix Model 555 oscilloscope was used for readout. Figure 3(a) shows the location of the gauges and figure 3(b) the

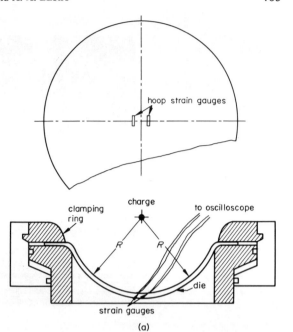

(a)

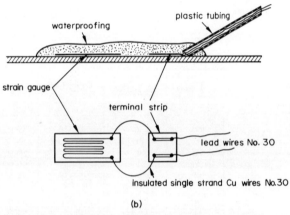

(b)

Figure 3(a). Location of strain gauges in test set-up. (b) Strain gauge installation.

installation. A schematic of the electronics is shown in figure 4.

Explosive charges of composition C-4, pentolite, and PETN varying in weight from 0.3 g to 6.0 g were

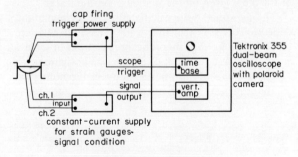

Figure 4. Schematic of electronics.

detonated by Dupont No. 6 blasting caps. The charges were placed at the centre of the die shell (figure 3(a)). The die was backed by water for all experiments.

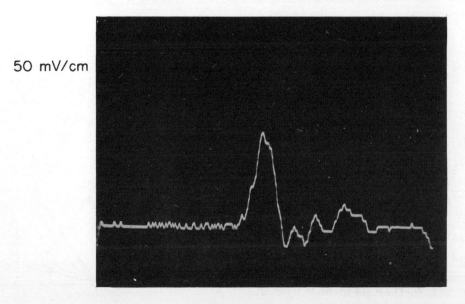

50 mV/cm

sweep rate 0.2 ms/cm m

Figure 5. Oscilloscope trace of output from strain gauge on aluminium die.

TABLE 1

Die material	Explosive	Weight of charge (g)	R/h	Maximum wall strain		Ratio Observed/Predicted
				Predicted	Observed	
Al	Cap (PETN)	0.316	59.0	4.43×10^{-4}	4.21×10^{-4}	0.95
Al	C4	1.0	59.0	0.896×10^{-3}	0.878×10^{-3}	0.98
Al	C4	2.0	59.0	1.41×10^{-3}	1.344×10^{-3}	0.954
MS	Cap (PETN)	0.316	54.0	4.54×10^{-4}	4.21×10^{-4}	0.93
MS	PETN	1.08	54.0	0.986×10^{-3}	0.98×10^{-3}	0.99
MS	PETN	2.16	54.0	1.485×10^{-3}	1.29×10^{-3}	0.87
MS	PETN	4.33	54.0	2.187×10^{-3}	1.779×10^{-3}	0.81
MS	C4	6.0	50.0	2.523×10^{-3}	2.24×10^{-3}	0.89

Acceptable results were obtained from the gauges during the tests. Figure 5 shows the actual recorded data from one of the tests. The maximum deflection of the die is given by the first peak on the trace.

The die strains were calculated from the maximum deflection w of the die wall. These strains are tabulated along with theoretical predictions in table 1 and plotted as a function of charge size in figure 6.

The die material properties used in the numerical computations are shown in table 2.

TABLE 2

Material	E (lb/in^2)	ρ (lb s^2/in^4)	Poisson's ratio
aluminium	10×10^6	2.59×10^{-4}	0.25
mild steel	30×10^6	7.25×10^{-4}	0.25

The water surrounding the die has properties $\rho_w = 9.362 \times 10^{-5}$ lb s^2/in^4 and $C_w = 6.0 \times 10^4$ in/s.

It is clear that, since deep draw domes were used to compare results with the spherical shell analysis,

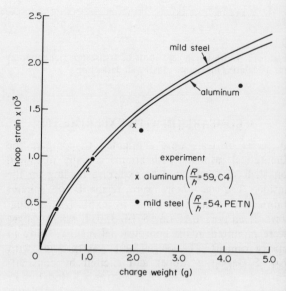

Figure 6. Maximum hoop strain versus charge weight comparison of theory and experiment.

there will be some error in the experimental maximum strains due to the existence of a clamping effect at the edge of the die. From the results of table 1, the good agreement between analytically and experimentally determined peak strain rates shows that the magnitude of this error is small. Further, the empirical formula for pressure given by equation (1) does not include the energy delivered in the reloading phase of the explosion. Because of the shallow depth of submergence used in the experiments to avoid the occurrence of this effect, it is not surprising, therefore, that the discrepancy between the experimental and analytical results is not significant.

Figure 7(a) and 7(b) show dimensionless plots, for steel and aluminium respectively, of the peak strain

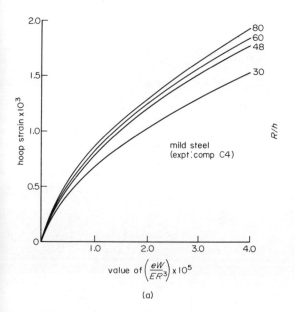

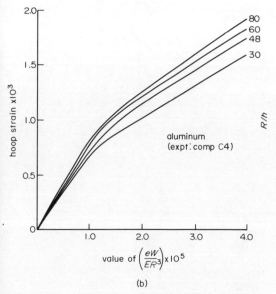

Figure 7. Parametric curves for die strains versus explosive energy.

as a function of the explosive energy for different die radius/thickness ratios. A value of 19.7×10^6 in lb/lb for the specific energy of C4 has been used in these plots. Alternative explosives and die materials could

be chosen resulting in obvious modifications to these parametric curves.

Of useful interest is the reduction in die stresses due to water immersion. If we define percentage alleviation as

percentage alleviation

$$= \frac{\text{max. displacement in air} - \text{max. displacement in water}}{\text{max. displacement in air}} \times 100$$

then figure 8 shows a plot of the percentage alleviation versus radius to thickness ratios, R/h, for mild steel and aluminium. An inspection of this figure indicates that for a R/h ratio of 80, alleviation of up to 46% for mild steel and 44% for aluminium is obtained. Therefore, a significant reduction in die stress is provided by the surrounding water.

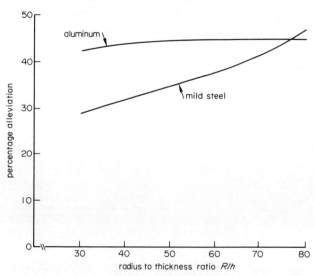

Figure 8. Percentage alleviation versus radius to thickness ratios.

Figure 8 also shows that alleviation increases with R/h. The explanation for this is simply that for the same amount of explosive charge, the velocity of the die wall increases as the ratio R/h is increased. Consequently, the damping force resisting the wall motion also increases. The alleviation increase, however, is greater for mild steel than for aluminium.

4. CONCLUSIONS

In this paper, the authors have presented a simple analysis to predict the peak strain (and hence peak stress) in a thin die subjected to known explosive pressures. These predictions, confirmed by experimental measurements, are necessary to prevent die failures in the explosive forming process.

The analysis has also been used to determine the effect of water immersion on the die stresses and shows that they are substantially reduced. It is interesting to note that the dynamic response of the die can be used to mobilise the resistance of the surrounding water and to reduce the stresses in the die walls. This important result can be effectively used to design, in a rational manner, thin shell dies for use in explosive forming, thereby substantially reducing the die cost.

ACKNOWLEDGEMENTS

This work was performed at the University of Denver Center for High Energy Forming, under the sponsorship of the Advanced Research Projects Agency of the U.S. Department of Defense, monitored by the U.S. Army Materials and Mechanics Research Center. The authors wish to acknowledge the able assistance of Mr. L. Brown and Mr. R. P. Marchese in performing the experiments.

REFERENCES

1. W. K. BEYER, 1964. 'The pool is the tool', Paper No. SP 64-124, *Creative Manufacturing Seminar,* American Society for Tool and Manufacturing Engineers. May.
2. H. REISMANN and W. H. JURNEY, 1966. 'Effect of water immersion on reduction of die stresses in explosive forming', *J. Engng. Ind.,* February, 111–16.

AN INVESTIGATION OF THE EDGE PULLIN IN EXPLOSIVELY-FORMED DOMES

by

MICHAEL A. KAPLAN* and SURESH B. KULKARNI*

SUMMARY

A simplified analysis for the prediction of edge pullin of explosively-formed domes is developed. The primary assumptions are that inertia effects are negligible and that the shape of the middle surface of the blank is known at all times in the forming process.

Markov's principle of minimum plastic work, in conjunction with the Rayleigh-Ritz method, is employed. The results of the analysis are used to compare thickness strain both in static dome formation and bulge testing and to predict the initiation of necking in the dome. There is good agreement in all cases.

Predicted values of edge pullin versus drawdepth also agree well with explosive dome forming data. The analysis, however, does not accurately predict the strain field in the dome, particularly in the region of the apex. By an extension of Markov's principle, it is shown that the error in the prediction of strain is not a result of inertia effects, but rather is caused by differences in loading in the static and dynamic forming processes. The reduced thinout which occurs dynamically is a result therefore of the interaction between charge location, energy transfer medium and the blank, and is not a property inherent in the explosive-forming process itself.

INTRODUCTION

In the explosive forming of domes a thin circular metal blank, supported by a suitable holddown mechanism near its outer edge, is subjected to a blast load when the explosive charge is detonated. The resulting pressure forces the central portion of the blank (the dome) into the die cavity. As the depth of the dome increases, the annular flange feeds material into the dome, thus preventing excessive thinout. The edge pullin, that is, the inward radial displacement at the edge of the blank, is appreciable for deep domes even with holddown pressures sufficient to prevent flange wrinkling. With shallow domes, edge pullin is insignificant.

The edge pullin is an important parameter in determining the mechanical state of an explosively-formed deep draw dome. Excessive pullin can produce flange wrinkling: insufficient pullin can result in tearing of the blank. Also, since the flange is usually discarded after the dome is formed, prediction of pullin is necessary for economic reasons.

In the present study, the edge pullin as a function of draw depth will be predicted for a draw-formed hemispherical dome.

The deformation of clamped metal blanks subjected to blast or impulsive type loading has been considered by Hudson[1], Wang[2] and Witmer et al.[3,4,5]. Hudson used a moving plastic hinge concept to examine an unloaded blank with a specified initial velocity field. Wang considered only bending stresses in his rigid plastic analysis, and Witmer et al. made a rigorous numerical analysis of a circular plate subjected to blast loading which gave good agreement with experimental results. Hill[6] used a quasistatic analysis to obtain an explicit solution for a clamped metal diaphragm which is bulged plastically by lateral pressure. These analyses do not include the flow of flange material into the die cavity and thus cannot be used to predict pullin.

A simpler formulation of the dynamic problem led to the analysis by Thurston[7] who extended the work of Boyd[8] to include specified amounts of edge pullin. This analysis however is restricted to shallow dish shapes.

DISCUSSION OF THE PROBLEM

A complete mathematical description of the process has not yet been achieved. A primary objective of this study is to provide a relatively simple analysis, based on realistic assumptions, that will yield reliable predictions of edge pullin. Accordingly, two simplifying assumptions are made.

(1) The shape of the dome is always a segment of a sphere.

(2) The dome is formed at a sufficiently slow rate for the flow process to be quasistatic, that is, the accelerations are negligible at all times.

Assumption (1) is based on an investigation conducted by Ezra[9] to determine optimal standoff distance with regard to uniformity of strain field. uniformity of draw, minimisation of thinout and minimisation of charge weight. He found that a standoff distance/die diameter ratio of 0.167 is most favourable. The domes that result from the use of this ratio are nearly spherical.

Assumption (2) was based on the result of experiments conducted for the purpose of comparing the edge pullin of statically and dynamically-formed domes. The blank material was 2014-0 aluminium. A die was used to assure the same final shape in both cases. Dynamic forming was achieved by forming the domes explosively in water. Static forming was accomplished by rubber pressing. A shaped rubber block, confined in a piston-cylinder arrangement, was pressed against a greased blank. The cylinder walls were attached to the blank holddown ring and pressure was applied to a steel piston on top of the rubber block.

Figure 1 shows the profiles of the statically and dynamically-formed domes and figures 2(a) and 2(b)

*Department of Mechanical Sciences and Environmental Engineering, University of Denver, Denver, Colorado 80210, USA.

show their circumferential and thickness strain distributions, respectively. The circumferential strains were determined from the change in length of the photographically-etched circumferential gridlines. The thickness strains were calculated from the change in thickness.

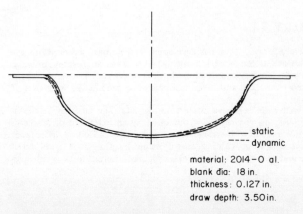

material: 2014–0 al.
blank dia: 18 in.
thickness: 0.127 in.
draw depth: 3.50 in.

Figure 1. Profiles of statically and dynamically die formed 2014-0 aluminium domes.

These results show that even though the strain distribution is somewhat different in the two domes, the edge pullins are nearly identical. This implies that an analysis which proceeds from assumptions (1) and (2) should be suitable for predicting pullin. It is clear from the experimental observations, however, that there will be some error in the predicted strain field, particularly in the neighbourhood of the apex of the dome.

Based on experimental observations, we shall use the usual membrane assumptions that bending effects are negligible and the shear stresses (and hence the shear strain rates) vanish across the thickness. In view

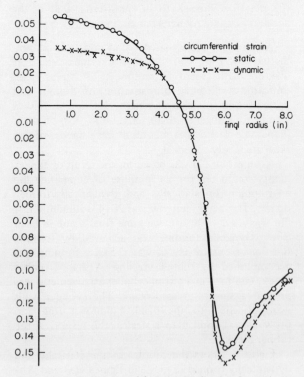

Figure 2(a). Circumferential strain distribution in statically and dynamically die formed 2014-0 aluminium domes.

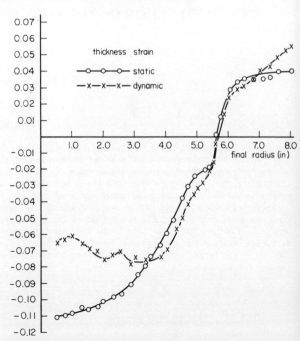

Figure 2(b). Thickness strain distribution in statically and dynamically die formed 2014-0 aluminium dies.

of (1) it is appropriate to use spherical coordinates (R,ϕ,θ) for the dome. Cylindrical coordinates (r,θ,z) will be used for the flange. The longitudinal Z-axis is assumed to coincide with the axis of symmetry of the dome and the flange (figure 3).

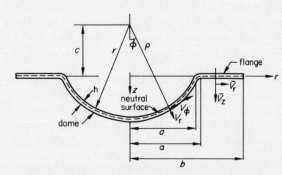

Figure 3. Nomenclature for dome and flange.

MATHEMATICAL MODEL

The analytical approach will be to use the Rayleigh-Ritz method in conjunction with Markov's principle which states that, among admissible solutions which satisfy the conditions of compatibility and incompressibility, as well as the geometrical boundary conditions, the actual solution renders the rate of work an absolute minimum.

The variational principle stated above has been developed for quasistatic problems in the flow theory of plasticity. Its use is restricted to rigid-plastic Prandtl-Reuss materials described by the equations

$$S_{ij} = \frac{\sqrt{2}}{\sqrt{3}} \frac{\sigma_0}{\sqrt{\dot{\epsilon}_{mn}\dot{\epsilon}_{mn}}} \dot{\epsilon}_{ij} \tag{1}$$

where S_{ij} and $\dot{\epsilon}_{ij}$ are the deviatoric stress and total strain rates respectively. σ_0 is the yield stress in simple tension.

The work done per unit time on an elemental volume dV is given by

$$W = \frac{\sqrt{2}}{\sqrt{3}} \sigma_0 \int_V \sqrt{\dot{\epsilon}_{ij}\dot{\epsilon}_{ij}} \, dV \qquad (2)$$

Equation (2) will be used to evaluate the rate of work for the dome and the flange.

Dome analysis

The physical components of the normal strain rates in spherical coordinates, assuming axial symmetry, are

$$\dot{\epsilon}_{RR} = v_{R,R}$$

$$\dot{\epsilon}_{\phi\phi} = \frac{v_R}{R} + \frac{1}{R} v_{\phi,\phi} \qquad (3)$$

$$\dot{\epsilon}_{\theta\theta} = \frac{v_R}{R} + \frac{v_\phi}{R} \cot\phi$$

and $\dot{\epsilon}_{R\phi} = 0$ because of the membrane assumption. In equations (3) the v_i are the components of velocity. The comma is used to denote partial differentiation with respect to the spatial coordinates.

Assumption (1) constrains the radial velocity field v_R at the neutral surface such that

$$v_R = \left(1 - \frac{\rho}{\sqrt{\rho^2 - a^2}} \cos\phi\right)\dot\rho \qquad (4)$$

where ρ is the radius to the neutral surface and a is the radius of the die. Because of (2), the time can be replaced by any other parameter which varies monotonically with it and is characteristic of the deformation. In the following, therefore, we shall replace real time by the pseudo time variable ρ. Noting that time differentiation with this procedure reduces to differentiation with respect to ρ gives

$$\dot\rho = 1 \quad \text{and} \quad ()_{,t} = ()_{,\rho} \qquad (5)$$

Thus, the expression for the radial velocity at the neutral surface reduces to

$$v_R = \left(1 - \frac{\rho}{\sqrt{\rho^2 - a^2}} \cos\phi\right) \qquad (6)$$

The first of relations (3), physically, is the thickness strain rate. Using Alamansi's definition of finite strain[10], the average thickness strain is given by

$$\epsilon_{RR} = \frac{1}{2}\left(1 - \frac{h_0{}^2}{h^2}\right) \qquad (7)$$

where h_0 and h are the original and current thickness respectively. For the velocity field associated with this problem the material derivatives D/Dt of the principal strains are equal to the strain rates (for the derivation refer to 11). Therefore the average thickness strain rate is given by

$$\dot{\epsilon}_{RR} = \frac{D\epsilon_{RR}}{Dt} = \frac{h_0{}^2}{h^3}\left(h_{,\rho} + \frac{v_\phi}{\rho}h_{,\phi}\right) \qquad (8)$$

The membrane assumption implies that the tangential strain rate $\dot{\epsilon}_{\phi\phi}$ and the hoop strain rate $\dot{\epsilon}_{\theta\theta}$ are constant across the thickness. Hence we can write the last two relations of equation (3), at the neutral surface, as

$$\dot{\epsilon}_{\phi\phi} = \frac{v_R}{\rho} + \frac{1}{\rho} v_{\phi,\phi}$$

$$\qquad (9)$$

$$\dot{\epsilon}_{\theta\theta} = \frac{v_R}{\rho} + \frac{v_\phi}{\rho} \cot\phi$$

Flange analysis

A bar will be used to distinguish the flange displacement, velocity, strain rate and stress components from similar quantities in the dome. Neglecting friction, the equilibrium equations in the radial and axial directions are

$$\bar\sigma_{rr,r} + \frac{1}{r}(\bar\sigma_{rr} - \bar\sigma_{\theta\theta}) = 0 \qquad (10)$$

$$\bar\sigma_{zz,z} = 0$$

and the equation in the tangential direction is identically zero. The boundary conditions are (figure 3)

$$\bar\sigma_{rr} = 0 \quad \text{at} \quad r = b,$$

$$\qquad (11)$$

$$\bar\sigma_{zz} = -P \quad \text{at} \quad z = \pm\frac{t}{2}$$

where P is the clamping pressure.

The von Mises yield condition, which represents an ellipse in the $\bar\sigma_{rr}$, $\bar\sigma_{\theta\theta}$ plane, can be approximated by the straight line

$$\bar\sigma_{rr} - \bar\sigma_{\theta\theta} = \sigma_0 \qquad (12)$$

for the stress condition in the flange, that is, $\bar\sigma_{rr} > 0$; $\bar\sigma_{\theta\theta} < 0$. The problem is statically determinate. Using (12) in (10) together with (11), the stress field in the flange is

$$\bar\sigma_{rr} = \sigma_0 \ln\frac{b}{r}$$

$$\bar\sigma_{\theta\theta} = \sigma_0(\ln\frac{b}{r} - 1) \qquad (13)$$

$$\bar\sigma_{zz} = -P$$

The stress–strain rate relations of equation (1) will be used to determine the velocities and strain rates. The ratio of the radial and hoop strain rates is

$$\frac{\dot{\bar\epsilon}_{rr}}{\dot{\bar\epsilon}_{\theta\theta}} = \frac{2\bar\sigma_{rr} - \bar\sigma_{\theta\theta} - \bar\sigma_{zz}}{2\bar\sigma_{\theta\theta} - \bar\sigma_{rr} - \bar\sigma_{zz}} \qquad (14)$$

Substituting the strain rate–velocity relations for the flange given by

$$\dot{\bar\epsilon}_{rr} = \bar v_{r,r}$$

$$\dot{\bar\epsilon}_{\theta\theta} = \frac{\bar v_r}{r} \qquad (15)$$

$$\dot{\bar\epsilon}_{zz} = \bar v_{z,z}$$

in (14) together with (13), gives

$$v_r = \frac{\dfrac{db}{d\rho} r \, (-2\sigma_0 + P)^3}{b \, [\sigma_0(\ln\dfrac{b}{r} - 2) + P]^3} \qquad (16)$$

after satisfying the boundary condition that $v_r = \dfrac{db}{d\rho}$ at $r = b$. It follows that the radial and hoop strain fields are

$$\dot{\bar{\epsilon}}_{rr} = \bar{v}_{r,r} = \frac{\dfrac{db}{d\rho}(-2\sigma_0 + P)^3 \, [\sigma_0(\ln\dfrac{b}{r} + 1) + P]}{b \, [\sigma_0(\ln\dfrac{b}{r} - 2) + P]^4}$$

and $\qquad\qquad (17)$

$$\dot{\bar{\epsilon}}_{\theta\theta} = \frac{\bar{v}_r}{r} = \frac{\dfrac{db}{d\rho}(-2\sigma_0 + P)^3}{b \, [\sigma_0(\ln\dfrac{b}{r} - 2) + P]^3}$$

respectively. The axial strain rate $\dot{\epsilon}_{zz}$ is obtained from the incompressibility condition $\dot{\bar{\epsilon}}_{rr} + \dot{\bar{\epsilon}}_{\theta\theta} + \dot{\bar{\epsilon}}_{zz} = 0$.

$$\dot{\bar{\epsilon}}_{zz} = \frac{-\dfrac{db}{d\rho}(-2\sigma_0 + P)^3 \, [\sigma_0(2\ln\dfrac{b}{r} - 1) + 2P]}{b \, [\sigma_0(\ln\dfrac{b}{r} - 2) + P]^4} \qquad (18)$$

Noting that $\dot{\bar{\epsilon}}_{zz} = \bar{v}_{z,z}$, (18) can be integrated to give the axial velocity as

$$\bar{v}_z = \frac{-\dfrac{db}{d\rho}(-2\sigma_0 + P)^3 \, [\sigma_0(2\ln\dfrac{b}{r} - 1) + 2P]}{b \, [\sigma_0(\ln\dfrac{b}{r} - 2) + P]^4} z \qquad (19)$$

after satisfying the boundary condition that $\bar{v}_z = 0$ at $z = 0$. The strain rate and velocity field for the flange have been found in terms of the outer radius b and its time rate of change.

Continuity conditions at the edge
The incompressibility of the blank material requires that the rate of mass flow at the inner edge of the flange be equal to that at the edge of the dome. Therefore

$$\alpha(\bar{v}_r \bar{h})_{r=\alpha} = a(v_\phi h)_{\phi=\phi_e} \qquad (20)$$

where the flange quantities are to be evaluated at $r = \alpha$ and the dome quantities at $\phi_e = \sin^{-1} a/\rho$ (figure 3). Experimental observation indicates that the change in blank thickness over the edge is quite small (about 0.2%). Equation (20) can then be simplified and expressed as two separate relations

$$\alpha(\bar{v}_r)_{r=\alpha} = a(v_\phi)_{\phi=\phi_e} \qquad (21)$$

and

$$(\bar{h})_{r=\alpha} = (h)_{\phi=\phi_e} \qquad (22)$$

Equations (21) and (22) provide two conditions for relating dome and flange quantities. In the preceding section, the flange velocities were found in terms of $b(\rho)$. Equation (21) is therefore a relationship between $b(\rho)$ and v_ϕ. Since both the radial and hoop velocities in the dome are known, the problem has been reduced to the determination of v_ϕ.

FORMULATION OF RATE OF WORK

The Rayleigh-Ritz method requires a representation of the velocity field in terms of arbitrary constants so chosen that the rate of work is minimised. A representation which satisfies the condition that the tangential velocity vanishes in the entire blank when it is flat ($\rho = \infty$) and also vanishes at the blank centre for all values of ρ can be written as

$$v_\phi = b_0 \left(\frac{a}{\rho}\right) \sin\phi + b_1 \left(\frac{a}{\rho}\right)^2 \sin2\phi + \dots \qquad (23)$$

where b_0, b_1 and so on are arbitrary constants to be determined by the minimisation procedure. The total rate of work is to be formulated and minimised with respect to these constants to determine the mechanical state in the deformed part.

The total rate of work is the sum of the rate of work in the dome, the rate of work in the flange and the rate of work at the lip of the die, as the material flows over the edge and into the die. The rate of work at the lip of the die is very small in comparison with that in the dome and flange and will therefore be neglected.

Rate of work in dome
The rate of work (2) in the dome, as a function of the principal strain rates and the thickness h, is

$$W_D = \frac{\sqrt{2}}{\sqrt{3}} \sigma_0 \int_0^{\sin^{-1}a/\rho} \int_0^{2\pi} \sqrt{\dot{\epsilon}_{RR}^2 + \dot{\epsilon}_{\phi\phi}^2 + \dot{\epsilon}_{\theta\theta}^2}$$
$$\times \rho^2 h \, \sin\phi \, d\phi \, d\theta \qquad (24)$$

The use of equations (23) and (6) in the strain rate–velocity relations (3) yields the following expressions for the strain rates

$$\dot{\epsilon}_{RR} = \frac{h_0^2}{h^3} \left[h_{,\rho} + \frac{1}{\rho} \left\{ b_0 \left(\frac{a}{\rho}\right) \sin\phi \right. \right.$$
$$\left. \left. + b_1 \left(\frac{a}{\rho}\right)^2 \sin2\phi + \dots \right\} h_{,\phi} \right]$$

$$\dot{\epsilon}_{\phi\phi} = \frac{1}{\rho} (1 - \frac{\rho}{\sqrt{\rho^2 - a^2}}\cos\phi) + \frac{1}{\rho}\left\{ b_0 \left(\frac{a}{\rho}\right)\cos\phi \right.$$
$$\left. + 2b_1 \left(\frac{a}{\rho}\right)^2 \cos2\phi + \dots \right\} \qquad (25)$$

$$\dot{\epsilon}_{\theta\theta} = \frac{1}{\rho} (1 - \frac{\rho}{\sqrt{\rho^2 - a^2}}\cos\phi) + \frac{1}{\rho}\left\{ b_0 \left(\frac{a}{\rho}\right)\cos\phi \right.$$
$$\left. + 2b_1 \left(\frac{a}{\rho}\right)^2 \cos^2\phi + \dots \right\}$$

The dome thickness h in (24) and (25) is a function of the coefficients $b_0, b_1, \ldots$. This follows from the recognition that h at any time depends on the previous history of deformation and that the specification of $b_0, b_1, \ldots$ determines the entire velocity field in the dome for all values of drawdepth. The necessary functional relationship can be obtained by using the incompressibility condition $\dot{\epsilon}_{RR} + \dot{\epsilon}_{\phi\phi} + \dot{\epsilon}_{\theta\theta} = 0$ and (25). The result is

$$\frac{h_0{}^2}{h^3} \left[h_{,\rho} + \frac{1}{\rho} \left\{ b_0 \left(\frac{a}{\rho}\right) \sin\phi + b_1 \left(\frac{a}{\rho}\right)^2 \sin 2\phi + \ldots \right\} h_{,\phi} \right]$$

$$= -\frac{2}{\rho} \left(1 - \frac{\rho}{\sqrt{\rho^2 - a^2}} \cos\phi\right) - \frac{1}{\rho} \left[b_0 \left(\frac{a}{\rho}\right) \cos\phi \right.$$

$$\left. + 2b_1 \left(\frac{a}{\rho}\right)^2 \cos 2\phi + \ldots \right] \quad (26)$$

$$- \frac{1}{\rho} \left[b_0 \left(\frac{a}{\rho}\right) \cos\phi + 2b_1 \left(\frac{a}{\rho}\right)^2 \cos^2\phi + \ldots \right]$$

The relation (26) is a nonlinear partial differential equation for the current thickness h in the dome. It is clear from an examination of the rate of work integral that a substantial simplification in the numerical procedure necessary for the determination of b_i will result if a closed-form solution for h in terms b_0, b_1 and so on can be obtained. We note that an exact close-formed solution for h can be determined at the centre ($\phi = 0$) and edge of the dome ($\phi = \sin^{-1} a/\rho$). For the case where $b_2, b_3, \ldots$ are zero, these solutions are given by

$$\frac{h_0^2}{h_{\text{edge}}^2} = 1 - 4 \int_0^{1/\rho} \frac{b_0 a\sqrt{1 - a^2 y^2}}{1 - \{b_0 a y + b_1 a^2 y^2} \, dy \quad (27)$$

$$\frac{h_0^2}{h_{\text{centre}}^2} = 0.5 - \ln 0.25 + 2\ln\left(\frac{1}{1 + \sqrt{1 - \frac{a^2}{\rho^2}}}\right)$$

$$- 2b_0 \left(\frac{a}{\rho}\right) - 2b_1 \left(\frac{a}{\rho}\right)^2 \quad (28)$$

An approximate solution for the entire dome is obtained by assuming a parabolic distribution for h between these two points. (Experimental values for h can be accurately fitted by parabolic curves). This solution is

$$h = (h_{\text{edge}} - h_{\text{centre}}) \frac{\phi^2}{\left(\sin^{-1}\frac{a}{\rho}\right)^2} + h_{\text{centre}} \quad (29)$$

The rate of work done in the dome can now be written as

$$W_D = \frac{4\pi\sigma_0}{\sqrt{3}} \int_0^{\sin^{-1} a/\rho} \sqrt{\dot{\epsilon}_{\phi\phi}^2 + \dot{\epsilon}_{\theta\theta}^2 + \dot{\epsilon}_{\phi\phi}\dot{\epsilon}_{\theta\theta}}$$

$$\times \rho^2 h \sin\phi \, d\phi \quad (30)$$

after integrating (24) with respect to θ and using the incompressibility condition, where the strain rates $\dot{\epsilon}_{\phi\phi}$ and $\dot{\epsilon}_{\theta\theta}$ and thickness h are known functions of the arbitrary constants $b_0, b_1, \ldots$.

Rate of work in flange

Proceeding as in the case of the dome, the rate of work in the flange is

$$W_F = \frac{4\pi\sigma_0}{\sqrt{3}} \int_\alpha^b \sqrt{\dot{\epsilon}_{rr}^2 + \dot{\epsilon}_{\theta\theta}^2 + \dot{\epsilon}_{rr}\dot{\epsilon}_{\theta\theta}} \; \bar{h} r \, dr \quad (31)$$

The current radius of the outer edge b is obtained as a function of ρ, b_0, b_1 and so on, from the continuity relationship (21). The result is

$$b = \alpha \exp\left[\frac{2\sigma_0 - P}{\sigma_0} - \frac{1}{\sigma_0} \sqrt{\frac{T}{b_0 U + b_1 S + \ldots - Q}}\right] \quad (32)$$

where

$$T = \frac{-\alpha(-2\sigma_0 + P)^3}{2\sigma_0}$$

$$U = -\frac{a}{\alpha}\left(\frac{a^2}{\rho}\right)$$

$$S = \frac{a}{\alpha}\left[\frac{2}{3}a\left(1 - \frac{a^2}{\rho^2}\right)\sqrt{1 - \frac{a^2}{\rho^2}} - \frac{2}{3}a\right]$$

$$Q = \frac{\alpha(-2\sigma_0 + P)^3}{2\sigma_0\left[\sigma_0\left\{\ln\left(\frac{B}{\alpha}\right) - 2\right\} + P\right]^2}$$

Equation (21) also enables the flange thickness strain to be expressed in terms of b_0, b_1 and so on. However, an approximate solution for $\bar{h}$, determined in a manner similar to the procedure adopted for the dome, is not possible. Therefore an expression for $\bar{h}$ derived by Boduroglu[12] will be used. This expression is given by

$$\bar{h} = \frac{h_0 \sigma_0 r^2}{\sigma_0 r^2 - b(b - B)(\sigma_0 \ln\frac{b}{r} - \frac{\sigma_0}{2} + P)} \quad (33)$$

where h_0 and B are the original thickness and original radius of the blank, respectively. Equation (33) has been found to agree closely with experimental results.

METHOD OF SOLUTION

With the results of the previous section, the total rate of work for the system can be written in the form

$$W_T = \int_0^{\sin^{-1} a/\rho} f(b_0, b_1) \rho^2 \sin\phi \, d\phi$$

$$+ \int_\alpha^{b(b_0, b_1)} g(b_0, b_1) r \, dr \quad (34)$$

where the integrals on the right hand side of (34) represent the rate of work in the dome (30) and flange (31), respectively. We are interested in the extremum of the function (34) under the subsidiary condition (22). This condition can be put in the framework of the minimisation procedure by the introduction of a Lagrange multiplier λ. Thus, the rate of work is minimised at a stationary value of the function.

$$I = \int_0^{\sin^{-1} a/\rho} f(b_0, b_1)\rho^2 \sin\phi \, d\phi$$

$$+ \int_\alpha^{b(b_0, b_1)} g(b_0, b_1)r dr + \lambda n(b_0, b_1) \qquad (35)$$

The values of b_0, b_1 and λ which render I stationary are found by solving the set

$$\frac{\partial I}{\partial b_0} = \int_0^{\sin^{-1} a/\rho} \frac{\partial f}{\partial b_0}(b_0, b_1)\rho^2 \sin\phi \, d\phi$$

$$+ \frac{\partial}{d b_0}\left[\int_\alpha^{b(b_0, b_1)} g(b_0, b_1)r dr\right] + \lambda\frac{\partial n}{\partial b_0}(b_0, b_1) = 0$$

$$\frac{\partial I}{\partial b_1} = \int_0^{\sin^{-1} a/\rho} \frac{\partial f}{\partial b_1}(b_0, b_1)\rho^2 \sin\phi \, d\phi$$

$$+ \frac{\partial}{\partial b_1}\left[\int_\alpha^{b(b_0, b_1)} g(b_0, b_1)r dr\right] + \lambda\frac{\partial n}{\partial b_1}(b_0, b_1) = 0$$

and

$$\frac{\partial I}{\partial \lambda} = n(b_0, b_1) = 0 \qquad (36)$$

The Newton–Raphson numerical iterative procedure is used to solve these equations.

In the solution of the present problem, only two arbitrary constants were included in the series representation for the tangential velocity in the dome. The solution obtained appeared to be sufficiently accurate without the inclusion of additional terms.

DISCUSSION OF RESULTS

The primary objective of this study was to predict the edge pullin. However, the analysis also enables the thickness strain distribution in the entire dome and flange to be determined. Because of assumption (2) these strains will be compared with experimental results obtained by static forming only.

The pullin at the outer edge of the blank is plotted as a function of the drawdepth w_0 in figure 4 for a B/a ratio (initial blank radius/die radius) of 1.5. The analytically determined pullin is in close agreement with experimental values. The numerical results indicate that the yield-stress has little effect on the pullin. The magnitude of the force field necessary to produce the deformation depends intimately on the yield-strength, however. Further, the assumption

that the blank is a membrane excludes the effect of blank thickness on the pullin.

Figure 5 shows a plot of the pullin as a function of drawdepth for different B/a ratios. It is evident that for a particular drawdepth, the pullin decreases as the ratio B/a increases. The explanation for this is

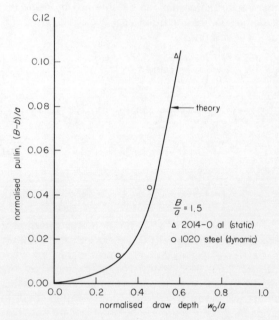

Figure 4. Blank pullin versus drawdepth: comparison of theory and experiment.

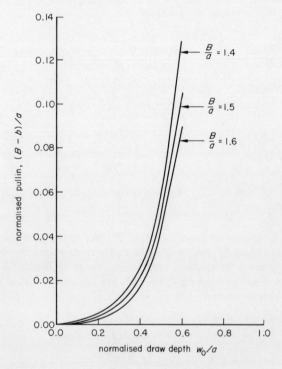

Figure 5. Parametric curves for blank pullin versus drawdepth.

simply that the surrounding annular flange restricts the flow of material into the die cavity. The larger the blank size, the greater the restriction. Figure 5 also shows that for a particular B/a ratio, the edge pullin increases as the drawdepth is increased. The curves have not been drawn beyond $w_0/a = 0.6$ due to instability at the apex of the dome. At this point the apex begins to neck (or thin rapidly). However, the pullin for very deep drawdepths can be obtained by linear extrapolation. This is demonstrated in figure 6 for a B/a ratio of 1.58 where the experimental points fall in the close neighbourhood of the extended curve. The experimental data for figure 6 have been obtained from the Martin–Marietta corporation.[13]

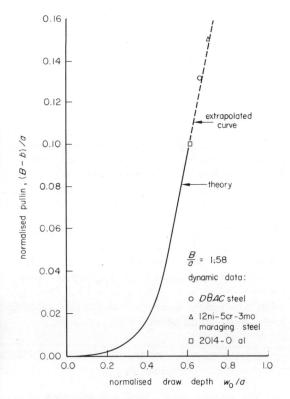

Figure 6. Comparison of linearly extrapolated theory with experimental values of pullin.

The analytically determined thickness strain field is in good agreement with thickness strain measurements obtained from a statically formed 2014-0 aluminium dome. This comparison is shown in figure 7. Additional verification of the formulation is provided by the prediction of flange instability and modelling of the bulge test.

The initiation of instability is assumed to be the point at which the flange becomes rigid so that further increase of drawdepth does not result in an increase in pullin. The predicted value of the apex thickness strain at the onset of necking for a 2014-0 aluminium dome is compared with experiment and Hill's analysis in table 1. The three values are within 3% of each other.

TABLE 1

Comparison of instability strains

	experiment	present analysis	Hill's analysis
$\bar{\varepsilon} = \ln \dfrac{h_0}{h}$	0.357	0.352	0.364

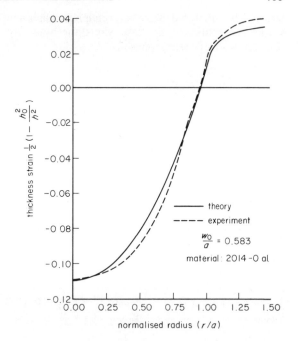

Figure 7. Comparison of theory and experiment for thickness strain distribution: static forming.

In the bulge test, hydrostatic pressure is applied to a thin circular diaphragm clamped at the edge. The clamping effect is introduced in the present analysis by making the flange radius large in comparison with the die radius. Figure 8 shows that the Thurston[14] and Hill[6] results are in the same range as the current numerical results.

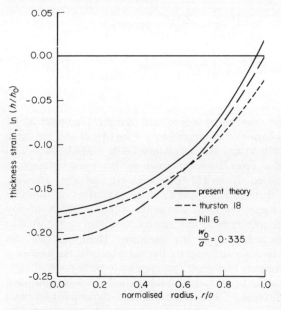

Figure 8. Comparison of different bulge test theories: thickness strain distribution.

Extension of Markov's principle to include inertia effects

The solution to the present problem was obtained under the assumption that the flow process is quasistatic. Inertia effects can be estimated by extending Markov's principle to include acceleration terms. The modified Markov principle, established in the Appendix for the case where the boundary conditions are entirely geometrical, is

$$\int_V \left(\frac{\sqrt{2}}{\sqrt{3}} \sigma_0 \sqrt{\dot{\epsilon}_{ij}^* \dot{\epsilon}_{ij}^*} - \rho_0 a_i v_i^* \right) dV$$

$$\geqslant \int_V \left(\frac{\sqrt{2}}{\sqrt{3}} \sigma_0 \sqrt{\dot{\epsilon}_{ij} \dot{\epsilon}_{ij}} - \rho_0 a_i v_i \right) dV \qquad (37)$$

where $\dot{\epsilon}_{ij}^*$ and v_i^* are arbitrary admissible quantities and the unstarred terms represent the correct solution to the problem. Therefore, a_i on the left-hand-side of (37) cannot be assumed in terms of arbitrary constants. The expansion of the inertia term in (37) yields

$$\int_{V_{dome}} -\rho_0(a_R v_R^* + a_\phi v_\phi^*)\,dv + \int_{V_{flange}} -\rho_0(\bar{a}_r \bar{v}_r^* + \bar{a}_z \bar{v}_z^*)dV \quad (38)$$

where the first integral is for the dome and the second for the flange.

Because of the nature of the explosive-forming process all but the $a_R v_R^*$ term in (38) is negligible compared to the term containing the strain rates in (37). This can be confirmed experimentally. Therefore, (38) reduces to

$$\int -\rho_0 a_R v_R^* \, dR \qquad (39)$$

and (37) becomes

$$\int_V \left(\frac{\sqrt{2}}{\sqrt{3}} \sigma_0 \sqrt{\dot{\epsilon}_{ij}^* \dot{\epsilon}_{ij}^*} - \rho_0 a_R v_R^* \right) dV$$

$$\qquad (40)$$

$$\geqslant \int_V \left(\frac{\sqrt{2}}{\sqrt{3}} \sigma_0 \sqrt{\dot{\epsilon}_{ij} \dot{\epsilon}_{ij}} - \rho_0 a_R v_R \right) dV$$

for the entire dome and flange. The method of solution is to prescribe $a_R = a_R(\rho,\phi,t)$ and minimise (40) at any instant of time (note that $D\rho/Dt \neq 1$ in this case). However, since v_R has been obtained in closed form (6) independently of the arbitrary constants b_0, b_1 and so on, minimisation of (40) with respect to these constants will not result in any contribution by the acceleration term to the present formulation of the problem. Therefore, we can conclude that inertia has no effect on the explosive forming of domes. This appears to be in contradiction to the experimental results which showed significant differences between static and dynamically-formed domes in the region of the apex of the dome. The reason for this contradiction is the assumption that the shape of the dome is a segment of a sphere at all times in the forming process.

This is not the case in the actual process, since the blank in the vicinity of the apex is nearly flat during a considerable portion of the forming operation. The favourable strain distribution produced dynamically is therefore due primarily to geometric rather than inertia effects.

CONCLUSIONS

The present study is part of a continuing investigation of the complex explosive-forming process. With the present state of knowledge, there is not enough information to predict all the required process parameters. Such prediction is necessary in order to derive maximum economic benefits, which lie in the limited production of very large parts.

In this study we have obtained satisfactory analytical results for predicting the edge pullin of a flat circular blank, formed into an axisymmetric dome. These can be used by designers to aid in the determination of initial blank size, given the final dimensions of the dome. The analysis also provides a rough estimate of blank thinout.

In view of the resulting simplicity of the formulation as compared with other methods for handling finite deformation plasticity problems, this approach appears to be suitable in all metalforming membrane problems in which the shape of the middle surface is specified during the forming process. Whether the forming is done statically or dynamically will, as indicated by the extended Markov's principle, have little effect on the final strain field although it will affect the forces necessary to form the part. Examples of problems of this type are the bulging and sizing of tubing (used in the plumbing systems of space-vehicles to carry fuel and oxygen to the engines), recessing of panels (used in heat exchangers) and the forming of other rotationally symmetrical shapes (conical, ellipsoidal, paraboloidal and so forth).

APPENDIX

Let the stresses, strain rates and velocities of the exact solution be denoted by σ_{ij}, $\dot{\epsilon}_{ij}$ and v_i, and the velocities and resulting strain rates of an admissible solution by v_i^* and $\dot{\epsilon}_{ij}^*$. An admissible solution is one which satisfies the conditions of incompressibility, compatibility and the geometric boundary conditions. By Schwarz's inequality,

$$S_{ij}\dot{\epsilon}_{ij}^* \leqslant \sqrt{S_{ij}S_{ij}} \sqrt{\dot{\epsilon}_{ij}^* \dot{\epsilon}_{ij}^*} \qquad (41)$$

Using the incompressibility condition $\dot{\epsilon}_{ii} = 0$ in

$$S_{ij} = \sigma_{ij} - \frac{1}{3}\sigma_{kk}\delta_{ij} \text{ gives}$$

$$S_{ij}\dot{\epsilon}_{ij}^* = \sigma_{ij}\dot{\epsilon}_{ij}^* . \qquad (42)$$

Substituting (42) and the yield condition $S_{ij}S_{ij} = \frac{2}{3}\sigma_0^2$ into (41), we obtain the relation

$$\sigma_{ij}\dot{\epsilon}_{ij}^* < \frac{\sqrt{2}}{\sqrt{3}}\sigma_0 \sqrt{\dot{\epsilon}_{ij}^* \dot{\epsilon}_{ij}^*} \qquad (43)$$

Combining the flow equation (1) into the incompressibility condition, we get

$$S_{ij}\dot{\epsilon}_{ij} = \sigma_{ij}\dot{\epsilon}_{ij} = \frac{\sqrt{2}}{\sqrt{3}}\sigma_0 \sqrt{\dot{\epsilon}_{ij}\dot{\epsilon}_{ij}} \qquad (44)$$

The subtraction of (44) from (43) yields

$$\frac{\sqrt{2}}{\sqrt{3}} \sigma_0 \left[\sqrt{\dot{\epsilon}_{ij}^* \dot{\epsilon}_{ij}} - \sqrt{\dot{\epsilon}_{ij}\dot{\epsilon}_{ij}}\right] \geqslant \sigma_{ij}(\dot{\epsilon}_{ij}^* - \dot{\epsilon}_{ij}). \qquad (45)$$

Now integrate both sides of (45) through the entire body. The integration of the right-hand-side of (45) gives

$$\int_V \frac{1}{2}\sigma_{ij}\left[(v_{i,j}^* + v_{j,i}^*) - (v_{i,j} + v_{j,i})\right] dV \qquad (46)$$

when the strain rate–velocity relation $2\epsilon_{ij} = v_{i,j} + v_{j,i}$ is considered. Since σ_{ij} is symmetric, we can write (46) as

$$\int_V \sigma_{ij}(v_{i,j}^* - v_{i,j}) dV \qquad (47)$$

or as

$$\int_V \left[\{(\sigma_{ij}v_i^*),_j - \sigma_{ij}v_i^*\} - \{(\sigma_{ij}v_i),_j - \sigma_{ij},_j v_i\}\right] dV \qquad (48)$$

Noting that $\sigma_{ij,j} = \rho_0 a_i$ where a_i is the acceleration and ρ_0 is the density of the continuum, the use of Green's theorem in (48) yields

$$\int_S \sigma_{ij}v_j v_i^* ds - \int_S \sigma_{ij}v_j v_i ds + \int_V \rho_0 a_i v_i^* dV - \int_V \rho_0 a_i v_i dV \qquad (49)$$

When only velocity boundary conditions are prescribed on the entire surface, $v_i = v_i^*$ on S, and (49) reduces to

$$\int_V \rho_0 a_i v_i^* dV - \int_V \rho_0 a_i v_i dV \qquad (50)$$

The integration of the left-hand-side of (50) through the entire body gives

$$\int_V \frac{\sqrt{2}}{\sqrt{3}} \sigma_0 \left[\sqrt{\dot{\epsilon}_{ij}^* \dot{\epsilon}_{ij}^*} - \sqrt{\dot{\epsilon}_{ij}\dot{\epsilon}_{ij}}\right] dV \qquad (51)$$

Combining (50) and (51) and rearranging terms, we obtain

$$\int \left(\frac{\sqrt{2}}{\sqrt{3}} \sigma_0 \sqrt{\dot{\epsilon}_{ij}^* \dot{\epsilon}_{ij}^*} - \rho_0 a_i v_i^*\right) dV$$

$$> \int \left(\frac{\sqrt{2}}{\sqrt{3}} \sigma_0 \sqrt{\dot{\epsilon}_{ij}\dot{\epsilon}_{ij}} - \rho_0 a_i v_i\right) dV \qquad (52)$$

In (52) the right-hand side consists of terms that involve actual quantities, that is, quantities which would be obtained by the correct solution of the problem. The left-hand side contains arbitrary admissible quantities (except for the a_i which are true acceleration components). The proof of the modified Markov's principle is completed if the a_i are prescribed

because the actual quantities are then separated entirely from the admissible quantities in equation (52).

REFERENCES

1. G. E. HUDSON. Theory of the dynamic plastic deformation of a thin diaphragm, *Journal of Applied Physics*, Vol. 22, No. 1, pp. 1–11, January 1951.
2. A. J. WANG. Permanent deflection of a plastic plate under blast loading, *Journal of Applied Mechanics*, Vol. 22, pp. 375–376, 1955.
3. E. A. WITMER, H. A. BALMER, J. W. LEECH and T. H. H. PIAN. Large dynamic deformations of beams, rings, plates and shells, *AIAA Journal*, Vol. 1, No. 8, pp. 1848–1857, 1963.
4. E. A. WITMER, H. A. BALMER and E. N. CLARK. Experimental and theoretical studies of explosive-induced large dynamic and permanent deformations of simple structures, presented at the SESA Spring Meeting, Denver, Colorado, May 5–7, 1965.
5. E. A. WITMER, J. W. LEECH and L. MORINO. PETROS 2: A new finite difference method and programme for the calculation of large elastic-plastic dynamically-induced deformations of general thin shells, prepared for Ballistic Research Laboratories, US Army Aberdeen Research and Development Center, Aberdeen Proving Ground, Maryland, Report No. ASRL TR 152-1, BRL Contract Report No. 12-Appendix D, December 1969.
6. R. HILL. A theory of the plastic bulging of a metal diaphragm by lateral pressure. *Phil Mag.*, Vol. 41, No. 1133, pp. 1133–1142, 1950.
7. G. A. THURSTON. On the effects of edge pullin on the explosive forming of domes. *Proceedings of the Seventh International Machine Tool Design and Research Conference*, Sept. 12–16, 1966. Pergamon Press.
8. D. BOYD. Dynamic deformation of circular membranes, *Journal of the Engineering Mechanics Division, Proceedings of the A.S.C.E.*, vol. 92, No. EM3, June 1966.
9. A. A. EZRA and M. A. MALCOLM. An investigation of the effect of explosive stand-off distance on the forming of metal blanks, *First International Conference of the Center for High Energy Forming*, Estes Park, Colorado, Vol. 2, pp. 6.4.1–6.4.20, June 19–23, 1967.
10. Y. C. FUNG. *Foundations of Solid Mechanics*, Prentice-Hall, Inc., New Jersey, 1965.
11. S. B. KULKARNI. The prediction of edge pullin in explosively-formed domes. Doctoral Dissertation, University of Denver, 1971.
12. M. A. KAPLAN and H. M. BODUROGLU. Flange buckling of explosively-formed domes, *Fifth Annual Report of the Center for High Energy Forming, US Army Materials and Mechanics Research Center*, Watertown, Massachusetts, Report No. AMRA CR-66-05/25, pp. 40–65, September 1970.
13. K. R. AGRICOLA, J. T. SNYDER, J. B. PATTON, *et al.* Explosive-forming processes, *Technical Report No. AFML-TR-67-70*, Martin Marietta Corporation, May 1967.
14. G. A. THURSTON and R. J. HARRIS. Plastic deformation of a circular membrane under pressure. *Developments in Mechanics, Proceedings of the 11th Midwestern Mechanics Conference*, Vol. 5, pp. 669–689, 1969.

THE RADIAL PISTON APPROACH TO THE EXPLOSIVE AUTOFRETTAGE OF THICKWALLED FORGING DIES

by

M. KAPLAN, H. GLICK, W. HOWELL and V. D'SOUZA*

SUMMARY

A new process for obtaining residual compressive hop stresses at the bore of thickwalled forging dies is presented. The process uses explosive energy in a controlled manner to produce plastic flow in the die. High residual stresses are generated as the applied pressures decay. A theoretical model of the process is developed which predicts residual stress and deformation as a function of the material properties and geometry of the system and the characteristics of the explosive. Experimental data is presented and compared with analytical results. The experiments demonstrate the validity of the analysis and the process itself. The application of explosive autofrettage to a specific production forging die is described. The process is compared with existing methods for producing residual stresses in thickwalled cylinders.

NOTATION

ρ, ρ_w, ρ^p	mass density of the outer cylinder, water and radial piston respectively
ρ_0	initial mass density of the water
p_g, p_w	gas and water pressures
p_0	initial explosion pressure
a, b, d	inner radius of the liner, radial piston and outer cylinder, respectively
a_0, b_0	initial inner radius of the liner and radial piston, respectively
c, e	outer radius of the radial piston and outer cylinder respectively
K^p, K	radial displacement functions for the radial piston and outer cylinder
e_{ij}	strain tensor
$\dot{e}_{ij}$	strain rate tensor
S_{ij}	deviatoric stress tensor
$\sigma_r, \sigma_\theta, \sigma_z$	radial, tangential and axial stress components
$\sigma_{rj}, \sigma_{\theta j}$	radial and tangential stress components at time t_j
$\sigma_{r\infty}, \sigma_{\theta\infty}$	residual radial and tangential stresses
k^L, k^p, k	$2/\sqrt{3}$ the tensile yield-strength of the liner, piston and outer cylinder, respectively
γ	specific heat ratio
E	specific energy of the explosive
η	radial position of the elastic-plastic interface
μ	positive scalar function

INTRODUCTION

There are many applications in which thickwalled cylindrical structures are subjected to large transient internal pressures. Failure, in these situations, generally occurs by the propagation of radial cracks from the inner radius of the cylinder. Structural life can often be extended by introducing residual compressive hoop stresses into the cylinder; they prevent premature failure by retarding radial crack growth.

Residual compressive hoop stress can be developed in a variety of ways. In the case of cannon barrels, it was empirically found many years ago that a single

* All of University of Denver, Denver, Colorado, U.S.A.

firing with a double charge of powder substantially increased the life of the cannon. More recently, cannon barrels have been strengthened by hydraulic or mechanical autofrettage. In the hydraulic autofrettage process a fluid is pumped into the cavity at high pressure and then released. Mechanical autofrettage involves forcing an oversized mandrel through the length of the bore. Both methods cause the cylinder wall to become partially or totally plastic during the loading phase. The resulting residual stress distribution is caused by the differences between the plastic loading stresses and the elastic stresses developed during unloading. Large and expensive facilities are required for either method, even for medium-sized cannon barrels.

In the case of cylindrical forging dies, a shrink ring on the outside diameter of the die produces compressive hoop stress at the inner radius. The method is not very satisfactory since only relatively small stresses can be achieved in practical applications. Mechanical autofrettage cannot be used on cylindrical forging dies because of die geometry, and the unit cost of hydraulic autofrettage appears to be too high to be practical.

The purpose of this paper is to describe a new method for producing large residual compressive hoop stresses in thick-walled cylinders. It is called the radial piston method of explosive autofrettage, and appears to be completely feasible for use on cylindrical forging dies.

THE RADIAL PISTON METHOD

Initial experiments in explosive autofrettage[1] used a line charge of explosive placed along the axis of a thick-walled tube immersed in water. Following detonation of the explosive, the inner wall of the tube was subjected to an intense pressure wave which decayed rapidly. Although the method produced substantial plastic flow in the wall, the residual hoop stresses were only about 30 percent as large as comparable stresses obtained with hydraulic autofrettage. It was demonstrated[1] that the loss of residual stress was due primarily to plastic flow during unloading of the tube, and that the development of full residual stress required some control of the magnitude and decay rate of the applied pressure.

These considerations led to the radial piston concept in which a ductile tube, called the radial piston, is placed coaxially with the thickwalled tube to be autofrettaged (figure 1). The radial piston is separated from the outer cylinder by water and is filled with explosive powder. The powder is ignited by a thin rod of high-detonation-velocity explosive so that the motion of the piston is essentially radial. A restraining end fixture is used to position the radial piston and outer tube and to prevent leakage of the water and gaseous products of the explosion.

The piston acts primarily as an attenuator of the rate of energy transfer. The explosive energy released

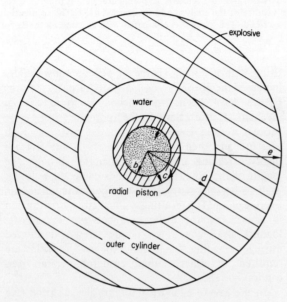

Figure 1. Radial piston autofrettage configuration.

at detonation is not transferred to the outer tube as a series of shock waves; rather, it is used to accelerate the piston, which in turn compresses the water sufficiently to produce plastic flow in the outer tube. A high rate of pressure decay is possible only if the piston is rapidly driven radially inward. This is prevented by the piston's inertia and yield-strength and the internal gas pressure in the piston. The gas pressure is maintained at relatively high values because the piston separates it from the surrounding water, thus preventing the sudden decrease in pressure produced by quenching.

MATHEMATICAL MODEL OF THE RADIAL PISTON CONFIGURATION

The purpose of the mathematical model is to predict residual stress and deformation as a function of system geometry and initial explosion pressure. Although explosive detonation generally produces significant wave effects, there is strong motivation for neglecting wave motion in this instance. The radial piston configuration is designed to smooth the pressure pulse produced by detonation. Because of the impedance mismatch between steel and water, only a small percentage of the initial detonation wave is transmitted to the water, and an even smaller percentage to the outer thickwalled tube. Although

wave motion in the radial piston is probably significant during its initial outward motion, a highly accurate description of piston motion is not necessary. The water gap acts as a weak nonlinear spring and moderate changes in piston position are reflected as small changes in the water pressure. Similarly, some error in the water pressure can be tolerated. The residual stress and deformation are functions of the integral of the pressure time history, not of the instantaneous pressure. Rapid time variations in the piston's internal gas pressure, for the same reasons, need not be modelled exactly. For these physical reasons, as well as for the substantial simplification produced in the mathematics, wave motions will not be considered.

The specific models used for the gas, radial piston, water and outer cylinder are described below.

(1) Outer cylinder

Typically, forging dies are composed of high yield-strength steels. Such materials exhibit little work-hardening and low strain rate sensitivity. Therefore, the die will be treated as an elastic—perfectly plastic strain rate insensitive body. Also, elastic incompressibility will be assumed. This produces a completely incompressible body and eliminates the possibility of wave solutions.

Under these conditions, the elasticity equations are written

$$\Delta S_{ij} = 2G\Delta e_{ij} \tag{1}$$

where the Δ's are used to represent the changes in deviatoric stress and strain from an initial stress state. The Prandtl-Reuss equations

$$\dot{e}_{ij} = \mu S_{ij} + \dot{S}_{ij}/2G \tag{2}$$

will be used to describe plastic flow, where μ is a positive scalar function, and $\dot{e}_{ij}$ is the strain rate tensor. $\dot{S}_{ij}$ is, for this particular problem, the material time derivative of S_{ij}. The Von-Mises yield condition

$$S_j^i S_i^j = \tfrac{1}{2}k^2 \tag{3}$$

is used as the condition for plastic flow. The constant k is equal to $2/\sqrt{3}$ times the yield-strength in tension.

(2) Water gap

The water between the radial piston and the outer cylinder is treated as a compressible fluid. Since the dynamic portion of the autofrettage process is generally completed in a time scale of microseconds, the compression and expansion of the water are assumed to be adiabatic. The pressure-specific volume adiabat for $p = 15$ lb/in^{2}a, $T = 80°$F, and $s = 0.0932$ Btu/lbm°R was obtained from[2]. This curve is closely approximated by the following equations.

$$p_w = 31.62Q \text{ for } 0 < Q \leqslant 0.079$$
$$p_w = -0.134 + 29.2Q + 82.1Q^2 + 104.9Q^3 \text{ for } \atop 0.079 < Q < 1.5 \tag{4}$$

where p_w is in kilobars and Q is a function of the ratio of the final and initial densities; that is,

$$Q = \frac{\rho_w}{\rho_0} - 1 \tag{5}$$

(3) Radial piston

The piston is treated as a rigid–perfectly plastic material. Its constitutive equation is

$$\dot{e}_{ij} = \mu S_{ij} \tag{6}$$

when the yield condition (3) is satisfied.

(4) Gas pressure

The gas pressure in the piston is computed from the isentropic expansion relations for a perfect gas. Thus

$$p_g = p_0 (b/b_0)^{2\gamma} \tag{7}$$

with p_0 the initial explosion pressure, γ the specific heat ratio and b_0 the initial radius of the piston. The value of $\bar{\gamma}$ is obtained from the ideal gas relation

$$p_0 = (\gamma - 1)E \tag{8}$$

where E is the specific energy of the explosive. The initial explosion pressure cannot be accurately computed from existing theory. It is estimated by correlating analytical predictions of residual deformation with experimental results: this is normally done only once for a given explosive.

Experimental evidence indicates that the flow process is essentially radial. Therefore, the axial and tangential velocities and displacements are zero, and the partial derivatives of all quantities with respect to these directions vanish; that is, a state of axial symmetry and plane strain exists. Under these conditions, the incompressibility equation in terms of velocities can be integrated to give the radial velocity v as

$$v = \dot{K}(t)/r \tag{9}$$

where $\dot{K}$ is an unknown time function and r is the radial position. Equation (9) is valid for a tube undergoing elastic, constrained plastic or total plastic flow. The radial acceleration, taking the material derivative of (9), is

$$a = \ddot{K}/r - \dot{K}^2/r \tag{10}$$

In the case of elastic or constrained plastic flow (when the tube is partially elastic and partially plastic) the convective term in the material derivative can be neglected. In this case, the radial displacement, velocity and acceleration in terms of K are simply

$$u = K/r; \quad v = \dot{K}/r; \quad a = \ddot{K}/r \tag{11}$$

The governing equations for the radial piston and outer cylinder are obtained when equations (1) to (11) are used with the radial equation of motion. With no body forces, this equation is

$$\frac{\partial \sigma_r}{\partial r} + (\sigma_r - \sigma_\theta)/r = \rho a \tag{12}$$

with ρ the mass density and σ_r and σ_θ the radial and tangential components of stress, respectively.

GOVERNING EQUATIONS FOR THE RADIAL PISTON

In order to avoid confusion between the piston and outer cylinder, all quantities associated with the piston will be superscripted with a p. The use of equation (9) in the strain-rate–velocity relations $\dot{e}_{ij} = \frac{1}{2}(v_{i,j} + v_{j,i})$ gives two nonvanishing components. These are

$$\dot{e}_r^p = \dot{e}_\theta^p = -\dot{K}^p/r^2 \tag{13}$$

Then, from equations (6), (3) and (13), it is found that

$$\sigma_\theta^p - \sigma_r^p = \pm k^p \tag{14}$$

and the axial stress is

$$\sigma_z^p = \frac{1}{2}(\sigma_r^p + \sigma_\theta^p) \tag{15}$$

The choice of sign in equation (14) is determined by the direction of the radial velocity. The sign is positive when the piston is moving outward; that is, when $\dot{K}^p > 0$.

The radial stress is found in terms of K^p by integration of the equation of motion using equations (14) and (10). The result is

$$\sigma_r = \rho^p [\ddot{K}^p \log r + (\dot{K}^p)^2/2r^2] - (\pm k^p \log r) \tag{16}$$

Application of the boundary conditions $\sigma_r = -p_g$ at $r = b$ and $-p_w$ at $r = c$ yields the Ricatti equation in K

$$-\ddot{K}^p = \frac{1}{2}\left(\frac{1}{b^2} - \frac{1}{c^2}\right)\frac{(\dot{K}^p)^2}{\log b/c} + \frac{1}{\rho^p}\left(\pm k^p + \frac{p_g - p_w}{\log b/c}\right) \tag{17}$$

The radii b and c and pressures p_g and p_w are functions of time in equation (17). The piston begins flowing outward plastically when

$$p_g - p_w > k^p \log c/b \tag{18}$$

When (18) is not satisfied, the solution to (17) violates the yield condition. The condition for plastic flow in the reverse direction is obtained by replacing the left hand side of (18) by $p_w - p_g$. The inequalities are applied only when the piston's velocity is zero. If the piston is moving, it continues flowing plastically until brought to rest by changes in the gas and water pressure.

When the wall thickness of the outer cylinder is not constant (as for a tapered die) a uniform charge in the radial piston produces nonuniform expansion of the cylinder. One method of correction is to use a stepped or tapered plastic liner in the piston. This allows the explosive energy per unit length of piston to be easily varied. Assuming the liner to be an incompressible material with yield stress k^L and density p^L, the governing equation for K^p becomes

$$-\ddot{K}^p = \frac{-\rho^L\left(\frac{1}{b^2} - \frac{1}{a^2}\right) + \rho^p\left(\frac{1}{b^2} - \frac{1}{c^2}\right)}{2(\rho^p \log b/c - \rho^L \log b/a)}(\dot{K}^p)^2$$

$$+ \frac{\pm(k^p \log b/c - k^L \log b/a) + p_g - p_w}{\rho^p \log b/c - \rho^L \log b/a} \tag{19}$$

where a is the inner radius of the liner. The use of a liner also requires b/b_0 in (7) to be replaced by a/a_0.

GOVERNING EQUATIONS FOR THE OUTER CYLINDER

As the water pressure increases, the outer cylinder first deforms elastically, then begins to yield at the inner surface $r = d$. With further increases in pressure, the extent of the plastic region grows until, at sufficiently high pressures, the tube is totally plastic. The elastic unloading phase of the process may, in some instances, be interrupted by additional plastic flow (reyielding). The governing equations for each of these conditions are given below. These equations were developed by following the same procedure used in obtaining the radial piston equations, the only differences being the use of equations (1) and (?) in place of (6), and (11) rather than (10) for elastic and constrained plastic flow.

(1) Elastic deformation

The equations are written for the time $t \geqslant t_j$ when the tube becomes completely elastic. The hoop stress and radial stress at t_j are $\sigma_{\theta j}(r)$ and $\sigma_{rj}(r)$ respectively. Similarly K at t_j is denoted by K_j. With the boundary conditions $\sigma_r = -p_w$ at $r = d$ and zero at $r = e$

$$\sigma_r = \rho\ddot{K}\log r/e - 2G(K - K_j)\left(\frac{1}{r^2} - \frac{1}{e^2}\right)$$

$$- \int_r^e \frac{1}{\lambda}(\sigma_{\theta j} - \sigma_{rj})\, d\lambda \tag{20}$$

$$\sigma_r - \sigma_\theta = \sigma_{rj} - \sigma_{\theta j} - 4G(K - K_j)/r^2 \tag{21}$$

$$\rho\ddot{K}\log e/d + 2G(K - K_j)\left(\frac{1}{d^2} - \frac{1}{e^2}\right)$$

$$= p_w - \int_d^e \frac{1}{\lambda}(\sigma_{\theta j} - \sigma_{rj})\, d\lambda \tag{22}$$

The stress field, (20) and (21), is determined from the solution to (22). The solution is valid until plastic flow begins; that is, until $|\sigma_\theta - \sigma_r| = k$.

(2) Elastic–plastic deformation

The tube is elastic in the region $\eta \leqslant r \leqslant e$ and plastic for $d \leqslant r \leqslant \eta$. The position of the elastic–plastic interface $r = \eta$ is a function of time. In addition to satisfying the boundary conditions, the stresses must also be continuous across $r = \eta$. The equations are written for time $t \geqslant t_j$ at which yielding begins, assuming a positive radial velocity.

In the plastic region,

$$\sigma_\theta - \sigma_r = K \tag{23}$$

$$\sigma_r = \rho\ddot{K}\log r/d + k\log r/d - p_w \tag{24}$$

In the elastic region, the equations for σ_r and $\sigma_r - \sigma_\theta$ are the same as (20) and (21). The equations for determining η and K come from application of the stress-continuity condition. The result is

$$K - K_j = (\eta^2/4G)(\sigma_{rj}(\eta) - \sigma_{\theta j}(\eta) + k) \tag{25}$$

$$\rho\ddot{K}\log e/d + 2G\left(\frac{1}{\eta^2} - \frac{1}{e^2}\right)(K - K_j) = k\log d/\eta + p_w$$

$$- \int_\eta^e \frac{1}{\lambda}(\sigma_{\theta j} - \sigma_{rj})\, d\lambda \tag{26}$$

For negative radial velocities, the sign of k is changed in the above equations. Plastic flow, once begun, continues until the radial velocity vanishes. The solutions for K and η are therefore valid until $\dot{K} = 0$ or the tube becomes totally plastic ($\eta = e$).

(3) Total plastic flow

Since the deformations are potentially large in this case, (10) is used. The final equations, for positive velocities, are

$$\sigma_\theta - \sigma_r = k \tag{27}$$

$$\sigma_r = \rho\ddot{K}\log r/e + k\log r/e + \tfrac{1}{2}\rho\dot{K}^2\left(\frac{1}{r^2} - \frac{1}{e^2}\right) \tag{28}$$

$$\rho\log\ddot{K}\log e/d = p_w - k\log e/d + \tfrac{1}{2}\rho\dot{K}^2\left(\frac{1}{d^2} - \frac{1}{e^2}\right) \tag{29}$$

Again, negative velocity requires a sign change in k. The equations represent the material's behaviour until $\dot{K}$ is zero.

The solution to the governing equations (14) to (29) is obtained by numerical integration with the initial values $p_w = K = \dot{K} = K^p = \dot{K}^p = 0$ and the initial gas pressure p_0. Typical results are given in the next section.

A similitude analysis can be conducted with the governing equations of the system. The results show that both model and prototype have the same residual stress and percentage residual deformation at the inner radius when they are geometrically similar and when the parameters γ, k^p/p_0, k/p_0, G/p_0 and ρ^p/ρ remain constant. This can be accomplished by using the same piston and cylinder material and the same explosive, packed at the same powder density, in both model and prototype.

NUMERICAL RESULTS

As an example of the dynamic response of the system following detonation, the time variations of cylinder velocity, piston velocity, water pressure and gas pressure are shown in figure 2 for a configuration used

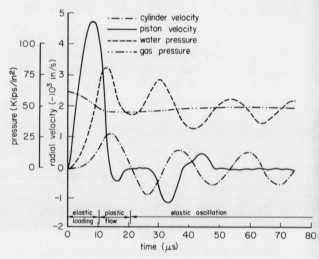

Figure 2. Predicted system response following detonation.

TABLE 1

radial piston: ID = 0.620 in,   OD = 0.750 in,   yield-strength = 69 kips/in²
outer cylinder: ID = 1.130 in,   OD = 1.984 in,   yield-strength = 160 kips/in²
initial explosion pressure = 62 kips/in²

in the experimental programme. The geometry, material properties and initial explosion pressure for this configuration are given in table 1.

The results show the oscillatory nature of the process. Following detonation, the piston's velocity rapidly increases. As the piston moves outward, volume changes produce an increase in the water pressure and a decrease in the gas pressure. When the water pressure exceeds the gas pressure, the piston begins slowing down. The point of zero velocity, that is, the point of maximum piston expansion, approximately corresponds to the initial pressure peak in the water. Because of the energy dissipated through plastic flow and the increase in the water gap produced by expansion of the outer cylinder, the oscillatory motion of the piston is highly damped. The piston and gas pressure become nearly stationary after a relatively short period of time. The outer cylinder, after deforming elastically, begins to yield slightly prior to the initial pressure peak. Plastic flow, extending through 70 per cent of the tube wall in this instance, ceases when the radial velocity vanishes. The tube then oscillates elastically in its fundamental breathing mode, producing an undamped oscillation of the water pressure.

Physically, end leakage gradually reduces the gas and water pressure to zero and internal dissipative mechanisms bring the outer cylinder to rest. The residual stresses are computed by assuming elastic unloading from time t_n to a rest state at $t = \infty$. The time t_n is any time beyond the cessation of all plastic flow. Thus, the final value of K is found by setting K and p_w to zero in (22) and replacing j by n. Using this result in (20) and (21), the residual stresses $\sigma_{r\infty}$ and $\sigma_{\theta\infty}$ are

$$\sigma_{r\infty} = \int_d^r \frac{1}{\lambda}(\sigma_{\theta n} - \sigma_{rn})d\lambda$$

$$- \left(\frac{1/e^2 - 1/r^2}{1/d^2 - 1/e^2}\right).\int_d^e \frac{1}{\lambda}(\sigma_{\theta n} - \sigma_{rn})d\lambda \qquad (30)$$

$$\sigma_{\theta\infty} = \sigma_{r\infty} + \sigma_{\theta n} - \sigma_{rn}$$

$$- \frac{2/r^2}{(1/d^2 - 1/e^2)}\int_d^e \frac{1}{\lambda}(\sigma_{\theta n} - \sigma_{rn})d\lambda \qquad (31)$$

The residual radial stress and hoopstress distributions for the previous example are exactly those obtained by hydraulic autofrettage under conditions producing the same degree of plastic flow. This result is typical of situations in which reyielding does not occur, reyielding being defined as plastic flow during contraction of the cylinder. Thus, with no reyielding, explosive and hydraulic autofrettage produce the same residual stress states. When reyielding does occur, it reduces the size of the residual hoop stress at the inner radius[1]. Reyielding is caused by excessive radial contraction following initial plastic

flow, a situation produced by rapid changes in the equilibrium position of the cylinder. With the radial piston method, reyielding can generally be avoided with reasonable choices for the radial piston and initial explosion pressure. The system substantially reduces the possibility of reyielding by reducing the rate of pressure decay. Also, the increase in water pressure during contraction of the outer cylinder limits the cylinder's radial displacement (see figure 2).

The effects of the variation of initial explosion pressure, piston material and piston geometry were studied with the analysis, and are discussed below. The basic configuration used for the study is given in table 1.

(1) Initial explosion pressure

The residual radial deformation at the inner diameter is shown in figure 3 as a function of initial explosion pressure. The residual stresses can be obtained from

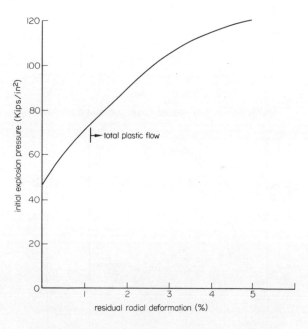

Figure 3. Effect of varying the initial explosion pressure.

the solid curve in figure 7. The cylinder begins yielding at an initial gas pressure of 46 kips/in², and total plastic flow is not developed until $p_0 = 74$ kips/in². This working range is large enough so that, in practice, substantial residual stress can be developed without the danger of the cylinder fracturing from excessive deformation. One of the advantages of explosive autofrettage (as compared to the hydraulic process) is that the rapid decay of pressure limits the amount of cylinder expansion without the use of constraining dies. Thus, total plastic flow at 74 kips/in² produces only a 1 percent expansion of the inner diameter. Even with a 25 percent overpressure of 92 kips/in², the permanent deformation is only about 2 percent.

(2) Piston geometry

The effect of varying the piston wall thickness, while keeping the outer diameter and initial explosion pressure constant, is shown in figure 4. The increase in stiffness produced by thickening the wall reduces the piston's response to internal pressure. The resulting lowering of the water pressure is reflected as a decrease in the residual stress and deformation in the outer cylinder. Extensive thinning of the wall increases the piston's response frequency. Although higher peak

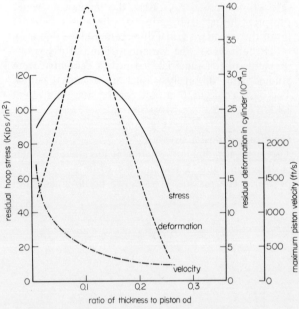

Figure 4. Effect of varying the piston wall thickness.

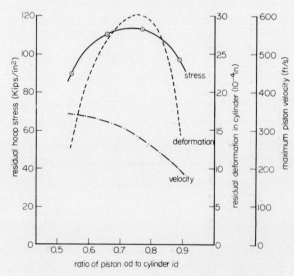

Figure 5. Effect of varying the size of the piston.

water pressures are developed, their application time is sufficiently reduced as to produce a decrease in plastic flow in the cylinder; thus, the resulting decrease in residual stress and deformation.

A similar effect occurs when the size of the water gap is varied by keeping the piston wall thickness constant and changing its outer diameter. Figure 5 shows that there is an optimal piston size for a given wall thickness. This is primarily due to changes in the stiffness of the nonlinear water spring. When the water

acts as a relatively soft spring (small piston), the piston expansion is not sufficient to produce high pressures. With small water gaps, the resulting high frequency system response reduces the time available for developing plastic flow.

(3) Material properties

The residual hoopstress was determined as a function of initial explosion pressure for four materials: 304 stainless, mild steel, 5056-0 aluminium and 1100-0 aluminium. These materials were chosen because of their high ductility. The piston, as shown in figure 6,

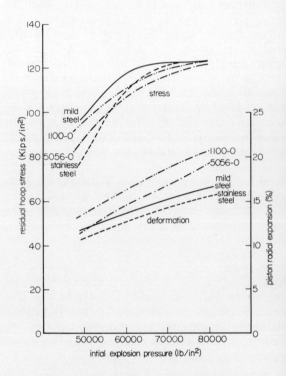

Figure 6. Effect of varying the piston material.

is typically required to expand 10 to 20 percent without fracture.

The effect of increased yield-strength in both aluminium and steel is to lower the residual stress in the outer cylinder. This result, as in the case of increased wall thickness, is caused by the change in the stiffness of the piston. The effect of material density can be seen from the mild steel and 5056-0 curves. Mild steel has a somewhat higher yield-stress, but it produces larger residual stresses than 5056-0 at equivalent initial explosion pressures. Although the results show mild steel to be the best of the four materials, the differences between them are small and all would be suitable in practice.

EXPERIMENTAL PROGRAMME

The experimental programme with the radial piston configuration was conducted using 4340 steel tubes, heat-treated to produce tensile yield-strengths of about 160,000 lb/in². The radial piston was standard 304 stainless steel seamless tubing. Measurements were taken of residual deformation as a function of axial and tangential position. Many of the tubes were bored out to produce the data necessary for residual stress determination by the Sachs method. The aim of the test programme was to demonstrate that explosive autofrettage would produce large residual hoopstresses. There was no attempt at optimisation, so the tests were conducted with only minor variations in the basic parameters.

The basic test sequence used a 1.98 in OD tube with a 0.425 in wall and a 3/4 in OD piston whose wall thickness was 1/16 in. Several tests were run using 34.9 g of an ammonium nitrate explosive with aluminium additive. The specific energy of the explosive is 1.9×10^6 ftlb/lb. A single axial strand of a 25 grain/ft detonating fuse, with PETN explosive as the core, was used for detonation.

The residual stresses were experimentally determined by the Sachs method[3] in which concentric rings are bored from the tube. Each cut relieves residual stresses at the inner diameter. The resulting change in strain is measured by a hoopgauge placed on the outside of the tube. The residual stresses are determined by relating the strain readings to the initial stress state through the standard elasticity equations. Because the technique is applied to short, open ended tube sections cut from the complete tube, the appropriate end condition is the vanishing of the axial stress (plane stress). The resulting residual hoopstress at the inner diameter is

$$\sigma_{\theta \infty}\Big|_{r=a} = (e^2 - d^2) Y \epsilon_{\theta 1}/(d_1^2 - d^2) \qquad (32)$$

where Y is Young's Modulus, and d_1 and $\epsilon_{\theta 1}$ are respectively the inner radius of the cylinder and the measured hoopstrain after the first cut.

The residual hoopstresses at the inner diameter for the basic test series are shown as the circular points in figure 7, plotted as a function of the

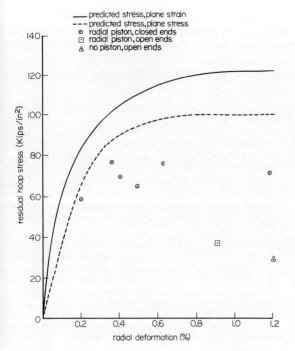

Figure 7. Comparison of experimental and predicted values of residual hoop stress.

percentage deformation of the inner radius. The solid curve is the residual stress predicted from the analysis (with no reyielding), and the dashed curve is the residual stress that would be predicted if the end conditions were changed from plane strain to open ended (plane stress). In computing this curve, the material was assumed to be elastically compressible. Initial plastic flow was governed by the Tresca rather than the Von Mises yield condition. The change in yield condition does not significantly effect the results.[4]

Although the analysis used the plane strain end condition for simplicity, in actual practice the end condition is closer to being open ended, since there is no restraint on the axial contraction of the cylinder. The expected residual stresses are therefore those given by the lower curve in figure 7. The experimental values of residual stress, typically about 70,000 lb/in², are considerably less than the predicted values. This phenomenon occurs in the hydraulic autofrettage process also and is generally attributed to reyielding produced by the Bauschinger effect[6]. Experimental results with a high strength 4330 steel show that the compressive yield-strength is reduced by as much as 50 percent following plastic flow in tension[5]. This also explains why the experimental value of residual stress at 1.2 percent deformation in figure 7 is no higher than at 0.4 percent deformation.

A check on the scaling law was made by doubling the size of the piston and outer cylinder and using four times as much explosive. The residual hoopstrains were essentially unchanged. The residual hoopstress was about 10 percent lower than was obtained with the original configuration. This could have been the result of a lower yield-strength in the larger cylinder; its hardness was 37 Rockwell C, while the smaller cylinder had a hardness of 43 Rockwell C.

One test was run using a radial piston with no end fixture. The explosive used was a type of detonating fuse with PETN core that is unaffected by water. With no way of preventing the water from leaking out of the ends, there was an increase in the rate of pressure decay. This apparently produced reyielding. The residual hoopstress was only 38 kips/in², despite the cylinder's nearly 1 percent residual deformation. In another test with no radial piston and open ends, the residual hoopstress was even lower, about 30 kips/in². These results are shown in figure 7.

The results of the experimental programme show that the radial piston method of dynamic autofrettage produces residual hoopstresses which are of the same order of magnitude as those developed in the static processes. It also demonstrates that the analysis can be used to predict accurately residual deformations and to provide reasonable estimates of residual stress.

THE AUTOFRETTAGE OF A PRODUCTION FORGING DIE

Two production forging dies, obtained from a manufacturing company, were autofrettaged explosively. The die material was H-13 steel hardened to 44-48 Rockwell C. The average diameter of the die cavity was 4.88 in and its minimum wall thickness 2.17 in. The autofrettage configuration, shown in figure 8, enabled both dies to be autofrettaged simultaneously. This eliminated end effects and reduced the time for the process. A steel seal was used to prevent leakage at the centre, where the two dies were in contact.

The residual deformation at the inner radius of the dies, following autofrettage, was 0.20 percent in one

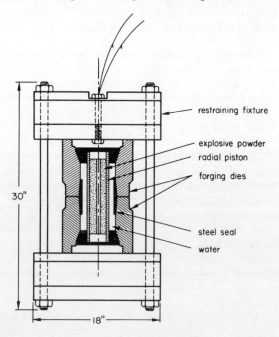

Figure 8. Autofrettage configuration: production forging dies.

die and 0.07 percent in the other. The theoretical residual stresses were 90 and 70 kips/in², respectively. The actual residual stresses were probably somewhat lower, but an accurate determination could not be made because the magnitude of the Bauschinger effect for the die material was not known. Sachs' method cannot of course be used on production dies.

The two dies were returned to the manufacturer and used in production. With no shrink ring, they failed by wear after producing an average of 3500 parts. This is about the same life and failure mode of nonautofrettaged dies when massive shrink rings are used.

CONCLUSIONS

It has been demonstrated that the radial piston method of explosive autofrettage produces large residual hoopstresses at the inner diameter of thick-walled, high strength cylinders. The primary advantage of the method over the conventional mechanical and hydraulic autofrettage processes is that it requires little in the way of capital investment: pumping equipment, external dies and hydraulic presses are eliminated. End sealing is not as critical as in hydraulic autofrettage because high pressures are maintained for extremely short times. Furthermore, the method can be readily applied to cylindrical forging dies. The hole

size in these dies is not generally uniform, so that the use of mechanical autofrettage would be difficult or impossible. As already noted, the unit cost of hydraulic autofrettage is probably excessive.

Stainless steel pistons were used in the experimental programme because of their ductility. It is expected, however, that in a production process a less expensive material like mild steel would be used. The piston represents the main portion of the material costs per shot, since the cost of explosive is minimal. The only other costs in the process are those associated with setup time and producing the end fixture. The end fixture is of course completely reusable. Also, if a water pool is not available, a small one must be built. All our experimental work was conducted in a 4 ft diameter by 4 ft deep pool. Rough cost estimates indicate that, *based on equal die life*, explosive autofrettage is competitive with the use of shrink rings.

The life of a forging die is generally maximised by using the hardest material that will fail by wear rather than cracking. Increasing the residual hoopstress reduces the possibility of fracture and permits the use of harder materials. For example, the H-13 production die described in the last section breaks after producing one part when it is used without a shrink ring. With a shrink ring, the die has a maximum compressive hoopstress of 7 kips/in² and fails by wear after 3,000–4,000 parts. The generation of higher residual stresses does not significantly affect wear. The autofrettaged dies, therefore, produced only 10 percent more parts on average than a typical die with a shrink ring. We believe, however, that major increases in die life are possible with autofrettage, since it allows the use of harder die materials than are feasible with shrink rings.

ACKNOWLEDGEMENTS

The authors wish to express their appreciation to A. Ezra and E. Knight for their important contributions towards the development of the process. The work was supported by an ARPA programme associated with the transfer of technology.

REFERENCES

1. J. MOTE, L. CHING, R. KNIGHT, R. FAY and M. KAPLAN. Explosive autofrettage of cannon barrels. *Army Materials and Research Centre Report* – AMMRC CR 70–25, February 1971.
2. 1967 ASME steam tables, ASME, New York, 1967.
3. G. SACHS and G. ESPEY. A new method for determination of stress distribution in thinwalled tubing, *Metals Tech.*, 1384, (1941).
4. T. DAVIDSON, D. KENDALL and A. REINER. Residual stresses in thickwalled cylinders resulting from mechanically induced overstrain, *Exp. Mech.*, 3, (1963), p. 253.
5. R. MILLIGAN, W. KOO and T. DAVIDSON. The Bauschinger effect in a high strength steel, *Journal of Basic Engineering*, 88 (1966), p. 480.
6. T. DAVIDSON and D. KENDALL. The design of pressure vessels for very high pressure operation, *Watervliet Arsenal Technical Report* WVT – 6917, May 1969.

INVESTIGATIONS ON THE ACCURACY OF REPRODUCTION OF ELECTROHYDRAULIC FORMING AND DEVELOPMENT OF AN ELECTROHYDRAULIC FORMING MACHINE

by

R. ZELLER*

SUMMARY

The accuracy of reproduction was investigated, by drawing conical hollow bodies with rotational symmetry, as dependent on the stored energy of the capacitors, the shape of the die cavity and the vacuum between sheet metal blank and die cavity. Also the limiting drawing ratio was investigated. The results follow. A cylindrical hollow body could not be produced, while there are certain limits for the cone angle. Optimum values were found for the stored energy and the vacuum. The long operational time in the water reservoir and the low efficiency led to the construction of a closed die. Therefore a machine for electrohydraulic forming with pneumatic sequence control was developed. By expanding tubes with this machine a cycle time of 40 seconds was achieved.

INTRODUCTION

The investigations discussed in this paper relate to the utility of the electrohydraulic forming process in production. This piece of research is the continuation of previous work done at the 'Institut für Umformtechnik' of the University of Stuttgart on the same process, which was concerned with the process of discharging, kinetics of deformation, energy balance, etc. Since the workpiece attains its final shape not by the simultaneous action of two die-halves but owing to the acceleration of the workpiece in the direction of the die cavity, this process seems to be interesting from the viewpoint of accuracy in reproducing desired contours. High pressures involved in the process might also permit higher amounts of deformation. The following points are investigated.

(1) The accuracy of reproducibility of hollow, axi-symmetric, conical bodies as dependent on the stored energy in the bank of condensers, the tool shape and the vacuum between the blank and the tool.

(2) The determination of the limiting drawing ratio (LDR) of pieces with different shapes and the causes of failures.

The very low efficiency of the process and the difficult handling conditions in the water tank are factors against a possible industrial application of this forming method. Cycle times that are more interesting in practice were achieved in a machine with closed dies developed for this purpose.

ACCURACY OF REPRODUCIBILITY

Test equipment
The tests were carried out in an impulse current plant comprised of three functional groups.

(1) Charging equipment (transformer, rectifier, load resistances).

(2) Energy store – capacitive (bank of condensers).

(3) Discharging equipment (cable collector, spark gap).

* Institut für Umformtechnik, Universität Stuttgart (TH)

The electrical details of the plant are as follows:

maximum capacitance	$C = 300\,\mu\text{F}$
load voltage	$U = 18\text{ kV}$
store energy	$E = 48.6\text{ kWs}$

The tests were performed in a water tank of 1500 mm diameter and 1500 mm depth to avoid reflections from the wall and for clear interpretation of the results. The tool set-up used is shown in figure 1. The hollow shapes to be formed were produced as inserts to make substitution easy. The parameters which remained unchanged throughout the series of experiments, like the height of water level above the electrode, the sparking distance and the space between electrodes, are also given in

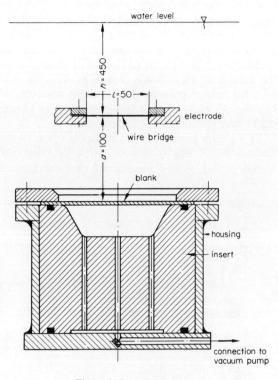

Figure 1. Experimental tool.

figure 1. The electrodes were connected by means of an aluminium wire (Al 99.5 − 0.8 mm thick).

The die inserts used to produce the conical, axi-symmetrical hollow bodies are shown in figure 2. For describing the propagation of shock waves a spherical wave is assumed in the literature[1]. Assuming that the formation of the hollow shape is achieved

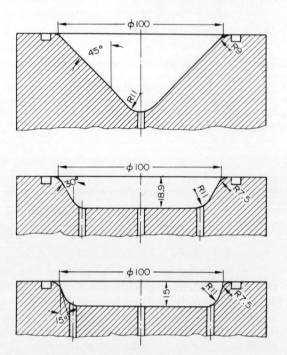

Figure 2. Tool inserts with semi-cone angles 45°, 30° and 15°.

by the acceleration of the sheet blank through the shock waves, one can deduce from the effective force field that the axisymmetrical conical shape can be easily obtained with truncated cones, while a relatively deep cylindrical cup (e.g. $h/d = 1$) with a flat bottom and a small transient radius from the side to the bottom is difficult to form. For these reasons the semi-cone angles were chosen as 45°, 30° and 15°; the order represents the increasing difficulty of forming. Al 99.5 (1.5 mm thick) and USt 14 (1 mm thick) were selected as test materials.

This energy-bound process naturally restricts the dimensions of the workpiece. In order that the reproducibility can be investigated as a function of

load energy, an energy reserve must be foreseen ever after completion of the deformation process. Work piece measurements obtained with 45° tool are shown in figure 2. A cylindrical die ring with a 100 mm diameter and a die entrance radius of 8 mm was used to determine LDR in free drawing.

The space between the blank and the tool must be evacuated, since the air present in this room cannot escape during the deformation process which takes place very quickly and it may lead to dents at some local points. The pressure could not be measured directly on the tool owing to the occurrence of shock waves and the sensitivity of the measuring probes.

Test programme

Table 1 shows the test programme. With reference to deep drawing, the LDR has been defined as the ratio between the blank diameter and the die entrance diameter. The LDR was investigated in the three different hollow forms and also in free drawing using a cylindrical die ring. A paste of zinc sulphide and mineral oil was used as a lubricant.

Interpretation of tests

The deviation of the dimensions of the workpiece formed from the measurements of the tool can be taken as a measure of error while forming (macro-analysis). Owing to local changes of wall thickness the measurements on the outer contour were the basis for evaluation. For cups with a 45° semi-cone angle the height of the cup measured as the distance between the flange portion and the lowest point of the drawn cup was a length that was dependent on the load energy in the evaluation of results. The outer contours of all the three types of cups were measured by using a contour projector in the region of the cone and transient radii (flange—mantle, mantle—bottom). This method could not be adopted for cups with flat bottoms (semi-cone angles 30° and 15°) since the bottom was cambered on the inner side. These contours were evaluated with a dial indicator in meridional sections. Wall thickness variations were determined from cups cut along the meridian plane. The springback of angle in 45° cups was also investigated. The blank diameter and the load energy at which the cups were torn during deformation were

TABLE 1
Test programme used to investigate the accuracy of reproducibility as a function of load energy, tool shape and vacuum.

Material	Semi-cone angle on tool (deg)	Drawing ratio $= d_0/d_{tool}$	Vacuum pressure (torr)	Stored energy (kW s)	Load voltage (kV)
Al 99.5 $s = 1.5$ mm	45 30 15	1.3	2×10^{-1}	$4.8 \leqslant E \leqslant 24.3$	18
USt 14 $s = 1$ mm	45 30 15	1.3	2×10^{-1}	$4.8 \leqslant E \leqslant 48.5$	18
Al 99.5 $s = 1.5$ mm	45 15	1.3	$1 \leqslant p \leqslant 760$	$E = 24.3$	18
USt 14 $s = 1$ mm	45 15	1.3	$1 \leqslant p \leqslant 760$	$E = 48.6$	18

other experimental parameters considered in the determination of LDR. Apart from these no other criteria were taken into account for the interpretation of test results.

Test results

Figure 3 shows the cup height as a function of load energy for the USt 14. Noteworthy in these experiments is the reduction of the flange diameter from 130 mm to 118 ± 0.2 mm for the various deformation

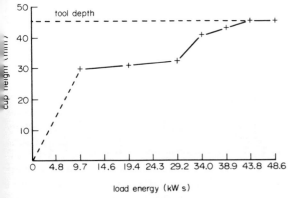

Figure 3. Height of cup versus load energy; 45° tool, material USt 14.

ratios. This phenomenon, that even with higher deformation ratios more material is not drawn in from the flange, can be explained as follows. The die used for drawing-in with spherical shock waves has a transition region in the form of periphery of a cone because of the given geometrical data. This results in very large normal forces and higher friction forces thereby preventing the flow of material. The inertia of the workpiece also plays a role. An increase of load energy therefore affects the blank thickness more. The influence of flow curve on the variation of cup height is shown in figure 3. A degressive tendency of the curve is to be expected from the viewpoint of the flow curve. The reduction of wall thickness occurring simultaneously brings about just the opposite effect. The pattern of curve obtained is

also influenced by the art of energy transmission. With progressive formation a greater portion of the energy contained in the blank as kinetic energy is levelled out through exchange of impulses. Hence the type of curve obtained is due to the simultaneous action of these three parameters.

Forming defects encountered are given in table 2 as the difference between the actual value and the nominal value. The minimum energy requirements tabulated denote the barest minimum required to see that the workpiece rests wholly on the tool during the deformation process. It is clear from this table that semi-cone angles of 45° and 30° could be achieved, but even with the maximum load energy a 15° semi-cone angle could not be produced. Reduction of the stand-off distance between the wire bridge and the blank made the formation feasible. The transient radii were also completely formed. The 45° cups showed the first signs of deviations from the tool geometry at one-third of the distance of the cone mantle, as measured from the flange. It was also interesting to analyse the inner cambering of cups with flat bottoms. Figure 4 shows how the deviation

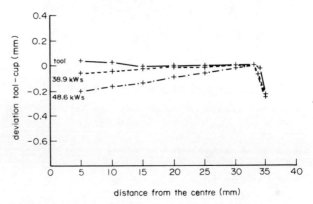

Figure 4. Deviation between cup bottom and tool; semi-cone angle 30°; material USt 14.

between the bottoms of 30° cups and the corresponding tool is dependent on the radius with load energy as parameter. The cup bottom and tool bottom were

TABLE 2
Error of formation as dependant on load energy: vacuum in tool space = 2 x 10⁻¹ torr.

Load energy (kW s)	material	Tool semi-cone angle (deg)	Measurements on workpiece			Maximum deviation – tool – workpiece – in the region of	
			semi-cone angle (deg)	transient flange – mantle (mm)	radii mantle – bottom (mm)	cone mantle (mm)	bottom (mm)
9.7	Al 99.5	45	45	9.0	11.0	0.15	–
19.4			45	8.6	10.5	0.15	–
43.8	USt 14		45	9.0	11.0	0.20	–
48.6		45	45	9.0	9.2	0.25	–
14.6	Al 99.5	30	30	7.4	10.8	–	0.08
19.4			30	7.4	10.9	–	0.13
38.9	USt 14	30	30	7.5	11.0	–	0.10
48.6			30	7.3	10.9	–	0.24
4.8	Al 99.5	15	20.5	7.3	10.9	–	0.12
19.4			17.5	7.5	10.8	–	0.07
19.4	USt 14	15	20	6.5	10.9	–	0.08
48.6			16.5	7.5	10.8	–	0.21

curved on opposite sides. No clear relation could be found between the load energy and the deviation of dimensions for Al 99.5 cups. In the case of cups produced from USt 14 an increase of load energy resulted in an increase in deviation from the nominal value. Measurements of radius carried out in two mutually perpendicular directions did not reveal any great changes. The three factors responsible for the deviation of cup bottom from the tool are the spring-back of workpiece, the mechanism of formation and the exchange of impulses between tool and workpiece[2].

The springback of the angle in the case of the 45° cups from USt 14 measured after sawing into two pieces was 15 min. This value was independent of the load energy. Such a determination was not possible for Al 99.5 cups owing to the fact that the residual stresses in the cups were still lower.

The shape of the tool influenced the reproducibility in the following way. The minimum cone angle that could be produced without any difficulty must be determined. It can be seen from figure 5 that an

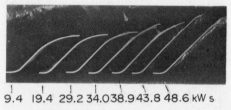

9.4 19.4 29.2 34.0 38.9 43.8 48.6 kW s

Figure 5. Variation of deformation ratio (45° cups — material USt 14) as a function of load energy.

increase of energy leads to a higher formability for 45° cups of USt 14. The workpiece rests more completely on the tool with increasing energy, a fact clearly recognisable from the figure. For all cups of the two different materials the 45° angle was obtained. The geometrical relationship between the tool and the stand-off distance makes this possible. Higher reductions of wall thicknesses involved cause necking at the cone peak in a few cups. Hence we found difficulty in forming a smaller radius at the cone peak.

A semi-cone angle of 30° was achieved for Al 99.5 for all load energies used (4.8, 9.7, 14.6, 19.4 and 24.3 kW s). The transient radius (side—bottom) was not wholly formed with the values mentioned first. The workpiece rested more smoothly on the surface of the die as reckoned from the transient radius and at the centre of the tool bottom, with the process of resting wholly on the surface increasing with load energy. The material required for forming at the entry of the die was drawn out from the blank thickness, since friction prevented the flow of material from the flange portion. This was also confirmed from the experiments. An angle of inclination of 49° due to bulging, which is partly present, resulted for USt 14 in a load energy of 4.7 kW s; this value was 37° for 9.7 kW s. The forming mechanism described above for Al 99.5 was found to be valid for USt 14 after an energy of 19.7 kW s. The cups with 45° and 30° angles are comparable within limits only. But the LDR and the transient radius are the same for both.

A comparison of the load energy required for 45° and 30° cups indicate clearly the higher amounts of energy required for 30° cups. It was not possible to obtain 15° cups for both the materials. In the case of USt 14 this desired cone angle is nearly obtainable (see table 2). This can be explained by the different types of bulging met with: Al 99.5 expands parabolically, whereas USt 14 expands spherically[1]. The build-up of cone in the area of free forming, i.e. to the bottom portion, is more truncated in the case of Al 99.5 compared with USt 14. Thus the USt 14 more closely approaches the angle.

The maximum reduction of a wall thickness of 57% was measured in 45° cups. In 15° and 30° cups a value of 15% was measured at the transient radius (mantle—bottom) and at, the centre of the cup, lower values were measured.

Vacuum played the following role in the accuracy of reproducibility. A vacuum of 2×10^{-1} torr was generated to investigate the parameters, load energy and the shape of the workpiece. The suction time involved was very high and the process may not therefore be economically sound. It is also important to determine the limiting value of the vacuum necessary to deform the blank completely. Initially the air trapped in the space between the blank and the die was left as such. This caused bulges and indentation in the workpiece (figure 6). These dents

Figure 6. Dents on 45° and 15° cups in tool not evacuated — material USt 14.

were formed by the resulting high pressure during deformation and the difficulty encountered by the trapped air escaping. Microscopic investigations did not confirm any melting away of the workpiece surface due to the remaining compressed air being heated up. The dents are dependent not only on the vacuum pressure but also on the position of holes provided to evacuate the container.

With the help of regulating valve and artificial leakage arrangement in the vacuum plant different values for vacuum pressure were obtained. A vacuum of 20 torrs was necessary for the $45°$ cups whereas the $15°$ cups needed 1 torr. The details given in table 2 are for a vacuum of 2×10^{-1} torr. For the limiting values already mentioned complete correspondence was found between the tool shape and blank shape. A higher value of vacuum seems, therefore, not worth while. The small amount of air remaining in the evacuated space prevents rebounding of the workpiece on the surface of the tool (damping effect). Various tool geometries gave rise to different optimum values for vacuum pressure.

With regard to investigations of the LDR, it should be pointed out that the failure of the cups in drawing with a cylindrical ring always occurred through tearing in the middle of the peak. $\beta_{max} = 1.82$ was obtained with an energy of 14.6 kW s; increasing the energy to 29.2 kW s reduced β_{max} to 1.5. A tool with a different form for $45°$ cups resulted in a value of $\beta_{max} = 1.3$. Independent of the energy used the flange diameter was reduced from 130 mm to 118 mm as already mentioned. Further drawing-in of the material in the flange was hindered by acceleration and friction forces. When the blank diameter was increased to values beyond those corresponding to $\beta = 1.3$, the blanks could not be fully deformed and failure occurred. Therefore it can be concluded that a generalisation of the results of the LDR is not possible, since it is dependent on the workpiece, shape of blank, load energy and the position of electrodes with respect to workpiece.

FORMING MACHINE

The technological and economical questions led to the development of this machine: for example the pressure chamber can be designed with closed dies of relatively simple material. Using the energy stored in the condenser banks in a better way and adopting easier and quicker handling possibilities would make this process more interesting from the point of view of economy. Vibrations and noise caused during the deformation process were kept to a minimum. The machine can be erected in a normal shop because of its convenient dimensions and a special housing facility for the high-voltage plant may be necessary. A line sketch of the machine is shown in figure 7 and a photograph of it in figure 8. The closed dies (for expansion of tubes) have a horizontal parting plane, the lower die being held tight in the lower portion of the machine and the upper die being actuated by an air piston. Locking of the two die halves is done by a pneumatically actuated wedge-shaped holder. The coaxial electrode used to produce the shock waves forms the most important part of the machine. Clamping spaces of the inner or outer electrodes lead to the cable collector through the copper wires.

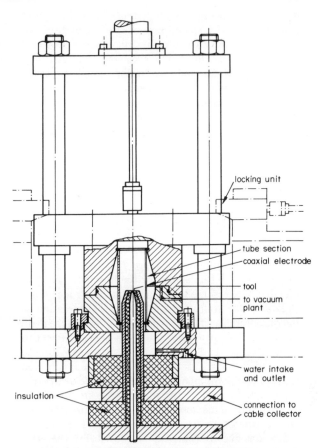

Figure 7. Line sketch of the working arrangement.

Figure 8. Partly mechanised electrohydraulic forming machine (front view).

Radial paths leading to a common round channel evacuate the tool space. The closing, regulation and opening of the die are by pneumatic sequence controls. Other functions like the filling-in and draining of water and evacuation could be easily manoeuvred with these controls. The following construction features deserve special mention. The die assembly is a thick hollow cylinder subjected to growing internal pressures. The regulators have to withstand a separating force of 1000 kN (approximately) arising out of shock waves on the two tool halves. Bearing in mind the probably high forces which might occur at the insulating plates under the machine table and at the end where current is supplied to the assembly, these portions are strengthened and bolted tightly with three bolts. The possible appearance of a leakage current should also be avoided. A minimum leakage distance of 60 mm is to be provided for lapping or binding the insulation together between the inner and outer electrode.

During the trial of the machine, the insulation in the ring clearance of the coaxial electrodes was the main source of difficulty. The insulation here is subject to both thermal and mechanical stresses. It can be observed that the exact stand-off distance is achieved for only one point along the circumference of the ring channel and therefore the shock waves are broken up eccentrically. This results in both compressive and tensile stresses on the insulation. The wear of the insulation due to burning is caused by the heat developed in the region of the spark gap. Hence the life period of the electrodes is reduced. A constructional method of the electrode head in which the insulation (the portion which wears away) is easily replaced was introduced to overcome this problem. It should also be seen that the dielectric strength of the insulation is not impaired by the surrounding water medium. The material of the electrode should not burn away easily and should possess high strength. Initial experiments revealed that nine diverse insulating materials did not withstand the thermal and mechanical stresses—for example, Teflon strengthened with glass fibres, Perlon, insulating compounds for cables used inside earth soil. With a soft and ductile polyurethane a life period of the insulation of about 130 shots could be achieved. Heat-treated hot working steel 55 NiCr MoV 6 was used as electrode material. Decarburisation with simultaneous increase of hard-

ness took place at the tip of the electrodes (both inner and outer electrodes) due to sparking. The small microcracks found on the outer microstructure may be probably due to the combined effect of steep temperature gradients and the impulse type of mechanical stressing. The propagation of such microcracks ended up in the failure of the outer electrode. The other steel used (for higher temperatures), namely hot working steel X 40 Cr MoV 51 stood the stressing with little erosion. Cycle times of 40 seconds were obtained during the bulging of cylindrical tubes with the above machine. In order to reduce the time needed for evacuating (this formed a greater portion of the cycle time), the tool was heated to about 80°C.

The present concept of the machine with coaxial electrodes has the following advantages and disadvantages. The inductance value of the whole unit is kept down to a minimum by the coaxial feeding-in of current in comparison with feeding-in through two opposing points. The coaxial electrodes can be used for both bulging of tubes and drawing of blanks, the system being more suitable for the latter case. Resetting the machine from the bulging operation to the drawing process is easily accomplished by changing the top half of the die. The vertical position of the point of emanation of the shock waves can be easily altered by placing in insulation, washers, etc.

The main disadvantage is that the stand-off distance is always eccentric to the tool axis. Since the outer electrode is subjected to an internal pressure, it cannot be made unduly small. The principle of using coaxial electrodes may not be applicable for tubes of less than 40 mm diameter. The non-symmetrical stressing accelerates the wear of the tip of the electrode.

REFERENCES

1. H. MÜLLER, 1969. 'Vorgänge beim elektrohydraulischen und elektromagnetischen Umformen von metallischen Werkstücken', *Berichte aus dem Institut für Umformtechnik,* Universität Stuttgart, Girardet, Essen, No. 11.
2. H. G. BARON, 1964. 'Spring-back and metal flow in forming shallow dishes by explosives', *Intern. J. Mech. Sci,* **6**, 435–44.
3. R. ZELLER, 1972. 'Untersuchungsergebnisse über die Abbildegenauigkeit bei der elektrohydraulischen Umformung', *Industrie-Anzeiger,* **94**, No. 2, 20–4.

POWDER COMPACTION
AND
SINTER FORGING

HIGH-SPEED COMPACTION OF METAL POWDERS

by

SHERIF ELWAKIL* and R. DAVIES*

SUMMARY

The work presented is the result of an investigation into the high-speed compaction of metal powders. The objective of the experiments was to compare the slow-speed and the high-speed compaction processes with regard to pressing characteristics, ejection loads, dimensional accuracy and friction produced at the die wall. The final properties of the sintered components compacted at slow and high speed were also compared.

INTRODUCTION

The most commonly used technique for compacting metal powders involves filling a form die with the required powder, pressing it by means of a hydraulic or a crank press and then sintering the compact at a high temperature in a controlled atmosphere furnace. The process is widely used to manufacture components which are difficult to form by forging or machining, as well as components made of powder mixtures. Good material utilisation and low machining costs are characteristics of the powder metallurgy process, since parts can be produced directly with good surface finish and close tolerances. The disadvantages of the process are the relatively high cost of the powders and the high capital cost of the pressing machines and tools. As a result, the process is generally uneconomic unless very large numbers of compacts are required. A further disadvantage is the limitation imposed by the size of the compact: the bigger it is, the larger is the physical size of the press used and the higher its capital cost. It is mainly for this reason that the high energy rate (HERF) process is being developed, since high-speed presses are generally of smaller physical size and lower capital cost than conventional presses of the same capacity.

Explosives were the first technique to be used for compacting metal powders at high speeds. This was achieved either by firing an explosive charge in direct contact with the powder[1-3] or by using the explosive to fire a piston at a very high speed[4,5]. The danger of handling explosives and the low cycling times impose serious limitations on these techniques in production. However, very high densities of over 99 per cent theoretical have been reported[2,4]. Moreover, it has been claimed that with this technique, less energy was required to attain a certain density, and the dimensional accuracy was higher than for compacts produced by conventional methods.

The early attempts to use HERF machines for compacting metal powders under conditions similar to those of production were carried out using Dynapak machines[6,7]. These machines were used to produce very large compacts (5 in diameter by 0.9 in thick) having very high densities (100 per cent of the theoretical for electrolytic iron powder: 97.5 per cent for sponge iron powder). The physical properties of the ultra high density samples were very good and the

tensile strengths comparable to those of cold worked wrought steels of similar chemical analysis.

Some preliminary experiments were performed using Petro-Forge to compact a mixture of electrolytic copper and molybdenum disulphide powders[8]. Although this use of Petro-Forge enabled compacts to be produced with densities close to the theoretical, distortion and cracking were observed after subsequent heat-treatment. These defects were attributed to air entrapped in the compact during the high-speed compaction process.

A later investigation was carried out using the Petro-Forge machine to determine the effects of high impact speeds on the compaction of sponge iron powder[9]. The physical and mechanical properties of green and sintered components compacted at both high and slow speeds were compared, and it was found that the properties of components compacted at high speed were generally better than for similar components compacted at slow speed. The Petro-Forge machine was proved suitable for powder compaction, especially for operations requiring conventional presses of very large capacities.

The aim of the present work was to add more detailed information on the effects of high speed. A comparison is given between high- and slow-speed compaction processes regarding pressing characteristics, ejection loads, dimensional accuracy and friction produced at the die wall, as well as the effects of changing the weight of powderfill on the above-mentioned factors. The physical and mechanical properties of the sintered components produced by both compaction processes are also compared.

EXPERIMENTAL EQUIPMENT AND MATERIALS USED

Compaction machines

High-speed compaction was carried out on a Petro-Forge Mk 1 machine having a maximum energy rating of 7000 ft lbf at an impact speed of about 60 ft/s at the end of a 9 in stroke[10]. In these experiments, the machine was operated to deliver energies up to about 4000 ft lbf. After each experiment, a 50 tonf Denison hydraulic testing machine was used to eject the compact from the die and the ejection load measured.

* Department of Mechanical Engineering, University of Birmingham

The slow-speed experiments were carried out with 50 tonf and 300 tonf Denison hydraulic testing machines. The components compacted at slow speed were ejected using the 50 tonf machine, under conditions similar to those for the high-speed compacts.

Tooling

A die of $1\frac{1}{2}$ in bore diameter was used for producing cylindrical compacts by single end pressing. Pressure pads having a close tolerance with the die bore were located on the powder mass and hit by the moving punch, which was made 0.1 in undersize to avoid the need for accurate alignment with the die bore. The maximum weight of powderfill was about 200 g, which was approximately the limit allowed by the height of the die.

Instrumentation

The instrumentation was designed to record the compaction load—displacement diagram. The corresponding load transmitted to the bottom of the powder mass was also recorded, the circuit used being illustrated in figure 1. The compaction load was

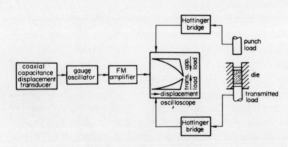

Figure 1. Block diagram of the instrumentation.

measured by four strain gauges mounted at the base of the punch, and the load transmitted to the base of the powder mass was measured by a loadcell located beneath the lower pressure pad. The displacement was measured by a capacitance-type displacement transducer. A plunger fixed to the upper moving platen moved along the bore of the transducer and the resulting change in capacitance was converted to a voltage by means of a gauge oscillator and an FM amplifier. A differential amplifier was used in the oscilloscope instead of the time base to enable the displacement signal to be indicated in the horizontal direction.

Sintering furnace

All sintering operations were carried out in a sintering furnace capable of providing sintering temperatures up to 1300°C with an accuracy of ±5°C. It consisted of a 4 in diameter furnace tube having a 9 in long uniform heat zone. Compacts were cooled to ambient temperatures in a watercooled jacket attached to one end of the furnace and having the same atmosphere. The atmosphere was cracked ammonia at a flow rate of 15 ft^3/h. The compacts were sintered for 40 min at a temperature of 1130°C.

Materials used

Hoganas sponge iron powder NC100-24 was used throughout the tests reported here. This was blended with 1 per cent of graphite powder and 1 per cent of zinc stearate as lubrication.

RESULTS

Pressing characteristics

Peak compaction pressures at slow and high speed are compared in figure 2 for a 200 g powderfill. For low green densities, less pressure is required to achieve

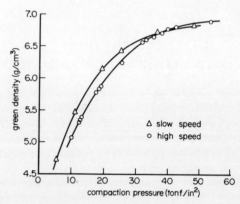

Figure 2. Variation of green density with compaction pressure for slow and high speeds. Material NC100−24 + 1% graphite + 1% zinc stearate: powderfill 200 g.

the same density in the case of slow-speed compaction, a trend that is reversed for green densities higher than about 6.8 g/cm^3.

The compaction energy was determined directly by measuring the area under the load—displacement curve using a planimeter. The compaction energy—green density relationship is illustrated in figure 3(a)

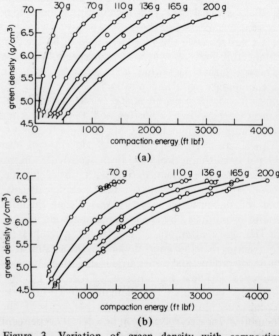

Figure 3. Variation of green density with compaction energy for different amounts of powderfill: (a) slow-speed compaction (b) high-speed compaction. Material NC100−24 + 1% graphite + 1% zinc stearate.

for various masses of powder compacted at slow speed. Similar curves for the same amounts of powder compacted at high speed are given in figure 3(b). From these figures it can be seen that for the same powderfill and the same density achieved, less energy is required for slow-speed compaction. However, the higher the density achieved, the less the difference between high-speed and slow-speed compaction processes.

A preliminary investigation was carried out to study the effects on pressing characteristics when compacting 200 g of powder by two successive equal blows. The peak compaction pressures were compared with those of slow-speed and high-speed single-blow compaction. These results show that the pressing characteristics of the two-blow compaction processes are intermediate between those of slow-speed and high-speed compaction processes. For low densities, the peak pressures required in the case of the two-blow compaction process are higher than those of the slow-speed process, but lower than peak pressures of the high-speed single-blow process. For high densities the trend is reversed.

Die wall friction

During a single end pressing process, the compaction load applied by the punch is not fully transmitted to the base of the powder mass. This is mainly due to the friction at the wall of the die bore. The variation of load distribution along the height of the compact is accompanied by similar variations in density and strength. The relation between the applied peak compaction load and the corresponding peak load transmitted to the base of the powder mass is given in figure 4 for both slow and high speeds. For

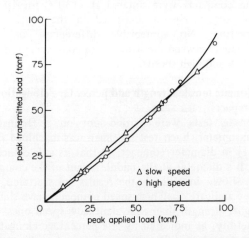

Figure 4. Variation of the peak load transmitted to the base of the powder with its simultaneous peak applied load. Material NC100−24 + 1% graphite + 1% zinc stearate: powderfill 200 g.

relatively small applied loads, the load transmitted to the base of the powder at slow speed is higher than that transmitted at high speed for the same applied load. This trend is reversed for high applied loads.

A better indication of the decay of the applied load along the height of the compact is given in figure 5. The ratio of the peak transmitted load to the simultaneous peak applied load is plotted against

green density for both slow and high speeds. For slow speed, this ratio is almost constant irrespective of the value of the green density achieved. For high speed, the ratio is lower than that for slow speed, should the density achieved be less than about

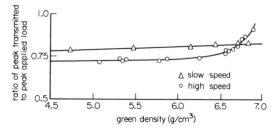

Figure 5. Variation of ratio of peak transmitted load to peak applied load with the green density. Material NC100−24 + 1% graphite + 1% zinc stearate: powderfill 200 g.

6.8 g/cm³. Conversely, high-speed compacts of densities higher than 6.8 g/cm³ would be expected to have a more even density distribution than those compacted at slow speed.

Ejection loads

In order to indicate the effect of compact density on ejection loads in a general form, irrespective of the powder mass compacted, the term *ejection pressure* is introduced, defined as the ejection load divided by the area of the cylindrical surface of the compact. The ejection pressures for both slow- and high-speed compacts are given in figure 6. The ejection pressures

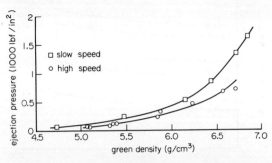

Figure 6. Variation of the ejection pressure with green density.

for high-speed compacts were always lower than those produced at slow speeds. It should be noted that high-speed compacts were not ejected immediately after being compacted, and thus any heat developed during compaction was dissipated before ejection.

Dimensional accuracy

It was found that the diameter of a green compact after ejection is not the same as the die bore diameter, but varied with the green density achieved. This is illustrated in figure 7 which shows the mean diameter of green compacts plotted against the corresponding green density for both slow and high speeds. The mean diameter was taken as the average of two readings taken across two perpendicular diameters at the middle height of the compact. It can be seen

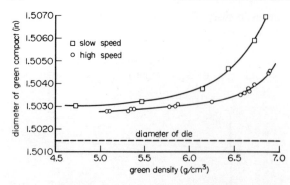

Figure 7. Variation of diameter of green compact with green density for both slow- and high-speed compacts. Material NC100–24 + 1% graphite + 1% zinc stearate: powderfill 200 g.

from figure 7 that the diameters of the compacts increase with increasing density for both slow- and high-speed compaction processes. Diameters of slow speed compacts are greater than those of high-speed compacts of the same density; the difference between both increases with increasing density.

Effect of varying the powderfill
In order to indicate the effects of varying the amount of powder on the compaction processes, the term *specific energy* is used, defined as the compaction energy per unit mass of powder. Specific energy would be expected to depend only on the density

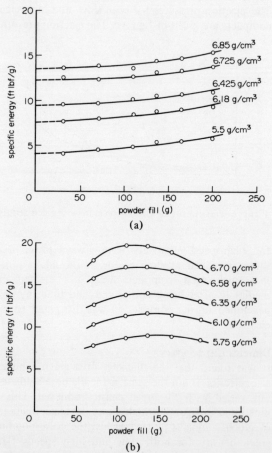

Figure 8. Variation of specific energy with powderfill for different green densities (a) slow speed (b) high speed. Material NC100–24 + 1% graphite + 1% zinc stearate.

achieved, and be independent of the amount of powder, but in practice part of the work done during the compaction process is used to overcome friction at the die wall, which increases with increasing weight of powder. The relationship between the specific energy and the weight of powder is given in figure 8(a) for different densities achieved with slow-speed compaction. The trend of the curves is in agreement with the above-mentioned expectations. The equivalent relationship for high-speed compaction is given in figure 8(b). Unlike the slow-speed curves, the specific energy first increases with increasing powder fill and then falls as the amount of powder is increased further. The higher the density achieved the smaller will be the powder weight at which the specific energy starts to decrease. This result agrees in general with the work of Gillemot[11].

Shrinkage of compacts during sintering
Shrinkage of slow- and high-speed compacts of pure iron with die wall lubrication has been studied. The compacts were sintered for 40 min at 1130°C. Mean values for the two measurements of diameter and four measurements of height were determined for each compact before and after sintering. It was found that shrinkage decreased with increasing density, varying from 2.4 per cent at a density of 5.31 g/cm³ to about 1 per cent at a density of 7.16 g/cm³ and that compaction speed has little effect on the shrinkage of compacts.

Microstructure
Micrographs were taken from pairs of samples, each pair consisting of samples made from a high- and a slow-speed compact of about the same density. The compacts were sintered at 1130°C for 40 min and the specimens sectioned in the longitudinal direction. No appreciable differences could be detected between the structure of compacts produced at low and high speeds.

Ultimate tensile strength and percentage elongation of compacts
Tensile tests were carried out on a Hounsfield Tensometer. Each test specimen was machined from a 1½ in diameter compact, so that its axis coincided with a diameter at the middle height of the compact. As shown in table 1, the results are scattered, but they indicate that high-speed compacts have slightly higher strength than equivalent slow-speed ones. The ductility, as indicated by the percentage elongation, was found to follow the same trend as the tensile strength.

DISCUSSION OF RESULTS

It was found from the experimental results that for high density components, improved pressing characteristics were obtained by the use of high speeds. High-speed compacts with densities higher than 6.8 g/cm³ require compaction loads lower than those of slow-speed compacts having the same densities. Moreover, the compacts produced at high speeds would have less variation of density and

TABLE 1
UTS and percentage elongation for compacts of different densities, sintered for 40 min at 1130°C.

slow speed			high speed		
sintered density (g/cm³)	ultimate tensile strength (tonf/in²)	percentage elongation	sintered density (g/cm³)	ultimate tensile strength (tonf/in²)	percentage elongation
5.72	9.2	4.82	6.16	17.5	7.50
6.24	16.5	6.27	6.35	15.25	6.27
6.47	14.5	5.80	6.50	17.0	6.52
6.57	15.5	5.54	6.63	14.50	5.55
6.75	15.0	5.80	6.69	14.25	6.27
—	—	—	6.57	14.0	5.55

properties, due to the more even pressure distribution in the pressing direction.

When analysing the specific energy — powder mass relationship at high speed shown in figure 8(b), it must be borne in mind that the higher the mass of powder used the higher will be the energy required to achieve a given density. Since the moving mass of the Petro-Forge machine is kept constant, this increase of energy is accompanied by an increase in impact velocity. The trend of the curves shown in figure 8(b) can be explained in terms of the increased velocity of compaction. The initial rise of these curves, which is at a faster rate than the equivalent slow-speed curves, is probably due mainly to inertia and strain-rate effects, which increase as the energy (and therefore the velocity) of compaction increases. Inertia effects result in less pressure being transmitted to the base of the compact as speeds increase, while strain-rate effects cause an increase in particle strength. Both these factors result in a greater energy requirement to achieve a given density. As the powder mass (and therefore the energy and velocity) increase in figure 8(b), the above trend is reversed and the specific energy decreases. This is probably due to two factors: firstly, both die wall and inter-particle friction decrease at the higher sliding speeds causing a decrease of energy required; secondly, any heat that is generated due to the deformation of particles, or generated to overcome friction between particles, remains concentrated near the surface because of the short time available for heat conduction, causing localised softening. These effects contribute towards a decrease of energy requirement with increasing compaction speed. The above factors suggest that optimum impact speeds should be somewhat higher than those used in the present work. The effects of increasing compaction speed also explain the variation of transmitted load under high-speed conditions as shown in figures 4 and 5.

The compaction of metal powders using several successive blows enables the useful working range of Petro-Forge to be extended considerably, since a relatively small machine can be used to compact very large components. This technique is not possible with conventional compaction process, due to their fixed stroke.

The high-speed compaction technique was also applied to other metal powders such as electrolytic, atomised and coarse iron powders, as well as stainless steel and prealloyed steel powders. The aim was to study the effects of high speed of compaction on a wide variety of metal powders used in industry. The early results indicate that high speed produces the same effects as those observed when compacting sponge iron powder.

The final properties of high-speed compacts are at least similar to those of slow-speed ones. Better dimensional accuracy and more-even property distribution at high densities are some additional advantages of the high-speed process. The fact that ejection loads might be higher if the components are immediately ejected, and the difficulty of controlling the component height in the pressing direction, are disadvantages of the high-speed process. However, it is clear that the high-speed technique can be adopted for the production of metal powder parts and that Petro-Forge is a suitable machine for achieving this.

CONCLUSIONS

(1) The Petro-Forge machine is suitable for the industrial production of metal powder components by the high-speed technique. The final properties of high-speed compacts were at least similar to those of slow-speed compacts. No traces of the effect of air, that might have been entrapped during the high-speed compaction process, were observed.

(2) For sponge iron powder, many advantages can be gained by using the high-speed technique to produce components with densities higher than about 6.8 g/cm³. Lower compaction loads and more-even load distribution along the height of the compact are some advantages of the high-speed compaction process. Similar trends were found with other metal powders.

(3) It is recommended that compaction should take place at higher speeds than for the present work, in order to widen the range of densities over which gains can be obtained.

(4) The feasibility of the multi-blow technique has been demonstrated. This technique can be employed to produce very large components having very high densities, which may be difficult or impossible to achieve by conventional methods. It can also be used to produce compacts of very high length-to-diameter ratios. The high-speed machines required are relatively small.

REFERENCES

1. R. W. LEONARD. Direct explosive compaction of powder materials. *Battelle Technical Review,* Nov.–Dec. (1968).

2. A. P. BOGDANOV and O. V. ROMAN. The structure and properties of pressing made from iron powder by dynamic pressing (in Russian), *Powder Metallurgy,* High Technical School, Minsk (1966).

3. W. T. MONTGOMERY and H. THOMAS. The compacting of metal powders by explosives. *Powder Metallurgy,* 6 (1960), p. 125.

4. E. M. STEIN, J. R. VAN ORSDEL and P. V. SCHNEIDER. High velocity compaction of iron powder. *Metal Progress,* 85, April (1964), p. 83.

5. G. GELTMAN. Explosive compacting of metal powders. *Progress in Powder Metallurgy,* 18 (1962), p. 17.

6. J. W. PRICE. Powder metal compaction by HERF. *Report No. MR-1159, Res. and Dev. Div.,* The National Cash Register Company, Ohio (1963).

7. D. W. BRITE and C. A. BURGESS. High energy rate pneumatically impacted UO_2–PUO_2 fuels. Presented to The Annual Meeting of The American Nuclear Society, San Francisco, California (1964).

8. R. C. FULLERTON-BATTEN. A preliminary investigation of high rate compaction of powders using the Petro-Forge. *Technical Report No. 66136,* Royal Aircraft Establishment, April (1966).

9. S. WANG and R. DAVIES. Some effects of high speeds in metal powder compaction. *Proc. 9th Int. Mach. Tool Des. Res. Conf.,* The University of Birmingham, September (1968).

10. L. T. CHAN, F. BAKHTAR and S. A. TOBIAS. Design and development of Petro-Forge high energy rate forming machines. *Proc. Inst. Mech.. Engrs.,* 180, (1965–66), p. 689.

11. L. GILLEMOT. The application of high energy rate densification in powder metallurgy. *Acta Technica Academial Scientiarum Hungaricae,* 64, March–April (1969), p. 427.

IMPULSE COMPACTING OF POWDER MATERIALS

by

P. A. VITYAZ* and O. V. ROMAN*

SUMMARY

The paper first classifies and reviews the various methods of impulse compacting of powders by 'gun' type units, hydrodynamic units, a high explosive and an electrical impulse. The principles of impulse compacting are discussed, and this is followed by a report on the influence of impulsive loading on physico-mechanical properties. Some experimental work is reviewed, and recommendations are given for the areas in which impulsive compaction is advantageous and where future developments will be directed.

INTRODUCTION

Recent years have seen intensified attention being paid to impulse methods of compacting powders. This has been prompted for two reasons. Firstly, owing to their inherent features, these methods are essential in some cases, particularly where no conventional method is available. Secondly, these methods can provide considerable economies where conventional presses of very large capacity would be required. Also, in some cases, certain types of component can only be manufactured satisfactorily by impulse methods, such as plated and multi-layered materials, large-size slabs and complex shapes.

Practical applications of impulse methods of compacting powders were first introduced during the 1960s, and have received attention in many countries including U.S.A., U.K., Germany, Japan, Canada and U.S.S.R. Developments of impulse compacting have mainly concentrated upon the production of high-density components, complicated shapes or large-size semi-finished articles which could be processed further by such methods as rolling, extrusion, drawing and forging.

METHODS OF IMPULSE COMPACTING

In spite of the availability of a large number of energy carriers and pressing methods, no attempt has previously been made to classify the methods, or the components which it is possible to produce by one of the methods. Below, a classification has been attempted on the basis of energy carriers and transmitting media. These are parameters which determine the velocities of wave propagation, and hence the pressure and impulse developed. It also takes into account the constructional details. Impulse compacting can be carried out by the following methods.

(1) Compacting in 'gun' type equipment.
(2) Compacting in hydrodynamic units.
(3) Compacting by a high explosive.
(4) Compacting by an electrical impulse.

(1) 'Gun' type units
Units utilising gun-powder (low explosive)[1,2] are either ballistic presses[3] or low explosive units for powder compaction[4]. For similar purposes high-speed

* Byelorussian Polytechnical Institute, Minsk, U.S.S.R.

pneumo-mechanical presses such as 'Dynapak'[5] or Petro-Forge machines[6], and also high-speed presses and hammers[7], are used. In general, these units involve the supply of kinetic energy directly or through some intermediate medium to the powder, resulting in its densification. The energy carrier may be a low explosive, compressed gases or inflammable mixtures. The powder can be pressed in solid dies.

The principle of this method is shown in figure 1 which shows an example of a low explosive unit[4].

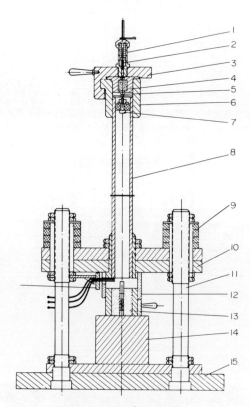

Figure 1. 'Gun' type unit.

The energy of the expanding explosive gases is given over to the impactor, 7, accelerating it along the bore, 8. The impactor impacts the punch without leaving the bore. The punch, in its turn, carries out the compaction of the powder in the die, 13. The speed of the impactor depends upon its mass and the amount of explosive charge, and varies between[4,8,9]

2 and 600 m/s. The large range of speed possible in such units make them very useful under laboratory conditions. Such units can be automated, and can be applied under industrial conditions for producing flat products. A large future application of units, such as hammers, explosive units, Petro-Forge and Dynapak machines, involves their use for the forging of powder preforms.

(2) Hydrodynamic units
The type of unit in which densification is carried out under impulse conditions in a completely closed chamber is generally referred to as a hydrodynamic unit. The design of these units resembles hydrostatic compaction presses, the difference being only the energy carrier for which low explosives are often used.

The literature refers to hydrodynamic units made by shortening the 14 inch barrels of naval guns[10]. In 1953 McKenna produced a patent on one type of hydrogen unit[11]. Schematically, the hydrodynamic unit is shown in figure 2. This type of unit depends

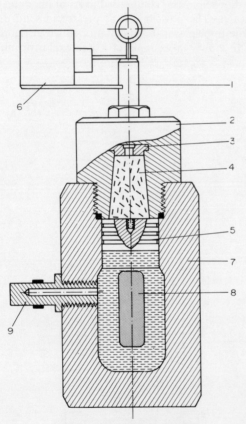

Figure 2. Hydrodynamic unit.

on the detonation of low explosive, 4, in a closed chamber, 7. This gives rise to high pressures, which can act either directly, or through a piston, 5, on the liquid. The powder, 8, inside the elastic bag is compressed by the pressure of the liquid.

From an analysis of the data given in the literature[12-14], it is evident that closed type hydrodynamic units preserve the advantages of hydrostatic units and have considerable industrial potential. The impulsive nature of pressure application makes it unnecessary to have complex constructions for hermetically

sealing working chambers filled with liquid. Also high-pressure pumps are no longer necessary. One must also take note of the fact that the design of the hydrodynamic units is simple, which reduces the cost of production.

At the Byelorussian Polytechnical Institute some experimental hydrodynamic units have been built with working pressures varying between 5 and 15 kbar and with the chamber diameters between 50 and 200 mm. Investigations were carried out on these units on the principles of densification, and on producing articles of complex shape from a range of powders.

Hydrodynamic units can be fully automated and used under industrial conditions for producing semi-finished articles of conventional or complex shape from various powders. The design and the construction must conform to the shape and size of the products, since pressure parameters may vary, depending upon the volume change of the powder mass during densification.

(3) Compacting by high-explosive energy
Explosive compacting refers to powder consolidation under the energy of high explosives. This can be further classified by the methods of applying the compacting load either by plane waves or by running waves. Methods vary according to the location of the explosive; compacting can take place with the charge in direct contact with the powder, through an intermediate plate or by a moving ram. The literature gives many methods and techniques for explosive compaction, the majority being for producing semi-finished parts such as slabs, tubes and rods[16-21]. A diagrammatic scheme for contact explosive compaction is shown in[16] figure 3. The high-explosive, 4, is in direct

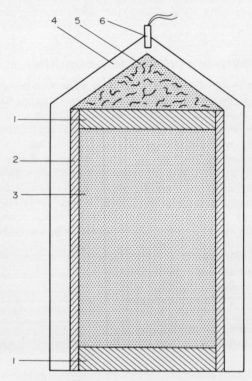

Figure 3. Scheme of contact explosive compaction.

contact with the powder 3, and the powder is compacted directly as a result of the explosive. 'Direct contact' also implies the presence of thin dividing containers, 2, the influence of which on compaction process can be neglected. However, contact explosive compaction has some disadvantages; firstly, the short direction of peak pressure, followed by rapid depressurisation to zero pressure, gives rise to non-uniformity of porosity along the cross section if insufficient charge is used[7]. On the other hand, an extremely large explosive charge results in the formation of innumerable fine cracks or central channels for asymmetrical articles.

Preparation of cylindrical or plate-shaped articles by explosive energy from plastic materials is also possible, but further mechanical treatment is very difficult to undertake because of defects in the microstructure. Thus, high-energy compaction of powder materials may more usefully be carried out at elevated temperatures. High-temperature high-speed compacting allows the possibility of compacting low-compressibility low-plasticity powders. This method may also be used to produce large-sized blanks for further mechanical treatment such as rolling and extrusion[22].

The use of high explosives reduces capital investments, and in many cases may completely replace high-power presses. This is particularly so because a high explosive drastically reduces the extent of the mechanical constructions necessary for generating high pressures, and makes it advantageous for producing very large semi-finished products. In order to obtain the optimum distribution and duration of pressure, the geometry of the charge, its distance from the powder and the properties of the intermediate medium have to be considered[22].

(4) Compaction by electrical impulse

The basic factor behind the use of electrical impulse energy of high voltage is the possibility of producing very high pressures. The 'mechanical' pressure can be generated by means of electrical impulses mainly in two ways; firstly, by high-voltage electric discharge in liquid and, secondly, by magnetic fields of high flux density. In spite of the appearance of these methods some years ago[23, 24] only recently have they been utilised in practice.

The block diagram showing the principles of electro-impulse units is shown in figure 4. It consists of two main parts: the impulse current generator (I.C.G.) and the converter of electrical energy into mechanical energy, called the working device (W.D.). There are also handling and control instruments, which are connected into a control block (C.B.), and various measuring gauges in the form of a measuring block (M.B.).

Depending upon the design of the W.D., the powder compaction may be carried out either by high-voltage discharge in a liquid[25] or by magnetic fields[26,27].

These devices can be readily mechanised and automated. The working element can be placed either inside or outside the powder container to give compaction by expansive or compressive fields. Compaction of complex shapes may be undertaken using plane or shaped inductors. If field concentrators are used, the applications of this method are considerably expanded, but with increasing loss of energy.

The main disadvantage of this method lies in the necessity of using a low-resistance electric conductor for the container shell. These shells are normally made from copper and aluminium tubes, which are difficult to use repeatedly, because of the large deformations involved. Obviously, the use of a new shell for each compact reduces the effectiveness of this method, and only when the cost of the shell becomes negligible in comparison with the total cost of the product will this method become an economic proposition. Multi-action shells, on the other hand, are more justifiable, and there are proposals for using liquid materials for this purpose. Production of bi-metals or special coatings constitute other possible variations.

SOME PRINCIPLES OF HIGH-SPEED DENSIFICATION OF POWDERS

The characteristic feature of high-speed compacting is its wave nature of loading and unloading. Obviously the actual process of powder densification presents a rather complex picture, particularly when reflected waves are involved. Under impulsive loading conditions both 'strong' and 'weak' shock waves may be generated in the powder mass. The mechanism of powder densification includes not only packing of particles, but also shear elastoplastic deformation. The complexity of the phenomena and the difficulty of measurement of the physio-mechanical properties during high-speed compaction do not allow for the description of the process by a single formula. However, within certain speed ranges, some relationships can be found which would describe the process of densification with sufficient accuracy.

The main problem of the physics of shock compression of matter is the determination of the functional relationship between Gruneisen parameters and volume. Unfortunately, when the substance is under the influence of high pressures and temperatures for a rather short period of time, the present theoretical method of determination of Gruneisen parameters is not applicable. The available semi-empirical relationships can only be used within very limited ranges.

Construction of Hugonoit curves involves complex technical problems. Hugonoit curves for metallic powders, as yet, do not exist, and thus their plotting by approximate methods is both essential and necessary.

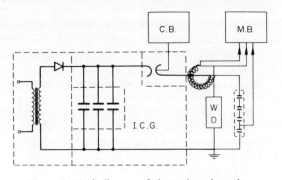

Figure 4. Block diagram of electro-impulse unit.

According to our postulation, such methods can be based upon the investigation of relationships between wave and mass velocities (u and v). As is known[28],

$$u = c_0 + \lambda v \qquad (1)$$

for compressible materials, where c_0 is the coefficient corresponding to the hydrodynamic value of velocity in still matter, and λ is the coefficient of dynamic compressibility.

For powder bodies, the function $u = f(v)$ in all ranges of velocity is of a non-linear nature, and can be expressed in the form of a stepped series. But, as the experiments show, on individual sections, the limits can be imposed by the first two members, i.e. to approximate it by analogous linear relationships. c_0 and λ are experimentally determined values, although they can be given a definite physical explanation, such as c_0 is the extrapolated value of wave velocity at infinite decrease of the wave amplitude and λ is the coefficient, taking into account the dynamic compressibility of the powder body.

Considering equation (1) and the laws of conservation of impulse and momentum,

$$p = \rho_0 u v \qquad (2)$$

and the law of conservation of mass,

$$v = u\left(1 - \frac{\rho_0}{\rho}\right) = u\left(1 - \frac{\gamma_0}{\gamma}\right) \qquad (3)$$

the approximate equation for the shock adiabatic becomes

$$\rho = \rho_0 c_0^2 \frac{1 - \gamma_0/\gamma}{1 - \lambda(1 - \gamma_0/\gamma)^2}$$

where ρ_0 or γ_0 and ρ or γ are material densities ahead of and on the wave front, respectively.

Thus to determine the functional relationship of equation (4) the values c_0 and λ must be known. They can be determined experimentally or by some approximated formulae.

Utilising the above method, the density distribution along the height of the compressed sample can be determined both with a contact explosive as well as with a moving plate on the powder mass. More details of this analysis are given in reference 22.

STUDIES OF THE PROCESS OF POWDER COMPACTION UNDER IMPULSIVE LOADING AND ITS INFLUENCE ON SOME PHYSICO-MECHANICAL PROPERTIES

The short process time is the basic feature of high-speed compacting, and must be taken into account in any analysis. The factors which are affected include that which involves the additional resistance of air entrapped between particles. Some preliminary estimations of the speed of escaping air from interparticle spaces were carried out on the basis of well-known laws of gas dynamics[29]. The results show that this speed is higher than the sound velocity in air at $\rho_0 = 1$ atm and $T_0 = 273$ °K, i.e. more than 330 m/s. Remembering that there is a large pressure in the 'air pockets', and that the powder particle mass velocity is less than the velocity of escaping air, it may be concluded that part of the entrapped air can

escape from the interparticle spaces. This means that entrapped air can only have an influence in high-density compaction.

The distribution of macrohardness and density along the height of the compact is one of the parameters which characterise a fundamental aspect of compacting and also the extent of mechanical deformation. The nature of the distribution along the height of an one-side-explosively-compacted compact depends upon the compacting conditions (weight of the explosive charge, the transmitting medium, the height of the powder layer, the base material, etc.)[30]. Such a density distribution is shown in figure 5, which

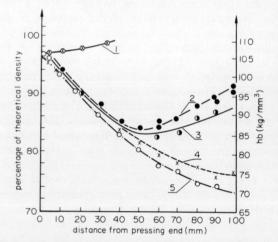

Figure 5. Variations of density and hardness along the section under explosive compaction of SAP: curves 1, 2, 4, density; curves 3, 5, hardness; curves 1, 2, 3, rigid base (steel plate); curves 4, 5 not rigid base (sand).

illustrates the large difference from that of the statically pressed compacts. The comparatively higher densities of the lower parts of the compacts (curve 1) can be explained by the generation of reflected waves from the rigid base.

In general, when the shock wave propagates from a medium of high compressibility to one of low compressibility, reflected and transmitted waves are generated at the boundary between the two media. Thus it follows that $A_1 < A_2$ ($A = c\rho$ is the acoustic resistance of the medium, where c is the sound velocity in the medium, and ρ is the density of the medium)[31]. When a shock wave is applied to the powder mass, the reflected wave from the steel base plate recompacts the already compacted material. When the shock wave passes from the low- to the high-compressibility medium, i.e. when $A_1 > A_2$, the reflected wave is one of rarefaction if the transmitted wave is compressive. In dynamically homogeneous media (i.e. the media are of the same compressibility) the reflected waves are not generated. The experiments confirm these principles.

In general, the properties of powder metallurgy products depend upon their densities, surface areas and quality of interparticle contacts. Static methods produce densities of the order of 80–90% of the theoretical one. Under such conditions the area of intermetallic contact is rather small and the properties (primarily tensile strength) depend mainly on mechanical interlocking of particles.

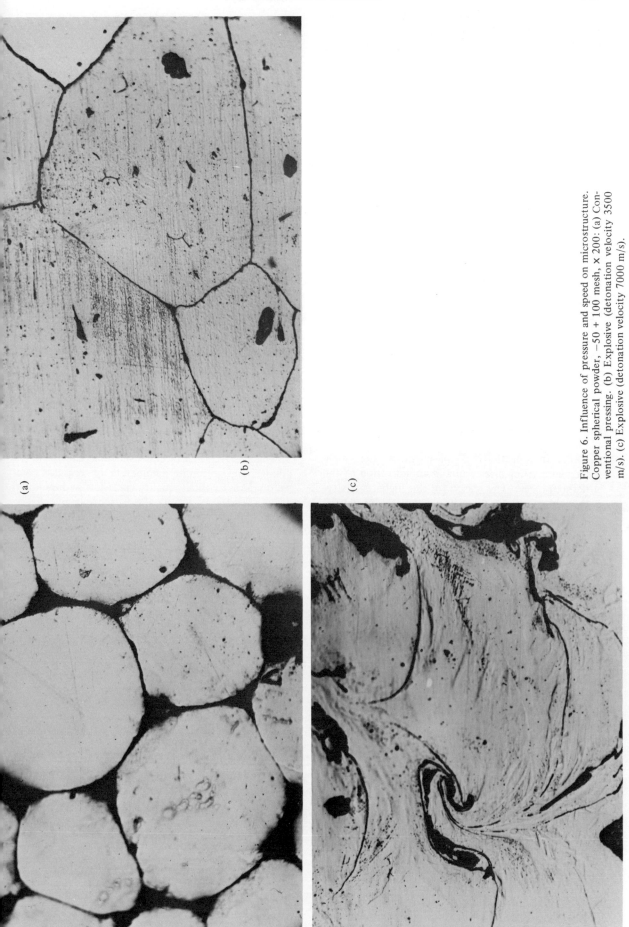

Figure 6. Influence of pressure and speed on microstructure. Copper spherical powder, −50 + 100 mesh, × 200: (a) Conventional pressing. (b) Explosive (detonation velocity 3500 m/s). (c) Explosive (detonation velocity 7000 m/s).

Previous work[6, 32, 33] has shown that iron powder compacts made by high-speed methods showed 10–15% higher tensile strength in comparison with those made statically. This difference increases with higher speeds of compacting, and with explosives this difference reaches a value of about 30%. The improvement of strength properties is mainly due to larger areas of intermetallic contact.

Samples prepared from spherical copper and silver powders pressed under different conditions showed a marked increase in metallic contact with increasing compacting speed, and also showed a different character of deformation (figure 6). The large compacting energy and localised deformation of the contact points cause the oxide film to be removed, resulting in the formation of large clean surfaces. High pressures on the contact points, which build up in microseconds, result in welding of the cleaned particle surfaces. Powder materials often show vortex type structure after being compacted under certain conditions by high explosives, because the removed oxide films are surrounded by the quickly deforming metal particles as shown in figure 6(c). Also, because of the high particle surface temperatures developed, the action of any entrapped air on the cleaned particle surfaces will cause an increase in the amount of oxides and nitrides present. These have a detrimental effect on the mechanical properties, and better results may be obtained by pressing in vacuum or in an inert atmosphere.

Other investigations[32,34] have shown that high-speed compacting also results in an activation of the sintering process. The reason for this higher 'activity' is thought to be due to larger quantities of structural defects in high-speed compacted samples in comparison with 'static' ones. Larger intermetallic contacts also play an important role and will accelerate the diffusion process during the early stages of sintering, when the oxide films have not yet fully formed. Investigations with 'Camca – MS – 76' powder confirmed the above phenomena at the contact zones[22].

During compaction of metal powders, as well as an increase of density, other effects take place in parallel; these are increases in the lattice distortion, which include both increased residual stresses and increased densities of dislocations.

X-ray methods for determining the residual stresses, block sizes and dislocation densities have been utilised[35], taking K_h for 110 and 220 lines. X-ray analysis showed that the residual stresses of the second order were 1.5 to 1.8 times greater for explosively compacted samples than for static ones[36]. Even larger values were expected, but, evidently, the increased temperature of contacts for explosive compacts reduced the value by annealing.

Increasing deformation results in reorientation of the lattice structure, both for explosive and for static samples, but, for the former, it takes place earlier. High temperatures are also developed during high-speed compaction resulting in some crystal growth. High-speed compaction also results in higher densities of dislocation; for explosive it is 2 to 3 times higher than 'static'[37].

CONCLUSIONS

Utilisation of impulse methods of loading opens wide technical possibilities for utilising new materials and obtaining special shapes of product.

Gun type units, such as Petro-Forge and Dynapak, can be used in industry for the production of comparatively small compacts with complex shapes, but with low height-to-diameter ratio. They can also be used for forging of powder preforms[37, 38].

Even larger industrial applications can be envisaged for hydrodynamic units using low explosive chambers which avoid the use of an expansive high pressure from 1 to 15 kbar. These units are cheaper than hydrostatic presses because of reduced rigidity required in the construction.

The following are some examples of the use of high explosive compacting.

(1) To obtain products from materials which are difficult or impossible to compact by conventional means.

(2) To obtain large size products or blanks for further processing.

(3) To improve the physico-mechanical properties, or to obtain new specific properties which are possible because of the special characteristics of high explosive compacting. A future development is the hot impulse compacting method.

Electrical impulse methods are specially suited for long asymmetrical products.

The researches now have the objective of determining the optimum parameters of impulse compacting so as to obtain quality compacts with required densities and configurations. A further objective is the wider application of newer methods in industrial fields.

ACKNOWLEDGEMENTS

The assistance of the staff of the Powder Metallurgy Laboratory of the Byelorussian Polytechnical Institute is gratefully acknowledged for their assistance in the preparation of this work. Thanks are also due to Dr. R. Davies for his assistance in the preparation of this manuscript and helpful discussions.

REFERENCES

1.   V. G. KONONENKO *et al.*, 1959. *Zh. Zavodskaya Lab.* (in Russian), No. 3.
2.   V. G. KONONENKO *et al.*, 1961. *Zh. Kuznechno Shtamovochoge Proiz.* (in Russian), No. 7.
3.   R. J. BREICHA and S. W. ME, 1962. *J. Am. Machinist*, **106**, 3.
4.   O. V. ROMAN and E. A. DOROSHKEVIR, 1968. *Poposhlovaya Metallurgiya, Rep. 9th Union Powder Metallurgy Conf.* (in Russian), Riga, U.S.S.R., p. 68.
5.   'High-speed deformation of metals' (in Russian), 1966. *Izd. Mashinostroyenie.*
6.   R. DAVIES and E. R. AUSTIN, 1970. *Developments in High-speed Metal Forming,* The Machinery Publishing Co. Ltd., London.
7.   Ju. DOROVEYEVE (Ed.), 1966. *Application of Dynamic Metal Ceramic Methods in Briquetting and Powder Metallurgy* (in Russian), Rostovsicoye Knizhnoye Izdatelstvo.
8.   *Metalworking Prod.*, 1967. 1, Feb., 48–53.

9. L. F. MAKSIMENKO *et al.*, 1970. *Poroshkovaya Metallurgiya.* No. 6 (in Russian).
10. *Iron Age*, 1952. 26th Sept., 70, 56.
11. *Patent U.S.A.*, 1953. No. 2648125.
12. L. N. AFANASEV, P. A. VITYAZ and O. V. ROMAN, 1966. *Reps. 8th Union Powder Metallurgiya Conf.* (in Russian), Minsk, U.S.S.R.
13. L. F. AFANASEV and O. V. ROMAN, 1967. *Collection of articles of Novocherkessk Polytechnical Institute* (in Russian), **173.**
14. L. N. AFANASEV and O. V. ROMAN, 1968. *Collection of Articles of Byelorussian Polytechnical Institute* (in Russian), No. 2.
15. P. A. VITYAZ and O. V. ROMAN, 1968. *Collection of Articles of Byelorussian Polytechnical Institute* (in Russian), No. 2.
16. R. W. LEONARD, 1968. *Battelle Tech. Rev.,* **7,** No. 10.
17. C. C. SIMONS, 1965. *Patent U.S.A.,* **No.** 3220103.
18. R. E. LENHARD, 1960. *Patent U.S.A.,* No. 2943933.
19. O. V. ROMAN, 1968. *J. Inst. Eng., India,* **48,** No. 11, Pt. MM3.
20. O. V. ROMAN, 1968. *Jerkont. Ann., Sweden,* **152.**
21. J. FELTMAN, 1967. 'Explosive compacting of metal powders', *New Methods For the Consolidation of Metal Powders,* Plenum Press, p. 75.
22. O. V. ROMAN (Ed.), 1971. *Modern Methods of Making Powder Products* (in Russian), Izdatelstvo 'Polymya', Minsk, p. 88.
23. C. Z. SUITS, 1936. *Cen. E.J. Rev.,* **39,** 430.
24. P. KAPITZA, 1923. *Proc. Cambr. Phil. Soc.,* **21,** 511.
25. I. M. KABELSKI *et al.,* 1966. *Zh. Promyshlennost of Byelorussian* (in Russian), No. 6.
26. TU. K. BARBANOVICH and T. POROSHKOVAYA, 1969. *Metallurgiya* (in Russian), No. 10.
27. T. DONALD, 1969. *Sandstram Metal Progr.,* **86,** No. 3, 91.
28. U. N. ZHARKOV and V. A. KALININ, 1961. *Equations of solid state under high pressure and temperature* (in Russian), *Izdatelstvo 'Nauka',* Moscow.
29. *Machien Builder Handbook,* Vol. II (in Russian), 1961. Izdatelstvo 'Mashgiz', Moscow.
30. A. P. BOGDANOV and O. V. ROMAN, 1966. *Rep. 8th Union Powder Metallurgiya Conf.* (in Russian), Minsk, U.S.S.R., p. 228.
31. L. P. ORLENKO, 1964. *Material Behaviour Under High Dynamic Stresses* (in Russian), Izdatelstvo 'Mashinosdroyenie', Moscow.
32. E. A. DORSHKEVICH and O. V. ROMAN, 1966. *Rep. 8th Union Powder Metallurgiya Conf.* (in Russian), Minsk, U.S.S.R., p. 161.
33. E. A. DOROSHKEVICH and O. V. ROMAN, 1969. *Novaye Technica and Progresivnaya Technologiya* (in Russian), Izdatelstvo 'Vyshaya shkola', Minsk, p. 276.
34. O. V. ROMAN and E. A. DORSHKEVICH, 1966. *Rep. 2nd Intern. Powder Metallurgy Conf.* (in Russian), Czechoslovakia.
35. A. IU. POGORASKI (Ed.), 1961. *X-rays in Physical Metallography* (in Russian), Izdatelstvo 'Metallurgiya', Moscow.
36. O. V. ROMAN, L. N. AFANOSEV, A. P. BOGDANOV and P. A. VIYAZ, 1970. *Rep. 3rd Intern. Powder Metallurgy Conf.* Czechoslovakia (in Russian), Vol. II, part I, p. 132.
37. R. DAVIES and R. H. T. DIXON, 1971. *Powder Metallurgy,* **19,** No. 28, 207.
38. J. B. MARX, R. DAVIES and T. L. GUEST, 1971. 'Some considerations of the hot forging of powder preforms', *Advan. Machine Tool Design Res.,* Pergamon Press, Oxford and New York.

PRELIMINARY INVESTIGATIONS OF THE COLD EXTRUSION OF POWDER PREFORMS

by

A. SINGH* and R. DAVIES*

SUMMARY

This paper reports on an investigation into the possibility of cold extruding billets which have been pressed and sintered from various metal powders. Powders selected for forward and backward extrusion were electrolytic iron, sponge iron, atomised iron and atomised–reduced iron. The extrusions were performed at slow and high speeds. The effect of speed on the extrudability of these powders has been studied.

INTRODUCTION

The forging of powder preforms has been considerably developed during the past few years. The main stress has been given to forgings produced at high temperatures, and only a small amount of information is available on the forging of powder preforms at room temperature. Some work has been carried out by Antes[1], who studied the formability and densification characteristics of powder preforms under compression for sponge iron and atomised iron at room temperature. He found that sponge iron required greater stress for a given strain than the atomised iron preforms. At higher densities the sponge iron preforms exhibited a greater tendency to crack under compression than atomised iron. Nakagawa et al.[2] extruded reduced iron at room temperature. The density of the forward and backward extrusions with a reduction ratio of 9 and 2.75 respectively was found to vary from 96 to 99% of theoretical, when the initial density of the preform was in the range of 83 to 86%.

With the development of powder forging techniques it is possible to achieve near full densities in hot forged parts, with properties comparable with those obtained in parts made from solid material. These developments have opened up possibilities of producing stressed parts from metal powders that were previously considered impossible. The aim of the present work is to extend the range of powder forging by forming the preforms in the cold condition. Extrusion has been chosen as the forming process in these preliminary investigations, but there is no reason to suppose why cold forming processes such as heading and upsetting could not also be carried out satisfactorily, provided the comparatively brittle nature of the preforms is taken into account in the design of dies and in the design of the preforms.

The possible advantages of forming powder preforms in the cold state when compared with hot forming are similar to those which apply to conventional cold forming of solid metals, namely improved surface finish, improved dimensional accuracy and improved strength due to work hardening. A further advantage which applies to cold forming of powders is that preforms are not exposed to air in their hot state, as in the hot forging of powder preforms, thus eliminating external and internal oxidation problems.

The present paper is mainly concerned with the feasibility and limitations of cold extrusion of preforms pressed from iron powders. Future work will concentrate on more detailed aspects of the subject and comparisons with hot formed components, including those produced from solid materials.

EQUIPMENT

Compaction and forging machines

The powder preforms were pressed on a 50 tonf hydraulic press. High-speed forward and backward extrusions were carried out on a MK IIA Petro-Forge machine. The velocity range of the Petro-Forge was 23 to 30 ft/s. The slow-speed extrusions were performed on a 200 tonf eccentric press. The velocity at the beginning of extrusion was 0.8 ft/s.

Tooling

Tooling for the compaction of powder preforms is shown in figure 1(a). The container bore and punch were of 1 in nominal diameter. Tooling for backward extrusion of sintered compacts is shown in figure 1(b). The container, A, was located in the main body, B,

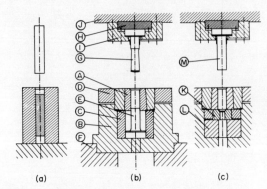

(a)   (b)   (c)

Figure 1. General layout of tooling.

backed by the ejector sleeve, C, and clamped by the ring, D. The ejector punch, E, guided by the ejector sleeve was operated by an hydraulic cylinder mounted under the fixed platen, F, of the machine. The backward extrusion punch, G, backed by the pressure

* Department of Mechanical Engineering, University of Birmingham

pad, H, was clamped by the holder, I, fixed to the moving platen, J, of the machine.

For forward extrusion, figure 1(c), the container, K, was located in the main body, C, and was backed by pressure pad, L. The forward extrusion punch, M, was clamped to the moving platen, J, in an identical manner to the backward extrusion punch.

Extrusion punches, container and die inserts and ejector punch were made of Carrs 69S high carbon, high chromium tool steel heat treated to 56–58 Rc. The forward extrusion dies were provided with 120° die entry angle. The forward extrusion dies gave reduction ratios of 1.78, 2.56, 3.30 and 4.00, and backward extrusion punches gave reduction ratios of 1.56, 1.90, 2.32 and 2.90.

Test materials and powder preforms

Materials selected for forward and backward extrusion tests were Sintrex – electrolytic iron 100 mesh, MH100 – Hoganas sponge iron, AHC100 – Hoganas atomised iron and MP32 – Domtar atomised iron powder.

Preliminary tests were conducted to select the initial density and sintering temperature of the powder preforms. Powders were compacted to various pressures in order to study the effect of initial density of the preform on the extrudability of these powders. The density of the preform was determined by weighing and measurement. The die wall was lubricated by a mixture of zinc stearate + acetone. For densities less than 6.5 g/cm³, surface cracks were observed on the extruded billets. Satisfactory results were obtained for densities higher than 6.5 g/cm³. For the present tests therefore, the powder was

compacted under a pressure of 38 tonf/in², which gave a sintered density range of 6.75 to 6.95 g/cm³.

Sintering of powder preforms

Preliminary tests showed that for sintering temperatures below 1000 °C surface cracks were observed on the extruded billets. For sintering temperatures above 1000 °C satisfactory results were obtained, so a sintering temperature of 1100 °C was selected for these tests. The powder preforms were sintered at this temperature for 50 min in a tube furnace provided with an atmosphere of cracked ammonia. The preforms were cooled to room temperature in the cooling zone of the furnace.

Lubrication

Graphite lubricant was used in all the extrusion tests. Graphite was sprayed uniformly on to the billet before extrusion.

EXPERIMENTAL RESULTS AND DISCUSSION

Extrusion pressures

Figure 2(a) and 2(b) show the curves for maximum extrusion pressure against reduction ratio for Sintrex, MH100, AHC100 and MP32 iron powders at slow and high speed. Extrusion pressures vary almost linearly with reduction. They are higher at high speed than at slow speed, probably because pure irons are highly strain rate sensitive. The difference in extrusion pressure is more noticeable in backward extrusions. This is possibly due to the effect of frictional forces between the container and the billet material, which

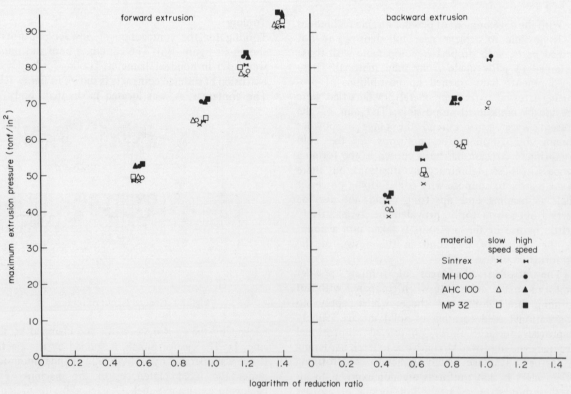

Figure 2. Variation of extrusion pressure with reduction ratio for Sintrex, MH100, AHC100 and MP32 powders at slow and high speeds.

are negligible in backward extrusion compared with those in forward extrusion. The effect of the increase of speed is to reduce the frictional forces, and this would have more effect in the reduction of pressure in forward extrusion than in backward extrusion. A further factor which will contribute towards an increase in extrusion pressure is poor lubrication, which was observed particularly in backward extrusions. The effect of lubrication is discussed later.

Extrusion pressures for Sintrex powder preforms are seen to be slightly lower than for the other materials. A small difference is also observed in the extrusion pressures for AHC100, MH100 and MP32 iron powders; there is a slight tendency for extrusion pressure to increase when these materials are considered in that order.

Effect of billet geometry

In the forward extrusion of cylindrical billets, surface cracks were observed, which are shown in figure 3. The cracks were found for all the reductions at slow speed and for 1.78 reduction ratio at high speed. Figure 3(c) shows a typical crack for a reduction ratio of 2.56 at slow speed. When the preforms were provided with $\frac{1}{8}$ in long and $120°$ taper at the leading end of the preform, figure 3(a), no crack was observed for reduction ratios of 2.56 and above. However, the

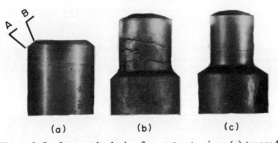

| (a) | (b) | (c) |

Figure 3. Surface cracks during forward extrusion: (a) tapered preform; (b) reduction ratio 1.78; (c) reduction ratio 2.56.

cracks were still present for a reduction ratio of 1.78, figure 3(b), irrespective of the use of tapered billets. Similar cracks were observed at high speed for a reduction ratio of 1.78. The geometry of the billet did not effect the extrusions at high speed. It therefore appears that in forward extrusion there is a limiting lower reduction below which it is difficult to extrude successfully.

The cracks observed in forward extrusion corresponded to the change of surface geometry at A and B as shown in figure 3(a). The cracks are probably due to a breakdown of lubricant film at these two points. At slow speed, cracking was more severe, which is due to high friction between the powder preform and dies.

Density of the extruded billet

The extruded billets were cut into two portions, the extruded and the undeformed parts. The density was determined by measuring weights in air and distilled water.

Figure 4 shows the variation in the density of the extruded and undeformed parts for forward and back-

ward extrusions of Sintrex iron powder at various reductions. The results show a tendency for the density at slow speed to be slightly higher than at high speed. This is possibly the effect of inertia and frictional forces acting during the process. Another factor which could effect the densities is the effect of speed on the material flow at the beginning of

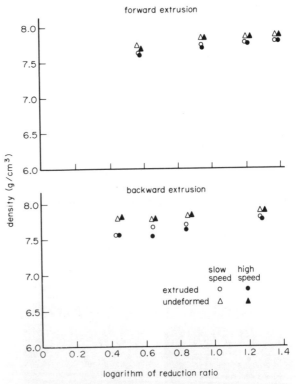

Figure 4. Variation of density with reduction for forward and backward extrusions of Sintrex iron powder.

extrusion. The first portion of material to be extruded undergoes a relatively small amount of deformation, which means that the material is extruded in a state close to its original form; this reduces the average density of the extruded part to some extent. The amount of this portion of the extruded part depends upon the friction level, which varies with speed. At high speed friction is decreased, so more undeformed metal passes the die in the beginning of extrusion than at slow speed.

TABLE 1.

Material	Slow speed		High speed	
	Extruded	Undeformed	Extruded	Undeformed
Sintrex	7.72	7.82	7.70	7.80
MH100	7.57	7.71	7.60	7.72
AHC100	7.70	7.76	7.67	7.71
MP32	7.70	7.80	7.70	7.80

Forward extrusion   Reduction ratio 2.56

Backward extrusion   Reduction ratio 1.9

Sintrex	7.72	7.80	7.61	7.80
MH100	7.57	7.76	7.40	7.75
AHC100	7.70	7.71	7.52	7.80
MP32	7.71	7.79	7.59	7.80

The density of the undeformed part at slow and high speeds were similar in both forward and backward extrusion, and was of the order of 99.2% at higher reductions. The density of the undeformed part was slightly higher than the extruded part. This is possibly due to the fact that the average density of the extruded part was lowered by the relatively undeformed part at the beginning of extrusion.

Table 1 compares the densities of the forward and backward extruded and undeformed parts at slow and high speeds for Sintrex, MH100, AHC100 and MP32 iron powders. MH100 preform extrusions gave the minimum density and Sintrex gave a slightly higher density for both extruded and undeformed parts.

Metallurgical examination

Extruded preforms were sectioned longitudinally, polished and etched for metallurgical examination.

Figure 5 shows some typical cracks observed in backward extrusions of Sintrex iron powder at slow and high speeds. These cracks are formed because of the formation of a dead metal zone or unsteady flow in the plastically deforming region. The cracks were present for a reduction ratio of 1.56 at both speeds, figure 5(a) and 5(b), and for a reduction ratio of 1.9

(a)　　　　　　(b)

(c)

Figure 5. Typical cracks in backward extrusions of Sintrex iron powder: (a) slow speed, reduction ratio 1.56; (b) high speed, reduction ratio 1.56; (c) high speed, reduction ratio 1.9.

at high speed, figure 5(c). The unsteady flow, figure 5(c), is the result of high friction at the surface. It was observed that graphite provided poor lubrication in backward extrusions, particularly at high speed. This is one of the reasons for the higher extrusion pressures in backward extrusions at high speed.

Figure 6 is a microstructure showing cracks observed in forward extrusion at slow speed for a reduction ratio of 1.78. These shear cracks on the

surface of the extruded billet are formed by the presence of high friction in the tapered zone AB (figure 3(a)) of the powder preform. Friction at the deforming region is considerably affected by the formation of a dead metal zone, which can be

Figure 6. A typical shear crack in forward extrusion at slow speed and a reduction ratio of 1.78.

reduced considerably by the correct choice of die entry angle. These cracks may be reduced to some extent by using better lubrication, a higher powder preform density and the optimum die entry angle.

The effect of impurities such as iron oxides on the material flow was studied. MH100 powder was chosen because of its relatively high oxide content compared with the other powders. A typical result is shown in figure 7 for backward extrusion at high

Figure 7. Material flow near the surface of a backward extrusion of MH100 at high speed and a reduction ratio of 1.9.

speed and a reduction ratio of 1.9. A similar flow pattern was observed for forward extrusion at both speeds. The impurities present in the iron powder only deform to a relatively small extent during extrusion and cause obstruction to the flow of material. The obstructed flow could cause a slight increase in extrusion pressure. Moreover these impurities reduce the density of the extrusion to some extent. It could be concluded that materials with impurities or oxides such as MH100 would be less suitable for extrusion purposes. No noticeable difference in the properties of AHC100 and MP32 was observed. Material flow in extrusions of Sintrex iron powder was uniform and no cracks were observed because of the absence of impurities. The densities of the extruded parts of Sintrex extrusions were higher than of other materials.

CONCLUSIONS

(1) Extrusion pressures were found to be higher at high speed than at slow speed; considerable differences were observed in backward extrusions. The lowest extrusion pressures were observed with Sintrex powder preforms at slow and high speeds.

(2) In forward extrusion, the geometry of the billet did not affect the properties of the extruded billets at high speed. At slow speed, billets which were tapered to fit the die angle were necessary to achieve satisfactory extrusions.

(3) In the forward extrusion at slow and high speed it was observed that there was a lower limiting reduction ratio, below which it would be difficult to extrude successfully.

(4) The density of the undeformed part was slightly higher than the extruded part. Highest densities were observed with Sintrex, which were of the order of 99%. MH100 powder gave the lowest densities of the extruded and undeformed parts, the lowest density being 95%.

(5) The effect of impurities in the iron powder was to make the flow less uniform. For extrusion purposes atomised or electrolytic powders would be most suitable.

ACKNOWLEDGEMENTS

The authors would like to thank Mr. T. L. Guest, Mr. M. K. Bhattacharjee and Mr. B. J. Beighton for their help in the experimental work and metallurgical assistance. This work is financed by a Science Research Council grant.

REFERENCES

1. H. W. ANTES, 1971. 'Cold forging iron and steel powder preforms', *Modern Developments in Powder Metallurgy, Vol. 4 – Processes, Plenum Press.*
2. T. NAKAGAWA *et al.*, 1971. 'Cold forging of ferrous M/P billet', *27th Symp. Japan Soc. Powder and Powder Metallurgy, Tokyo.*

ON THE COLD FORGING OF SINTERED IRON POWDER PREFORMS

by

T. NAKAGAWA*, T. AMANO*, K. OBARA†, Y. NISHINO† and Y. MAEDA‡

SUMMARY

In order to increase the toughness of sintered ferrous products and to produce more complicated shapes, cylindrical sintered billets prepared from reduced iron powder were forged at room temperature. Generally it is considered impossible to cold form a powder even after sintering, but according to our experiments it was found that sintered billets can be forged successfully and various kinds of products can be obtained. Cold forgeability of sintered iron powder and the effects of shape and reduction are discussed in this paper. Also density and hardness distributions of the forged products were investigated in detail, and their microscopic material flow was observed.

1. INTRODUCTION

Recently P/M hot forging, one method of producing fully dense sintered products, has been developed. By this method products with a nearly theoretical density and excellent mechanical properties can be prepared. On the other hand, it should be noted that there are problems similar to those occurring in the ordinary hot forging of bulk metal; these are die wear, lubrication, dimensional accuracy and surface quality of products, production speed, etc. These problems would be avoided by using cold forging instead of hot forging. However, sintered powder preforms were considered to be too brittle to be formed, and there was no experience of whether forging is possible or not at room temperature. Therefore, cold forging of sintered iron powder preforms was examined in this paper and its possibility was discussed.

2. SINTERED IRON POWDER BILLETS

The sintered billets were produced from the reduced iron powder KIP-270 without any other alloy element. KIP-270 is the millscale reduced iron powder with high purity. Table 1 shows the properties of this iron powder.

TABLE 1
Details of reduced iron powder KIP-270 used in this experiment

	Fe	bal.
Chemical composition %	Si	0.09
	C	---
	Mn	---
	H −loss	0.27
Screen	−100/+325	73
Analysis (%)	−325	27
Apparent density (g/cm³)		2.6

Cylindrical billets for cold forging have been employed with a diameter of 30 mm and a height of 30 mm; the microstructure of the billet in cross section is shown in figure 1. Billets were compacted by the floating die method with a lubricant of 0.7 wt. %

(a)

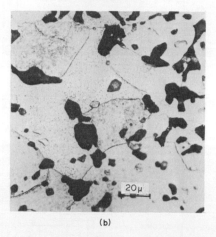

(b)

Figure 1. Size of the sintered cylindrical billet and its metallurgical structure compacted at a pressure of 6 ton/cm² from the millscale reduced iron powder KIP-270: (a) size of sintered billet; (b) photomicrograph of metallurgical structure of sintered iron powder billet.

TABLE 2
Mechanical properties of sintered compacts
(compacting pressure = 6 ton/cm²)

Density (g/cm³)	6.75
Hardness (Hv)	53.2
Yield point (kg/cm²)	9.10
Tensile strength (kg/cm²)	16.3
Elongation (%)	14.1

* Institute of Industrial Science, University of Tokyo, Roppongi, Minatokiu, Tokyo, Japan
† Central Research Laboratory, Mitsubishi Metal Mining Co. Ltd., Omiva. Saitama, Japan
‡ Miyamoto Kogyo Co. Ltd., Takinogawa, Kita-ku, Tokyo, Japan

zinc stealite and sintered at 1150 °C for 30 min in an atmosphere of ammonium cracking gas in a pusher type furnace. The compacting pressure of billet varies from 4 ton/cm² to 8 ton/cm², and the average density and mechanical properties of compact are shown in table 2.

The compacting and sintering conditions of these billets do not differ from those of ordinary powder forming.

3. COLD FORGING OF IRON POWDER PREFORMS

In order to investigate the restrictions of the shape in the cold forging of sintered iron powder preforms, seven kinds of standard forging have been examined; these are upsetting, upsetting with forward extrusion, forward extrusion, backward extrusion, two kinds of complex extrusion and tube extrusion. Among these forgings, the effects of reduction has been investigated

in the cases of forward and backward extrusions which are considered as an elementary forging. All of the experimental conditions are illustrated in detail in table 3. Billets for tube extrusion, backward and complex extrusion-A have been prepared by machining from original billets.

The tool construction is the usual one; the punch material is high-speed steel (JIS-SKH9) and the die material is special tool steel (JIS-SKS4). A 400 ton mechanical toggle press was used for cold forging. The forging load was measured by two strain gauges stuck on the punch. As a lubricant for cold forging Boderite—Bonderlube was used, and billets after lubrication were dried in order to absorb the moisture contained in the billet pores.

4. RESULTS AND DISCUSSION

4.1. Cold forgeability of sintered iron powder preforms

Cold forged products should be assessed from every

TABLE 3
Details of cold forging experiment

Kinds of forging	Forming shapes		Reduction (%)	Size of billet (mm)
	Shapes	Size		
forward extrusion		1 $d_1 = 30.5^\phi$ $d_2 = 20.0^\phi$ $O = 120°$	38.5	cylindrical
		2 $d_1 = 30.5^\phi$ $d_2 = 10.0^\phi$ $O = 120°$	88.9	$30.5^\phi \times 29.6$
backward extrusion		1 $d_1 = 29.6^\phi$ $d_2 = 15.6^\phi$ $h = 6$	26.4	cylindrical $29.3^\phi \times 20$
		2 $d_1 = 29.6^\phi$ $d_2 = 23.6^\phi$ $h = 5$	63.6	cylindrical $29.3^\phi \times 20$
		3 $d_1 = 29.6^\phi$ $d_2 = 26.6^\phi$ $h = 6$	88.3	cylindrical $29.3^\phi \times 12$
tube extrusion		$d_1 = 30.5^\phi$ $d_2 = 12.0^\phi$ $d_3 = 16.0^\phi$ $\theta = 90°$	85.8	hollow cylindrical inner diam. 12.0^ϕ outer diam. 30.5^ϕ height 29.6
complex extrusion – A		1 $d_1 = 29.6^\phi$ $d_2 = 26.6^\phi$ $d_3 = 12.2^\phi$ $h = 2.8$	80.3 (forward) 83.0 (backward)	cylindrical $29.3^\phi \times 12$
complex extrusion – B		2 $d_1 = 30.5^\phi$ $d_2 = 24.5^\phi$ $d_3 = 27.5^\phi$ $\theta = 20°$	64.5 (backward)	
		3 $d_1 = 30.5^\phi$ $d_2 = 24.5^\phi$ $d_3 = 16.5^\phi$ $\theta = 20°$		
upsetting				cylindrical $30.5^\phi \times 29.6$
upsetting extrusion		1 $h = 7.2$ $d = 20^\phi$		
		2 $h = 7.2$ $d = 10^\phi$		

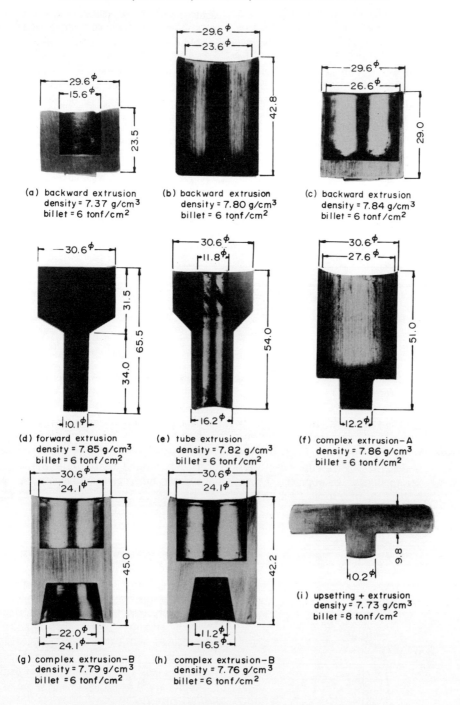

(a) backward extrusion
density = 7.37 g/cm³
billet = 6 tonf/cm²

(b) backward extrusion
density = 7.80 g/cm³
billet = 6 tonf/cm²

(c) backward extrusion
density = 7.84 g/cm³
billet = 6 tonf/cm²

(d) forward extrusion
density = 7.85 g/cm³
billet = 6 tonf/cm²

(e) tube extrusion
density = 7.82 g/cm³
billet = 6 tonf/cm²

(f) complex extrusion-A
density = 7.86 g/cm³
billet = 6 tonf/cm²

(g) complex extrusion-B
density = 7.79 g/cm³
billet = 6 tonf/cm²

(h) complex extrusion-B
density = 7.76 g/cm³
billet = 6 tonf/cm²

(i) upsetting + extrusion
density = 7.73 g/cm³
billet = 8 tonf/cm²

Figure 2. Successful cold forged products of sintered iron powder billets.

(a) forward extrusion
reduction = 38.5%
billet = 4 tonf/cm²

(b) forward extrusion
reduction = 38.5%
billet = 8 tonf/cm²

(c) upsetting + extrusion
extrusive diameter = 17ϕ
billet = 4 tonf/cm²

(d) upsetting by flat die
lubricant: lanolin, anhyd
billet = 6 tonf/cm²

Figure 3. Unsuccessful cold forged products of sintered iron powder billets.

industrial point of view; however, they were evaluated in this paper from the crack formation observed in or on the forged products. Figure 2 shows several successful forged products and figure 3 shows the unsuccessful ones in which some cracks or defects were observed. All the experimental results are classified into four groups and summarised in table 4.

the origin and propagation of cracks is most probably explained by the existence of pores and weak grain boundaries. Besides, cracks take place exclusively in forged products with low reduction. This tendency can also be observed in ordinary cold forging of conventional steel which has rather little deformability. From the point of view of crack formation it

TABLE 4
Classified results of cold forging of sintered iron powder billets

Kinds of forging		Forward extrusion		Backward extrusion			Tube extrusion	Complex extrusion			Upsetting + extrusion	
No.		1	2	1	2	3		A	B–1	B–2	1	2
Reduction (%)		38.5	88.9	26.4	63.6	80.3	85.8					
Billet	4 ton/cm²	✕	◎	✕	◎	◎	○	◎	◎	◎	✕	△
	6 ton/cm²	✕	◎	△	◎	◎	○	◎	◎	◎	✕	△
	8 ton/cm²	✕	◎	◎	◎	/	/	/	◎	◎	/	○

Note. The results of forging are classified into four groups by following marks: ◎, perfect; ○, no crack, but rough surface; △, microcrack; ✕, crack; ╱, no experiment.

Judging from these results, it was proved that the sintered iron powder preforms have sufficient plastic deformability for forging at room temperature. Such a surprisingly good cold forgeability cannot be explained by the extension of the tensile test, but it seems that the sintered powder has enough deformability in itself under the high compressive stress state; also this deformability is supported by the coining process which exists prior to the cold forging.

Most of the cracks are observed in unsuccessful products whose billets are compacted at low pressures;

can be concluded that the forgeability of sintered iron powder preforms is equivalent to that of conventional medium carbon steel.

Another important point on which to judge the cold forgeability seems to be the forging load. Figure 4 shows the comparison of load–time curves measured in the complex extrusion-A between a sintered iron powder billet and a conventional pure iron billet (annealed at 850 °C for 1 hour). From this figure it is found that the forging load of sintered iron powder billet is lower than that of conventional iron billet.

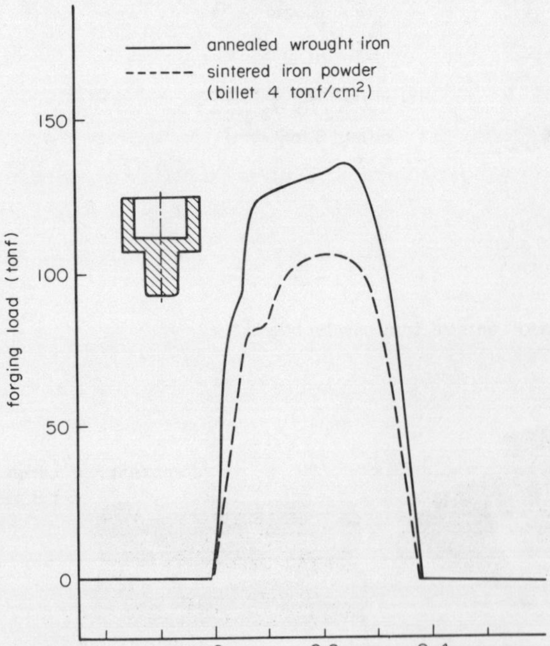

4.2. Material flow and hardness distribution of the cold forged products

Figure 5 shows the schematic flow patterns and the hardness distributions on typical forged products. In this figure, flow patterns were observed by microscope after etching, and hardness distributions were measured on the cross-sectional surface which did not include any pores. It can be seen that the flow pattern corresponds well to the hardness distribution. These results are almost the same as those of cold forging of bulk material.

For an example of the material flow and fracture behaviour in cold forging, the photographs of microstructure in forward extruded products with low reduction are shown in figure 6. At F, extruded at the first stage, the deformation is small and the porosity remains large, similar to that of the original billet shown in figure 1. At A, which is exposed by the surrounding high pressure, no deformation can be seen but porosity decreases remarkably. The size of the pore becomes smaller with increase of reduction, and only a few traces can be seen in the case of forward extrusion with 88.9% reduction. The obvious fibre flow can be observed at C, D, E and the pores become longer and smaller as the degree of fibre flow becomes larger. It can be estimated that the porosity becomes small also because of the effect of plastic

Figure 4. Comparison of the forging load between the annealed wrought iron and sintered iron powder in complex extrusion-A.

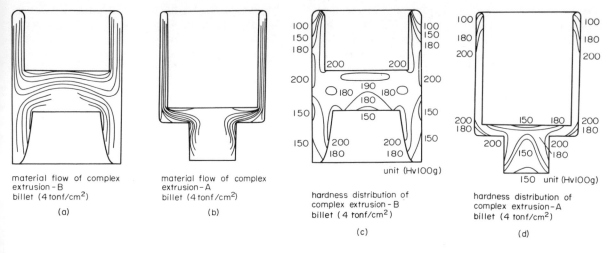

Figure 5. Material flow and Vickers hardness distribution of complex extrusion-A and B.

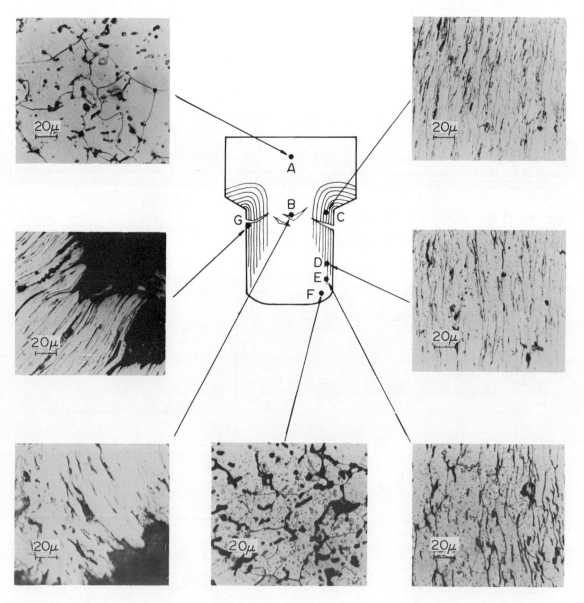

Figure 6. Photomicrographs of metallurgical structure at seven portions of forward extrusion with 38.5% reduction from the billet compacted at a pressure of 4 ton/cm².

flow. G and B are the microphotographs showing the cracks on the outer surface and at the centre of extruded portion. The outside crack is due to the tensile stress at the exit of the die hole followed by large plastic flow. On the other hand, the central crack originates in the difference of flow speeds between the outside and the inside. The deformation at the portion of central crack is as small as that of general sintered products at the time of fracture.

measured by the Archimedean principle. The forged products undergoing higher reduction possess an extremely high density over 7.7 g/cm³ or 7.8 g/cm³. Both high pressure and plastic flow are reasons for such a high density. However, in the case of forward extrusion with 88.5% reduction, the average density is reached at 99.9% of the theoretical density of iron, in spite of the existence of a large undeformed portion. From this fact, the main reason for the

(a) material flow around
an inclusion

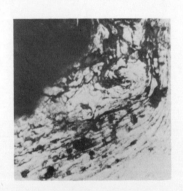

(b) vortex flow of material
at the inside corner of
backward extrusion

(c) turbulent flow of material
at the inside corner of
complete extrusion

Figure 7. Some examples of material flow observed in forged products.

Furthermore, detailed observations on material flow have been made, and these are shown in figure 7. In these photographs, it is found that the material flow seems to be similar to liquid flow. This means that the sintered iron powder has a large deformability in the high compressive stress state.

4.3. Average density and density distribution of forged products

The large decrease of porosity could be confirmed by the above-mentioned observations of microstructure of forged products. The average densities of the cold forged products shown in figure 2 have been

density increase can be attributed to the extremely high pressure, approximately 250 kg/mm².

The density distributions (shown in figure 8) have been measured by using the change of volume and weight by gradual machining of forged products. The portion extruded at the earlier stage has a comparatively low density, but other portions have as high a density as the theoretical density. It should be remarked that the difference of average density and density distribution between the high-density billet and the low-density billet becomes small after these billets are cold forged. This is attributed to the coining process which occurs before the forging process accompanying the material flow.

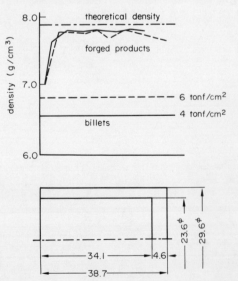

(a)

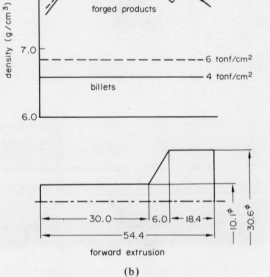

(b)

Figure 8. Density distributions of backward and forward extrusion.

4.4. Characteristics of the cold forging process of sintered powdered preforms

From the experimental results described above, the characteristics of the cold forging process of sintered powder preforms can be summarised as follows.

At the earlier stage of forging, billet pores are squeezed; their density increases as the punch pressure increases. Taking into account the facts that the maximum forging loads of billets of different density give almost the same values, and that it takes a longer time for the lower-density billet to reach the maximum point of forging load than it does for the higher-density billet, it can be concluded that the actual density of different-density billets is almost the theoretical density in the case of higher reduction forging. Where the plastic flow stage follows the coining stage, the grains are deformed and work hardened to a large extent.

On the other hand in the case of smaller reduction, the pressure increment does not reach a sufficient value, and the material flow starts without much increment in density. Therefore many pores still remain and they are the source of cracks and also promote crack propagation.

5.   CONCLUSIONS

Cold forging of sintered iron powder preforms has been considered to be impossible because of their small ductility; however, contrary to expectation it was revealed that cold forging of sintered iron powder preforms was easily possible. Comparisons between ordinary methods and the cold forging process are as follows.

5.1. Comparison with ordinary cold forging

The weight of billet can be controlled more precisely and closed die forging without producing any burr is easy owing to the changeability of volume during the forging operation. The billet is preformed in any shape by powder compacting and the products with a complicated shape can be produced in a single stamping. The preformed billet can be forged partially on the portion of the product where higher strength or density is necessary. The machinability of forged products is considered to be better than that of ordinary bulk annealed material. As the sintering time is not long, a continuous process from the first compacting to the final forging can possibly be developed.

As to the defects of cold forging of powder preforms, the cold forgeability is inferior to that of bulk material and the application range of this process is restricted to the high-ductility powder such as iron powders. The main problem of introducing this into actual production is economical; that is, the higher costs of powder material and sintering.

5.2. Comparison with ordinary powder forming

Because of the fact that products with an almost theoretical density can be obtained by this process, the mechanical properties such as the impact resistance and fatigue resistance and machinability can be improved, the heat treatment becomes easy and the electrical and magnetic properties are also improved. Billet shape can be simplified and it becomes possible to obtain special shaped products which cannot be formed by ordinary powder forming. The defects of this process are the limitation of powder material with high ductility, the breakage risk of forging dies and the increase of cost by additional cold forging.

5.3. Comparison with hot forging of powder preforms

As this process is performed at room temperature, the preheating of sintered billet can be eliminated and the difficulty of temperature control at the time of transformation from the furnace to the forging press is also overcome. There are several other merits for cold forging; these are easier lubrication, longer tool life, better dimensional accuracy and surface quality, and improved strength by work hardening. The disadvantages of this process, compared with hot forging, are that the limitation of powder material is severe, that the forging load is high and that forging of a complicated shape cannot be done in a single operation.

ACKNOWLEDGEMENTS

The authors would like to thank Mr. M. Nagase and Mr. M. Oshima for their assistance and cooperation in this experiment. They also wish to thank Mr. K. Miyamoto, the president of Miyamoto Kogyo Co. Ltd., and Mr. Y. Saito, senior researcher of Mitsubishi Metal Mining Co. Ltd., for their encouragement and valuable suggestions for this work.

REFERENCES

1. Japan Society of Powder and Powder Metallurgy, 1967. *Handbook of Design for Mechanical Sintered Parts* (in Japanese) Gijutsushoin, Tokyo.
2. Y. ISHIMARU, 1971–75. 'Recent review of mechanical sintered parts' (in Japanese), *Text of Lecture to Japan Soc. Mech. Engng.*, No. 341, 63.

THE PRODUCTION OF COMPONENTS BY FORGING OF POWDER PREFORMS

by

R. DAVIES* and J. B. MARX*

SUMMARY

This paper is a report of part of a long-term investigation into the various aspects of the powder forging process. The parts of the work covered in the paper are concentrated on the production of some of the components which have been manufactured, with particular reference to the design of the most suitable shape of powder preform to produce a sound final forging. Recommendations are given on types of powder, preform density and preform design.

INTRODUCTION

Powder preform forging, sometimes referred to as sinter forging, is a process that has received some considerable attention in the last five or six years. It involves the application of both pressure and heat. Pressure is applied to the cold powder in order to form the preform which is subsequently heated and then forged whilst still hot. The heating stage of the process is often prolonged in order to ensure that the grains of the preform are sufficiently well bonded to permit the preform to be hot worked, and is referred to as sintering.

A preform can of course be made from solid material, but this can be expensive as regards labour, time and material wastage. The advantages of compacting the preform from powder lie in the excellent surface finish, accurate dimensions and weight control, ease of alloying, materials savings, subsequent machining, savings and homogeneous structures that can be readily achieved.

Although a considerable amount of work has been recently published on powder forging[1-4], no information is available on the detailed considerations that apply to the design of preforms for forging, although the forging of many components has been reported by investigators[5-7]. Thus, it is intended that this paper will give an appreciation of the extent to which powder forging may be successfully applied and establish some basic design criteria for the process.

EQUIPMENT

Compaction machines

Slow- and high-speed machines were used to compact the preforms for the experiments reported here. The slow speed machine was a Denison T1A/MC hydraulically powered 300 tonf press with a maximum die closure rate of 0.05 in/s. The same tooling was also used on the Petro-Forge Mk I high-speed press. Although more powerful Petro-Forge machines are available[8] the nominal capacity of 5000 ft lbf was adequate for the production of the preforms used in these experiments. This energy level corresponds to a platen impact velocity of 50 ft/s.

Sintering furnace

Sintering was performed in an Efco-Royce 12 kW molybdenum wound furnace with a heating zone 9 in

in length. The tube diameter was 4 in, and a platinum/rhodium thermocouple was employed for use up to 1300 °C. A sintering atmosphere of cracked ammonia gas was passed through the furnace at a flow rate of about 14.5 ft³/h. The furnace was fitted with a water cooled section which enabled the furnace charge to be cooled in the cracked ammonia atmosphere, should this be required.

Forging machines

Forging was carried out on a Petro-Forge Mk II machine. It is capable of providing a range of impact energies from 3000 to 20 000 ft lbf with corresponding impact speeds of about 20—50 ft/s. For low-energy working the combustion process may be dispensed with and the machine operated solely by compressed air at pressures up to 200 lbf/in². Under these conditions the machine is capable of providing energies up to 1400 ft lbf at impact speeds up to 13 ft/s.

Tooling

Compaction tooling was designed on the basis of previous work on high-speed compaction[9]. The only significant difference between this tooling and that used in industrial compaction is the use of an accurately fitted pressure pad placed between the powder and the impacting punch. This avoided the necessity of an accurate alignment between punch and die.

Some previous experience[10] in HERF die assemblies was used in the design of the forging tooling, although modifications for the forging of powder preforms were developed during these experiments. The die insert technique was used in all the die sets. The inserts were made in Jessop Saville H50 steel heat treated to 45—47 Rc.

Powder

The principal powder used in these experiments was MH100 sponge iron. Commercial production of this powder is by crushing, disintegrating and screening of the reduced oxide ores. The MH100 powder was selected because of its current relative cheapness and wide usage in industrial compaction, although it suffers from a rather high oxide content when compared with other varieties.

Other powders used in these experiments were Domtar MP32 and MP35 iron powders and MP55 pre-alloyed steel powder (2% Ni + 0.6% Mo). These

*Department of Mechanical Engineering, University of Birmingham

powders were produced by atomisation of molten metal by high-pressure water, followed by drying, ball milling, blending with mill scale, decarburising and finally sizing.

In order to add more carbon to the powders, where desired, Rocol Graphite X7119 was blended using a mixing unit. In assessing the correct graphite addition to be made to the mix, it must be remembered that not all of the graphite will be converted to combined carbon in the steel. Graphite loss occurs through reaction with oxides in the powder and the oxygen in the pores of the compacted preform.

Lubrication

As far as the compaction of the preforms was concerned, 1% zinc stearate was mixed with the powder prior to compaction as this simulates normal industrial practice.

The lubricant used in the forging stage of the process was Acheson Delta 31, a water-based graphite lubricant, which was brushed onto the dies prior to each forging operation.

EXPERIMENTAL RESULTS AND DISCUSSIONS

As there exists no published information on the design of preforms for powder forging, the technique had to be acquired without any prior knowledge. Theoretical considerations of powder flow and densification during the actual forging operation in complex shaped dies have not yet been fully developed, so a method of trial and error was adopted. The experience and information gained from forging a particular shaped preform was used to assist in the design of the following preform and thus to build up a picture of the nature of powder forging in general, and in these forgings in particular.

This work constitutes a continuation of previous experiments[11] in which several other types of components were produced.

Gear blank

This component shown in section in figure 1(a), is a flash and gutter type forging of $2\frac{1}{2}$ in overall diameter,

originally intended to be produced from solid material. A variety of preform shapes were attempted, but none was successful owing to the faults that were induced from allowing the metal to proceed across the flash lands. The situation was probably aggravated by the fact that the forging operation was carried out at high speed, thus not giving the flash lands time to cool the metal flowing through them and hence obstruct further flow. Cracks at corners 'u' and 't' as well as tears on surfaces 'j' and 'g' were manifestations of these faults.

Owing to the failure of the flash and gutter die configuration, the component was re-designed so as to be suitable for closed die forging and is shown in figure 1(b). The only difference is the elimination of flash and the removal of the taper on the outside diameter of the forging. When the preform shown in figure 2(a) was forged, severe cracks on forging

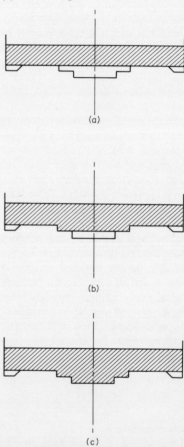

Figure 2. Preform shapes and position in die prior to forging of gear blank.

surface 'a' and poor filling at corners 'q' and 'r' were evident. No surface tears were produced. The preform in figure 2(b) also produced cracks at surface 'a' and poor filling of corner 'r', these faults being attributable to the lack of metal preformed into the centre of the forging. Thus the preform in figure 2(c) was attempted and a fault-free component produced. A full theoretical determination of this preform's dimensions in a similar manner to that previously described[12] is not feasible. If it were applied the resulting outside diameter of the preform would be far greater than the outside diameter of the forging die cavity.

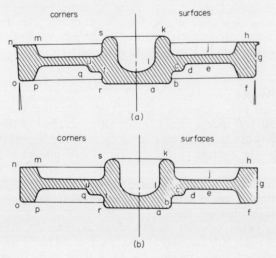

Figure 1. Gear blank forging: (a) flash and gutter, (b) closed die.

The initial preform density was about 82% of theoretical and the final density achieved in the forging was 99.8%. The material used was MH100, and the preform was sintered for 20 min at 1100 °C and then forged at the same temperature. A very minimal amount of lubrication was used with this forging, since any excess of lubricant caused cracking and folds associated with gross metal movement around the corners of the forging.

As mentioned in the introduction, two of the main advantages of powder preform forging are the elimination of subsequent machining operations and the saving of material. Both of these advantages can be used to good effect by endeavouring to forge a hole through the centre of the gear blank under consideration here. Thus the closed die set configuration was modified to permit the forging of a $\frac{1}{2}$ in diameter central hole. The preforms attempted are shown in figure 3.

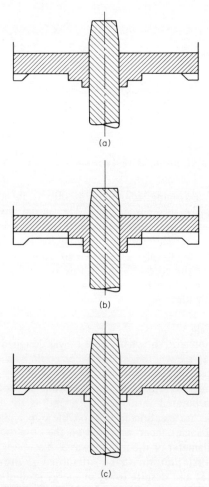

(a)

(b)

(c)

Figure 3. Preform shapes and position in die prior to forging of gear blank with central hole.

Using the experience derived from the previous gear blank forgings it was anticipated that the preform shown in figure 3(a) would forge satisfactorily. There was, however, a slight fault on surface 'b' caused by corner 't' being moved downwards as the forging consolidation proceeded. This was perhaps to be expected as the large-scale radial movement of metal in the forging in figure 1(b) no longer existed now that a hole had been produced, in the preform.

The preform shown in figure 3(b) failed to eliminate the fault and the preform was simplified to that shown in figure 3(c), and a satisfactory forging was produced. A parallel-sided core rod was used, thus ensuring that the central hole in the forging was to an accurate dimension. The surface finish in the hole itself was very good, and the final forging density achieved was 99.9%. As with the previous forging only a minimal amount of lubricant was applied, except on the core-rod where liberal quantities were used.

Spur gear

The die-set that was used to produce the forging shown in figure 4(a) was originally intended for the high-speed forging of solid material. Hence the

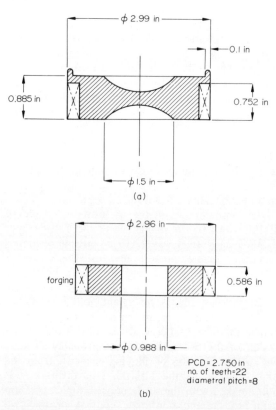

(a)

forging

PCD = 2.750 in
no. of teeth = 22
diametral pitch = 8

(b)

Figure 4. Spur gear forging: (a) with flash, (b) closed die with central hole.

flash allowance and absence of central hole. No attempt was made to produce an accurate involute gear profile on the lower die as the experiment was conducted merely to examine the feasibility of forging such a shape from powder preforms. The overall diameter of the forging was 3 in.

A simple cylindrical billet was used as the preform with an outside diameter of just less than the root diameter of the teeth on the forging die. This was to allow for expansion when raised to 1100 °C and also for ease of insertion into the forging die just prior to forging. Using an energy of 14 000 ft lbf it was found possible to just fill the cavities of the die, with lubricant applied. Forgings were produced in MH100, MP35 and MP55 + 0.5%C with final densities of 99.5%, 99.3% and 99.2% respectively. Tensile test specimens were taken from the centre of each forging and yielded the following results in the normalised condition:

MH100, upper tensile strength 20.3 tonf/in^2, elongation 25.9%

MP 35, upper tensile strength 22.5 tonf/in^2, elongation 31.3%

MP 55, +0.5%, upper tensile strength 59.7 tonf/in^2, elongation 12.7%

Owing to the far cleaner nature of the MP35 iron powder the properties are considerably superior to those obtained from the MH100. The MH100 and MP35 preforms were sintered for 20 minutes at 1100 °C and forged at 1100 °C, whereas the MP55 + 5% C preform was sintered at 1130 °C for 40 min prior to forging at 1130 °C.

A spur gear is a good example of where the benefits of duplex preform forging may be applied. The gear teeth are often required to withstand severe wear and fatigue as well as possess high mechanical strength. The bulk of the spur gear, inside the root diameter, is not generally required to perform under such exacting conditions so there would be a considerable cost saving if the central section of the gear could be made from a cheaper, less highly alloyed material. Thus duplex preforms were pressed, using the single press technique where both powders were pressed simultaneously as well as the double press technique where the powders were pressed separately but sintered and forged together.

The central core of the preform was made of MH100 iron 2.055 in diameter and the outside layer of MP55 alloy steel was 2.416 in outside diameter. With the double pressed preform the clearance between the outer ring and the inner billet was varied from 0 to 0.005 in. All the preforms were forged satisfactorily as far as the join of the two materials was concerned, and subsequent tests demonstrated that the join was at least as strong as the weaker of the two materials, and probably stronger due to the diffusion of alloying elements from the MP55 inwards to the MH100. This diffusion was more marked with the single pressed duplex preform owing to the more intimate contact of the two materials during the sintering stage of the process. A duplex forging is shown etched in cross-section in figure 5. The small zones of decarburisation that occurred during sintering at 1130 °C are visible. Forging was carried out at 1130 °C and the forgings allowed to cool in a nitrogen atmosphere to prevent external scaling.

As far as production of these forgings is concerned, it is suggested that the double pressed preform be employed as this is closest to current industrial experience and presents no difficulties as regards powder fill. A clearance of 0.001 in between the outer ring and the inner billet is the most satisfactory as regards ease of assembly prior to sintering, and transference of the hot duplex preform from the furnace to the forging die prior to the forging operation. When a clearance of greater than 0.001 in was employed, the inner billet slipped out of the outer ring during the transfer movement.

In order to utilise fully the surface finish and material saving advantages of the powder forging process it was decided to redesign the spur gear die set in order to incorporate a central hole in the forging. Furthermore, all flash was eliminated by making the punch to a gear profile which could mate with the lower die. The forging, shown in figure 4(b),

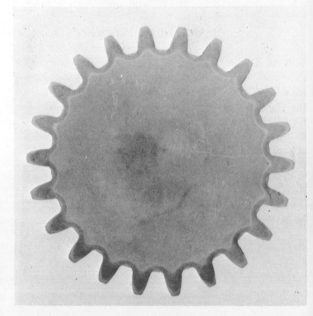

Figure 5. Duplex gear forging showing the use of different materials for teeth and central portion.

required a forging energy of 12 000 ft lbf to fill satisfactorily the die cavities. The initial preform density of 82% was converted to a final forging density of 98.7% after sintering and forging at 1100 °C. The preform was simply a ring-shaped compact which fitted over the core-rod, and whose outside diameter was just less than the root diameter of the teeth on the forging die.

Cup forging

This component, shown in figure 6, was designed as an exercise to see whether such a shape could be successfully forged in fixed tooling from a powder preform. Unlike previous forging die configurations the core-rod was incorporated into the punch and as the Petro-Forge has no ejection facility on the upper platen a 5° (semi-angle) taper was used on the core-rod. The overall diameter of the forging was 3 in.

The first preform shape attempted is shown in figure 7(a) and despite using an energy level of about 6500 ft lbf it was not found possible to fill the top section of the forging at corners 'n' and 'm'. As a considerable surface area of the forging is involved in the backward extrusion, the heat losses from the metal in this section of the forging act in a similar nature to those in the flash lands of an open die forging, namely to restrict the flow of material in this direction with the result that it becomes increasingly difficult to fill corners 'n' and 'm'. Thus it was felt that by endeavouring to forge with a higher energy level than 6500 ft lbf it was more likely that the die set would fracture before the die could be filled.

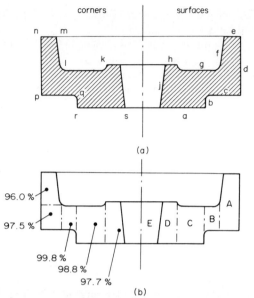

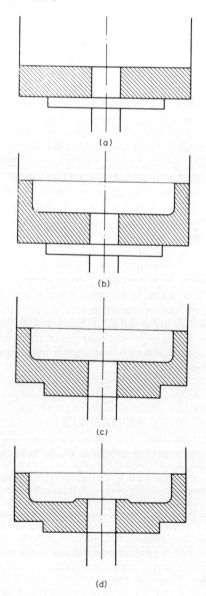

Figure 6. (a) Cup-shaped forging, (b) density variation in cup-shaped forging.

Figure 7. Preform shapes and position in die prior to forging of cup-shaped component.

Thus the preform in figure 7(b) was attempted. It was found that, if an energy level in excess of 4500 ft lbf was employed, then cracking and blistering was apparent on surface 'e' of the forging. This was considered due to an excess of energy used in forging the outer rim. Corners 'r' and 'k' were unfilled at this energy level so the preform was modified to that shown in figure 7(c). This also proved unsatisfactory owing to poor filling of corner 'k' if cracking and blistering was to be avoided. The preform in figure 7(d), which closely resembles the final forging, was forged to produce a sound component at an energy level of 3000 ft lbf. The final density that could be achieved was only 97.4% of theoretical owing to the cracking and blistering of surface 'e' that occurred, when energy in excess of 3000 ft lbf was used. The nature of the density distribution in the final component is shown in figure 6(b).

As the high-speed process prohibited the use of high forging energies with this component the preform shown in figure 7(d) was forged on an eccentric press using the same die-set as before. With a peak forging load of 181 tonf it was possible to attain a density of 99.8% of theoretical. Thus, should it be desired to achieve a high density in this component, then the slow-speed eccentric press is more suitable than the Petro-Forge.

The material used was MH100 at a sintering and forging temperature of 1100 °C.

Pivotal bracket forging

This component is currently being manufactured by conventional drop forging, and a request was made by the manufacturer to produce the component by powder forging. The outline shape of the bracket can be seen from the die profile of figure 8, and consisted of a $1\frac{1}{2}$ in diameter head and a reduced portion of $\frac{7}{8}$ in diameter. In the manufacture of this component by conventional forging, flash and gutter dies were used, and a 7° draft was incorporated on the forging. The forging was then trimmed to remove flash, and the $\frac{7}{8}$ in portion was milled to produce $\frac{5}{8}$ in wide flats. In the redesign of the part for powder forging, the draft was removed, and the flats were incorporated into the die in order to eliminate the milling operation.

The first preform to be attempted is shown in figure 8(a). The smaller diameter of the preform was designed to fit just between the flats of the forging die. Although the upper rim of the forging forged satisfactorily, there was a severe shortage of metal in the barrel of the forging, caused by insufficient material being preformed into this section. This was evident by the poor filling and bad cracking that occurred in this region.

The preform in figure 8(b) was an improvement, but was also unsuccessful for the same reasons as the previous forging. Thus it became apparent that, in order to fill the die cavity, the flats would have to be incorporated into the preform.

Hence the preform shown in figure 8(c) was forged successfully. The distance across the flats of the preform, when heated, was made to be a close fit with the corresponding distance in the forging die. This was to avoid misalignment of the preform in the die which would lead to poor filling in two opposite corners of the die cavity. The diameter of the top section of the preform was designed so that the mass of metal in the rim of the preform was equal to the mass in the rim of the forging. This was to avoid the folds and cracks that are associated with gross metal movement

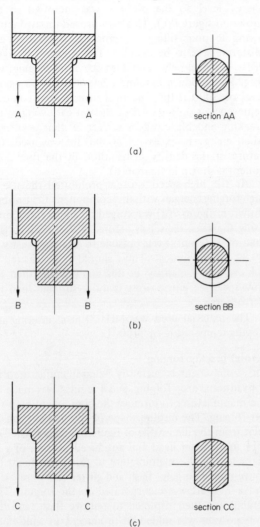

Figure 8. Preform shapes and position in die prior to forging of pivotal bracket.

around corners as densification proceeds[11]. The overall component density was 99.7% of theoretical produced from a preform density of 81.5% of theoretical. The material used was MH100 + 0.2% C and the preform was sintered and forged at 1130 °C.

RECOMMENDATIONS

The following remarks are presented as some of the major points to be borne in mind when designing preforms for powder forging.

(a) Selection of powder. The type of powder chosen is dependent upon the properties required in the forged component. If fatigue, impact and elongation properties approaching those of normal wrought material are required, then an electrolytic or atomised powder will be necessary. If, however, tensile strength is the main requirement, then the poorer quality

reduced powders will suffice. The actual mesh size of the powder is not critical.

(b) The preform should be designed in such a way as to eliminate, wherever possible, the flow of metal from one depth of the forging to another. In other words flow around corners in the vertical plane is to be avoided, but flow around corners in the horizontal plane may be considered beneficial if grain flow is required into, say, a gear tooth.

(c) The preform should be designed such that all sections of the forging densify to the same final density simultaneously. With fixed tooling, as used in the work reported here, it may not always be possible to achieve this so movable type die sets similar to those used in conventional compaction may be necessary.

(d) Small radii on the preform can be forged and are often essential if small radii are to be achieved on the forging.

(e) Forgings with central holes can readily be forged from preforms with corresponding holes. The hole diameter of the preform should be made slightly less than the hole on the forging.

(f) The production of components made up of different materials in various sections is possible by the powder forging process, and could result in significant material cost savings.

(g) In order that the process make best use of current industrial experience in compaction operations it is suggested that a pressing pressure of 30 tonf/in^2 be used. This will result in preform densities of about 80% depending on the type of powder used.

(h) The design of the preform must allow for expansion of the preform when heated to forging temperature and also must incorporate a clearance between the die and the hot preform if air is likely to be trapped under the preform when it is inserted into the die.

(i) Adequate location of the preform in the lower die must be provided to ensure a sound forging.

(j) Where large changes in cross-section are being forged it is desirable to forge the preform in both parts of the section simultaneously.

(k) Lubricant should be applied with care, as excessive lubrication will cause the metal to flow easily around corners where such a movement may not be desirable.

REFERENCES

1. R. T. CUNDILL *et al.*, 1970. 'Mechanical properties of sinter/forged low-alloy steels', *Powder Met.*, **13**, No. 26, 130.
2. H. F. FISCHMEISTER *et al.*, 1971. 'Deformation and densification of porous preforms in hot forging', *Powder Met.*, **14**, No. 27, 144.
3. G. W. CULL, 1970. 'Mechanical and metallurgical properties of powder forgings', *Powder Met.*, **13**, No. 26, 156.
4. G. ZAPF, 1970. 'The mechanical properties of hot-precompacted iron−nickel sintered alloys', *Powder Met.*, **13**, No. 26, 130.
5. G. T. BROWN and P. K. JONES, 1970. 'Experimental and practical aspects of the powder forging process', *Intern. J. Powder Met.*, **6**, 4, 29.

6.   Y. ISHIMARU *et al., 1969*, 'On the properties of the forged P/M ferrous alloys', *Mitsubishi Metal Mining Co.* (Tokyo).

7.   P. K. JONES, 1970. 'The technical and economic advantages of powder forged products', *Powder Met.*, **13,** No. 26, 114.

8.   L. T. CHAN and S. A. TOBIAS, 1968. 'Performance characteristics of Petro-Forge Mk I and Mk II machines', *Proc. 9th Intern. M.T.D.R. Conf., Birmingham.*

9.   S. WANG and R. DAVIES, 1968. 'Some effects of high speeds in metal powder compaction', *9th Intern. M.T.D.R. Conf., Birmingham.*

10.   S. C. JAIN and A. N. BRAMLEY, 1968. 'Characteristics of the high-speed hot forging process', *Proc. 9th Intern. M.T.D.R. Conf., Birmingham.*

11.   J. B. MARX, R. DAVIES and T. L. GUEST, 1970. 'Some considerations of the hot forging of powder preforms', *Proc. 11th Intern. M.T.D.R. Conf., Birmingham.*

12.   J. B. MARX, 1971. 'The forging of powder preforms', *Ph.D. Thesis*, University of Birmingham.

THE INTERRELATION OF DENSITY AND HARDNESS IN THE ISOSTATIC COMPACTION OF POWDERS

by

S. SHIMA* and J. M. ALEXANDER*

SUMMARY

Stress—strain and hardness—strain curves were determined for sintered iron powder in the annealed state. These results were used to explain the behaviour of isostatically compacted green compacts and the density changes observed due to either particle deformation and/or sliding.

By introducing the concept of a reduced pressure, that is, pressure divided by instantaneous flow stress, it was found possible to develop a generalised curve of relative density versus reduced pressure which may have some general applicability to powders of ductile materials.

INTRODUCTION

Recently a technique called 'forging of powder preforms' has been attracting much attention[1], because it produces powder products reaching nearly 100% of the theoretical density as well as developing mechanical properties competitive with conventional pore-free products. The workpiece in the case of the powder forging process undergoes significant density increase or volume decrease during deformation, which does not allow conventional plasticity theories to be applied in this case. Although there have been proposed new plasticity theories for porous metals[2,3], they do not give enough information for the behaviour of green compacts.

It is important to investigate how sintered porous metals and nonsintered porous compacts behave, or how they are strengthened during densification, in order to generalise the compaction and powder forging processes.

In the past, the overall hardness has been considered an important measure for representing the strength of the porous metals, both sintered and nonsintered. However, it is not a proper parameter as in the case of pore-free metals, unless the combined effect of the apparent density and the material strength on the apparent strength is determined, as will be discussed later. The strength of a porous metal can be increased by raising its density. It can also be increased by an increase in strength of the matrix metal, due to workhardening, without an increase in density. In fact the density increase is always accompanied by deformation of the matrix metal. In other words, there exists an interrelation between the increase in the apparent density and the effective deformation in the matrix metal. In the case of nonsintered compacts the problem is more complicated, because not only deformation of particles, but also their sliding, is likely to occur.

In this investigation an attempt was made to investigate the effective deformation in the matrix metal due to increase of density. Furthermore, on the basis of these results, a curve of relative density versus reduced pressure (= compacting pressure/instantaneous flow stress of the matrix metal at the particular density concerned) was produced for isostatic compaction of powders. This method will be explained in the next section in more detail.

There have been many investigations into the relationship between density and pressure for the compaction of metal powders[4], but these relationships are not capable of being generalised, because they have not been related to a basic property of the material (such, for example, as the basic stress—strain curve) due to lack of investigation of the interrelation mentioned above. On the other hand, the relative density versus reduced pressure curve obtained in this experiment could be applied to other materials, provided the effects of other powder characteristics on the curve are determined. In this experiment the curve was produced for iron powder and was applied to Cu powder, to predict its relative density versus pressure relationship. Fairly good agreement with experimental results was obtained.

EXPERIMENTAL PROCEDURE

Method of estimating the effective deformation in the matrix metal due to increase in the apparent density

In this investigation the apparent stress condition was limited to the hydrostatic state of stress, that is, isostatic compressing. To investigate the amount of effective deformation in the matrix due to apparent density increase, the following assumptions were made:

1. The hardness of the matrix metal is always uniform throughout a porous specimen.

2. The apparent hardness of a nonsintered compact is identical to that of a sintered metal if the density and the matrix material hardness respectively are the same.

The following is a method of estimating the effective deformation in the matrix metal due to an increase in the apparent density.

(i) Determine the stress—strain curve and hardness—strain curve for the iron matrix (that is, pore-free).

*Department of Mechanical Engineering, Imperial College of Science and Technology, London, S.W.7.

(ii) Determine the relationship between the apparent hardness and the relative density (that is, $H_A - \rho$) for sintered iron.

(iii) Estimate the hardness of annealed iron (pore-free) H_M from the $H_A - \rho$ curve by extrapolation to 100% theoretical density.

(iv) Determine the relationship between the hardness ratio H_A/H_M and the relative density by combining the H_M value and the curve obtained in (ii).

The curves obtained in (i) and (ii) were used as basic data to determine the effective deformation in the matrix metal.

(v) Obtain the apparent hardness versus relative density relationship for sintered iron after recompressing isostatically.

(vi) Find the matrix metal hardness (or flow stress) versus relative density relationship after recompressing by combining the curves obtained in (iv) and (v).

(vii) Finally, obtain the relationship between the effective strain in the matrix metal and the relative density from the curves obtained in (i) and (v).

For the case of green compacts, a similar method was employed on the basis of the second assumption given above.

Isostatic compaction and preparation of specimens

Isostatic compaction was carried out on a standard 250-ton Avery Testing Machine. The apparatus in

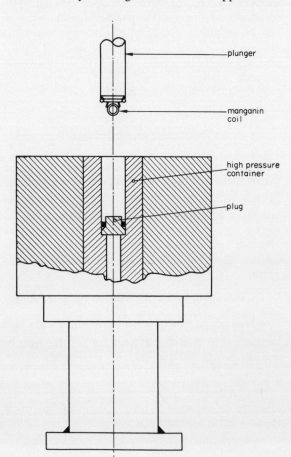

which compacts were pressurised is shown diagrammatically in figure 1. Powders were put into PVC flexible tubing with two rubber stoppers at the ends. 'Evostik' was used as a sealant at the interfaces of the PVC tubing and rubber stoppers to prevent the high pressure castor oil used as the pressurising medium from penetrating into the powder compacts.

Sintered specimens were recompressed isostatically in a similar manner.

Green compacts were then turned into cylindrical specimens $\frac{1}{2}$ in diameter by 1 in long.

Sintering was carried out at 1100°C for iron powder compacts and at 900°C for Cu, for 1 h in vacuum ($\approx 10^{-4}$ torr).

Hardness testing

Hardness was measured on a Vickers hardness testing machine with a 5 kg load, although in some cases a 10 kg load was used, on transverse cross-sections. Six to ten readings were made for each specimen and the scatter was within ± 4.5%.

Stress–strain curve and hardness–strain curve for iron

Although it was intended to obtain a stress–strain curve for the sintered iron with 100% of the theoretical density for the basic property of the matrix metal, this did not prove possible. Therefore, as will be shown in the next section, compressive specimens with 97.5% of the theoretical density were used. Friction at the specimen–tool interfaces was mostly eliminated; thus, up to a strain of 0.25 (natural strain), appreciable barrelling was not observed. The hardness–strain curve was similarly obtained by measuring hardness at each step of strain.

Density measurement

Density was obtained simply by measuring specimen dimensions and weight, and in order to confirm this, the hydrostatic (or Archimedes) method was also employed in some cases.

Powder

The iron powder used in this experiment was Hoganas NC 100, as previously used by Alexander and Quainton[5], and Cu powder was supplied by McKechnie

Figure 1. Apparatus for isostatic compaction.

plunger

manganin coil

high pressure container

plug

TABLE 1
Details of Hoganas iron powder NC100

Chemical analysis	approx. %
iron	98+
carbon	0.1
oxygen (loss in weight in hydrogen)	0.6
silica	0.3
sulphur	0.015
phosphorous	0.015

Physical properties	
Sieve analysis (BS sieves)	
+100	1% Max.
−100 + 150	15~25%
−150 + 200	20~30%
−200 + 240	5~15%
−240 + 350	15~30%
−350	15~30%

apparent density 2.4 g/cm³
theoretical density 7.87 g/cm³

TABLE 2
Details of copper powder

Chemical analysis

    copper 99.50%
    oxygen 0.15%

Physical properties

    sieve analysis (BS sieves)

+ 100 mesh	1.0 max.
−100 + 200 mesh	25~40%
−200 + 350 mesh	30~40%
−350 mesh	30~45%

    apparent density (untapped) 2.50~2.90 g/cm³
    theoretical density 8.96 g/cm³

Chemicals, their analysis being shown in tables 1 and 2 respectively.

It was observed that iron powder tended to strain-age, therefore all the tests were carried out carefully to eliminate this effect.

EXPERIMENTAL RESULTS

Determination of stress–strain curve and dependent variables for the iron

Figures 2(a) and (b) show the stress–strain and hardness–strain curves respectively for sintered iron with 97.5% of the theoretical density. They are described by the following expressions:

$$\sigma = 8.618 \, (1 + 160.511 \, \epsilon)^{0.3241} \quad (\text{tonf/in}^2) \quad (1)$$

$$H_M = 77.876 \, (1 + 143.172 \, \epsilon)^{0.1772}$$
$$(\text{VHN} \equiv \text{Vickers hardness number}) \quad (2)$$

which were determined by a least squares method. These equations will be used later for estimating strain increment due to density change. Although the stress–strain curve in figure 2 is for the iron with 2.5% porosity, it was assumed to be identical with that of the pore-free iron under compressive load for the following reasons:

● As will be shown later (figure 3(a)), there is no increase in hardness beyond 96% of the theoretical density.

● It is very likely that, if fractures had taken place from pores, they would have lowered the density of the specimens, and consequently the stress–strain curve obtained would have given a misleading result. However, density measured at each step of straining did not show an appreciable decrease.

Curve *a* in figure 3(a) shows apparent hardness (H_A) against relative density (= apparent density/the density of the pore-free metal) for sintered iron. Curve *b* shows the change in apparent hardness after

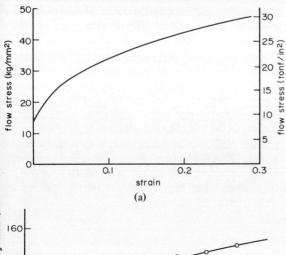

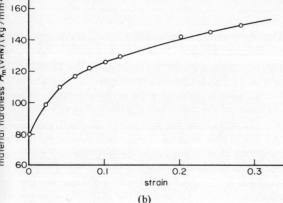

Figure 2. (a) Stress–strain curve for the iron (b) Hardness–strain curve for the iron.

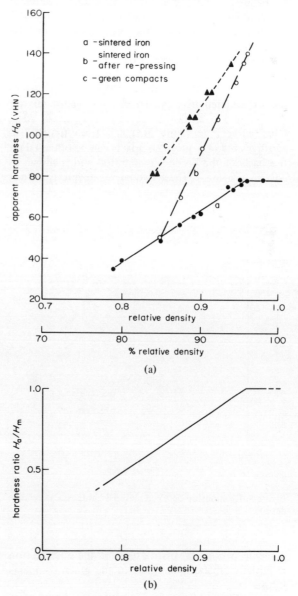

Figure 3. (a) Hardness–density relationship (b) Hardness ratio–density relationship.

recompressing (isostatically) sintered iron with an initial relative density of 85%. It is seen from curve a that there is no significant increase in apparent hardness above 96% of the theoretical density.

Since it can be assumed that the matrix metal is fully annealed for the case of sintered iron, the extrapolation of curve a in figure 3(a) to 100% of the theoretical density suggests that the actual material hardness of the annealed iron would be 78 (VHN). Curve c shows a relationship between hardness and density for green compacts of iron powder. It is clearly seen by comparing the curves b and c, that the sintered iron (after repressing) with a relative density of 85%, where the matrix metal is annealed, possesses a lower apparent hardness than the corresponding green compact. However, the hardness for the former case increases more rapidly with increasing density than for the latter.

The hardness of the annealed iron can be now estimated from the above; thus H_A/H_M can be plotted against relative density (figure 3(b)), where H_A denotes overall (apparent) hardness in the sintered iron and H_M the matrix material hardness (= 78 VHN for the present case). This relationship should be applicable to any porous metal, provided the matrix material is ductile like iron. From this figure, together with curve b in figure 3(a) the instantaneous hardness or strength of the matrix metal for recompressed sintered iron can be plotted against relative density and is shown as curve a in figure 4. Also, from curve c in figure 3(a) and from figure 3(b) another curve for green compacts can be obtained on the basis of the second assumption and is shown as curve b in figure 4. These two curves obviously do not

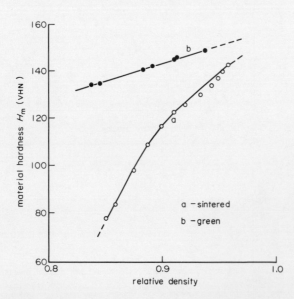

Figure 4. Relationship between material hardness and relative density.

coincide, because the matrix metal in the sintered iron is initially annealed but that in the green compacts is not.

Combining this figure with the hardness strain curve (figure 2(b)), the effective strain in the matrix

metal ϵ due to densification is plotted against relative density and is shown in figure (5a), curve a showing the relationship for sintered iron and b for green compacts. Since the matrix metal is fully annealed at the density of 85% the initial effective strain is zero, whereas green compacts possess some initial strain. In order to give a better comparison the gradients $d\epsilon/d\rho$ of both curves in figure 5(a) are plotted against relative density in figure 5(b), curve a for sintered iron and curve b for green compacts. It is seen that sintered iron possesses larger values of $d\epsilon/d\rho$ than do the green compacts, which implies that in the latter not only deformation but also particle sliding plays an important part in densification. It is more clearly seen by comparing the total external work with the energy consumption due to effective deformation of the particles during densification. The external work was obtained from the density—pressure curves a and b for sintered and nonsintered compacts respectively shown in figure 7(a) and the plastic work from figure 2(a) and figure 5(a). For convenience of comparison W_{ext}, the total external work done during densification from 85% to 96% of the theoretical density, was taken as unity. Thus, in figure 5(c) W_P/W_{ext} was plotted against fractional external work W_{ext}, where W_P the fractional plastic energy was obtained in a similar way to the case of W_{ext}. In green compacts (curve b in figure 5(c)), the plastic deformation energy is much smaller than the total external work, probably due to particle sliding, whereas in sintered iron (curve a in figure 5(c)) both are nearly the same. The broken line in the figure shows the case where the external work and deformation energy are exactly the same. Figure 5(d) shows $d\overline{W}_P/d\overline{W}_{ext}$ against relative density for green compacts, where $d\overline{W}_P/d\rho$ and $d\overline{W}_{ext}/d\rho$ are the rates of change of plastic deformation energy and external work W_P to changes of density respectively. It is seen from this figure that the ratio of the rates of plastic energy to external work can be seen to increase and deformation of particles is therefore dominant in the vicinity of 90% of the theoretical density. Presumably, particle sliding is at a minimum in this region (that is, around 90% theoretical density for green compacts). However, above this density it tends to decrease and it appears that the energy required for particle sliding becomes dominant.

Density-pressure relationship for isostatic compaction
Combining figures 2 and 5(a) together with the pressure values p corresponding to the density for isostatic compaction and recompressing (figure 7(a)), a relationship between relative density and reduced pressure p/σ is obtained, (curve a, figure 6), where σ is the instantaneous strength of the matrix metal at the particular density concerned. Although it was expected that the relative density-reduced pressure curve for the isostatic compaction of green compacts would not coincide with that for the recompressing of sintered iron, because of the different patterns of energy consumption, both curves surprisingly coincide with one another.

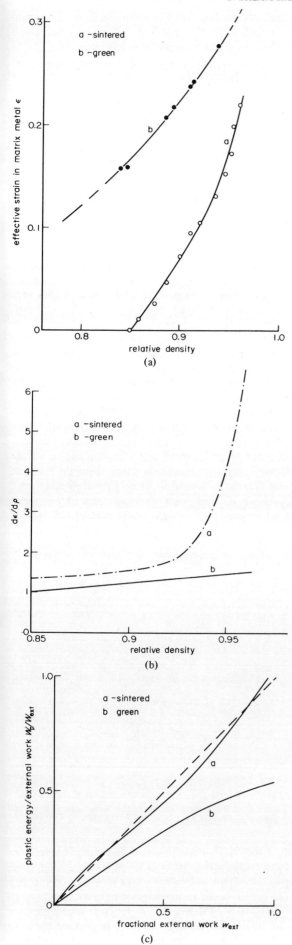

(a)

(b)

(c)

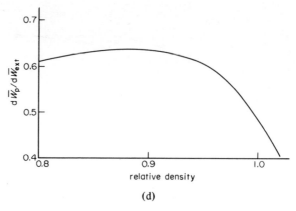

(d)

Figure 5. (a) Dependence of effective strain in matrix metal on relative density (b) Dependence of $d\epsilon/d\rho$ on relative density (c) Comparison between plastic energy and external work (d) Ratio of plastic energy rate to total external work rate against relative density.

As mentioned above this curve might also be applied to other materials, provided they are sufficiently ductile. Therefore, from this curve, coupled with figure 5(a), which may also be assumed to apply generally and a stress–strain curve for the particle material concerned, a density–pressure relationship for isostatic compaction may be predicted for other materials. It should be noted that here the effect of other powder properties such as particle size, particle

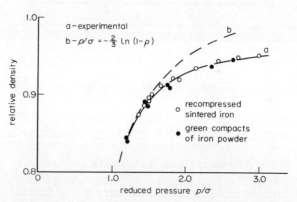

Figure 6. Relative density-reduced pressure curve for iron compacts.

size distribution, particle shape, surface condition and so on is not taken into account. The stress–strain curve for Cu was taken from an experimentally determined curve for commercially pure Cu metal[6] (curve a, figure 7(b)). For confirmation, a stress–strain curve for sintered Cu (with 98.3% of the theoretical density) was produced (curve b, figure 7(b)) in a similar manner to that for iron powder. As had been expected, the predicted curve (curve d, figure 7(a)), using the latter stress–strain curve is very close to the curve c in figure 7(a) for the former one. Although other characteristics of the powders were neglected, the predicted curves show good agreement with the experimental results for the copper powder, also shown in figure 7(a).

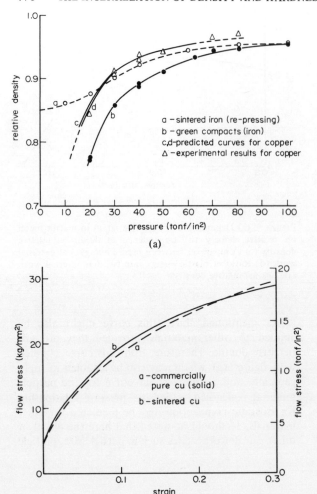

a – sintered iron (re-pressing)
b – green compacts (iron)
c,d-predicted curves for copper
△ – experimental results for copper

(a)

a – commercially
pure cu (solid)
b – sintered cu

(b)

Figure 7. (a) Relative density–pressure relationship (b) Stress–strain curve for copper.

DISCUSSION

Plastic deformation in the matrix metal due to density increase

From figure 5(c) it can be seen that the total external work is nearly equal to the total plastic energy consumption in the matrix metal for sintered iron, which verifies the estimation of effective strain due to density increase.

By means of the curves in figures 2 and 3 and a in figure 5(a), the two interrelating factors which influence the apparent strength of a sintered metal (namely the relative density and the strength of the matrix metal) can always be separated. In other words, if the initial density and density increase in a porous metal, which has been cold-worked, are assumed to be uniform throughout the body, the increase in strain (and consequently in strength) can be estimated.

In the case of green compacts, on the other hand, it is difficult to confirm whether or not the second assumption is correct. However, hardness tests on green compacts indicate that it is likely to be true. For the indentations from the pyramidal indenter were of good, square shape and not distorted,

especially in the higher density region, which would be expected if separation of the bond and/or some collapse of the surface particles had occurred.

It was at first considered that the higher the isostatic pressure to which a green compact was subjected, the more it would behave like a sintered body, that is, with less particle sliding. This was because at high hydrostatic pressure it was thought that the interfacial pressure between particles would be so high that the frictional stress at the interfaces would overcome the shear strength of the underlying particle material. Therefore deformation in the particles would be more likely to occur than interfacial sliding. It is interesting to see that $d\epsilon/d\rho$ in the sintered body (curve a, figure 5(b)) in fact increases rapidly above 93% of the theoretical density. However, in the green compact (curve b, figure 5(b)) although it still increases, it increases only gradually over the whole density range. By comparing the two curves, for green compacts particle sliding still plays a very important role, even predominating at very high pressures.

Under an isostatic pressure of p, the hydrostatic component of stress at the particle interfaces can be assumed to be p/ρ (where ρ is the corresponding relative density), if stress concentrations can be neglected. Considering a very simple case, where two particles move relative to one another with an interfacial pressure of p/ρ in a direction parallel to the interface, in this case the frictional stress at the interface will be $\mu p/\rho$, where μ is a friction coefficient. In figure 8 the instantaneous shear strength τ of the iron powder (assumed to be half the flow stress) and $\mu p/\rho$ are plotted, with various values of μ, against relative density. Although it can be seen from figure 8

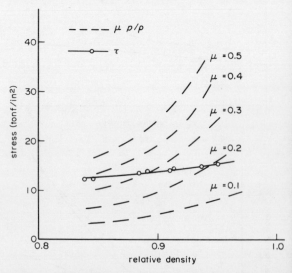

Figure 8. Comparison between the shear strength of the particles and friction stress.

that the friction coefficient at the particle interfaces apparently varies from 0.19 to 0.38 during compaction, there is too little data to verify this; it is a factor that could be investigated in more detail.

Figure 5(d) shows that under hydrostatic pressure the higher the density the more important becomes particle sliding as far as densification is concerned.

This implies that a state of stress deviating from the hydrostatic one would cause sliding more easily and a more rapid increase in density could be expected.

Curve b in figure 5(a) would possibly be influenced by the initial density in such a way that the curve might be shifted horizontally. It is also thought that particle size and surface condition would affect the gradient $d\epsilon/d\rho$ of this curve. Consequently, these factors would possibly make some difference to the curve of relative density versus reduced pressure, which should be examined further in detail.

Relative density–pressure curve for isostatic compaction

As mentioned in the previous section, there have been proposed some equations[4] describing the relationship between density and pressure in the compaction of powders. Among them Heckel's equation is now widely accepted.

$$\ln \frac{1}{1-\rho} = Kp + A \tag{3}$$

or

$$p/\frac{1}{K} = -\ln(1-\rho) - A \tag{3}'$$

where K and A are constants characterising powders. The constant A is considered to be related to an initial density and also to densification which takes place at low pressures before appreciable interparticle bonding occurs. K is related to a measure of the strength of the particle materials. However, it is not clear how K is described in terms of strength during the compacting process. In this work the charac- teristics of powders were neglected in order to investigate the effect of the stress–strain curve of a material on the density–pressure relationship, since this was thought to be the most important factor.

As is seen in figure 7(a) a relative density–pressure curve for Cu powder, which has been predicted on the basis of the results on iron powder (curve b in figure 5(a) and curve a in figure 6) agrees well with the experimental results. Curve b in figure 5 would be influenced by powder characteristics and hence the relative density versus reduced pressure curve would be changed as mentioned previously. However, an attempt will be made here tentatively to discuss the density–pressure relationship on the basis of curve a in figure 6. In the case of sintered metal it can be assumed that no sliding of particles takes place but only deformation of the matrix metal during densifi- cation (from curve a in figure 5(c)). Therefore, a very simple model of a spherical shell subjected to external pressure can be used to discuss the behaviour of a sintered metal under hydrostatic pressure. Using basic plasticity theory the relationship between an external pressure p and a relative density ρ, assuming there is no internal pressure, is described by the following equations:

$$p/\sigma = 2 \ln (R_o/R_i) \tag{4}$$

where σ is the yield stress or the instantaneous flow stress of the material, R_i and R_o are the internal and external radii of the shell respectively. The relative density can be described in terms of R_i and R_o as

$$\rho = (R_o^3 - R_i^3)/R_o^3$$

which leads to

$$(R_o/R_i) = 1/(1-\rho)^{1/3}$$

Substituting in equation (4) gives

$$p/\sigma = -\frac{2}{3} \ln (1 - \rho) \tag{5}$$

which is plotted as curve b in figure 6.

Although this equation does not fit curve a in figure 6 very well, it is interesting to compare it with equation (3)'. If we assume the constant A in equation (3') is zero, then comparing the two equations (3)' and (5), $1/K$ is apparently equivalent to $\frac{2}{3}\sigma$. K values for isostatic compaction have been proposed by some investigators[4,7] and range from 1.1 to 2.3 x 10^{-5} lbf/in^2 for iron powder and from 3.4 to 12 x 10^{-5} lbf/in^2 for Cu powder. Values of $1/K$ derived from these figures are compared with values of $\frac{2}{3}\sigma$ for the present experiment and summarised in table 3.

TABLE 3
Comparison of $1/K$ and flow stress values

	$1/K$ 10^4 lbf/in^2	$\frac{2}{3}\sigma$ 10^4 lbf/in^2
iron powder	4.35~9.1	3.7~4.5 above 85%
copper powder	0.83~2.94	2.2~2.86 above 85%

Although it is not possible to make a direct com- parison between the two parameters, since it is not clear whether $1/K$ corresponds to the same density range as for the present experiments, they appear to agree fairly well with each other.

K values have been considered by other workers[4,7] as constants for powder materials, except in the initial stage of compaction. It is obvious from what has previously been discussed that K is a function of the instantaneous flow stress at the particular density concerned.

In view of this a further investigation of the friction coefficient mentioned in the previous section and the $d\epsilon/d\rho - \rho$ relationship may lead to a better understand- ing of the compaction process.

CONCLUSION

(1) For the case of sintered iron the close interrelation between density increase and strain increment in the matrix metal was examined and verified by the fact that the external work and plastic energy consumption are nearly the same. Hence, it is always possible to separate the two factors, namely, the increase in strength of the matrix metal and the increase in density, both of which influence the apparent strength of a sintered body.

(2) In the case of the green compacts of iron powder, strain increment in the particles is always smaller, because of particle sliding, than in the sintered iron, for the same amount of density increase. Particle

478 THE INTERRELATION OF DENSITY AND HARDNESS IN THE ISOSTATIC COMPACTION OF POWDERS

sliding plays an important role even under very high pressure. As a result of this the plastic energy consumption in the particles is much smaller than the total external work.

(3) A relative density versus reduced pressure curve was obtained for both sintered and green compacts, which can apparently be applied to other materials. A relative density versus pressure relationship for copper was predicted by the use of this curve obtained from iron and shows good agreement with the experimental results. However, it is necessary to investigate the effect of other powder characteristics on the compaction process, possibly by means of investigating the coefficient of friction between the particles.

REFERENCES

1. For example,
G. T. BROWN and P. K. JONES, 1971. Experimental and practical aspects of the powder forging process, *Modern Developments in Powder Metallurgy* (Ed. by H. Hausner), Plenum Press, New York, p. 369.
H. W. ANTES. Cold forging of iron and steel preforms, *ibid*, p. 415.
R. DAVIES and R. H. T. DIXON, 1971. The forging of powder preforms by using petro-forge machines, *Powder Metallurgy*, 14, No. 28, p. 207.
2. H. A. KUHN and C. L. DOWNEY, 1971. Deformation characteristics and plasticity theory of sintered powder materials. *Int. J. of Powder Metallurgy*, 7, p. 15.
3. M. OYANE, S. SHIMA and Y. KOHNO. Plasticity theory for sintered metals, to be published.
4. For example,
R. W. HECKEL, 1961. *Trans. Metl. Soc. A.I.M.E.*, 221, p. 1001.
W. P. JONES, 1960. Fundamental Principles of Powder *Metallurgy*, Arnold, London.
M. F. BURR and M. J. DONACHIE, 1962. *J. Metals*, 151, p. 849.
5. J. M. ALEXANDER and D. QUAINTON, 1971. On the isostatic compaction and hydrostatic extrusion of iron powder. *Proc. 12th Int. MTDR Conf.*, Macmillan, p. 51.
6. S. KAMYAB and J. M. ALEXANDER, 1972. Pressure distribution in hydrostatic extrusion dies, *J. of Strain Analysis*, 7, p. 205.
7. W. R. MORGAN and R. L. SANDS, 1969. Isostatic compaction of metal powders. *Metals and Materials*, 3, 5, p. 85.

METAL FORMING:
PROCESSES AND MACHINES

PLASTIC COMPRESSION OF RECTANGULAR BLOCKS BETWEEN TWO PARALLEL PLATENS

by

F. KANACRI,* C. H. LEE,* L. R. BECK* and SHIRO KOBAYASHI*

SUMMARY

Hill's analysis for compression of a long rectangular strip was modified so that the analysis is applicable to the compression of a rectangular block having any length—width ratio. The predictions were compared with experimental results obtained with annealed pure aluminium in terms of load—displacement relationships, changes in specimen geometry, flowlines, and grid distortions. Also, a comparison was made for the load—displacement curves by the modified analysis and by the sandheap analogy.

INTRODUCTION

In forging operations, metals are deformed into various shapes and sizes. An estimation of forging loads can be made with some success by approximate methods, such as the slab method[1] and the bounding method[2]. Several investigators in the past have attempted an analytical treatment of the problem of free-surface barrelling in forging. However, little effort has been devoted to an understanding of the deformation characteristics in forging where simplifying factors, such as plane-strain and axial symmetry, do not exist.

The present investigation deals with plastic compression of rectangular blocks between parallel platens. The prediction of changes in shape as well as the determination of forging loads during compression is of primary concern in this study.

A few investigators proposed theoretical solutions to problems of this type. Hill[3] formulated a theory for the calculation of stresses and displacements in a thin sheet of arbitrary shape that is being plastically compressed between parallel plates. The theory provided a means of estimating the coefficient of friction in a compression test from observations of the change in shape of a rectangular lamina. Hill[4] also proposed a general method of analysis, combining flexibility with rigorous principles, for any technological forming process. The details of procedure were illustrated by a preliminary analysis of the compression of a prismatic bar, with an arbitrary cross-section, flat-tool forging and bar drawing.

Baraya and Johnson[5] conducted experiments with flat-bar forging to obtain experimentally the co-efficients of sideways spread, the maximum amount of bulge and the bulge profiles. They concluded that the coefficient of spread as evaluated by Hill fits the experimental results fairly well. Johnson et al.[6] applied the sandheap analogy[7] to provide an over-estimation of the load required for the quasi-state compression of a rectangular block. They further indicated the usefulness of the sandheap analogy in predicting gross flow during compression. While Johnson drew attention only to a similarity between two-dimensional pressure distribution during the compression and the sandheap analogy, Eder[8,9] established the sandheap analogy by deriving the differential expression for the function involved in the stress distribution.

In this study, Hill's analysis for the compression of a long rectangular strip was modified so that the analysis is applicable to the compression of a rectangular block having any length—width ratio. The predictions were compared with experimental results obtained with annealed pure aluminium in terms of load—displacement relationships, changes in specimen geometry, flowlines and grid distortions. Also, a comparison was made for the load—displacement curves by the modified Hill's analysis and by the sandheap analogy.

MODIFICATION OF HILL'S ANALYSIS

A rectangular block having a thickness h small in comparison to the other dimensions (width $2W$, length $2B$) is compressed between parallel platens. The restriction of small thickness is made for the sake of simplicity, since the variations of stress and strain through the thickness can then be neglected.

(a) If the width $2W$ is greater than about ten times the length $2B$, and if B/h is of order unity, the variation in stress across the length may be negligible in comparison with the variations along the width. The only stress components are then the compressive stress q in the width direction, the pressure p and the friction. It is assumed that the frictional stress can be represented by a constant fraction of the shear strength of the material (λk). The equation of equilibrium becomes

$$h\frac{dq}{dy} = -2\lambda k, \qquad \text{where } 0 \leqslant \lambda \leqslant 1.0. \qquad (1)$$

The coordinate system is shown in figure 1.

Linearizing the von Mises yield criterion,

$$p - \frac{1}{2}q = mY, \qquad (2)$$

where Y is the flow stress and m takes a value between 0.866 and 1.0.

* Mechanical Design, University of California, Berkeley, California

Integrating equation 1 and using the boundary condition that $q = 0$ at $y = W$,

$$q = \frac{2\lambda k}{h}(W - y). \tag{3}$$

From (2) and (3),

$$p = mY + \frac{\lambda k}{h}(W - y). \tag{4}$$

The Levy–von Mises plastic stress–strain relations are

$$\dot{\epsilon}_y/\dot{\epsilon}_z = \frac{2q - p}{2p - q}, \tag{5}$$

where

$$\dot{\epsilon}_y = \frac{\partial v}{\partial y} \qquad \text{and} \qquad \dot{\epsilon}_z = \frac{\dot{h}}{h}.$$

Substituting (3) and (4) in (5),

$$\frac{\partial v}{\partial y} = -\frac{\dot{h}}{h}\left[\frac{1}{2} - \frac{3}{2}\frac{v}{h}(W - y)\right] \tag{6}$$

where $v = \lambda/\sqrt{3m}$, taking $k = Y/\sqrt{3}$.

Note that when friction is small, v corresponds to the coefficient of friction defined by the ratio of frictional stress and the pressure.

(i) When $vW/h \leqslant \frac{1}{3}$, then $v \geqslant 0$ for $0 \leqslant y \leqslant W$ and the velocity component in the y-direction is obtained by integrating (6) as

$$v = -\frac{\dot{h}}{h}\left[\frac{1}{2}y - \frac{3v}{2h}\left(Wy - \frac{1}{2}y^2\right)\right] \tag{7}$$

The velocity component u in the x-direction can be found from the incompressibility condition

$$u = -\frac{\dot{h}}{h}\left[\frac{1}{2} + \frac{3v}{2h}(W - y)\right]x. \tag{8}$$

(ii) When $\frac{vW}{h} > \frac{1}{3}$

$$v = 0 \qquad \text{and} \qquad u = -\frac{\dot{h}}{h}x \tag{9}$$

for the range of y given by $0 \leqslant y \leqslant W - \frac{h}{3v}$.

For $W - \frac{h}{3v} < y < W$,

$$v = \int_{W - \frac{h}{3v}}^{y} -\frac{\dot{h}}{h}\left[\frac{1}{2} - \frac{3v}{2h}(W - y)\right]dy$$

$$= -\frac{\dot{h}}{2h}\left[y - \frac{3v}{2h}(2Wy - y^2) + \frac{3v}{2h}\left(W - \frac{h}{3v}\right)^2\right] \tag{10}$$

and

$$u = -\frac{\dot{h}}{h}\left[\frac{1}{2} + \frac{3v}{2h}(W - y)\right]x$$

(b) If the length of the rectangular block is much greater than the width, the foregoing analysis can be carried out along the length. Corresponding velocity distributions are obtained by replacing W by B and interchanging x and y, and u and v, in the expressions given by equations (7) to (10).

(c) In order to obtain the velocity field for rectangular blocks having various width and length ratios, the velocity fields for the above two extreme cases are combined according to

$$u = c_B u_B + c_W u_W$$
$$v = c_B v_B + c_W v_W \tag{11}$$

where (u_B, v_B) and (u_W, v_W) are the velocity fields for the cases where $B \gg W$ and $W \gg B$, respectively, and c_B and c_W are weight factors defined by

$$c_B = \frac{B}{W + B} \qquad \text{and} \qquad c_W = \frac{W}{W + B} \tag{12}$$

It must be noted that the dimensions W and B represent gross flow and are the average values of actual width (w) and length (b) respectively, at any stage during compression as shown in figure 1.

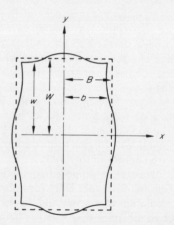

Figure 1 Deformed rectangular block.

In the following, the velocity components for various friction conditions and specimen dimensions $(W \geqslant B)$ are summarized.

(i) When $vW/h \leqslant 1/3$ and $vB/h \leqslant 1/3$

$$u = -\frac{\dot{h}}{h}\left[c_B\left\{\frac{1}{2}x - \frac{3}{2h}\left(Bx - \frac{1}{2}x^2\right)\right\}\right.$$
$$\left. + c_W\left\{\frac{1}{2} + \frac{3v}{2h}(W - y)\right\}x\right]$$

$$v = -\frac{\dot{h}}{h}\left[c_B\left\{\frac{1}{2} + \frac{3v}{2h}(B - x)\right\}y\right.$$
$$\left. + c_W\left\{\frac{1}{2}y - \frac{3}{2h}\left(Wy - \frac{1}{2}y^2\right)\right\}\right] \tag{13}$$

for the entire region given by $0 \leqslant x \leqslant b$ and $0 \leqslant y \leqslant w$.

(ii) When $vW/h > 1/3$ and $vB/h \leqslant 1/3$

$$u = -\frac{\dot{h}}{h}\left[c_B\left\{\frac{1}{2}x - \frac{3v}{2h}\left(Bx - \frac{1}{2}x^2\right)\right\} + c_W x\right]$$
(14a)

$$v = -\frac{\dot{h}}{h}\left[c_B\left\{\frac{1}{2} + \frac{3v}{2h}(B-x)\right\}y\right]$$

are valid in the region defined by $0 \leqslant x \leqslant b$ and $0 \leqslant y \leqslant W - h/3v$, while the velocity field for the region given by $0 \leqslant x \leqslant b$ and $W - h/3v \leqslant y \leqslant w$ is expressed by

$$u = -\frac{\dot{h}}{h}\left[c_B\left\{\frac{1}{2}x - \frac{3v}{2h}\left(Bx - \frac{1}{2}x^2\right)\right\}\right.$$

$$\left. + c_W\left\{\frac{1}{2} + \frac{3v}{2h}(W-y)\right\}x\right]$$

$$v = -\frac{\dot{h}}{h}\left[c_B\left\{\frac{1}{2} + \frac{3v}{2h}(B-x)\right\}y\right.$$

$$\left. + c_W\left\{\frac{1}{2}y - \frac{3v}{2h}\left(Wy - \frac{1}{2}y^2\right) + \frac{3v}{4h}\left(W - \frac{h}{3}\right)^2\right\}\right]$$
(14b)

(iii) When $vW/h > 1/3$ and $vB/h > 1/3$ the velocity components are defined in the following regions:

In the region given by $0 \leqslant x \leqslant B - h/3v$ and $0 \leqslant y \leqslant W - h/3v$,

$$u = -\frac{\dot{h}}{h}[c_W x]$$

$$v = -\frac{\dot{h}}{h}[c_B y].$$
(15a)

In the region given by $0 \leqslant x \leqslant B - h/3v$ and $W - h/3v \leqslant y \leqslant w$,

$$u = -\frac{\dot{h}}{h}\left[c_W\left\{\frac{1}{2} + \frac{3v}{2h}(W-y)\right\}x\right]$$

$$v = -\frac{\dot{h}}{h}\left[c_B y + c_W\left\{\frac{1}{2}y - \frac{3v}{2h}\left(Wy - \frac{1}{2}y^2\right)\right.\right.$$

$$\left.\left. + \frac{3v}{4h}\left(W - \frac{h}{3v}\right)^2\right\}\right]$$
(15b)

In the region $B - h/3v \leqslant x \leqslant b$ and $0 \leqslant y \leqslant W - h/3v$,

$$u = -\frac{\dot{h}}{h}\left[c_B\left\{\frac{1}{2}x - \frac{3v}{2h}\left(Bx - \frac{1}{2}x^2\right)\right.\right.$$

$$\left.\left. + \frac{3v}{4h}\left(B - \frac{h}{3v}\right)^2\right\} + c_W x\right]$$

$$v = -\frac{\dot{h}}{h}\left[c_B\left\{\frac{1}{2} + \frac{3v}{2h}(B-x)\right\}y\right].$$
(15c)

Finally, in the region defined by $B - h/3v \leqslant x \leqslant b$ and $W - h/3v \leqslant y \leqslant w$,

$$u = -\frac{\dot{h}}{h}\left[c_B\left\{\frac{1}{2}x - \frac{3v}{2h}\left(Bx - \frac{1}{2}x^2\right) + \frac{3v}{4h}\left(B - \frac{h}{3v}\right)^2\right\}\right.$$

$$\left. + c_W\left\{\frac{1}{2} + \frac{3v}{2h}(W-y)\right\}x\right]$$

$$v = -\frac{\dot{h}}{h}\left[c_B\left\{\frac{1}{2} + \frac{3v}{2h}(B-x)\right\}y\right.$$

$$\left. + c_W\left\{\frac{1}{2}y - \frac{3v}{2h}\left(Wy - \frac{1}{2}y^2\right) + \frac{3v}{4h}\left(W - \frac{h}{3}\right)^2\right\}\right]$$
(15d)

Equations (13) to (15) define the velocity distributions over the entire tool—workpiece interface during compression. Thus, the step-by-step calculations of incremental displacements by these equations permit us to obtain the spread contours, flowlines and grid distortions at any stage of compression.

(d) The forging load can be determined by equation (4). Combining the two extreme cases for the pressure,

$$p = mY + \frac{\lambda k}{h}\{c_B(B-x) + c_W(W-y)\}. \quad (16)$$

Integrating the pressure over the contact surface, the total forging load P is expressed by

$$\frac{P}{4BW} = mY + \frac{\lambda k}{2h}(c_B B + c_W W)$$

$$= Y\left(\frac{\lambda}{2\sqrt{3}h}\right)\left(\frac{B^2 + W^2}{B + W}\right)$$
(17)

The way in which the pressures for the two extreme cases are combined is, by all means, arbitrary. Therefore, the approximation of the expression given by equation (17) for the forging load is not known. However, the experiments later show that this simple expression predicts the load—displacement curves very well for a certain range of length—width ratios.

SANDHEAP ANALOGY

The sandheap analogy for the solution of torsion of prismatic bars of noncircular section is well known. Under certain assumptions the analogy is also possible with the problem of compression of noncircular prismatic blocks. When the angle between the x-direction and the direction of frictional stress, λk, is given by $\theta(\leqslant 90°)$, the equilibrium equations are

$$\frac{\partial q_x}{\partial h} = -\frac{2\lambda k}{h}\cos\theta,$$

$$\frac{\partial q_y}{\partial y} = -\frac{2\lambda k}{h}\sin\theta, \quad (18)$$

where q_x and q_y are the compressive stresses in the x and y directions respectively.

It is now assumed that the direction along which the frictional stress acts coincides with the direction of the maximum slope of the distribution of the pressure p. Then,

$$\frac{p}{y} = \frac{p}{x}\tan\theta. \quad (19)$$

Combining (18) and (19)

$$\frac{\partial p/\partial y}{\partial p/\partial x} = \frac{\partial q_y/\partial y}{\partial q_x/\partial x}, \quad (20)$$

and noting the relationship of equation (20), the following differential expression for the pressure p is derived:

$$\left(\frac{\partial p}{\partial x}\right)^2 + \left(\frac{\partial p}{\partial y}\right)^2 = \left(\frac{\partial q_x}{\partial x}\right)^2 + \left(\frac{\partial q_y}{\partial y}\right)^2 = \left(\frac{2\lambda k}{h}\right)^2 \quad (22)$$

or

$$|\operatorname{grad} p|^2 = \left(\frac{2\lambda k}{g}\right)^2.$$

With the property given by equation (22), the condition that p = constant along the edge determines the pressure distribution. Thus, the sandheap analogy to the compression problem is established. The pressure distribution roof is obtained by constructing a constant-slope roof starting a wall around the boundary of height Y. The total forging load is represented by the volume beneath the pressure distribution roof.

For a rectangular block the calculation of this volume is simple and results in the forging load expressed by

$$\frac{P}{4BW} = Y + \frac{\lambda k}{h}\left(B - \frac{1}{3}\frac{B^2}{W}\right)$$
$$= Y\left|1 + \frac{\lambda}{\sqrt{3}}\left(\frac{B}{h} - \frac{1}{3}\frac{B}{h}\frac{B}{W}\right)\right| \quad (23)$$

By comparing (17) and (23), it can be seen that the average forging pressures by both methods become identical for all friction conditions when $B/W = 0.5$ (that is $W/B = 2$) and m is taken as unity. It is also noted that the average pressure increases with increasing width according to (17), while (23) indicates that the effect of width on the average pressure diminishes when the width is very large in relation to the length. This point was not examined in the present study, and the average pressures by both equations are more or less the same within the range of investigation. Therefore, the theoretical load—displacement relationships were calculated using (17) for comparison with experiments in the later section.

EXPERIMENTS

Experiments were conducted with a compression apparatus placed on a 160 000-lb Tinius Olson testing machine. Quasi-state compression of rectangular blocks of annealed pure aluminium was performed under two friction conditions at the tool—workpiece interface for several specimen dimensions. The load—displacement curves were obtained by a recorder of the testing machine using a deflectometer for displacement measurements. Spread profiles and grid distortions on the plane perpendicular to the compression axis were examined.

Rectangular blocks having various width—length ratios were prepared from a 3-in diameter bar of commercially pure aluminium with the compression axis coincident to the bar axis. The dimensions of the specimens are summarised in table 1.

TABLE 1
Specimen dimensions
(subscript 0 indicates the initial values)

specimen no.	$2W_0$ (in)	$2B_0$ (in)	W_0/B_0	h_0 (in)
1	0.75	0.75	1	0.375
2	1.00	0.75	1.33	0.375
3	1.50	0.75	2	0.375
4	2.25	0.75	3	0.375

The specimens were split along the central plane perpendicular to the forging direction, and square grids were printed on this central plane by the photo-resist method.

The compression dies were made of hardened and lapped Graph-Mo. Two halves of the specimen, for each experiment, were placed together and joined with a strip of masking tape for alignment. The forging was performed with a speed of 0.004 in/min. Two types of loading were used: the specimen was compressed up to 50 per cent reduction in height continuously in one; in the other, loading was interrupted for each 10 per cent reduction in height. In the latter, specimens were unloaded after each 10 per cent reduction and the grid patterns on the central plane were photographed in order to examine the spread profiles and grid distortions at 10 per cent reduction increments. The two friction conditions at the interface were achieved by cleaning the specimens and dies with acetone for the dry condition, and by applying a thin film of lanoline to the specimen for the case of lubrication. The procedure for ensuring the two friction conditions was repeated, for the tests of interrupted loading, before each loading commenced.

The coordinates of grid points were measured on the photographs of grid patterns by a microscope and were punched on IBM cards. Then the grid distortions, spread profiles and flowlines were plotted by a computer.

RESULTS AND DISCUSSION

Continued-loading and interrupted-loading
The load—displacement curves obtained experimentally in continued-loading and interrupted-loading of the specimen of $W_0/B_0 = 2$ are shown for two friction conditions in figure 2. In interrupted-loading

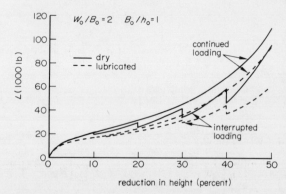

Figure 2 Experimental load—displacement curves in continued loading and interrupted loading of the specimen, $W_0/B_0 = 2$, for two friction conditions.

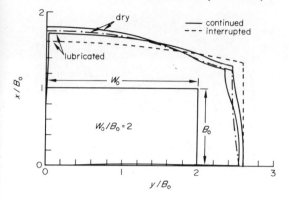

Figure 3 Comparison of spread contours at 50 per cent reduction in height in continued and interrupted-loading tests.

the forging load required to deform the specimen after unloading is less than that before unloading. Therefore, the load—displacement curves in interrupted-loading are lower than those recorded in continued-loading. This observation is true for all the specimen dimensions regardless of whether the inter-

face was dry or lubricated. The reason for this difference appears to be the difference in actual friction in the two loading types. The examination of the appearance of the die-specimen contact surfaces indicates the difference in the actual state of friction. Apparently, the frictional constraints, in both dry and lubricated cases, are less in interrupted tests than those under the continued-loading condition. This argument is indeed confirmed qualitatively in figure 3 by comparing the spread contours at the 50 per cent reduction in height obtained under the two types of tests. First, observing the spread contours for dry and lubricated cases, it is concluded that the amount of total spread along the x-axis is larger, and the spread along the y-axis tends to become smaller with increased frictional constraint. Furthermore, the curvatures of the contour are more severe for larger frictional constraints. With these observations it is obvious that the frictional constraints are less in the interrupted test than in the continued-loading tests when the spread contours for both loading conditions are compared. It appears that the load—displacement

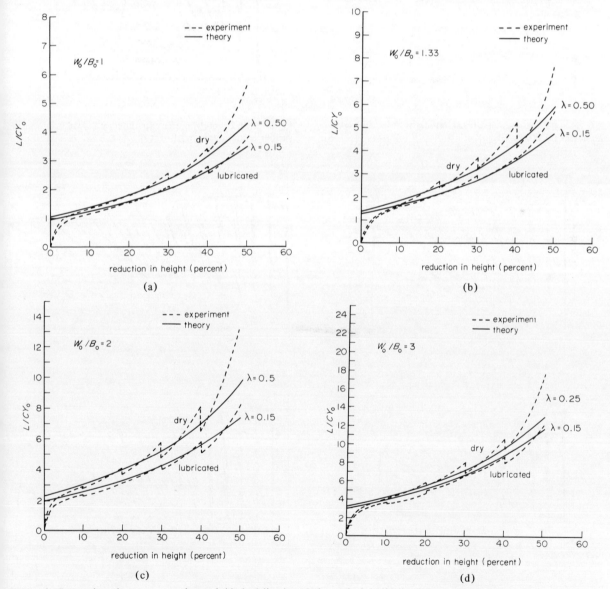

Figure 4. Comparison between experimental (dashed lines) and theoretical (solid lines) load—displacement curves for various specimen dimensions under dry and lubricated conditions; (a) $W_0/B_0 = 1$, (b) $W_0/B_0 = 1.33$, (c) $W_0/B_0 = 2$, (d) $W_0/B_0 = 3$.

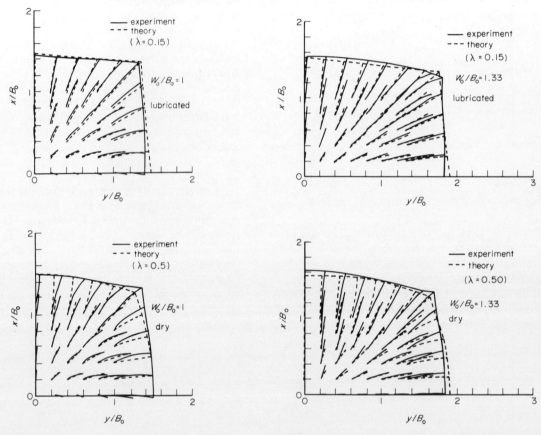

Figure 5(a) Experimental spread contours at various reductions in height for $W_0/B_0 = 1$.

Figure 5(b) Theoretical and experimental spread contours at various reductions in height for $W_0/B_0 = 1.33$.

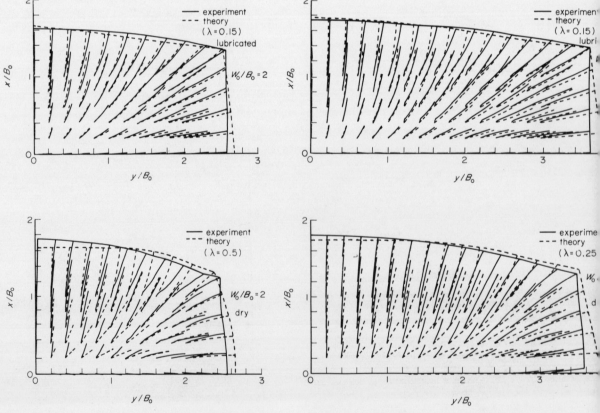

Figure 5(c) Theoretical and experimental spread contours at various reductions in height for $W_0/B_0 = 2$.

Figure 5(d) Theoretical and experimental spread contours at various reductions in height for $W_0/B_0 = 3$.

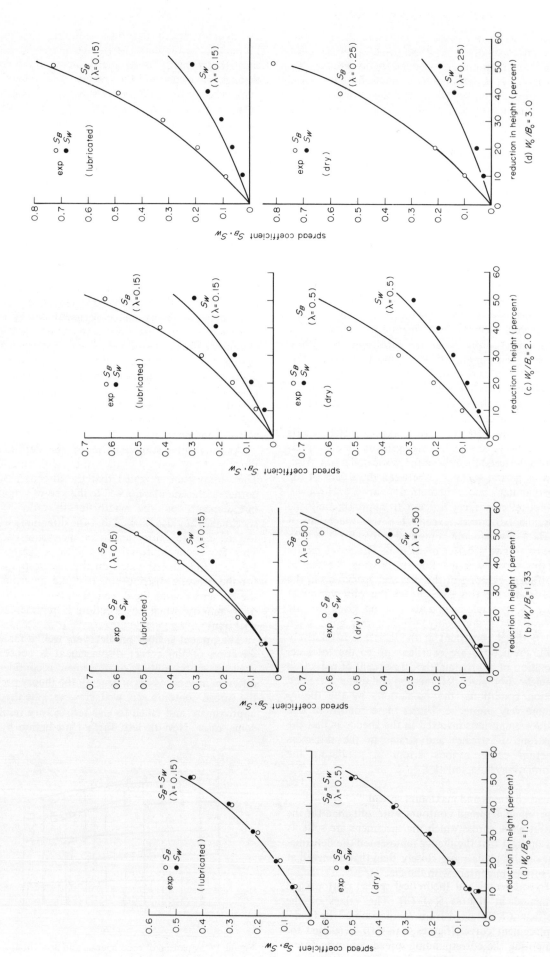

Figure 6 The maximum spreads in the x and y-directions for various specimen sizes under dry and lubricated conditions. (a) $W_0/B_0 = 1$, (b) $W_0/B_0 = 1.33$, (c) $W_0/B_0 = 2$, (d) $W_0/B_0 = 3$.

curves in interrupted loading represent the relationships under relatively uniform frictional constraint during the compression. Since the theory assumes the friction stress to be a constant fraction of the shear yield-stress in the following, the theoretical calculations were compared with the results obtained from the interrupted-loading tests.

Load—displacement relationship

The experimental and theoretical load—displacement relationships were compared in figure 4(a)—(d). In the figure, the load L (lb) is normalised by dividing by cY_0 where $Y_0 = 13\ 000$ p.s.i. and $c = 9/16$ in^2. In computing the load—displacement relationships, the current yield stress Y was estimated from the stress—strain curve of the material at the strain determined from the current reduction in height assuming homogeneous deformation. The value of m in equation (17) was assumed to be 0.9. Figure 4(a)—(d) reveal that the theory with the coefficient $\lambda = 0.15$ predict very well the experimental behaviour for all the specimen sizes under the lubricated condition. However, the slope of the experimental curve differs from that of the computed curve even during one-step deformation of 10 per cent reduction in height. Furthermore, deviation of the slopes tends to be more pronounced as the total deformation increases.

Under the dry condition experimental results show more variations in forging loads (due to interrupting the tests) than those for lubricated cases. Also, there is greater difference between the slopes of the experimental and theoretical curves. However, theoretical curves for $\lambda = 0.5$ fit approximately the experimental curves except for the specimen of $W_0/B_0 = 3$ where the computed curve for $\lambda = 0.25$ appears to give a better prediction. The exact reason for this anomaly is not known, but may be due to insufficient cleaning of the dies and specimens in the experiment for this case. From the observation of figure 4, general comments can be given to the applicability of the theory, which is very good when the frictional constraint at the interface is relatively small. Predictions are excellent up to the total deformation of approximately 30 per cent of reduction in height. For larger deformation and under increased friction conditions, the predictions by the theory become very imprecise. Under these conditions one of the assumptions involved in the theory is that the variations of stresses and strains in the thickness direction cause serious errors in predicting the deformation characteristics.

Spread contours and maximum spread

Experimental spread contours were obtained on the central plane across which the specimens are split. It was thought that the theory represented the deformation on this plane more closely than that observed at the surface in contact with the die.

Experimental and theoretical spread contours are compared in figures 5(a)—(d). The values of the coefficient λ obtained from comparison of the load—displacement curves (figures 4(a)—(d)) were used for computing the corresponding spread contours. For a square block given in figure 5(a), the predictions are

excellent for both friction conditions even at the 50 per cent reduction in height. This can be expected since the geometry is symmetric about the 45° line through the origin. For the specimen of $W_0/B_0 = 1.33$ comparison in figure 5(b) still shows good agreement between theory and experiment. However, with increasing W_0, theoretical contours begin to deviate from the observed contours. Particularly, the results shown in figure 5(d) for the specimen of $W_0/B_0 = 3$ under the dry condition are not encouraging.

In order to express the spread quantitatively, the maximum spreads along the x and y directions are plotted in figure 6. The maximum spreads along the two axes are defined by

$$S_x = \frac{b - B_0}{B_0} \quad \text{and} \quad S_y = \frac{w - W_0}{W_0}$$

where b and w are the half-length and half-width measured along the x and y axes, respectively. Agreement between theory and experiment is very good, particularly under the lubricated condition for the maximum spread in the x-direction. The predictions of the maximum spread in the y-direction, however, deviate more from the measurements, and the amount of deviation increases with increasing reduction in height. This is because the maximum spread in the y-direction reflects sensitively the distribution of the velocity components.

Another interesting feature of the deformation mode is the movement of the corner of the specimen. The observations revealed that the direction of the corner displacement is not 45° to the axis of symmetry but depends on the width—length ratio of the specimen. They also showed that the direction changes toward the 45° direction with increasing friction. This fact is in conformity with the suggestion by Johnson and others[6] based on the sandheap analogy for the pressure distribution, that the movement of the corner be along the direction with 45° to the axis of symmetry when the friction is represented by sticktion.

The present analogy predicts very well as far as the direction of the corner displacement is concerned, although the magnitude of movement is considerably different in some cases. In general, the theory predicts the spread contours very well. However, the theory is approximate and failed to give satisfactory results in some cases. Nevertheless, further prediction by this

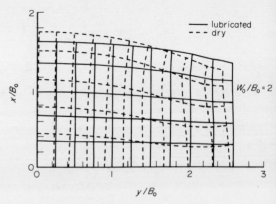

Figure 7 Comparison of experimental grid distortions for two friction conditions; $W_0/B_0 = 2$.

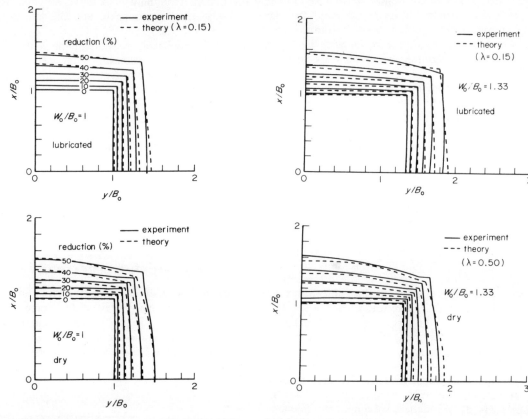

Figure 8(a) Theoretical and experimental flowlines for two friction conditions, $W_0/B_0 = 1$.

Figure 8(b) Theoretical and experimental flowlines for two friction conditions, $W_0/B_0 = 1.33$.

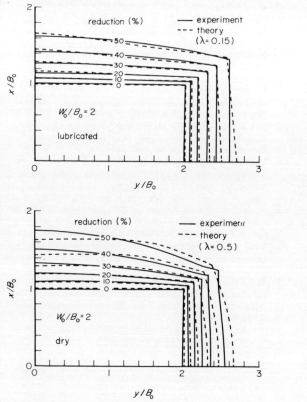

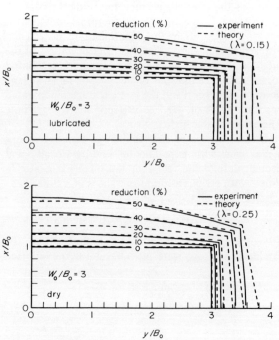

Figure 8(c) Theoretical and experimental flowlines for two friction conditions, $W_0/B_0 = 2$.

Figure 8(d) Theoretical and experimental flowlines for two friction conditions, $W_0/B_0 = 3$.

theory was made for the theoretical spread contours for $\lambda = 1.0$. It may be of interest to note that under extremely high friction (seizure or sticktion) the corner moves in the direction of $45°$ to the axis and that the sharp corner of the rectangular specimen tends to disappear, resulting in a rounded specimen.

Grid distortions, flowlines

Experimental grid distortions at 50 per cent reduction in height are compared for the two friction conditions. The square specimen was compressed, grids distorted more and the spreads along the axes were larger for increased friction.

As the width of the specimen increases, the effect of friction on grid distortion is characterised by less spread in the width direction and more spread in the length direction under the dry condition. These characteristics become clearer with the specimens of larger widths. A typical example is shown in figure 7 for $W_0/B_0 = 2$. It is of interest to note that the frictional constraint in restricting the displacement appears to be more effective in the y-direction than that in the x-direction. Therefore, when the width is very large relative to the length, even a slight trace of friction at the interface may affect the spread in the width direction.

Experimental flowlines were obtained by tracing the grid points at various reductions in height. They are compared with the theoretical flowlines in figures 8(a)–(d). For square specimens the flowlines computed are in excellent agreement with those observed for both friction conditions (figure 8(a)). The flowlines for other specimens show good agreement between theory and experiment for the cases where the interfaces are lubricated. However, the deviations of the prediction become appreciable when the frictional constraint is large under the dry condition. Most differences are observed in the central region of the quarter specimen and along the edges of free surface. In the theory the frictional constraint is not as effective as in the experiment to restrict the displacement in the y-direction. The deviation of the predicted flowlines from those observed is attributed to the fact that the split plane did not remain plane during compression and the two halves were separated near the edges of the free surfaces due to barrelling when friction was high.

It is surprising to find that in general the theory agrees well with experimental findings locally and overall, despite the fact that the theory is relatively simple with various assumptions and approximations.

SUMMARY AND CONCLUSIONS

Two theories, the modified Hill theory and the sandheap analogy, were introduced for the solution of the three-dimensional deformation problem. The modified Hill theory predicted not only the load–displacement relationship but also the detailed deformation characteristics, while the sandheap analogy is only applicable to finding the load. The load–displacement relationships obtained by the two theories were almost identical.

The theoretical results were computed by the modified Hill theory in terms of load–displacement curves, spread profiles, grid distortion and flowlines. Experiments performed with commercially-pure aluminium revealed that, in general, the theoretical predictions are in good agreement with the experimental results locally and overall despite the assumptions and approximations involved in the development of the theory.

ACKNOWLEDGEMENTS

The authors wish to thank the Air Force Materials Laboratory, Wright-Patterson Air Force Base, for its contract No. F33615-70-C-1320, and the National Science Foundation for its grant NSF GK–14946, under which the present work was possible.

REFERENCES

1. T. ALTAN and R. J. FIORENTINO, 1971, Predictions of loads and stresses in closed-die forging. *Journal of Engineering for Industry, ASME Trans.*, pp. 477–484.
2. H. TAKAHASHI and T. MURAKAMI, 1971, Effect of tool angle and friction in an open-die forging in axisymmetry: Part 2, Theory. *Journal of Japan Society for Technology of Plasticity*, vol. 12, no. 121, pp. 122–133.
3. R. HILL, 1950, On the inhomogeneous deformation of a plastic lamina in a compression test. *Philosophical Magazine*, vol. 41, p. 733.
4. R. HILL, 1963, A general method of analysis for metal-working processes. *Journal of Mechanics and Physics of Solids*, vol. 11, pp. 305–326.
5. G. L BARAYA and W. JOHNSON, 1964, Flat-bar forging. *Proceedings of the Fifth Machine Tool Design and Research Conference*, pp. 449–469.
6. W. JOHNSON, R. A. C. SLATER and A. S. YU, 1966, The quasi-static compression of non-circular prismatic blocks between very rough platens using the friction Hill concept. *Int. J. Mech. Sci.*, vol. 8, pp. 731–738.
7. A. NADAI, 1950, *Theory of flow and fracture of solids*, vol. 1, McGraw-Hill Book Co., p. 494.
8. E. EDER, 1967, Der Stofffluß und die spezifische Normalkraft in der Grenzschicht plastisch uniformbarer Körper beim freien Stauchen zwischen planparallelen Preßflächen. *Eisenhüttenwesen*, S.821/29.
9. E. EDER, 1968, Anwendung des Sandhügelgleichnisses auf das freie Stauchen plastisch deformierbarer Körper zwischen planparallelen Preß flächen. *Eisenhüttenwesen*, S.511/14.

MECHANICS OF PLANE-STRAIN DEEP INDENTATION WITH FLAT PUNCHES

by

S. SOHRABPOUR* and SHIRO KOBAYASHI†

SUMMARY

The detailed mechanics of flat-punch indentation into specimens of finite dimensions under plane-strain conditions were obtained, based on the observed flow patterns of commercially-pure, hard copper specimens. Slip-line fields were constructed and a stress calculation was made for some cases. The effects were examined of friction at the interface, and initial dimensions of the specimens, on the deformation characteristics.

INTRODUCTION

In recent years, the process of indentation has attracted special attention as one of the basic operations involved in advanced forging techniques, such as incremental forging. The pattern of metalflow during the relatively deep penetration of the punch is of primary interest in incremental forging. The present investigation deals with this aspect of indentation under plane-strain conditions.

For plane-strain flat-punch indentation, a number of investigators obtained solutions at the yield-point using the slip-line method for rigid-plastic materials with no workhardening[1-4]. The problem of indentation into specimens of finite dimensions was treated by Lee and Kobayashi[5]. Using actual materials, they computed the detailed mechanics of indentation beyond the yield-point. However, the range of punch penetration considered in the investigation was such that the overall geometrical change of the workpiece was insignificant. Johnson and Kudo[6] employed the upper-bound method to estimate the pressure for indentation of a semi-infinite body in their investigation of plane-strain deep indentation. In spite of these extensive studies, there is still a dearth of information on metalflow during the forging operation which takes into account the workhardening property of the material.

Oxley[7] has shown how the basic equations for the stress variation along the slip-lines are modified if the yield-stress of the material varies throughout the field. He constructed a slip-line field in the machining process from the observed boundary of the plastic zone. The hodograph for this slip-line field was then constructed and compared with the observed velocity distribution. The initial shape of the plastic zone was adjusted until a more acceptable hodograph was obtained.

Inasmuch as experimental observations of velocities must be made for constructing a slip-line field, these velocities may then be used directly to complete the corresponding slip-line field, thus obtaining a complete solution to the problem. This is the procedure exactly followed in the visioplasticity method[8-12], which has been used extensively to solve extrusion problems[13,14,15]. Usui[16,17] applied the visioplasticity technique to the analysis of discontinuous chip formation. He used the equilibrium equations referring to the slip-line directions for deriving the stresses. Using a new technique for observing the flow pattern in chip formation, Childs[18] made a stress analysis based both on the slip-line and principal-stress fields.

In our investigation, the detailed mechanics of flat-punch indentation into specimens of finite dimensions under plane-strain conditions were obtained, based on the observed flow patterns of commercially-pure, hard copper specimens. Slip-line fields were constructed and a stress calculation made for some cases. The effects of friction at the interface and initial dimensions of the specimens on the deformation characteristics were examined.

METHOD OF ANALYSIS

The process of indentation is a non-steady-state problem. The plastic zone size and the specimen geometry vary from moment-to-moment. For rigid-plastic materials with workhardening, the flow stress $\bar{\sigma}$ is expressed as a function of the effective strain $\bar{\epsilon}$, where $\bar{\epsilon}$ and $\bar{\sigma}$ are defined by

$$\bar{\epsilon} = \int d\bar{\epsilon}$$

$$= \frac{\sqrt{2}}{3} \int \sqrt{(d\epsilon_1 - d\epsilon_2)^2 + (d\epsilon_2 - d\epsilon_3)^2 + (d\epsilon_3 - d\epsilon_1)^2}$$

and

$$\bar{\sigma} = \sqrt{\tfrac{1}{2}\left[(\sigma_1 - \sigma_2)^2 + (\sigma_2 - \sigma_3)^2 + (\sigma_3 - \sigma_1)^2\right]}$$

with the principal stresses σ_1, σ_2, σ_3 and the principal incremental strains $d\epsilon_1$, $d\epsilon_2$, $d\epsilon_3$. The effective strain distribution, and thus the yield stress distribution, can be obtained at any instant by integrating the incremental effective strain for each element from the beginning of deformation to that instant. This necessitates a continuous observation of flow patterns during the process. At any stage of deformation, the incremental displacement or velocity distribution can be obtained by the superposition of two consecutive flow patterns having a short time-interval. The incremental strain components or strain-rate components are then calculated from the observed displacement field. The known strain-rate components can be used not only for finding the yield-stress distribution due to workhardening, but also for calculating the stresses.

* Presently in Pahlavi University, Shiraz, Iran.
† Mechanical Design, University of California, Berkeley, California.

This is a principle of the visioplasticity method. Since the workhardening characteristics of the material are expressed by $\bar{\sigma} = H(\bar{\epsilon})$, we assume the materials are incompressible, obeying the von Mises yield criterion and its associated flow rule.

With reference to the Cartesian coordinate system, x, y, z, the governing equations under plane-strain conditions ($\dot{\epsilon}_z = 0$) are

$$\frac{\partial \sigma_x}{\partial_x} + \frac{\partial \tau_{xy}}{\partial_y} = 0 \qquad \frac{\partial \tau_{xy}}{\partial_x} + \frac{\partial \sigma_y}{\partial_y} = 0 \qquad (1)$$

$$\sqrt{\tfrac{3}{4}(\sigma_x - \sigma_y)^2 + 3\tau_{xy}^2} = \sigma \qquad (2)$$

and

$$\frac{\dot{\epsilon}_x}{\sigma_x - \sigma_m} = \frac{\dot{\epsilon}_y}{\sigma_y - \sigma_m} = \frac{\dot{\gamma}_{xy}}{2\tau_{xy}} \qquad (3)$$

where σ_x, σ_y, σ_{xy} are stress components and $\dot{\epsilon}_x$, $\dot{\epsilon}_y$, $\dot{\gamma}_{xy}$ are strain-rate components. Equations (1) and (2) can be solved for three stress components by the method of characteristics, namely, the slip-line method. Since the construction of a slip-line field is extremely difficult, particularly when the yield stress varies throughout a deforming body, the known strain-rate components are used in various ways for the stress analysis. In one method[12] by combining (2) and (3),

$$\dot{\epsilon}_x = -\dot{\epsilon}_y = \dot{\lambda}(\sigma_x - \sigma_m),$$
$$\dot{\gamma}_{xy} = 2\dot{\lambda}\tau_{xy}, \qquad (4)$$

where

$$\dot{\lambda} = \frac{3\,\dot{\epsilon}}{2\,\bar{\sigma}} \qquad \text{and} \qquad \sigma_m = \sigma_z = \frac{1}{2}(\sigma_x + \sigma_y)$$

Equation (4) provides two relationships for three stress components and the third will be one of the two equilibrium equations given in (1). The remaining equilibrium equation can be used partially for sub-stantiating insufficient boundary conditions and also for examining the accuracy of the computed solution.

In a second method[18], the principal stress field is constructed from the strain-rate components according to

$$\tan 2\theta = \frac{\dot{\gamma}_{xy}}{2\dot{\epsilon}_x} \qquad (5)$$

where θ is the angle which is measured anticlockwise from the x-axis to the σ_1-axis ($\sigma_1 > \sigma_2$). Then, the principal stresses are calculated from the equilibrium equations expressed along the principal stress directions as

$$\frac{\partial \sigma_1}{\partial s_1} + (\sigma_1 - \sigma_2)\frac{\partial \theta}{\partial s_2} = 0$$
$$\frac{\partial \sigma_2}{\partial s_2} + (\sigma_1 - \sigma_2)\frac{\partial \theta}{\partial s_1} = 0. \qquad (6)$$

The overall accuracy of the field construction and stress computation is examined with the aid of the yield condition.

The third method consists of performing the stress

calculation, based on the equilibrium equation along the slip-lines, the fields of which can be constructed from the known strain-rate distribution[7,16]. The equilibrium equations along the slip-lines become

$$\frac{\partial \sigma}{\partial s_\alpha} - 2k\frac{\partial \phi}{\partial s_\alpha} + \frac{\partial k}{\partial s_\beta} = 0 \qquad \text{along } \alpha\text{-line}$$

$$\frac{\partial \sigma}{\partial s_\beta} + 2k\frac{\partial \phi}{\partial s_\beta} + \frac{\partial k}{\partial s_\alpha} = 0 \qquad \text{along } \beta\text{-line} \qquad (7)$$

where σ and k are the normal and shear stresses, respectively, along the slip-line, and s_α and s_β are the distances along the slip-lines of each family. The two families of slip-lines α and β are distinguished in such a way that if they are regarded as a pair of right-handed curvilinear axes, the line of action of the algebraically-greatest principal stress falls in the first and third quadrant[19] The angle ϕ is the anticlockwise angular rotation of the α-line from the x-axis. In equation (7), the shear stress k is given by $k = \bar{\sigma}/\sqrt{3}$ according to the yield condition, and its distribution is known throughout a deforming body. The accuracy of the computation as well as the construction of the slip-line field can be checked by the fact that the stress value at a point is independent of the integration paths, or by an examination of the orthogonality relation between the slip-line field and the hodograph. This orthogonality condition should hold for work-hardening materials also, so long as the materials are assumed to be incompressible.

In the present investigation the slip-line method was used for the stress analysis. Integrating (7), we obtain

$$\sigma - 2\int k\frac{\partial \phi}{\partial s_\alpha}\,ds_\alpha + \int\frac{\partial k}{\partial s_\beta}\,ds_\alpha = \text{constant, along } \alpha\text{-lines}$$

$$\sigma + 2\int k\frac{\partial \phi}{\partial s_\beta}\,ds_\beta + \int\frac{\partial k}{\partial s_\alpha}\,ds_\beta = \text{constant, along } \beta\text{-lines}$$

$$(8)$$

Equation (8) is the basis for determining the stress field from the slip-line field. Once the value of σ is known, the stress components can be deduced from the relationships given by

$$\sigma_x = \sigma - k\sin 2\phi \qquad \tau_{xy} = k\cos 2\phi$$
$$\sigma_y = \sigma + k\sin 2\phi \qquad (9)$$

This same method was applied previously to the analysis of plane-strain side-pressing[20].

EXPERIMENTS

The experiments consisted of indenting, at room temperature, rectangular blocks of commercially-pure copper with flat punches under the plane-strain condition. The test methods included the observation of grid distortion and load—displacement measure-ments for various specimen dimensions with smooth and rough punches.

Rectangular blocks 1 and 1.5 in wide and 0.25, 0.50, 0.75 and 1.0 in high were prepared from commercially-pure, hard copper. The punches and the bottom dies were made of Graph-Mo and heat-treated to a hardness of Rockwell C50. Four punch widths were selected so that a variety of specimen width to punch and specimen height to punch half-width ratios were obtained in the experiments. The values of these ratios are

W/w	1.33	1.5	2	3	4	6			
h_0/w	1		1.33	1.5	2	2.66	3	4	6	8

$2W$ = specimen width; $2w$ = punch width; h_0 = specimen height

Two interface friction conditions were used in the tests. Low friction was achieved by using smooth dies and punches lapped to an 8 μin surface finish and by applying fluorocarbon spray lubricant. For high friction, rough dies and punches having machined serrations 0.005 in deep were used, and specimens and die and punch surfaces were cleaned with ethylene dichloride and dried before each test.

The experiments were performed with an apparatus placed on a 160 000 lb Tinius Olson testing machine with a speed of 0.05 in/min. This apparatus was also used for a previous study of plane-strain side-pressing[20]. A Nikon camera was mounted in front of the apparatus to photograph the grid-patterns behind a glass plate.

Grid lines with 0.050 in spacing were printed on one end-surface of the specimen, using the Kodak Photo-Resist method. The copper specimens were coated with tin before grid printing to give a better contrast in the photographs. A series of photographs of the grid-patterns was taken at an interval of 0.010 in of punch displacement.

The grid distortions were traced by a scanning machine at the Lawrence Radiation Laboratory, and the current coordinates of each grid point in these patterns were punched on IBM cards. These coordinates constitute the input data for the analysis.

Fracturing is one of the limiting factors for deformation in metalworking processes, although copper did not exhibit any crack within the range of investigation. Therefore, an additional study was made to observe fracturing during indentation using an aluminium alloy 7075-Ty. Two specimen sizes were selected with the punch of 0.5 in width. One was 0.9 in wide and 0.25 in thick; the other 0.6 in wide and 0.6 in thick. The two friction conditions (smooth and lubricated, rough and dry) were used for the bottom die, while the punch was smooth and lubricated.

RESULTS AND DISCUSSION

From the observed flow patterns, modes of deformation were examined. Using the measurements of grid distortion, the distributions of strain components were calculated and contours of constant effective strain $\bar{\epsilon}$ throughout the deformation zone were determined. Based on the experimentally-determined slip-line directions, approximate slip-line fields were constructed and approximate stresses computed for some cases. The effect of workpiece dimensions on the pressure—displacement relationship and on the yield behaviour was also discussed.

Furthermore, fracturing, which was observed with an aluminium alloy, was analysed by the slip-line theory.

Deformation mode

Contrary to the problem of indentation of an infinite strip, the assumption of the rigid portions moving apart horizontally along the foundation is not valid here. A considerable change in geometry takes place during the process and the velocity boundary conditions are more complex. As far as the h/w ratio (height of the workpiece under the punch/punch half-width) is concerned, all the tests performed were in the range $1 < h/w < 8$ except two cases where h/w was equal to unity initially. The type of local deformation in the area just between the punch and the foundation was different for ratios of h/w less than unity, as was apparent from the load values. The load—displacement curves for these cases deviated considerably from the general trend of the curves for the other cases, as shown in the next section.

In general, two completely different modes of deformation with visibly distinct flow characteristics were observed during the tests for the range of punch and specimen sizes investigated. The first mode, which occurs for large initial workpiece thicknesses and for small workpiece widths, is characterised by some bulging of the side surfaces of the workpiece while the bottom surface rests on the foundation.

As the punch moves down, the two upper corners and the bottom of the workpiece remain rigid and the two sides move apart with a velocity distribution which depends on the geometry. As the indentation proceeds, the tensile horizontal stresses in the lower part of the workpiece increase and the plastic zone spreads downward until the entire workpiece, except the two upper corners and possibly the tip of the two lower corners, becomes plastic. The second mode involves the phenomenon called *edge lifting*. The plastic zone is a narrow zone under the punch which extends to the foundation and becomes wider at the bottom of the workpiece while the rest of the workpiece remains rigid.

The phenomenon of edge-lifting can be explained qualitatively as follows. At small workpiece heights, the rigid wedge of material under the punch, in its downward movement, comes close to the foundation: the material in-between is squeezed outward near the foundation, whereas the material near the punch corners is not subjected to this condition. Consequently, a bending moment is induced and causes the end-section to rotate upwards. For small values of initial workpiece thickness to punch half-width ratio h_0/w and large values of workpiece width-punch width ratio W/w, this mode is present from the beginning of indentation. Also, mode 1 changes to mode 2 after a certain amount of deformation at a critical h_c/w ratio which depends on h_0/w and W/w. In figure 1, the normalised penetration at the start of edge lifting, $\Delta = (h_0 - h_c)/w$, was plotted against W/w for various values of h_0/w. The arrows indictae where edge-lifting did not occur in the range of penetrations achieved in the test and it will occur at a higher value of Δ. As can be seen from the curves, for $h_0/w = 1$, the edge-lifting mode is present from the beginning and W/w has no effect on the Δ for the range being investigated. For larger values of h_0/w, however, W/w becomes important (in the range of W/w where it is comparable with h_0/w). This is due to the fact that at small h_0/w, the edge-lifting mode is present and the two side ends of the workpiece are rigid and have no role in the mechanics of the process. Thus any increase in W/w

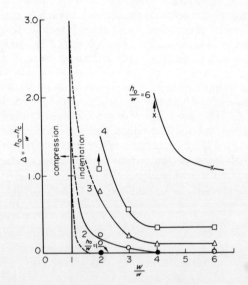

Figure 1 Critical penetration as a function of relative specimen width for various specimen heights.

will have no effect. But, for larger values of h_0/w, the plastic zone extends all across the workpiece (mode 1) and W/w becomes an influential parameter in the deformation mechanics. As W/w decreases, the duration of mode 1 at the beginning increases and thus Δ increases. This is because more deformation is required before the plastic zone extends to the foundation.

At $W/w = 1$, the process is similar to plane-strain compression. In this case, edge-lifting (except when h_0/w is very small and after large deformation) will not occur and maximum normalised compression (equivalent to penetration Δ) will be $\Delta = h_0/w$. At large values of h_0/w, plastic buckling will set a limit to this process. Another interesting point is the effect of frictional conditions on the deformation modes. It was observed that the friction on the punch in the range $h/w > 1$ has practically no effect on the deformation pattern. Friction on the foundation, however, is very important and prevents edge-lifting by resisting the horizontal movement of the material in contact with the foundation. Of course the amount of friction may be decisive in preventing the edge-lifting. With the rough foundation used in the experiments in this work, it was possible to prevent edge-lifting completely, whereas, for the same geometry, using a smooth foundation, edge-lifting occurred from the very beginning.

Punch pressure—displacement curves
The experimental results obtained for punch average pressure versus punch penetration half-punch-width ratios d/w are shown in figures 2(a) to 2(d). As can be observed from figure 2(a), at small workpiece

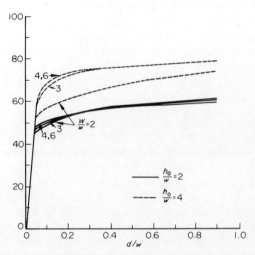

Figure 2(a) Punch pressure—displacement curves for various values of W/w and h_0/w.

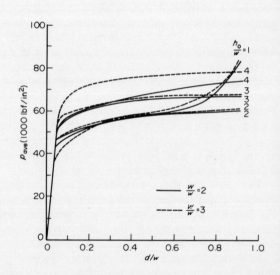

Figure 2(b) The effects of specimen height on punch pressure—displacement curves for small values of W/w.

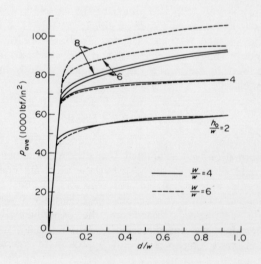

Figure 2(c) The effects of specimen height on punch pressure—displacement curves for large values of W/w.

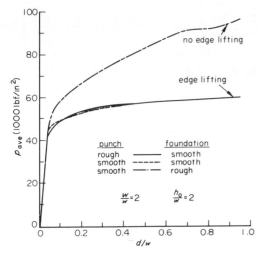

Figure 2(d) The effects of friction conditions at the punch face and the foundation on punch pressure—displacement curves.

heights the increase in width of the workpiece has no effect on the pressure. This is due to the fact that at small heights the edge-lifting mode is present and the two sides of the workpiece are essentially stress-free so that adding to the width of the workpiece does not add to the constraint and thus has no effect on the pressure. On the other hand, at larger heights the pressure becomes sensitive to (and increases with) the workpiece width because the plastic zone spreads throughout the workpiece (mode 1) and the amount of constraint increases with the workpiece dimensions.

The effect of relative workpiece height on the pressure—displacement curves is shown in figures 2(b) and (c). As can be expected, the increase in height increases the constraint and thus the pressure. However, there seems to be a limit to this increase for large values of height and the pressure 'saturates' as the height increases. This limit depends on the value of workpiece width: for smaller workpiece widths the 'saturation' of pressure is faster and the limiting values are smaller. This is because for a tall (but narrow) workpiece the plastic zone does not spread much downward due to small horizontal constraint at the top of the workpiece. However, even for large values of workpiece width, as the height increases, the pressure curve seems to be approaching a limit which, for sufficiently large values of height and width, will be the pressure—displacement curve for indentation of a semi-infinite body. For the largest values of workpiece initial height and width, namely, $h_0/w = 8$ and $W/w = 6$, the limit was still not achieved. Figures 2(b) and (c) also show the different types of pressure—displacement curve for $h/w < 1$. This difference is consistent with the predictions of slip-line theory[21].

The effect of frictional conditions on the pressure—displacement curves is shown in figure 2(d). It is observed that friction on the punch in the examined range of $h/w > 1$ has negligible effect on pressure, whereas the friction on the foundation increases the pressure and changes the pressure—displacement behaviour considerably.

Yield-point load

Although this work is primarily concerned with large plastic indentations, some observations were made regarding the yield-point behaviour of workpieces with finite dimensions. The yield-points of specimens were determined by using the method employed by Watts and Ford[22]. In this method, the elastic and fully-plastic parts of the load—displacement curve are extrapolated to intersect at a point which will be taken as the yield-point. The values of normalised average punch pressure $p/2k$ (k is the shear yield-strength) were plotted against workpiece height to punch half-width ratio h/w and compared with the theoretical curve for the infinite strip in figure 3(a). It is seen that nearly all the experimental values fall below the theoretical curve and the reason is the deviation from the assumption of infinite strip and the assumed velocity boundary conditions in the slip-line solution. Two parameters are important as can be seen in figure 3(a).

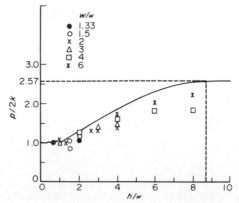

Figure 3(a) Comparison of experimental yield-pressures with the theoretical solution for an infinitely wide strip.

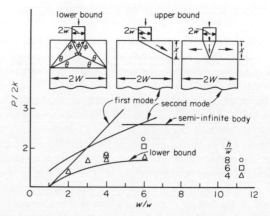

Figure 3(b) Comparison of experimental yield-pressures with lower and upper bound solutions for specimens with infinite height.

(1) For a constant h_0/w ratio as the workpiece width to punch ratio W/w increases, the pressures become closer to the theoretical values. This is because the assumption of infinite strip tends to become more realistic and the increased constraint of the surrounding elastic material increases the pressure.

(2) For a constant workpiece width to punch-width ratio, the deviation between theoretical and experimental pressures increases as the h_0/w ratio increases

because the effect of foundation on the constraint decreases as the distance between the punch and the foundation increases.

Inasmuch as there is no theory available for indentation of workpieces of finite width and finite height, the yield-pressures for some larger values of h/w, namely, 8, 6, 4, were compared with an upper bound–lower bound solution for workpieces of finite width but infinite height[23]. The effect of specimen width on the yield-pressure is shown in figure 3(b). As expected, for large h/w ratios, the theory is in very good agreement with the experiment, but as h/w decreases, the pressures are closer to the lower bound.

Strains and stresses

Strains and maximum shear strain rate directions were computed from the deformed gridlines. Depending on the mode of deformation, strain distributions differ considerably. Figure 4 shows ϵ_x along the workpiece

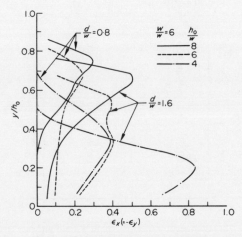

Figure 4 Strain distribution along the vertical axis at two stages of deformation for three specimen sizes.

vertical axis at two different stages of deformation for three types of specimen all having the same width but having different heights. It is seen that for both stages of $h_0/w = 8$ and for the first stage of $h_0/w = 6$, the strain is mostly concentrated in the area just beneath the punch. These are the cases where edge-lifting has not occurred and the deformation is of mode 1. For the case of $h_0/w = 6$, edge-lifting occurs at some stage and another zone of high straining is formed in the lower portion of the workpiece. For $h_0/w = 4$, edge-lifting takes place from the beginning and the point of maximum strain is always near the bottom of the workpiece. Figures 5(a) and (b) show the contours of effective strain $\bar{\epsilon}$ for the cases corresponding to those in figure 4. These contours clearly demonstrate the difference between the types of straining in the two modes. Maximum $\bar{\epsilon}$ is always located at the corner of the punch. When there is edge-lifting, another peak is formed in the lower part of the vertical axis and moves downward as deformation continues.

Figure 6 gives the contours of constant effective strain $\bar{\epsilon}$, for two different conditions of friction on the foundation. As seen from these figures, the high friction on the foundation, by preventing edge-lifting,

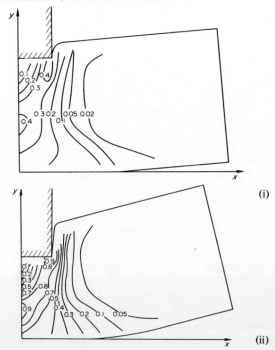

Figure 5(a) Contours of constant effective strain for $W/w = 6$ and $h_0/w = 4$ at (i) displacement of $d/w = 0.8$, and (ii) displacement of $d/w = 1.6$.

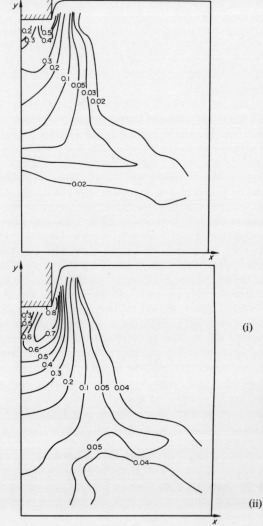

Figure 5(b) Contours of constant effective strain for $W/w = 6$ and $h_0/w = 8$ at (i) $d/w = 0.8$, and (ii) $d/w = 1.6$.

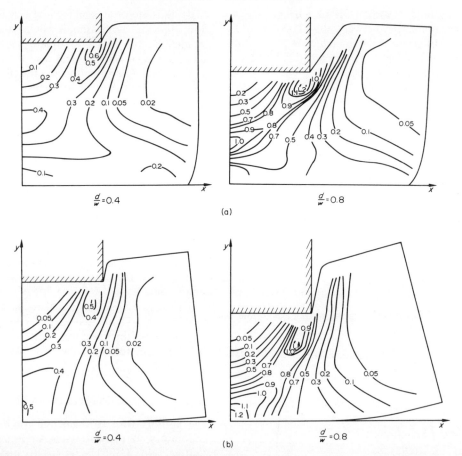

Figure 6 Contours of constant effective strain for $W/w = 2$ and $h_0/w = 2$ at $d/w = 0.4$, and $d/w = 0.8$ with (a) rough foundation and (b) smooth foundation.

reduces the high straining at the bottom of the work-piece and causes a much more uniform strain distribution along the vertical axis.

Slip-lines were drawn using the calculated directions and known boundary conditions. Since the velocity boundary conditions were somewhat complicated, it was not possible to draw a hodograph and correct the slip-lines. Thus, the slip-lines and also the calculated stresses are approximate. Figures 7(a)

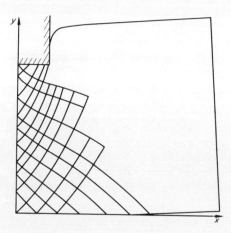

Figure 7(b) Slip-line field for mode 2; $W/w = 6$, $h_0/w = 6$, and $d/w = 1.6$.

and (b) show the approximate slip lines for mode 1 and mode 2 (edge-lifting) respectively. Figure 8(a) gives σ_x and σ_y values along the vertical axis for the case of edge-lifting while figure 8(b) shows the contours of constant σ_z for the same case. Figure 8(b) can be compared with the observation since the material exerts high pressure on the glass plate and breaks down the lubricant in the region of negative σ_z and separates from the glass plate in the positive region. In general, agreement is good.

A comparison of slip-lines, velocity boundary conditions and σ_x distribution along the vertical axis

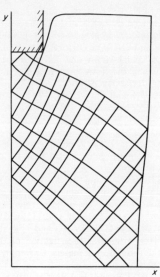

Figure 7(a) Slip-line field for mode 1; $W/w = 4$, $h_0/w = 8$, and $d/w = 1.6$.

for edge-lifting modes, with those of pure bending, reveals certain similarities between the two cases. Of course, the existence of the punch with its compressive force and downward movement makes the

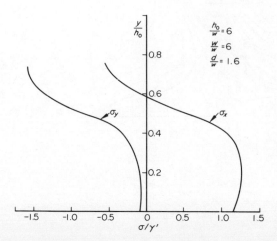

Figure 8(a) Stress distributions along the vertical axis for the edge-lifting mode.

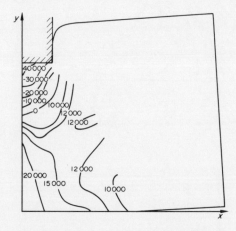

Figure 8(b) Contours of constant σ_z for edge-lifting; $W/w = 6$, $h_0/w = 6$, and $d/w = 1.6$.

problem asymmetric and boundary conditions complicated. Neglecting the punch force and velocity, the boundary conditions are similar to the problem of pure bending of beams with rectangular notches[24]. An examination of σ_x distribution in figure 8(a) reveals that

● The integral of σ_x along the vertical axis will yield a net force which will be resisted by friction on the foundation.
● The stress distribution indicates that a bending moment is induced at the central plane producing edge-lifting.

SUMMARY AND CONCLUSIONS

The detailed mechanics were derived in plane-strain indentation based on the observed grid distortions using commercially-pure copper specimens. Deformation modes, yield-point load, pressure–displacement relationships, and stress and strain distributions were investigated. In particular, the effects of specimen dimensions and interface friction on deformation

characteristics were examined. It was found that the flow models with simple boundary conditions, assumed in the previous slip-line solutions, are not applicable. For the range of $h/w > 1$, there are two basic modes of deformation and the edge-lifting mode starts at a critical h/w ratio which depends on W/w and h_0/w ratios. Also, it was found that, while friction on the punch face has negligible effect on the deformation, friction on the foundation prevents edge-lifting.

In the comparison of punch pressure–displacement curves, it was noted that at small thicknesses the width of the workpiece has no effect on the pressure, whilst at larger thicknesses the pressure increases with the width. The increase in thickness increases the pressure to a certain limit and then the pressure 'saturates' with thickness and this limiting value depends on the width. The saturation is faster for small widths: for large widths it approaches that for a semi-infinite body.

The strains were computed to show the different deformation characteristics in the two modes. Approximate slip-lines for both modes and stress distributions for the case of edge-lifting were obtained.

ACKNOWLEDGEMENTS

The authors wish to thank the Air Force Materials Laboratory, Wright-Patterson Air Force Base, for its contract No. F33615-70-C-1320, under which the present work was performed.

REFERENCES

1. L. PRANDTL, 1920. Uber die Haerte Plastischer Koerper, *Nachlichten der Akademie der Wissenschaften,* Gottingen, Mathematisch-Physikalische Klasse, p. 74.
2. R. HILL, 1950. The Mathematical Theory of Plasticity, *Clarendon Press,* Oxford.
3. R. HILL, E. H. LEE and S. J. TUPPER, 1951. A method of numerical analysis of plastic flow in plane strain and its application to the compression of a ductile material between rough plates. *Trans. ASME, J. Appl. Mech.,* vol. 18, p. 46.
4. A. P. GREEN, 1951. A theoretical investigation of the compression of a ductile material between smooth flat dies. *Philosophical Magazine,* vol. 42, p. 400.
5. C. H. LEE and SHIRO KOBAYASHI, 1970. Elastoplastic analysis of plane-strain and axisymmetric flat-punch indentation by the finite element method. *International Journal of Mechanical Sciences,* vol. 12, pp. 349–370.
6. W. JOHNSON and H. KUDO, 1964. Plane-strain deep indentation. *Proceedings of the Fifth International Machine Tool Design and Research Conference,* Pergamon, pp. 441–447.
7. P. L. B. OXLEY, 1957. A photographic investigation of the metal cutting process. Ph.D. thesis for the University of Leeds, England.
8. E. G. THOMSEN and J. T. LAPSLEY Jr., 1954. Experimental stress determination within a metal during plastic flow. *Proceedings of the American Society of Experimental Stress Analysis,* vol. 11, pp. 59–68.
9. E. G. THOMSEN, C. T. YANG and J. B. BIERBOWER, 1954. An experimental investigation of the mechanics of plastic deformation of metals. University of California Press, Berkeley.
10. E. G. THOMSEN and J. FRISCH, 1955. Stresses and strains in cold extrusion of 2SO aluminium. *Trans. ASME,* vol. 77, p. 1 344.

11. E. G. THOMSEN, 1963/4. Visioplasticity. *CIRP Annalen,* vol. 12, p. 127.
12. E. G. THOMSEN, C. T. YANG and SHIRO KOBAYASHI, 1964. Mechanics of Plastic Deformation in Metal Processing, Macmillan.
13. A. H. SHABAIK and E. G. THOMSEN, 1967. Investigation of the application of visioplasticity methods of analysis to metal deformation processes. *Final Report prepared for the Navy,* Bureau of Naval Weapons.
14. A. H. SHABAIK and E. G. THOMSEN, 1968. Investigation of the application of visioplasticity methods of analysis to metal deformation processes. *Final Report prepared for the Navy,* Bureau of Naval Weapons.
15. H. S. MEHTA, A. H. SHABAIK and SHIRO KOBAYASHI, 1970. Analysis of tube extrusion. *Trans. ASME,* Journal of Engineering for Industry, vol. 92, pp. 403–411.
16. E. USUI, 1967. Plasticity analysis of discontinuous chip formation (Part 1): deformation mode, slip-line field and hodograph. Seimitsu-Kikai, vol. 33, no. 2, pp. 77–85 (in Japanese).
17. E. USUI, 1967. Plasticity analysis of discontinuous chip formation (Part 2): deformation mode, slip-line field and hodograph. Seimitsu-Kikai, vol. 33, no. 2, pp. 77–85 (in Japanese).
18. T. H. C. CHILDS, 1970. A new experimental technique for observing deformation. *Proceedings of the 11th International Machine Tool Design and Research Conference.*
19. R. HILL, 1950. The mathematical theory of plasticity, Oxford, Clarendon Press.
20. SHIRO KOBAYASHI, S. SOHRABPOUR and M. M. SEHGAL, 1971. Study of deformation and defects occurring in advanced forging technique. *Technical Report AFML-TR-71-69.*
21. J. F. W. BISHOP, 1958. On the effect of friction on compression and indentation between flat dies. *J. of Mech. Phys. Solids,* vol. 6, p. 132.
22. A. B. WATTS and H. FORD, 1952. An experimental investigation of the yielding of strip between smooth dies. *Proc. Inst. Mech. Engrs.,* IB, p. 448.
23. C. R. CALLADINE, 1969. Engineering Plasticity, Pergamon Press.
24. G. LIANIS nad H. FORD, 1957. The yielding of notched bars due to bending, *9th Int. Congr. Appl. Mech.,* Brussels, vol. 8, p. 235.
25. S. C. JAIN and SHIRO KOBAYASHI, 1970. Deformation and fracture of an alum: *um* alloy in plane-strain side-pressing. *Proceedings of the 11th International Conference on Machine Tool Design and Research.*

A THEORETICAL STUDY OF TUBE DRAWING WITH A FLOATING PLUG

by

D. J. SMITH* and A. N. BRAMLEY*

SUMMARY

Stimulated by the reported observation of recovery and recrystallisation occurring in heavily drawn copper tube, a theoretical study has been made to establish the relative importance of the various parameters of the floating plug tube-drawing process. As far as the authors know, this is the first solution of the process using the upper-bound technique. A kinematically admissible velocity field is assumed for the deformation process. The solution takes into account redundant work and friction but assumes a constant yield-stress. Circumferential and thickness straining are calculated using a hodograph derived from the velocity field, and solutions are optimised for varying plug die angles, plug lands and frictional conditions. Some preliminary experimental results obtained on a hydraulic drawbench are compared with theoretical predictions.

NOTATION

R_0 = initial outside radius of tube
R_i = initial inside radius of tube
R_{of} = final outside radius of tube
R_{if} = final inside radius of tube
α = die taper angle
β = plug taper angle
L_p = length of plug land
L_d = length of die land
m = friction factor
V_1 = initial velocity of tube
V_4 = final velocity of tube
ϵ_R = thickness strain
ϵ_θ = circumferential strain
$\bar{\epsilon}$ = equivalent strain
k = shear yield-stress
$\bar{\sigma}$ = mean effective stress
x = parameter determining plug position
σ_x = drawing stress
L = applied load in plane strain compression test
W = width of specimen in P.S. compression test
t_0 = initial thickness of specimen P.S. compression test
b = breadth of die in P.S. compression test
$\dot{w}$ = rate of working
A = area of velocity discontinuity
V = volume of deformed metal.

INTRODUCTION

Tube drawing with a floating plug has been practised for many years in this country and abroad. The advantages of the floating plug process over other means of tube manufacture are: better control of tube dimensions and surface finish, ability to draw on bull-blocks and (in some circumstances) the drawing force required may be less[4].

The floating plug is drawn in to the deformation zone, and, depending on the geometry of the zone, will stabilise itself in its equilibrium position. Frictional forces acting on the plug, drag the plug forward and the normal forces acting on the conical surfaces of the plug tend to force it backward, out of the deformation zone. Thus the plug is maintained in its equilibrium position. Incorrect choice of plug

* Department of Metallurgy, Leeds University.

geometry may cause instability resulting in excessive oscillation about the equilibrium position, which will produce dimensional inconsistency, ring formation and possible tube fracture.

Thin-walled tubes may be coiled on bull-blocks before and after drawing, so long lengths of tube can be drawn at high speeds. Electron metallographic studies of drawn copper tubes have shown[1] that recovery and recrystallisation can occur as a result of large amounts of deformation. Total deformation of 98% reduction-in-area can be achieved without interstage annealing as a result of this recovery and recrystallisation. However, as the amount of deformation is increased the workhardening exponent n in the general stress strain relation $\sigma = k\epsilon^n$ decreases, and then increases at very high deformations. The susceptibility to tube failure is high if the deformation is such that the value of n is a minimum.

One aim of the present work is to correlate recovery and recrystallisation with the parameters of the process so that ultimately it will be possible to exploit its occurrence. For example, it may be desirable to promote recovery and recrystallisation by maximising the work done per pass and thus extend the drawing schedule.

Theoretical investigations of the geometry of the deformation zone in the floating plug tube-drawing process have been mainly carried out overseas. Perlin[2], Bisk et al.[3,5], Orro and Savin[4], Schneider et al.[6,7] have developed criteria for the position of the plug in the deformation zone and for calculating the dimensions of the floating plug, based on the equilibrium of forces acting on the plug. Meadows and Lawrence[8], assuming the maximum reduction per pass was achieved such that the strain disparity (and hence the residual circumferential stress) is zero, showed how the required plug and die sizes can be determined graphically.

All the early theories of tube drawing were mainly concerned with tube sinking (Sachs et al.[9,10,11,12,13]) and assumed plane strain conditions.

In 1960 Green[14] reviewed the existing plane strain theories and adapted them to accommodate friction, workhardening and redundant work.

It is fairly recently that theoretical solutions to

axi-symmetrical drawing processes have been developed. Avitzur[15] analysed axi-symmetrical tube sinking assuming a kinematically admissable velocity field in order to determine a lower upper-bound solution, and it is possible to adapt this solution for parallel mandrel drawing.

UPPER-BOUND SOLUTION

A pattern of velocity discontinuities is shown in figure 1. Plastic work is dissipated by shearing across the velocity discontinuities and by friction at the tool—workpiece interface. Further plastic work is dissipated in regions 2 and 3 due to circumferential and thickness straining causing an increase in velocity in the deformation zone.

It is assumed that during sinking, prior to contact with the plug, the wall thickness remains constant. In practice there may be some wall thickening during sinking but it is assumed to be small compared with the total reduction in thickness.

In region 1 the undrawn tube travels with unit velocity in a direction parallel to the drawing axis. In region 2 the material flows parallel to the die force at an angle α to the drawing axis. In region 3 the material flows parallel to the conical face of the plug at an angle β to the drawing axis. In region 4 the drawn tube travels with a velocity V_4 given by $(R_0^2 - R_i^2)/(R_{of}^2 - R_{if}^2)$ parallel to the drawing axis.

In regions 2 and 3 the magnitude of the velocity is not constant but increases because of circumferential and thickness straining.

The hodograph associated with figure 1 is shown in figure 2.

The rate of plastic working across a velocity discontinuity is given by:

$$\dot{w} = kAu \tag{1}$$

Thus the rate of working across each velocity in figure 1 can be computed by using figure 2 and equation (1). The rate of working across the tool—workpiece interfaces as a result of friction is computed in a similar way.

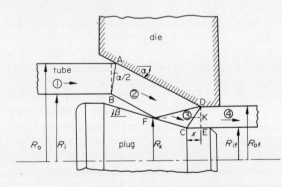

Figure 1. Pattern of velocity discontinuities.

In addition to plastic work being dissipated across velocity discontinuities, and as a result of friction, there is also the work dissipated due to circumferential straining. This is computed using the work formula

$$w = V\sigma\epsilon \tag{2}$$

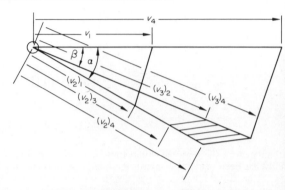

Figure 2. Hodograph associated with figure 1.

The power of the external pull is given by

$$P = \pi(R_0^2 - R_i^2)\sigma_x \tag{3}$$

When the power given by (3) is equated to the total power given by (1) and (2) we arrive at an expression for the drawing stress, given by equation (4).

$$\frac{\sigma_x}{2k} = \frac{\sin\alpha}{2\cos^2\alpha/2} + \frac{\sin\beta}{2\sin\mathrm{DCE}\,\sin(\beta + \mathrm{DCE})}$$

$$+ M\left[\frac{R_0(R_0 - R_{of})}{(R_0^2 - R_1^2)\sin\alpha} + \frac{R_{if}(R_x - R_{if})\sin\mathrm{DCE}}{\sin\beta\,\sin(\beta + \mathrm{DCE})(R_{of}^2 - R_{if}^2)}\right.$$

$$\left. + \frac{R_{of}L_d}{(R_{of}^2 - R_{if}^2)} + \frac{R_{if}L_p}{(R_{of}^2 - R_{if}^2)}\right]$$

$$+ \frac{R_{of}(R_{of} - R_x)\sin(\alpha - \beta)\sin\mathrm{DCE}}{(R_{of}^2 - R_{if}^2)\sin\mathrm{DFK}\,\sin(\alpha + \mathrm{DFK})\sin(\beta + \mathrm{DCE})}$$

$$+ [\epsilon_R^2 + \epsilon_R\epsilon_\theta + \epsilon_\theta^2]^{\frac{1}{2}} \tag{4}$$

Where $\sin\mathrm{DCE} = \dfrac{(R_{of} - R_{if})}{[(R_{of} - R_{if})^2 + x^2]^{\frac{1}{2}}}$

$$\sin\mathrm{DFK} = \frac{R_{of} - R_x}{\mathrm{FD}}$$

$$R_x = (FC\sin\beta + R_{if})$$

$$FC = \frac{(R_0 - R_i) - \cos\alpha(R_{of} - R_{if} + x\tan\beta)}{\sin(\alpha + \beta)}$$
$$- \frac{x}{\cos\beta}$$

$$\epsilon_R = \ln\frac{(R_0 - R_i)}{(R_{of} - R_{if})} \qquad \epsilon_\theta = \ln\frac{R_0}{R_{of}}$$

The full derivation of equation (4) is given in the Appendix. It is totally defined by the variable x and the geometry of the pass, for a constant yield stress.

Evaluation of equation (4) was carried out using an appropriate programme for an ICL computer. Variations of $\sigma_x/2k$ with x and the process variables were determined using this programme.

The value of x, which determines the position of the plug in the deformation zone, must lie within a range of values determined by the geometry of the zone.

According to Perlin[2], the minimum value of x is given by

$$x_{\min} = (R_{of} - R_{if})\tan(\beta/2) \tag{5}$$

According to references (3, 4, 5, 6) the minimum value of x is zero. However, if x equals zero the current wall thickness of the tube in the working zone of the pass, say CD in figure 1, would be less than the thickness of the drawn tube, DE, which is clearly impossible when drawing on a conical plug. However, in practice, the value of x_{min} as given by equation (5) is usually very small and in some circumstances may be taken as zero.

The maximum value of x is when the floating plug is in such a backward position that there is no contact between the conical portion of the plug and the tube.

$$x_{max} = \frac{(R_o - R_i)}{\sin\alpha} - (R_{of} - R_{if})\cot\alpha \quad (6)$$

As x approaches its maximum value, the length of contact between the plug and the tube, FC in figure 1, approaches zero and all the reduction in wall thickness of the tube is produced by the cylindrical land of the plug, in a similar way to parallel mandrel drawing.

The passes under investigation were taken from part of an actual drawing schedule performed on an industrial scale. The dimensions of each pass are given in table 1.

coefficient is to raise the value of $\sigma_x/2k$ for a given value of x due to the increased resistance along the tool–workpiece interface. Also increasing the friction coefficient between the tools and the tube will tend to pull the plug further into the deformation zone thus decreasing the value of x which causes an increase in the drawing stress.

The effect of increasing the length of the plug land is the same as increasing the friction coefficient. Therefore from a consideration of minimum drawing stress the length of the plug land should be as small as possible. However, a short plug land would produce instability of the plug leading to dimensional inconsistency in the drawn tube and possible ring formation. Therefore there is a minimum length of plug land which is necessary for a given pass in order to maintain stability.

Figure 4 shows the variation of $\sigma_x/2k$ with x for varying values of the plug taper angle β for a given value of α, the die taper angle. The value of β has no effect on the minimum value of $\sigma_x/2k$ because as pointed out above, this minimum occurs when $x = x_{max}$ and the conical portion of the plug plays no part in the deformation.

However, for x less than x_{max} the taper of the plug does have an effect on the drawing stress. For a

TABLE 1
Dimensions of each pass (in)

pass	α	β	R_o	R_i	R_{of}	R_{if}	L_p	L_d
1	15	13	0.913	0.850	0.775	0.725	0.45	0.25
2	12	10	0.775	0.725	0.670	0.626	0.30	0.15
3	10	10	0.670	0.626	0.550	0.513	0.24	0.35
4	11	10	0.550	0.513	0.456	0.423	0.25	0.15
5	11	10	0.456	0.462	0.383	0.354	0.20	0.20
6	10	10	0.383	0.354	0.324	0.297	0.20	0.20

Figure 3 shows the variation of $\sigma_x/2k$ with x, for pass 1. The value of x corresponding to the minimum value of $\sigma_x/2k$ is equal to x_{max}, given by equation (6), that is, the minimum drawing stress corresponds to the plug being in its extreme backward position. Not only is the frictional contribution to the drawing a minimum when $x = x_{max}$ but the shear power across velocity discontinuity FD is zero and hence the redundant work contribution is also a minimum.

The effect of increasing the value of the friction

particular value of x the lower the value of β the lower the drawing stress.

Figure 5 shows the variation of $\sigma_x/2k$ with die taper angle α for pass 1 and for $x = x_{max}$. The curve shows a minimum, indicating that there is an optimum die angle for a particular pass. This curve is independent of β the plug taper angle, as x was chosen

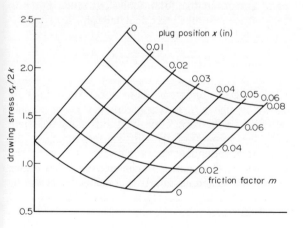

Figure 3. $\sigma_x/2k$ against x for pass 1, varying friction factor.

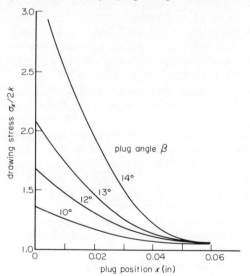

Figure 4. $\sigma_x/2k$ against x for pass 1, varying $\beta(\alpha = 15°)$.

to be equal to x_{max}. At low values of α the frictional contribution to the drawing stress is high, because low values of α correspond to large areas of contact between the die and the tube. At high values of α the area of contact between the die and the tube is low and consequently the frictional contribution to the total drawing stress is low. However, friction losses can never be eliminated because obviously there will always be some contact between the die face and the tube; also, the areas of contact between the tube and the die and plug lands are independent of the die taper angle.

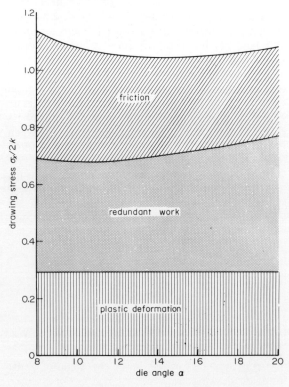

Figure 5. Contributions to the total drawing stress for pass 1.

At low values of α the redundant work caused by shearing across the velocity discontinuities is low as a result of the small distortions at these values of α. As the value of α increases, the distortion of the material in the deformation zone increases and the redundant work contributions increase. However, in some cases, as shown in figure 5 the contribution by redundant work shows a minimum at low α. The occurrence of this minimum depends on the particular reduction being considered and is due to the fact that as α decreases, x_{max} increases and the area of velocity discontinuity CD increases sufficiently to produce an increase in the redundant work contribution. As mentioned above this minimum depends on the geometry of the pass, and graphs of $\sigma_x/2k$ versus α for passes 2, 4, and 6 did not show any minimum in the redundant work contribution.

For rod-drawing[16] the redundant work contribution does not pass through a minimum because the limitations are not the same as those for floating plug tube drawing. In rod-drawing both the velocity discontinuities and their areas decrease with decreasing α.

As can be seen from figure 5 the resistance to overcome plastic deformation depends only on the total strain put into the material, which depends on the initial and final dimensions of the tube and consequently is independent of the die taper angle α.

EXPERIMENTAL

The schedule in table 1 was drawn on a 5-ton hydraulic drawbench using Esso IL 1709 full flood lubricant. The tubes and the tools were supplied by Yorkshire Imperial Metals Ltd. Each pass was drawn at two speeds, 6 ft/min and 60 ft/min. The drawing stress was recorded using strain gauges attached to the ram of the drawbench and connected to an amplifier and UV recorder.

Over the range used it was found that the drawing stress was independent of the drawing speed.

In order to determine the value of the true mean yield-stress k in equation (4), samples of drawn tube were taken from each pass and plane strain compression tests were carried out using a subpress of a design based on that developed by Watts and Ford[16]. The test samples needed no lengthy preparation for the surface finish was good as a result of the drawing process; however, they were degreased and polished. The thickness measurements were made using a hand micrometer with reduced points, a mean value of five equispaced measurements being taken. The proportions of die width to specimen dimensions were kept at $b/t = 2$ to 4 and W/b not less than 6. The tests were carried out on a 500 kN Denison hydraulic testing machine and the lubricant was graphited grease. Each test was performed several times and the mean curve constructed in each case.

The equivalent stress was calculated using the expression

$$\bar{\sigma} = \frac{\sqrt{3}L}{2wb}$$

and the equivalent strain by

$$\bar{\epsilon} = \frac{2}{\sqrt{3}} \ln \frac{t_0}{t}$$

The undrawn tube from pass 1 was taken as the basic curve and the total equivalent strain put into each subsequent tube was calculated using the relation

$$\bar{\epsilon}_2 = \frac{2}{\sqrt{3}} [\epsilon_R{}^2 + \epsilon_R \epsilon_\theta + \epsilon_\theta{}^2]^{\frac{1}{2}}$$

which neglects any redundant work. The results are shown in figure 6. It is apparent that curve 2 could be fitted to the basic curve by a shift along the strain axis, the amount of the shift corresponding to the amount of redundant strain. However, this is not the case for 3, 4, 5 and 6 obtained from tubes drawn later in the schedule, and it is concluded that during these passes some recovery and recrystallisation must have occurred.

The value of k used in the theoretical calculations was calculated from the mean effective stress $\bar{\sigma}$

obtained graphically from the appropriate curve in figure 6 using the relation

$$k = \frac{\bar{\sigma}}{\sqrt{3}}$$

The tube from the final pass was not tested, as its thickness was too small and accurate results could not be attained using the tools available. However, the tubes drawn in this investigation were taken from a schedule in which the original tube size was 2.0 in bore 2.24 in OD. No interstage annealing has been carried out and consequently the tubes had already undergone large amounts of deformation and their yield-stress was approximately constant.

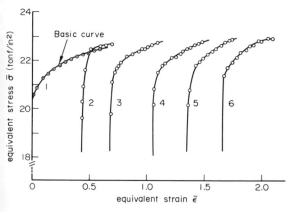

Figure 6. Plane strain compression curves.

TABLE 2

Calculated values of drawing stress and those determined experimentally

pass	σ_x (tonf/in²) experimental	σ_x (tonf/in²) calculated
1	19.65	25.7
2	13.95	20.7
3	19.72	27.5
4	20.00	23.0
5	20.70	23.4
6	17.35	21.3

Examination of sections taken of the deformation zone when the drawing operation was stopped indicated that there was only a small amount of contact between the conical surface of the plug and the tube. Consequently the maximum value of x was used in the calculations. An arbitrary value of $M = 0.03$ was also used in the calculations.

In order to investigate the variation of drawing stress with percentage reduction in area, annealed tubes of different dimensions were drawn down to

the same final dimensions using the same die and plug.

The tubes were all in the annealed condition initially and the basic stress–strain curve was determined using the plane strain compression test. Values of the true mean yield-stress were calculated from the curve using the values of the total equivalent strain shown in table 3.

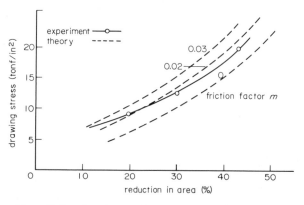

Figure 7. Drawing stress against percentage reduction in area.

Figure 7 shows the experimentally determined drawing stress and the calculated values using equation (4) plotted against percentage reduction in area. The calculated values of the drawing stress show good agreement with those found experimentally for $m = 0.02$.

CONCLUSIONS

(1) It is possible, using the upper-bound technique using velocity fields and hodographs, to derive a solution for the floating-plug tube-drawing process which incorporates terms accommodating redundant work, friction and circumferential straining. With the aid of the computer it is possible to investigate the effect of the various process parameters on the drawing force needed to accomplish the reduction.

(2) The drawing stresses calculated give reasonable agreement with those found experimentally for the reductions investigated.

(3) For the drawing force to be a minimum, the floating plug should position itself in its most backward position possible, so that $x = x_{max}$, and the conical portion of the plug serves only to prevent the plug being pulled through the die with the tube. However, the position of the plug will depend on the value of the friction factor between the plug and the tube, and this is being investigated at the present time, together with the effect of the plug taper angle. The plug taper angles for four of the passes investigated were either equal to (or only one degree

TABLE 3
Dimensions (in) of each of the tubes drawn

α	β	R_o	R_i	R_{if}	R_{of}	% red. in area	$\bar{\epsilon}$	k
15	12.5	0.570	0.531	0.457	0.493	19	0.21	7.2
15	12.5	0.576	0.532	0.457	0.493	30	0.344	9.1
15	12.5	0.608	0.558	0.457	0.493	43	0.542	10.5

less than) the die taper angle and consequently the plug could not be expected to penetrate the deformation zone.

(4) From the analysis described it can be concluded that drawing speed does not affect the drawing stress. However, drawing at high speeds, beyond the capability of the drawbench used in the study, will alter the mechanical properties of the material and no provision is made for this in the solution.

(5) For any combination of reduction and friction factor there exists an optimum die angle, for which the drawing stress is a minimum. With too small a die angle, the length of contact between the tube and the tools is high, causing significantly-high friction losses. With too large a die angle, the distortion becomes a predominant factor.

(6) The solution described does not account for strainhardening and in this respect the solution is not complete. The occurrence of recovery and recrystallisation at large strains cannot be accommodated into the solution and has to be determined experimentally.

REFERENCES

1. J. M. CAIRNS, J. CLOUGH, P. DEWEY and J. NUTTING, 1971. The structure and mechanical properties of heavily deformed copper. *JIM* 99, p. 93.
2. I. L. PERLIN, 1958. Consideration of stability in the floating plug drawing process for tubes. *Tsvetyne Metally*, 31, p. 463.
3. M. B. BISK, Z. A. SOMINSKI and V. V. SHVEIKIN, 1963. Drawing tubes on a self adjusting mandrel on straight drawing mills. *Stal.*, 6, p. 463.
4. P. I. ORRO and G. A. SAVIN, 1963. Drawing medium-grade carbon steel tubes on a floating mandrel. *Stal.*, 6, p. 466.
5. M. B. BISK and V. V. SHVEIKIN, 1964. Relation between the shape and position of the self-stopping mandrel in the zone of deformation and the parameters of the tube drawing operation. *Stal.*, 7, p. 894.
6. M. SCHNEIDER and J. PIWOWARSKI, 1966. Parameters of the floating plug tube drawing process. *Archeiwum Hutnictwa*, 11(i), p. 40.
7. M. SCHNEIDER, L. GABLANKOWSKI and L. SADOK, 1968. Optimisation of plug and die design for the floating plug tube drawing process. *Archeiwum Hutnictwa*, 13(i), p. 77.
8. B. J. MEADOWS and A. G. LAWRENCE, 1970. Theoretical approach to the rationalisation of tube drawing schedules. *JIM*, 98, p. 102.
9. G. SACHS, J. N. LUBAHN and D. P. TRACY, 1944. Drawing thin-walled tubing with a moving mandrel through a stationary die. *Trans. ASME, Journal of Applied Mechanics*, 11, p. 119.
10. G. SACHS and G. ESPEY, 1942. Residual stress in sunk cartridge brass tubing. *Transactions A.IMME.*, 147, p. 74.
11. G. SACHS and G. ESPEY. A new method of determining stress distribution in thin-walled tubing. *Ibid.*, p. 67.
12. G. SACHS and G. ESPEY. Strength distribution in sunk brass tubes.*Ibid.*, p. 348.
13. G. SACHS and W. M. BALDWIN. Stress analysis in tube sinking. *Trans. ASME*, 68, p. 655.
14. A. P. GREEN, 1960. Plane strain theories of drawing. *Proc. Inst. Mech. Engrs.*, 174, p. 847.
15. B. AVITZUR, 1968. Metal forming: process and analysis. McGraw-Mill Book Co., p. 327.
16. A. B. WATTS and M. FORD, 1952/3. An experimental investigation of the yielding of strip between smooth dies. *Proc. Inst. Mech. Engrs.*, 1B, p. 448.

APPENDIX

Plastic work across discontinuities

The rate of plastic working across a discontinuity is given by equation (1).

$$\dot{w} = kAu$$

Referring to figure 1 the rate of working across discontinuity AB is given by

$$\dot{w}_{AB} = k\ area_{AB}\ V_{12} \qquad (A.1)$$

It is assumed that there is no thickness change immediately after passing discontinuity AB and therefore from figure 1 we have

$$A\hat{B}F = \frac{\alpha}{2} \quad \text{and}\ area_{AB} = \frac{\pi(R_o{}^2 - R_i{}^2)}{\cos\alpha/2}$$

From figure 2

$$V_{12} = \frac{\sin\alpha}{\cos\alpha/2} \quad (V_1 = \text{unity})$$

so (A.1) becomes

$$\dot{w}_{AB} = \frac{k\pi(R_o{}^2 - R_i{}^2)\sin\alpha}{\cos^2\alpha/2} \qquad (A.2)$$

The rate of working across discontinuity CD is given by

$$\dot{w}_{CD} = k\pi\ area_{CD}\ V_{34} \qquad (A.3)$$

From figure 1 we have

$$area_{CD} = \frac{\pi(R_{of}{}^2 - R_{if}{}^2)}{\sin DCE}$$

From figure 2 we have

$$V_{34} = \frac{\sin\beta V_4}{\sin(\beta + DCE)} = \frac{\sin\beta(R_o{}^2 - R_i{}^2)}{\sin(\beta + DCE)(R_{of}{}^2 - R_{if}{}^2)}$$

therefore

$$\dot{w}_{CD} = \frac{k\pi(R_o{}^2 - R_i{}^2)\sin\beta}{\sin DCE\ \sin(\beta + DCE)} \qquad (A.4)$$

The rate of working across velocity discontinuity FD is given by

$$\dot{w}_{FD} = k\ area_{FD}\ V_{23} \qquad (A.5)$$

From figure 1 we have

$$area_{FD} = \frac{2\pi R\delta R}{\sin DFK}$$

From figure 2 we have

$$V_{23} = (V_{23})_4 \frac{R_{of}}{R} \quad \text{and}\ (V_{23})_4 = \frac{\sin(\alpha - \beta)(V_3)_4}{\sin(\alpha + DFK)}$$

therefore

$$(V_{23})_4 = \frac{\sin(\alpha - \beta)\ \sin DCE(R_o{}^2 - R_i{}^2)}{\sin(\beta + DCE)\ \sin(\alpha + DFK)(R_{of}{}^2 - R_{if}{}^2)}$$

R varies from R_x to R_{of}, therefore

$$\dot{w}_{FD} =$$

$$\frac{2\pi k R_{of}(R_{of} - R_x)(R_o^2 - R_i^2)\sin(\alpha - \beta)\sin DCE}{(R_{of}^2 - R_{if}^2)\sin DFE \sin(\alpha \div DFK)\sin(\beta + DCE)}$$

$$(A.6)$$

Allowance for friction

The surface of the velocity discontinuity AD is a boundary between the die and the tube. The shear stress acting along this surface is assumed to be a constant given by

$$\tau = mk \qquad (A.7)$$

The rate of working across discontinuity AD is given by

$$\dot{w}_{AD} = mk\,area_{AD}\,V_2 \qquad (A.8)$$

The velocity V_2 along the die face changes due to acceleration in region 2 of figure 1. If $(V_2)_1$ is the velocity along the die–tube interface at point A, the velocity at any point whose distance from the drawing axis is R, is given by

$$V_2 R = (V_2)_1 R_0 \qquad \text{assuming volume constancy.}$$

For a small change δR in R, the incremental area of the die face δA is given by

$$\delta A = \frac{2\pi R \delta R}{\sin\alpha}$$

Therefore the rate of working against friction over this incremental area is given by

$$d\dot{w}_{AD} = \frac{2\pi mk dR\,R_0 (V_2)_1}{\sin\alpha}$$

Integrating from $R = R_{of}$ to $R = R_0$ we have the total rate of working against friction along the die face,

$$\dot{w}_{AD} = \frac{2\pi mk\,R_0(R_0 - R_{of})(V_2)_1}{\sin\alpha}$$

from figure 2

$$(V_2)_1 = V_1 = 1$$

therefore

$$\dot{w}_{AD} = \frac{2\pi mk(R_0 - R_{of})R_0}{\sin\alpha} \qquad (A.9)$$

The surface of the discontinuity FC is a boundary between the plug surface and the tube. The shear stress acting along this surface is assumed to be equal to that acting along the die – tube interface.

Therefore the rate of working across discontinuity FC is given by

$$\dot{w}_{FC} = mk\,area_{FC}\,V_3 \qquad (A.10)$$

Following a similar procedure used for the die–tube interface, we have

$$V_3 R = (V_3)_4 R_{if}$$

$$\delta A = \frac{2\pi R \delta R}{\sin\beta}$$

therefore

$$d\dot{w}_{FC} = \frac{2\pi mk\,R_{if}\,dR(V_3)_4}{\sin\beta}$$

R ranges from R_{if} to R_x, therefore

$$\dot{w}_{FC} = \frac{2\pi mk\,R_{if}(R_x - R_{if})(V_3)_4}{\sin\beta}$$

From figure 2

$$(V_3)_4 = \frac{\sin DCE\,V_4}{\sin(\beta + DCE)}$$

therefore

$$\dot{w}_{FC} = \frac{2\pi mk\,R_{if}(R_x - R_{if})(R_o^2 - R_i^2)\sin DCE}{\sin\beta(R_{of}^2 - R_{of}^2)\sin(\beta + DCE)} \qquad (A.11)$$

Allowances for friction between the die land and the tube, and the plug land and the tube are made in a similar way.

The rate of working against friction along the die–land-tube interface is given by

$$\dot{w}_{DL} = 2\pi mk\,R_{of}L_d\,V_4$$

$$= \frac{2\pi mk\,R_{of}(R_o^2 - R_i^2)L_d}{(R_{of}^2 - R_{if}^2)} \qquad (A.12)$$

Similarly the rate of working against friction along the plug–land-tube interface is given by

$$\dot{w}_{PL} = 2\pi mk\,R_{if}\frac{(R_o^2 - R_i^2)L_p}{(R_{of}^2 - R_{if}^2)} \qquad (A.13)$$

Plastic work due to circumferential straining

From the work formula, the work done in plastic deformation is given by

$$w = V\bar{\sigma}\bar{\epsilon}$$

and

$$\dot{w} = \dot{V}\bar{\sigma}\bar{\epsilon} \qquad (A.14)$$

According to the von Mises yield criterion

$$\bar{\sigma} = \sqrt{3}\,k$$

and

$$\bar{\epsilon} = \frac{2}{\sqrt{3}}\left[\epsilon_R^2 + \epsilon_R\epsilon_\theta + \epsilon_\theta^2\right]^{\frac{1}{2}}$$

where $\epsilon_R = \ln\dfrac{(R_0 - R_i)}{(R_{of} - R_{if})}$ $\epsilon_\theta = \ln\dfrac{R_0}{R_{of}}$

The volume of deformed metal per unit time is given by

$$\dot{V} = \pi(R_o^2 - R_i^2)$$

Therefore the rate of working due to circumferential straining is given by

$$\dot{w}_P = 2\pi k(R_o^2 - R_i^2)[\epsilon_R^2 + \epsilon_R\epsilon_\theta + \epsilon_\theta^2]^{\frac{1}{2}} \qquad (A.15)$$

The rate of working of the external pull is given by

$$P = \pi(R_o^2 - R_i^2)\sigma_x \qquad (A.16)$$

Power balance

Equating the power P given by (A.16) with the power given by the summation of equations (A.2), (A.4), (A.6), (A.9), (A.11), (A.12), (A.13) and (A.15), we have

$$\frac{\sigma_x}{2k} = \frac{\sin x}{2\cos^2 \alpha/2} + \frac{\sin\beta}{2\sin DCE \, \sin(\beta + DCE)}$$

$$+ \frac{R_{of}(R_{of} - R_x) \sin(\alpha - \beta) \sin DCE}{(R_{of}^2 - R_{if}^2) \sin DFK \, \sin(\alpha + DFK) \, \sin(\beta + DCE)}$$

$$+ M\left[\frac{R_o(R_o - R_{of})}{(R_o^2 - R_i^2)\sin\alpha} + \frac{R_{if}(R_x - R_{if})\sin DCE}{\sin\beta \, \sin(\beta + DCE)(R_{of}^2 - R_{if}^2)} \right.$$

$$\left. + \frac{R_{of}L_d}{(R_{of}^2 - R_{if}^2)} + \frac{R_{if}L_p}{(R_{of}^2 - R_{if}^2)} \right]$$

$$+ [\epsilon_R^2 + \epsilon_R \epsilon_\theta + \epsilon_\theta^2]^{\frac{1}{2}} \tag{A.17}$$

From figure 1

$$\sin DCE = \frac{R_{of} - R_{if}}{[(R_{of} - R_{if})^2 + x^2]^{\frac{1}{2}}}$$

$$\sin DFK = \frac{R_{of} - R_x}{FD}$$

$$R_x = (FC \sin\beta + R_{if})$$

$$FC = \frac{(R_o - R_i) - \cos\alpha(R_{of} - R_{if} + x\tan\beta)}{\sin(\alpha + \beta)} - \frac{x}{\cos\beta}$$

$$FD = [FC^2 + CD^2 + 2FC \, CD \, \cos(\beta + DCE)]^{\frac{1}{2}}$$

$$CD = [(R_{of} - R_{if})^2 + x^2]^{\frac{1}{2}}$$

The right-hand side of equation (A.17) is totally determined by the geometry of the deformation zone, friction coefficient and the variable x which determines the position of the floating plug in the deformation zone.

DEFORMATION AND ITS RATE AS TWO CONCEPTS OF DESIGN OF TOOLS FOR THE SECONDARY TUBE-PIERCING OPERATION

by

P. V. VAIDYANATHAN* and T. Z. BLAZYNSKI*

SUMMARY

Tool designs based on the concept of the variation of homogeneous strains or that of strain-rate referred to the physical boundary are discussed. The pattern of roll separating forces and roll torques ascertained when using lead billets in a suitably-instrumented experimental tube mill are given. The variation of non-homogeneous strains using wax billets is also discussed. Comparison is made between the performance of the two types of tool.

NOTATION

A	tools designed on the basis of the concept of strain
B	tools designed on the basis of the concept of strain-rate
D	outer diameter of billet
d	inner diameter of billet
e_0	homogeneous generalised strain
$\dot{e}_m$	strain-rate referred to physical boundary
e_1, e_2, e_3	principal strains: strains at sections 1,2,3
e_1', e_2', e_3'	non-homogeneous strains
e_n	homogeneous strain at section 'n'
m	rate of deformation
t	tube thickness
T	time
V	velocity of billet
x	distance along workpiece
Z	parameter which depends on the geometry of the workpiece

INTRODUCTION

The initial development of the rotary secondary piercing process, as an extension of the Assel tube elongating operation, was described earlier by Jubb and Blazynski[1] and Blazynski[2]. Further experimental and analytical work carried out in this connection indicated the need for a careful investigation into the concepts involved in the design of tools required in the process.

As originally pointed out by one of the authors[3], the 'traditional' method of trial-and-error could be eliminated by using the concept of preservation of the constancy of strain ratios along the working zone of a metal forming pass. The use of this concept will account for the variation in the pattern of flow of the metal, without omitting from the consideration the effect of any homogeneous strain.

This approach was in fact used by Blazynski in the combined rotary piercing-elongating[4] and in the forward extrusion[5] processes. In both cases a reduction in non-homogeneous or redundant strains was observed. The severity of the secondary piercing operation calls for an attempt to reduce unnecessary (non-homogeneous) deformation. This paper deals with the development of tool profiles based, on the one hand, on the concept of the constancy of strain ratios and, on the other, on the more general concept of the constancy of strain-rate. The strain-rate is in this case referred to the physical boundary of the pass.

The results of experimental investigations involving both ideas are discussed in some detail. The changes in the pattern and in the incidence of the non-homogeneous strains were studied by means of internally marked wax billets and the load and torque requirements using lead specimens.

THE INVESTIGATED PROCESS

In the secondary rotary tube-piercing process (figure 1), the tubes which have been initially rotary pierced, are repierced to their final size in a three-roll mill. The three rolls are situated at 120° to each

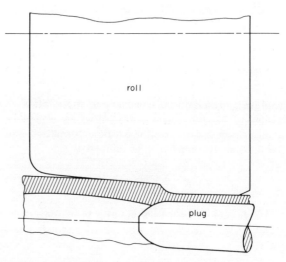

Figure 1. Diagrammatic representation of the secondary piercing process.

other and are so inclined to the mill axis that an axial pull is applied to the billet providing the necessary feed. As in an Assel mill, the rolls have a hump for elongating the tube. Unlike an Assel process, however, the secondary piercing operation

* Department of Mechanical Engineering, University of Leeds.

uses a piercing plug introduced centrally between the rolls. The plug is located in such a way that its shoulder lies in the gorge formed by the roll humps.

The tube first meets the rolls and is partially sunk before it touches the profiled plug. As the tube passes over the plug it experiences elongation and repiercing simultaneously till it reaches the gorge.

The main advantage of the process is that large elongations can be obtained in a single pass: in the Assel process the amount of reduction is restricted by the hump height.

TOOL DESIGN

Theoretical approach to the design of tools

Basically the adopted principle of design should be able to produce a correctly-designed tool acceptable to the industry, and one that leads to the reduction of the non-homogeneous strain.

The concept of the total generalised strain provides a common basis equally applicable to all metal-forming processes and includes all the essentials of the process by ensuring that the effect of the variation of the three homogeneous strains is accounted for.

One approach to the basic principle of design, as already mentioned, uses the concept of maintaining a constant ratio of successive increments of total generalised strain[3]. The analytical development of this concept produces a mathematical function which is directly related to the change in physical dimensions of the workpiece and consequently defines that shape of the boundary given by the tool profiles. In addition, the rate-of-change of homogeneous deformation, which is an important parameter, can be varied by the designer.

In the second approach, the concept of maintaining the strain-rate (referred to the physical boundary) as a constant is used, and this fulfils automatically the condition imposed by the strain approach.

Tool design based on the constancy of strain ratios

Assuming that the elastic deformations are negligibly small in comparison with the plastic deformation, the plastic strain increment can be defined as

$$de = \frac{\sqrt{2}}{3}(de_{ij}de_{ij})^{\frac{1}{2}} \tag{1}$$

Developing this equation in terms of homogeneous strains, the homogeneous generalised strain is given by

$$de_0 = \frac{\sqrt{2}}{3}[(de_1 - de_2)^2 + (de_2 - de_3)^2 +$$

$$(de_3 - de_1)^2]^{\frac{1}{2}} \tag{2}$$

Solving equation (2) between the section (0) of the pass, that is, the transverse section in which plastic deformation begins, and any section (n) along the pass,

$$e_n = \ln(Z) \tag{3}$$

where function Z is defined by the physical dimensions of the workpiece.

The condition of equal strain ratios along the working zone of the pass can be stated as follows:

$$\frac{e_2 - e_1}{e_1 - e_0} = \frac{e_3 - e_2}{e_2 - e_1} = \frac{e_n - e_{n-1}}{e_{n-1} - e_{n-2}} = m \tag{4}$$

Combining (3) and (4)

$$\frac{Z_n}{Z_{n-1}} = (Z_1)^{m^{(n-1)}} \tag{5}$$

Equation (5) provides the solution to the problem of tool design since it is established on the basis of both the concept of generalised strain and the principle of constancy of ratios of strain increments.

In the secondary piercing process, the physical dimensions which affect the value of the function Z are the workpiece diameter D, the plug diameter d and the workpiece thickness t. Since all the three quantities vary, particularly in the repiercing zone, it is not possible to write a simple expression relating the variables. A possible method of solution would be to relate D, t and d to the axial distance x along the pass. An expression for strain derived in this manner becomes rather complex, however.

One condition that can reasonably be satisfied in many secondary piercing operations is that the mean circumference of the workpiece remains constant, that is,

$$d + t = \text{constant} \tag{6}$$

This condition leads to the value of Z given by

$$Z_n = \left(\frac{t_n}{t_0}\right)^{2\sqrt{2}} \tag{7}$$

where t_n and t_0 are the tube thicknesses at two considered sections.

In equation (5) the value of constant m determines the rate-of-change of the homogeneous deformation in the pass. The basic rates possible are the decelerated, uniform and accelerated respectively: these can be varied by the designer at any stage of the design.

Tool design based on the strain-rate

The idea of using the rate-of-deformation as a basis of tool design is of recent origin, though the interest in the methods of calculation of strain-rates in metal forming processes has been growing since the development of high-energy rate forming techniques and the use of model materials for the investigation of conventional metal-forming processes. The mean strain-rate often used is that based on the time required for the material element to travel through the working zone of the pass: it does not therefore take into account the variation in velocity of the strained elements. Consequently, the mean strain-rate calculated on this basis will refer solely to the element under consideration and not to the total volume of material undergoing change at the given instant.

This limitation can be overcome by referring the mean strain-rate to the boundary of the zone of deformation as indicated by, among others, Hodierne[6] and Blazynski[7]. The strain-rate is referred to the horizontal projection of the length of contact between the tool and the workpiece. The method therefore gives the advantage of choosing more reliable values of mean yield-stress, especially at high strain-rates, than is generally possible when considering strain-rate values based on the concept of total time required for a given deformation to take place.

Mathematically, the definition of the mean strain-rate referred to the physical boundary, is given by

$$\bar{\dot{e}}_m = \frac{1}{x} \int_{x_1}^{x_2} \dot{e}_0 \, dx \qquad (8)$$

where x is the length along the tool and $\dot{e}_0$ the instantaneous strain-rate defined by

$$\dot{e}_0 = \frac{de_0}{dT} \qquad (9)$$

From equation (2), the increment of the generalised strain in the secondary piercing process is given by

$$de_0 = 2\sqrt{2} \frac{dt}{t} \qquad (10)$$

Considering the condition of constancy of volume

$$(D_0^2 - d_0^2) \, V_0 = (D^2 - d^2) \, V \qquad (11)$$

where D_0 and d_0 refer to the external and internal diameters of the original bloom, D and d to the diameters at a given section, and V_0 and V are the corresponding velocities.

Applying equations (6) and (11), the velocity V is found to be

$$V = \frac{(D_0 - d_0)}{2t} V_0 \qquad (12)$$

Putting (12) in (8), the value of the mean strain-rate referred to the physical boundary for any two consecutive sections (1) and (2) is given by

$$\bar{\dot{e}}_m = \sqrt{2} \frac{(D_0 - d_0)}{x} V_0 \left(\frac{1}{t_1} - \frac{1}{t_2} \right) \qquad (13)$$

With the values of t_1 and t_2 obtained from equation (7), the value of x can be determined from (13). It is seen therefore that both design concepts are incorporated in the B-type tools.

Evaluation of non-homogeneous strains

In the rotary process, the non-homogeneous strains are caused by the general non-uniformity of deformation obtained in each part of the rotation of the workpiece. Measurements of distortions and deformations were made by means of two types of internally marked wax bloom. Three basic types of non-homogeneous, shearing strain can be isolated in a rotary tubemaking operation. These are (i) longitudinal, (ii) circumferential and (iii) strain due to twist. Of the three, the strain due to twisting of the workpiece between the rolls, caused by the mismatch of tangential velocities, is the least important.

The longitudinal and circumferential shears are mainly responsible for the distortions and the overstraining of the worked material.

For the purpose of this paper it is not necessary to discuss the method of measurement of these strains. A detailed discussion is provided in a paper by Blazynski and Cole[8].

EXPERIMENTAL INVESTIGATION

Experimental mill

The experiments were all performed on a quarter-linear-scale model mill having 3 rolls disposed at 120° to each other. Each roll was driven by a separate hydraulic motor at a constant speed of 67 rev/min. The mill was equipped to measure roll-separating force and torque on all three rolls. The signals from the load and torque transducers were all recorded on a UV recorder.

The hydraulic motors were trunnion-mounted and torque was measured by means of strain gauge torque-reaction arms, screwed into the motor bodies. The full bridge arrangement of the strain gauges gave a sufficiently large output signal. Measurement of roll-separating force was achieved by quartz crystal load cells in combination with charge amplifiers.

The mill was also provided with a pneumatic quick-release mechanism which enabled the rolls to be retracted and the rolling to be stopped at any instant. An examination of the state of strain in any section of the pass could thus be carried out.

Tools

The rolls were manufactured from 0.4% carbon steel and were a quarter-scale version of typical industrial rolls with a height of hump of 0.063 in. Since the coefficient of friction between lead and steel is small, it was necessary to knurl the surface of the rolls in order to provide sufficient grip on the workpiece. Similar Tufnol rolls, without knurl, were used when testing wax specimens.

In order to test the validity of the theoretical methods of designing tools, two basic sets of plug were used. The design of one set was based on the strain: the other on the strain-rate concept. To examine the influence of the rate of deformation (m), on the performance of the tools, each set of plugs was designed to give three basic rates: decelerated, uniform and accelerated. The rates were 0.8, 1.0 and 1.2 respectively.

Plugs were made of both steel and Tufnol in order to measure the forces and non-homogeneous strains. Steel plugs and rolls were used when testing lead

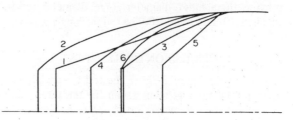

Figure 2. Representative plug profiles.

TABLE 1
Plug details

plug number	design concept	design condition	value of m	minimum diameter (in)	maximum diameter (in)	plug length (in)
1	A	decelerated	0.8	0.652	1.000	0.670
2	B	decelerated	0.8	0.652	1.000	0.741
3	A	uniform	1.0	0.664	1.000	0.394
4	B	uniform	1.0	0.664	1.000	0.513
5	A	accelerated	1.2	0.679	1.000	0.236
6	B	accelerated	1.2	0.679	1.000	0.388

A – tool profile with strain-increment-ratio constant; B – tool profile with strain-rate constant.

specimens, while Tufnol tools were used for testing wax specimens. All plugs were designed to merge with a 1 in diameter plugbar which was held stationary by the plugbar holder.

A representative set of plug profiles is shown in figure 2.

Specimens

Lead was used as a model material for measuring the roll-separating force and roll torque. The tubing was manufactured from commercially pure lead. Any ovality or eccentricity present was not considered to be sufficiently great to affect the measurements made.

TABLE 2
Sizes of lead tube used in the experiments

nominal outside diameter (in)	nominal wall thickness (in)	D/t ratio
1.366	0.183	7.464
1.428	0.214	6.673
1.486	0.243	6.115
1.678	0.339	4.950

Wax was used as the model material for measuring the non-homogeneous strains. Wax billets were made from a mixture of four waxes: 8% carnauba wax, 8% beeswax, 40% ceresin wax and 44% paraffin wax. Rolling of the billets was stopped by retracting the rolls and the partly-rolled billet was sectioned to reveal the distortions of the respective internal markings. The non-homogeneous strains were determined by making suitable measurements on the wax billets, using the technique of 8. The billets used were machined to 1.678 in OD, 1.00 in ID and 2 in long.

Rolling conditions

In all the tests the gorge diameter was varied to provide three billet diameter/gorge diameter ratios, that is, 1.10, 1.15 and 1.20. The roll feedangle was 6°.

DISCUSSION OF RESULTS

The graphs in figure 3 show the roll torque plotted against total homogeneous strain for the three billet diameter/gorge diameter ratios chosen. Figure 4 shows a similar set of curves for the roll-separating

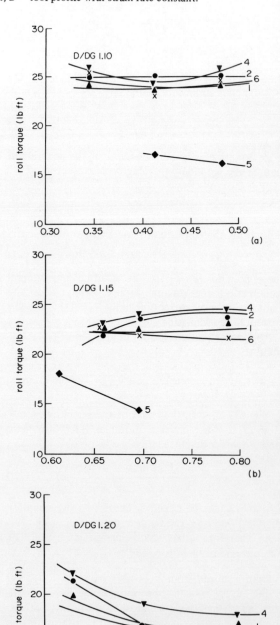

Figure 3(a), (b), (c). Variation of roll torque with strain for billet diameter/gorge diameter ratios of 1.10, 1.15 and 1.20.

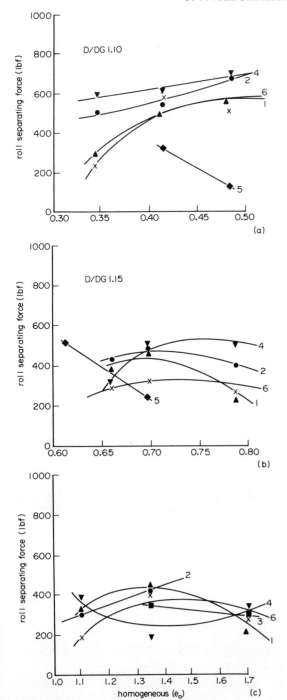

Figure 4(a), (b), (c). Variation of roll-separating force with strain for billet diameter/gorge diameter ratios of 1.10, 1.15 and 1.20.

force. It will be noticed that these curves have almost the same form as those of torque.

A general observation that can be made, on the evidence provided by these figures, is that the torque and roll-separating force increase if (i) the strain imposed is increased, (ii) the billet diameter/gorge diameter is reduced. Both torque and roll-separating force increase with increasing strain, there being a tendency to level-off (or in some cases fall-away) at the higher values of strain.

Though plugs (3) and (5) appear to give the minimum values of torque and roll-separating force,

it will be observed that their performance could not be recorded satisfactorily in certain cases. All tubes could be satisfactorily repierced in all conditions, with plugs (2), (4) and (6). It is obvious that plug (6) produces the conditions of minimum torque and roll-separating force in all the cases. It also gives the best performance for billet diameter/gorge diameter ratio 1.20 and at high strains.

Figure 5 shows that variation of non-homogeneous strain with distance along the pass, figure 5(a) representing the circumferential shear strain, figure 5(b) the longitudinal shear strain and figure 5(c) the shear strain due to twist. In figure 5(a), plugs

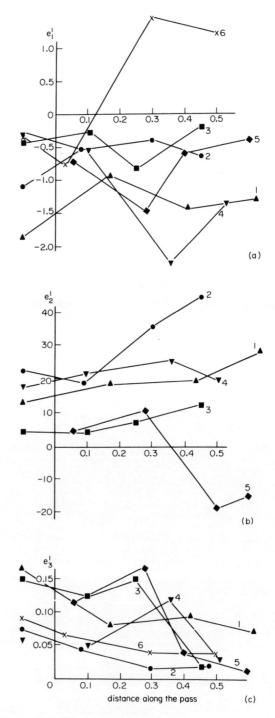

Figure 5(a), (b), (c). Variation of non-homogeneous strain with distance along the pass.

(2) and (3) give the least amount of reversals of strain while in figure 5(c), plugs (2) and (6) produce the least value of non-homogeneous strain. Further, the variation of shear strain due to twist is well defined for plug (6), while plug (3) gives a well-defined variation of longitudinal shear strain. Though plug (5) appears to give the minimum non-homogeneous strain, it is not suitable for all conditions of repiercing.

Considering the geometry of the plugs, it is seen that plug (5) has the shortest length. The effect of friction can therefore be expected to be small since the area of contact between the tool and the workpiece is relatively low. This is probably the main reason for the observed reduction in the roll torque, roll-separating force and in the non-homogeneous strains. Since, however, the required reduction in the wall thickness of the tube has, in this case, been achieved in a short pass, the plug profile is comparatively steep. As a result of this a number of tubes could not be repierced with this plug. From the industrial point of view an important factor in the selection of a plug profile is that of wear. Plug (5), with a sharp profile, would be expected to wear quickly, and is therefore the least attractive. In comparison plug (6), although of short length, does not suffer from the disadvantages of plug (5). Its profile is not sharp enough to produce undue wear, and it is, as already pointed out, eminently suitable for repiercing the whole range of tube sizes.

CONCLUSIONS

In the light of the results obtained from the experiments discussed, the tool profiles designed on the basis of the concept of strain-rate, referred to the physical boundary, seem to be industrially attractive when large deformations are required, particularly when an accelerated rate of deformation is employed. Although such tools do not eliminate the non-homogeneous strains completely, they tend to reduce the severity of these redundant strains.

ACKNOWLEDGEMENT

The authors wish to thank the Association of Commonwealth Universities for the financial support given to P. V. Vaidyanathan.

REFERENCES

1. C. JUBB and T. Z. BLAZYNSKI, 1969. Development of the Assel tube-elongating process into a secondary-piercing operation. *Proc. 9th International MTDR Conference,* Pergamon Press, p. 369.
2. T. Z. BLAZYNSKI, 1971. Development of high-ratio elongating processes. *Metal Forming,* **38,** p. 50.
3. T. Z. BLAZYNSKI, 1967. Theoretical method of designing tools for metal-forming processes. *Metal Forming,* **34,** p. 50.
4. T. Z. BLAZYNSKI, 1969. Design of tools for the combined piercing–elongating tube making process. *Proc. 9th International MTDR Conference,* Pergamon Press, p. 351.
5. T. Z. BLAZYNSKI, 1971. Optimisation of die design in the extrusion of rod using model materials. *Int. J. Mech. Sci.,* **13,** p. 113.
6. F. A. HODIERNE, 1961. University of London, Ph.D. Thesis. (1961).
7. T. Z. BLAZYNSKI and I. M. COLE, 1963/4. An analysis of redundant deformations in rotary piercing. *Proc. Inst. Mech. Engrs.,* **178,** p. 867.

AN EXPERIMENTAL INVESTIGATION OF THE SANDWICH ROLLING OF THIN HARD SHEETS

by

A. A. AFONJA* and D. H. SANSOME†

SUMMARY

The rolling of hard metals, particularly to thin gauges, presents many sheet manufacturers with considerable difficulties. The high forces to which the rolling mill is subjected may cause excessive elastic distortion of the rolls, thereby imposing a limit on the reduction that can be achieved, or even resulting in roll breakage. The use of small diameter rolls reduces the roll force substantially but the need to support the rolls to prevent them from bending makes such a mill expensive and uneconomic unless the level of production is high. However, a considerable reduction in roll separating force can be achieved on a conventional mill at an economic level by rolling the hard metal sandwiched between softer metal. Also, the rolling of clad metals is necessary to meet particular technological requirements. An experimental investigation of the rolling of high carbon steels sandwiched between the softer metals, mild steel, copper and aluminium, has been carried out on a laboratory 2-high mill instrumented for the measurement of roll force and torque. The effect of various parameters (such as interface friction, relative metal hardness ratios and thickness ratios) on the roll force is considered and an outline of the theoretical analysis of the process given. It is concluded that a reduction of more than 50 per cent in roll force can be achieved with a consequent increase in the maximum reduction possible, and that the process has considerable potentiality particularly for the rolling of thin, hard sheets on a small or medium scale.

NOTATION

p normal stress

τ shear stress

σ horizontal stress

R roll radius

h half-instantaneous thickness of strip

θ angle of bite

K mean yield-stress in plane strain

μ coefficient of friction at roll-clad interface

SUFFIX

i entry of deformation zone

o exit of deformation zone

c clad

m matrix

INTRODUCTION

The problem of cold rolling high strength metals in strip form to thin gauges has been solved largely by the introduction of the Sendzimir-type multi-roll mill. However, this type of mill is expensive and often requires a relatively high level of production to operate economically. The need to apply high tensions, to ensure that the product is of good shape, often imposes a limit on the width of strip that can be rolled and requires the use of powerful coiler and decoiler units. The main advantage of the multi-roll mill is that it employs small diameter workrolls. Consequently, the roll separating force is reduced and higher reductions can be achieved between anneals. However, the small workrolls have to be supported to reduce bending and this has led to the development of mills with as many as eighteen back-up rolls. Comparable reductions in roll separating force can be achieved much more cheaply on a conventional mill by rolling the hard metal sandwiched between sheets of softer metal. This technique has been applied to the rolling of titanium alloy sheets[1] but, as far as the authors are aware, no work has been done on its applicability to the rolling of high strength steels. In view of this, an investigation has been carried out on the rolling of stainless steel and high carbon steels between layers of aluminium, copper or mild steel: the results are presented in this paper.

THEORETICAL CONSIDERATIONS

Figure 1(a) shows the stresses acting on an elemental vertical section of the sandwich between the rolls. A magnified view of the element is shown in figure 1(b). The softer clad has a lower mean yield-stress and will tend to elongate in preference to the harder metal constituting the matrix, but its movement will be restricted by the frictional shear stress (τ_m) at the

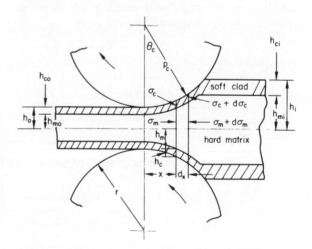

Figure 1(a). Section through the rolls and workpiece showing the usual arrangement with the softer layers outermost.

*Lecturer in the Dept. of Chemical Engineering Technology, University of Ife, Nigeria.
†Reader in the Dept. of Mechanical Engineering, University of Aston in Birmingham.

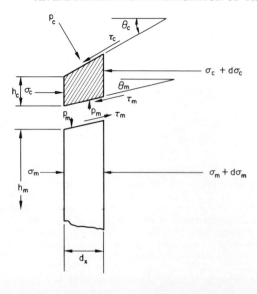

Figure 1(b). Magnified view of the elemental length of the compound workpiece in the entry zone.

clad-matrix interface. As a result, tensile compressive stresses are induced in the matrix and clad respectively. It can be shown[2] by applying the yield criterion that in these circumstances of applied compressive stresses, the clad will become more difficult to deform, that is, the values of the principal stresses are increased. On the contrary, the matrix will be easier to deform because of the tensile stresses. Clearly the overall effect of τ_m on the total roll separating force depends on the relative yield stresses of the component metals, the proportion of clad in the sandwich, and the frictional conditions at the clad-matrix interface. By careful selection of conditions and parameters it is possible to obtain an overall reduction in the roll separating force. It is well known that roll-flattening and consequently the roll force increases as the strip thickness decreases and vice versa. The increase in thickness of the strip due to cladding will therefore result in a further decrease of the roll force.

EXPERIMENTAL TECHNIQUES

Equipment
The investigation was carried out on a laboratory 2-high mill having hardened steel rolls, 4.5 in diameter and 4 in long (see Plate 1). The rolls were finish ground to a surface roughness of approximately 14 μin. Roll force and torque meters incorporating straingauge bridges were specially designed for the mill. The signals from the bridges were fed into a 10-channel UV recorder. Hand micrometers were used for width and thickness measurements. The maximum safe load on the mill was about 20 tonf.

Materials
Details of the materials used in the tests are shown in table 1.

Preparation of specimens
Strips 5 or 12 in long were cut from 1 in wide coils of hard and soft metal respectively. Oxide films were removed from the surfaces of the specimens by rubbing with fine emery paper and all were thoroughly cleaned with trichloroethylene. The soft metal was folded and the hard strip sandwiched between the soft layers. All three layers were riveted together at the free ends to prevent them from slipping sideways. Preliminary measurements showed that the riveting had no effect on the roll separating force except near the rivets. First-pass reductions of 10–40 percent were made on single strips as well as sandwiched metals without using any rolling lubricant. Measurements of the roll force, the roll torque and the initial and final thickness of the strips were recorded.

DISCUSSION OF RESULTS

Figures 2 to 4 show that there is a decrease in the roll force when high strength steels are rolled by the sandwich method. In all cases, the maximum reduction in strip thickness possible without exceeding the maximum safe load of the mill is more than doubled. The load economy is considerably greater for the

TABLE 1

position in sandwich	material	thickness (in)	hardness (DPN) in annealed state
matrix (hard)	austenitic stainless steel	0.047	135
	0.71% carbon steel	0.026	160
	1.25% carbon steel	0.036	280
clad (soft)	commercial purity aluminium	0.028–0.05	17–20
	high conductivity copper	0.01–0.04	60–70
	mild steel	0.018–0.05	90–100

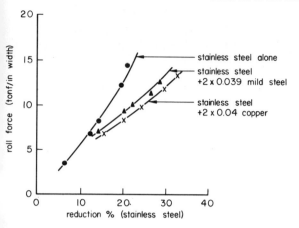

Figure 2. Effect of cladding on the roll separating force developed during the rolling of austenitic stainless steel.

0.71 percent carbon steel strip than for the harder but thicker 1.25 percent carbon steel strip (see figures 3 and 4). Apparently, the contribution to the load reduction by the increase in the initial thickness is higher in the former case. In all cases, the roll force is lower when copper is used as the clad than when mild steel is used, other parameters remaining unchanged. This appears to be reasonable since the copper is softer and tends to deform more readily

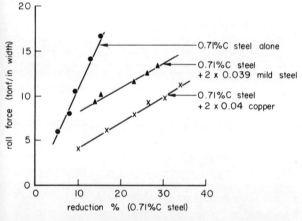

Figure 3. Effect of cladding on the roll separating force developed during the rolling of 0.71% C steel.

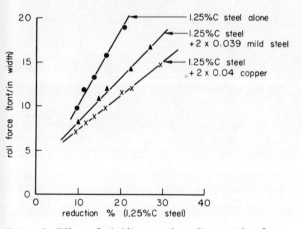

Figure 4. Effect of cladding on the roll separating force developed during the rolling of 1.25% C steel.

than the mild steel, thereby inducing a higher interface shear stress τ_m.

Figure 5 shows that the roll torque is higher when copper is rolled between aluminium layers than when it is rolled alone. In figure 6 the opposite is true; the roll torque is higher for the stainless steel strip than for the sandwich. This is not surprising. The reduction

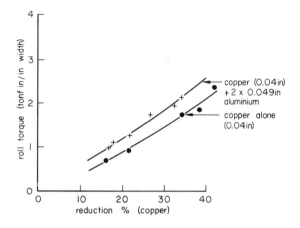

Figure 5. Effect of cladding on the roll torque developed during the rolling of copper.

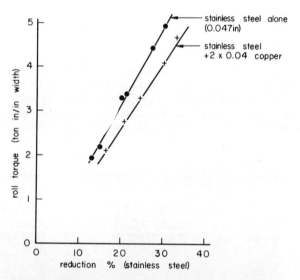

Figure 6. Effect of cladding on the roll torque developed during the rolling of austenitic stainless steel.

in roll force due to cladding will cause a reduction in the roll torque, but the increase in thickness due to cladding will result in an increase in the roll torque. The overall effect of cladding on the roll torque therefore depends on the clad-sandwich thickness ratio and the clad-matrix yield-stress ratio. In figure 5 the thickness of the clad is the dominant factor whereas in figure 6 the effect of the load reduction due to cladding is more marked.

In figure 7 the roll force decreases with increasing clad-sandwich thickness ratio B until B attains a value between 0.5 and 0.6 when the roll force starts to rise. This means in effect that there is an optimum value of B that gives a maximum reduction in the roll force. At this value, the frictional shear stress τ_m has reached the maximum value possible for the particular set of

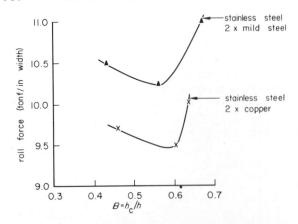

Figure 7. Effect of clad thickness on the roll separating force developed during the rolling of austenitic stainless steel (25% reduction).

process parameters. Beyond this point, the increase in the area of contact between the rolls and the sandwich, as a result of an increase in B, becomes the dominant factor.

SUMMARY OF CONCLUSIONS

The foregoing results show that a substantial reduction in the roll force can be achieved by rolling thin, high strength metal between layers of softer metal. Consequently, higher reductions in gauge can be achieved than is possible without cladding. From the point of view of load reduction, the softer clad gives a higher reduction in roll force, the limit being when deformation becomes restricted to the softer metal alone and cannot penetrate into the matrix. However, the choice of clad will be greatly influenced too by economic considerations, since the clad may be discarded after use.

The roll torque may increase or decrease as a result of cladding, depending on the clad-sandwich thickness ratio, the clad-matrix yield-stress ratio and the reduction in gauge achieved. In any case, the roll torque is usually not a limiting factor in normal cold rolling practice. The choice of clad material and thickness influences the reduction in roll force that can be achieved. Experiments indicate that the optimum value of clad-sandwich thickness ratio B is between 0.5 and about 0.6.

ACKNOWLEDGEMENTS

The authors would like to thank Mr. R. R. Arnold of the Research Department of Imperial Metal Industries Limited, Birmingham for his helpful comments, and Messrs. I.M.I., J.B. and S. Lees, Arthur Lee & Sons, the Steel Company of Wales and Samuel Fox for supplying the materials. Also, their thanks are due to Mr. G. M. Jones and Mr. J. Hirons for assistance with the manufacture and instrumentation of the experimental mill.

REFERENCES

1. R. R. ARNOLD and P. W. WHITTON, Proc. Inst. Mech. Eng. 173 (8), 241 (1959).
2. A. A. AFONJA. The sandwich rolling of thin hard strip. Ph.D. Thesis, University of Aston, Birmingham (1970).

Plate 1. The rolling mill and instrumentation.

FLATNESS OF SHEET AND STRIP IN COLD ROLLING

by

G. W. ROWE* and A. S. FEDOSIENKO†

SUMMARY

A short investigation is reported on the initiation of buckling in strip produced by rolls with excessive camber. A simple elastic analysis gives qualitative agreement with the experimental results, but still requires elaboration for quantitative predictions.

INTRODUCTION

For all practical purposes it is desirable for rolled sheet and strip to be flat. The camber of rolls is carefully chosen so that the thickness of strip is uniform across the width, despite the inevitable bending of the rolls. Strip which is thinner at the edges tends to develop 'loose edges' because of the additional elongation, while over-camber tends to produce 'loose middle' with a wavy central zone as shown in figure 1. The permissible range between these extremes is usually fairly easily determined

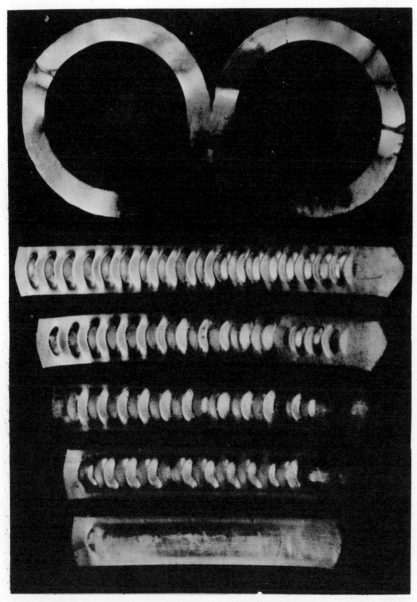

Figure 1. A photograph of six aluminium strips, each initially 2 in wide, rolled with increasing nominal reductions of area (24%, 29%, 33%, 39%, 45% and 50%), using severely over-cambered rolls. Buckling, or loose middle, starts with reductions over 20%, and at 50% the strip is cut into two.

*Department of Industrial Metallurgy, University of Birmingham.
† Chelyabinsk Polytechnic Institute, U.S.S.R.

empirically but the conditions for flatness are not clearly understood. A few investigations have been reported, but it has always been necessary to introduce rather broad assumptions. The aim of the present work is to examine possible theories and to undertake accurate experiments with deliberately over-cambered rolls to check the validity of the theoretical approach and to assist in formulating realistic assumptions.

The basic problem can conveniently be stated in two ways; firstly, to formulate an expression for the conditions under which buckling will occur, in terms of the ingoing and outgoing thicknesses of the strip, its width and its elastic and plastic properties; secondly, to describe the shape of a buckled strip, for example by its amplitude and wavelength. The first problem is clearly the important one in practice, but the second is essential for developing the theory.

THEORETICAL STUDY

Consideration of elastic buckling of strip

To ensure that central buckling or loose middle will occur, the rolls were severely over-cambered. The central region of the strip will thus be elongated appreciably more than the edges, and the emergent strip can be considered, to a first approximation, to consist of three zones. As the leading end emerges from the rolls it will be flat but longer in the middle than at the edges, so that the central zone is subject to a restraining longitudinal compression. At some stage, as the rolled length increases, this compression will cause a collapse of the central region into a buckle, in much the same way as a strut collapses at a critical load. The compressive stress is thereby released, but immediately begins to build up as rolling continues, producing a cyclic behaviour, as shown in figure 1.

As a first approach, therefore, it is possible to consider the buckling to be governed by strut

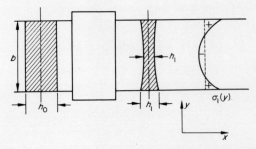

Figure 2. A diagram indicating the strip thickness h_0 before rolling, the profile from $h_{1\,max}$ to $h_{1\,min}$, after rolling, and the stress distribution in unbuckled strip. The central region is subject to compression, which eventually causes buckling.

collapse theory with appropriate boundary conditions. The imposed stresses result from elastic strains, indicated in figure 2, until they disappear or diminish at the moment of collapse. Euler's theory of elastic collapse of a simple strut can therefore be used as a basis for calculation.

For a simple freely supported strut the critical buckling stress is given by

$$\sigma_c = \pi^2 \frac{EI}{l^2} \quad \text{or} \quad \pi^2 E \left(\frac{k}{l}\right)^2 \qquad (1)$$

More generally, account can be taken of end constraint by introduction of a constant κ, so that, applying the criterion to a wide sheet of thickness h,

$$\sigma_c = \kappa E \left(\frac{h}{l}\right)^2 \qquad (2)$$

For a long sheet the length l becomes the half-wavelength l_w of the regular buckles.

The problem is then to determine the form of κ, which is likely to depend upon the dimensionless ratios of b/h or b/l_w. If the longitudinal compressive stress is less than σ_c the strip will be expected to stay flat, though it will contain finite residual stresses.

Application of Euler theory of elastic buckling of a long thin strut

Since the thickness of the strip varies across the width, we may conveniently define the average thickness of the strip after rolling as

$$h_{1a} = \frac{1}{b} \int_{-b/2}^{+b/2} h_1(y)\,dy \qquad (3)$$

The simple buckling condition for long strip is then given by the following equation for the maximum compressive stress after rolling:

$$(\sigma_1)_{max} = \sigma_c = \kappa E \left(\frac{h_{1a}}{l_w}\right)^2 \qquad (4)$$

We can now make the common simplifying assumption that all straight lines parallel before rolling remain parallel after the process, which implies homogeneous deformation. Figure 3 shows diagrammatically the typical lateral profile.

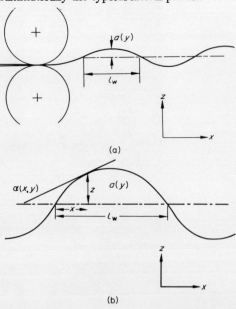

(a)

(b)

Figure 3. A diagram of the profile of the strip, seen from the side, after buckling has occurred. The amplitude a is a function of the lateral position y as well as being an assumed sinusoidal function of x.

Since volume is conserved in the plastic deformation we may consider the length of an element after making allowance for elastic recovery, and write the following equation for a portion of length l_0 before rolling and l_1 after, as in figure 3.

$$h_1(y) \int_{l_0}^{l_1} \frac{1 - \sigma_1(y)/E}{\cos\alpha(x,y)} \, dx = h_0 l_0 \qquad (5)$$

Further information can be obtained from the condition that the resultant force arising from the residual stresses is zero, so that

$$\int_{-b/2}^{+b/2} \sigma_1(y) \, h_1(y) \, dy = 0 \qquad (6)$$

Condition prior to buckling
Up to the instant that buckling commences the strip remains flat, so that

$$\alpha(x,y) = 0 \qquad (7)$$

Consequently for this flat strip equation (5) may be written

$$1 - \frac{\sigma_1(y)}{E} = \frac{l_0}{l_1} \frac{h_0}{h_1(y)} \qquad (8)$$

Thus, using the volume constancy condition that $l_0 h_0 = l_1 h_{1a}$

$$\sigma_1(y) = E \left\{ 1 - \frac{h_{1a}}{h_1(y)} \right\} \qquad (9)$$

and the maximum value of the compressive stress

$$(\sigma_1)_{\overline{max}} = E \left\{ \frac{h_{1a}}{(h_1)_{min}} - 1 \right\} \qquad (10)$$

But we have already the condition for buckling in equation (4), so combining this with equation (10) we see that buckling will occur when

$$(\sigma_1)_{\overline{max}} = \kappa E \left(\frac{h_{1a}}{l_w} \right)^2 = E \left\{ \frac{h_{1a}}{(h_1)_{min}} - 1 \right\}$$

Thus the strip will stay flat when

$$\left\{ \frac{h_{1a}}{(h_1)_{min}} - 1 \right\} < \kappa \left(\frac{h_{1a}}{l_w} \right)^2 \qquad (11)$$

It will be seen that this relationship does not contain the modulus of elasticity E, so that according to these assumptions the condition for the strip to remain flat depends only on the geometry. This arises because the maximum residual stress and the stress to cause buckling are both directly proportional to E for elastic buckling.

Shape of strip after buckling
Let us assume a simple distribution of residual stress such that once buckling has occurred all stresses are relaxed; that is, for the buckled strip

$$\sigma_1(y) \equiv 0 \qquad (12)$$

Then we can rewrite equation (5), using a specific value of l, equal to the half-wavelength of the buckling. l_{0w} is the corresponding initial length.

$$\int_{l_0}^{l_w} \frac{dx}{\cos\alpha(x,y)} = \frac{l_{0w} h_0}{h_1 y} \qquad (13)$$

Referring again to figure 3, for a sinusoidal wave form the height $Z(x)$

$$Z(x) = a(y) \sin\frac{\pi x}{l_w} \qquad (14)$$

Hence

$$\tan\alpha(x,y) = \frac{dz}{dx} = \frac{\pi a(y)}{l_w} \cos\frac{\pi x}{l_w} \qquad (15)$$

or

$$\alpha(x,y) = \tan^{-1} \left\{ \frac{\pi a(y)}{l_w} \cos\frac{\pi x}{l_w} \right\}$$

$$= \cos^{-1} \frac{1}{\left[1 + \left\{ \frac{\pi a(y)}{l_w} \cos\frac{\pi x}{l_w} \right\}^2 \right]^{1/2}} \qquad (16)$$

Substituting this into the integral of equation (13)

$$2 \int_0^{\frac{1}{2}l_w} \left[1 + \left\{ \frac{\pi a(y)}{l_w} \cos\frac{\pi x}{l_w} \right\}^2 \right]^{1/2} dx = l_{0w} \frac{h_0}{h_1(y)} \qquad (17)$$

This can be transformed to

$$2 \frac{l_w}{\pi} \left[1 + \pi^2 \left\{ \frac{a(y)}{l_w} \right\}^2 \right]^{1/2} \times$$

$$E^* \left[\frac{\left\{ \frac{\pi a(y)}{l_w} \right\}^2}{1 + \left\{ \frac{\pi a(y)}{l_w} \right\}^2} \right]^{1/2} = l_{0w} \frac{h_0}{h_1(y)} \qquad (18)$$

where $E^*(x)$ is the full elliptical integral of the second class over the variable x,

$$E^*(x) = \int_0^{\pi/2} (1 - x^2 \sin^2\chi \, d\chi)^{1/2}$$

values of which are given in standard tables of integrals.

Now, for a simple distribution of residual stress

$$\frac{l_{0w}}{l_w} = \frac{(h_1)_{max}}{h_0} \qquad (19)$$

so we may write

$$\left[1 + \pi^2 \left\{ \frac{a(y)}{l_w} \right\}^2 \right]^{1/2} E^* \left(\left[\frac{\left\{ \frac{\pi a(y)}{l_w} \right\}^2}{1 + \left\{ \frac{\pi a(y)}{l_w} \right\}^2} \right]^{1/2} \right)$$

$$= \frac{\pi}{2} \frac{(h_1)_{max}}{h_1(y)} \qquad (20)$$

TABLE 1

Experimental results (Thickness measurements in 0.001 in units, length in 1 in units)

Strip No.	h_0	h_1 max.	h_1 min.	$(a)_{max}$ meas.	$(a)_{max}$ calc.	meas./calc.	l_w meas.	l_w calc.
(1) Width b = 1 in								
1.1	66	18 (18, 18)	16 (16, 16)				cut along centre	
(2) Width b = 1.5 in								
2.1	66	35	30	0	0		∞	∞
2.2	66	31 (32, 30)	26 (27, 25)	99 (99, 92)	36	2.75	0.72 (0.75, 0.72)	0.36
2.3	66	27 (27, 27)	22 (23, 22)	104 (109, 101)	41	2.53	0.67 (0.72, 0.64)	0.34
2.4	66	25 (26, 24)	20 (21, 20)	106 (110, 101)	43	2.46	0.67 (0.70, 0.62)	0.33
2.5	66	23 (24, 23)	19 (20, 18)	103 (106, 97)	42	2.45	0.66 (0.67, 0.62)	0.33
2.6	66	22 (23, 22)	18 (19, 17)	108 (112, 104)	45	2.40	0.61 (0.62, 0.56)	0.30
(3) Width b = 2 in								
3.1	66	50 (51, 50)	45 (46, 44)	0	0		∞	∞
3.2	66	47 (48, 47)	42 (43, 42)	179 (182, 175)	68	2.63	0.94 (0.95, 0.89)	0.47
3.3	66	44 (44, 43)	38 (38, 37)	185 (188, 179)	73	2.53	0.89 (0.92, 0.86)	0.45
3.4	66	40 (40, 39)	33 (33, 33)	179 (183, 173)	73	2.45	0.83 (0.89, 0.78)	0.42
3.5	66	36 (37, 36)	29 (31, 29)	175 (181, 172)	73	2.39	0.80 (0.83, 0.76)	0.40
3.6	66	33	23				cut	

TABLE 1 (*continued*)

(4) Width b = 2.5 in	4.1	66	61	62 / 61	55	56 / 55	263	265 / 262	104	2.52	1.19	1.15 / 1.19	0.57
	4.2	66	59	59 / 58	52	53 / 51	253	259 / 250	100	2.53	1.15	1.20 / 1.09	0.57
	4.3	66	55	56 / 54	47	48 / 46	248	249 / 247	100	2.48	1.08	1.11 / 1.06	0.54
	4.	66	51	52 / 51	43	44 / 43	244	249 / 237	100	2.44	1.04	1.08 / 1.02	0.52
	4.5	66	47	48 / 47	39	39 / 38	244	248 / 242	102	2.39	1.00	1.05 / 0.95	0.50
	4.6	66	44	44 / 43	35	36 / 39	231	238 / 224	98	2.35	0.89	0.92 / 0.83	0.45
	4.7	66	44	45 / 44	35	35 / 34	232	235 / 228	98	2.36	0.88	0.91 / 0.86	0.44
	4.8	66	41		29							cut	
(5) Width b = 3 in*	5.1	66	60	61 / 59	51	51 / 50	306	314 / 299	128	2.39	1.39	1.41 / 1.37	0.70
	5.2	66	58	59 / 57	48	49 / 46	309	316 / 296	130	2.37	1.30	1.34 / 1.25	0.65
	5.3	66	54	55 / 53	42	44 / 41	301	304 / 298	130	2.31	1.23	1.30 / 1.17	0.62
	5.4	66	52		37							cut	
(6) Width b = 3.5 in	6.1	66	61	62 / 60	48	48 / 47	359	371 / 351	155	2.31	1.37	1.42 / 1.33	0.68
	6.2	66	61	62 / 60	47	48 / 46	357	364 / 347	155	2.30	1.41	1.44 / 1.37	0.70
	6.3	66	60		41							cut	

*Negative spread was observed.

Thus, if the distribution $h_1(y)$ is known, the amplitude function $a(y)/l_w$ can be calculated. For all practical purposes (though not necessarily for our exaggerated experiments)

$$\frac{a(y)}{l_w} \ll 1 \qquad (21)$$

so that

$$E^* \left(\left[\frac{\left\{ \frac{\pi a(y)}{l_w} \right\}^2}{1 + \left\{ \frac{\pi a(y)}{l_w} \right\}^2} \right]^{1/2} \right) = \frac{\pi}{2} \frac{(h_1)_{max}}{(h_1)y} \qquad (22)$$

This can be simplified and solved to give

$$\frac{a(y)}{l_w} \approx \frac{1}{\pi} \left[\left\{ \frac{(h_1)_{max}}{h_1(y)} \right\}^2 - 1 \right]^{1/2} \qquad (23)$$

The maximum value of a is then

$$\frac{(a)_{max}}{l_w} = \frac{1}{\pi} \left[\left\{ \frac{(h_1)_{max}}{(h_1)_{min}} \right\}^2 - 1 \right]^{1/2} \qquad (24)$$

The same result is given by Nalcha, and a similar formula appears in the paper by Pearson[1] (possibly with a mistake). It is found that since $2h_{1a} \approx (h_1)_{max} + (h_1)_{min}$ this equation (24) can be combined with the buckling condition (11) to show that at the moment of buckling the maximum wave amplitude is given by

$$\frac{(a)_{max}}{l_w} = \frac{2\sqrt{\kappa}}{\pi} \qquad (25)$$

Thus, if this approximation is valid, it is a simple matter to obtain empirical values for κ.

Non-sinusoidal buckling of a strut

It is observed practically that the waveform is not sinusoidal, but more nearly triangular (equal saw-tooth). The corresponding analysis leads to an equation closely resembling equation (24), namely

$$\frac{(a)_{max}}{l_w} = \frac{1}{2} \left[\left\{ \frac{(h_1)_{max}}{(h_1)_{min}} \right\}^2 - 1 \right]^{1/2} \qquad (26)$$

This differs only in the constant $\frac{1}{2}$ in place of $\frac{1}{\pi}$.

Consideration of buckling theory applied to wide plates

The above analysis does not include variation of amplitude $a(y)$ across the width y of the strip in deriving the buckling condition. This deficiency of unidimensional theory has already been noted in the paper by Wistreich[3].

Current theory of buckling can be extended to a flat plate loaded uniformly along two sides, with various load distributions on the other two, for example both being elastically restrained, or one being elastically restrained and the other free.

Under the conditions of strip rolling both edges are free and the stress is not uniformly distributed across the width.

If we simplify this to assume free edges and uniform stress across the strip, the condition reverts to that for the simple unidimensional strut, and the strip width plays no part.

It seems most appropriate for future extension of the work to apply the Ritz method[6], considering the total strain energy in unit length of strip. This has not yet been applied.

Plastic buckling

A further approach worth considering is based on the plastic collapse theory, in which the moment applied to a plastic hinge is considered by the methods of Green[7] and of Baker et al.[8]. The form of the observed buckling strongly suggests that plastic collapse is important.

EXPERIMENTAL RESULTS

The experiments are simple in principle. A Stanat–Mann 2-high rolling mill was used with 5 in diameter, 8 in wide rolls. Special rolls were made with an exaggerated conical profile rising centrally to 0.050 in above the edge diameter. The stock aluminium was 0.066 in thick and strips were cut to appropriate widths, all passes being taken with fresh specimens. It is necessary to restrain the strip on the run-in and run-out tables to prevent curling, but the forces involved have no influence on the buckling.

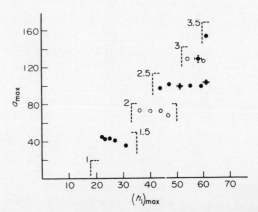

Figure 4. The relationship between calculated values of a_{max} and the reduction in cross-section. For any given strip width the amplitude is little influenced by the reduction, but there is an approximately linear relationship between amplitude and width.

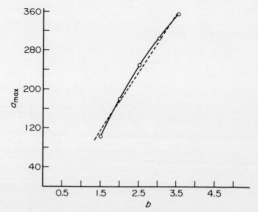

Figure 5. The experimental relationship between the amplitude of buckle a and the strip width b.

TABLE 2

Correlation in terms of non-dimensional ratios

No.	h_{1a}	$\dfrac{b}{h_{1a}}$	$\dfrac{(h_1)_{max}}{(h_1)_{min}}$	a_{max}		$\dfrac{l_w}{b}$
				$\dfrac{a_{max}}{l_w}$	$\dfrac{a_{max}}{h_{1a}}$	
				$b = 1.5$ in		
1	32	47	1.17	–	–	–
2	28	54	1.19	0.1	1.28	0.24
3	24	62	1.23	0.12	1.71	0.23
4	22	68	1.25	0 13	1.55	0.22
5	21	71	1.21	0.13	2.00	0.22
6	20	75	1.22	0.15	2.25	0.20
				$b = 2.0$ in		
1	47	43.	1.11	–	–	–
2	44	45	1.12	0.14	1.55	0.23
3	41	49	1.16	0.16	1.78	0.22
4	36	56	1.21	0.17	2.03	0.21
5	32	62	1.24	0.18	2.28	0.20
6	28	71	1.43	–	–	–
				$b = 2.5$ in		
1	58	43	1.11	0.18	1.80	0.23
2	55	45	1.13	0.18	1.82	0.23
3	51	49	1.17	0.19	1.96	0.22
4	47	53	1.19	0.19	2.13	0.21
5	43	58	1.20	0.20	2.38	0.20
6	39	64	1.26	0.22	2.52	0.18
7	39	64	1.26	0.22	2.52	0.18
8	35	71	1.41	–	–	–
				$b = 3.0$ in		
1	55	55	1.18	0.18	2.33	0.23
2	53	57	1.21	0.20	2.46	0.22
3	48	62	1.29	0.21	2.71	0.21
4	44	68	1.41	–	–	–
1	54	64	1.27	0.23	2.87	0.19
2	54	64	1.30	0.22	2.87	0.20
3	50	70	1.46	–	–	–

Numerous tests have been conducted using partially strain-hardened aluminium stock of normal commercial

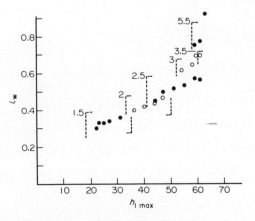

Figure 6. The relationship between the calculated wavelength of buckle and the reduction in cross-section.

purity. Figure 1 shows a set of narrow strips in which the development of buckling can be followed as a function of percentage reduction in cross-sectional area in single passes. The details of this particular set

of rolling tests are given in table 1 for 2 in wide specimens. The parameters $(a)_{max}$ describing the maximum amplitude and l_w the wavelength of the buckled strip are shown in these tables as functions of the average strip thickness, before and after rolling. The amplitudes were measured with a dial gauge reading in 0.0001 in units and mounted on a rigid C frame. Wavelengths were measured between peaks located by light scratching against a steel straight-edge.

It is convenient to present the results in the form of graphical relationships between amplitude or wavelength and the outgoing thickness, bearing in mind that the ingoing thickness was maintained constant at 0.066 in. Figure 4 shows quite clearly that for any given strip width the amplitude of the buckle (when a regular cyclic buckling is established) is almost independent of h_1 and therefore of the reduction in cross-sectional area. There is, however, almost a linear dependence of the amplitude upon the width of the strip, as shown in figure 5. Figure 6 shows the relationship between the wavelength and the reduction in cross-section. This is approximately linear, but with very wide strips the possible range of reductions is limited and the relationship less clear.

Calculations were then performed in an attempt to correlate the non-dimensional groups. Table 2 shows the data. It appears that l_w is related to the width b by an almost constant multiplier, whereas the ratio $(a)_{max}/b$ increases steadily with both increasing reduction and increasing width. The value of κ is approximately constant, as required by equation (25), but is numerically too large.

CONCLUSION

The results appear to be reproducible and reliable. The theory is not sufficiently advanced to allow useful comparison, and it appears to be necessary to introduce plastic buckling.

ACKNOWLEDGEMENTS

We thank Professor E. C. Rollason for provision of facilities in the Aitchison Laboratory.

REFERENCES

1. W. K. J. PEARSON, 1964/5. 'Shape measurement and control', *Jnl. Inst. Metals*, **93**, 169–78.
2. B. SABATINI and K. A. YEOMANS, 1968. 'An algebra of strip shape and its application to mill scheduling', *Jnl. Iron and Steel Inst.*, **206**, 1207–13.
3. M. D. STONE, R. GRAY and R. A. SOMERVILLE, 1965. 'Theory and practical aspects in crown control', *Iron and Steel Engr.*, **42B**, 73–90.
4. K. N. SHOHET and N. A. TOWNSEND, 1971. 'Flatness control in plate rolling', *Jnl. Iron and Steel Inst.*, **209**, 769–75.
5. J. G. WISTREICH, 1968. 'Control of strip shape during cold rolling', *Jnl. Iron and Steel Inst.*, **206**, 1203–6.
6. W. RITZ, 1909. 'Über eine neue Methode zur Lösung gewisser Variationsprobleme der Mathematische Physik', *Jnl. für reine u. angew. Math. Phys.*, **135**, 1–61.
7. A. P. GREEN, 1953. 'The plastic yielding of notched bars due to bending', *Q.J. Mech. Appl. Maths.*, **6**, 223–39.
8. J. F. BAKER, M. R. HORNE and J. HEYMAN, 1956. *The steel skeleton*, vol. II: *Plastic behaviour and design*, Cambridge U.P.

A NEW CONCEPT IN SHEET METAL FORMING LUBRICATION

by

B. FOGG*

SUMMARY

General requirements of lubricants for sheet forming are presented together with a discussion on how far the most recent developments satisfy these requirements. In this context a new concept of lubrication based on the provision of externally pressurised films using orifices with restrictors within the tooling is introduced.

Test results presented in this paper demonstrate drastic changes in final strains in various modes of stretch forming and hence the feasibility of the new system. Suggestions for further work are put forward.

INTRODUCTION

Perhaps the most significant development in sheet metal forming lubrication over the past decade or so, has been the introduction of precoated 'dry' films. The use of methacrylate resins in volatile solvents, the water-soluble soap and alkali films and the films of fatty acids and resins which are also water soluble led to a new philosophy and outlook in the press-working industry.

With the conventional 'wet' lubricants, such as oils, emulsions and pastes, it is commonplace to use hand application especially on large blanks, and, even though there is some advantage to be gained from a judicious application of the lubricant by a skilled operator, the process is time consuming and is sometimes considered inconvenient for use in automated press lines.

The 'dry' films, on the other hand, are applied by dipping or roller coating and dried prior to delivery to the press line. A high degree of consistency of performance is claimed and the film characteristics are such that even rough handling prior to pressing does not seem to damage the film. The major disadvantage is that, in achieving this consistency, it is necessary to install expensive application plant, which demands high production rates to do full justice to such an installation. This need, however, is synonymous with that of an automated line of presses because the pre-coated sheets can be fed and handled with the necessary speed and because the initial coating will normally accommodate subsequent forming stages without breakdown.

Despite these advantages, two schools of thought remain. There are those who feel that for effective control of metal flow in the more complicated pressings, differential lubrication rather than all-over lubricant coating is still essential and that considerable advantages accrue from the use of a sheet having a closely prescribed surface texture in conjunction with a good conventional 'wet' or paste type lubricant.

Other developments include the use of discrete polymer films such as polythene, and there is now a tendency to supply stainless steel sheet with at least one surface protected by a thin polymer film attached to the metal by a suitable adhesive. One of the major purposes of this film is to protect the surface from scratching and other forms of damage likely to occur during handling and forming; however, the film will certainly play the role of a lubricant. It has also been noted that simple forms of synergistic effect are achieved by applying a 'dry' film over the plastic coating or by using oil between the workpiece surface and polythene film.

It is acknowledged that whilst these 'all-over' pre-coating lubricants and discrete films are extremely effective, they are also expensive. There is also a feeling that perhaps a compromise situation is desirable in that it is not always necessary to provide an 'all-over' dry coat and that it would be useful to be able to apply this type of lubricant in certain specific areas of a blank, but still on an automatic basis and prior to delivery to the press line. In this way differential lubrication would be achieved on a consistent and repeatable basis. More recently it has been reported[1] that low and high friction films of MoS_2 can be deposited in discrete areas of a blank in deep drawing and, if cost is not prohibitive and production rates are high enough, this could also provide an extremely useful and consistent method of lubrication for difficult conditions.

TRIBOLOGICAL CONDITIONS IN SHEET METAL FORMING

The range of tribological situations arising over the wide spectrum of sheet metal forming operations makes a simple generalisation of criteria for the actual choice of lubricant virtually impossible. Even in the most straightforward case of producing a cylindrical cup from a circular blank, where the deformation modes and associated tribological situations are readily isolated and quantified[2], the choice of lubricant might depend on factors other than the predictable tool pressures, temperatures and sliding velocities. If product accuracy and surface finish are important, or if the blank is thin relative to its planar dimensions, then a low-viscosity carrier oil or emulsion would probably be prescribed. The choice of additive would depend on the chemical or physical compatibility between the tool and workpiece materials. Other factors affecting choice include the method of application and removal, the type of press, component size, volume of production or production rates, and the effect the lubricant residuals might have on the

*Department of Mechanical Engineering, University of Salford

annealing and finishing of the product; lubricant cost might also be important in certain classes of pressing.

It is not yet possible to predict quantitatively the precise tribological conditions existing in irregular shaped pressings. The problem is complicated by the fact that sharp re-entrant areas of irregular pressings may be susceptible to the 'crowding' of material from adjacent regions, thus giving rise to body wrinkles or folds with subsequent locking and tearing. For this reason it is necessary to have lubricants capable of providing adequate film strength but offering sufficient frictional resistance to restrict material flow into these regions of potential crowding.

Even though the basic deformation modes such as radial drawing, uniaxial and biaxial stretching, bending and unbending, etc., can be recognised, it is difficult to specify the exact proportion of each mode in a pressing; hence it is difficult to predict the lubricant requirements. Lubrication in these circumstances is still something of an intuitive process and until recently the problem of rationalisation seemed intractable. However, Keeler in the U.S.A.[3] and Pearce in the U.K.[4] pioneered the use of the electromarked circular grid to examine the influence of the strain state at localised regions in pressings on the largest principal strain at fracture. This technique should contribute indirectly to a basis of rationalising the lubrication effects in irregular pressings. Using this technique it is now possible to explore the influence of lubrication on local critical strains in specific and neighbouring regimes. Perhaps the most significant feature emerging from this work is the possibility of recognising and eliminating strain conditions which give rise to low critical strains in pressings, as demonstrated by the Keeler type forming limit diagrams, e.g. plane strain deformation. By using a lubricant of low interfacial shear stress at strategic points, metal will be allowed to flow towards regions exhibiting the plane strain condition, hence displacing the strain state towards uniaxial tension. Alternatively by using a lubricant containing rosin, for example, the interfacial shear stress will be increased and metal flow into these regions of plane strain will be even more restricted. This would be expected to produce a higher state of biaxiality and presumably larger principal strains to fracture. This type of differential lubrication has been well known to press setters and operatives for many years, although the reasons for its success have not been understood. The technique of lubricating only that side of the blank which comes into contact with the die, and sometimes with the blankholder, has led to deeper pressings due to higher friction at the punch nose and at the punch wall when ironing. However, dry conditions lead to 'pick-up' and scoring, especially when ironing, so that high film strength and high shear strength lubricants on the punch side could provide the solution.

Whilst it is difficult to quantify the frictional conditions in irregular shaped pressings, it is a fairly simple exercise to estimate the potential increase in drawing limit if it were possible to eliminate completely friction in the flange region when drawing a circular blank into a cylindrical cup. Assuming the ratio of the wall stress with and without friction is equal to the ratio of the corresponding drawing loads, then, as shown in an earlier paper[5], the increase in limiting drawing ratio for stainless steel is around 10%, and 15% for soft aluminium. The estimated effects are based on the assumption that the frictional conditions over the punch nose are unchanged.

In components where stretching is the predominant mode of deformation, it is desirable to take advantage of the full intrinsic 'tensile' ductility of a metal. However, with punches having sharp variations in profile radius, extensive regions may exist where the final thinning strains are low relative to the potential uniform thinning strains. This is merely because these areas of low strain are adjacent to regions where the tribological situations are more severe and frictional 'locking' of the material occurs in a way which prevents transmission of stresses to those regions where extra thinning would be more economical in terms of material utilisation and depth of pressing.

It is clear that a need exists for predictable and consistent methods of controlling friction in the general field of pressworking and even its complete elimination is beneficial in certain regions as one fact of this control.

THICK FILM LUBRICATION

Several attempts have been made to achieve almost frictionless conditions in sheet metal forming operations by providing thick fluid films between the tool and workpiece surfaces. Phase change and specific melting point thick films which create this condition after the change of state have been used for some time. For example, Wallace[6] utilised the phase change brought about by the effect of pressure on ice to produce extremely low friction during the plane stretch wrapping and forming of 'difficult' sheet materials. This technique was completely effective for its original purpose but it does not appear to have been tried in press forming operations, as was suggested by Wallace; perhaps this was not a practical proposition for high production rates and normal shop floor conditions.

The methacrylate resin film referred to earlier is apparently 'dry' under ambient conditions but it is suggested by the manufacturer that the film melts under the conditions prevailing at the tool—workpiece interface during forming, to produce a low-friction fluid film. Tests have shown that this lubricant does exhibit the relatively low friction expected of such a mechanism but there is still scope for further reduction in friction as will be seen later.

Pressurised fluid films have been tried by Kasuga[7], by the present author[8], using tandem dies in redrawing and ironing, and by P.E.R.A.[9] using a special case of lubricant entrapment.

The Kasuga method, figure 1(a), utilises a die cavity filled with oil into which the punch forces the base of the cup and displaces the oil. Since in the simplest case the displaced oil must be forced between the die and deforming blank surfaces, the workpiece is virtually supported by a pressurised thick film and the friction on the die side is negligible. The pressure building up in the fluid in the cavity tends to clamp the base of the cup onto the punch, thus reducing thinning. The pressure could be controlled by providing a relief valve in the supply line to the cavity.

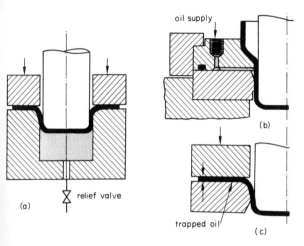

Figure 1. Various methods used for providing a pressurised film of lubricant.

This process is mainly applicable to the production of deep cups and extremely large increases in limiting drawing ratio can be achieved (2.77) if the oil pressure is augmented by an external source. The major disadvantages of this process are that it cannot be used for operations where the mode of deformation is predominantly stretching and that punch loads much greater than those attained in conventional pressing are involved. A process based on this principle now exists in a highly developed state and is marketed by SMG under the trade name 'Hydro-Mec'[10].

In the present author's original scheme for pressurised lubrication, oil was introduced under pressure between a redrawing ring and an ironing ring, figure 1(b). However, because the combined redrawing and ironing of cylindrical cups is a transient process, considerable loss of oil occurred at each cycle of the press; there was also a tendency to separate the rings at the higher pressures. However, with adequate die separation and efficient sealing and clamping, considerable pressure could be exerted between the inner wall of the cup and the punch body and this also assisted the ironing process in addition to the improved die lubrication. Obviously there is still scope for development.

In the P.E.R.A. method, on the other hand, it was found that by using fairly viscous fluids and greater than normal blank holding loads, lubricant entrapment occurred in the flange region during conventional first stage drawing, figure 1(c). This resulted in a small increase in the limiting drawing ratio but it cannot strictly be classed as a process involving externally pressurised lubrication and again there are size and shape limitations associated with this method.

TOOLING WITH EXTERNALLY PRESSURISED LUBRICANT FILMS

The purpose of the present paper is to present a new concept in the field of sheet forming lubrication[6]. The technique is based on the principle of the externally pressurised pad bearing commonly used in machine tools and nuclear reactors, etc. In these applications oil is supplied from an externally pressurised source via a flow restrictor to a cavity in one of the bearing faces. The oil attains some desirable fraction of the supply pressure and is allowed to flow from the cavity between the bearing surfaces to provide a pressurised film capable of carrying high normal loads and allowing tangential movement with very low friction.

In applying this technique to sheet metal forming problems it is only necessary to provide orifices at strategic points in the press tool and a supply pressure capable of overcoming the local tool contact pressures, thereby producing a thick pressurised film and hence extremely low frictional resistance to metal movement during deformation.

Even with a pressure of 10 tonf/in^2 (154 MN/m^2) leakage from a single restricted orifice can be made extremely small. The significance of this feature is obvious in that when the sheet material being deformed, or about to be deformed, comes over an orifice outlet there is an almost instantaneous rise in pressure and a thick film of pressurised oil is formed. The other important feature is that, when the deformed sheet moves clear of the outlet, the pressure at the surface drops to atmospheric and there is very little leakage; the tool is then ready to receive the next blank which does not require pretreatment.

FEASIBILITY STUDIES

To study the feasibility of this lubrication concept, four commonly occurring tool configurations were adopted; figure 2. The first three conditions are normally associated with the nose of a punch, whether drawing a circular or irregular pressing, and consequently represent plane strain or biaxial stretching modes with very low sliding velocities. The fourth condition occurs in radial drawing processes such as redrawing, tube drawing, etc., where tool pressures and sliding velocities are much higher.

There is considerable evidence to suggest that friction has a particularly marked influence on strain distribution in stretching operations, and furthermore, Kaftanoglu and Alexander[11] have shown that when stretching a blank over a hemispherical punch under quasi-static conditions, the friction coefficient is three or four times greater than that normally expected in radial drawing regions. Consequently the initial emphasis in this study has been devoted to stretch forming over the various shaped punches illustrated in figure 2.

The first tests were carried out on blanks of aluminium, mild steel and 18/8 stainless steel, all nominally 0.030 in thick, stretch formed over a 4 in diameter hemispherically ended punch. The punch

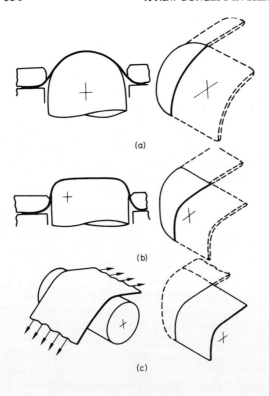

(a)

(b)

(c)

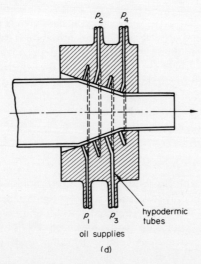

(d)

Figure 2. Commonly occurring modes of deformation and tooling configurations in sheet forming.

was made from 'Jewellite' low melting point alloy into which were cast four restricted orifices and arranged as shown in figure 3. Shell Tellus 27 hydraulic fluid was used to provide the film, but almost any mineral oil, or even water, could be used as the lubricating fluid.

In the early stages of the work, oil was fed directly to the restrictors from a hand pump but later it was found necessary to include a Greer-Mercier accumulator in the circuit to provide the necessary pressure control. It was also found necessary to have a reasonable amount of control over the flow, even though the gross oil flow was generally small. The present arrangement contains a permanent restriction in the tool which can, if necessary, be adjusted in steps, although this means dismantling the tool to

make the adjustment; consequently an infinitely variable restrictor is provided in the supply circuit. In this way the system can be made to respond rapidly when the restrictor outlets in the punch face are covered by the deforming metal. The arrangement

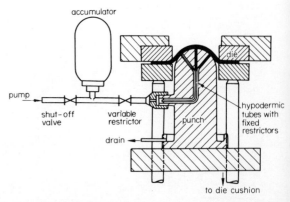

Figure 3. Diagrammatic arrangement of new externally pressurised lubrication system used in stretching over a solid punch.

is shown diagrammatically in figure 3. The hemispherical and other tools were mounted in a 200 tonf capacity hydraulic press and blank-clamping was achieved by a Worson die cushion via four marquette pins.

LUBRICATION ASSESSMENT IN BIAXIAL STRETCHING

To assess the effectiveness of the new lubricating technique it was necessary to compare its behaviour with conventional methods. Lloyd[12] proposed the use of the Olsen stretching test as a means of assessing lubricants but obviously this can only be done on a comparative basis using a discrete polymer film as the standardised lubrication condition. Unfortunately the end-point based on fracture is obscure for certain materials and this can lead to considerable scatter in cup depth at fracture. Bastien[13] has described the Parkinson test, which is similar to the test suggested by Lloyd except that the lubricant is applied after a prescribed depth of draw and then the position of fracture relative to the die edge is used as the criterion of lubricant performance.

In the present tooling it is impossible to guarantee control on the amount of 'draw-in' of the flange and certainly the controls of a machine of this size are too insensitive to predict accurately an end-point based on fracture, both of which affect the cup depth. Consequently it was considered more desirable to use either the surface strains in the vicinity of the pole or the variations in thickness over the region between the 'neck' and pole. Clearly an improvement in lubrication will lead to greater thinning at the polar cap and, in the limit, the frictionless conditions associated with the hydraulic bulge test will provide an indication of the potential thinning that could be achieved under ideal conditions. Consequently the amount of thinning in the bulge test can be used to provide an absolute index of the effectiveness of a lubricant under biaxial stretching conditions over a solid punch. The performance index to be used in this, and perhaps future work in 'stretching' lubricant assessment, is the ratio of the 'polar thickness strain in stretching over a solid hemispherical punch' to the 'polar thickness strain in the corresponding bulge test on the same material', i.e.

$$\eta_s = \frac{e_t \text{ (solid punch)}}{e_t \text{ (bulge test)}}$$

FLUID PRESSURE REQUIREMENTS

It is obviously desirable to operate the system with the smallest possible supply pressure, or more precisely with a pressure drop between the supply and film which is capable of providing a small but adequate flow rate. To determine these conditions it is necessary to have some knowledge of the mean tool contact pressure at various pole heights for a range of materials and blank thicknesses. These data were obtained by measuring the punch load at various pole heights and by making use of the current projected area of contact with the punch to determine the mean pressure, at each pole height. Corresponding values of maximum mean tool contact pressure for different thicknesses are plotted in figure 4.

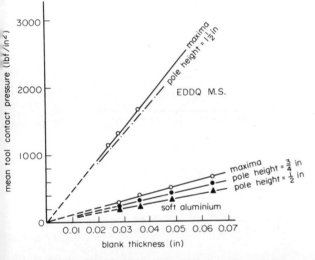

Figure 4. Variation of maximum mean tool contact pressure with thickness of blank in stretching over a hemispherical punch.

Intuitively, from the conditions for yielding under biaxial stretching, one feels that the variation of pressure across the current contact region should be negligible, especially if the friction is small. In fact the analysis of Woo[14] supports this assumption and he also shows that the pressure should be proportional to the metal thickness, figure 4. The measured values should obviously be fairly reliable.

The tests on the aluminium blanks included an examination of the response of the performance index to changes in supply pressure around and beyond the predicted critical value.

Special attention was also given to the effect of flow rate, although at this stage it was impossible to quantify the amount of flow apart from stating whether it was small or moderate — certainly the

pressures used were well below those required to make the fluid flow as a jet. Attempts were also made to establish the extent of the pressurised film in relation to the restrictor outlet by using the fact that non-existence of a film led to asperity junction growth and burnishing of the workpiece surface due to direct contact with the punch.

Tests on the mild steel and stainless steel were designed to establish the effectiveness of an externally pressurised film on these higher yield strength materials and where the friction coefficient is not so high.

RESULTS FROM TESTS USING HEMISPHERICAL PUNCH

The curves showing the distribution of final thickness across the deformed zone for aluminium at pressures around the critical value, and with adequate fluid flow, are given in figure 5. The curves for the 'dry'

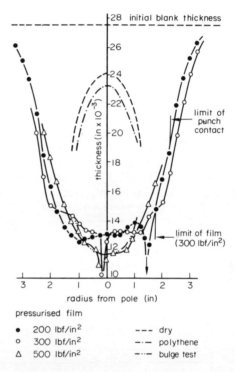

Figure 5. Thickness distributions in stretching for near-critical supply pressure — soft aluminium.

condition, for polythene film and for hydraulic bulging are also included to provide a direct comparison. It is obvious that a pressure of 200 lbf/in² with full and adequate flow is not sufficient to produce the full effect of the technique. The fracture is away from the maximum stress location at the pole and the final thickness, whilst still well below those produced by conventional methods, is greater than the potential final thickness indicated by the bulge test. If the pressure is increased to 300 lbf/in² which is just above the maximum mean tool contact pressure of 280 lbf/in², then by maintaining adequate flow, via the external restrictor, full hydrostatic lubrication is ensured.

The effect on the pole strain, or performance index, of increasing the supply pressure beyond

TABLE 1
Lubricant performance factors in stretching over a solid
hemispherical punch

	Lubricant	Performance factor η_S
Aluminium (soft)	dry	0.134
	polythene	0.197
	externally pressurised	
	200 lb/in²	0.890
	350 lb/in²	0.890
	600 lb/in²	1.000
	bulge test	1.000
EDDQ mild steel	dry	0.117
	TSD 996 (Esso)	0.117
	7618 (Castrol)	0.192
	polythene	0.376
	polythene + oil	0.394
	trilac	0.404
	externally pressurised	
	5000 lb/in²	0.695
	bulge test	1.000
Stainless steel 18/8	dry	
	TSD 996 (Esso)	0.196
	plastic coating	0.355
	externally pressurised	
	6000 lb/in²	0.688
	bulge test	1.000

300 lbf/in² is shown in figure 6. The flow rate was maintained at an adequate level throughout the range and it can be seen that the maximum performance index and hence the least final thickness occurs at a pressure of 650 lbf/in². Clearly there appears to be an optimum ratio of supply pressure to maximum mean tool contact pressure which in the present case is around 2.3:1; but further work is required on this aspect. The effect of an inadequate flow rate at 200 and 300 lbf/in² is clearly indicated in figure 6.

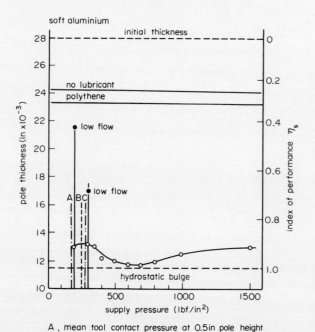

A, mean tool contact pressure at 0.5in pole height
B, maximum bulge pressure
C, mean tool contact pressure at fracture

Figure 6. Effect of supply pressures greater than critical – soft aluminium.

The extent of the film in relation to the location of the boundary where the deforming material leaves the punch is also dependent on the supply pressure as shown in figure 5. It is also necessary to balance the flow from the orifices in order to achieve an effective thick film. In the present experiments, the four orifices were supplied from a common pressure source and it was necessary to reduce the amount of fixed restriction at the pole outlet.

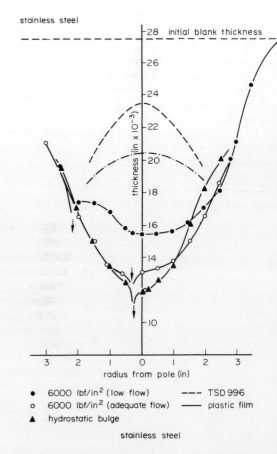

● 6000 lbf/in² (low flow)   --- TSD 996
○ 6000 lbf/in² (adequate flow)   — plastic film
▲ hydrostatic bulge

stainless steel

Figure 7. Thickness distributions for 18/8 stainless steel for different methods of lubrication.

The preliminary results for the stainless steel blanks are shown in figure 7 and are representative of the most difficult tribological condition. However, provided adequate flow is achieved, full benefits from the new method of lubrication can be obtained.

TESTS ON THE FLAT NOSED PUNCH

It is well known that because of frictional 'locking' at the profile radius of a flat nosed punch, the final thickness of the material within the flat central region is well above its potential limiting value; although the amount of thinning in this region obviously depends on the magnitude of the profile radius.

The effect of lubricant choice on base thickness was examined by stretching nominally 0.028 in thick soft aluminium blanks over a 4 in diameter punch having a 0.625 in profile radius. It is clear from figure 8 that, when lubricating conditions ranging from 'dry' to polythene film were used, the amount of thinning only changed from 10% to 20%; based on the initial blank thickness.

However, when an externally pressurised film was introduced at the profile radius, the base thinning increased drastically to values shown in figure 8, i.e. a

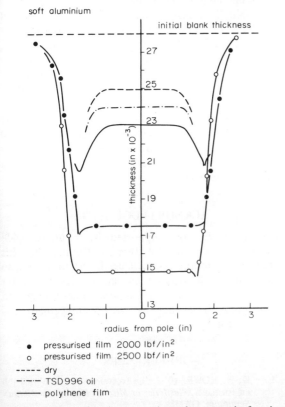

soft aluminium

* pressurised film 2000 lbf/in²
o pressurised film 2500 lbf/in²
----- dry
-·-·- TSD 996 oil
——— polythene film

Figure 8. Base thinning occurring when stretch forming aluminium over a flat nosed punch using various lubricants.

reduction of 46% at a supply pressure of 2500 lbf/in². The location of fracture moved to the point of intersection between radius and flat base, and strains over the whole stretching region were much more uniform.

The depth of cup at fracture, which is the alternative measure of lubricant effectiveness, as mentioned earlier, was measured for a wider range of conventional lubricants and for externally pressurised films. The results are given in table 2, and, whilst the conventional lubricants only increase the depth by

TABLE 2

Effect of lubrication on cup depth when stretching over a flat nosed punch

Aluminium (Soft)

Lubricant	Cup height (in)	Increase in height as % of dry cup
dry	1.0	—
oil (TSD 996)	1.06	6
polythene film	1.09	9
polythene + oil	1.13	13
PTFE film	1.18	18
dupont Vydax	1.15	15
pressurised film		
1750 lbf/in²	1.38	38
2000 lbf/in²	1.50	50

15% relative to the dry conditions, the introduction of the pressurised film produces an increase in depth of 50%.

A precautionary note must be injected at this stage in that improvements in performance in stretching could obviously lead to decreases in depth when drawing circular cups; because of a lower cup wall strength. However, regions exist in complex pressing where local stretching occurs, and drastic reductions in friction could, in these circumstances, remove the dangers of localised necking. The forming limit criterion in these latter cases is not maximum gross fracture load but local limiting strain.

STRETCHING OVER A PLANE, SINGLE CURVATURE FORMER

This mode of deformation, figure 2(c) is fairly representative of the conditions at the punch profile along the straight sides of rectangular or even irregular pressings. In positions away from the corner radii, plane strain modes exist which usually give rise to small strains to fracture[3].

In the present tests, 4 in wide strips of 0.028 in thick aluminium and mild steel were stretched over a cylindrical former using the tooling arrangement shown in figure 9 and the previous range of lubricants.

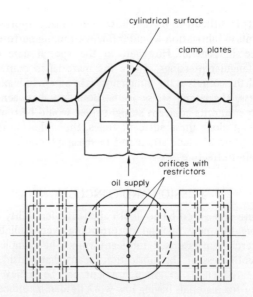

Figure 9. Tooling configuration used for stretching wide strip over a cylindrical former.

Perhaps the most significant feature of this test is the overall lateral strain of the material in contact with the former.

Under 'dry' conditions, the reductions in width for aluminium and mild steel are 3% and 11% respectively and the deformation in this region is therefore virtually plane strain. Even the best conventional lubricant used in the present series of tests was only capable of allowing a reduction in width of 8% for aluminium and 20% in the case of mild steel, and fracture still occurs in the free zone between the blank holder and punch. Again the introduction of the pressurised film drastically reduces friction giving reductions in width of 13% and 32% for aluminium and mild steel respectively, with fracture along the axis of the punch. The full results are summarised in table 3.

TABLE 3
Lateral frictional restraint when stretching over a cylindrical former

(a) Aluminium (soft)

Lubricant	Reduction in width of strip (%)
Dry	3
PTFE film	6
polythene + oil	8
pressurised film	
600 lbf/in²	5
1000 lbf/in²	12

(b) EDDQ mild steel

Lubricant	Reduction in width of strip (%)
Dry	10
Oil (TSD) 996	11
PTFE film	20
Polythene + oil	20
pressurised film	
2000 lbf/in²	28
3000 lbf/in²	31

It is difficult to predict how these drastic improvements in lubrication would influence tooling performance in general. However, in the special case of rectangular pressings, where the material in contact with the corners at the punch nose is in a state of biaxial stress, the lateral stress at the boundary between a corner and its adjacent straight sides would be transmitted along these adjacent zones, thus relieving the condition of plane strain and increasing the principal strain to fracture.

DISCUSSION AND CONCLUSIONS

There appears to be no doubt about the feasibility of using this new method of providing a controllable externally pressurised lubricant film. The complete effectiveness of the lubrication means that considerable reductions in strain gradient can be achieved, even for materials having low work hardening indices. With these improvements, the required depth of pressing could be obtained without the development of narrow zones of high strain, as discussed by Keeler[3] and others. Consequently more uniform general straining should be possible.

There is also the advantage that, by strategically placing the orifices and by providing adequate pressure and flow control external to the tool, repeatable differential lubrication could be achieved without the reduction in cycle time of the press normally associated with hand application of the lubricant; as mentioned earlier. However, for a successful applica-

tion of this technique many decisions about lubrication will have to be taken at the tool design stage.

Developments are well underway for the application of this system of lubrication to the flange area in deep drawing and in redrawing (which always presents a problem of re-lubrication) and to geometries where frictional 'locking' can occur.

Emphasis is also being placed on the practical problems of embodying orifices in conventionally designed tooling, rather than using the low melting point alloys, and this latter activity has met with a measure of success. It has also been found possible to build the pressure source into the tooling for operation by the press ram rather than by using a pump unit outside the press. The problem of providing adequate rapidity of build-up of film pressure when this method of lubrication is adapted for use on crank presses is also being examined but in the context of a real industrial problem.

ACKNOWLEDGEMENTS

This work is now supported by a grant made available by the Science Research Council and this support is gratefully acknowledged. The author is also indebted to Mr. D. Bretnall and Mr. K. Barber for their enthusiasm and careful development work and to Professor J. Halling for his encouragement and many useful discussions.

REFERENCES

1. G. W. ROWE, 1971. 'Le frottement et la lubrication en formage', *Plasticité et Mise en Forme des Métaux Seminaire*, GAMI.
2. B. FOGG, 1967. 'The relationship between the blank and product surface finish and lubrication in deep drawing and stretching operations', *Sheet Metal Ind.*, **44**, 478.
3. S. P. KEELER, 1965. 'Determination of forming limits in automotive stampings', *SAE 1965*, 650365.
4. R. PEARCE and J. WOODTHORPE, 1969. 'The effect of *r* and *n* upon forming limit diagrams of sheet metal', *Sheet Metal Ind.*, **46**, 12.
5. B. FOGG, 1971. 'A new concept in sheet metal forming lubricants', *Brit. Deep Drawing Res. Group Conf. Swansea.*
6. J. F. WALLACE, 1960. 'Phase change lubrication in stretch forming', *Aircraft Prod.*, **22**, 395–401.
7. Y. KASUGA, N. NOZADKI and K. KONDO, 1961. 'Pressure lubrication in deep drawing', *Bull. J.S.M.E.*, **4**, 14.
8. B. FOGG, unpublished.
9. P.E.R.A., 1958. *Rept.*, No. 60.
10. E. BUERK, 1967. 'Hydro-mechanical drawing', *Sheet Metal Ind.*, **44**.
11. B. KAFTANOGLU and J. M. ALEXANDER, 1970. 'On quasi-static axisymmetrical stretch forming', *Intern. J. Mech. Sci.*, **12**, 1065.
12. D. H. LLOYD, 1966. 'Lubrication for press forming', *Sheet Metal Ind.*, **43**.
13. E. L. H. BASTIEN, 1951. *Metal Working Lubricants*, McGraw-Hill, New York.
14. D. M. WOO, 1965. 'The stretch-forming test', *The Engineer*, November 26.

CHARACTERISTICS OF FORGING PRESSES: DETERMINATION AND COMPARISON

by

J. R. DOUGLAS* and T. ALTAN*

SUMMARY

This study discusses the important characteristics of forging presses and describes methods for determining them. A 700 ton hydraulic press, a newly installed 500 ton mechanical press of scotch yoke design, and a 400 ton screw press were used in the study. In all three presses high-temperature ring-compression tests were used to determine the practical effect of press speed on die chilling, metalflow and load requirements. In the mechanical press, room temperature copper upset tests were employed to determine the dynamic press stiffness, the offcentre loading ability, the flatness of ram and bolster surfaces under load and the available energy capacity at various production rates.

The methods used in the present study can be applied for the evaluation, comparison and standardisation of forging presses under practical production conditions.

INTRODUCTION

The purchase of new forging equipment, and the efficient use of various types of existing equipment, require a thorough understanding of the effect of equipment characteristics on the forging operations, of load and energy requirements for the specific forging operation and of the capabilities and characteristics of the specific forging machine used for that operation. Today the trend is to install presses, especially mechanical or screw presses, instead of hammers, except for very large capacities. The hammer, although it is the least expensive forging machine, has several disadvantages such as limited accuracy, noise pollution and difficulty in automation. The mechanical forging press is most effectively used for large production series in which required tool changes and setups are infrequent. The screw press competes with hammers and with mechanical presses, especially in forging relatively thin parts with great accuracy. Hydraulic presses are used mostly for open-die forging. For closed-die forging operations, the hydraulic press is in general too slow, gives long contact times and causes die chilling. Consequently, the hydraulic press is only practical for forging aluminium and magnesium alloys, where no die chilling is present, and for forging very large parts requiring forging loads above 6000 or 8000 tons.

The principal process and equipment variables and their interactions in hot forging under presses are shown schematically in figure 1, where a line between two blocks indicates that one variable influences the other[1, 2]. As seen on the left-hand side of figure 1, the flowstress $\bar{\sigma}$, the interface friction conditions and the forging geometry (dimensions, shape) determine both the load L_p at each position of the stroke and

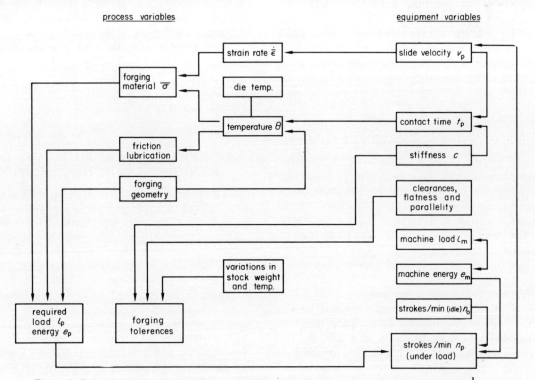

Figure 1. Relationships between process and equipment variables in closed-die forging in presses[1].

*Metalworking Division, Battelle Columbus Laboratories, Columbus, Ohio, USA.

the energy E_p required by the forging process. The flowstress $\bar{\sigma}$ increases with increasing deformation rate $\dot{\bar{\epsilon}}$ and with decreasing temperature θ. The magnitude of these variations depends on the specific forging material. The friction at the tool material interface increases with increased die chilling.

As indicated by the lines connected to the temperature block, for a given initial stock temperature, the temperature variations in the forging are largely influenced by the surface area of contact between dies and forging, the part thickness of volume, die temperature, the amount of heat generated by deformation and friction, and the contact time under pressure. During deformation, the heat transfer from the hot forging to the colder dies is nearly perfect with graphite-base lubricants. With glass-base lubricants, however, the heat transfer is greatly reduced, depending on the interface temperature and the thickness and type of glass coating.

The velocity of the slide under pressure V_p determines mainly the contact time under pressure t_p and the deformation rate $\dot{\bar{\epsilon}}$. The number of strokes per minute n_0 under no load conditions, the machine energy E_M and the deformation energy E_p required by the process influence the slide velocity under load V_p and the number of strokes n_p under load: n_p determines the maximum number of parts forged per minute (that is the production rate) provided the feeding and unloading of the machine can be carried out at that speed.

As indicated in figure 1, the stiffness of the press influences the contact time under pressure. The stiffness also influences the thickness tolerance of forged parts in mechanical press forging, where usually the upper and lower dies do not touch or kiss during a forging stroke. In hydraulic and screw presses, the stiffness does not affect forging tolerances. The flatness and parallelity of upper and lower die surfaces determine the degree of skewness and mismatch in the forged part. Therefore, these characteristics too are important for evaluating the overall performance of a forging press.

SIGNIFICANT CHARACTERISTICS OF FORGING PRESSES

Characteristics of a machine consist of all design and performance data on that machine pertinent to its economical use[3]. These data are necessary for an optimum selection of equipment for a given operation. The characteristic data for forging presses can be divided into three groups[3,4]; data for load and energy, data that are time-dependent and data for accuracy.

Characteristic data for load and energy

Available energy E_M is the energy supplied by the machine for carrying out the deformation but does not include the energy necessary to overcome frictional, inertial and deflection losses. *The available load* L_M is the load available at the ram to carry out the deformation process. The following conditions must be satisfied to complete a forging operation[1,2]: the available machine load $L_M >$ the load required

by the process L_p at any time during the working stroke and the available machine energy $E_M >$ the energy required by the process E_p for an entire stroke.

Time-dependent characteristic data

The *number of strokes per minutes under pressure* n_p is the most important characteristic of any machine since it determines the production rate. The *contact time under pressure* t_p determines the heat transfer between the hotter forged part and the cooler dies, and greatly influences the flowstress of the forged material and the die wear.

The *velocity under pressure* V_p, determines the contact time under pressure and the rate of deformation or the strain-rate. Thus, it influences the flow stress of the forged material and affects the load and energy requirements.

Characteristic data for accuracy

For *unloaded* machine conditions, the stationary surfaces and their relative positions are established by clearance in the gibs, parallelism of upper and lower beds, flatness of upper and lower beds, perpendicularity of slide motion with respect to lower bed, and concentricity of tool holders[3]. All these machine characteristics affect the tolerances in the forged part. Much more significant however are the quantities obtained under *load* and under *dynamic* conditions.

The stiffness of a press C (the ratio of load to the total elastic deflection between upper and lower dies) influences the energy lost in press deflection, the velocity versus time curve under load and the contact time. In mechanical presses, variations in forging thickness, due to volume or temperature changes in the stock, are also smaller in a stiffer press.

Very often the stiffness of a press (ton/in) is measured under static loading conditions, but such measurements are misleading: for practical use the stiffness of a press must be determined under dynamic loading conditions.

DESCRIPTION OF EXPERIMENTAL WORK, EQUIPMENT AND INSTRUMENTATION

The purpose of the present study is to develop test methods and to demonstrate their application in order to determine the most significant characteristics of forging presses. These tests must be practical and relatively simple so that they can be conducted easily by the press manufacturers as well as by the press users in a forge shop.

Three different presses used in the experimentation are a 700 ton hydraulic press and a 500 ton mechanical press with scotch yoke design installed at the Metalworking Laboratory of Battelle's Columbus Laboratories, and a 400 ton production screw press in the Brave, Pennsylvania, plant of the Accurate Brass Corporation.

The following significant press characteristics were evaluated for these three presses:

investigated presses	nominal capacity (in ton)	dynamic stiffness	production rate/energy	ram tilting, load increase in offcentre loading	ram speed and contact time
Hydraulic	700				X
Mechanical	500	X	X	X	X
Screw	400	X			X

The vertical 700-ton hydraulic press is direct driven, uses hydraulic oil as a working medium and has an adjustable ram speed of 0.1 to 80 in/min. In operating this press, there is a certain dwell time after the ram touches the workpiece and before the deformation proceeds: there is another dwell time after the stroke is completed and before the ram is lifted.

The mechanical press is a newly installed Erie press with scotch yoke design, illustrated in figure 2. The

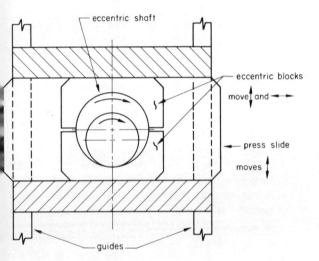

Figure 2. Principle of the scotch yoke type drive for Erie's mechanical forging presses[1].

ram contains a top and bottom eccentric block which retains the eccentric shaft. As the shaft rotates the eccentric blocks move in both horizontal, front-to-back and vertical directions and the ram is actuated by the eccentric blocks only in a vertical direction. This mechanical press, rated at 500 ton at 0.25 in before bottom deadcentre, is a high-speed forging press. It has a stroke of 10 in and a nominal idle speed of 90 strokes/min.

The screw press is a Weingarten PSS 255 with a nominal rating of 400 metric ton, 2250 m kg (97.2 in ton) energy and a stroking rate of about 30 per min. The manufacturer specifies a maximum ram velocity of 0.64 m/s (2.1 ft/s) at a maximum stroke of 15.8 in. The static stiffness of the press is given as 300 ton/mm (8400 ton/in). The press uses an electric drive to accelerate the flywheel and the screw assembly, and converts the angular kinetic energy into the linear energy of the ram. As illustrated in figure 3 the reversible electric motor, which accelerates the screw up or down, is built directly on the screw and on the frame, above the flywheel[7].

In all three presses the displacement versus time

was measured by a linear variable-displacement transducer LVDT. In the hydraulic and mechanical presses, the load was measured by strain bars attached on the press frame[9]. In the screw press, a loadcell was placed directly under the lower die. To measure the flywheel speed in the mechanical press, two magnetically operated switches were placed on the housing of the

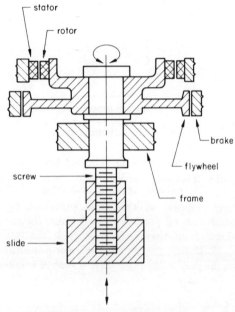

Figure 3. Schematic of the direct electric drive screw press with reversing motor.

flywheel, and four magnets were attached to the flywheel. With this simple setup, a signal on the oscillograph would be obtained at every 1/8 revolution of the flywheel. The distance between two signals allowed the determination of the flywheel speed. The principle of this technique is seen in figure 4.

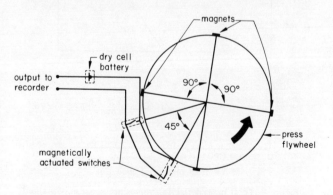

Figure 4. Illustration of the technique used for measuring flywheel velocity in the mechanical press.

DYNAMIC STIFFNESS IN THE MECHANICAL PRESS

To obtain the dynamic stiffness of the mechanical press, copper samples of various diameters, but of the same height, were forged under oncentre conditions (table 1). The samples of wrought pure electrolytic copper were annealed for 1 h at 900 F.[8,9] The press setup was not changed throughout the tests. Lead samples about 1 in square and 1.5 in high were placed near the forged copper sample, about 5 in to the side. As indicated in table 1, with increasing

The linear portion represents the actual elastic deflection of the press components. The slope of the linear curve is the dynamic stiffness: it was determined as 5800 ton/in for the 500 ton Erie forging press.

The method, described above, requires the measurement of the load in forging annealed copper samples. If instrumentation for load and displacement would be impractical for forgeshop measurements, the flow-stress of the copper can be used for estimating the load and the energy for a given reduction in height. The details of this approximate approach are given elsewhere[9].

TABLE 1
Copper samples forged under oncentre conditions in the 500 ton mechanical press

sample	sample size (in) height	sample size (in) diameter	predicted* load (tons)	measured load (tons)	predicted† energy (in tons)	measured energy (in tons)
1	2.00	1.102	48	45	24	29
2	2.00	1.560	96	106	48	60
3	2.00	2.241	197	210	98	120
4	2.00	2.510	247	253	124	140
5	2.00	2.715	289	290	144	163
6	2.00	2.995	352	350	176	175

* Based on an estimate of 50 000 lb/in² flowstress for copper at 50 per cent reduction in height.
† Estimated by assuming that the load–displacement curve has a triangular shape, that is, energy = 0.5 load x displacement.

sample diameter the load required for forging also increases. The press deflection is measured by the difference in heights, of lead samples forged with and without copper, at the same press setting. The variation of total press deflection versus forging load, obtained from these experiments is illustrated in figure 5. During the initial nonlinear portion of the curve, the play in the press driving system is taken up.

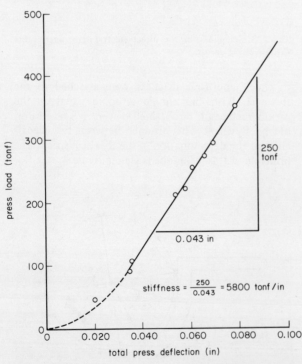

Figure 5. Total press deflection versus press loading obtained under dynamic forging conditions for a 500 ton Erie scotch yoke press.

DYNAMIC STIFFNESS IN THE SCREW PRESS

Usually, in forging in a screw press, the dies kiss during each blow. Thus, the stiffness of the press influences not the thickness tolerances of the forged part, but the contact time and determines the amount of energy lost in press deflection.

The static stiffness of the 400 ton Weingarten screw press, as given by the manufacturer, is 8400 ton/in. This stiffness does not include the torsional stiffness of the screw which occurs under dynamic conditions[10]. As pointed out by Watermann[11], who conducted an extensive study on the efficiency of screw presses, the torsional deflection of the screw may contribute up to 30 per cent of the total losses at maximum load (about 2.5 times nominal load). Based on experiments conducted in a smaller Weingarten press (model P160, nominal load 180 metric ton, energy 800 kg m), Watermann concluded that the dynamic stiffness was 0.7 times the static stiffness[11]. Assuming that this ratio is approximately valid for the 400 ton press, the dynamic stiffness is 0.7 x 8400 ≈ 5900 ton/in.

During a downstroke the total energy supplied by a screw press E_T is equal to the sum total of the machine energy used for the deformation process E_P, the energy necessary to overcome friction in the press drive E_F and energy necessary elastically to deflect the press E_D. Thus

$$E_T = E_P + E_F + E_D \qquad (1)$$

Expressing E_D in terms of press stiffness C, equation (1) can be rewritten as

$$E_T - E_F = E_P + L_M^2/2C \qquad (2)$$

In a forging test, the energy used for the process E_P (surface area under the load–displacement curve) and the maximum forging load L_P can be obtained from oscillograph recordings. By considering simultaneously two tests, conducted at the same energy setting, and by assuming that E_F remains constant during the tests, one equation with one unknown C can be derived from equation (2). But, in order to obtain reasonable accuracy, it is necessary that in both tests considerable press deflection is obtained, that is, high loads L_P and low deformation energies E_P are measured. Thus, the errors in calculating E_P do not impair the accuracy of the stiffness calculations. This practical procedure could not be used in the study described here because the capacity of the load cell used in screw press experiments was limited to 500 ton and the press should be loaded in the 800 to 1000 ton range in order to obtain reasonably accurate stiffness values from equation (2).

MAXIMUM PRODUCTION RATE VERSUS ENERGY IN THE MECHANICAL PRESS

In mechanical presses, the flywheel slows down during each stroke and supplies the energy available for the process E_P, the energy necessary to overcome the inertial and frictional resistance in the drive mechanism E_F and the energy necessary to deflect the press E_D. The time necessary for the flywheel to reach its idle speed, before starting the next stroke, will be prolonged if too much energy is lost from the flywheel. Therefore, the energy capacity of a mechanical press is known only if a curve describing the actual stroking rate versus available energy per stroke is available[9, 12].

For this purpose, the results of the oncentre copper forging tests of table 1 were used. During each stroke, the time required by the flywheel to recover 99 per cent of its original velocity was measured. The deformation energy and the maximum load required by each sample were obtained from the oscillograph recordings. This information was sufficient to obtain the possible production rate versus energy curve of figure 6. In this curve the energy used includes also

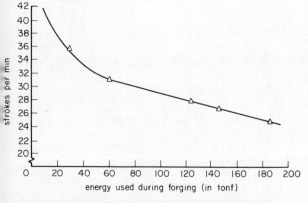

Figure 6. Maximum strokes/min versus energy available in the 500 ton mechanical Erie press.

the energy needed to deflect the press. During a forging stroke, in order to overcome inertial and frictional resistance, the flywheel slows down before the ram touches the part. This slowdown corresponds

to an energy loss of about 7 per cent of the total flywheel energy of 1170 in ton (calculated for a flywheel rpm of 92 from the moment of inertia of rotating parts). Thus, about 70 in ton of energy is required to overcome the frictional and inertial resistance during a downstroke.

RAM TILTING IN OFFCENTRE FORGING

Offcentre loading conditions occur often in mechanical press forging when several operations are performed in the same press. Especially in automated mechanical presses, the finish blow (which requires the highest load) occurs on one side of the press. Consequently, the investigation of offcentre forging is particularly significant in mechanical press forging.

The offcentre loading characteristics of the 500 ton mechanical press were evaluated as follows. During each test, a copper specimen, which requires 220 ton to forge, was placed 5 in from the press centre in one of the four directions, that is, left, right, front or back. A lead specimen, which requires not more than 5 ton, was placed an equal distance on the opposite side of the centre. On repeating the test for three other directions, the comparison of the final height of the copper and lead forged during the same blow gave a good indication of the nonparallelity of the ram and bolster surfaces over a 10 in span. In conducting this comparison, the local elastic deflection of the dies in forging copper must be considered. Therefore, the final thickness of the copper samples were corrected to counteract this local die deflection.

In offcentre loading with 220 ton (or 44 per cent of nominal capacity) an average ram-bed nonparallelity of 0.038 in/ft was measured in both directions, front-to-back and left-to-right. In comparison, the nonparallelity under unloaded conditions was about 0.002 in/ft. Before conducting the experiments described above, the clearance in the press gibs was set to 0.010 in. The nonparallelity in offcentre forging would be expected to increase with increasing gib clearance.

EVALUATION OF RAM SPEED AND CONTACT TIME

In order to evaulate the effects of ram speed and contact time on the forging process, the well-known ring-compression test was used.

Principles and application of the ring test

The ring test is primarily used for investigating the friction in forging[15, 16] and for evaluating forging lubricants under various forging conditions[17]. Recently however it is also being used for determining the flowstress of forged materials[16, 18]. In the ring test, a flat ring-shaped specimen is upset forged at a known reduction. The change in internal and external diameters of the forged ring depends very much on the friction at the tool—specimen interface. The internal diameter of the ring is reduced if the interface friction is large, and is increased if the friction is low. Thus, the change in the internal ring diameter,

540

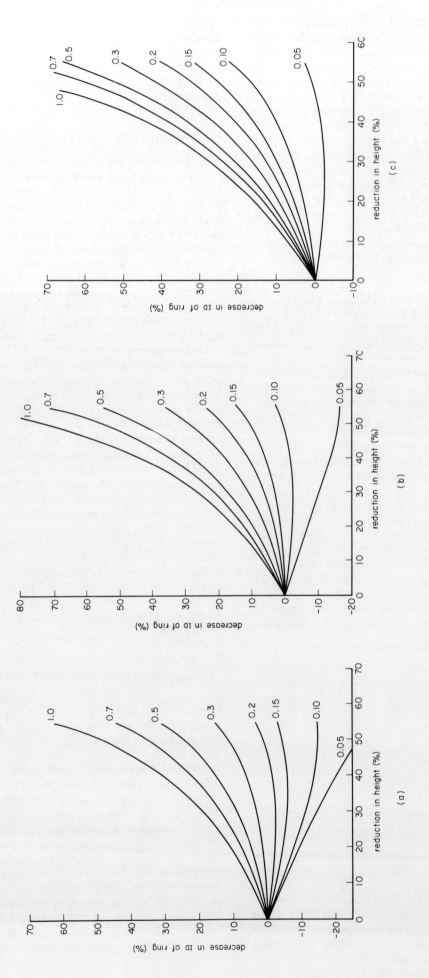

Figure 7. Theoretical calibration curves for upsetting rings with different OD:ID thickness ratios.
(a) ring ratio 6:3:2
(b) ring ratio 6:3:1
(c) ring ratio 6:3:0.5

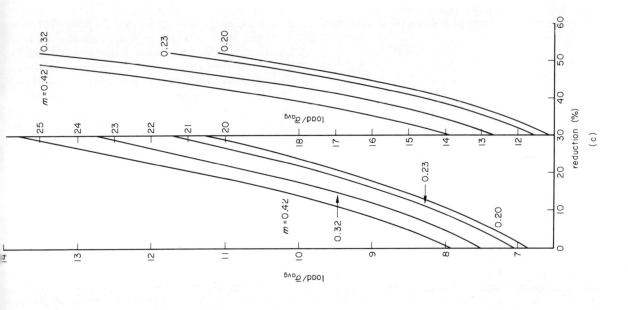

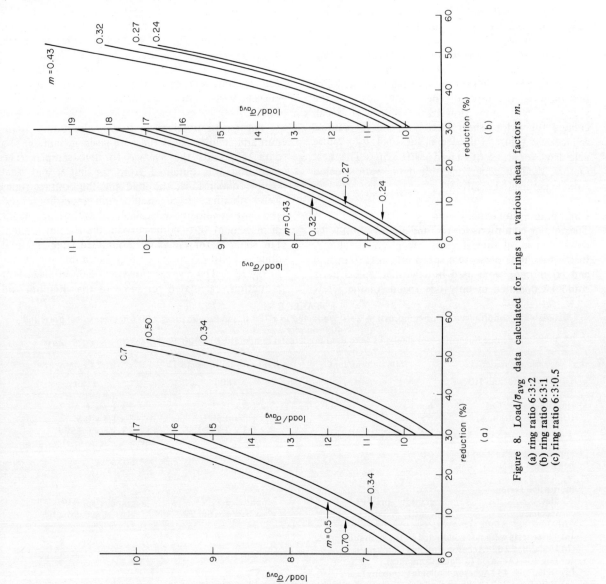

Figure 8. Load/$\bar{\sigma}_{avg}$ data calculated for rings at various shear factors m.

(a) ring ratio 6:3:2
(b) ring ratio 6:3:1
(c) ring ratio 6:3:0.5

before and after upsetting, represents a simple method for evaluating interface friction.

The interface friction shear stress (τ) is expressed by

$$\tau = f\bar{\sigma} = \frac{m\bar{\sigma}}{\sqrt{3}} \qquad (3)$$

where f = friction factor = $\dfrac{m}{\sqrt{3}}$, $0 \leqslant f \leqslant 0.577$

m = shear factor, $0 \leqslant m \leqslant 1$

$\bar{\sigma}$ = flowstress of forged material

In order to obtain the magnitude of the friction factor, the internal diameter of the upset ring must be compared with values predicted by using various friction factors. In this study, a computer programme, developed at Battelle's Columbus Laboratories, was used[16]. This computer programme simulates the ring-upsetting process for given shear factors m by including the bulging of the free surfaces. Thus, ring dimensions for various reductions and shear factors can be determined and theoretical calibration curves generated for rings of different dimensions. Calibration curves are given in figure 7 for the rings having OD x ID x thickness ratios of 6:3:2, 6:3:1 and 6:3:0.5. These rings were used in the present study.

The computer programme, simulating the ring-upsetting test, can be used for predicting the average flowstress of the forged ring material, provided the forging load has been measured during the test.[16] For this purpose, the ratio of the forging load to the average instantaneous flowstress (load/$\bar{\sigma}_{avg}$) is calculated for various shear factors m and for various reductions. Figure 8 illustrates the load/$\bar{\sigma}_{avg}$ values calculated for 6:3:2, 6:3:1 and 6:3:0.5 rings. The shear factors, used in generating this data, were selected from experimental results.

Ring-compression experiments

The forging conditions for the ring-compression tests, conducted in all three presses, are given in table 2. In the hydraulic press, two speed settings (20 in/min and 80 in/min) were used but the mechanical press could be operated at only one speed. In the screw

press, the maximum energy setting and (due to tooling height) a maximum stroke length of 13.5 in were used (the maximum press stroke was 15.8 in). The ring materials were selected so that variations in heat conductivity (titanium versus aluminium) and various degrees of flowstress versus temperature and strain-rate dependencies (titanium versus stainless steel and aluminium) could be obtained. The dimensions of the samples were so selected that the maximum forging load in forging the thinnest samples (6 in OD x 3 in ID x 0.5 in thick for aluminium 6061: 3 in OD x 1.5 in ID x 0.25 in thick for 403 stainless and Ti-7A1-4Mo) would be in the range of 500 ton. Thus, the capacity of all three presses would be sufficient to perform the experiments. The lubrication and the temperatures of sample and die were kept unchanged in tests conducted in all three presses. Lubricants, recommended by one supplier, were selected without any effort to optimise lubrication conditions, and usual commercial practice was followed in lubricating the samples and the dies.

All the samples were placed on the lower die, on top of a 0.030 in diameter, mild steel wire, in order to prevent die chilling before deformation. The transfer time (time necessary to take a sample from the furnace and place it on the lower die) was about 4 s in mechanical and hydraulic press trials, but in screw press trials, conducted in a forging plant, the transfer time was about 10 s. The effect of this difference in transfer times on the test results is discussed later.

The results obtained in forging aluminium, titanium and stainless steel rings are given in table 3. The data represents the average for two samples. The reduction was obtained from the initial and final sample dimensions; the load and the contact times were obtained from oscillograph recordings, and the contact velocity was calculated as the slope of the displacement versus time curve at the start of forging. The shear factor values were obtained from the final dimensions of the rings by using the calibration curves in figure 7. The average flowstress data at maximum reduction, calculated for some of the titanium and

TABLE 2

Materials and conditions for ring-compression tests conducted in a 700 ton hydraulic press, 500 ton mechanical press and 400 ton screw press

(The same hot work tool steel dies were used in all tests; temperature $\cong$ 300 F, surface finish $\cong$ 25 μin)

sample material	6061 aluminium	AISI 403 stainless steel	Ti-7A1-4Mo
temperatures (F)	800	1800	1750
ring dimensions (in) (OD x ID x thickness)	6 x 3 x 2 6 x 3 x 1 6 x 3 x 0.5[a]	3 x 1.5 x 1[a] 3 x 1.5 x 0.5[a] 3 x 1.5 x 0.25[a]	3 x 1.5 x 1 3 x 1.5 x 0.5 3 x 1.5 x 0.25[a]
approximate reduction in height (per cent)	50	50	50
lubrication system dies specimen	graphite spray[b] caustic precoat + graphite[c]	graphite spray[d] glass base coating[e]	graphite spray[d] glass base coating[e]

(a) These tests were not conducted in the hydraulic press.
(b) Deltaforge 105 (Acheson Colloids Company).
(c) Dag 137 (Acheson Colloids Company).
(d) Deltaforge 43 (Acheson Colloids Company).
(e) Deltaforge 347 (Acheson Colloids Company), glass-bare lubricant diluted in isopropanol with solid content of 15 per cent by weight. The samples were dipped in the solution before heating.

TABLE 3

Data obtained while forging rings in a hydraulic press, mechanical press and screw press. Experimental conditions are given in table 2; the data represent the average for two samples.

No.	press	reduction (per cent)	load (tons)	shear factor m	contact time (s) dwell(a)	loading	unloading	total	contact velocity (in/s)	flow stress (10³ lb/in²)
				6061 Aluminium						
				6:3:2 ring ratio						
A-1	hydraulic(b, e)	51.0	285	0.63	1.00	3.60	0.32	4.92	(d)	
A-2	hydraulic(c, e)	51.2	250	0.65	0.44	0.83	0.29	1.56	(d)	
A-3	mechanical(e)	51.0	260	0.53	–	0.079	0.018	0.097	26	
A-4	screw(f)	34.9	160	0.49	–	0.051	0.035	0.086	19	
				6:3:1 ring ratio						
A-5	hydraulic(b, e)	50.2	465	0.48	0.92	2.56	0.42	3.90	(d)	
A-6	hydraulic(c, e)	51.0	386	0.42	0.47	0.53	0.35	1.35	(d)	
A-7	mechanical(e)	49.8	338	0.31	–	0.047	0.029	0.076	19	
A-8	screw(f)	47.0	275	0.35	–	0.041	0.021	0.062	22	
				6:3:0.5 ring ratio						
A-9	mechanical(e)	45.7	520	0.40	–	0.038	0.027	0.065	14	
A-10	screw(f)	45.6	400	0.34	–	0.023	0.017	0.040	22	
				Ti-7A1-4Mo alloy						
				6:3:2 ring ratio						
T-1	hydraulic(b,e)	51.0	255	0.72	1.03	2.16	0.31	3.50	(d)	
T-2	hydraulic(c, e)	52.2	215	0.40	0.43	0.49	0.29	1.21	(d)	
T-3	mechanical(e)	49.0	225	0.70	–	0.056	0.018	0.074	22	29.1
T-4	screw(f)	40.0	250	0.64	–	0.043	0.019	0.062	20	41.0
				6:3:1 ring ratio						
T-5	hydraulic(b, e)	47.7	690	0.26	1.57	0.99	0.69	3.25	(d)	92.0
T-6	hydraulic(c, e)	48.8	335	0.28	0.50	0.33	0.37	1.20	(d)	43.1
T-7	mechanical(e)	46.2	275	0.42	–	0.044	0.017	0.061	17	33.6
T-8	screw(f)	44.8	365	0.44	–	0.024	0.021	0.045	22	46.5
				6:3:0.5 ring ratio						
T-9	mechanical(e)	30.8	400	0.42	–	0.033	0.020	0.053	15	56.7
T-10	screw(f)	37.6	445	0.20	–	0.019	0.018	0.037	22	68.0
				AISI 403 stainless steel						
				6:3:2 ring ratio						
S-1	mechanical(e)	51.1	225	0.34	–	0.047	0.028	0.075	20	31.3
S-2	screw(f)	47.3	270	0.50	–	0.040	0.022	0.062	21	38.2
				6:3:1 ring ratio						
S-3	mechanical(e)	51.2	275	0.24	–	0.037	0.025	0.062	16	34.2
S-4	screw(f)	45.6	330	0.32	–	0.026	0.015	0.041	22	44.6
				6:3:0.5 ring ratio						
S-5	mechanical(e)	46.1	330	0.23	–	0.029	0.023	0.052	11	38.0
S-6	screw(f)	40.0	390	0.32	–	0.020	0.016	0.036	23	47.5

(a) This dwell, which occurs only in the hydraulic press, is the length of time that the ram is in contact with the specimen before deformation begins.
(b) Nominal ram speed 20 in/min.
(c) Nominal ram speed 80 in/min.
(d) In the hydraulic press, the contact velocity depends on the free fall of the ram. On average, this velocity was about 20 in/s.
(e) The transfer time from furnace to press was about 4 s for the hydraulic and mechanical presses.
(f) The transfer time from furnace to screw press was about 10 s, due to the greater distance of the furnace from the press.

stainless steel samples, were obtained by using the measured forging load and the curves given in figure 8.

In tests conducted with stainless steel rings, having 3 in OD and 1.5 in ID, the flowstress and the shear factor are larger in the screw press than in the mechanical press. This is due to longer transfer and cooling times in air (10 s for the screw press and 4 s for the mechanical press). In order to verify this observation, additional experiments were conducted in the mechanical and screw presses. In these tests a scale-pre-

venting coating (Turco Pretreat) was used instead of the glass-base coating (Acheson 347) and the transfer

ring ratio	press	reduction (per cent)	shear factor m
6:3:2	mechanical	49.9	0.37
6:3:2	screw	45.8	0.42
6:3:1	mechanical	44.4	0.26
6:3:1	screw	44.0	0.25
6:3:0.5	mechanical	44.0	0.30
6:3:0.5	screw	37.6	0.20

times were kept at 10 s. The shear factor values are given below:

These results indicate that there is no significant difference in shear factor, provided the transfer time is the same in both presses. (The data for 6:3:0.5 rings cannot be compared since the reductions are different in two presses).

DISCUSSION AND CONCLUSIONS

The purpose of the work described here was to develop simple and useful methods for evaluating presses for closed-die forging. The effect of equipment behaviour on the forging process has been reviewed and the most significant characteristics of hydraulic, mechanical and screw presses investigated.

For evaluating and comparing the performance of mechanical presses, practical test methods have been developed for determining the dynamic stiffness under forging conditions, determining the maximum possible production rate versus available energy for deformation, and evaluating the ram and bed parallelity in offcentre loading conditions. The application of these methods has been demonstrated on a high-speed 500 ton mechanical press. These tests can be used for comparing mechanical presses of different suppliers, or for evaluating the performance of used presses.

The dynamic stiffness of a screw press has been discussed and a practical method for its determination suggested. However, due to capacity limitation of the loadcell used in screw press trials, the practical application of the suggested technique could not be carried out.

The principles of the ring-compression test have been reviewed and its application for determining friction shear factors and flowstress values discussed. 6061 aluminium, Ti-7A1-4Mo and 403 stainless steel rings of various dimensions with 6:3:2, 6:3:1 and 6:3:0.5 ratios have been forged in a hydraulic, a mechanical and a screw press of comparable capacities. For all these tests, ram speeds, contact times and forging loads have been measured. The shear factors and flowstress values have been calculated by using the experimental results and the theoretical calibration curves which give the shear factor versus reduction and load/average flowstress versus reduction. These curves have been obtained from computer programmes which simulate the ring-compression test by taking bulging of free surfaces into account.

The results of the ring tests give useful practical information for comparing the time-dependent characteristics of the hydraulic, screw and mechanical presses investigated. From these results the following observations and conclusions are made:

● For actual forging strokes of 1 in or less, the screw press exhibits shorter contact times than the mechanical and hydraulic presses.

● During unloading, the ram of the screw press is lifted through elastic energy stored in the frame and in the screw. Therefore, in the screw press the unloading time decreases with increasing forging load. A similar trend is not recognisable in the mechanical press.

● The influence of transfer time (here 4s in the mechanical and hydraulic presses and 10 s in the screw press) on friction and load is not significant in large aluminium rings (6 in OD x 3 in ID). However, this influence is considerable in small titanium and stainless steel rings. In the screw press, calculated flowstress and measured load values are larger than in the mechanical press, for all ring thicknesses. These differences are due to larger transfer times in screw-press trials and cannot be explained by strain-rate effects, because the strain-rate does not differ significantly in these two presses.

● The differences in contact times and deformation speeds, in mechanical and screw presses, do not affect the friction shear factor m in aluminium, titanium and stainless steel ring compression tests. (For stainless steel the trials conducted with 'Turco Pretreat' and 10s transfer time must be used for comparison).

● In aluminium rings, the load decreases with decreasing contact time and increasing deformation speed. Thus, in forging aluminium (which is an excellent heat conductor) with graphite-base lubricants (which ensure good heat transfer at the die-material interface) the effect of die chilling is significant while the effect of strain-rate is negligible. This observation explains the extensive use of screw presses (30 per cent of all Weingarten presses built[7] are used in hot forging aluminium and copper alloys).

● A combined effect of increasing strain-rate and decreasing temperature on forging load is seen by comparing different thickness rings forged in the same press. The increase in flowstress with decreasing ring thickness is particularly significant between 6:3:1 and 6:3:0.5 rings. These results, for stainless steel and titanium rings, indicate that die chilling must have a far more significant effect in titanium than in stainless steel. This is valid for both mechanical and screw presses and is explained by the fact that the flow stress of titanium is more temperature-dependent than that of stainless steel.

In summary, the ring tests indicate that, except for aluminium alloys, the differences in contact time and deformation rate, which existed in the mechanical and screw press used in the study, are not expected to influence appreciably the metalflow and forging load. However, shorter contact times in the screw press may result in lower die wear in practical forging operations. This point was not further investigated in the present study.

ACKNOWLEDGEMENTS

The present work was supported by the Army Materials and Mechanics Research Centre, Watertown, Massachusetts, and by the U.S. Army Aviation Material Command, St. Louis, Missouri. The authors wish to thank Messrs. R. Gagne of AMMRC, Mr. J. S. Willison of USAAMC and Mr. R. J. Fiorentino, Chief of the Metalworking Division at Battelle's Columbus Laboratories for their valuable suggestions at various parts of this study.

REFERENCES

1. T. ALTAN and A. M. SABROFF. Comparison of mechanical presses and screw presses for closed-die forging. *SME Technical Paper* MF70-125.

2. J. STOTER. Investigation of the forging process in hammer and press, particularly as related to die fill (in German). *Doctoral Dissertation,* Technical University, Hanover, 1959.

3. O. KIENZLE. Characteristics of data in machine tools for closed-die forging (in German). *Werkstattstechnik,* **55,** 1965, p. 509.

4. T. ALTAN and A. M. SABROFF. Important factors in the selection and use of equipment for forging. *Precision Metal,* **28,** Part I, II, III, and IV, June, p. 54, July, August, and September, 1970.

5. U. KLAFS. On the determination of temperature distribution in dies and workpiece during warmforming (in German). *Doctoral Dissertation,* Technical University, Hanover, 1969.

6. O. KIENZLE. Development trends in forming equipment (in German). *Werkstattstechnik,* **49,** 1959, p. 479.

7. P. KNAUSS. The design and application of the modern percussion press. *Sheet Metal Industries,* February, 1970, p. 137.

8. D. WATERMANN. Determination of available energy in hammers and presses with copper cylinders (in German). *Werkstattstechnik,* **52,** 1962, p. 95.

9. T. ALTAN and D. E. NICHOLS. Use of standardised copper cylinders for determining load and energy in forging equipment. *ASME Paper 71-WA/Prod-3.*

10. H. BOHRINGER and K. H. KILP. The significant characteristics of percussion presses and their measurement. *Sheet Metal Industries,* May, 1968, p. 335.

11. H. D. WATERMAN. The blow efficiency in hammers and screw presses (in German). *Industrie-Anzeiger,* No. 77, September 24, 1963, p. 53.

12. Th. KLAPRODT. Comparison of characteristics of screw and crank presses for die forging (in German). *Industrie-Anzeiger,* **90,** 1968, p. 1423.

13. H. D. WATERMANN. The work accuracy of hammers and screw presses in offcentre loading (in German). *Werkstattstechnik,* **53,** 1963, No. 8, p. 413.

14. H. MAKELT. *The Mechanical Presses* (in German). Carl Hanser Verlag, Munich, 1961.

15. A. T. MALE and V. DE PIERRE. The validity of mathematical solutions for determining friction from the ring compression test. *ASME Paper No. 69-WA/Lub-8.*

16. C. H. LEE and T. ALTAN. Influence of flowstress and friction upon metal flow in upset forging of rings and cylinders. *ASME Paper 71-WA/Prod-9.*

17. S. C. JAIN and A. N. BRAMLEY. Speed and frictional effects in hot forming. *Proceedings of the Institution of Mechanical Engineers,* Vo. 182, 1967–68, Part I, Number 39, p. 783.

18. G. SAUL, A. T. MALE and V. DE PIERRE. A new method for the determination of material flowstress values under metalworking conditions. *Technical Report AFML-TR-70-19,* January, 1970.

AN EXPERIMENTAL WIDE RING ROLLING MILL OF NOVEL DESIGN†

by

J. B. HAWKYARD*, E. APPLETON* and W. JOHNSON*

SUMMARY

An account is given of the design and initial testing of an experimental ring rolling mill having a somewhat novel form of construction. The mill is intended for basic studies into the mechanics of ring rolling, which appears to have received relatively little attention in the past, and the capabilities of this particular design will be investigated. The novel feature of the mill is in the mounting of the mandrel which operates within the ring during rolling. Commercial ring mills have a cantilevered mandrel, the end of which may be restrained by a pivotting arm during forming, whereas the present mill has a mandrel which retracts axially for loading and unloading and is symmetrically supported between roller bearings within a four-column frame during rolling.

It is believed that this arrangement provides better support to the mandrel, allowing the use of larger length-to-diameter ratios, for the rolling of cylinders and for multi-stage form rolling, for example.

INTRODUCTION

Ring rolling mills have been in use in steelworks for many years, engaged principally in the production of tyres for railway rolling stock, and the process is now finding increasing application in wider fields, for example in the production of pipe flanges, ring gear blanks, bearing races and gas turbine rings. Complex sections can be produced in two or three stages and there are reports of ring components up to about 100 tons in weight being manufactured. Despite the increased employment of the process, there appears to be little published information on the mechanics of ring rolling, or on systematic investigations even into simpler aspects of the process, and the operation of mills still appears to be based largely on experience and trial and error.

Ring rolling differs from the more conventional rolling operations in several respects: the reduction per pass is relatively small, the rolls are of unequal diameters and one is undriven. Also, special means are needed to maintain the circularity of the ring during expansion.

Some experimental work on ring rolling was carried out in the Mechanical Engineering Department of U.M.I.S.T. about five years ago, using a small experimental mill[1-3] and the present paper describes further work on a new experimental mill with increased capacity and a somewhat novel form of construction, which is believed to offer advantages in certain respects in relation to most conventional commercial mills.

MILL DESIGN

General principles

The arrangement of rolls adopted for the experimental mill is shown in figure 1(a). The ring is rolled between an internal mandrel which is freely rotating but undriven, and a driven main roll. Reduction in ring thickness is achieved by moving the mandrel towards the driven roll.

To keep the ring in a central position during rolling and to maintain circularity, small guide rolls attached to synchronised arms impose stabilising forces. The roll arrangement is generally similar to that used on many commercial ring rolling mills, although other arrangements are possible, and of particular interest is that shown in figure 1(b), where the internal roll is powered and the external roll rotates freely and is

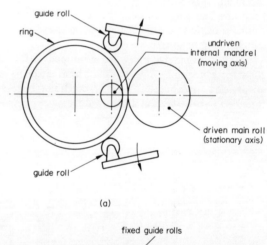

(a)

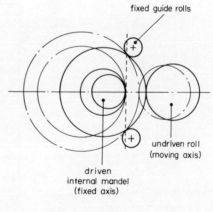

(b)

Figure 1. (a) Arrangement of rolls in experimental machine. (b) Arrangement of work and guide rolls, described in reference 4.

* Department of Mechanical Engineering, The University of Manchester Institute of Science and Technology.
† Provisional patent application No. UKPA 26174/72, assigned to the National Research Development Corporation.

moved towards the fixed-axis internal roll. The guide rolls in that mill are arranged so that the points of contact with the ring lie on a line tangential to the inner roll, and in principle the axes of the guide rolls can remain fixed during a rolling operation. An account of the machine, with an analysis of the guide roll arrangement, is given in reference 4.

The majority of modern machines roll the ring up from a forged hollow billet in a single operation, in which case the diameter of the internal mandrel is generally made as small as reasonably practicable to minimise forging and piercing work on the billet. To allow for loading and unloading of the workpiece the mandrel is free at one end[1-3], although the machines may have a support to the end of the mandrel during rolling. High bending stresses are produced in the mandrel, which, in combination with the thermal stresses induced in hot rolling by the enclosing ring, create severe operating conditions. This is particularly so for small ring mills, where the heat can fully penetrate the mandrel during an operation.

The design of the present experimental mill differs from known commercial mills mainly in the mounting of the mandrel and in the support offered by the bearing system. The mandrel and roll are supported equally at each end, in the manner of a normal rolling mill and to allow loading and unloading of the ring the mandrel is withdrawn axially from the operating position. The ring can then be fed to and from the rolling position across a flat worktable, which is adjustable in height.

Many commercial mills are equipped with edge rolls for controlling ring width. This feature is not at present incorporated in this machine, but it is intended to provide it at a future stage.

Design details

Details of construction of the mill are given in the sectional view of figure 2 and the photograph of figure 3. The rolls are situated symmetrically within a four column main frame, the main roll being carried between bearings supported on a vertical cross member and the mandrel in the throat of a C frame, which slides within the main frame members on plain brass bearings.

In the rolling position, the top end of the mandrel engages in a tapered sleeve in the top bearing, the bearing housing being spring loaded axially to produce a controlled engagement force at the taper. The mandrel is moved vertically, between the rolling and retracted positions, by a hydraulically operated lever mechanism as indicated in figure 2.

In the raised position the main arms of the linkage are opened out into an 'in-line' situation, thereby locking the mandrel in the rolling position. The lower bearing housing, which moves with the mandrel, is then pressed against a stop with a force of about 3 tons by means of a pre-stressed disc–spring system. The mechanism allows the mandrel height to be adjusted, for alignment with the main roll. The main roll is driven by a 10 hp induction motor through a variable-speed gearbox and final reduction box, to give a speed range continuously variable between 22 and 200 rev/min. Roll force is applied by a hydraulic cylinder giving 15 tonf capacity at 1500 lbf/in^2.

The mill is designed to accommodate various sizes of main roll and mandrel; the arrangement shown in figure 3 has a main roll diameter of $8\frac{1}{4}$ in and a mandrel diameter of $2\frac{3}{4}$ in. Spherical roller bearings are used to support the roll and mandrel.

The worktable is adjustable in height to enable

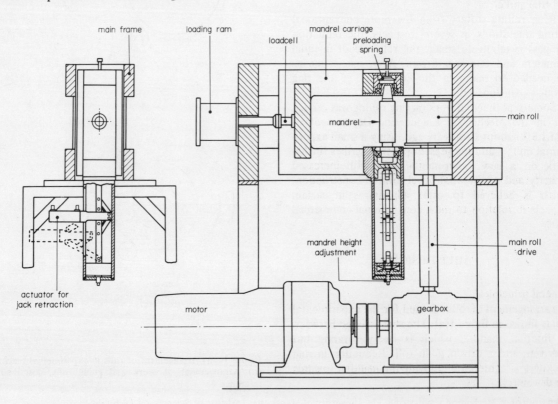

Figure 2. Sectional view of experimental ring rolling mill (excluding guide rolls, guide arms and worktable).

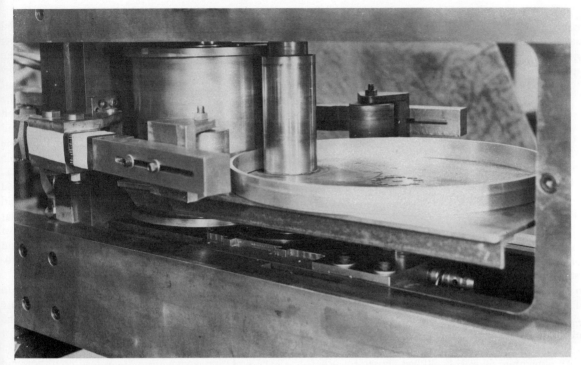

Figure 3. View of ring rolling mill showing guide rolls and ring supported on worktable.

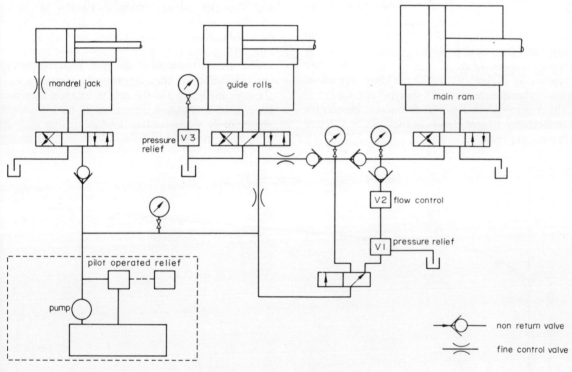

Figure 4. Ring rolling mill hydraulic circuit.

the workpiece to be formed on any desired part of the roll. The guide rolls serve to keep the ring central during rolling, thereby maintaining circularity. A linkage mechanism ensures synchronous movement of the arms and a hydraulic cylinder coupled to the linkage moves the guide rolls inwards prior to rolling and provides a controlled centralising force during rolling.

The hydraulic circuit controlling the main ram, guide rolls and mandrel retracting jack is shown schematically in figure 4.

Instrumentation
Rolling force is determined from a strain gauge load cell at the end of the ram shaft, the output being fed to a multi-channel ultraviolet recorder, together with signals of main roll torque, roll closure and rotation of the main roll. Main roll torque is determined from the reaction torque on the final reduction gearbox which is mounted on a turntable with rolling contact bearings. A small load cell on the side of the turntable transmits the torque to the fixed baseplate and provides a sensitive means of torque measurement.

Motion of the mandrel relative to the main roll is recorded by means of a simple slide-wire resistance gauge and rotation of the main roll by a cam-operated switch which is arranged to impose a small positive impulse on the load trace at each revolution.

PRELIMINARY INVESTIGATIONS

A description is given of some of the investigations carried out to explore the capabilities of the mill in producing various ring products, with observations on the accuracy achieved. A simple theory for predicting rolling loads is also presented and compared with experimental load measurements.

Test materials

For preparatory work on the rolling of hot metals, tests were conducted using lead billets in diameters up to $4\frac{1}{2}$ in and lengths between 2 and $7\frac{1}{2}$ in, giving billet weights between 8 and 29 lb. These were made by casting solid ingots in thin metal moulds, machining out the bore and removing a small amount from the ends and outer diameter by machining.

An aluminium alloy H.E. 30 was chosen for other tests in which comparisons were made with theoretical rolling forces. The alloy in its fully heat treated condition exhibits a relatively small degree of strain hardening and is also insensitive to strain rate, so that it approaches an ideal perfectly plastic condition. A uniaxial compressive stress strain curve for the alloy is shown in figure 5. Initial yielding occurs at 20 tonf/in² and increases to 25 tonf/in² at $\epsilon = 1.0$.

Other materials tested include copper, chosen for its malleability to give strains as high as $\epsilon = 1.7$, and mild steel and brass.

Operating conditions

Initial tests indicated that it was necessary to exceed a critical rolling load for a particular ring, depending on wall thickness, length and yield stress. As the critical load was approached, some lateral flow occurred near the inner and outer surfaces, without any increase in ring diameter, indicating that the plastic zones under the rolls had not penetrated sufficiently to join across the centre which remained elastic. A relatively small increase in load above the critical was sufficient to achieve an appreciable reduction rate.

In the region of the critical load it was observed that an unstable yielding condition could be produced resulting in distortion of the ring into fairly regular straight-sided forms; see figure 6(e). The number of sides developed was dependent on ring thickness and diameter and the shapes were reasonably reproducible.

The hydraulic circuit for the mill, as shown in figure 4 allows for the control of main ram pressure by the pressure relief valve V1 and feedrate by the flow control valve V2. The procedure adopted for preliminary tests was to set the pressure to a value in excess of that giving the critical load condition and adjust the feedrate above that at which the instability and distortion could develop, and below the stalling condition where the bite was too great and slipping occurred between the main roll and the ring. The feedrate setting and the main roll rotational speed were kept constant throughout the test. Thus the feedrate per revolution of the ring actually increased as its diameter increased and the rate of rotation decreased. To achieve a constant bite it would be necessary to progressively decrease the flowrate to the cylinder, or increase the roll speed.

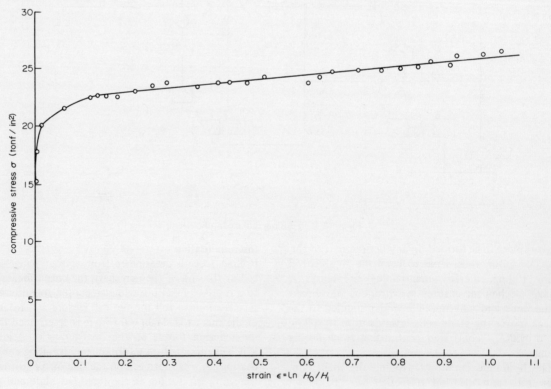

Figure 5. Quasi-static compressive stress–strain curve for aluminium alloy (H.E. 30).

Figure 6. Ring rolled products: (a) aluminium rings (H.E. 30); (b) lead rings; (c) copper belt; (d) composite brass–steel ring; (e) distorted rings produced with insufficient rolling loads

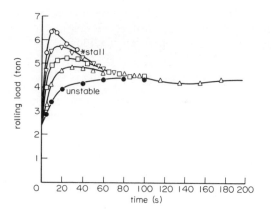

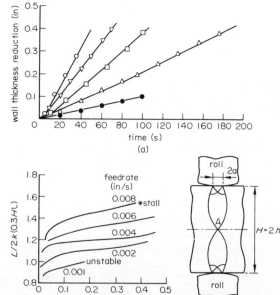

Figure 7. Results of rolling tests on aluminium alloy rings (initially $4\frac{2}{8}$ in outer diameter, $\frac{3}{4}$ in length, $\frac{7}{8}$ in wall thickness). Main roll speed 30 rev/min.

Guide roll force

It was found that the guide roll force necessary to maintain stability of the ring was generally low and throughout the tests a setting of 100 lbf/in^2 was used on the guide roll cylinder relief valve (V3, figure 4) corresponding to a total guide roll force in the region of 120 lbf. Increase in guide roll force to about 360 lbf did not reduce the tendency for instability to develop under the critical rolling conditions described above.

ANALYSIS FOR CRITICAL ROLLING LOAD

It is possible to derive a simple expression for critical rolling load, based on a slip line analysis for the indentation of a slab of finite thickness[5,6]. Figure 7(b) shows the slip line field, and it is assumed that with increasing roll force the widths (2a) of the contact regions increase until this field is created at a ratio of $h/a = 8.74$, as given in reference 5, where h is the half-thickness of the slab.

At lower roll forces and narrower contact widths, the plastic regions would be relatively shallow under each roll and the slip line fields would resemble that for the indentation of a semi-infinite slab as proposed by Prandtl.

At the critical condition the indentation pressure p may be shown to be

$$p = 2(1 + \pi/2) = 2k \, 2.57$$

and the indentation or rolling load is then

$$L = 2a \times l \times p = 2k \times 5.14 \times a \times l$$

$$= 2k \times 0.295 \, H \times l \qquad (1)$$

$$\approx 2k \times 0.3 \, H \times l$$

where l is the axial length of the ring, $H = 2h$ the ring thickness and $2k$ the compressive yield stress of the material in plane strain.

This analysis is obviously a considerable simplification of actual conditions in the roll gap — for example the effects of motion of the ring through the roll gap are ignored, curvature of the rolls and ring is not considered, nor the restraint imposed by the ring on lateral spread in the roll gap[1]. However, comparison with experiments described below indicate that equation 1 does give a useful estimate of critical rolling load. It should be noted also that the reduction per pass in ring rolling is generally small in relation to most other rolling processes and the critical rolling condition as outlined above may not differ greatly from a normal operation condition.

RESULTS

Figure 6 shows a selection of ring products made on the mill at room temperature. The aluminium alloy rings, figure 6(a), were given wall thickness reductions of up to 76% and the lead rings, figure 6(b), of axial lengths between 2 in and $7\frac{1}{2}$ in, were reduced in wall thickness by about 50%; higher reductions would have made the lead rings too thin for subsequent handling during measurement.

Figure 6(c) shows a flexible copper belt of final thickness 0.022 in produced from a 5 in diameter ring of $\frac{1}{8}$ wall thickness and figure 6(d) shows a composite cylinder consisting of a brass lining in a mild steel outer member.

Rolling load

Figure 7(a) shows results of load and displacement measurements taken during the rolling of aluminium alloy rings at different feedrates. At the lowest rate (0.001 in/s, or 0.002 in per revolution of the main roll) rolling became unstable and the ring was distorted, while at the highest rate of 0.008 in/s the ring slipped and stalled owing to the excessive bite, estimated at about 0.016 in. A greater reduction rate would have been achieved with a higher friction coefficient between the ring and roll.

The load curves generally show an initial peak and subsequent fall, which would be expected with reduction in wall thickness.

In figure 7(b) the results are shown replotted in terms of

$$L' = L \big/ 2k(0.3\,Hl)$$

and thickness strains $ln(H_0/H)$, using current values of $2k$, derived from the stress–strain curve, and thickness H assuming plane strain conditions. The lowest stable feed rate gives values of L' between 1.05 and 1.10 over most of the operation, which suggests that equation (1) does give a satisfactory estimate of critical rolling load. The yield stress value used is of course for static loading but it would be expected to be within a few per cent of the relevant dynamic condition because of the low strain rate sensitivity. At higher feedrates the curves rise to about $L' = 1.5$, at which the stalling condition occurs.

Axial spread and edge shape

In commercial ring rolling operations the edge shape is generally controlled by the use of edge rolls, or flanges on the main roll. Problems can occur owing to edge distortion, resulting in laps and fins.

In the present tests the rings are unconstrained axially and the results of distortion and changes in length are presented. Figure 8(a) shows profiles of rolled aluminium alloy rings rolled at various feedrates. The lowest rate results in pronounced spread at the inner and outer surfaces suggesting that the plastic regions under the rolls are barely penetrating to the centre of the section. With increasing feedrate the

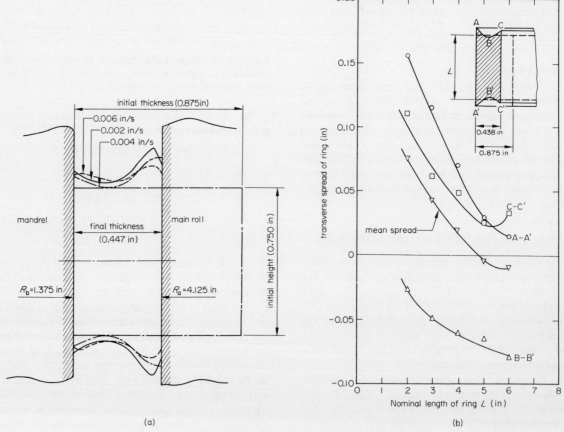

(a)   (b)

Figure 8. Showing axial spread of rings: (a) effect of feedrate (aluminium alloy); (b) effect of axial length (lead).

profile becomes more regular. The influence of axial length of ring on axial spread can be seen from the results of figure 8(b) for lead cylinders reduced in wall thickness by about 50%. Length changes are shown for the centre and edges of the section, and a mean value is given. As would be expected, there is a general reduction in spread with increase in length, the maximum being only 0.032 in for the 6 in long cylinder. It is interesting to note that the central sections of the cylinders actually undergo a reduction in length, by about 0.075 in for the 6 in cylinder. This can be explained by reference to the slip line field given in figure 7(b). It can be readily shown that a triaxial tensile system exists about point A, and the effect of an axial tensile stress at the central section would be to modify the plane strain situation to produce axial contraction.

Accuracy

Observations were made on roundness, wall thickness, and taper over the lengths of the rings to provide some

and cylinders up to a relatively high length-to-diameter ratio. Future work will be directed towards developing the analysis for more accurate prediction of rolling load, torque and stability conditions. It is intended to investigate cold precision rolling and hot rolling.

The mill controls are being developed to enable a rolling operation to be performed automatically, after placing a billet on the worktable, and it is hoped to achieve rolling times of between 5 and 10 s. Short rolling times are desirable in hot ring rolling to minimise heating of the mandrel, which can be a serious problem on small mills when rolling steel.

ACKNOWLEDGMENTS

The authors wish to thank the technical staff engaged in the manufacture of the equipment and the experimental work. The project is supported by a Science Research Council grant, which is gratefully acknowledged, and the continued interest of BISRA

TABLE 1
Rolling of pure lead (dimensions are in inches)

Billet dimensions			Final diameter (outer)				Final wall thickness		
Axial length	Outer diameter	Wall thickness	Top		Bottom		Top	Bottom	Centre
			(Max.)	(Min.)	(Max.)	(Min.)			
6	4.500	0.812	7.322	7.318	7.310	7.225	0.437	0.438	0.436
5	4.500	0.813	7.275	7.265	7.260	7.250	0.438	0.439	0.438
4	4.501	0.815	7.273	7.264	7.264	7.257	0.438	0.439	0.437
3	4.501	0.815	7.268	7.248	7.281	7.269	0.435	0.436	0.434
2	4.500	0.815	7.110	7.095	7.102	7.088	0.438	0.438	0.438

indication of the accuracy of the mill. No special adjustments were made to achieve optimum conditions and the results relate to single test specimens. Measurements from the lead rings are given in table 1. Roundness is generally to within 0.020 in, taper on mean diameter is about the same and wall thickness varies by a maximum of 0.002 in over any ring. This uniformity of wall thickness is at least partly attributable to the method used for limiting the motion of the carriage, by means of stops between the bearing blocks. Further tests performed upon lead rings using the main ram solenoid valve as the switch to stop forward motion of the carriage have also produced rings of good dimensional quality. For a ring of 4 in length the tolerance on ring thickness was found to be ±0.002 in for a reduction in wall thickness from 0.815 to 0.363 in.

CONCLUDING REMARKS

The mill has performed satisfactorily in the preliminary tests, in producing reasonably accurate rings

(now The Corporate Laboratories, British Steel Corporation) in this field of research is also greatly appreciated.

REFERENCES

1. W. JOHNSON and G. NEEDHAM, 1968. 'Experiments on ring rolling', *Intern. J. Mech. Sci.,* **10**, 95.
2. W. JOHNSON, I. MACLEOD and G. NEEDHAM, 1968. 'An experimental investigation into the process of ring rolling or metal type rolling', *Intern. J. Mech. Sci.,* **10**, 455.
3. W. JOHNSON and G. NEEDHAM, 1968. 'Plastic hinges in ring indentation in relation to ring rolling', *Intern. J. Mech. Sci.,* **10**, 487.
4. D. M. POTTER, 1960. 'Ring rolling', *Aircraft Prod.,* **22**, 468.
5. R. HILL, 1950. *Mathematical Theory of Plasticity,* Clarendon Press, Oxford.
6. R. VENTER, W. JOHNSON and M. C. de MALHERBE, 1971. 'The plane strain indentation of anisotropic aluminium using a frictionless, flat rectangular punch', *J. Mech. Engng. Sci., Inst. Mech. Engrs,* **13**, 416–428.

INVESTIGATION INTO THE POSSIBILITIES OF TESTING LUBRICANTS FOR COLD AND WARM EXTRUSION OF STEEL

by

H. KAISER*

SUMMARY

Lubrication is one of the most important criteria for the extrusion of steel, but the choice of a proper lubricant is not very easy in practice. Most of the investigations now published are about the coefficient of friction, but there is little about the applicability of lubricants. In this report a possibility is indicated as to how lubricants may be tested by using the can extrusion of steel. Lubricant requisites for this forming process are extremely high, because the inner can surface increases enormously. At the same time the lubricant has to withstand a very severe stressing. In this paper it will be shown that the relation between inner length and inner diameter of the extruded can is a measure of the applicability of a lubricant. Results of investigations at room temperature, $300°$ C, $500°$ C and $700°$ C are given.

NOTATION

a = punch land (mm)
A_0 = initial cross-sectional area (mm^2)
A_1 = final cross-sectional area (mm^2)
d_i = inner diameter of can (mm)
d_o = billet diameter ≈ outer diameter of can (mm)
e_a = fractional reduction of area = $(d_i/d_o)^2$
h_i = inner height of can (mm)
h_o = height of billet (mm)
$\bar{p}$ = average punch pressure (N/mm^2)
r_p = punch nose radius (mm)
R = deformation ratio = A_0/A_1
θ = billet temperature ($°$C)
θ_d = die temperature ($°$C)

INTRODUCTION

Lubrication is one of the biggest problems in the cold extrusion of steel — in actual fact the cold extrusion of steel was made possible only after suitable lubricants and lubricant carriers were discovered. The demands on a lubricant used in deformation processes and especially in cold extrusion of steel are very severe. A lubricant (1) should avoid direct contact between the tool and the workpiece, since this may lead to cold welding (scuffing), (2) should result in reduced friction thereby bringing down the force requirements, and (3) should stretch itself out as the surface area enlarges itself during deformation. In addition to the qualities mentioned above a lubricant should adhere to the surface of the workpiece and should be easily removable after the completion of deformation. A decision on the quality and applicability of a lubricant can be arrived at only after a careful consideration of these factors.

It becomes, therefore, increasingly difficult for the user to decide on a particular lubricant based on these requisites. The industry offers a wide variety of lubricants for extrusion and test results are available for most of them[1-3]. These experimental results are concentrated around the determination of friction coefficients by various methods. These coefficient of friction values give little evidence about the probable behaviour of such lubricants in deformation

* Institut für Umformtechnik, Universität Stuttgart (TH).

processes other than the one for which the values were established.

Based on all these considerations ring compression tests with different lubricants and diverse workpiece—tool combinations were conducted in a temperature range from room temperature to 700 $°$C at the Institut für Umformtechnik of the University of Stuttgart[4,5]. The ring compression test has proved itself to be a reasonably good test to check the suitability of lubricants. Average friction coefficients are determined from the change in dimensions of the upset cylindrical rings.

These investigations gave sufficient information on the nature and value of coefficient of friction for a few normal lubricants and workpiece—tool pairs in the temperature range tested. But the question now arises about the validity of applying these coefficient of friction values to the cold extrusion process. One encounters in these ring compression tests punch pressures around 1000 N/mm^2 and a maximum deformation ratio of $R = 2$. The relative movement between the workpiece and the tool is also not very high and the maximum increase of surface of the ring that one can achieve is approximately three times the initial value. This means that one cannot draw definite conclusions from the values obtained in ring compression tests about the behaviour of a lubricant in question in some other deformation process in which larger deformations, higher pressures and enormous increase of surfaces are involved.

Can extrusion (backward) is a classical example of this viewpoint. Pressures up to the maximum load-carrying capacity of tools — in laboratory tests up to about 3500 N/mm^2 — are possible. At the same time the surface of the workpiece increases greatly — in the extrusion of a can with a ratio of $h_i/d_i = 3.5$, the inner surface increases by about 17 times — at some local places this value may even be greater.

AIM OF THE INVESTIGATIONS

Can extrusion — in this case backward extrusion — was selected as the process by the use of which the quality and applicability of the lubricants would be tested. In order to effect a classification of lubricants tested later on, a suitable reproducible parameter was aimed at. As the coefficient of friction is quite difficult to determine by this process, this possibility was ruled out and another characteristic value was sought for. The suitability of lubricants formed the second aspect of these experimental investigations.

TEST SET-UP AND TEST CONDITIONS

The tool set-up used is shown in figure 1. Heating coils heated the die to about 160 $^{\circ}$C (θ_d) to obtain

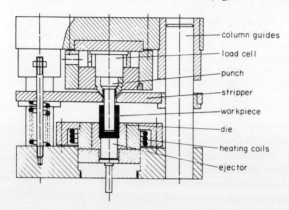

Figure 1. Tool set-up (schematic).

conditions normally adopted in the usual manufacture. Billets of materials Ma8 and Ck15 were subjected to a coating of zinc phosphate after the initial setting operation; this was done to ensure that the billet completely fills the die hole. The billet diameter and height were 28 mm (d_o) and 30 mm (h_o) respectively. Cans with various inner diameters were produced with punches of different diameters. The area reductions thus obtained by the use of these punches were e_a = 64%, 72% and 80% which correspond to the deformation ratios of $R \approx 3, 4$ and 5. Six lubricants in fine powder form as

Force—travel diagrams recorded during the experiments served for this purpose. Some cups were sawn longitudinally into two portions to study the build-up of the inner surface.

The force—travel diagrams did not give sufficient information regarding the characteristics of the lubricant employed. A typical diagram is shown in figure 2. After attaining a maximum value the force

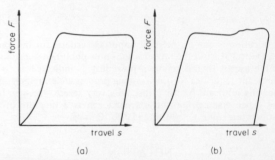

Figure 2. Typical force—travel diagrams (schematic): (a) faultless lubrication; (b) occurrence of scuffing.

reduces uniformly. When severe scuffing takes place, especially in cold extrusion, the force increases again after a definite amount of travel of the punch; this can be seen on the right-hand side of the same figure. Even for small scratches on the inner surface a force—travel diagram as on the left-hand side was obtained, leading to difficulties in predicting failures.

Although the force necessary for deformation with a particular lubricant can be read off these diagrams, a definite conclusion regarding the applicability of the lubricant is difficult to arrive at. The lubricants, for which higher friction coefficients were measured in ring compression tests, resulted in higher force requirements in these tests also. A lubricant having a lower coefficient of friction as obtained from ring compression tests showed a smaller maximum

TABLE 1
Lubricants investigated

Lubricant number	Type	Manufacturer's name	Mixing proportion (lubricant/water) (volume ratio)
1	colloidal graphite in water	Dag 1713 k	2 : 3
2	graphite suspension in water	Delta 31	1 : 2
3	MoS$_2$-base lubricant suspension in water	Molydag 15	1 : 2
4	solid lubricants (graphite, inorganic phosphates) in synthetic oils soluble in water	VN 4096	1 : 10
5	solid lubricants (graphite, zinc sulphide) in synthetic oils soluble in water	VN 4112	1 : 10
6	graphite suspension in water	Hydrokollag 300A	2 : 3

suspensions in water or in synthetic oils soluble in water (see table 1) were distributed on the tool surface by spraying.

PRELIMINARY TESTS

During the first part of the tests a method of deciding whether a lubricant has failed to function and has not withstood the load was evaluated.

force but the value of a lubricant cannot be described from this one phenomenon only. In addition to this, the difference in force requirements using various lubricants was found to be small — generally less than 10%, with a maximum of about 15%. So, grading of lubricants from the viewpoint of force requirements only is not advantageous. Figure 3 shows the narrow band in which the force variations fall. The deformation force decreases up to a temperature of 300 °C and remains more or less constant in the blue brittleness region (300 °C to 600 °C). A further increase of temperature leads to a reduction of force.

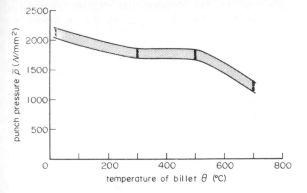

Figure 3. Punch pressure versus billet temperature (workpiece Ck 15, $e_a = 0.72$).

It should be pointed out at this stage that the observations made in the last few paragraphs were to find out the extent to which the lubricants can be loaded without any trace of cold welding, and the forces refer to the maximum values occurring at the beginning of the deformation process. Malfunctioning of the lubricant generally occurred with cups of greater height. It can be stated, therefore, that the force–travel plots and the maximum deformation force as determined from such diagrams are not the criteria in deciding the usefulness of a lubricant. For this reason a careful visual examination was undertaken, which revealed any eating away of material either on the inner wall of the cup or on the outer contour of the punch, if present. Small particles of test billets sticking onto the shoulder (land) of the punch and longitudinal scratches along the entire length of the inner cup wall were some of the points establishing the above-mentioned fact. The longitudinal scratches or flutes were better identified when the cups were cut along the longitudinal axis and

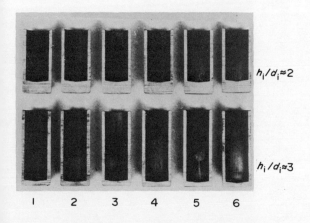

Figure 4. Longitudinally cut cups for the appraisal of inner surface obtained with diverse lubricants.

viewed under illumination. Figure 4 shows a row of cups for which such observations were made. The formation of the surfaces using different lubricants can be observed in this figure.

MAIN INVESTIGATIONS

Preliminary investigations gave information on the failure of a lubricant in general. The main tests were aimed at finding out when a lubricant actually failed to function. By adjusting the lower return point of the punch so that cups of various height could be produced, experiments were carried out with the lubricants at desired temperatures. Such an experimental series is shown in figure 5.

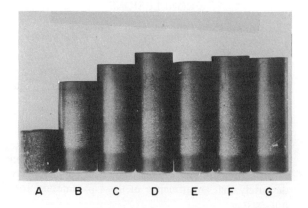

Figure 5. Experimental procedure to determine $(h_i/d_i)_{max}$.

Starting with a cup of small height, cups of larger heights were produced till the first signs of scuffing appeared. By bracketing a definite height of the cup at which no scuffing occurred, cups with different heights were extruded. By this method one could arrive at the maximum possible ratio of h_i/d_i without any longitudinal cracks, or in other words with a good surface. This ratio was assumed as the characteristic parameter of the lubricant investigated. In the example illustrated in figure 5 wear was observed at position D and the cup E was faultless; cup F showed slight flutes; G was again faultless. This method leads to the determination of $(h_i/d_i)_{max}$ as measured from cup G.

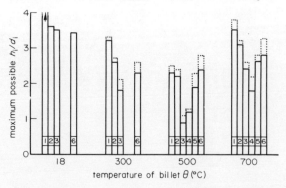

Figure 6. $(h_i/d_i)_{max}$ as a function of billet temperature and different lubricants.

Summarising the observations made on the above tests, the following can be said (figure 6). Maximum possible h_i/d_i ratios are plotted as a function of temperature. No failure of lubricant 1 at room temperature was observed, but limitations of producing a cup with higher ratios of h_i/d_i, greater than 3.9, prevented the determination of a ratio at which the failure might probably have occurred. Lubricants 4 and 5 can be used at higher temperatures only. The results presented here are reproducible within limits. Test results with the two workpiece materials and at the three deformation ratios mentioned above conformed to the control region shown in figure 6. The influence of the geometry of the punch through the use of diverse punch lands ($a = 3, 5$ and 7 mm) and different punch radii ($r_p = 0.5, 1, 1.5$ and 2 mm) could not be demonstrated in these tests.

CONCLUSIONS

The aim of testing the suitability of a lubricant for cold extrusion process can be considered to have been achieved. $(h_i/d_i)_{max}$ values give an idea of the possibility of producing cans without any scratches with a lubricant in question. Based on the investigations carried out, the influence of the workpiece material and deformation ratio on the applicability of a lubricant should be answered in the negative. However, it should be borne in mind that the difference in the materials used and the deformation ratios employed were not large enough to say definitely that an influence of these parameters does not exist.

As a by-product of these laboratory tests punch pressure variations with temperatures were obtained. A classification of the lubricants used was also made possible. Graphite (No. 1) proved itself to be the best lubricant in the temperature region considered, whereas MoS$_2$ (No. 3) was not found to be suitable.

REFERENCES

1. M. BURGDORF, 1967. 'Über die Ermittlung des Reibwertes für Verfahren der Massivformung durch den Ringstauchversuch', *Industrie-Anzeiger*, 89, 799–804.
2. O. PAWELSKI, 1966. 'Probleme der Reibung und Schmierung in der Umformtechnik', *Schmiertechnik*, 13, 267–73.
3. G. GRAUE, W. LÜCKERATH, G. GEBAUER and W. GRADTKE, 1962. 'Die Schmierung bei der Warmumformung von Metallen', *Schmiertechnik*, 9, 245–53.
4. R. GEIGER, 1970. 'Untersuchung von Schmierstoffen zum Warmumformen von Stahl', *Industrie-Anzeiger*, 92, 623–31.
5. R. GEIGER, 1970. 'Einfluss der Temperatur auf die Grösse des Reibwertes verschiedener Schmierstoffe beim Umformen von Stahl', *Industrie-Anzeiger*, 92, 1553–4.
6. H. KAISER, 1971. 'Möglichkeiten zur Prüfung von Schmierstoffen zum Kalt- und Warm-fliesspressen von Stahl', *Industrie-Anzeiger*, 93, 2311–13.

OBSERVATION OF NEW PHYSICAL PROPERTIES OF ZINC–ALUMINIUM AND TIN–LEAD SUPERPLASTIC ALLOYS

by

SHYAM KINKAR SAMANTA*

SUMMARY

The paper reports the initial results of efforts to utilise superplastic alloys in conventional metal forming operations, by exploring strain rate sensitivity (SRS) of the eutectoid zinc–aluminium alloy (78 wt % Zn, 22 wt % Al) and the eutectic tin–lead alloy (63 wt % Sn, 37 wt % Pb) at high strain rates and temperatures. Compression tests were conducted at strain rates from (i) $10^{-2} - 3 \times 10^{-1}$ s^{-1} and (ii) $8 \times 10^{1} - 2.2 \times 10^{2}$ s^{-1} at 250 °C, 270 °C and 290 °C for the Zn–Al alloy and at room temperature and 170 °C for the Sn–Pb alloy.

The true stress versus strain data from the compression tests at all loading rates show negligible strain hardening. Stress versus strain rate plotting from the dynamic tests produced a very good linear distribution of the data, which indicates that the parabolic stress–strain rate characteristic $\sigma = K\dot{\epsilon}^m$ could be used. However, the low strain rate data exhibited the characteristic variation of m with strain rate, as was previously observed by Backofen and many others. The important features are that high SRS ($m \sim 0.5$) is observed at high strain rates and that total deformation without fracture were also high. Furthermore, the results indicate that superplastic deformation can be obtained above the eutectoid invariant at high strain rates.

The high m values and relatively large deformations at moderately high strain rates suggest the possibility of industrial utilisation of superplastic alloys in conventional forming.

INTRODUCTION

Materials that experience extraordinary elongations before failure are said to be *superplastic*. Superplastic behaviour is found in a variety of materials. Its distinguishing characteristic is the ability of the material to undergo very large tensile deformation (> 1000%) without fracture. Many of the superplasticity studies are surveyed by Underwood[1] and Chaudhari[2]. An excellent review article has been given recently by Johnson[3] on this subject. Some of the more recent and most comprehensive studies of superplasticity are the investigations of Backofen and his colleagues[4]. Interpretations of the superplasticity phenomenon are also given by Sherby[5] and his group and Weiss[6] and his associates.

Hence, on the metallurgical front there has been a rapidly growing interest and information over the last few years in the so-called superplastic materials. A number of alloy systems have been reported[2, 3] which exhibit this effect.

There are two different ways of obtaining these large elongations. One involves stressing the material and simultaneously cycling it about a transformation temperature; this is described[6] as transformation superplasticity and, although metallurgically interesting, the industrial utilisation of this phenomenon would seem to be remote. The other is 'micrograin superplasticity'. Micrograin superplastic alloys have a very fine and stable microstructure and exhibit large total elongations, when deformed slowly at a temperature of about half the absolute melting point. At low deformation speeds, the flow stress of these materials is extremely low but it increases rapidly with increasing strain rate. In so far as one can characterise the degree of superplasticity, the more nearly the material behaves as a viscous liquid the greater is its superplasticity[7].

The mechanical properties of these alloys have been investigated almost exclusively by means of simple tension tests and in the literature there is extensive information on the results of such tests on the common superplastic alloys. There are certain features of the properties which are always observed in these materials.

(1) Strain hardening is negligible.
(2) The flow stress at low strain rates is a small fraction of the flow stress in the same alloy in the non-superplastic state.
(3) The flow stress increases rapidly with strain rate (most experiments were performed at initial strain rates of 10^{-5} and 10^{-2} s^{-1}).

The properties of these alloys at the superplastic deformation temperatures is described by the familiar exponential relationship

$$\sigma = K\dot{\epsilon}^m$$

where σ is the true flow stress in uniaxial tension and $\dot{\epsilon}$ is the strain rate associated with that value of stress. K is a constant for that material and m is called the index of strain rate sensitivity which for normal materials usually is less than 0.2. In normal materials, m usually decreases slightly with strain rate; in superplastic materials, m typically increases. The initial value of m may be about twice that for a normal material and will increase with strain rate up to a certain level and will decrease thereafter. Typical initial or peak values of m are usually in the range of 0.4 to 0.8. Peak values of m occur at strain rates which are frequently in the range 10^{-4} and 10^{-2} s^{-1}. When $m = 1$, which characterizes Newtonian flow, the deformation is uniform, i.e. neck free, and the stress is directly proportional to the strain rate. When $m = 0$ necking will occur and the metal is not sensitive to strain rate. The large value of m for superplastic metals indicates there is a reduced tendency for necking.

A material with such unusual properties has some

*Department of Mechanical Engineering, Carnegie-Mellon University rather obvious technological applications but so far these properties have only been exploited to a limited extent in any commercial metal forming. The difficulty has arisen principally because the alloys are superplastic either at high temperatures[2] and/or at strain rates too low[4] to be of practical value.

Processes for the forming of superplastic alloys have been recently investigated by Al-Naib and Duncan[7], Fields and Stewart[8], and Thomson et al[9]. The forming times in these processes are longer than those in industrial processing and therefore are at present commercially unattractive. From the point of view of practical application the superplastic alloys must be formable at reasonable temperatures, exhibit superplasticity at practical forming rates and have adequate properties when cooled to room temperature.

Although the literature is extensive on the strain rate dependence of stress, σ, strain rate sensitivity index, m, grain size and temperature of superplastic alloys, little has been reported on the mechanical behaviour of superplastic alloys at strain rates comparable with conventional metal forming operations. Moreover, most tests have geen performed under simple tension while behaviour resulting from compressive state of stress is largely unexplored.

If the phenomenon of superplasticity is not to remain a laboratory curiosity, research efforts must be devoted to the development of economic production of components in conventional forming processes. In processes such as closed die forming and stretch forming the forming properties of superplastic materials could be utilised provided the materials can be processed at a reasonable deformation rate and temperature, and the finished product has desirable mechanical properties. In closed-die forging, superplastic alloys would not only form at lower loads, but the ability of the material to creep rapidly in the die should result in excellent die filling and an enhanced ability to pick up fine details. In stretch forming conventional metals, the maximum deformation between annealing operations is limited because of the danger of necking. Much larger deformations in a single stage can be expected with superplastic materials.

Neck-free elongation of very large total deformation is remarkable in superplastic alloys when compared with ductile materials such as deep-drawing brass and soft copper. This phenomenon is obviously of interest in the metal-forming industry but it must be remembered that total elongation in the tensile test is not, by itself, a particularly useful criterion of formability. In a recent paper, Thomson et al.[9] pointed out that many shapes of practical interest do not require so much strain. In forming a hemisphere from a flat blank the total elongation along a meridian plane is only about 57%.

In the present investigation, attention is given to the mechanical behaviour of two common superplastic alloys, e.g. the eutectoid zinc–aluminium alloy (78 wt % Zn, 22 wt % Al) and the eutectic tin–lead alloy (62 wt % Sn, 38 wt % Pb) under high compression rate tests. The Zn–Al alloy was tested at 250 °C, 270 °C and 290 °C, while the Sn–Pb alloy was tested at room temperature and at 170 °C. The true compressive strain rates were in the range of $2 \times 10^{-2} < \dot{\varepsilon} < 2 \times 10^2 \ s^{-1}$.

EXPERIMENTAL TECHNIQUES

Specimens

Zinc and aluminium of 99.99% purity were used to prepare the Zn–Al superplastic alloy by melting in an induction furnace under argon and chill casting. The ingot was then machined to remove surface defects, and hot extruded at 250°C to $\frac{5}{8}$ and $\frac{3}{8}$ in diameter rods. Cylindrical specimens $\frac{1}{2}$ in long were then cut from the extruded rods. The specimens were heated to 375 °C for 50 h, followed by a rapid quench into iced brine. The specimens were then heat treated at 250 °C for 0.2 h and were brine quenched and stored in a refrigerator. The grain size obtained after heat treatment at 250 °C was L (mean linear intercept) = 0.6 μm.

The second superplastic material was eutectic Sn–Pb alloy. Each ingot of the alloy was given a 30:1 area reduction by extrusion at room temperature to rod of $\frac{3}{4}$ in diameter and then stored in a refrigerator. Ingots were machined all over to remove surface defects before extrusion. From the extruded product the cylindrical specimens of $\frac{5}{8}$ in diameter x $\frac{1}{2}$ in long and $\frac{3}{8}$ in diameter x $\frac{1}{2}$ long were prepared. The initial grain size obtained was L = 2.0 μm. The specimens were preserved in a refrigerator.

Dynamic tests

The basic apparatus of the present investigation was developed and used by Lindholm[10]. In Lindholm's

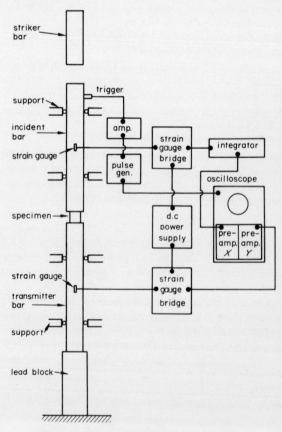

Figure 1. The split Hopkinson pressure bar.

apparatus, the loading pulse was obtained by longitudinal impact with a steel striker bar which has been accelerated by means of a spring mechanism. For the present study a gravity-accelerated striker bar was used. This apparatus consists of elastic steel pressure bars between which the short compression specimens are sandwiched as shown schematically in figure 1. The striker bar and the pressure bars were made of $\frac{7}{8}$ in diameter centreless ground stainless steel rod. The pressure bars were supported in Teflon O rings mounted in steel support blocks. In order to ensure uniform compression of specimens, the bars were carefully aligned so that the contact faces with the specimen would be parallel. The pressure bar system was mounted on a cylindrical lead block, ends of which were carefully turned in a lathe in order to obtain perfect alignment with the pressure bar, to absorb the energy of the pulse. The ends of the lead block were machined at regular intervals.

The impact end of the striker bar was rounded slightly to assure axial impact. The faces of the pressure bars were carefully turned on a lathe to a flat surface and then finished with a fine emery paper.

The pressure bar arrangement outlined in figure 1 is exactly similar to the one used by Lindholm[10]; hence, for the details of the principal of the Hopkinson bar arrangement, theory of measurements, recording system and calibration, the interested reader is referred to the above-mentioned paper.

From X–Y oscilloscope records stress–strain histories and from dual beam oscilloscope records strain-time histories were obtained. The slope of the strain–time records, which is very nearly linear, is taken as the strain rate of the test.

Low strain rate tests

Low strain rate tests were performed on a constant crosshead velocity hydraulic press. The upper and lower platen assembly, together with the test piece, was enclosed in a cylindrical furnace. As the strain rate was changing during the deformation, the stress–strain histories for large deformation were not obtained, the specimens were deformed only up a to true strain of 0.104 and the true stress and strain rate at that strain is used.

Teflon spray was used as a lubricant for room temperature tests and molybdenum disulphide spray at elevated temperatures.

EXPERIMENTAL RESULTS AND DISCUSSIONS

For the dynamic compression tests, two types of tests were conducted.

In the first set of experiments the nominal dimensions of the specimens were maintained at $\frac{3}{8}$ diameter and $\frac{1}{2}$ in length. Three specimens at the same strain rate and at the same temperature were tested in the Hopkinson pressure-bar. Strain rates were 80, 100, 130, 155 and 220 s^{-1} and temperatures were 250 °C, 270 °C and 290 °C for Zn–Al alloy and room temperature and at 170°C for Sn–Pb alloy. Stress–strain histories were not recorded, because of the inaccuracy of the measurements,

particularly in the beginning of the deformation, when the specimen diameter is much smaller than that of the bar[11]. At all strain rates the specimens were deformed at least up to a height reduction of 80% and fracture was never observed in any test piece. This observation

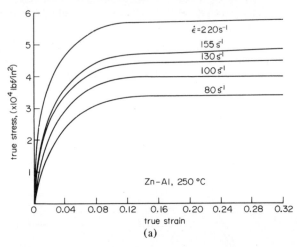

(a)

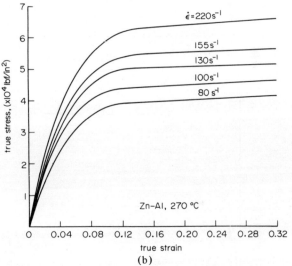

(b)

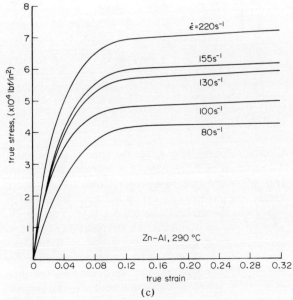

(c)

Figure 2. Stress–strain curves for Zn–Al superplastic alloy at different strain rates and temperatures: (a) 250 °C, (b) 270 °C, (c) 290 °C.

is contrary to the findings of Saller and Duncan[12]. About one-third of their quenched Zn–Al alloy test pieces failed in a brittle manner in the drop-hammer tests. Our experiments demonstrate that the Zn–Al alloy has sufficient ductility at high strain rates ($\dot{\epsilon} = 10^2$ s^{-1}), so we feel that the failures of Saller and Duncan's specimens may be due to the defects which appeared during specimen preparation.

In the group second of tests, the height of the specimens was $\frac{1}{2}$ in and the diameter was $\frac{5}{8}$ in. The continuous records of strain versus time and stress versus strain were recorded simultaneously. Figures 2 and 3 present stress–strain curves from the dynamic test for Zn–Al and Sn–Pb alloy. The curves

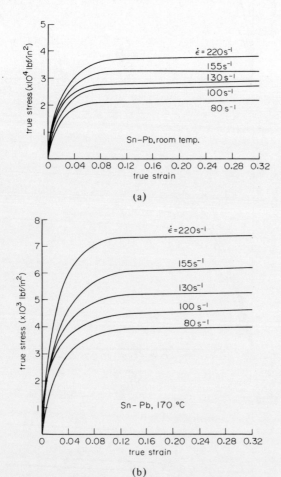

Figure 3. Stress–strain curves for Sn–Pb superplastic alloy at different strain rates and temperatures: (a) room temperature, (b) 170 °C.

shown are average curves for the same group of tests. All the stress and strain data are presented in terms of true stress and true strain. Corrections to true stress values were made to account for adiabatic temperature rise, giving the stress at true temperatures. For Sn–Pb alloy, it is seen that increased temperature has a very pronounced effect on the magnitude of stress. For Zn–Al alloy the flow stress increased slightly with the increase of temperature. For both the materials the flow stress increased markedly with the increase of strain rate at all temperatures.

Strain hardening is negligible for both the materials above $\epsilon = 0.12$ and at all temperatures.

The true stress versus strain rate relationships of the materials are shown in figures 4 and 5. The flow stresses at lower strain rates were plotted along with the data of Zehr and Backofen[13] and Holt[14]. Our results agree well with their results. Above $\dot{\epsilon} = 10^{-2}$ s^{-1} the flow stress of the Sn–Pb alloy at room temperature and the Zn–Al alloy at all temperatures increases very slowly with the strain rate but high SRS ($m \sim 0.48$

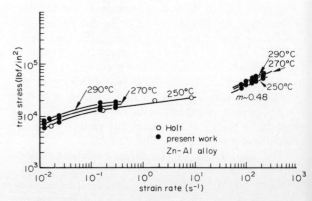

Figure 4. Stress–strain rate relation for Zn–Al superplastic alloy.

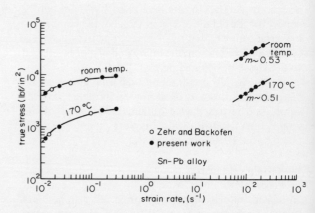

Figure 5. Stress–strain rate relation for Sn–Pb superplastic alloy.

for Zn–Al alloy, $m = 0.53$ Sn–Pb alloy at room temperature, and $m \sim 0.51$ at 170 °C) is observed over a strain rate range $8 \times 10^1 < \dot{\epsilon} < 2 \times 10^2$ s^{-1}. Earlier work[13,14] had shown that the high SRS ($m \sim 0.5$) was observed only over a limited range of strain rates, $10^{-4} < \dot{\epsilon} < 10^{-2}$ s^{-1}, and the technological role to be filled by superplastic materials is anticipated to be limited. The large m value of these alloys at high strain rates accounts for ductility and the experiments clearly demonstrate that these materials can even be formed by conventional forming technique.

Another important feature is that high SRS is obtained for Zn–Al alloy at 290 °C, i.e. above the eutectoid invariant, at high strain rates. Although the flow stress increased slightly at 270 °C and 290 °C, relatively large deformation was obtained at high rates of deformation. It is probable that the grain size of Zn–Al eutectoid alloy was greater the higher the temperature and this may be the reason for the increase in flow stress with increase in temperature.

CONCLUSIONS

Compression tests with the Sn—Pb and Zn—Al superplastic alloys, at more practical rates of deformation, revealed the following.

(1) The alloys possess a level of ductility which is significant for metal forming with conventional metal-deformation processing methods.

(2) The SRS factor is significantly high ($m \sim 0.5$) at higher strain rates and the resistance to plastic deformation is moderate.

(3) For the Zn—Al alloy superplastic deformation can be obtained above the invariant temperature.

A material which maintains a high SRS while undergoing deformation is in a superplastic state irrespective of the mode of deformation and strain rate. The ductility of superplastic alloys can sometimes approach elongation of 2,000 per cent but usually at deformation rates that are too slow to be practical. Our results clearly demonstrate the engineering potential of superplastic alloys at practical rates of deformation and further work in this area would be fruitful.

REFERENCES

1. E. E. UNDERWOOD, 1962. 'A review of superplasticity and related phenomena', *J. Metals*, **14**, 914.

2. P. CHAUDHARI, 1968. 'Superplasticity,' *Science and Technology*, 42.

3. R. H. JOHNSON, 1970. 'Superplasticity', *Metals Rev.*, **15**, 115.

4. W. A. BACKOFEN, F. J. AZZARTO, G. S. MURTY and S. W. ZEHR, 1968. 'Superplasticity', *Ductility*, A.S.M., Metals Park, Ohio, p. 279.

5. C. M. PACKER and O. D. SHERBY, 1967. 'An interpretation of the superplasticity phenomenon in two-phase alloys', *Trans. A.S.M.*, **60**, 21.

6. V. WEISS and R. KOT, 1967. 'Superplasticity', *Intern. Conf. Manuf. Tech.*, A.S.T.M.E., Ann Arbor, p. 1031.

7. T. Y. M. AL—NAIB and J. L. DUNCAN, 1970. 'Superplastic metal forming', *Int. J. Mech. Sci.*, **12**, 463.

8. D. S. FIELDS and T. H. STEWART, 1971. 'Strain effects in the superplastic deformation of 78 Zn—22 Al', *Intern. J. Mech. Sci.*, **13**, 63.

9. T. H. THOMSON, D. L. HOLT and W. A. BACKOFEN, 1970. 'Forming superplastic sheet metal in bulging dies', *Metal Engng, Quart.*, **10**, 1.

10. U. S. LINDHOLM, 1964. 'Some experiments with the split Hopkinson pressure bar', *J. Mech. Phys. Solids*, **12**, 317.

11. S. K. SAMANTA, 1971. 'Dynamic deformation of aluminium and copper at elevated temperatures', *J. Mech. Phys. Solids*, **19**, 117.

12. R. A. SALLER and J. L. DUNCAN, 1971. 'Stamping experiments with superplastic alloys', *J. Inst. Metals*, **99**, 173.

13. S. W. ZEHR and W. A. BACKOFEN, 1968. 'Superplasticity in lead—tin alloys', *Trans. A.S.M.*, **61**, 300.

14. D. L. HOLT, 1968. 'The relation between superplasticity and grain boundary shear in the aluminium—zinc eutectoid alloy', *Trans. A.I.M.E.*, **22**, 25.

THE NECKING OF CYLINDRICAL BARS UNDER LATERAL FLUID PRESSURE

by

J. CHAKRABARTY*

SUMMARY

A cylindrical specimen is observed to neck when subjected to fluid pressure on the lateral surface only. Moreover, the uniform elongation of the specimen at the onset of necking is found to be the same as that in simple tension. A theoretical analysis for the necking problem is presented as an attempt to explain this phenomenon. The experimental investigation is also fully described.

1. INTRODUCTION

Suppose that a tensile test is being performed under an all-round fluid pressure σ_0, where σ_0 is the tensile stress at the initiation of necking in the conventional tensile test. Since an all-round fluid pressure is known to have no effect on tensile instability, the specimen will begin to neck when the tensile stress reaches σ_0. But at this stage the resultant longitudinal stress is zero and the specimen is left with a lateral pressure σ_0 alone. If the specimen is initially work-hardened, so that its yield stress before the experiment is σ_0, the deformation that takes place before instability is purely elastic. As the manner in which the final state is reached during an elastic deformation is immaterial in relation to the ensuing plastic behaviour, the specimen will also neck when subjected to lateral fluid pressure alone when the intensity of the pressure becomes σ_0.

The fact that the hydrostatic part of the stress has no effect on yielding has nothing to do with the preceding argument which merely makes use of the experimental fact that an all-round fluid pressure has no influence on plastic instability of a specimen subjected to longitudinal tension. The argument given by Pugh[1] in support of the above conclusion is erroneous, because it makes no distinction between a fluid and a rigid die in transmitting the pressure. It can be shown that a rigid die will completely suppress plastic instability.

It is reasonable to expect that plastic instability will occur under lateral fluid pressure even in an annealed specimen, although an argument simiiar to above will not be convincing in that case. A theoretical attempt is therefore made in the present paper to solve the instability problem for general strain-hardening materials. The specimen is taken to be in the form of a cylindrical bar which is supported in such a way that the ends of the specimen can move freely during the deformation. It is assumed at the outset that, out of all possible instability m ndes, we need to consider only those which are axially symmetrical. The configuration of the bar at the instant when the stability is examined will be referred to as the initial state. It is convenient to use cylindrical co-ordinates in which the z axis coincides with the axis of the bar, the specimen being assumed to extend from $z = 0$ to $z = l$.

Hill[2] has shown that application of uniform fluid pressure on the lateral surface of a specimen subjected to longitudinal tension does not enhance the uniform elongation. It should be noted that the boundary conditions at the ends of the specimen in the present problem are different from those assumed by Hill. The result derived here does not therefore follow from Hill's conclusion.

2. BOUNDARY CONDITIONS

At the onset of plastic instability, the deformation mode ceases to be uniform and quasi-static deformation of the bar continues temporarily under constant pressure. Since the future position of an unconstrained boundary is not known in advance, it is convenient to formulate the boundary condition in terms of the rate of change of the nominal traction (defined below). Let dP_r, dP_z, be the components of the load vector acting on a final surface element

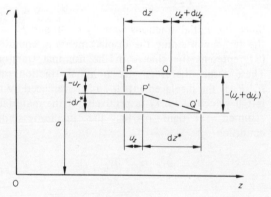

Figure 1.

which was initially of area dS. If F_r, F_z are the radial and axial components of the nominal traction, then

$$dP_r = F_r \, dS, \qquad dP_z = F_z \, dS \qquad (1)$$

Figure 1 shows a line element PQ (solid line) on the generator of the initial surface of the bar. $P'Q'$ is the corresponding line element (broken line) in the final state, so that the components of the displacement of P are (u_r, u_z) and those of Q are $(u_r + du_r,$

* Department of Mechanical Engineering, University of Birmingham.

$u_z + du_z$). The vertical and horizontal projections of $P'Q'$ are

$$
\left.\begin{aligned}
-dr^* &= -du_r = -\frac{\partial u_r}{\partial z}\,dz \\[2mm]
dz^* &= dz + du_z = \left(1 + \frac{\partial u_z}{\partial z}\right)dz
\end{aligned}\right\} \tag{2}
$$

The elemental surface dS is assumed to be generated by a rotation of PQ about the z axis by an amount $d\theta/2$ on either side of the meridian plane. Since the elemental arc $a\,d\theta$ becomes $(a + u_r)\,d\theta$ in the final state, the projected components of the final surface element normal to r and z directions are $(a + u_r)\,d\theta\,dz^*$ and $(a + u_r)\,d\theta\,dr^*$ respectively. Hence the radial and axial components of the load vector acting on the final surface element are

$$
dP_r = -pa\,d\theta\,\left(1 + \frac{u_r}{a}\right)dz^*
$$

$$
\approx -pa\,d\theta\,\left(1 + \frac{u_r}{a} + \frac{\partial u_z}{\partial z}\right)dz
$$

$$
dP_z = pa\,d\theta\,\left(1 + \frac{u_r}{a}\right)dr^*
$$

$$
\approx pa\,d\theta\,\frac{\partial u_r}{\partial z}\,dz
$$

where the last steps follow from equations (2), higher-order terms being neglected. Substituting for dP_r, dP_z from (1) and observing that $dS = a\,d\theta\,dz$, we obtain the nominal tractions as

$$
\left.\begin{aligned}
F_r &= -p\left(1 + \frac{u_r}{a} + \frac{\partial u_z}{\partial z}\right) \\[2mm]
F_z &= p\,\frac{\partial u_r}{\partial z}
\end{aligned}\right\} \tag{3}
$$

Since the initial tractions were $F_r = -p$ and $F_z = 0$, the terms containing the displacements in equations (3) represent the change in the nominal tractions. These terms therefore give nominal traction rates $\dot{F}_r$, $\dot{F}_z$ if the displacements u_r, u_z are replaced by the initial velocities v_r, v_z respectively. If the material is assumed as rigid plastic, the incompressibility condition

$$
\frac{\partial v_r}{\partial r} + \frac{v_r}{r} + \frac{\partial v_z}{\partial z} = 0 \tag{4}
$$

must be satisfied everywhere in the bar. Using this equation, the nominal traction rates are expressed as

$$
\dot{F}_r = p\,\frac{\partial v_r}{\partial r}, \qquad \dot{F}_z = p\,\frac{\partial v_r}{\partial z} \tag{5}
$$

where the derivatives are to be evaluated at the surface $r = a$. These expressions take into account possible axisymmetric distortions of the cylindrical surface on which the fluid pressure is applied. The nominal tractions obviously vanish on the stress-free ends of the specimen.

3. THE CONDITIONS FOR STABILITY

A body (initially at rest) is said to be stable if its motion following an arbitrarily small displacement from the equilibrium position remains as small as we please. Evidently, a sufficient condition for stability is that, in any geometrically possible small displacement from the equilibrium position, the internal energy dissipated must exceed the work done by the external loads. Since these two quantities are always equal when calculated to the first order, second-order terms must be taken into account for the investigation of stability. It is important to remember that the body need not be in continuing equilibrium during the additional virtual displacement.

Let the displacements of a generic particle in the bar be (u_r, u_z) during an interval of time δt. Since the nominal tractions increase (under constant pressure) from $(-p, 0)$ at the initial state by the amounts $\dot{F}_r\delta t$ and $\dot{F}_z\delta t$ at the final state, the work done by the surface forces, to second order, is

$$
\begin{aligned}
W &= \int\{(-p + \tfrac{1}{2}\dot{F}_r\delta t)u_r + \tfrac{1}{2}\dot{F}_z\,\delta t u_z\}\,dS \\
&= \int(-pu_r)\,dS + \tfrac{1}{2}(\delta t)^2\int(\dot{F}_r v_r + \dot{F}_z v_z)\,dS
\end{aligned}
$$

where the integrals extend over the cylindrical surface only. The first integral (denoted by W_1) represents the first-order term, while the second integral (denoted by W_2) represents the second-order term. In view of (5), we have

$$
\left.\begin{aligned}
W_2 &= \tfrac{1}{2}p(\delta t)^2\int\left(\frac{\partial v_r}{\partial r}\,v_r + \frac{\partial v_r}{\partial z}\,v_z\right)dS \\[2mm]
W_1 &= \int(-pu_r)\,dS
\end{aligned}\right\} \tag{6}
$$

If E is the internal energy dissipated per unit volume during the additional displacement, a sufficient condition for stability is

$$
\int E\,dV - W > 0 \tag{7}
$$

where the volume integral extends throughout the volume of the bar. Since the initial state of stress in the bar is given by

$$
\sigma_r = \sigma_\theta = -p, \qquad \sigma_z = \tau_{rz} = 0 \tag{8}
$$

the first-order part for the energy of distortion per unit volume is

$$
E_1 = -p\left(\frac{\partial u_r}{\partial r} + \frac{u_r}{r}\right) = -\frac{p}{r}\frac{\partial}{\partial r}(ru_r) \tag{9}
$$

It is easily verified by direct integration, after substitution for the surface and volume elements

$$
dS = 2\pi a\,dz, \qquad dV = 2\pi r\,dr\,dz
$$

that the first-order quantities E_1 and W_1 satisfy the equation

$$
\int E_1\,dV = W_1
$$

The stability condition (7) therefore becomes

$$
\int E_2\,dV - W_2 > 0 \tag{10}
$$

where the second-order part E_2 for the energy of distortion per unit volume[3,4] is

$$
E_2 = \tfrac{1}{2}(\delta t)^2\left[H\dot{\bar{\varepsilon}}^2 + p\left\{\left(\frac{\partial v_r}{\partial r}\right)^2 + \left(\frac{v_r}{r}\right)^2 + \frac{\partial v_r}{\partial z}\frac{\partial v_z}{\partial r}\right\}\right] \tag{11}
$$

where H is the slope of the uniaxial stress–strain curve and $\dot{\bar{\epsilon}}$ the equivalent strain rate at the initial state of hardness. Substituting for W_2 and E_2 from (6) and (11) into the stability criterion (10), we obtain the condition for stability as

$$\int_0^l \int_0^a \left[H\dot{\bar{\epsilon}}^2 + p\left\{\left(\frac{\partial v_r}{\partial r}\right)^2 + \left(\frac{v_r}{r}\right)^2 + \frac{\partial v_r}{\partial z}\frac{\partial v_z}{\partial r}\right\}\right] r\, dr\, dz$$

$$-pa\int_0^l \left(\frac{\partial v_r}{\partial r} v_r + \frac{\partial v_r}{\partial z} v_z\right) dz > 0 \qquad (12)$$

where a is the current radius and l the current length of the bar.

According to the Lévy–Mises flow rule, the strain rates associated with the stress state given by (8) satisfy the equations

$$\dot{\epsilon}_r = \dot{\epsilon}_\theta = -\tfrac{1}{2}\dot{\epsilon}_z, \quad \dot{\gamma}_{rz} = 0 \qquad (13)$$

Hence the virtual velocity field must satisfy

$$\frac{\partial v_r}{\partial r} = \frac{v_r}{r} = -\tfrac{1}{2}\frac{\partial v_z}{\partial z}, \quad \frac{\partial v_r}{\partial z} + \frac{\partial v_z}{\partial r} = 0$$

These equations have the general solution

$$\left.\begin{aligned} v_r &= -r(Az + B) \\ v_z &= A(z^2 + \tfrac{1}{2}r^2) + 2Bz \end{aligned}\right\} \qquad (14)$$

where A and B are arbitrary constants. If these constants are now so adjusted that one end of the specimen is just at the point of unloading, the velocity components become

$$\left.\begin{aligned} v_r &= -Br\left(1 - \frac{z}{l}\right) \\ v_z &= B\left(2z - \frac{r^2 + 2z^2}{2l}\right) \end{aligned}\right\} \qquad (15)$$

This velocity field furnishes

$$\left.\begin{aligned} \frac{\partial v_r}{\partial r} &= \frac{v_r}{r} = -B\left(1 - \frac{z}{l}\right) \\ \frac{\partial v_r}{\partial z} &= -\frac{\partial v_z}{\partial r} = \frac{Br}{l} \\ \dot{\bar{\epsilon}} &= \frac{\partial v_z}{\partial z} = 2B\left(1 - \frac{z}{l}\right) \end{aligned}\right\} \qquad (16)$$

where B must be positive for the rate of plastic work to be so. It is easily verified that the strain rates do not change sign anywhere in the field and they vanish at $z = l$. Substituting from (16) into (12) and integrating, we obtain

$$H > p\left(1 - \frac{3a^2}{8l^2}\right)$$

On the generalised stress–strain curve, the onset of plastic instability is therefore given by

$$\frac{1}{\bar{\sigma}}\frac{d\bar{\sigma}}{d\bar{\epsilon}} = 1 - \frac{3}{8}\left(\frac{a}{l}\right)^2 \qquad (17)$$

If the second term on the right-hand side is neglected,

as would be justified for normal specimen sizes, we recover the well-known condition for plastic instability in uniaxial tension. The additional term, however, does not arise when the elastic deformation of the bar is taken into account[5].

4. EXPERIMENTAL INVESTIGATION

Carefully machined cylindrical specimens, $\frac{3}{8}$ in in diameter and 4.75 in long, were held between two plungers as shown in figure 2. The plungers were bored along their lengths to allow the specimens to move freely during the elongation caused by the lateral fluid pressure. Neoprene O-ring seals, backed by beryllium–copper anti-extrusion rings, were used both inside and outside the plungers.

The experiment was performed in a Fielding and Platt 300 ton hydraulic extrusion press (figure 3) described in detail by Alexander and Lengyel[6].

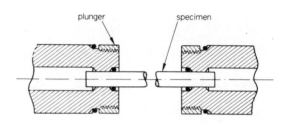

Figure 2. Plungers holding the specimen.

Briefly, it consists of a downstroking main ram fitted to the press main frame and three auxiliary rams fitted in a separate yoke. Two of the auxiliary rams are diametrically opposed to one another and the third has its axis perpendicular to the common axis of the other two. The subpress consists of a container block resting on a high-pressure block with adequate sealing. The container block can be lifted by four hydraulically operated rams. The high-pressure block can be rotated about a vertical axis and contains horizontal bores to line up with the opposing auxiliary rams. The plungers were pushed into the horizontal bores of the machine and held in position by a pair of opposing auxiliary rams through spherical seats. The cavity of the high-pressure block was filled with a mixture of castor oil and methyl alcohol, care being taken to ensure that there were no air bubbles trapped inside. The pressure was applied by a hand-operated pump, which in turn pushed down a plunger fitted to the main ram. The pressure steadily increased as the specimen continued to deform, until a stage was reached when it was not possible to build up the pressure. The specimen broke into pieces (which were expelled with violence) immediately after the pressure started to decrease. The broken pieces were taken out and the diameter of the uniform part was measured with a micrometer.

The experiment was repeated with several specimens. On one occasion the pressure was applied on an incremental basis, the specimen being withdrawn each time and the diameter measured. The applied pressure was recorded at each stage from a pressure gauge fitted to the hand pump. This was

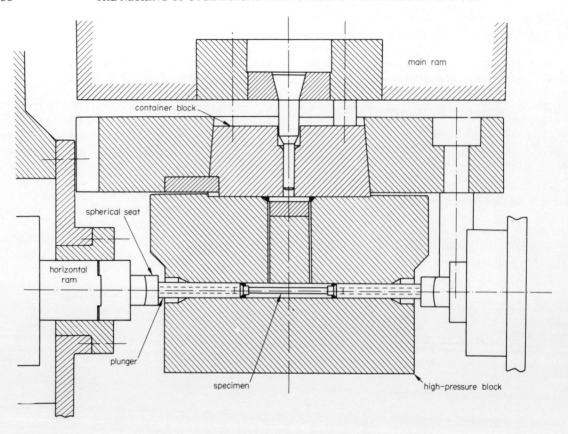

Figure 3. Section through the subpress.

done to assess the effect of frictional losses in raising the strain-hardening curve above the actual one (obtained by conventional methods). The actual curve would have been traced if the pressure were measured in the region surrounding the specimen (for instance, from the change in resistance of a manganin wire).

5. RESULTS AND DISCUSSION

Specimens made of commercially pure aluminium and an alloy of 90% aluminium and 10% silicon were used in the above experimental work. The stress—strain curves for these materials, obtained by the Cook and Larke compression test, are shown in figures 4 and 5. The point of plastic instability under uniaxial tension (given by $d\sigma/d\epsilon = \sigma$) is also shown in each diagram. A tensile test for the aluminium—silicon alloy showed that the specimen fractured in a brittle manner with a uniform strain of about half the theoretical instability strain.

The tests under lateral fluid pressure showed that necking began in the aluminium specimens with exactly the same uniform elongation as that in simple tension, but the cross-sectional area at the neck was almost reduced to a point. This was obviously because of increased ductility caused by the compressive hydrostatic part of the stress, in contrast with that in simple tension. A notable difference in behaviour from that in simple tension was observed for the aluminium—silicon alloy specimens. Distinct necks were formed under the lateral fluid pressure and considerable reduction in

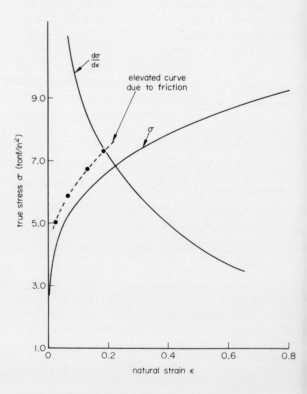

Figure 4. Stress—strain curve for commercial aluminium.

area took place before fracture. The instability strain was, however, the same as the theoretical value (indicated in figure 5).

For extremely brittle materials such as glass, the fracture mode would remain brittle even under lateral

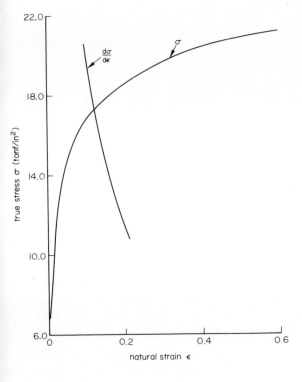

Figure 5. Stress—strain curve for aluminium—10% Si alloy.

fluid pressure. Bridgman[7] observed that a specimen of glass-hard tool-steel broke at right angles to the axis without any necking, when subjected to lateral fluid pressure. It should be noted that in the case of ductile materials, as soon as necking begins, there is a tensile stress tending to pull the specimen apart. This, however, does not account for the beginning of necking and is therefore not the true cause of instability. Materials which are ordinarily brittle apparently behave in a ductile manner under fluid pressure, but the instability strain (given by the generalised stress—strain curve) remains unchanged.

ACKNOWLEDGEMENTS

The author wishes to thank Professor J. M. Alexander, under whose supervision the experimental work was carried out at Imperial College, London (1964).

REFERENCES

1. H. LL. D. PUGH, 1964. *Nat. Engng. Lab. Rept.*, No. 192, D.S.I.R.
2. R. HILL, 1957. *J. Mech. Phys. Solids*, **5**, 153.
3. R. HILL, 1957. *J. Mech. Phys. Solids*, **6**, 1.
4. J. CHAKRABARTY, 1969. *Intern. J. Mech. Sci.*, **11**, 732.
5. J. CHAKRABARTY, 1969. *J. Appl. Math Phys.* (ZAMP), **20**, 696.
6. J. M. ALEXANDER and B. LENGYL, 1965. *J. Inst. Metals*, **93**, 137.
7. P. W. BRIDGMAN, 1912. *Phil. Mag.*, July, 63.

AUTHORS' INDEX

SUBJECT INDEX